AF694533

Millets and Millet Starch: Chemical Composition, Functional Modification, and Applications

Jaspal Singh • Mayank Kumar Malik
Editors

Millets and Millet Starch: Chemical Composition, Functional Modification, and Applications

Nutri-cereals from Farm to Fork and Beyond

Springer

Editors
Jaspal Singh
Department of Chemistry
Gurukula Kangri (Deemed to be University)
Haridwar, Uttarakhand, India

Mayank Kumar Malik
Department of Chemistry
Gurukula Kangri (Deemed to be University)
Haridwar, Uttarakhand, India

ISBN 978-3-032-09501-5 ISBN 978-3-032-09502-2 (eBook)
https://doi.org/10.1007/978-3-032-09502-2

© The Editor(s) (if applicable) and The Author(s), under exclusive license to Springer Nature Switzerland AG 2026

This work is subject to copyright. All rights are solely and exclusively licensed by the Publisher, whether the whole or part of the material is concerned, specifically the rights of translation, reprinting, reuse of illustrations, recitation, broadcasting, reproduction on microfilms or in any other physical way, and transmission or information storage and retrieval, electronic adaptation, computer software, or by similar or dissimilar methodology now known or hereafter developed.
The use of general descriptive names, registered names, trademarks, service marks, etc. in this publication does not imply, even in the absence of a specific statement, that such names are exempt from the relevant protective laws and regulations and therefore free for general use.
The publisher, the authors, and the editors are safe to assume that the advice and information in this book are believed to be true and accurate at the date of publication. Neither the publisher nor the authors or the editors give a warranty, expressed or implied, with respect to the material contained herein or for any errors or omissions that may have been made. The publisher remains neutral with regard to jurisdictional claims in published maps and institutional affiliations.

This Springer imprint is published by the registered company Springer Nature Switzerland AG
The registered company address is: Gewerbestrasse 11, 6330 Cham, Switzerland

If disposing of this product, please recycle the paper.

For the hands that till the soil and the hearts that nurture us

Preface

Millets (Shree Anna), considered as “Miracle grain,” are not merely a compendium of nutritional facts. It is a celebration of the wisdom inherent in these small seeded marvels. These small grained minor cereal crops are water efficient and resistant to pests, diseases, and climatic stress. Despite their diverse benefits, millets have gained limited attention from farm to table, research, policies, and markets. The narrowing of diversity in crop species contributing to the world’s food supplies has been considered a potential threat to global food security. Recently, interest in the potential of millet has increased to contribute to inclusive and sustainable agriculture development. Millets can, therefore, be viewed as a future crop that will help agriculture overcome obstacles. In recent years, millet-based meals have become more and more advocated for a balanced diet and treating numerous health conditions. There is a strong need to include “millet” as a staple food in addition to wheat and rice in the current condition of population blast, shrinking land, food shortage, and nutritional deficiency. The importance of millets as a crop has been recognized by various organizations and researchers. With an aim to create awareness of consumption and increase production of millets, the United Nations at the behest of the Government of India declared the year 2023 as “International Year of Millets.” The Food and Agriculture Organization of the United Nations (FAO) has emphasized millets’ potential to support food and nutrition security, particularly in areas with high malnutrition rates and poverty.

Thus, the book *Millets and Millet Starch: Chemical Composition, Functional Modification, and Applications* is an attempt to put forward various applications of millets by which millets can be proposed as nutraceutical crops from their status as orphan crops. The book, comprising 23 chapters, will cover the current research and development in the field of millet nutrition, processing, value-added products, human health, and millet-based entrepreneurship development. Millets are nutritionally similar or superior to some major cereal grains. The additional benefits of the millets like gluten-free proteins, high fiber content, low glycemic index, and richness in bioactive compounds make them a suitable healthy food. Similarly, millet starch has very high medicinal and commercial value. The slow digestion of millet starch results in low glycemic load on the human body; therefore, it is an ideal food for people suffering from diabetes and obesity. Chapters 1, 2, 3, 4, 5, 6, 7, 8, 9, 11 and 12 cover this aspect and attempt to compile the latest advances in understanding millet starch's structural and functional properties. These chapters provide thorough information of the origin, geographical distribution, cultivation, storage, processing, and starch isolation methods. These further highlight the approaches used for tailoring the functionality of the millet starch and how these approaches impact the functionality of millet starch. These give a cursory impression of the way millet starch in native and modified form has been utilized in industries. Chapter 10 explores the nutritional composition and traditional uses of lesser known browntop millet. Chapter 13 is devoted to pseudocereals. Increasing obesity and other lifestyle disorders forced researchers to explore new sources of resistant starch and dietary fiber. Millets are a rich source of dietary fiber and slowly digestible starch. Chapter 14 is focused on comprehending the fraction of resistant starch in various millets. Further, health benefits of millets' resistant starch are highlighted. Millets are ancient grains that have served as a foundational traditional food for thousands of years in cultures across Asia and Africa, featuring prominently in diets and rituals as a staple food and even an ingredient in traditional beverages. Chapter 15 is devoted to the cultural significance and traditional food recipes of the millets. The role of nutricereals in weight management and controlling the obesity is covered in Chap. 16. The unique and unequalled potential and versatility of millets starch in native and modified form for nanotechnology applications is highlighted in Chap. 17. It is considered that millets are not only the potential sources of energy but also possess a therapeutic role in preventing metabolic disorders and decreasing the risk factors for lifestyle disorders and cardiovascular and renal disease. Chapter 18 and the last two chapters of the book cover the therapeutic potential of Indian nutricereals and the medicinal importance of several millets. Fermented foods are of great importance because they provide and preserve great amounts of nutritious and healthy foods in a wide diversity of flavors, aromas, and textures, which enrich the human diet. Chapter 19 delves into the expansive realm of fermented millet cuisines prevalent across African and Asian regions, elucidating their cultural significance, nutritional attributes, and prospective health advantages. Molecular biology and nutritional attributes of millet have been discussed in depth in Chap. 20. Strengthening food and nutrient security is crucial to feeding the ever-growing world population. Promotion of millet cultivation could help to meet the sustainable development goals (attaining zero hunger) of the UN. The objective of this goal is "to end hunger, achieve food security and improve the nutrition and promote sustainable agriculture." The role of millets in attaining food security is explained in Chap. 21.

The chapters of the book have been authored by various experts keeping in view the syllabi of different universities, researchers, students, as well as the requirements of the industry. This book will serve as an instructional material for researchers in millet science, biopolymer industry, food science, and process engineering and as a reference material for those working in industry and R&D labs. Efforts have been made to avoid overlapping in contents; however, some overlapping in isolated spots is unavoidable. We express our sincere thanks to all the contributors of the 23 chapters. We sincerely appreciate their continuous cooperation, starting from the first submission of drafts to the revision of the chapters as suggested by the editors. We also thank our family members for bearing with us throughout the process of editing of this book. We are thankful to the Springer Nature team, especially Daniel Falatko and Sofia Priya Dharshini Valsendur, for their support and suggestions.

Haridwar, Uttarakhand, India

Jaspal Singh
Mayank Kumar Malik

Contents

About the Authors

Jaspal Singh is currently working as Associate Professor in the Department of Chemistry, Gurukula Kangri (Deemed to be University), Haridwar, India. He earned his Doctorate of Philosophy in Chemistry from the Gurukula Kangri (Deemed to be University), Haridwar. Dr. Singh is a double postgraduate: M.Sc. Chemistry from IIT, Roorkee, Uttarakhand, India, and M.E. in Polymer Chemistry from Delhi College of Engineering, Delhi. He is engaged in research work related to chemical kinetics, synthesis of drug, and polymer science, basically in emulsion and emulsion paint technology. He is a recipient of GATE 96, GATE 98, and NET (UGC-CSIR) 2000. He has over 15 years of research/academic experience and over 66 publications in reputed international/national peer-reviewed scientific journals and has supervised 6 successful Ph.D. scholars. Dr. Singh has visited several research laboratories and industries. He has chaired various technical sessions, organized several conferences, delivered invited talks, and presented more than 60 research papers in national/international conferences. Dr. Singh has authored four books credited for his research work and edited two books. He is actively working as convener of the Indian Science Congress (Haridwar Chapter). He is running a research project funded by Uttarakhand Science Education Research Centre as Principal Investigator.

Mayank Kumar Malik is presently working as Assistant Professor (C) in the Department of Chemistry, Gurukula Kangri (Deemed to be University), Haridwar, India. He is an associate member of the Royal Society of Chemistry, London. He received his Doctorate of Philosophy in Chemistry from Gurukula Kangri (Deemed to be University), Haridwar, India. Dr. Malik has also done Postgraduate Diploma in GIS and Remote Sensing with distinction from C.C.S. University, Meerut, and PGDBA from Annamalai University, India. Dr. Malik has also worked as Research Assistant (Project) in Uttarakhand Council of Science and Technology sponsored research project, at the Department of Pharmaceutical Sciences, Gurukula Kangri (Deemed to be University), India. He has over 5 years of research/academic experience and over 13 national and internationally published book chapters. He has 16 publications in reputed international peer-reviewed scientific journals, presented over 25 research papers in national/international conferences, published many articles in newspapers, delivered radio talks, acted as resource person, and delivered 3 invited talks. He has one Indian patent published in his research kitty. In 2025, he was awarded the Vigyan Ratna Puraskaar by the AACCRS. He received best paper presentation award conferred by the Association of Pharmacy Professional (Uttar Pradesh and Uttarakhand Chapter) and excellence award for outstanding performance in research conferred by IPGA (Uttarakhand State Branch). His research interests include millet science; starch structure, functionalities, and modifications; starch transitions in pharmaceutical and food/non-food industry; and the thermal, rheological properties of bio-polymers.

Chapter 1
Finger Millet: Starch Isolation, Modification, and Food/Nonfood Applications

Mayank Kumar Malik , Anita Rawat, Vipin Kumar, and C. P. Singh

1.1 Introduction

Millet is a broad collection of small gluten-free grain, short-growing crop (Paschapur et al. 2021). Millets include finger millet (*Ragi*), sorghum millet (*Jowar*), foxtail millet (*Kakum*), pearl millet (*Bajra*), little millet, proso millet (*Chena*), kodo millet, barnyard millet (*Sanwa*), and brown top millet (Shobana et al. 2013). Among these millets, one of India's oldest crops, finger millet (*Eleusine coracana* L.), is known in ancient Indian Sanskrit literature as "nrtta-kondaka," or "dancing grain." It was also referred to as "rajika," or "markataka, Ragi, Mandua" (Malik et al. 2023a, b; Achaya 2009; Shobana et al. 2013). It is also known as Koracan in Sri Lanka and by numerous other names in Africa. It has historically been a significant millet staple food in Central and Eastern Africa and India (FAO 1995). It is enriched in protein, fiber, and minerals, and plays a key role in contributing to food security in semi-arid

M. K. Malik (✉)
Department of Chemistry, Faculty of Science, Gurukula Kangri (Deemed to be University), Haridwar, India
e-mail: Mayank.malik@gkv.ac.in

A. Rawat
Former Director, Uttarakhand Science Education and Research Center, Uttarakhand, Dehradun, India
e-mail: dr.anitaarawat@gmail.com

V. Kumar
Department of Pharmaceutical Sciences, Gurukula Kangri (Deemed to be University), Haridwar, India
e-mail: vipin@gkv.ac.in

C. P. Singh
Former Professor, Department of Chemistry, Sahu Jain (P.G) College, Najibabad, Uttar Pradesh, India

© The Author(s), under exclusive license to Springer Nature Switzerland AG 2026
J. Singh, M. Kumar Malik (eds.), *Millets and Millet Starch: Chemical Composition, Functional Modification, and Applications*,
https://doi.org/10.1007/978-3-032-09502-2_1

regions. Globally, finger millet secures fourth rank in importance among other millets (Malik et al. 2022a, b). It is high in carbs, resistant starch, and other nutritious components. The presence of resistant starch allows it as an ideal superfood for people suffering from type II diabetes. Ayurveda also signifies its importance in the treatment and cure of numerous diseases (Devi et al. 2014; Roopa and Premavalli 2008).

Based on their functional characteristics, including paste clarity, viscosity, gel strength, and degree of syneresis, starches often have a wide range of industrial uses as thickening, stabilizing, emulsifying, and cross-linking agents (Makroo et al. 2021; Babu et al. 2019). Starch is considered the key constituent of the millets (Mahajan et al. 2021). Researchers explored finger millet to isolate starch and characterize its structural and functional characteristics (Roopa and Premavalli 2008; Malik et al. 2023a, b; Yadav et al. 2024; Dey et al. 2024). Studies supported that finger millet is rich in starch content (65–70%)(Yadav et al. 2024) and granules of finger millet starch are irregular, spherical, and 5–15 μm in size, whereas polygonal finger millet starch granules are 2.5–17 μm in size (Malik et al. 2022a, b; Yasir et al. 2024).

Yasir et al. (2024) and Shobana et al. (2013), offered a comprehensive account of finger millet nutritional value, processing, and health attributes. Nevertheless, it was proven that more meticulous information on isolation of starch from finger millet, modification of finger millet starch, and recent advances in its application in food and nonfood industries is lacking. Our objective is to block this breach of information by performing a comprehensive study that emphasizes the research available on the finger millet and the probable approaches used for the isolation of starch from it. Further, potential benefits of modifying the functionality of finger millet starch using different chemical and physical methods will be explored thoroughly. This chapter aims to deliver considerable insight for employing the native starch and modified starch in diverse pharmaceutical and food industries applications.

1.2 Origin, Domestication, and Geographical Distribution

Finger millet, an allotetraploid, is commonly called Ragi/Mandua in India and Wimbe in East Africa, and was developed in the Ethiopian region of Africa from *E. coracana* ssp. *Africana*. It is considered that it is attained from the wild progenitors *E. floccifolia* and *E. indica* (Krishna et al. 2020; Babu et al. 2013). *E. coracana* showed resemblance to *E. indica* (2n = 18) and *E. coracana* subsp. *Africana* (2n = 36) (Sood et al. 2016; Hilu and De Wet 1976). Because of its great variety throughout Africa and India, as well as its physical resemblance to both *E. indica* and *E. africana*, the origin of *E. coracana* has lingered a matter of controversy. Nonetheless, it is generally acknowledged that Africa is the primary origin and its domestication started nearly 5000 years ago in the Ethiopian highland and Western Uganda. However, it was domesticated in India around 3000 BC (Hilu and de Wet

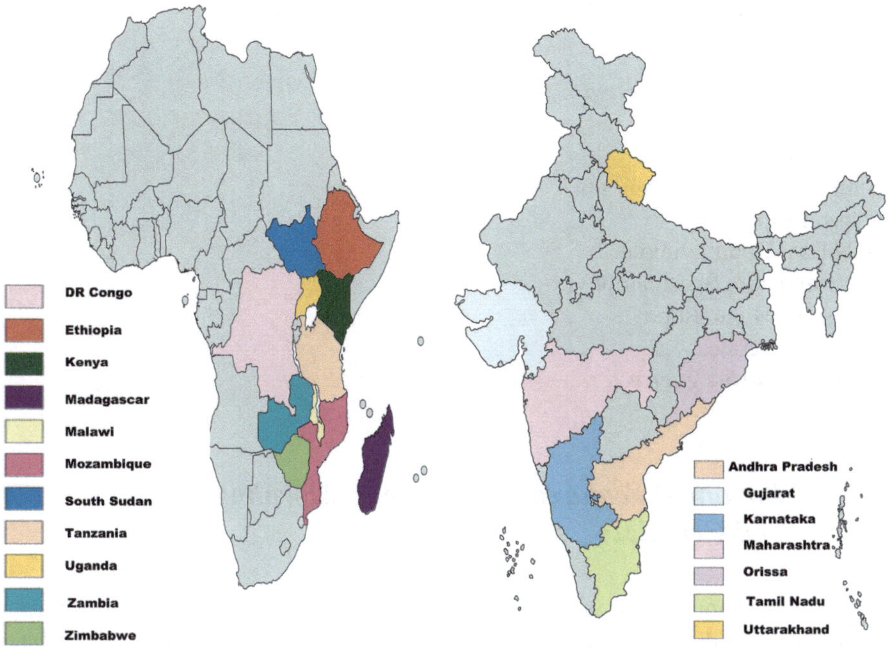

Fig. 1.1 Geographical distribution of finger millet in Africa and India

1976; Vetriventhan et al. 2016). In Eastern Africa, it is commonly distributed in Uganda, Kenya, and the United Republic of Tanzania while in southern Asia it is usually distributed in India and Nepal. However, it has been testified that India is the chief cultivar of finger millet. It is also farmed in Rwanda, Zambia, Malawi, Iran, Jordan, Burundi, Sri Lanka, Pakistan, Myanmar, and Sudan to a lesser extent (Maharajan et al. 2021). Figure 1.1 represents the global distribution of finger millet. Besides the African and Asian regions, it is also grown in Europe and the America, specifically in Germany, Italy, Ukraine, Mexico, and the USA (Mirza and Marla 2019; Radchuk et al. 2012; Backiyalakshmi et al. 2021; Gebreyohannes et al. 2024). Globally, it has >34,1600 cultivable genotypes out of them 22,583 genotypes available in India and covers 2.7 million hectares of agriculture fields globally (Gebreyohannes et al. 2021; Ceasar et al. 2023). In India, it is grown abundantly in Rajasthan, Uttarakhand, Karnataka, Odisha, Tamil Nadu, Jharkhand, Andhra Pradesh, Maharashtra, Telangana, Assam, Kerala, Arunachal Pradesh, Meghalaya, Nagaland, and Tripura (KSDA 2018; APEDA 2025; ICAR-IIMR 2018; Ganapathy 2017). India is the leading cultivar of finger millet (1.19 million hectares of agriculture land). Ethiopia stands second in finger millet (32,554,127 hectares of agriculture land) production (Gebreyohannes et al. 2021).

1.3 Botanical Classification and Description

Species: *Eleusine coracana* (L.) *Gaertn.*
Family: *Poaceae*
Kingdom: Plantae
Subkingdom: Tracheobionta
Class: Liliopsida
Subclass: Commelinidae
Superdivision: Spermatophyta
Division: Magnoliophyta
Order: Cyperales
Genus: *Eleusine gaertn*

1.3.1 Botanical Features (Appearance, Morphology, and Genetics)

It is a robust annual grass with fibrous roots and can stretch up to 1.7 m in height. It needs 90–180 days to complete its seed cycle. It has smooth, dark green linear leaves (Fig. 1.2). The leaf blades show no cross-veination with narrow blades. The inflorescence branches are curved at the tip when robust and reflexed when slender (Chandrashekar 2010; Gomez and Gupta 2003). The inflorescence branches of finger millet are 24 cm long, slender to robust, and have a cluster of 3–26 fingers known as digital spikes with dense spikelet's containing seeds (Fig. 1.2). Its spikelets are 3–4 mm wide and 5–8 mm long with 4–10 florets on each spikelet (Sood et al. 2019; Rambhai 2009; DeWet et al. 1984). Lower-flying glumes called lemma and a pair of scales called palea cover the florets. Lemma and palea both are awnless. However, lemma is hairless and pointed with three nerved and palea is two nerved. Close to the base of the ovary are binary tiny scales called lodicules. Whereas the androecium contains three hypogynous stamens, the unilocular bicarpellary gynoecium of flowers has a superior ovary and a single ovule. Self-pollination is ensured by the stigma being fully encircled by the androecium. The grain shape contrasts from oval and oblong to round. Finger millet seed is a utricle, which is why the grain can be effortlessly detached from the seed coat and seed pericarp is independent of the kernel (McDonough et al. 1986). Seeds vary in weight from 1 to 3 mg in diameter (Mohan et al. 2018) and from white to reddish brown in color (Kayastha et al. 2024). A study conducted on 92 finger millet accessions showed that in the seeds majority of genotypes had brown color (55.43%), rough surface (52.17%) enclosed glume cover (58.69%), round shape (42.39%), and nonpersistent pericarp (76.06%) (Kumar et al. 2019).

Fig. 1.2 Leaf, inflorescence, roots, and grain of *Eleusine coracana* depicted. (**a**) Stem part with leaves; (**b**) roots with stem part; (**c**) inflorescence (close view of the flower); (**d**) part of inflorescence branch; (**e**) spikelet; (**f**) harvested finger millet grains. (Photo courtesy: Mr. Balkrishna)

1.4 Cultivation Technology

Finger millet is a highly self-pollinated millet. Finger millet has a significant growth flexibility as it can be grown in numerous soil types from high fertile to poor fertile. Besides these characteristics, it has a high adaptation due to its short, flexible growing season. It can thrive even in specific degree of alkalinity, under alluvial soil, loamy soil, and in sandy soil with well drainage (Sood et al. 2019). It yields well and cultivates best in the mean temperature range of 26–29 °C. It thrives in soil with a pH between 5.0 and 8.2 with moderate precipitation (Duke 1978; Ceasar et al. 2023). However, it was reported that the crop grows well in tropical region of Africa and India where the temperature ranges between 18 and 27 °C (Sood et al. 2019). It survives well in areas with a yearly rainfall of around 1000 mm.

Although it can thrive in poor shallow to rich loam soil but better water-holding capacity, high organic content and proper porosity are preferable for high yield. The best time for cultivation is Kharif start with the onset of monsoon between June and July and Rabi from October to September. However, Kharif is the most considerable season for cultivation. During the growing season, the field should be kept well-drained and clear of weeds at the time of sowing. Ploughing is required to create a tilth, and then 3–4 harrowing are required to create ridges on a leveled field that are 75 cm apart. The seeds should be pretreated with a disinfectant to avoid infection. The appropriate seed rate for transplanting a one hectare field is 4–5 kg and 8–10 kg seeds based on variety. For the purpose of transplanting, 3–4-week-old seedlings are considered ideal (Ceasar et al. 2023; Upadhyaya et al. 2008; ICAR-IIMR 2018; Warisa et al. 2025). Figure 1.3 represents the growth of finger millet in agriculture

Fig. 1.3 Finger millet field. (**a**) Finger millet saplings after the transplantation. (**b**) Full-grown finger millet in natural field in 2025 at Village Jaikhal, Almora District of Uttarakhand, India. (**c**) Growth of finger millet along with *Amaranthus* in natural field in 2025 at Village Jaikhal, Almora District of Uttarakhand, India. (Photo courtesy: Mr. Balkrishna, B.Sc II Student at Gurukula Kangri (Deemed to be University), Haridwar, India)

field. Diammonium phosphate (DAP) (100 kg/ha) must be utilized as a basal dose before sowing the seed and urea (100 kg/ha) to be supplied to field as top dressing after 3 weeks of sowing. The weeding should be performed after 3 weeks of planting via hands before providing the top dressing of urea, and alien species should be removed for better growth and yield (Ceasar et al. 2023; Upadhyaya et al. 2008). Despite being resilient and able to withstand drought circumstances better than main cereals, extended dry spells result in a notable decrease in both grain yield and fodder. Hence, supplemental irrigation should be performed to maintain the desired soil moisture until flowering. It is considered that it is cherished for its superior nitrogen use efficacy and ability to consume rock phosphate (Sood et al. 2019; Gupta et al. 2012; Goron et al. 2015; Neshamba 2010). Each finger millet cultivar may have a different maturity duration, ranging from 4–5 months. In order to facilitate speedy threshing, the ear heads are typically harvested 40 days after blooming. Then, after cutting the earheads, it is sundried for 2–4 days. The dried earheads need to be threshed to obtain seeds. On a rainfed system, finger millet yields between 1.2 and 1.5 t/ha, but on an irrigated crop, it varies between 4.0 and 4.5 t/ha (Upadhyaya et al. 2008). A variety of techniques, including milling, popping, puffing, malting, flaking, and debranning, are used to process the seeds depending on their intended use.

1.5 Traditional and Modern Methods of Storage of Finger Millet

In the realm of food preservation, appropriate storage is required to hinder spoilage, conserve nutritional content, and restrict the growth of toxic, food-spoilage microorganism. There are both conventional and contemporary ways to preserve finger millet. Briefly, mud or earthen constructions like "hagevu" (underground pits) or "kothis" (earthen bins) are examples of traditional methods (Srusti et al. 2023). Modern techniques make use of metal bins, vacuum-sealed bags, and airtight containers. Both techniques need proper drying of seeds before storage to a moisture level of 12–13%, and long-term storage also requires temperature regulation. The walls of the mud houses are made by mixing the mud with paddy. Mud houses are used to preserve 10–25 quintal of finger millet. The standard average size of the mud house is 4 m * 4 m * 3 m for 25 quintal storage (Ghode 2023; Nickhil et al. 2025). Bamboo house is another kind of mud house used to store finger millet. It is commonly called Dhukar and Doke-Doke and can store 50–100 kg of finger millet (Negi and Solanki 2015). Figure 1.4 shows bamboo house (Dhukar) used for finger millet storage used by farmers of Uttarakhand state of India. Plants with insecticidal activities and available locally have been used by farmers to protect the finger millet grains from insects (Swamy and Wesley 2020). Bhakaar and Tumri is another traditional finger millet storage system. Bhakaar is a rectangular wooden box made from the pine or deodar wood while tumri is round or oval-shaped dried bottle gourd. Tumri can store 10–15 kg of grains but bhakaar can be of different storage sizes (Nickhil et al. 2025; Negi and Solanki 2015). In Nigeria, people used Rumbu and Sacks for storage of finger millet (Damar et al. 2016). Kothis and earthen pots are also used to store finger millet in regions with dry and hot climates. These pots allow some moisture to escape from them and keep the grains cool, moisture free (Nickhil et al. 2025; Mobolade et al. 2019). Besides, easy availability and eco-friendly nature of all these traditional storage system face a significant challenge

Fig. 1.4 Traditional storage bamboo house (Dhukar) for finger millet storage used in Uttarakhand State of India. (Photo courtesy: Mr. Balkrishna)

today. In the past few years, to attain food security and preserve the grains, significant innovations have been made by researchers in finger millet storage. In the past few years, hermetic storages have gained significant attention due to its uniqueness in preserving the quality of finger millet and extending its shelf life. It is a unique way to store finger millet in airtight containers (Nimbkar et al. 2022; Meenatchi and Loganathan 2022). Storage of finger millet in modularly controlled atmosphere chambers is resolving the drawbacks of conventional techniques. Their modular design enables flexible storage conditions, which are essential for different climates and harvests. In addition to extending shelf life, controlled atmospheres lower the danger of microbiological development and insect infestations. These chambers allow users to maintain finger millet storage-friendly temperature and humidity (Juliano and Reyes-De-Corcuera 2022).

1.6 Post-Harvest Processing of Finger Millet and Their Impact

To remove the inedible portions, increase their shelf life, enhance their nutritional and sensory qualities, finger millet must be processed before consumption. The principal processing following harvesting is cleaning, destaining, and storing. The secondary or post-harvest processing techniques for finger millet include germination, fermentation, malting, roasting, puffing, extrusion, and baking. These techniques are used to create tertiary products, which require minor processing before consumption.

1.6.1 Milling

Milling is considered the primary processing technique for finger millet. It was reported that drying of grains is needed before milling (Abioye et al. 2022). Due to the delicate endosperm and intact seed coat of finger millet kernels, the grain cannot be polished and boiled like rice or other cereals (Shobana et al. 2013). Further, the high tannin content in seed coat of finger millet is responsible for the less attractive, darker food (Devi et al. 2014). For this purpose, researchers indicate that laboratory milling be used to develop refined flour from finger millet (Kurien and Desikachar 1966). Although the flour becomes whitish after treatment, it also tailors the digestion behavior of the flour (Shobana et al. 2013). When finger millet is processed by milling and then sieving, the bioavailability of the nutrient is increased but the amount of nutrients and antinutrients decreased significantly (Oghbaei and Prakash 2012). The impact of milling techniques (attrition, abrasion, comminuting, and compression-cum-shearing) on the quality of the finger millet flour was investigated by Katti et al. 2008. The study concluded that the flours from the hammer and pin mills had intact starch granules with particles that had sharp cut edges, whereas the flours from the emery, plate, and roller mills contained particles with many exposed starch granules (Katti et al. 2008).

1.6.2 Soaking

Soaking is a typical method for reducing the fraction of antinutrients in food. Soaking of finger millet includes the steeping of finger millet seeds in distilled water for 12 h at 60–30 °C. Further, the grains are drained to remove the excess water and washed through (Abioye et al. 2022; Banusha and Vasantharuba 2013). Shigihalli et al. 2018 reported that soaking of finger millet grains for 48 h curtailed the phytic acid content (15–18%) and boosted the bioavailability of zinc. The impact of soaking duration (0, 18, 16, 14, and 12 h) and water-to-grain ratio (1:3 and 1:2) on the proximate composition of finger millet grains was investigated by Dey et al. 2025. Dey et al. 2025 concluded that an increase in soaking time significantly increased crude fiber, total ash, energy, and total carbohydrate and decreased total protein and fat content. Arzoo et al. 2024 observed that soaking finger millet (3 g) for 6 h at 45 °C in distilled water (11 ml) significantly changed water absorption capacity (18.83% to 21.86%), oil absorption capacity (15.68% to 23.48%), total phenol content (7.36 mg/g to 6.05 mg/g), and total tannins (1.69 mg/g to 0.46 mg/g) (Arzoo et al. 2024). According to Patel and Dutta 2018, soaking of finger millet caused a decrease in the trypsin (13.51%) and phytate (13.22%) inhibition activity.

1.6.3 Germination

Germination is a natural process for healthy seed and the primary step in the life cycle of the next-generation plant. However, germination requires oxygen, appropriate moisture, and specific temperature for the improvement of nutritional composition and functionality (Makhaye et al. 2021; Abioye et al. 2022; Pushparaj and Urooj 2011). Germination is considered a superior choice for the preparation of nutrient-rich finger millet products (Abioye et al. 2022). It was described that germination of finger millet diminished the phytate content and tannins in finger millet flour to half of the initial value (Abioye et al. 2018; Patel and Dutta 2018). Azeez et al. 2022 investigated the impact of germination on digestibility and functionality of finger millet and reported that germination increased the protein content (9.03% to 14.73%), resistant starch content by 14.32%, decreased the total starch content by 6.1% and oxalate content by 54.46%. According to Mbithi-Mwikya et al. 2000, germination reduced total starch content to 35.1% from 71.3%. Sripriya et al. 1997 also reported that germination (24 h) caused a reduction in total starch content (71.3% from 81.1%) and total phenols (1.43% to 1.28%). It was reported that germination caused a decrease in phosphorus and increase in iron and calcium (Chauhan and Sarita 2018). These findings are advantageous for achieving food security and nourishing human health. Further, reduction in antinutrients significantly augmented the acceptability, nutrient bioavailability, and digestibility behavior. Hence, such a household technique can be easily employed for tailoring the nutritional quality to promote finger millet utilization.

1.6.4 Malting

Finger millet is considered a super millet for malting due to its good malting characteristics. Malting of finger millet was first performed in 1939 by Sastri 1939. It was concluded in the study that the minor size of finger millet grains is worthy for germination and kilning (Sastri 1939). Malting involves a number of steps. Soaking, germination, kilning, and milling are some of the unit operations utilized for malting grains. Pathare et al. 2025 utilized malted finger millet as a substitute for barley in beer production and reported that a higher content of malted finger millet in beer increased the content of ethanol, color, total flavonoids (18.51–30.58 mg QE/L), and total phenolic (194.45–271.21 mg GAE/L). According to Nayak et al. 2024, malting of finger millet flour caused a significant reduction in the commonly applied pesticides such as chlorpyrifos and lindane. Thus, malted finger millet can be used for generating a variety of food items. Chaudhary et al. 2025 explored malted finger millet for the preparation of pasta and optimized its content (11% malted finger millet flour, 70% durum wheat semolina, 3% gluten, and 16% malted lentil) in pasta for improvement in nutritional profile and cooking time. Krishnan et al. 2012 concluded that malted finger millet showed greater bioaccessibility of iron, zinc, and calcium and reduction in polyphenols, total phytic acid, and dietary fiber. The impact of malting on antioxidant and microstructural characteristics of the finger millet was assessed by Murungweni et al. 2024. It was reported that the total color differences, hue angle, protein, DPPH activity, total phenolic, and total fiber content increased after malting (Murungweni et al. 2024).

1.6.5 Popping

Puffing or popping is one of the widely applied processing techniques to develop ready-to-eat foodstuffs. This technique comprises treatment of finger millet seeds at a high temperature for a short time duration using sand as a medium of heat transfer (Ushakumari et al. 2007). Pooped finger millet is extremely rich in aroma and flavor. In Karnataka, India, popped finger millet is named *hurihittu* (Shobana et al. 2013). Popping increased alcoholic activity in finger millet (Dharshini and Meera 2023). The bulk density of popped finger millet (0.47 g/ml) is lower than native finger millet (0.63 g/ml) (Arzoo et al. 2024). The energy, crude protein, alkaloids, and antioxidants have been found to be increased and crude fiber content decreased after popping the finger millet (Kaur et al. 2023).

1.6.6 Roasting

Roasting, known as dry frying, is a traditional technique utilized in rural areas and household to enhance storage life. Further, roasting of millet also reduced the toxic component and antinutritional components. According to Singh et al. 2018, roasting of finger millet caused a decrease in protein, moisture, and fat content and enhancement in bioavailability of iron calcium, total carbohydrate, and ash content. Significant enhancement in the tannin content (threefold), total polyphenols (by 17%), and salicylic acid concentration (182–1822 μg/g) was observed for roasted finger millet (Hithamani and Srinivasan 2014). Amino acids are the key components necessary for the biosynthesis of body tissue, metabolic activity, secretion of hormones, and neurotransmission. The composition of essential amino acid and nonessential amino acids was impacted by roasting and germination. It was reported that roasting decreased the essential amino acid in white finger millet (23.10% to 16.10%) but increased nonessential amino acids (76.90% to 83.90%) (Navyashree et al. 2022). The rotti and pittu developed from roasted finger millet showed a higher total phenolic content. Such a rise in phenolic content after roasting is possibly due to the degradation of cell structure (Chandrasekara et al. 2022).

1.6.7 Fermentation

It is a process used to generate beverages by breaking of carbohydrates under aerobic or anaerobic conditions using microorganisms. The utilization of microorganism for this purpose brings significant changes in digestion behavior, flavor, and other functional phytochemicals. A traditional mild-alcoholic beverage called Kodo ka Jaanr is used in Sikkim and Darjeeling hills in India, Bhutan, and Nepal (Thapa and Tamang 2004). Antony et al. 1996 investigated the impact of fermentation on chief nutrients available in finger millet and reported that fermentation significantly lowered the total fat content (diminished by 42.9%), long-chain fatty acids, and starch content. Antony and Chandra also investigated the change in mineral availability and antinutrients compositions and concluded that fermentation for 24 h significantly lowered phenols by 20%, trypsin inhibitor activity by 32%, and phytate by 20% (Antony and Chandra 1998). Fermentation of finger millet via yoghurt as starter for 8–36 h significantly enhanced the acidity of finger millet slurry and lowered the protein content, suggesting that fermentation can be used to develop low-protein food products (Gowthamraj et al. 2021). Finger millet grains have been utilized as functional grain to develop nonalcoholic beverages. For this purpose, researchers used *Lactobacillus fermentum* for fermentation and reported a reduction in viscosity and citric acid content in prepared fermented nonalcoholic beverage. Further, it was noticed that 24 h fermented beverages secure better amount of antioxidants (Udeh et al. 2025).

1.6.8 Extrusion

Extrusion is a crucial hydrothermal process that alters the functioning of flour by causing starch to degrade and gelatinize, dietary fiber to dissolve, and protein to aggregate. In this sense, extruded flours provide intriguing substitutes for hydrocolloids and pre-gelatinized starch in order to boost food materials such as bread production in the baking sector. Considering the advantages of extrusion, researchers used extruded finger millet to prepare Uji (Onyango et al. 2005; Onyango et al. 2004), bread (Patil et al. 2016), and snacks (Sukumar and Athmaselvi 2019). It was testified that breads prepared from extruded finger millets had superior textural, technological (loaf height and higher specific volume) characteristics compared with breads developed from unextruded finger millet (Patil et al. 2016). Similar findings were reported by Kumar et al. 2020. According to Kumar et al. 2020, the addition of extruded finger millet in pearl millet-based unleavened flatbread significantly tailored the dough extensibility, puffing, nutritional, and textural quality.

1.7 Isolation of Starch from Finger Millet

Researchers reported numerous approaches for the isolation of starch from finger millet grains. However, the ultimate goal of all the extraction approaches was to isolate the starch with the highest yield and purity. The commonly applied methods for the extraction of starch from finger millet are alkaline stepping method, acid stepping method, and distilled water stepping method. Different researchers applied alkali isolation approach for the isolation of starch from finger millet. Briefly, finger millet seeds or seed powder was used and treated with sodium hydroxide of different concentrations for a prolonged time. The alkali suspension containing finger millet seeds or powder was centrifuged. The resultant slurry was repeatedly treated with alkali and centrifuged till achieving a white layer of starch. Finally, the white starch layer was scraped carefully and the pH of the starch was adjusted to 7 (Gautam et al. 2021; Malik et al. 2022a, b; Umar et al. 2025; Malleshi et al. 1986). The isolated starch was stored in an airtight container. Yadav et al. 2024 compared the impact of ultrasound-assisted isolation approach on the conventional isolation method on characteristics of the finger millet starch. For this purpose, the operating frequency, input voltage, and power of the bath-type sonicator were maintained at 40 kHz, 220 V, and 100 W, respectively. Powdered and sieved finger millet flour was treated with sodium hydroxide (0.25%) solution (1:6) in ultrasound bath system for 15 min sonication time. The suspension was filtered and then centrifuged for 15 min. The starch was collected, dried, and stored in zip-lock pouch. Distilled water soaking approach was also applied to extract starch from the finger millet. Finger millet flour was soaked in distilled water (5:1 solvent-to-flour ratio) and blended. The blended slurry was centrifuged and a white starchy residue was collected. The white starchy residue was washed twice with distilled water to obtain a pure white starch.

Table 1.1 Impact of isolation methods on yield (%) and purity characteristics of finger millet starch

Isolation method	Yield%	Protein%	Fat%	Ash%	Swelling power	References
Alkali isolation	59.0	0.6	0.7	0.4	4.4 g/g at 70°C	Umar et al. (2025)
Alkali isolation	60.37	2.71	–	2.70	3.53 g/g	Balasubramanian et al. (2011)
Alkali isolation	37.67	1.08	1.54	1.73	–	Adebowale et al. (2005)
Alkali isolation	35.740	–	–	0.185	226.51% at 37°C	Malik et al. (2022a, b, 2023a, b)
Alkali isolation	63.5	–	–	0.0031	1.3 g/g at 60°C	Gautam et al. (2021)
Aq SO_2 steeping	52.4	1.45	0.24	0.32	–	Lawal (2009)
Water steeping with $HgCl_2$	42.0	0.14	0.20	0.11	1.4 g/g at 30°C	Mohan et al. (2005)
Ultrasound-assisted extraction (ultrasound time 15 min; flour to solvent ratio (4:5); power density (0.1 W/ cm^3) using sodium hydroxide (0.25%)	60.2	–	–	–	4.53 g/g at 80°C	Mishra et al. (2025)
Alkali soaking and pulverization	–	0.81	0.04	0.15	2.85 g/g at 60°C	Dey et al. (2024)
Dry milling followed by Na_2SO_4 (0.5%) stepping	25.95	–	–	–	–	Panwar et al. (2024)
Wet milling followed by Na_2SO_4 (0.5%)	30.83	–	–	–	–	Panwar et al. (2024)

The thoroughly washed starch was stored in a zip-lock pouch (Yadav et al. 2024). Table 1.1 shows the yield of starch obtained by various researchers via alkali isolation approach and other methods and their impact on purity characteristics.

In another investigation, Lawal (2009) extracted starch from finger millet by steeping the seeds in 0.2% aq. SO_2 at 30 °C for 28 h. The water-washed swollen seeds were blended for 30 min to obtain slurry. The slurry thus generated was centrifuged for 30 min after distilled water (5 L) resuspension. Resuspension in water and centrifugation step is repeated, and the slurry thus obtained was treated with toluene to remove protein. The protein layer formed at the interface of water and toluene was removed. Finally, the slurry of starch was washed with acetone and air-dried for 48 h at 30 ± 2 °C. Mohan et al. 2005 isolated starch from finger millet by steeping the deglumed ragi in distilled water containing $HgCl_2$ (100 ppm) for 16 h. The steeped material was screened from 85-micron sieve after blending with 5 volume distilled water. The process of water blending and sieving was repeated till the residue gave a negative iodine test. The slurry thus obtained was washed via alkali, followed by sodium chloride and toluene washing. The starch thus obtained was dried over methanol and kept in an air-proof container.

1.8 Modification of Finger Millet Starch

The functionality of the native starch depends on the botanical origin from which the starch is extracted. There are numerous restrictions such as poor thermal stability, high rate of retrogradation, syneresis, insolubility in water, and shear strength of native starch that hinder its utilization in pharmaceutical, food, and other industries. Finger millet starch also showed these limitations in its native form (Malik et al. 2023a, b). To overcome these limitations and endow native finger millet starch with industrially sound characteristics, numerous modification approaches have been employed, classified as physical, chemical, and enzymatic modification. Specific examples of how these modifications tailored the functionality of alkali starch are discussed in this section.

1.8.1 Physical Modifications of Finger Millet starch and Their Impact on Properties

Physical method of tailoring the functionality of starch has garnered significant attention of researchers due to their advantages, which include no use of toxic chemicals, that is, environment-friendly approach, better handling, and less processing cost. Keeping in view the advantages of these methods, researchers explored various physical treatment methods for changing the characteristics of finger millet starch. Some of these are discussed in this section.

Hydrothermal treatments such as annealing and heat moisture are the most frequently applied physical treatment in starch modification that tailor the physicochemical characteristics of the finger millet starch, without altering the structure of starch granule (Adebowale et al. 2005). It was stated that finger millet starch on heat moisture processing at (20%, 25%, and 30%) moisture level, 100 °C, and 16 h significantly tailored the pasting property of the finger millet starch. The paste viscosity of the native finger millet starch changes from 286.75 RVU to 243.88 RVU (Adebowale et al. 2005). In another study, the impact of heat moisture processing (110 °C, 6 h and 25% moisture content) on in vitro digestibility of finger millet starch and flour was assessed (Umar et al. 2025). It was concluded in the study that treatment significantly increased the rapidly digestible starch (14.3–22.4%) and slowly digestive starch (28.0–60.9%) content and decreased the resistant starch content (55.7–30.9%). Further, dramatic increment in the swelling power (3.4–18.4 g g^{-1}), solubility index (0.14–0.44%), and oil absorption capacity (1.9–2.5 g g^{-1}) was observed for the heat moisture-treated finger millet starch (Umar et al. 2025). Hydrothermal treatment of white finger millet starch was investigated by Balakumaran et al. 2023. It was confirmed in the study that hydrothermal modification induced significant changes in least gelation concentration (13.63% to 17.74%), the relative crystallinity, and short-range order of starch (Balakumaran et al. 2023).

Plasma treatment is an atmosphere-sensible, unique, and nonthermal method used for the modification of starch. The influence of atmospheric pressure cold plasma on the rheology and gelling characteristics was investigated by Kaur and Annapure 2024a, b. The study concluded considerable variations in the rheological characteristics of finger millet starch, the peak viscosity decreased (3.35 to 0.553 Pa.s), and paste clarity increased due to depolymerization of starch (Kaur and Annapure 2024a). In an another study, Kaur and Annapure studied the mechanism of atmospheric cold plasma-modified finger millet starch and suggested that cold plasma treatment initiated for 20 to 30 min at the highest voltage incorporated chemical (hydroxyl group oxidized to carbonyl followed by carbonyl group at C6 position) and physical modification in finger millet starch. Further, remarkable increase in the water solubility (6.7 times) of plasma-modified finger millets starch was observed at 230 voltage and 30 min time (Kaur and Annapure 2024a, b).

Ultrasonication is another method of physical modification of starch to tailor functional characteristics. Ultrasonication can be employed alone or in combination with other physical methods of modifications. Resistant starch was prepared from the alkali-isolated finger millet starch using ultrasonication and ultrasonication-autoclaving modification. For this purpose, the 25% starch suspension prepared in distilled water was treated with 600 W ultrasonic power for 10 min. However, for the ultrasonic-autoclaving purpose, the ultrasonicated starch was gelatinized and autoclaved at 121°C for 45 min. It was concluded in the study that ultrasonic-autoclaving treatment of finger millet starch caused lamellar debris on the starch granule surface. Further, the sedimentation rate of ultrasonic-autoclaving finger millet starch (16.67%) was higher than ultrasonicated finger millet starch (13.43%). The resistant starch content was 7.60 ± 0.29%, 5.44 ± 0.63% in ultrasonicated finger millet starch and ultrasonic-autoclaving finger millet starch, respectively (Wang et al. 2022).

High-pressure processing (HPP) has attracted considerable attention in tailoring the starch functionality as it can be operated easily without using toxic reagents. HPP originated changes in physical and functional characteristics like the color of the food, swelling power, and water-holding capacity of various food products. Numerous researchers have explored the possible applications of HPP modification as a substitute for traditional thermal modifications for food processing and conservation (Zhang et al. 2019; Yang et al. 2017;Nath et al. 2023; Liu et al. 2016). White finger millet starch was explored to study the influence of HPP modification on the characteristics of the alkali-isolated starch. The HPP was executed by placing the starch suspension containing PET bottles in the high-pressure chamber. For the modification purpose, 600, 400, and 200 MPa pressure was applied for 10 min using water as a pressure-transmitting medium in the instrument. It was claimed in the study that apparent amylose content and water absorption capacity in HPP-modified finger millet starch increased to 21% from 18%, 4.39 g/g from 1.92 g/g, respectively, with an increase in treatment pressure to 600 from 200 MPa (Sherin et al. 2024). Besides all these thermal and nonthermal physical modifications, finger millet starch has also been tailored via microwave modification technique. During microwave modification, starch is tailored by dielectric heating and electromagnetic polarization (Rao et al. 2021). White finger millet starch (20% moisture content) was treated for 10 min through microwave power of 510 W, and the treated starch was dried at 40°C for 24 h and stored at 4°C. It was claimed in the study that relative crystallinity (57.27% to 52.41%) and oil absorption capacity (1.06% to 0.98%) get reduced after microwave treatment. Further, microwave-tailored white finger millet starch showed agglomeration and irregular shape due to incomplete gelatinization of starch molecules (Balakumaran et al. 2023). Ananthu et al. 2023 explored the impact of microwave treatment at 720, 840, 900 W power level for 2 to 6 min exposure time on finger millet starch functional and pasting characteristics. The findings of the study suggested that microwave power level and exposure time both significantly reduced the free peak viscosity, free fatty acid, and oil absorption index (Ananthu et al. 2023). Table 1.2 represents the impact of different chemical & physical modifications on various characteristics of the finger millet starch.

1.8.2 Chemical Modifications of Finger Millet Starch

Hydroxypropylation of finger millet (Jeero, African finger millet species) starch was done using propylene oxide to investigate any change in functionality. It was found that hydoxypropylation significantly reduced turbidity (1264 ± 2.0 to 21.7 ± 1.6 NTU) for modified finger millet starch. Further, significant changes were observed in the peak viscosity from 365 to 436 RVU after modification. It was reported in the investigation that the introduction of bulky hydroxypropyl groups restricted the reorganization of molecular chains to the orderly structure after gelatinization (Lawal 2009). Acetylation and oxidation of finger millet starch have been performed using acetic anhydride and NaOCl, respectively, to tailor the tableting characteristics of native finger millet starch. The swelling power of the acetylated finger millet starch amplified significantly while for the oxidized starch the highest solubility was noticed. Further, both the modifications augmented the gelatinization profile with a significant decrease in the enthalpy (9.64 J/g to 8.76 J/g) (acetylated finger millet starch) and 3.88 J/g (oxidized finger millet starch) (Afolabi et al. 2012). Esterification of finger millet starch was also done by using acetic anhydride in glacial acetic acid medium, and the toxicity analysis was performed for the esterified finger millet starch. It was established in the study that no adverse effects were noticed in the treated animals at a dose of 2000 mg/kg body weight of mice, and no substantial changes in the liver and kidney parameters were observed. Besides, significant changes in the morphology (granules lost their shape) and degree of crystallinity have been reported after modification (Malik et al. 2023a, b). Balasubramanian et al. 2011 extracted starch from finger millet via alkali approach, and the extracted finger millet starch was subjected to acidic, enzymatic modifications. For this purpose, hydrochloric acid and fungal amylase were used as acidic and enzymatic modification reagents. The modified starches were characterized for various physicochemical characteristics. It was concluded in the study that the treatment of starch with fungal amylase caused a reduction in swelling power (3.18% from 3.45%) and

Table 1.2 Effects of chemical and physical modifications on finger millet starch characteristics

Modifications	Chemical reagents/ methodology used	Major findings	References
Chemical modifications			
Esterification	Glacial acetic acid and acetic anhydride (1:2) with 2.5N H_2SO_4	Low decomposition rate and degree of crystallinity, better thermal stability, increase in paste clarity, high swelling power, increase in moisture content, loss in smoothness and geometry of modified starch granules	Malik et al. (2022a, b)
	10.2g of acetic anhydride with 1M HCl	Increase in gelatinization temperature $T_{p,}$ and Tc values; decrease in enthalpy value; increase pasting property; high swelling power, rupture in starch granules, increase in bulk and tap density	Afolabi et al. (2012)
Crosslinking	Epichlorohydrin	Low swelling, solubility, moisture, paste clarity, and pH; an increase in water binding capacity; rough surface with groove formation on the starch granules	Malik et al. (2022a, b)
Etherification	Propylene oxide	Increase in molar substitution improves swelling capacity, turbidity is reduced, decrease in pasting temperature and syneresis, increased peak viscosity, reduced starch retrogradation	Lawal (2009)
Phosphorylation	STPP/STMP	Decrease in water-holding capacity, paste clarity, swelling power, paste clarity; high oil absorption capacity, enthalpy and phase transition temperature, decrease in crystalline nature of starch granules	Malik et al. (2023a, b)
Acid modified (hydrolysis)	0.14 mol eq./L of aq. HCl	Increase in water-binding capacity, slightly high pasting property, increase in L-value; low solubility and low swelling power	Balasubramanian et al. (2011)
Enzymatic modification	0.1% of fungal amylase (*Aspergillus oryzae*)	Decrease in water-binding capacity, low gel consistency and pasting property, increase in L-value, high solubility, show low swelling power	Balasubramanian et al. (2011)

(continued)

Table 1.2 (continued)

Modifications	Chemical reagents/ methodology used	Major findings	References
Oxidation	NaOCl	High ΔT_{gel}, $T_{p,}$ and Tc values; increase in pasting property; high solubility; irregularity in starch granular surface; high bulk and tap density; high flow property	Afolabi et al. (2012)
Physical modifications			
Hydrothermally modification	Native starch was heated at 110 ± 2°C for 8 hours and modified starch was dried at 40 °C for 24 hours and stored at 4 °C for later analysis	Gel hydration property was enhanced, gelation concentration was increased up to 17.74%, higher change in enthalpy, and change in b-value	Balakumaran et al. (2023)
Cold plasma treatment	Starch was subjected to partially ionized gas under different voltage levels (170, 200, and 230 volts) for varying durations of 10, 20, and 30 minutes	Significant decrease in degree of polymerization, zeta potential, pH, amylose content, Tp; decrease in water absorption capacity with increase in voltage and time; water solubility increases with increase in voltage and time; initially no change noted but increase in voltage and time shows increase in % of syneresis; oil absorption shows no significant changes	Kaur and Annapure (2024a, b)
Microwave-assisted modification	Native starch was microwave-treated at 510 W for 10 minutes, then the modified starch was dried at 40 °C for 24 hours and stored at 4 °C for later analysis	Gelation concentration was increased up to 16.41%, increase in the agglomeration of starch granules	Balakumaran et al. (2023)
Ultrasound treatment	The starch slurry was treated in an ultrasound bath at 33 kHz and 50 °C for 30 minutes; the mixture was filtered, dried at 40 °C for 6 hours, and then ground into a fine powder	Increase in water absorption capacity and swelling power, increase in L* value, high ΔH, T_o, T_p, and T_c values, and crack formation in modified starch	Amarnath et al. (2023)

(continued)

Table 1.2 (continued)

Modifications	Chemical reagents/ methodology used	Major findings	References
High-pressure modification	Starch was modified at 200, 400, and 600 MPa for 10 minutes	5% increase in amylose content; increase in pressure increases alkaline water retention, pasting temperature, solubility, syneresis, water and oil absorption capacity; paste clarity decreases with an increase in storage time; high color difference seen at 600MPa	Sherin et al. (2024)
Annealing	Recrystallized starch was then dried at 40 °C for 6 hours	66% increase in syneresis, decrease in L-value, pore formation in modified starch, high gelatinization temperature and enhanced crystallinity	Amarnath et al. (2023)

water-binding capacity (199.9 ± 0.42% from 256.9 ± 1.34%) while an increase in the gel consistency from 120 ± 0.70 mm to 92.5 ± 3.6 mm was observed. Wang et al. 2022 also studied the impact of enzymatic treatment on resistant starch content of the finger millet starch. Suspension of the alkali-isolated finger millet starch in phosphate buffer (pH 6.8) was prepared and treated with α-amylase for 10 min. It was concluded in the study that treatment with α-amylase yields 5.82 ± 0.51% resistant starch (Wang et al. 2022). Ozonation of the finger millet starch was done using ozone gas liquid mixing machine. Finger millet starch was placed in a closed catalytic chamber infused with ozone gas liquid machine and treated with ozone gas for 10 min exposure time. The study claimed that ozonation of finger millet starch caused an enormous increase in water holding rate (926.47 ± 103.26%) compared with enzyme-modified (360.19 ± 45.77%) and ultrasonicated finger millet starch (335.18 ± 52.89%) (Wang et al. 2022). In another investigation, crosslinking of the finger millet starch was done using epichlorohydrin (1.0% w/w, solid dry basis). The starch slurry was treated with epichlorohydrin at pH 10.5, reaction temperature 50 °C for 5 h, and the in vitro digestion resistibility was determined for the cross-linked starch. It was reported in the investigation that on crosslinking with epichlorohydrin-resistant starch content increased to 27% w/w from 9.4% w/w and swelling capacity decreased from 226.510% to 10.3.140% (Malik et al. 2022a, b). Phosphorylation of the finger millet starch was done to enhance the digestion resistibility and functionality of the finger millet starch. Finger millet starch was treated with the combination of the sodium tri metaphosphate and the sodium tripolyphosphate (19:1) at pH 12, and it was reported that digestibility of the finger millet starch was decreased after phosphorylation (32.64 ± 1.98% from 61.12 ± 2.54% in simulated intestinal fluid). Moreover, the paste clarity changed from 6.566 ± 0.321% to 0.604 ± 0.0089%, and the water-binding capacity changed from 1.080 ± 0.011%

w/w to 0.962 ± 0.014% w/w (Malik et al. 2023a). The major changes reported by various researchers after chemical modifications of finger millet starch are represented in Table 1.2.

1.9 Applications of Finger Millet, and its Starch

1.9.1 Industrial Applications of Finger Millet Starch

Modified finger millet starch has numerous applications in food, pharmaceuticals, and other industries. Esterified finger millet starch was utilized for the development of highly foldable, sustainable, eco-friendly film. For this purpose, the glycerol was used as a plasticizer. The study concluded that esterification of finger millet starch using acetic anhydride in acetic acid medium improved the folding endurance of the developed films. Further, the esterified mandua starch film was crack free and has improved % elongation. Based on the study's findings, researchers suggested that these films can be used as a substitute for developing the biodegradable, low-cost packaging, and other products (Malik et al. 2022a, b). In another study, researchers developed finger millet (variety: VL- 347, VL-315 and VL-324) starch (3% w/w) film using sorbitol (25% w/w) and solvent casting method. The developed film was explored as a substitute for the thin food packaging film. It was concluded in the study that films were free from cracks, translucent with film thickness ranging 0.03 mm to 0.04 mm. Further, the solubility and swelling index of the films developed from finger millet starch were superior compared to corn starch film. Researchers claimed that water vapor permeability of the finger millet starch film was almost similar to the film developed by the Amarantha starch ($2.76 \pm 0.05 \times 1$ $0^{-10}gm^{-1}\ s^{-1}\ Pa^{-1}$) and mango kernel starch ($0.98 \pm 0.04 \times 10^{-10}gm^{-1}\ s^{-1}\ Pa^{-1}$) (Gautam et al. 2021). The study revealed that thin film fabricated from dry heat-treated finger millet starch showed homogeneity, and was free from holes and cracks. Further, mechanical characteristics, water vapor permeability, and water solubility of the dry heat-treated finger millet starch film showed significant improvement compared to unmodified finger millet starch (Gautam et al. 2024). Sharma et al. 2023 assessed the impact of succinic acid-based crosslinking and plasticizer (sorbitol and glycerol) on the film-forming ability of the finger millet starch. It was confirmed in the study that the addition of sorbitol and glycerol significantly improved the peeling efficiency, brittleness, and flexibility. It was concluded that glycerol-based finger millet starch-based film showed better compactness, homogeneity, and thermal stability. The study claimed that finger millet starch can be used for the production of edible films for packaging and enhancing the shelf life of the food materials (Sharma et al. 2023).

Finger millet polyphenols have been incorporated in gelatin–chitosan composite film for the preservation of pork. It was reported that film containing 1.0% polyphenols showed significant antioxidant and antibacterial activity. The decrease in elongation at break by 36.45% and increase in tensile strength by 2.67 MPa were noticed when 1.5% finger millet polyphenol was incorporated into chitosan gelatin composite film (Fang et al. 2025).

Finger millet starch has also been explored for its potential application in solar cells. Yogananda et al. 2021 developed finger millet starch-based gel electrolyte for dye-sensitized solar cells. Guanidinium thiocyanate and N-methylbenzimidazole were incorporated in finger millet starch gel prepared by gelatinization in DMSO. By adding N-methylbenzimidazole to pure finger millet starch gel, a dramatic increments in open-circuit voltage was observed (Yogananda et al. 2021).

Afolabi et al. 2012 investigated the tableting and functional characteristics of the finger millet starch in native and modified forms (oxidized and acetylated) using Heckel and Kawakita's analysis. It was reported that oxidized finger millet starch showed lower Pk value and higher Py value, indicating that during the compression process the oxidation lowers the plastic deformation of starch. Tablets developed using acetylated finger millet starch showed the lowest disintegration time compared to oxidized finger millet starch and alkali-isolated finger millet starch. Thus, acetylated finger millet starch could be employed as a pharmaceutical excipient when fast disintegration is needed, whereas oxidized finger millet starch could be used as fillers in capsule formulations (Afolobi et al. 2012). Polymeric film coatings are commonly explored to acquire the sustained, controlled, or pH-dependent release of bioactive agents. Phosphorylated finger millet starch was explored as a controlled drug release coating dispersion agent to control the release of mesalamine. Phosphorylated finger millet starch aqueous dispersion-coated mesalamine matrix tablets when assessed in healthy human volunteer showed that drug release was controlled with 10% w/w coating level. It was reported that only 6% drug release occurred in the initial 2 h, which reached 59.51% in simulated colonic fluid after 14 h of administration (Malik et al. 2024). In another study, researchers explored the potential of phosphorylated finger millet starch as a carrier material and claimed that phosphorylated finger millet starch could be applied as a novel drug carrier for colon-specific delivery of bioactives. The mesalamine matrix tablets were developed using the phosphorylated finger millet starch as matrix material and coated with pH-responsive polymer Eudragit S 100. It was suggested in the study that only 30.93% cumulative drug release occurred with 5% w/w Eudragit S 100 coating. In vivo gamma scintigraphic investigation confirmed the gastric emptying time of the coated phosphorylated tablet was 90–120 min (Malik et al. 2023a, b).

1.9.2 Application of Finger Millet in Food Industries

Finger millet is processed through popping, fermentation, malting, and decortication milling. It is employed to develop diverse food products, including pancakes, pasta, vermicelli, noodles, Indian sweets (Ladoo), and other bakery items (Shobana et al. 2013: Mbithi-Mwikya et al. 2002). Nontraditional items such as papads (rolled and dried) (Vidyavati et al. 2004), pasta (Krishnan and Prabhasankar 2010), vermicelli (Sudha et al. 1998), biscuits (Saha et al. 2011), and noodles (Shukla and Srivastava 2014) are also emerging. Alshehri et al. 2024 assessed potential health benefits of mixture of finger millet flour and banana peel and determined its applicability in developing pancakes. Interestingly, finger millet flour and banana peel flour mixture (70:30) successfully improved the liver and lipid profile in rats and is also acceptable for the development of pancakes. Thus, finger millet and banana peel mixture can be used by obese individuals (Alshehri et al. 2024). Recently, noodles were developed using finger millet flour and assessed for nutritional and hypoglycemic activity. It was reported that 30% finger millet-incorporated noodles showed significantly lower glycemic index (45.13) than control (62.59) and were best fit on sensory criteria (Shukla and Srivastava 2014). Jyotsna et al. 2011 explored the potential of finger millet flour in replacement of wheat flour for the preparation of muffins. Biscuits have been developed from finger millet seed coat-based flour. It was concluded in the study that compared to control biscuits (1560 g), the biscuits made with the composite flour (finger millet seed coat based) had a crisp texture and a breaking strength of 1480–1690 g. The biscuits had higher levels of calcium, dietary fiber, and protein and had a mild gray color (DE = 40–50) (Krishnan et al. 2011).

1.10 Future Perspectives of Finger Millet

Finger millets, despite their exceptional nutritional content and utilization in daily diet as food, are still restricted to the rural population, especially to the geographical area of origin. This is associated with poor awareness regarding finger millets' nutritional benefits and poor availability of the ready-to-eat products. Recently, efforts have been made to commercialize the utility of the finger millet and other millets by producing ready-to-eat value-added food materials. Starch is the key ingredient of finger millet. Its native starch and modified starch can be utilized as a key ingredient for diverse industrial utilizations apart from its application in feed and food applications. Finger millet starch modification is achieved by processing it with approaches such as acetylation, cross-linking, etherification, oxidation, heat-moisture treatment, high-pressure treatment, sonication, gamma irradiation, enzymatic treatment, or their combinations. These approaches will improve the industrial suitability, structural and functional characteristics of finger millet starch for specified utilization as edible coating on foods for self-life enhancement, packaging film, resistant starch-rich products as dietary supplements, carrier material in drug delivery, coating material as pH-responsive delivery of drug, disintegrating agent in tablets, etc. Since its modification was restricted to very few investigations, the prospect of tailored finger millet starch should be reconnoitered more in a diverse way as the call for functional foods is rising. Furthermore, exploration should be attentive on producing genetically altered finger millet and green chemical reagent-based modification of its starch instead of using conventional approaches of modifications.

1.11 Conclusion

Finger millet is gaining considerable attention in food industries. However, finger millet starch is still underutilized at a commercial stage compared to tuber starches. Investigation of the alteration of finger millet starch using green chemicals, physical, and other combined methods still has areas that require the attention of researchers for investigation. However, finger millet is explored for the development of bakery items and other ready-to-eat products; there are still areas for further investigation of its starch and other components in the food and nonfood industries. Beyond the food industries, finger millet and its starch can be employed in other industries, and a proportional investigation with the traditionally used commercial starches will undoubtedly provide new areas of its utilization. Considering the need of the era, this chapter ascertained a comprehensive overview of the various processing techniques, origin, and geographical distribution of finger millet, various methods used for the isolation of starch, and their impact on yield and other qualities. Further, the chapter provided an in-depth literature of various methods used to tailor the functionality of finger millet starch and how these modifications enable the finger millet starch suitable for various industries.

References

Abioye VF, Ogunlakin GO, Taiwo G (2018) Effect of germination on anti-oxidant activity, total phenols, flavonoids and anti-nutritional content of finger millet flour. J Food Process Technol 9:719

Abioye VF, Babarinde GO, Ogunlakin GO et al (2022) Varietal and processing influence on nutritional and phytochemical properties of finger millet: a review. Heliyon 8(12):e12310

Achaya KT (2009) The illustrated foods of India: A-Z. Oxford University Press, New Delhi

Adebowale KO, Afolabi TA, Olu-Owolabi BI (2005) Hydrothermal treatments of finger millet (*Eleusine coracana*) starch. Food Hydrocoll 19(6):974–983

Afolabi TA, Olu-Owolabi BI, Adebowale KO (2012) Functional and tableting properties of acetylated and oxidised finger millet (*Eleusine coracana*) starch. Starch-Starke 64(4):326–337

Alshehri AA, Younes NM, Kamel R, Shawir SM (2024) Characterization and potential health benefits of millet flour and banana peel mixtures on rats fed with a high-fat diet. Heliyon 10(20):e39424

Amarnath MS, Muhammed A, Antony AK et al (2023) White finger millet starch: physical modification (annealing and ultrasound), and its impact on physicochemical, functional, thermal and structural properties. Food Humanit 1:599–606

Ananthu R, Buvaneswaran M, Meena L, Sunil CK (2023) Microwave treatment effect on functional and pasting properties, and storage stability of white finger millet. J Food Process Eng 46(7):e14341

Antony U, Chandra TS (1998) Antinutrient reduction and enhancement in protein, starch, and mineral availability in fermented flour of finger millet (*Eleusine coracana*). J Agric Food Chem 46(7):2578–2582

Antony U, Sripriya G, Chandra TS (1996) Effect of fermentation on the primary nutrients in finger millet (*Eleusine coracana*). J Agric Food Chem 44(9):2616–2618

APEDA. Millet Portal. Agricultural and Processed Products Export Development Authority, Government of India. Available from: https://apeda.gov.in/IndianMillets. Accessed 7 July 2025

Arzoo N, Shukla AK et al (2024) Effect of processing techniques on functional, antioxidants, and structural properties of finger millet (*Eleusine coracana*) flour. Food and Humanity 2:100305

Azeez SO, Chinma CE, Bassey SO et al (2022) Impact of germination alone or in combination with solid-state fermentation on the physicochemical, antioxidant, in vitro digestibility, functional and thermal properties of brown finger millet flours. LWT-Food Sci Technol 154:112734

Babu K, Thakur RP, Upadhyaya HD et al (2013) Resistance to blast (*Magnaporthe grisea*) in a mini-core collection of finger millet germplasm. Eur J Plant Pathol 135(2):299–311

Babu AS, Mohan RJ, Parimalavalli R (2019) Effect of single and dual-modifications on stability and structural characteristics of foxtail millet starch. Food Chem 271:457–465

Backiyalakshmi C, Vetriventhan M, Deshpande S et al (2021) Genome-wide assessment of population structure and genetic diversity of the global finger millet germplasm panel conserved at the ICRISAT genebank. Front Plant Sci 12:692463

Balakumaran M, Nath KG, Giridharan B et al (2023) White finger millet starch:Hydrothermal and microwave modification and its characterisation. Int J Biol Macromol 242:124619

Balasubramanian S, Sharma R, Kaur J, Bhardwaj N (2011) Isolation, modification and characterization of finger millet (Eleucine coracana) starch. J Food Sci Eng 1(5):339

Banusha S, Vasantharuba S (2013) Effect of malting on nutritional contents of finger millet and mungbean. Am-Eurasian J Agric Environ Sci 13(12):1642–1646

Ceasar SA, Maharajan T, Krishna TA, Ignacimuthu S (2023) Finger millet (*Eleusine coracana (L.) Gaertn*). In: Farooq M, Siddique KHM (eds) Neglected and underutilized crops. Academic Press, pp 137–149

Chandrasekara A, Senanayake I, Kumari D, Shahidi F (2022) Effect of processing on the antioxidant activities of porridges and Pittu prepared from finger millets (*Eleusine coracana*). Food Prod Process Nutr 4(1):18

Chandrashekar A (2010) Finger millet: *Eleusine coracana*. In: Taylor SL (ed) Advances in food and nutrition research, vol 59. Academic, pp 215–262

Chaudhary S, Singh R, Mishra S et al (2025) Optimization and characterization of multigrain pasta enriched with malted finger millet and lentil flour. Food and Humanity 5:100652

Chauhan ES, Sarita (2018) Effects of processing (germination and popping) on the nutritional and anti-nutritional properties of finger millet (*Eleusine coracana*). Curr Res Nutr Food Sci 6(2):566–572

Dey BB, Choudhury M, Das M (2025) Impacts of soaking conditions on proximate composition of finger millet. J Adv Biol Biotechnol 28(7):364–370

Damar WK, Jwanya A, Haggai PT, Manggoel W, Damar I (2016) Finger millet (*Eleusine coracana L. Gaertn*): production, management and utilization status in geo-political northern plateau state, Nigeria. Greener J. Agron. Forest. Horticult. 4(1):012–018

Devi PB, Vijayabharathi R, Sathyabama S et al (2014) Health benefits of finger millet (*Eleusine coracana L.*) polyphenols and dietary fiber: a review. J Food Sci Technol 51(6):1021–1040

DeWet JMJ, Rao KP, Brink DE, Mengesha MH (1984) Systematics and evolution of *Eleusine coracana* (Gramineae). Am J Bot 71(4):550–557

Dey A, Dutta D, Singh A, Sit N (2024) Physicochemical and functional properties of starches isolated from different millets. Meas Food 15:100188

Dharshini SS, Meera M (2023) Effect of popping and milling on physical, chemical, structural, thermal properties and angle of repose of amaranth seed (*Amaranthus cruentus L.*) and finger millet (*Eleusine coracana L. Gaertn*) from Udhagamandalam. Appl Food Res 3(2):100306

Duke JA (1978) The quest for tolerant germplasm. In: Jung GA (ed) Crop tolerance to suboptimal land conditions, vol 39. American Society of Agronomy, Crop Science Society of America, and Soil Science Society of America, pp 1–61

Fang F, Li H, Cai Z et al (2025) Preparation and characterization of edible gelatin-chitosan films incorporated with finger millet (*Eleusine coracana L.)* polyphenols and its application in pork. Int J Biol Macromol 307:142178

Food and Agricultural Organisation (FAO) of the United Nations (1995)

Ganapathy KN (2017) Improvement in finger millet: status and future prospects. In: Patil JV (ed) Millets and sorghum: biology and genetic improvement. Wiley, pp 87–111

Gautam N, Garg S, Yadav S (2021) Underutilized finger millet crop for starch extraction, characterization, and utilization in the development of flexible thin film. J Food Sci Technol 58(11):4411–4419

Gautam N, Garg S, Yadav S (2024) Dry heat treatment of finger millet starch for thin films-studies on physicochemical, mechanical, and barrier properties. Starch-Stärke 76(11–12):2300097

Gebreyohannes A, Shimelis H, Laing M (2021) Finger millet production in Ethiopia: opportunities, problem diagnosis, key challenges and recommendations for breeding. Sustainability 13(23):13463

Gebreyohannes A, Shimelis H, Mashilo J et al (2024) Finger millet (*Eleusine coracana*) improvement: challenges and prospects—a review. Plant Breed 143(3):350–374

Ghode N (2023) Tribal methods for millets preservation and storage in Bastar District of Chhattisgarh. Pharma Innovation 12(2):3749–3753

Gomez MI, Gupta SC (2003) Millets. In: Caballero B (ed) Encyclopedia of food sciences and nutrition, vol 2, 2nd edn. Academic, pp 3974–3979

Goron TL, Bhosekar VK, Shearer CR (2015) Whole plant acclimation responses by finger millet to low nitrogen stress. Front Plant Sci 6:652

Gowthamraj G, Raasmika M, Narayanasamy S (2021) Efficacy of fermentation parameters on protein quality and microstructural properties of processed finger millet flour. J Food Sci Technol 58:3223–3234

Gupta N, Gupta AK, Gaur VS, Kumar A (2012) Relationship of nitrogen use efficiency with the activities of enzymes involved in nitrogen uptake and assimilation of finger millet genotypes grown under different nitrogen inputs. Sci World J 2012(1):625731

Hilu KW, De Wet JMJ (1976) Domestication of *Eleusine coracana*. Econ Bot 30(3):199–208

Hithamani G, Srinivasan K (2014) Effect of domestic processing on the polyphenol content and bioaccessibility in finger millet (*Eleusine coracana*) and pearl millet (*Pennisetum glaucum*). Food Chem 164:55–62

ICAR-IIMR (2018) Finger Millet Assessed from https://www.millet.res.in/technologies/brochures/Finger_Millet_Brochure.pdf

Juliano P, Reyes-De-Corcuera JI (2022) Food engineering innovations across the food supply chain: debrief and learning's from the ICEF13 congress and the future of food engineering. In: Juliano P, Knoerzer K, Sellahewa J, Nguyen MH, Buckow R (eds) Food engineering innovations across the food supply chain. Academic, pp 431–476

Jyotsna R, Soumya C, Indrani D, Venkateswara Rao G (2011) Effect of replacement of wheat flour with finger millet flour (*Eleusine corcana*) on the batter microscopy, rheology and quality characteristics of muffins. J Texture Stud 42(6):478–489

Katti VS, Kumar S, Malleshi NG (2008) Studies on the effect of milling finger millet in different pulverisers on physico-chemical properties of the flour. J Food Sci Technol 45(5):398–405

Kaur P, Annapure US (2024a) Rheological and gelling properties of atmospheric pressure cold plasma treated finger millet (*Eleusine coracana*) starch. Food Res Int 187:114418

Kaur P, Annapure US (2024b) Understanding the atmospheric cold plasma-induced modification of finger millet (*Eleusine coracana*) starch and its related mechanisms. Int J Biol Macromol 268:131615

Kaur H, Oberoi HK, Ganapathy KN, Bhardwaj R (2023) Effect of popping and malting processing techniques on physiochemical, antinutrients and antioxidant properties of millets flour. J Food Sci Technol 60(9):2370–2384

Kayastha S, Sahoo JP, Mahapatra M, Sharma SS (2024) Finger millet *(Eleusine coracana)* enhancement through genomic resources and breeding methods: current implications and potential future interventions. Planta 259(6):139

Krishna TA, Maharajan T, Roch GV et al (2020) Hybridization and hybrid detection through molecular markers in finger millet [*Eleusine coracana (L.) Gaertn.*]. J Crop Improv 34(3):335–355

Krishnan M, Prabhasankar P (2010) Studies on pasting, microstructure, sensory, and nutritional profile of pasta influenced by sprouted finger millet (*Eleusina coracana*) and green banana (*Musa paradisiaca*) flours. J Texture Stud 41(6):825–841

Krishnan R, Dharmaraj U, Manohar RS, Malleshi NG (2011) Quality characteristics of biscuits prepared from finger millet seed coat based composite flour. Food Chem 129(2):499–506

Krishnan R, Dharmaraj U, Malleshi NG (2012) Influence of decortication, popping and malting on bioaccessibility of calcium, iron and zinc in finger millet. LWT-Food Sci Technol 48(2):169–174

KSDA, ICAR-IIMR (2018) The story of millets. Karnataka State Department of Agriculture in association with ICAR-Indian Institute of Millets Research

Kumar A, Chawla HS, Jeena AS, Rohit (2019) Morphological characterization of finger millet germplasm collected from Uttrakhand hills for qualitative traits. J Curr Microbiol App Sci 8(9):2105–2109

Kumar P, Kaur C, Sethi S, Jambh KH (2020) Effect of extruded finger millet on dough rheology and functional quality of pearl millet-based unleavened flatbread. Cereal Chem 97(5):991–998

Kurien PP, Desikachar HSR (1966) Preparation of a refined white flour from Ragi (*Eleusine coracana*) using a laboratory mill. J Food Sci Technol 3(2):56–58

Lawal OS (2009) Starch hydroxyalkylation: Physicochemical properties and enzymatic digestibility of native and hydroxypropylated finger millet (Eleusine coracana) starch Food Hydrocolloids 23(2) 415–425. https://doi.org/10.1016/j.foodhyd.2008.02.013

Liu H, Fan H, Cao R et al (2016) Physicochemical properties and in vitro digestibility of sorghum starch altered by high hydrostatic pressure. Int J Biol Macromol 92:753–760

Mahajan P, Bera MB, Panesar PS, Chauhan A (2021) Millet starch: a review. Int J Biol Macromol 180:61–79

Maharajan T, Ceasar S, Krishna TPA, Ignacimuthu S (2021) Finger millet [*Eleusine coracana (L.)* Gaertn]: an orphan crop with a potential to alleviate the calcium deficiency in the semi-arid tropics of Asia and Africa. Front Sustain Food Syst 5:684447

Makhaye G, Mofokeng MM, Tesfay S et al (2021) Influence of plant biostimulant application on seed germination. In: Gupta S, Staden JV (eds) Biostimulants for crops from seed germination to plant development. Academic, pp 109–135

Makroo HA, Naqash S, Saxena J, Sharma S, Majid D, Dar BN (2021) Recovery and characteristics of starches from unconventional sources and their potential applications: a review. Appl Food Res 1(1):100001

Malik MK, Kumar T, Kumar V et al (2022a) Sustainable, highly foldable, eco-friendly films from Mandua starch derivative. Sustain Energy Technol Assess 53:102398

Malik MK, Kumar V, Sharma PP (2022b) Improvement in digestion resistibility of mandua starch (*Eleusine coracana*) after cross-linking with epichlorohydrin. ACS Omega 7(31):27334–27346

Malik MK, Kumar V, Singh J et al (2023a) Phosphorylation of alkali extracted mandua starch by STPP/STMP for improving digestion resistibility. ACS Omega 8(13):11750–11767

Malik MK, Kumar V, Singh J, Kumar P (2023b) Efficiency of phosphorylated mandua starch in matrix tablet for targeted release of mesalamine in colon. J Drug Deliv Sci Technol 81:104251

Malik MK, Kumar V, Kumarasamy V, Singh OP (2024) Film coating of phosphorylated Mandua starch on matrix tablets for pH-sensitive release of Mesalamine. Molecules 29(13):3208

Malleshi NG, Desikachar HSR, Tharanathan RN (1986) Physico-chemical properties of native and malted finger millet, pearl millet and foxtail millet starches. Starch-Stärke 38(6):202–205

Mbithi-Mwikya S, Van Camp J, Yiru Y, Huyghebaert A (2000) Nutrient and antinutrient changes in finger millet (*Eleusine coracana*) during sprouting. LWT-Food Sci Technol 33(1):9–14

Mbithi-Mwikya S, VanCamp J, Mamiro PR et al (2002) Evaluation of the nutritional characteristics of a finger millet based complementary food. J Agric Food Chem 50(10):3030–3036

McDonough CM, Rooney LW, Earp CF (1986) Structural characteristics of *Eleusine Corocana* (finger millet) using scanning electron and fluorescence microscopy. Food Struct 5(2):9

Meenatchi R, Loganathan M (2022) Millet storage and pest management. In: Anandharamakrishnan C, Rawson A, Sunil CK (eds) Handbook of millets-processing, quality, and nutrition status. Springer, Singapore, pp 49–61

Mirza N, Marla SS (2019) Finger Millet (*Eleusine coracana* L. Gartn.) Breeding. In: Al-Khayri JM, Johnson DV, Jain SM (eds) Advances in plant breeding strategies: cereals. Springer, pp 83–132

Mishra S, Yadav S, Vivek K et al (2025) Ultrasound-assisted isolation, quality characterization of finger millet (*Eleusine coracana*) starch, and comparison with conventional isolation methods. J Food Sci Technol 62:1173–1183

Mobolade AJ, Bunindro N, Sahoo D, Rajashekar Y (2019) Traditional methods of food grains preservation and storage in Nigeria and India. Ann Agric Sci 64(2):196–205

Mohan BH, Gopal A, Malleshi NG, Tharanathan RN (2005) Characteristics of native and enzymatically hydrolyzed ragi (*Eleusine coracana*) and rice (*Oryza sativa*) starches. Carbohydr Polym 59(1):43–50

Mohan H, Arya L, Verma M et al (2018) Morpho-molecular variation in Indian finger millet (*Eleusine coracana (L.)* Gaertn.) varieties and landraces. Indian J Plant Genet Resour 31(3):276–285

Murungweni KT, Ramashia SE, Mashau ME (2024) Effect of malting on physicochemical, antioxidant, and microstructural properties of finger millet (*Eleusine coracana*) flours. Food Sci Nutr 12(1):547–563

Nath KG, Pandiselvam R, Sunil CK (2023) High-pressure processing: effect on textural properties of food-a review. J Food Eng 351:111521

Navyashree N, Sengar AS, Sunil CK, Venkatachalapathy N (2022) White finger millet (KMR-340): a comparative study to determine the effect of processing and their characterisation. Food Chem 374:131665

Nayak PP, Das RP, Mahanta SK, Singh A (2024) Simultaneous folate fortification and pesticide residue degradation in finger millet (*Eleusine coracana*) via malting and *Lactiplantibacillus plantarum*-mediated fermentation. Food Biosci 1(62):105429

Negi T, Solanki D (2015) Tradition grain storage structures and practices followed by farm families of Kumaon region in Uttarakhand. Indian Res J Ext Educ 15:137–141

Neshamba SM (2010) Variability for drought tolerance in finger millet (*Eleusine coracana (L.) Gaertn.*) M.Sc. thesis, University of Zambia

Nickhil C, Singh R, Deka SC et al (2025) Exploring finger millet storage: an in-depth review of challenges, innovations, and sustainable practices. Cereal Res Commun 53:57–79

Nimbkar S, Raja V, Shanmugasundaram S et al (2022) Post-harvest treatments and storage of millets. In: Anandharamakrishnan C, Rawson A, Sunil CK (eds) Handbook of millets-processing, quality, and nutrition status. Springer, Singapore, pp 151–171

Oghbaei M, Prakash J (2012) Bioaccessible nutrients and bioactive components from fortified products prepared using finger millet (*Eleusine coracana*). J Sci Food Agric 92(11):2281–2290

Onyango C, Noetzold H, Bley T, Henle T (2004) Proximate composition and digestibility of fermented and extruded uji from maize–finger millet blend. LWT-Food Sci Technol 37(8):827–832

Onyango C, Noetzold H, Ziems A, Hofmann T (2005) Digestibility and antinutrient properties of acidified and extruded maize–finger millet blend in the production of uji. LWT-Food Sci Technol 38(7):697–707

Panwar S, Sindhu R, Gehlot R, Rekha et al (2024) Finger millet starch extraction: comparative analysis of solvent efficiency, yield, and colour attributes. Int J Adv Biochem Res 8(8):779–782

Paschapur AU, Joshi D, Mishra KK et al (2021) Millets for life: a brief introduction. In: Kumar A, Tripathi MK, Joshi D, Kumar V (eds) Millets and millets technology. Springer, Singapore, pp 1–32

Patel S, Dutta S (2018) Effect of soaking and germination on anti-nutritional factors of garden cress, wheat and finger millet. Int J Pure App Biosci 6(5):1076–1081

Pathare AM, Vyawhare IR, Singhal RS, Gokhale JS (2025) Malted finger millet is an effective and economical adjunct in brewing. Food Biosci 68:106367

Patil SS, Rudra SG, Varghese E, Kaur C (2016) Effect of extruded finger millet (Eleusine coracan L.) on textural properties and sensory acceptability of composite bread. Food Biosci 14:62–69

Pushparaj FS, Urooj A (2011) Influence of processing on dietary fiber, tannin and in vitro protein digestibility of pearl millet. Food Nutr Sci 2(8):895–900

Radchuk V, Radchuk R, Pirko Y et al (2012) A somaclonal line SE7 of finger millet (*Eleusine coracana*) exhibits modified cytokinin homeostasis and increased grain yield. J Exp Bot 63(15):5497–5506

Rambhai MA (2009) A contribution to ethnobotany of Mehsana District, North Gujarat (Doctoral dissertation, Ph. D. Thesis, Saurashtra University. http://etheses. saurashtrauniversity. edu)

Rao MV, Akhil KG, Sunil CK (2021) Effect of microwave treatment on physical and functional properties of foxtail millet flour. Int J Chem Stud 9(1):2762–2767

Roopa S, Premavalli KS (2008) Effect of processing on starch fractions in different varieties of finger millet. Food Chem 106(3):875–882

Saha S, Gupta A, Singh SRK, Bharti N (2011) Compositional and varietal influence of finger millet flour on rheological properties of dough and quality of biscuit. LWT-Food Sci Technol 44(3):616–621

Sastri BN (1939) Ragi, *Eleucine coracana* Gaertn. – a new raw material for the malting industry. Curr Sci 8(1):34–35

Sharma R, Gautam N, Giribabu G et al (2023) Synergistic effect of plasticizer and cross-linking agent on the properties of finger millet starch-based film. Starch-Stärke 75(9–10):2200251

Sherin AJ, Sunil CK, Chidanand DV, Venkatachalapathy N (2024) Structural, physicochemical and functional properties of high-pressure modified white finger millet starch. Int J Biol Macromol 261:129919

Shigihalli S, Ravindra U, Ravishankar P (2018) Effect of processing methods on phytic acid content in selected white finger millet varieties. Int J Curr Microbiol App Sci 7(2):1829–1835

Shobana S, Krishnaswamy K, Sudha V, Malleshi NG et al (2013) Finger millet (Ragi, Eleusine coracana L.): a review of its nutritional properties, processing, and plausible health benefits. In: Henry J (ed) Advances in food and nutrition research, vol 69. Academic, pp 1–39

Shukla K, Srivastava S (2014) Evaluation of finger millet incorporated noodles for nutritive value and glycemic index. J Food Sci Technol 51(3):527–534

Singh N, David J, Thompkinson DK et al (2018) Effect of roasting on functional and phytochemical constituents of finger millet (*Eleusine coracana L.*). Pharma Innovation 7(4):414–418

Sood S, Kumar A, Babu BK, Gaur VS, Pandey D, Kant L, Pattanayak A (2016) Gene discovery and advances in finger millet [*Eleusine coracana* (L.) Gaertn.] genomics—an important nutri-cereal of future. Front Plant Sci 7:1634. Sorghum and millets in human nutrition. In FAO Food and Nutrition Series, 27. Assessed 31 July 2025 https://www.fao.org/4/t0818e/t0818e00.htm#Contents

Sood S, Joshi DC, Chandra AK et al (2019) Phenomics and genomics of finger millet: current status and future prospects. Planta 250:731–751

Sripriya G, Antony U, Chandra TS (1997) Changes in carbohydrate, free amino acids, organic acids, phytate and HCl extractability of minerals during germination and fermentation of finger millet (*Eleusinecoracana*). Food Chem 58(4):345–350

Srusti SNR, Babu HNR, Rajeshwari N (2023) Survey study on traditional storage practices of finger millet. Indian J Appl Pure Biol 38(3):1160–1166

Sudha ML, Vetrimani R, Rahim A (1998) Quality of vermicelli from finger millet (*Eleusine coracana*) and its blend with different milled wheat fractions. Food Res Int 31(2):99–104

Sukumar A, Athmaselvi KA (2019) Optimization of process parameters for the development of finger millet based multigrain extruded snack food fortified with banana powder using RSM. J Food Sci Technol 56(2):705–712

Swamy SVS, Wesley BJ (2020) Traditional knowledge of post-harvest crop handling by tribal farmers of northern Andhra Pradesh. Ind J Ecol 47(2):383–389

Thapa S, Tamang JP (2004) Product characterization of kodo ko jaanr: fermented finger millet beverage of the Himalayas. Food Microbiol 21(5):617–622

Udeh HO, Duodu KG, Jideani AIO (2025) Effect of fermentation time on the physicochemical properties and phenolic composition of malted finger millet beverages. Food Sci Nutr 13(3):e4659

Umar M, Hassan M, Rungraeng N (2025) In vitro digestibility and physicochemical analysis of heat-moisture treated finger millet flour and starch. Sustainable Food Technol 3(1):215–226

Upadhyaya HD, Reddy VG, Sastry DVSSR (2008) Regeneration guidelines finger millet. International crops research Institute for the Semi-Arid Tropics (ICRISAT), Andhra Pradesh, India. Assessed 31 July 2025 from https://genebanks.cgiar.org/resources/publications/finger-millet-eng/

Ushakumari SR, Rastogi NK, Malleshi NG (2007) Optimization of process variables for the preparation of expanded finger millet using response surface methodology. J Food Eng 82(1):35–42

Vetriventhan M, Upadhyaya HD, Dwived SL et al (2016) Finger and foxtail millets. In: Singh M, Upadhyaya HD (eds) Genetic and genomic resources for grain cereals improvement. Academic, San Diego, pp 291–319

Vidyavati HG, Begum JM, Vijayakumari J (2004) Utilization of finger millet in the preparation of papad. J Food Sci Technol 41:379–382

Wang L, Chen J, Lu S et al (2022) Structural characterization, physicochemical properties and in vitro digestion of finger millet–resistant starch prepared by different methods. Int J Food Sci Technol 57(1):416–425

Warisa S, Debbarma A, Hazarika A, Nath AJ (2025) Millet cultivation in India: geographical distribution, nutrient management practices, and policy implications. Discov Soil 2(1):12

Yadav S, Mishra S, Vivek K, Mishra S (2024) Characterisation of finger millet (*Eleusine coracana*) starch isolated by conventional and ultrasound-assisted starch isolation methods. Int J Food Sci Technol 59(8):5710–5720

Yang Z, Chaib S, Gu Q, Hemar Y (2017) Impact of pressure on physicochemical properties of starch dispersions. Food Hydrocoll 68:164–177

Yasir KM, Bhasin A, Maibam BD, Sharma M (2024) Review on finger millet (*Eleusine coracana* L.): nutritional composition, modification, its effect on physicochemical, structural, functional properties, and its applications. J Food Compos Anal 135:106623

Yogananda KC, Ramasamy E, Vasantha Kumar S, Rangappa D (2021) Ragi (finger millet) starch-based gel electrolytes for dye-sensitized solar cell application. Bull Mater Sci 44(3):211

Zhang D, Mu T, Sun H, He J (2019) Effects of ifferent high hydrostatic pressure-treated potato starch on the processing performance of dough-like model systems. Food Res Int 120:456–463

Chapter 2
Foxtail Millet: Chemical Composition, Starch Isolation, Modification, Functionality, and Food/Nonfood Application

Dibya Ranjan Sahoo, Aparajita Priyadarshini, Monika Mishra, and Pradeep Kumar Naik

Abbreviations

WHO	World Health Organization
FAO	Food and Agricultural Organization
IYM	International Year of Millets
Vit	Vitamins
UHP	Ultra high pressure
AC	Amylose content
RDS	Rapidly digestible starch
H_2SO_4	Sulphuric acid
HNO_3	Nitric acid
BD	Breakdown
DHT	Dry heat treatment
LDL	Low-density lipoprotein
HDL	High-density lipoprotein
NaOH	Sodium hydroxide
HCl	Hydrochloric acid
WUE	Water use efficiency
RVA	Rapid viscosity analyzer
HMT	Heat–Moisture Treatment
H_3PO_4	ortho-phosphoric acid

D. R. Sahoo · M. Mishra · P. K. Naik (✉)
Centre of Excellence in Natural Products and Therapeutics, Department of Biotechnology and Bioinformatics, Sambalpur University, Sambalpur, Odisha, India

A. Priyadarshini
P.G. Department of Food Science Technology and Nutrition, Sambalpur University, Sambalpur, Odisha, India

© The Author(s), under exclusive license to Springer Nature Switzerland AG 2026
J. Singh, M. Kumar Malik (eds.), *Millets and Millet Starch: Chemical Composition, Functional Modification, and Applications*,
https://doi.org/10.1007/978-3-032-09502-2_2

PV	Peak viscosity
SB	Setback
CP	Cold plasma

2.1 Introduction

Millets, often called "ancient grains," have been integral to human diets for thousands of years. These small-seeded, hardy cereal crops have been a staple in many cuisines and have played a significant role in traditional medicine systems worldwide. Various cultures have recognized and utilized their medicinal properties for their numerous health benefits. In this exploration of millets in conventional medicine, we will delve into the rich history of these grains and how they have been harnessed for their therapeutic qualities. Among the most widely consumed cereal grains globally are millets, particularly in Asia and Africa's semi-arid and desert regions (India and China). Their excellent nutritional value and agro-industrial significance interest them.

Figure 2.1 shows there are seven main varieties of millet such as finger millet (*Eleusine coracana*), foxtail millet (*Setaria italic*), pearl millet (*Pennisetum glaucum*), little millet (*Panicum sumatrense*), porso millet (*Panicum miliaceum*), barnyard millet (*Echinochloa esculenta*), kodo millet (*Paspalum scrobiculatum*), each with distinct hues, forms, sizes, and growing regions. These small-seeded,

Fig. 2.1 Seven different types of millet and sorghum millet

round-shaped cereals, part of the Poaceae family, are the earliest and most likely the first cereal grains humans have used for domestic purposes (Sharma and Niranjan 2018). The sixth highest yielding grain in the world is millet. Approximately 815 million people globally suffer from hunger and malnutrition, according to a World Bank report. One of the main problems in developing nations is that they lack enough food, leading to sicknesses and fatalities worldwide (Vila-Real et al. 2017). A significant portion of human nutrition comes from millets, whose output has risen over the past few decades to keep up with the growing global population's dietary needs. All essential nutrients, including protein, fat, carbohydrates, minerals, vitamins, and bioactive substances, are abundant in millet. However, food preparation methods, including decortications/dehulling, soaking, germination/malting, milling, and fermentation, might affect cereal grains' nutrients, bioactive chemicals, and uses.

2.1.1 History of Millets and Foxtail Millet

A diverse group of small-seeded grasses, millet is grown all over the world as a cereal crop or grain for human consumption and fodder. The cultivation of millet has a long history; evidence indicates that it was one of the first domesticated crops, having been grown in East Asia and West Africa thousands of years ago (Ingle et al. 2023). Foxtail millet (*Setaria italica*) and broomcorn millet (*Panicum miliaceum*) have been cultivated in East Asia, especially China, since at least 6000 BC (Huang et al. 2016). These grains were among the first plants to be domesticated in this region. Archaeological evidence, including carbonized seeds and imprints found in pottery, indicates that millets were a staple crop of ancient Chinese civilizations. They played a crucial role in the diets of these populations long before the widespread cultivation of rice. The Yellow River region is often cited as a center of early millet cultivation (Doust et al. 2009). For thousands of years, pearl millet (*Pennisetum glaucum*) has been grown in West Africa, especially in the Sahel region. Its cultivation has been documented since at least 2500 BC. Because of its exceptional resistance to drought, pearl millet is a priceless grain in arid and semi-arid regions of Africa. For millions of people in the area, it continues to be a staple diet, offering a substantial supply of minerals, protein, and carbohydrates (Kumari et al. 2024). From their centers of origin, millets spread across the globe. Their cultivation did not require complex technologies, facilitating their spread and adoption among various cultures and civilizations. Millets were introduced into Europe by various migrations and trade connections. They became an important crop during the Middle Ages in some regions, although they later declined in importance with the rise of more productive cereals. Whereas millets were introduced to the Americas in pre-Columbian times, but they never became as important as in Asia and Africa (Turner 2022).

Millets have gained popularity again in recent years because of their sustainability, gluten-free qualities, and nutritional advantages. They are regarded as crops that, particularly in developing nations, can help fight malnutrition and ensure food security. In addition to its historic beginnings, millet cultivation is noteworthy for its contributions to agriculture and human civilization. Millets have thrived in conditions where other crops fail, making them a cornerstone of food security in many parts of the world. Their rich history underscores the importance of preserving and promoting diverse agricultural practices to ensure sustainable and resilient food systems. Indian mythological texts such as the Samhita, Purana, and Veda highlighted the nutritional value and cultural significance of millets. These grains were praised for their extraordinary capacity to flourish in unfavorable environmental circumstances, which made them a dependable option for farming. Andhra Pradesh, Tamil Nadu, Karnataka, and Maharashtra are the leading states for foxtail millet production.

2.1.2 Nutritional Importance of Foxtail Millet

Scientifically known as "*Setaria italica*", foxtail millet has been a beloved mainstay in Indian homes for centuries, along with Kangni or Kakum. It has an important role in Ayurveda, India's traditional medical system, in addition to being a dietary choice. Foxtail millet is revered in Ayurveda, sometimes referred to as the "science of life," for its many health advantages. This investigation dispels common misconceptions about this ancient grain and examines its historical significance, nutritional profile, Ayurvedic qualities, health advantages, culinary integration, and sustainability aspects (Tiwari et al. 2023a, b). The other vernacular names of foxtail millet are Kangni, Kakum (Hindi), Kanguni (Sanskrit), Tenai (Tamil), Navane (Kannada), Kangu (Odia), Korra, Korralu (Telugu), Thina (Malayalam), Kang, Rala (Marathi), Kang (Gujarati), Kaon (Bengali), and Kangni (Punjabi).

By preventing common digestive problems and guaranteeing effective nutrient absorption, the fiber content of foxtail millet supports healthy digestion. Research indicates that it contributes to cardiovascular health by lowering cholesterol and promoting heart health. A low-glycemic index option that helps with blood sugar regulation, foxtail millets raise blood sugar levels. Because it is gluten-free, it is a great option for people with celiac disease or gluten sensitivity and fits in with contemporary dietary preferences (Maurya et al. 2023).

The nutritional security of foxtail millet can be greatly increased since it contains various components, particularly starch, protein, vitamins, and minerals, which is appropriate for vegans and vegetarians trying to get enough protein. Vitamins like niacin, riboflavin, thiamine, and vitamin B6 are necessary for overall health and energy metabolism. High concentrations of vital minerals like calcium, magnesium, phosphorus, and iron are found in foxtail millets, which support a variety of bodily functions like bone health, oxygen transport, and energy production. Because of its higher concentration of vitamins and minerals than other cereals, it is a simple and affordable solution for nutritional adversity. Because of its nutritional makeup, foxtail millet has a number of health advantages. Its high fiber and low saturated fat content support cardiovascular health. It helps in the management of bodyweight because it contains dietary fibers and complex carbohydrates, supplying vital nutrients for good health, such as B vitamins for energy metabolism and iron to prevent anemia (Harish and Maraddi 2023). Phenolic acids, flavonoids, and tannins, in addition to macro- and micronutrients, were known to have antioxidant, antimutagenic, antiviral, and anti-inflammatory properties (Neeraja et al. 2017). Given its nutritional advantages, Pasricha et al. (2021) noted that foxtail millet is commonly used as an energy source by children, diabetics, and expectant and nursing mothers. The purpose of this chapter is to review the characteristics of foxtail millet, the processing methods to use it in food products, and also its health attributes.

2.2 Geographical Distribution

A staple crop in many parts of the world, including arid and semi-arid regions of China, parts of India, and Japan, foxtail millet (*Setaria*) is grown for silage and hay in South and North America. The United Nations General Assembly declared 2023 to be the International Year of Millets (Mohapatra et al. 2024) in March 2021 during its 75th session. FAO is the primary organization responsible for commemorating the year in collaboration with other relevant stakeholders (Mohapatra et al. 2024). Millets can be grown on arid fields with little help and are resilient to climate change. They are therefore the ideal means for countries to reduce their reliance on imported cereal grains and increase their level of self-sufficiency (www.fao.org). According to the latest data from the US Department of Agriculture, global millet production will total 30.802 million mt in 2023/2024. India is the largest producer of millet in the world, with an average annual production of around 12 million tons. It is followed by Niger, China, and Mali in terms of millet production in the world (https://ipad.fas.usda.gov/). Figure 2.2 represents the foxtail millet cultivation across the world.

The Economic Survey of India 2023 emphasized that India produces 80% of Asia's millets and contributes 20% to worldwide production. The average millets harvested in India are 1239 kg/ha, while the average international yield is 1229 kg/ha. In 2022–2023, the nation produced 17.32 million tons of millets (Shree Anna), with Rajasthan leading the list of states that make the most millet, accounting for 27% of the nation's total millet production. Maharashtra is followed by and contributes 15%, and Uttar Pradesh is the next with a contribution of 14% (https://pib.gov.in). From the various small millets grown in India, foxtail millet, kodo millet, and finger millet account for about 80% of production. Grown in dry lands with rainfed irrigation, foxtail millet, along with other small millets, is a highly drought-tolerant crop belonging to the Poaceae family. Tamil Nadu, Karnataka, and Andhra Pradesh are the Indian states that are central locations for foxtail millet cultivation. The nutritional value of foxtail millet is very noteworthy as 100 g of the grain contains 8 g of fiber, 12.3 g of protein, 60.9 g of total carbohydrates, 4.3 g of fat, 31 mg of calcium, 2.8 mg of iron, 290 mg of phosphorus as minerals, 3.3 g of vitamins, and 323–350 kcal of food energy (Pine 2020).

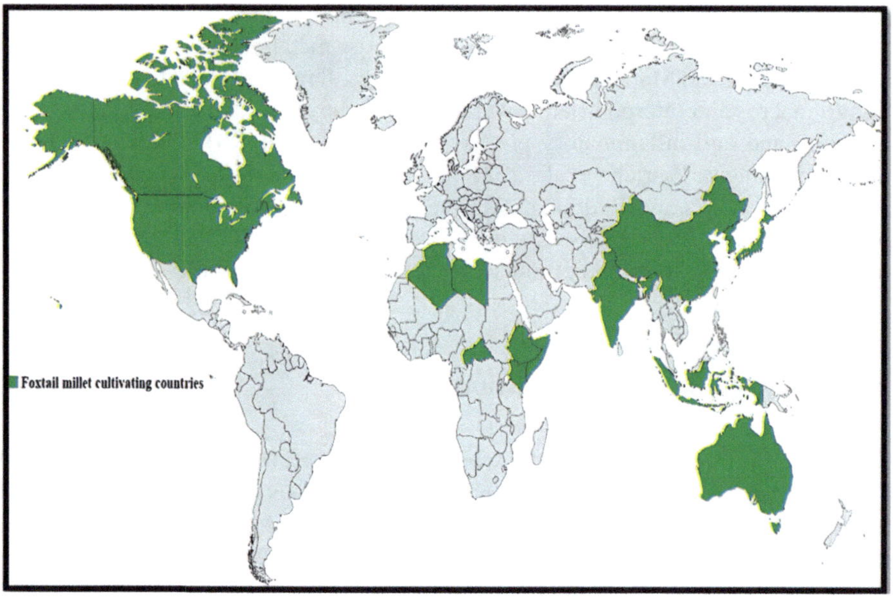

Fig. 2.2 Geographical distribution of Foxtail millet

The foxtail millet is a drought-tolerant crop that grows well in a wide range of solid conditions and climates and is highly esteemed as a short-duration crop that serves as a valuable food, feed, and fodder source. The production of this crop is viable in regions with little rainfall and at elevations ranging from sea level to 2000 meters above mean sea level (Diao and Jia 2017). It is primarily cultivated as a rain-fed crop but has favorable performance when provided with irrigation (He et al. 2015). A wide scale of foxtail millet cultivation was also seen in India and presented in Fig. 2.3.

2.3 Cultivation

The three decades leading up to 1960 saw the Green Revolution, during which rice and wheat were favored as the main staple food grain crops for consumption globally due to their large yields and immediate ability to alleviate food poverty. With an eye toward improving rice and wheat production worldwide, this concentration led to significant investments in irrigation infrastructure, the upgrading of management practices, the creation of hybridized seeds, artificial fertilizers, and targeted insecticides (Paarlberg 2013). Millets like foxtail millet, which can flourish with very minimal agricultural inputs in contrast to any other major cereal that would almost likely fail to survive, were favored over major cereals like rice and wheat in small-scale farming (Zhang et al. 2015; Padulosi et al. 2015). Because of these

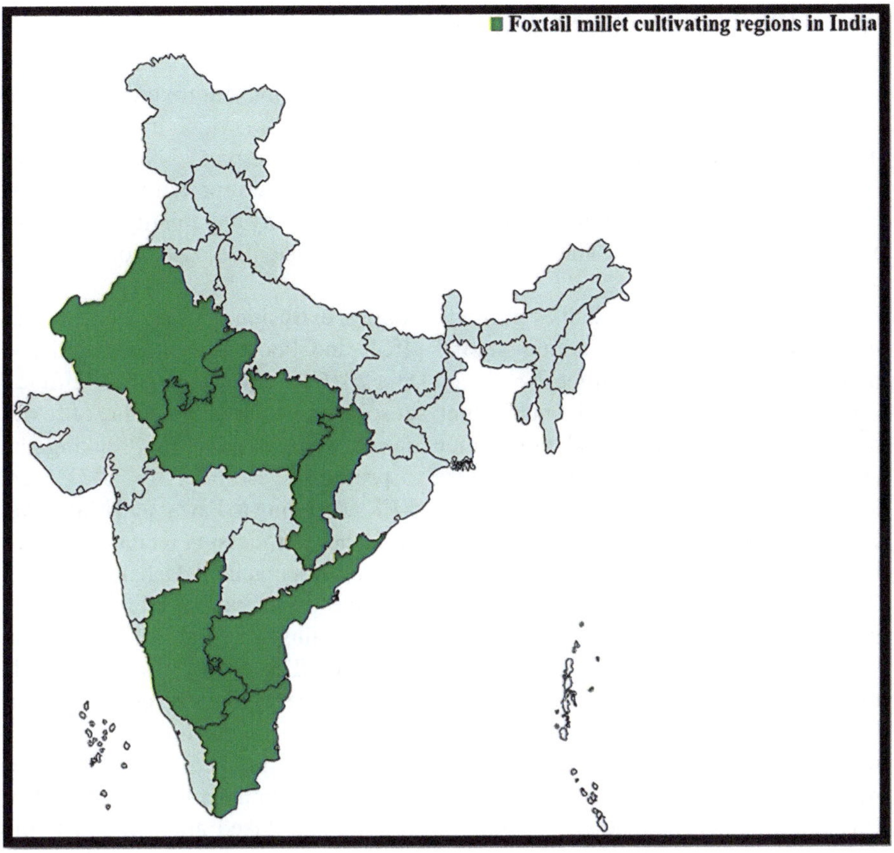

Fig. 2.3 Map of India representing the regions highlighted in green where the foxtail millet cultivation in practice

circumstances, the wealthier tribes exclusively cultivate foxtail millet seeds for bird feed and fodder, while the lower groups mostly consume the seeds (Chandrasekara et al. 2012). Therefore, the commercial development of foxtail millet has been hampered by underinvestment and perceived as a feed element. It is widely acknowledged that using millets, such as foxtail millet, could reduce malnutrition at a reasonable expense. The conversion of foxtail millet into food products with added value could potentially significantly boost the economic development of several Asian and African countries. Ironically, foxtail millet and other millets are grown in large quantities in the countries with the highest incidence of malnutrition, suggesting a significant underutilization of their potential as a food source. Foxtail millet is produced in large quantities in China and India. With 4.6 million tons produced yearly, China dominates the world in foxtail millet production, accounting for 15–17% of China's overall dietary intake.

2.4 Storage and Processing

Foxtail millet must be dehulled before processing or eating. But foxtail millet without a shell ages poorly, fades quickly, and smells bad, all of which have an adverse effect on the grain's appearance and nutritional value (Sharma and Niranjan 2018). Certain phytochemicals present in foxtail millet have been reported to have strong antinutrient effects (1), which may lower the bioavailability and quality of nutrients (Devisetti et al. 2014). The grain has an inconsistent distribution of these antinutrients.

Processing techniques such as milling, cooking, extrusion, germination, and fermentation can reduce the antinutritional effect and boost the bioavailability of micronutrients in foxtail millet (Nazni and Devi 2016). An overview of the different techniques used to improve the nutritional, sensory, and commercial value of foxtail millet is given in Table 2.1. The fundamental procedures of dehulling, soaking, and heating, for example, increased the in vitro protein digestibility by 32.71% and decreased the antinutrient content by almost 50%, according to Pawar and Machewad (2006). Numerous experts have looked into how these processing techniques affect the characteristics of foxtail millet. Using a rice milling machine, Ushakumari et al. (2004) decorticated the foxtail millet grains. Native grains were used for the roasting and popping process, while the decorticated grains were subjected to further processing by roller drying, extrusion, and flaking. The results showed that the popped grains had the best nutritious composition since the bran layer did not burn during roasting. Nazni and Devi (2016) obtained findings on nutritional characteristics that were similar. Functional characteristics of both popped and flaked grains included equal swelling powers, solubility, and capacity for absorbing water and oil. This could be because the starch granules in burst and flaked grains are changing structurally. Processed items made from foxtail millet can be successfully made using any of these processing techniques.

2.4.1 *Milling*

Cereal grains are polished and dehulled during the post-harvest processing stage. A large amount of husk as well as bran on the outer pericarp layer is present in foxtail millet grain. The pericarp layer is typically removed with a wooden mortar and pestle (Chandrasekara et al. 2012) or a sheller and polisher (Devisetti et al. 2014). It was discovered that the nutritious, antinutritious, and bioactive components of different milled fractions of foxtail millet, such as whole, brown, and polished grain flours, varied noticeably. It was found that the brown grain flour fraction had a higher total protein content and the whole brown grain flour was revealed to be extremely rich in minerals. Whole millet flours had higher levels of insoluble dietary fiber, which is made up of lignin, cellulose, and hemicellulose, while brown grain flours had higher levels of soluble dietary fiber. Proanthocyanidin, total flavonoid

Table 2.1 The impact of various processing techniques on characteristics of Foxtail millet

Sl. no.	Method	Objective	Results	References
1	Roasting, popping, and flaking	Evaluation of the possible changes in the volatile compounds due to roasting	↑ Aldehydes and heterocyclics; ↓ hydrocarbons and benzene derivatives; improved aroma feature	Jingke et al. (2014)
		Optimization of the popping conditions for FM grains	Finest popping yield at 230 °C with higher grain improvement and doneness quality	Choudhury et al. (2011)
2	Fermentation	Combined fermentation and HMT on FM flour	↑ Crude protein; ↓ carbohydrates; better starch quality and pasting properties	Amadou et al. (2014)
		FM fermentation using *Lactobacillus paracasei* FN032	↑ Antioxidant and antimicrobial actions; supplemented with bioactive peptides	Amadou et al. (2013)
		Effect on macronutrient profile	Better protein quality; appropriate for varied food product development	Antony et al. (1996)
		Microbial fermentation using probiotic yeast and LAB	Co-inoculation (*S. boulardii* + *L. acidophilus*) improved physicochemical characters; *L. acidophilus* alone ↑ mineral bioavailability and ↓ phytic acid	Pampangouda et al. (2014)
3	Germination	Calibration of malting procedure	Max. sprouting at 20h soaking + 72h germination (25–30°C); ↓ protein and fat; ↑ starch digestibility	Choudhury et al. (2011)
		Assessment of nutrient shifts post-germination	↓ Fat, FFA, carbs; ↑ TSS, TA, antioxidant capacity, minerals, vitamins	Coulibaly and Chen (2011)
		Result of controlled malting	Improved nutritional importance at 12h soaking + 48h sprouting (25°C)	Laxmi et al. (2015)
		Optimization of sprouting parameters	Soaking (15.84h) + sprouting (40h at 25°C) → ↑ phenolics, flavonoids, antioxidants, protein, minerals; ↓ antinutrients	Sharma et al. (2015)
4	Extrusion	Development of FM-based extruded snacks	Ideal extrusion at 118.23 °C, 400 rpm, 15.88% feed moisture → formed high-quality extrudates	Kharat et al. (2015)
		Structural integrity of modified FM flours throughout extrusion	Fiber, fat, and protein elimination condensed product roundness, wall width, and exact volume	Zhao and Wei (2014)

(continued)

Table 2.1 (continued)

Sl. no.	Method	Objective	Results	References
5	Dehulling, soaking, and cooking	Effect of alkaline cooking on nutritional profile	↑ Protein, ash, carbs, polyphenols; ↓ fat and antioxidant activity	Rajeswari et al. (2015)
		Influence of dehulling, soaking, and thermal action	↓ Protein, phytate phosphorus, polyphenols, iron, and zinc; ↑ bioavailability of ionizable iron and soluble zinc	Pawar and Machewad (2006)

*FM, foxtail millet.

content, and total phenolics content were found to be higher in the whole grain fraction of the antinutritional components, but phytic acid was found to be higher in the brown grain fraction. Devisetti et al. (2014) found that foxtail millet flour has unique functions that may be applied to the creation of new products. Numerous other studies have shown that the milled fraction with a higher proportion of hulls may be used as a potential source of antioxidants due to its significant phenolic content (Chandrasekara et al. 2012; Shejawale et al. 2016).

2.4.2 *Fermentation*

Many Asian and African nations have processed different cereals using traditional fermentation techniques. In many of these cultures, these extremely nutrient-dense fermented cereal products are staple diets (Antony et al. 1996). Fermentation is a practical, simple, and affordable way to improve the nutritional value, as well as the functional and sensory qualities, of foxtail millet (Amadou et al. 2014). A variety of microbial strains, including *Saccharomyces boulardii* and *Lactobacillus acidophilus, Lactobacillus paracasei, Lactobacillus acidophilus, Lactobacillus rhamnosus, Bifidobacterium bifidum,* and *Bifidobacterium longum* (Farooq et al. 2013), have been used to ferment foxtail millet flour.

Amadou et al. (2014) also reported that the phytochemical components, thermal properties, starch digestibility, and pasting qualities of foxtail millet flour were significantly impacted by the combination of fermentation and high-moisture treatment. These scientists came to the conclusion that high-moisture treatment had a substantial impact on rapidly digesting starch, slowly digesting starch, and resistant starch. It also affected the tryptophan and riboflavin spectra as demonstrated by fluorescence spectroscopy. The increased physicochemical and nutritional qualities, as well as antioxidant and antibacterial activity, have been demonstrated through the fermentation of foxtail millet employing different microbial strains. The presence of numerous antioxidative antimicrobial peptides has demonstrated that this technology can be successfully employed to manufacture alternative dietary supplements and food preservatives.

2.4.3 Germination

The biochemical process of germination in seeds produces new chemicals and interconverts existing ones, enhancing the cereals' useful qualities. Because germination produces naturally occurring, lightly processed, extremely healthy goods that may also help minimize the hazard of illnesses, likely hyperglycemia and hyperlipidemia, it has recently gained interest among the general public. The most frequent result of germination on millet is a higher output of water-soluble protein that obstructs hydroxyl radicals (Bai et al. 2009).

Because foxtail millet germination can greatly increase the bioactive components of the grain, researchers are becoming increasingly interested in this process. In Sharma et al. (2015), the germination of foxtail millet seeds improved and changed the composition of the seed by making more bioactive substances like dietary fiber, proteins, minerals, antioxidants, total flavonoids, and total phenolics available and lowering antinutritional factors. According to Laxmi et al. (2015), foxtail millet's nutritious qualities were greatly enhanced by soaking it for around 12 hours and then letting it germinate for 48 hours. Based on the phytochemical and functional characteristics of foxtail millet, several researchers have developed ideal germination conditions. Choudhury et al. (2011) suggested that soaking for 20 hours and germination for 72 hours at temperatures ranging from 25 to 30^0 C produced foxtail millet with significantly lower fat and carbohydrate content, increased starch and protein digestibility, and negligibly reduced protein content. There have also been reports on the impact of germination, dry heat treatment (roasting), and wet heat treatment (boiling and pressure cooking) on different characteristics of foxtail millet flour (Nazni and Devi 2016). While the solubility of germinated foxtail millet flours was found to be the highest, there was no discernible difference between the raw and processed flours in terms of swelling power or solid loss.

According to the aforementioned research, soaking duration, germination temperature, and germination time all affect the functional qualities of foxtail millet. When all three of these factors are combined in the right ways, phytochemicals can result in major health advantages. To maximize germination, more investigation is required.

2.5 Appearance, Morphology, and Genetics

Approximately 125 species of the genus *Setaria* are found in warm-temperate zones across the globe, including *Setaria italica*, also known as foxtail millet. It is classified under the subfamily *Panicoideae*; this genus is part of the tribe *Paniceae*. It includes species of grains, wildflowers, and weeds with various life cycles, breeding strategies, and ploidy levels (Lata et al. 2023). Foxtail millet (*Setaria italica* (L.) P. Beauv.) is recognized as one of the earliest domesticated crops. This versatile plant, with significant agricultural and nutritional potential, possesses nine chromosome pairs ($2n = 2x = 18$) (Babele et al. 2022; Antony Ceasar and Maharajan 2022).

It ranks as the second-oldest cultivated millet species. Its wild ancestor, green foxtail (*Setaria viridis* (L.) P. Beauv.), is believed to have given rise to cultivated foxtail millet through domestication approximately 7400 years ago in northern China. Additionally, archaeological evidence of its early use has also been uncovered in parts of Europe and the Middle East (Sakamoto 1987; Doust et al. 2009).

A study involving 62 varieties of the *Setaria* genus, classified into five groups using RFLP analysis (with EcoR1V digestion), revealed significant genetic diversity. Using inter simple sequence repeat (ISSR) markers, researchers identified a high degree of polymorphism among 81 accessions from nine *Setaria* species across China (Wei et al. 2012). These findings support the idea that South China served as a major center of domestication for *Setaria* and also indicate substantial gene flow between accessions from diverse eco-geographical regions. The availability of a draft reference genome for foxtail millet, approximately 423 Mb in size, is a key resource for identifying functional genetic variants and advancing future breeding efforts. Furthermore, resequencing of 312 foxtail millet accessions has yielded a comprehensive genomic variation dataset (Li et al. 2021) (Fig. 2.4).

The taxonomic hierarchy of foxtail millet is as follows:

Kingdom:	Plantae
Subkingdom:	Tracheobionta
Super division:	Spermatophyta
Division:	Magnoliophyta
Class	Liliopsida
Subclass:	Commelinidae
Order	Cyperales
Family:	Poaceae
Genus:	*Setaria*
Species:	*italica*

Fig. 2.4 Foxtail millet panicle and grains

The architectural form of foxtail millet is typical of domesticated plants, with large inflorescences that mature almost simultaneously, and either a single stalk or several tillers. A fully grown foxtail millet plant has thin, upright stems with many leaves and can reach a height of 120–200 cm (about 4 to 6.5 feet). The stems are straight and slender with hollow sections between the joints. Its leaves are broad, smooth, and have no hairs. At the top of the plant, it produces bristly flower clusters (panicles) that are usually reddish or purplish and measure between 5 and 30 cm in length.

A schematic diagram of the longitudinal section of foxtail millet is described in Fig. 2.5 (Kalsi and Bhasin 2023). Cultivated millets in the *Setaria* genus are commonly called "foxtail" because their flower clusters (panicles) look like a fox's tail. The flower head is a compact panicle that often bends at the top. Its short branches make it look similar to a spike. The stiff bristles, which indicate branches that no spikelets have developed, are mixed in and crowded with the spikelets. There is only one yellow-pistilled flower per spikelet. It can produce hundreds of seeds from each flower cluster and has a short growth cycle—taking about 5–8 weeks to start flowering and 8–15 weeks to fully mature, depending on the variety (Singh et al. 2017).

Hundreds of tiny, convex seeds with a diameter of 2 mm can be produced by a single inflorescence. The seeds have a thin, papery covering that comes off easily during threshing. Seed color varies a lot between different types. In both greenhouses (with up to 100 plants per square meter) and in fields in warm or mild climates, the seeds—if not dormant—germinate easily (Dekker 2003). It is harvested in the summer and usually planted in late spring. However, after 65–70 days, harvesting for hay or silage (with a typical yield of 15,000–20,000 kg/ha of green matter or 3000–4000 kg/ha of hay), and after 75–90 days, was done. Its quick growth and efficient use of water make it well-suited for cultivation in dry and arid regions (Singh et al. 2017).

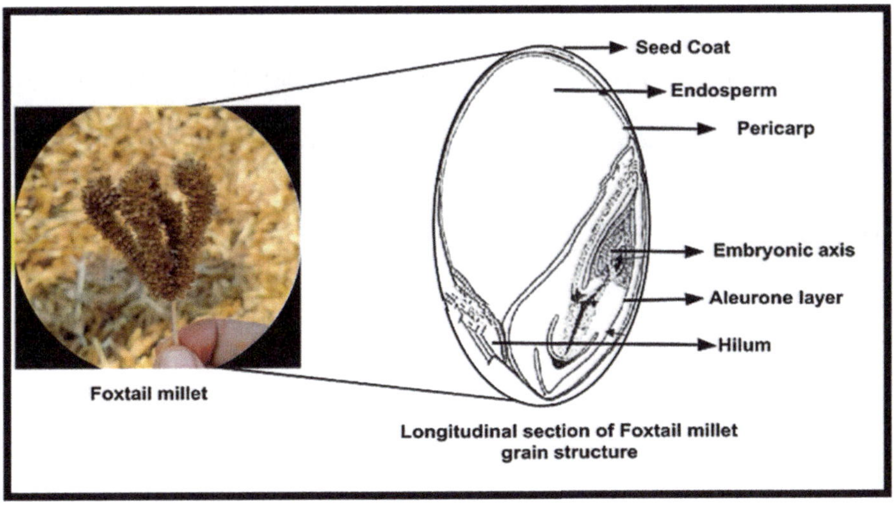

Fig. 2.5 Longitudinal section of a foxtail millet grain

2.6 Phytochemical Composition and the Health Attributes

Millets are easily digested, free of gluten, and do not add to acidity; they provide excellent nutritional value, offering high levels of calories, protein, healthy fats, essential vitamins, minerals, dietary fiber, and polyphenols. Therefore, it is also known as the powerhouse of nutrients. As a C4 photosynthetic crop, foxtail millet naturally exhibits high water use efficiency and nitrogen use efficiency. Diao et al. (2014) reported it also possesses a number of morpho-physiological traits, such as reduced leaf area, thicker cell walls, and compact and deep root systems, which were thought to provide long-lasting resistance to a variety of abiotic stresses, primarily heat, salinity, and drought (Lata et al. 2023). Moreover, unlike major cereals such as rice, wheat, and maize that need a minimum of 45% of their grain weight in water for germination, foxtail millet requires only 26%, making it significantly more water-efficient. Similar to this, foxtail millet needs the least amount of water 257 g compared to other cereals, as principal cereal crops like wheat and maize need 470 and 510 g, respectively, to produce 1 g dry biomass (Diao et al. 2014). These foxtail millet properties highlight its important agronomical traits and climate resilience. The nutritional composition of different millets (pearl millet, finger millet, porso millet, foxtail millet, kodo millet, little millet, and barnyard millet), rice, and wheat is shown in Table 2.2.

The morphological features of this millet (foxtail millet) are similar to those of other millet grain varieties, including the presence of pericarp layer (husk and bran). The husk makes up 13.5% (w/w), while the bran and germ combined make up only 1.5–2% (w/w). The plant grows rapidly, typically taking 5–8 weeks to flower after seeding and 8–15 weeks for its seeds to fully mature. Moreover, hundreds of seeds may be produced by each inflorescence. Table 2.2 lists a number of essential macronutrients that foxtail millet is high in, including protein, carbohydrates, and minerals. About 79% of foxtail millet grains are digestible, while the remaining

Table 2.2 Nutritional composition of different millets, rice, and wheat per 100 g (Longvah et al. 2017)

Millet type	Carbohydrate (g)	Protein (g)	Fats (g)	Crude fiber (g)	Mineral matter (g)	Energy (kcal)
Pearl millet	67.5	11.6	5.0	1.2	2.3	363
Finger millet	72.0	7.3	1.3	3.6	2.7	336
Porso millet	70.4	2.5	1.1	2.2	1.9	364
Foxtail millet	60.9	12.3	4.3	8.0	3.3	351
Kodo millet	65.9	8.3	1.4	9.0	2.6	353
Little millet	75.7	8.7	5.3	8.6	1.7	329
Barnyard millet	74.3	11.6	5.8	14.7	4.7	300
Rice	76.0	7.9	2.7	1.0	1.3	362
Wheat	71.0	11.6	2.0	2.0	1.6	348

Source: Indian Food Composition Tables, NIN—2017

indigestible portion of the grain contains a number of antinutritional ingredients and a comparatively high amount of fiber because of its coarse texture.

According to Muthamilarasan et al. (2016), foxtail millet grains are an excellent source of minerals (Fe, Ca, and Zn), dietary fiber (6–8%), crude fat (7–8%), and protein (10–15%). Similar to other millets, foxtail millet is a rich source of natural, unrefined dietary fiber. It has a favorable laxative effect on the digestive system because it facilitates bowel movement and aids in digestion (Sharma and Niranjan 2018). Because of its nutritional qualities, foxtail millet is used extensively in China to make noodles, fermented beverages, sweet dishes, and bakery products (Krishna et al. 2018; Yang et al. 2013). According to Zhang and Liu (2015), foxtail millet has many health benefits in addition to its nutritional qualities (Table 2.3), such as cancer prevention, hypoglycemic, and hypolipidemic effects.

2.6.1 Protein

Due to the diverse biochemical compositions across different millet varieties, the protein content in millet seeds constitutes a significant share of their total dry weight. For instance, the main storage proteins in foxtail millet are alcohol-soluble prolamines (Monteiro et al. 1982; Mohamed et al. 2009). Numerous researchers have also thoroughly examined the amino acid composition and other protein components of foxtail millet, proving that it is an excellent and reasonably priced source of protein for enhancing the flavor of food products. Protein constitutes approximately 11–15% of the dry weight of foxtail millet. It is clear from comparing this type of millet to other popular varieties that it contains more protein in addition to more essential amino acids and sulphur-containing amino acids like cysteine and methionine. (Tiwari et al. 2023a) reported the essential amino acids present are methionine (180 mg), cystine (100 mg), threonine (190 mg), leucine (1040 mg), isoleucine (480 mg), arginine (220 mg), histidine (130 mg), lysine (140 mg), **L**-tryptophan (60 mg), phenylalanine (420 mg), and valine (430 mg) per 100 g of millet.

Jingke et al. (2014) analyzed the protein composition of foxtail millet and identified the isoelectric points of its protein components across different varieties using the Osborne method, adapted from the modified approach by Lookhart and Bean (1995). Among the various protein types, albumin was the most abundant, followed by gliadin, globulin, glutein, and others. The study reported an average total protein content of 11.54 g per 100 g. The main protein stored in foxtail millet is alcohol-soluble prolamin (Monteiro et al. 1982) and reported as a primary source of protein by Kamara et al. (2009) in foxtail millet. The latter also stated that for various foxtail millet cultivars, the amino acid, prolamine, ranged from 41 to 77.5% of the total protein. Other scientists had analyzed the amino acid composition of foxtail millet and conducted an in-depth analysis of protein components. They found a high-quality, reasonably priced source of protein that can be added to food products to improve their nutritional content.

Table 2.3 Macro and micro nutritive values of foxtail millet compared to rice and wheat (Longvah et al. 2017)

Nutrients	Foxtail millet	Wheat (whole)	Rice (raw, milled)
Carbohydrates (g/100 g)	60.9	71.2	78.2
Protein (g/100 g)	12.3	11.8	6.8
Fat (g/100 g)	4.3	1.5	0.5
Energy (kcal)	331	346	345
Crude fiber (g/100 g)	8	1.2	0.2
Mineral matter (g/100 g)	3.3	1.5	0.6
Starch Composition			
Amylose (%)	17.5	25	–
Amylopectin (%)	82.5	75	88
Macro minerals (mg/100 g)			
Ca (mg)	31	41	10
P (mg)	290	306	160
Essential amino acid (mg/g of N)			
Arginine	220	290	480
Histidine	130	130	130
Lysine	140	170	230
Tryptophan	60	70	80
Phenyl alanine	420	280	280
Tyrosine	–	180	290
Methionine	180	90	150
Cystine	100	140	90
Threonine	190	180	230
Leucine	1040	410	500
Isoleucine	480	220	300
Valine	430	280	380
Vitamins (mg/100 g)			
Thiamin	0.59	0.41	0.41
Niacin	3.2	5.1	4.3
Riboflavin	0.111	0.1	0.04
Vitamin A (carotene) (μg/100 g)	32	64	0
Vitamin B6	–	0.57	–
Folic acid	15	36.6	8
Vitamin B5	0.82		–
Vitamin E	31	–	–
Micronutrients (mg/100 g)			
Fe	2.8	5.3	0.7
Zn	2.4	2.7	1.4
Mg	81	138	90
Na	4.6	17.1	–
K	250	284	–
Cu	1.4	0.68	0.14

(continued)

Table 2.3 (continued)

Nutrients	Foxtail millet	Wheat (whole)	Rice (raw, milled)
Mn	0.6	2.29	0.59
Mb	0.07	0.051	0.058
Cr	0.03	0.012	0.004
S	171	128	–
Cl	37	47	–
Fatty acids (%)			
Palmitic	6.4	24.5	15
Palmoleic	–	0.8	–
Stearic	6.3	1	1.9
Oleic	13	11.5	42.5
Linoleic	66.5	56.3	39.1
Linolenic	–	3.7	1.1

Source: Nutritive value of Indian foods (2017), National Institute of Nutrition, India

The amino acid study of foxtail millet protein by Kamara et al. (2009) and Mohamed et al. (2009) revealed that it was high in essential (lysine and leucine) and nonessential amino acids (glutamic acid, **L**-alanine, and aspartic acid). The essential amino acid composition of foxtail millet demonstrated that its amino acid content met FAO, WHO, and UNU standards and was even on par with soy protein concentrate (Mohamed et al. 2009). Another study by Kasaoka et al. (1999) detailed the amino acid composition of foxtail millet protein concentrate after enzymatic digestion. The findings showed that the protein concentrate was made from foxtail millet, which was digested using a heat-stable α-amylase, that contained comparatively more amino acids than other millets. The biological value, protein consumption, and digestibility of proteins were also covered by these writers. The authors demonstrated that foxtail millet protein concentrate served as a superior dietary supplement compared to other millet protein concentrates by comparing the outcomes in rats fed a millet-based diet with those fed a millet-free diet.

Therefore, the sum of total protein content in a cereal grain played a vital part in its suitability for human feeding (Sharma and Niranjan 2018). The protein in foxtail millet is easily digestible and efficiently utilized by the body.

2.6.2 *Crude Fat*

People primarily get their energy from fat, which has twice as many calories as carbs. Over 80% of the fat in millet is unsaturated fatty acid. According to Zhang et al. (2015) and Singh et al. (2010), distinct foxtail millet varieties planted in five different regions of China had fat contents ranging from 3.38 to 6.49%. Linoleic acid was the most prevalent fatty acid detected, followed by oleic acid, palmitic acid, stearic acid, and linolenic acid. According to Liang et al. (2010), the amounts of α-tocopherol, ɣ-tocopherol, and δ-tocopherol in crude millet oil are 15.53 ± 0.31 mg/100 g oil, 48.79 ± 0.46 mg/100 g oil, and 0.51 ± 0.06 mg/100 g oil, respectively.

The study concluded that breeding programs could effectively improve the fatty acid profile of foxtail millet as a correlation was observed between regional varieties and their fat content.

2.6.3 Carbohydrate

Numerous epidemiological studies have demonstrated that including more foxtail millet in regular meals reduces the likelihood of acquiring chronic illnesses, including type 2 diabetes mellitus and cholesterol metabolism. Foxtail millet demonstrated a dose-dependent decrease in fasting blood glucose levels in diabetic rats up to a dosage of 300 mg/kg body weight, beyond which the reduction plateaued at approximately 41% (Sireesha et al. 2011). These beneficial effects were largely attributed to its rich content of antioxidants, vitamins, minerals, phytochemicals, and other bioactive compounds, making foxtail millet a promising candidate as a functional food (Zhang and Liu 2015).

Complementary low glycemic index scores indicate that carbohydrates that are ingested and digested slowly are linked to a decreased risk of type 2 diabetes. Foxtail millet's high content and variety of fatty acids are mostly to blame for the starches' continued poor digestion. For instance, in foxtail millet, oleic acid significantly reduces the rate of starch hydrolysis. Additionally, complexes containing both starch and oleic acid exhibit greater resistance to enzymatic hydrolysis than those containing elaidic acid. Additionally, foxtail millet helps regulate hyperglycemia by preventing conformational changes near the trans form and minimizing fatty acid loss during processing (Sharma and Niranjan 2018).

2.6.4 Minerals

Studies show that the following mineral contents are found in 100 grams of foxtail: 290 mg of phosphorus, 31 mg of calcium, 2.8 mg of iron, 2.4 mg of zinc, 81 mg of magnesium, 4.6 mg of sodium, 250 mg of potassium, 1.4 mg of copper, and 171 mg of sulphur (Moharil et al. 2019; Nazni and Devi 2016). The functional quality of the cereal is influenced by the millets' high mineral content. Foxtail millet is a powerhouse of macrominerals such as calcium, phosphorus, magnesium, and sulphur, all of which aid in building stronger bones and muscles.

2.6.5 Vitamins

In general, the cereals are high in vitamins A, B, and E. The bran layer of cereals contains the majority of the vitamin E content, which is lost during grain preparation. Vitamin B12, which is abundant in foxtail millet, is essential for maintaining healthy skin and hair development, as well as for supporting neurological system function and cardiovascular health. In 100 grams of foxtail millet, the other vitamins that are present in a good amount are tocopherol (vit E) 31 mg, carotene (vit A) 32 mg, folic acid (vit B9) 15 mg, pantothenic acid (vit B5) 0.82 mg, thiamine (vit B1) 0.59 mg, niacin (vit B3) 3.20 mg, riboflavin (vit B2) 0.11 mg, and tocopherol (vit A) 32 mg (Bhatt et al. 2022; Yang et al. 2022).

2.6.6 Dietary Fiber

Food samples' dietary fiber is divided into two categories: soluble and insoluble fibers, which comprise polymers of carbohydrates such as lignin, polysaccharides, and oligosaccharides. In recent years, cereals rich in dietary fiber have been widely reviewed for their numerous health-promoting properties. According to Sharma and Sharma (2022), the study reveals that foxtail millet contains a handsome amount of total dietary fiber (15.20%), of which 10.52% are insoluble fibers and 4.68% are soluble fibers. The millet's arabinoxylan percentage is determined to be 1.52%, of which 1.26% are insoluble arabinoxylan. This demonstrates how a high insoluble fiber content might be advantageous when manufacturing bread because of its potent water-binding ability. Adequate consumption of insoluble dietary fiber can also benefit the health of the digestive system and cardiovascular system.

Bangoura et al. (2013) also showed that insoluble dietary fibers derived from foxtail millet could slow the diffusion of **D**-glucose and enhance its absorption in the digestive tract. These fibers may additionally inhibit α-amylase activity, delay carbohydrate breakdown, and reduce the release of **D**-glucose. This has been shown to increase insulin sensitivity and reduce the chance of developing diabetes mellitus (Kim et al. 2011). The dietary fibers in foxtail millet are not the only components that have a noticeable impact on plasma levels of lipid, **D**-glucose, insulin, and adiponectin. These levels are associated with risk factors, including high blood sugar (hyperglycemia), increased low-density lipoprotein (LDL) cholesterol, reduced high-density lipoprotein (HDL) cholesterol, and abnormal lipid profiles (dyslipidemia).

2.6.7 Antinutritional Compounds

A significant number of polyphenols, flavonoids, proanthocyanidins, and phytic acid were found in the pericarp of cereals. The majority grains' phytonutrient content may subsequently decline as a result of grain processing techniques. The phytonutrient availability of grains decreases as a result of dehulling and dehusking. It is commonly recognized that phenolic compounds interrelate with the proteins to form complexes, which decrease the proteins' solubility and make protein complexes less vulnerable to proteolytic attack than if the proteins were attacked alone (Sharma and Niranjan 2018). They also interact with proteolytic enzymes to reduce their activity and obstruct the digestion of starch and disaccharides.

The grains' phytotic acid concentration acts as an antinutritional factor, reducing the absorption of zinc, iron, and calcium. Additionally, during digestion, the tannins and saponins in the grains may prevent the body from absorbing minerals. The enzymes that are essential to digestion, such as pepsin, trypsin, and amylases, are likewise blocked by phytates. According to published research, foxtail millet contains 5.4–11.7% of phytic acid per gram of grain (Devisetti et al. 2014). Grain processing lowers the antinutritional elements. Grain phytonutrient content decreases upon cooking (Table 2.4).

The phenolic extract of foxtail millet has been shown to inhibit DNA scission by neutralizing hydroxyl radicals. Moreover, at a concentration of 0.5 mg/ml, it reduced the proliferation of HT-29 human colon cancer cells by 30–40% over a period of 72–96 hours. In a separate study, Zhang and Liu (2015) identified various phenolic acids in foxtail millet, including ferulic acid, p-coumaric acid, syringic acid, caffeic acid, and chlorogenic acid. These compounds exhibited dose-dependent antiproliferative effects against human HepG2 liver cancer cells and breast cancer cells.

Therefore, this millet was utilized to make a variety of low-GI foods that could result in a sensible improvement in the long-term management of cholesterol and diabetes. Before initiating any intervention trials with foxtail millet, it is essential to clearly determine its effects on hyperglycemia and hyperlipidemia.

Table 2.4 Antinutritional factors of different varieties of foxtail millet

Sl. no.	Foxtail millet variety fraction	Total phenolic content (TPC) (mg/g)	Total flavonoid content (TFC) (mg/g)	Proanthocyanidin (PC) (mg/g)	Phytic acid (mg/g)
1	PS-4 whole	1.70 ± 0.20	0.71 ± 0.01	0.59 ± 0.06	5.4 ± 0.3
2	PS-4 brown	0.81 ± 0.03	0.75 ± 0.06	0.16 ± 0.02	11.7 ± 0.1
3	PS-4 polished	0.50 ± 0.04	0.14 ± 0.00	0.06 ± 0.01	1.9 ± 0.1
4	SIA-3126 whole	1.83 ± 0.14	0.78 ± 0.03	0.79 ± 0.06	9.9 ± 0.1
5	IA-3126 brown	0.79 ± 0.03	0.65 ± 0.04	0.11 ± 0.01	11.0 ± 0.2
6	SSIA-3126 polished	0.58 ± 0.08	0.16 ± 0.00	0.08 ± 0.01	4.9 ± 0.1

Source: Devisetti et al. (2014)

2.7 Starch

In addition to being used in food products, starch is one of the valuable base materials in the paper, textile, and pharmaceutical sectors. It has been discovered that the chemical configuration, morphology, and physicochemical features of millet starch vary naturally (Dey and Sit 2017). Foxtail millet starch is a primary carbohydrate component, comprising roughly 61% of the dry mass and two major carbohydrate components in the ratio of 25:75, that is, amylose and amylopectin (Ushakumari et al. 2004). Amylose is a linear chain structure with a few branches arising from backbone, and amylopectin is a rather branched structure. The two vital molecular modules play a vital role in determining the biosynthesis, granular structure, functionality, and potential applications of starch. The common techniques that are used for the analysis of the surface morphology and structural dimensions of foxtail millet starch are scanning electron microscopy (SEM) and X-ray diffraction. The foxtail millet starch granules were mostly small or large polygonal (especially pentagonal), and rarely elliptical in shape with round edges and some indentations at the surface caused by the protein bodies. The structure of the starch granules resembles that of the maize starch (Fujita et al.1996). Among all other millets, foxtail millet has the largest starch granules (Annor et al. 2014).

This millet is categorized into three major types based on its amylose level: nonwaxy, low amylose concentration, and waxy. While the nonwaxy species is found worldwide, the waxy variety, which accounts for more than half of Taiwan's foxtail millet yield, is exclusive to East and Southeast Asia (Babu et al. 2019). One major aspect of cereal food processing is starch viscosity, which can be measured with the rapid viscosity analyzer (RVA). It is considered to be one of the useful tools for the assessment of the physicochemical properties of starch, which provides a vision into the eating and cooking quality of cereal grains (Cozzolino et al. 2016).

Granules of starch extracted from different plant sources can have a variety of forms, such as spherical, oval, or polygonal (Singh et al. 2003). Rice, sorghum, maize, oats, and other cereal crops are frequently found to have polygonal shapes. Including foxtail millets, several millets had starch granules with smaller spherical and small polygonal forms, measuring less than 10 μm in diameter (Krishna Kumari and Thayumanavan 1998). The shape of the starch granules is polygonal, with a small number of spherical ones. The average diameter of the starch granules in 53 different Chinese types ranged from 6.8 to 11.8 μm (Fujita et al. 1996; Annor et al. 2014). The internal cavity of the starch granule has a waxy accession and is identical to that of rice, maize, sorghum, and millet (Yin et al. 2019). The polygonal shape of the starch granules nonchalky endosperm has nonchalky endosperm with a high amylose content. There is a notable negative correlation between the cavity area and the amylose content. The amylose content (AC) in foxtail millet grains varies from 0 to 31.9%, while starch accounts for 60–70% of the grain's dry matter (Regina and Guzmán 2020). According to Yin et al. (2019), the amylose, amylopectin, rapid digestible starch, slow digestible starch, and resistant starch contents of foxtail millet are 35, 87.65–98.62, 13.10–26.80, 32.20–43.10, and 35.2–51.2%, respectively (Table 2.5).

Table 2.5 Characteristics and amylose content of foxtail millet starch

Sl. no.	Method	Amylose content (%)	Size (μm)	Shape	References
1	SEM and Amylose determination method	18.08	0.80–9.60	Small spherical; small polygonal (pentagonal)	Krishna Kumari and Thayumanavan (1998)
2	SEM and amylose determination method	3.30–11.40	L = 10.10–25.0 D = 4.70–12.50	Smooth with some fissures; A-type crystalline	Kim et al. (2009)
3	SEM, particle size analyzer, X-ray diffractometry, spectrophotometry	32.50	3.0–24.0	Polygonal and spherical; A-type crystalline	Annor et al. (2014)
4	SEM	11.40–27.10	6.80–11.80	Similar to maize starch granule	Fujita et al. (1996)

The digestion and absorption process of any functional component of food by the human body is crucial for the health-promoting characteristics of that component. Although heat processing enhances the digestion of starch, a portion is thought to become resistant to enzymes involved in the digestion process of mammals. The starch component, referred to as resistant starch, has a notable beneficial impact on human health, namely in terms of glucose metabolism, the treatment of chronic diet-related disorders, mental function, and weight control. Foxtail millet flour contains a very good proportion of resistant starch, ranging from 13% to 15%, which has particularly beneficial effects in the treatment of diabetes compared to other commonly consumed cereal grains like rice and wheat. According to Sharma and Gujral 2019, foxtail millet has the lowest rapidly digestible starch (RDS) response that can be due to the presence of high phenolic acid content that has a negative correlation with the postprandial glycemic response.

2.8 Modification of Starch

The starch component of the proso millet was found to be the highest, followed by foxtail millet, kodo millet, barnyard millet, and little millet, that is, 73.41 ± 3.69%, 66.20 ± 0.25%, 57.23 ± 0.37%, 52.12 ± 0.23%, and 50.97 ± 0.16%, whereas the browntop millet has the least starch content of 45.51 ± 0.46% with $p \leq 0.01$ (Akshitha et al. 2022).

2.8.1 *Isolation of Starch*

The wet milling process is typically used to separate millet starch. Depending on its chemical makeup and qualities, the grain is moistened for a specific amount of time in water (water, alkali, or acid). The specific extraction technique (enzymatic, alkaline, or acidic) has a big impact on the amount of starch produced. The processing and functionality of the grain are determined by the basic starch content and the soaking condition, that is, the neutral, alkaline, or acidic conditions (Fig. 2.6).

2.8.1.1 Alkaline Extraction Method

Using 0.3% sodium hydroxide (NaOH) solution, the grains are soaked for 24 hours at 4 °C in the alkaline steeping process. After the grains have been soaked, they are ground in a mill to a wet slurry and sieved through a sieve with a mesh size of 100. Subsequently, the samples undergo a 15-minute centrifugation at 3000 rpm, the supernatant is extracted, and the residual material is reconstituted in water. After three cycles of this washing process, HCl is used to nullify the slurry. The starch cake is dehydrated for 48 hours at 40 °C after being cleaned and neutralized (Sun et al. 2019).

2.8.1.2 Acid Hydrolysis Method

Using the acid steeping method, grains are soaked for 48 hours at 52 °C in a solution containing 0.15% sulphur dioxide (SO_2). After being soaked, the grains are ground in a blender, put through a 40-mesh sieve, and then given a water wash. After being smashed with a mortar and pestle, the leftover material is filtered first using a 270-mesh sieve, followed by 200-mesh sieve. Water is used for the washing of the residues, and the residues are further filtered with a No. 2 Whatman filter paper, followed by centrifugation and overnight drying at 45 °C (Zhu et al. 2014).

2.8.1.3 Enzymatic Hydrolysis Method

The extraction of starch by enzymatic method primarily requires the hydrolysis of proteins and cellulose. The enzymatic removal protein is carried out by protease. The enzymatic removal of protein is better compared to the thermal denaturation of proteins. The raw materials had to be chopped and ground, the cellulase enzyme and distilled water added, the solution had to be mixed, heated, filtered, and dried. The enzymes mainly required in the digestion of the starch are α-amylase, β-amylases, and glucoamylases. *Aspergillus oryzae* is the organism that is approved by the Food and Drug Administration to obtain amylases. *Aspergillus oryzae* is treated as the most efficient amylase in the scarification (Machida et al. 2008).

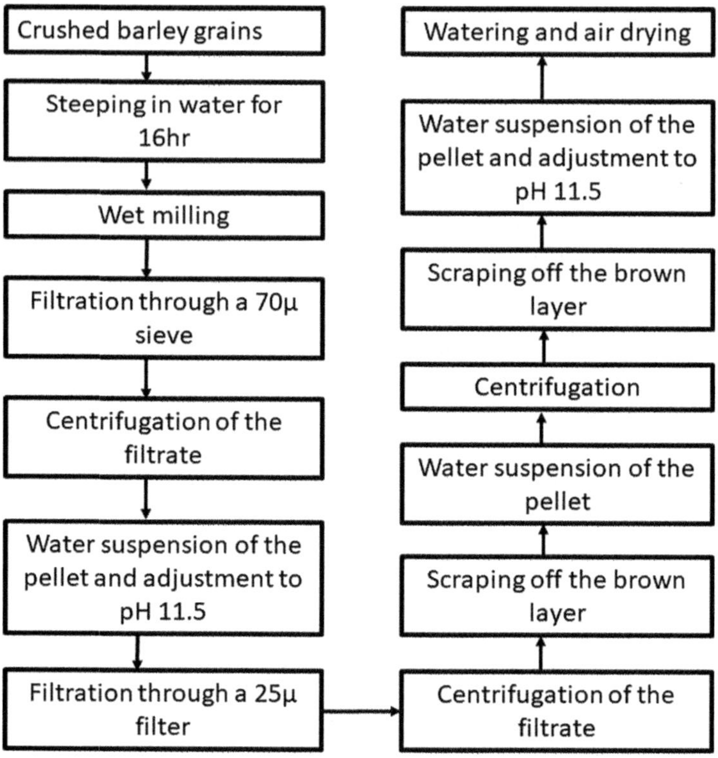

Fig. 2.6 General method of millet starch isolation

The starch content of millet is comparable to that of other cereal grains. Usually, 20–30% of the millet starch is amylose and 70–80% is amylopectin. Achieving certain intended functional goals is significantly impacted by the presence of contaminants in millet starch grains. For instance, the majority of the lipids in millet starch are polar and nonpolar. Polar phospholipids account for 89% of the total lipid composition, with nonpolar triglycerides making up the remaining portion. These lipids have the ability to form compounds with the starch's amylose component, which may reduce the starch's ability to expand and flow. The cohesive nature and hydrophobic bonding of the lipids are the source of this.

2.8.2 *Starch Structure Alteration*

The plant from which native starch is derived determines its composition and uses. Native starch has many drawbacks that prevent it from being used directly in the food and nonfood industries, chief among them being its high degree of retrogradation, low shear strength, and low thermal stability (Jyothi et al. 2006). Despite their pastes' weak gelling qualities, native starches may also be susceptible to syneresis, which increases coagulation and further lowers water content. The food, paper, and textile industries frequently use physical, chemical, and enzymatic methods to stimulate and increase certain functional qualities of native starch, such as PMS, in order to overcome this starch deficiency (Punia et al. 2021). Derivatization (such as esterification, etherification, and crosslinking), fragmentation (via acid hydrolysis and starch oxidation), treatments using enzymes (e.g., fermentation), and modifications using heat–moisture treatment (HMT) and ultra-high-pressure treatment (UHP) are the frequently used approaches for starch alteration.

2.8.2.1 Acid Treatment

Acid modification is a chemical procedure that involves using mineral acids to hydrolyze the α-glucan chains in starch. The structure and characteristics of native starch are changed as a result of this process, which breaks glycosidic linkages (Hoover 2000). An extensively used method for starch modification is acid hydrolysis. It includes subjecting starch to mineral acids, like HCl, H_2SO_4, H_3PO_4, and HNO_3, which promotes the formation of short starch chains—like those in amylose—to rise. The main drawback of this technique, despite its widespread use, is its strong dependence on chemical agents, which may have negative effects on the environment (Mahajan et al. 2021).

Acid modification not only modifies the arrangement of natural starch but also modifies its physicochemical qualities, enabling its application in a variety of industries, including food and textiles. Among other things, this modified starch is used to make paper, gum candy, cationic and amphoteric starches, and gum (Punia et al. 2021). Acid-modified starch has a lower water-binding ability than native PMS. According to some theories, this decrease results from acid alteration, shrinking the amorphous region's size and making the binding sites less accessible (Singh and Adedeji 2017).

2.8.2.2 Hydrothermal Treatment

The hydrothermal processing technique known as "heat-moisture treatment" (HMT) is used at high temperatures (90–120 °C) and subsequently lower moisture content (35% or less). In order to obtain a low moisture content, starch is heated to a measured temperature beyond its glass transition temperature for a predetermined amount of time—typically 1–24 hours. The intended result and particular treatment plan will determine how long the procedure takes (Hoover 2000). When hydrothermally the starch is heated with a water content of 40–65% at temperatures below the start of gelatinization, it is known as annealing (ANN). The crystalline portion's structural flaws are repaired as a result of this treatment's facilitation of the interaction and reassociation of the amylose and amylopectin chains inside the starch granules (Zhu et al. 2014). Heat-moisture treatment (HMT) upturns the pasting and gelatinization temperatures of proso millet starch (PMS), foxtail millet starch, while considerably dropping PMS swelling power, solubility, peak viscosity (PV), breakdown (BD), setback (SB) values, and gelatinization enthalpy (Kaur et al. 2026; Kumar et al. 2023; Zheng et al. 2020). The decrease in breakdown and setback standards donates to a better hot and cold paste firmness. Moreover, a reduction in swelling potential is helpful for certain uses, such as noodle manufacture. According to Kumar et al. (2023), HMT improves the in vitro digestibility of PMS by collectively increasing the scopes of rapidly digestible starch (RDS) and slowly digestible starch (SDS), while dropping the resistant starch (RS). The influence of HMT on starch digestibility is through several aspects, comprising (i) the type of starch source, (ii) moisture content, (iii) treatment temperature and period, and (iv) relations between different starch components, such as amylopectin–amylopectin, amylose–amylose, and amylose–amylopectin interactions (Pratiwi et al. 2018; Zeng et al. 2015).

2.8.2.3 Dry Heat Treatment

A physical change known as dry heat treatment (DHT) modifies the physicochemical characteristics of starch without affecting its granular structure. An easy, safe, and healthy alternative to chemical-based methods is dry heat. The new characteristics of thermally treated starch are comparable to those obtained through chemical crosslinking. Starch's pasting and functional qualities are enhanced by dry heat just as much as by the chemical substitute. Wet milling is the most common method for modifying the starch; however, DHT, having a moisture content of 9%, temperature at 130 °C, and for a period of 2 and 4 h, decreased the breakdown (BD) and pasting temperature (PT) and increased the final (FV) and setback (SB) viscosities compared to the native starch, with the effects increasing with treatment time. DHT-modified starch showed increased hot paste stability similar to chemically cross-linked starch by becoming more resistant to mechanical and thermal shear. Products requiring a higher ultimate viscosity and hot paste stability may employ DHT starch. Moreover, DHT caused a drop in the gelatinization enthalpy (ΔH) and an increase in the onset (To) and peak (Tp) gelatinization temperatures. A drop in crystallinity in the starch granule following DHT treatment may be the cause of a drop in ΔH (Sun et al. 2014).

2.8.2.4 Ultra-High-Pressure (UHP) Treatment

A nonthermal modification technique called UHP can be used to physically alter or cause gelatinization in a variety of starch varieties. When using UHP instead of traditional thermal processing, a number of variables affect the degree of gelatinization that can be achieved. These variables include the type of starch, pressure levels, water content, temperature ranges, and treatment duration. At specific ambient temperatures and appropriate pressure, UHP totally gelatinizes starch (Li et al. 2018).

There is not much research done on using UHP to alter the molecular structure or content of millet starch. Application of ultra-high pressure (UHP) action at pressures stretching from 150 to 450 MPa led to a growth in trough and final thicknesses as well as pasting temperature, while declining peak and breakdown viscosities associated to natural starch. However, proso millet starch preserved at the highest pressure of 600 MPa for 15 minutes showed a decrease in all paste viscosities during pasting (Li et al. 2018).

The applied pressure may have facilitated the penetration of liquid molecules into the starch particles, broken hydrogen bonds, and caused modifications to the starch structure. The starch treated at 600 MPa showed different pasting qualities than the starch treated at 150–450 MPa, suggesting that the starch underwent full gelatinization at greater pressure (Błaszczak et al. 2007; Li et al. 2018).

2.8.2.5 Fermentation

Modifying the chemical and physical properties of starch by means of fermentation is possible. This process enhances the starch's structural properties. A study on the effects of lactic acid bacteria (LAB) fermentation on the structure and physical characteristics of glutinous (waxy) PMS was carried out (Bianet al 2022). Microbial acids and enzymes hydrolyze the starch molecules, which are mostly found in the unstructured region, into smaller molecules during fermentation compared to the nonfermented starch, and the fermented starch has a reduced molecular weight, swelling power, and solubility, but a higher amylose concentration and crystallinity. After fermentation, all of the pasting characteristics of starch decreased, with the exception of the setback viscosity (SB), which increased. The gelatinization enthalpy (ΔH) of the fermented starch increased compared to that of the nonfermented starch, reflecting the rise in crystallinity after fermentation (Bian et al. 2022).

2.8.2.6 Dual Modifications

In contemporary science, dual modification of starches is becoming more and more common. This is due to the possibility that individual modification techniques will not always meet the functional specifications needed for industrial and culinary applications (Babu et al. 2019). The range of uses for starch can be expanded using dual modification, which allows for the customization of starch for particular claims and the improvement of single-modified starch's functionality. The impact of dual physical modification using cold plasma (CP) and ultra-high-pressure (UHP) on the characteristics and digestibility of starch (Sun et al. 2019). Cold plasma is a nonthermal, ecologically benign process that uses gas ionization to create a variety of free radicals that respond with it and cause both depolymerization and crosslinking of the starch without the need for chemicals (Chaiwat et al. 2016). Gelatinization and depolymerization of PMS as well as an increased concentration of resistant starch (RS) were the outcomes of applying dual modification with UHP at 600 MPa and CP. Furthermore, the dual modified starch showed reduced breakdown viscosity, suggesting that the paste was more resistant to heat and mechanical shear (Sun et al. 2022).

2.9 Food and Nonfood Applications

2.9.1 *Foxtail Millet Used as a Food Source*

Aside from Asia and Africa, foxtail millet is currently gaining more attention in other regions of the world where it can be consumed every day in a variety of forms such as drinks, porridge, and ready-to-eat snacks.

Functional components that are concentrated from milling foxtail millet can subsequently be used to produce infant foods, snack foods, dietary foods, and traditional sweets on a large scale (Subramanian and Viswanathan 2007). Moreover, you can use this gluten-free flour to partially replace wheat flour in recipes (Devisetti et al. 2014). Table 2.6 enumerates a few conventional and commercially manufactured food products made from foxtail millet that have been made up to this point using different unit operations. It is clear that a lot of studies have shown that dishes made with foxtail millet have extremely positive nutritional, physical, rheological, and sensory qualities. The growing consumer demand for gluten-free, high-fiber goods motivates stakeholders to concentrate more on food commodities derived from millet. The food industry has recently focused more on low-calorie, gluten-free cereal, which can be substituted for wheat with ease in baked goods. Foxtail millet contains a significant amount of slowly digesting starch, and it can be a key component of baked products and confections.

Increased physicochemical and nutritional qualities, as well as antioxidant and antibacterial activity, have been demonstrated through the fermentation of foxtail millet employing different microbial strains. The presence of numerous antioxidative antimicrobial peptides has demonstrated that this technology can be effectively employed to manufacture substitute dietary supplements and food additives (Amadou et al. 2014).

According to Sharma et al. (2015), the germination of foxtail millet seeds improved and changed the composition of the seed by making more bioactive substances, such as antioxidants, total flavonoids, dietary fiber, proteins, and minerals, available and lowering antinutritional factors. Soaking duration, germination temperature, and germination time all affect the functional qualities of foxtail millet, and a suitable combination of these factors can result in major health benefits in the form of phytochemicals.

In order to enhance the germination process, more studies are required.

Table 2.6 Food applications of foxtail millet

Sl. no.	Product	Properties	References
1	Biscuits and burfi	The product showed low glycemic index that is highly nutritious and acceptable	Thathola et al. (2011)
2	Semolina	Highly nutritious and acceptable based on the texture	Dharmaraj et al. (2016)
3	Probiotic dahi	Better firmness and syneresis; increased acidity; better sensory properties	Syama et al. (2015)
4	Yakju beverage	Better shelf-life as related to thermal treated drink; highly acceptable	Kang et al. (2005)
5	Sikhye beverage	Highly nutritious, rich source of antioxidants and acceptable	Jeong et al. (2014)
6	Foxtail millet bran oil	Linoleic acid was identified and it is considered to be a good source of tocopherols; fiber production can be done through brans	Liang et al. (2010)
7	Ready-to-eat flaked snack	↓ Carbohydrate; ↓ oil and water absorption; reduced unsaturated fatty acids	Ushakumari et al. (2004)
8	Ready-to-eat extruded snack	Highly nutritious and acceptable	Deshpande and Poshadri (2011)
9	Composite bread	Highly nutritious and ↓glycemic index	Chhavi and Sarita (2012)
10	Unfortified weaning mix	Highly nutritious; satisfactory rheological and sensory properties	Thathola and Srivastava (2002)

2.9.2 Foxtail Millet Used as a Nonfood Source

Egypt, the Middle East, South America, India, and the United States all produce millets for green fodder. Foxtail millet forage cultivars and hybrids are planted in India to generate superior quality fodder for feeding animals. Sorghum and foxtail millet crops come in a variety of multicut hybrid varieties that guarantee a steady supply of green fodder throughout the summer and kharif seasons. It is also popular in dairy farms across the nation. The interspecific pearl millet and Napier grass hybrid, well known as Napier bajra hybrid in India, is a perennial fodder crop. One of the main components of cow, pig, and poultry feed in the Western Hemisphere, China, and Australia is sorghum grain and foxtail millet. For forage purposes, sorghum is also widely farmed; in northern India, it is fed to animals either raw or in the form of silage or hay. Foxtail millet starch can serve as an ideal starch replacement for conventional starch sources. Modified foxtail millet starch can be used in pharmaceutical and food industries.

2.10 Conclusion

The food processing industries are concentrating on underutilized substances that have the potential to improve health in any way. So, recently, the rate of global population growth, feeding the entire population with the available cereals, will become increasingly challenging. Millets, along with foxtail millet, thrive under unfavorable agroclimatic circumstances like hardiness, shorter growth season, and resistance to drought. However, due to the impending threat of climate change, these qualities make them favorable crops. This cereal is appropriate for addressing food and nutrition security since it can yield food with significant health benefits. You can make a lot of different items with processed whole grain flours, such as porridge, biscuits, and breads. Knowledge of foxtail millet's nutritional makeup and possible uses that are now available in the literature is sufficient, but further investigation and advancement are needed to fully realize this plant's potential. Foxtail millet starch, can be used to generate films and other materials. One promising crop to address the current climate change challenges is foxtail millet, which can be utilized for biofortifying micronutrients. Antinutrients components are found in foxtail millet by nature but have the potential to impede nutrient absorption. However, a number of processing methods have been shown to reduce these adverse effects. Additionally, raising awareness of the importance of these millets for human well-being is strongly encouraged. Thus, foxtail millet value addition needs to be promoted in order to create incredibly nutrient-dense food products.

References

Akshitha R, Puskuri JS, Devi TS et al (2022) Comparative study of starch content in foxtail, little, Kodo, Proso, Barnyard and Brown top millets processed by traditional and modern methods. Biol Forum 14(3):1265–1270

Amadou I, Le GW, Shi YH (2013) Evaluation of antimicrobial, antioxidant activities, and nutritional values of fermented foxtail millet extracts by *lactobacillus paracasei* Fn032. Int J Food Prop 16(6):1179–1190

Amadou I, Gounga ME, Shi YH, Le GW (2014) Fermentation and heat-moisture treatment induced changes on the physicochemical properties of foxtail millet (*Setariaitalica*) flour. Food Bioprod Process 92(1):38–45

Annor GA, Marcone M, Bertoft E, Seetharaman K (2014) Unit and internal chain profile of millet amylopectin. Cereal Chem 91(1):29–34

Antony Ceasar S, Maharajan T (2022) The role of millets in attaining United Nation's sustainable developmental goals. Plants People Planet 4(4):345–349

Antony U, Sripriya G, Chandra TS (1996) Effect of fermentation on the primary nutrients in finger millet (*Eleusine coracana*). J Agric Food Chem 44(9):2616–2618

Babele PK, Kudapa H, Singh Y et al (2022) Mainstreaming orphan millets for advancing climate smart agriculture to secure nutrition and health. Front Plant Sci 13:902536

Babu AS, Mohan RJ, Parimalavalli R (2019) Effect of single and dual-modifications on stability and structural characteristics of foxtail millet starch. Food Chem 271:457–465

Bai Q, Chai M, Gu Z et al (2009) Effects of components in culture medium on glutamate decarboxylase activity and γ-aminobutyric acid accumulation in foxtail millet (*Setariaitalica* L.) during germination. Food Chem 116(1):152–157

Bangoura ML, Nsor-Atindana J, Zhu K et al (2013) Potential hypoglycaemic effects of insoluble fibers isolated from foxtail millets [*Setariaitalica* (L.) P. B eauvois]. Int J Food Sci Technol 48(3):496–502

Bhatt D, Fairos M, Mazumdar A (2022) Millets: nutritional composition, production and significance: a review. J Pharm Innov 11:1577–1582

Bian X, Chen JR, Yang Y et al (2022) Effects of fermentation on the structure and physical properties of glutinous proso millet starch. Food Hydrocoll 123:107144

Błaszczak W, Fornal J, Kiseleva VI et al (2007) Effect of high pressure on thermal, structural and osmotic properties of waxy maize and Hylon VII starch blends. Carbohydrpolym 68(3):387–396

Chaiwat W, Wongsagonsup R, Tangpanichyanon N et al (2016) Argon plasma treatment of tapioca starch using a semi-continuous downer reactor. Food Bioprocess Technol 9(7):1125–1134

Chandrasekara A, Naczk M, Shahidi F (2012) Effect of processing on the antioxidant activity of millet grains. Food Chem 133(1):1–9

Chhavi A, Sarita S (2012) Evaluation of composite millet breads for sensory and nutritional qualities and glycemic response. Malays J Nutr 18(1):89–101

Choudhury M, Das P, Baroova B (2011) Nutritional evaluation of popped and malted indigenous millet of Assam. J Food Sci Technol 48:706–711

Coulibaly A, Chen J (2011) Evolution of energetic compounds, antioxidant capacity, some vitamins and minerals, phytase and amylase activity during the germination of foxtail millet. Am J Food Technol 6(1):40–51

Cozzolino D, Degner S, Eglinton J (2016) The effect of the addition of emulsifiers on the pasting properties of barley grain and malt. Food Anal Methods 9:664–669

Dekker J (2003) The foxtail (*Setaria*) species-group. Weed Sci 51(5):641–656

Deshpande HW, Poshadri A (2011) Physical and sensory characteristics of extruded snacks prepared from Foxtail millet based composite flours. Int Food Res J 18(2)

Devisetti R, Yadahally SN, Bhattacharya S (2014) Nutrients and antinutrients in foxtail and proso millet milled fractions: evaluation of their flour functionality. LWT- Food Sci Technol 59(2):889–895

Dey A, Sit N (2017) Modification of foxtail millet starch by combining physical, chemical and enzymatic methods. Int J Biol Macromol 95:314–320

Dharmaraj U, Rao BS, Sakhare SD, Inamdar AA (2016) Preparation of semolina from foxtail millet (*Setariaitalica*) and evaluation of its quality characteristics. JCereal Sci 68:1–7

Diao X, Jia G (2017) Origin and domestication of foxtail millet. In: Genetics and genomics of Setaria, pp 61–72

Diao X, Schnable J, Bennetzen JL, Li J (2014) Initiation of Setaria as a model plant. Front Agric Sci Eng 1(1):16–20

Doust AN, Kellogg EA, Devos KM, Bennetzen JL (2009) Foxtail millet: a sequence-driven grass model system. Plant Physiol 149(1):137–141

Farooq U, Mohsin M, Liu X, Zhang H (2013) Enhancement of short chain fatty acid production from millet fibers by pure cultures of probiotic fermentation. Trop J Pharm Res 12(2):189–194

Fujita S, Sugimoto Y, Yamashita Y, Fuwa H (1996) Physicochemical studies of starch from foxtail millet (*Setariaitalica* Beauv.). Food Chem 55(3):209–213

Harish BP, Maraddi GN (2023) A study on adoption of foxtail millet production technology by the farmers in North Karnataka. Seed 65:54-16

He L, Zhang B, Wang X et al (2015) Foxtail millet: nutritional and eating quality, and prospects for genetic improvement. Front Agric Sci Eng 2(2):124–133

Hoover R (2000) Acid-treated starches. Food Rev Int 16(3):369–392

Huang P, Shyu C, Coelho CP et al (2016) *Setariaviridis* as a model system to advance millet genetics and genomics. Front Plant Sci 7:231871

Ingle KP, Suprasanna P, Narkhede GW et al (2023) Biofortified foxtail millet: towards a more nourishing future. J Plant Growth Regul 99(1):25–34

Jeong MS, Ko JY, Song SB et al (2014) Physicochemical characteristics of Sikhye (Korean traditional rice beverage) using foxtail millet, proso millet, and sorghum. J Korean Soc Food Sci Nutr 43(11):1785–1790

Jingke L, Wei Z, Shao-hui L et al (2014) Effect of roast on the volatile compounds in foxtail millet. Food Sci Technol 39:181–185

Jyothi AN, Moorthy SN, Rajasekharan KN (2006) Effect of cross-linking with epichlorohydrin on the properties of cassava (ManihotesculentaCrantz) starch. Starch-Stärke 58(6):292–299

Kalsi R, Bhasin JK (2023) Nutritional exploration of foxtail millet (*Setariaitalica*) in addressing food security and its utilization trends in food system. eFood 4(5):e111

Kamara MT, Ming ZH, Kexue Z (2009) Extraction, characterization and nutritional properties of two varieties of defatted foxtail millet flour (*Setariaitalica* L.) grown in China. Asian J Biochem 4(3):88–98

Kang YJ, Oh YJ, Koh JS (2005) Non thermal process and quality changes of foxtail millet Yakju by micro filtration. J Korean Soc Food Sci Nutr 34(2):277–284

Kasaoka S, Oh-hashi A, Morita T, Kiriyama S (1999) Nutritional characterization of millet protein concentrates produced by a heat-stable α-amylase digestion. Nutr Res 19(6):899–910

Kaur T, Panesar PS, Riar CS (2026) Effect of controlled hydrothermal treatment on the structural, functional and in vitro digestibility characteristics of foxtail millet grain starch. Eur Food Res Technol 252:42

Kharat S, Hiregoudar S, Beladhadi RV et al (2015) Optimization of extrusion process parameters for the development of foxtail millet based extruded snacks. Karnataka J Agric Sci 28(2):301–303

Kim SK, Sohn EY, Lee IJ (2009) Starch properties of native foxtail millet, *Setariaitalica*Beauv. J Crop Sci Biotechnol 12:59–62

Kim JS, Hyun TK, Kim MJ (2011) The inhibitory effects of ethanol extracts from sorghum, foxtail millet and proso millet on α-glucosidase and α-amylase activities. Food Chem 124(4):1647–1651

Krishna Kumari S, Thayumanavan B (1998) Characterization of starches of proso, foxtail, barnyard, kodo, and little millets. Plant Foods Hum Nutr 53:47–56

Krishna TA, Maharajan T, David RHA et al (2018) Microsatellite markers of finger millet (*Eleusine coracana* (L.) Gaertn) and foxtail millet (*Setariaitalica* (L.) Beauv) provide resources for cross-genome transferability and genetic diversity analyses in other millets. Biocatal Agric Biotechnol 16:493–501

Kumar SR, Tangsrianugul N, Sriprablom J et al (2023) Effect of heat-moisture treatment on the physicochemical properties and digestibility of proso millet flour and starch. CarbohydrPolym 307:120630

Kumari N, Prakash A, Tiwari P et al (2024) Origin, diversity, floral biology, pollination, and genetics in foxtail millet. In: Mishra S, Kumar S, Srivastava RC (eds) Genetic improvement of small millets. Springer, Singapore

Lata C, Gupta S, Prasad M (2023) Foxtail millet: a model crop for genetic and genomic studies in bioenergy grasses. Crit Rev Biotechnol 33(3):328–343

Laxmi G, Chaturvedi N, Richa S (2015) The impact of malting on nutritional composition of foxtail millet, wheat and chickpea. J Nutr Food Sci 5:407

Li W, Gao J, Saleh AS et al (2018) The modifications in physicochemical and functional properties of proso millet starch after ultra-high pressure (UHP) process. Starch-Starke 70(5–6):1700235

Li S, Zhao W, Liu S et al (2021) Characterization of nutritional properties and aroma compounds in different colored kernel varieties of foxtail millet (*Setariaitalica*). J Cereal Sci 100:103248

Liang S, Yang G, Ma Y (2010) Chemical characteristics and fatty acid profile of foxtail millet bran oil. J Am Oil Chem Soc 87(1):63–67

Longvah T, Ananthan R, Bhaskarachary K et al (2017) Indian food composition tables 2017, Indian Council of Medical Research, Hyderabad, Telangana, India.

Lookhart GE, Bean S (1995) Separation and characterization of wheat protein fractions by high-performance capillary electrophoresis. Cereal Chem 72(6):527–532

Machida M, Yamada O, Gomi K (2008) Genomics of *Aspergillus oryzae*: learning from the history of Koji mold and exploration of its future. DNA Res 15(4):173–183

Mahajan P, Bera MB, Panesar PS, Chauhan A (2021) Millet starch: a review. Int J Biol Macromol 180:61–79

Maurya R, Boini T, Misro L et al (2023) Comprehensive review on millets: nutritional values, effect of food processing and dietary aspects. J Drug Res Ayurvedic Sci 8(Suppl 1):S82–S98

Mohamed TK, Zhu K, Issoufou A et al (2009) Functionality, in vitro digestibility and physicochemical properties of two varieties of defatted foxtail millet protein concentrates. Int J Mol Sci 10(12):5224–5238

Mohapatra M, Mahapatra S, Das S, Jena AK, Mahapatra M, Sahoor JP (2024) International Year of Millets-Unleashing its Potential in Making a Sustainable World. In Millets and Other Potential Crops (pp. 1-9). CRC Press

Moharil MP, Ingle KP, Jadhav PV et al (2019) Foxtail millet (*Setariaitalica* L.): potential of smaller millet for future breeding. In: Al-Khayri JM, Jain SM, Johnson DV (eds) Advances in plant breeding strategies: cereals. Springer, Singapore

Monteiro PV, Virupaksha TK, Rao DR (1982) Proteins of Italian millet: amino acid composition, solubility fractionation and electrophoresis of protein fractions. J Sci Food Agric 33(11):1072–1079

Muthamilarasan M, Mangu VR, Zandkarimi H et al (2016) Structure, organization and evolution of ADP-ribosylation factors in rice and foxtail millet and their expression in rice. Sci Rep 6(1):24008

Nazni P, Devi RS (2016) Effect of processing on the characteristics changes in barnyard and foxtail millet. J Food Process Underw Technol 7(3):566

Neeraja CN, Babu VR, Ram S et al (2017) Biofortification in cereals: progress and prospects. Curr Sci 25:1050–1057

Paarlberg R (2013) Food politics: what everyone needs to know. Oxford University Press, Oxford

Padulosi S, Mal B, King OI, Gotor E (2015) Minor millets as a central element for sustainably enhanced incomes, empowerment, and nutrition in rural India. Sustainability 7(7):8904–8933

Pampangouda P, Munishamanna KB, Khanagoudar S, Niveditha SB (2014) Nutritional improvement of foxtail millet [*Setariaitalica* (L) Beauv] through fermentation using probiotic yeast and lactic acid bacteria. Biochem Cell Arch 14(2):329–333

Pasricha SR, Tye-Din J, Muckenthaler MU, Swinkels DW (2021) Iron deficiency. Lancet 397(10270):233–248

Pawar VD, Machewad GM (2006) Processing of foxtail millet for improved nutrient availability. J Food Process Preserv 30(3):269–279

Pine S (2020) Nutrient management in foxtail millet: a review. IJONS 10(60):23156–23160

Pratiwi M, Faridah DN, Lioe HN (2018) Structural changes to starch after acid hydrolysis, debranching, autoclaving-cooling cycles, and heat moisture treatment (HMT): a review. Starch-Stärke 70(1–2):1700028

Punia S, Kumar M, Siroha AK et al (2021) Pearl millet grain as an emerging source of starch: a review on its structure, physicochemical properties, functionalization, and industrial applications. Carbohydr Polym 260:117776

Rajeswari JR, Guha M, Jayadeep A, Rao BS (2015) Effect of alkaline cooking on proximate, phenolics and antioxidant activity of foxtail millet (*Setariaitalica*). World Appl Sci J 33(1):146–152

Regina A, Guzmán C (2020) Starch and starch-associated proteins: impacts on wheat grain quality. In: Igrejas G, Ikeda TM, Guzmán C (eds) Wheat quality for improving processing and human health. Springer, Singapore

Sakamoto S (1987) Origin and dispersal of common millet and foxtail millet. JARQ 21(22):84–89

Sharma B, Gujral HS (2019) Characterization of thermo-mechanical behavior of dough and starch digestibility profile of minor millet flat breads. J Cereal Sci 90:102842

Sharma N, & Niranjan K (2018) Foxtail millet: Properties, processing, health benefits, and uses. Food Rev Int 34(4):329–363

Sharma R, Sharma S (2022) Anti-nutrient & bioactive profile, in vitro nutrient digestibility, techno-functionality, molecular and structural interactions of foxtail millet (*Setariaitalica* L.) as influenced by biological processing techniques. Food Chem 368:130815

Sharma S, Saxena DC, Riar CS (2015) Antioxidant activity, total phenolics, flavonoids and antinutritional characteristics of germinated foxtail millet (*Setariaitalica*). Cogent Food Agric 1(1):1081728

Shejawale DD, Hymavathi TV, Manorama K, Zabeen F (2016) Effect of processing on nutraceutical properties of foxtail millet (*Setariaitalica*) varieties grown in India. J Food Meas Charact 10(1):16–23

Singh M, Adedeji AA (2017) Characterization of hydrothermal and acid modified proso millet starch. LWT Food Sci Technol 79:21–26

Singh N, Singh J, Kaur L et al (2003) Morphological, thermal and rheological properties of starches from different botanical sources. Food Chem 81(2):219–231

Singh KP, Mishra HN, Saha S (2010) Moisture-dependent properties of barnyard millet grain and kernel. J Food Eng 96(4):598–606

Singh RK, Muthamilarasan M, Prasad M (2017) Foxtail millet: an introduction. In: Prasad M (ed) The foxtail millet genome. Springer, Singapore

Sireesha Y, Kasetti RB, Nabi SA et al (2011) Antihyperglycemic and hypolipidemic activities of *Setariaitalica* seeds in STZ diabetic rats. Pathophysiology 18(2):159–164

Subramanian S, Viswanathan R (2007) Bulk density and friction coefficients of selected minor millet grains and flours. J Food Eng 81(1):118–126

Sun Q, Gong M, Li Y, Xiong L (2014) Effect of dry heat treatment on the physicochemical properties and structure of proso millet flour and starch. Carbohydr Polym 110:128–134

Sun J, Luu NS, Chen Z et al (2019) Generation and characterization of a foxtail millet (*Setariaitalica*) mutant library. Front Plant Sci 10:369

Sun X, Saleh AS, Lu Y et al (2022) Effects of ultra-high pressure combined with cold plasma on structural, physicochemical, and digestive properties of proso millet starch. Int J Biol Macromol 212:146–154

Syama A, Hiremath JP, Kempanna C (2015) Sensory and physico-chemical attributes of probiotic dahi incorporated with foxtail millet (*Setariaitalica*) flour. Indian J Nutr Diet 52:177–183

Thathola A, Srivastava S (2002) Physicochemical properties and nutritional traits of millet-based weaning food suitable for infants of the Kumaon hills, Northern India. Asia Pac J Clin Nutr 11(1):28–32

Thathola A, Srivastava S, Singh G (2011) Effect of foxtail millet (*Setariaitalica*) supplementation on serum glucose, serum lipids and glycosylated hemoglobin in type 2 diabetics. Diabetol Croat 40(1):23–29

Tiwari H, Singh PK, Naresh RK, Islam A, Kumar S, Singh KV & Shukla A (2023a) Millets based integrated farming system for food and nutritional security, constraints and agro-diversification strategies to fight global hidden hunger: a review. Int J Plant Soil Sci 35(19):630–643

Tiwari A, Misra A, Srivastava S (2023b) Trinadhanya as a new era super-food: revival of interest in traditional millet for their nutritional and therapeutic benefits. J Drug Res Ayurvedic Sci 8(Suppl 1):S42–S49

Turner PM (2022) Millet. DigiCat

Ushakumari SR, Latha S, Malleshi NG (2004) The functional properties of popped, flaked, extruded and roller-dried foxtail millet (*Setariaitalica*). Int J Food Sci Technol 39(9):907–915

Vila-Real C, Pimenta-Martins A, Maina N et al (2017) Nutritional value of indigenous whole grain cereals millet and sorghum. Int J Nutr Food Sci 4(1):1–5

Wei LI, Hui ZH, Yong-fang WA (2012) Assessment of genetic relationship of foxtail millet with its wild ancestor and close relatives by ISSR markers. J Integr Agric 11(4):556–566

Yang XS, Wang LL, Zhou XR et al (2013) Determination of protein, fat, starch, and amino acids in foxtail millet [*Setariaitalica* (L.) Beauv.] by Fourier transform near-infrared reflectance spectroscopy. Food Sci Biotechnol 22(6):1495–1500
Yang X, Tian Q, Yan J, Chen Y (2022) Characterizing root morphological traits in 65 genotypes of foxtail millet (*Setariaitalica* L.) from four different ecological regions in China. Agronomy 12(6):1472
Yin SY, Kuo SM, Chen YR et al (2019) Genetic variation of physicochemical properties and digestibility of foxtail millet (*Setariaitalica*) landraces of Taiwan. Molecules 24(23):4323
Zeng F, Ma F, Kong F et al (2015) Physicochemical properties and digestibility of hydrothermally treated waxy rice starch. Food Chem 172:92–98
Zhang LZ, Liu RH (2015) Phenolic and carotenoid profiles and antiproliferative activity of foxtail millet. Food Chem 174:495–501
Zhang A, Liu X, Wang G et al (2015) Crude fat content and fatty acid profile and their correlations in foxtail millet. Cereal Chem 92(5):455–459
Zhao X, Wei Y (2014) Effect of sequent extracting major components in foxtail millet on the properties of its expanded extrudates. Cereal Foods World 8:13–17
Zheng MZ, Xiao Y, Yang S et al (2020) Effects of heat–moisture, autoclaving, and microwave treatments on physicochemical properties of proso millet starch. Food Sci Nutr 8(2):735–743
Zhu C, Ming C, Zhao-shi X et al (2014) Characteristics and expression patterns of the aldehyde dehydrogenase (ALDH) gene superfamily of foxtail millet (*Setariaitalica* L.). PLoS One 9(7):e101136

Chapter 3
Fonio Millet: Chemical Composition, Starch Isolation, Modification, Functionality, and Food/Nonfood Applications

Rohit Bhardwaj, Vishnu D. Rajput, and Satyendra Kumar Rajput

Abbreviations

FM	Fonio millet
WF	White FM
BF	Black FM
Mb	megabase pairs
NFE	Nitrogen-free extract
NPK	Nitrogen, phosphorous, and potassium
D	Digitaria

3.1 Introduction

Nutrition is becoming increasingly popular today, yet many simple foods are processed to prioritize taste over nutritional value (Touvier et al. 2023). This shift is a major factor behind the growing weight gain challenge and numerous health-related ailments. Research has shown that healthy dietary habits, such as consuming

R. Bhardwaj · S. K. Rajput (✉)
Department of Pharmaceutical Sciences, Faculty of Pharmaceutical Sciences, Gurukula Kangri (Deemed to be University) Haridwar, Haridwar, Uttrakhand, India
e-mail: rohit.bhardwaj@gkv.ac.in; rohitbhardwaj.aip@gmail.com; Satyendra.rajput@gkv.ac.in; satyendraj@gmail.com

V. D. Rajput
Academy of Biology and Biotechnology, Southern Federal University, Rostov-on-Don, Russia
e-mail: rvishnu@sfedu.ru

© The Author(s), under exclusive license to Springer Nature Switzerland AG 2026
J. Singh, M. Kumar Malik (eds.), *Millets and Millet Starch: Chemical Composition, Functional Modification, and Applications*,
https://doi.org/10.1007/978-3-032-09502-2_3

various fruits, grains, legumes, and cereals, could mitigate approximately 71% of the diabetes cases and related fatalities (Popkin and Ng 2022). In this context, millets stand out as a particularly beneficial grain as they possess a high dietary content, a low glycemic index, and the ability to support sustainable agricultural practices, and thus improving human health and combating diet-related diseases (Jacob et al. 2024). Millets are associated with the *Poaceae* family and are among the earliest cultivated crops in human history (Kheya et al. 2023). One such traditional millet is fonio millet (FM), a small-sized cereal from the genus *Digitaria*, cultivated in West Africa. Unlike major cereals (rice and wheat), FM is a dormant grain with significant ability for food utilization (Zhu 2020). It has gained global interest for its use in food products and is widely advertised in the United States as a "nutritious" cereal. Currently, FM has gained approval of unique food grain by the European Union (Yadav and Upsana 2024). *Digitaria exilis* [white fonio millet (WFM)] and *Digitaria iburua* [black fonio millet (BFM)] are the two principal forms of FM. The former is known as fundi, hungry rice, and acha while the latter is referred to as iburu (Zhu 2020). Other associated types like *Digitaria compacta* (Raishan), *Digitaria sanguinalis* (hairy crabgrass), and *Digitaria eriantha* (Pangola) can enhance FM's agronomic traits, including yield, seed size, and diseases resistance through genetic improvements (Aguyoh and Masiunas 2003). FM's yield is relatively low, primarily because it is often grown on marginal lands with minimal inputs. However, ongoing breeding programs show promise for enhancing productivity, and fertilizers can also be employed to increase yields (Barnaud et al. 2022). Understanding FM's nutritional properties, dietary value, and culinary uses is essential for developing these breeding programs. Recent studies have covered FM breeding and established its genome (Abdul and Jideani 2019). FM thrives on surplus land and can withstand drought and less fertile soils. Its production has been increasing, reaching over 6 lacs tonnes per year (Animasaun et al. 2023). FM is involved in food safety, particularly climate change makes more lands unsuitable for major cereal crops. It may be adapted for large-scale cultivation in different areas of the world to enhance dietary safety and crop varieties expansion (Porcuna-Ferrer et al. 2024). Traditionally, FM is used in dishes like couscous, thin porridge, stiff porridge, and alcoholic beverages (Koreissi-Dembélé et al. 2013). There is a rising concern in incorporating FM into diverse food practices. Apart from the grains, different parts of the FM plant (husk, stalk, and leaves) could be utilized comprehensively. However, information on its chemical applications and nutritional benefit is scarce, with conflicting outcomes from different studies creating confusion among researchers and hindering FM's development as an innovative and resilient food crop. This chapter provides FMs' chemical, physicochemical, and dietary roles and applications. The implementation of these findings is discussed, and the applications of FM are reviewed. Further, this chapter will also explore the role of starch collected from the FM starch in pharma and other industries.

3.2 Geographical Distribution

FM, derived from the Malinke patois of Upper Niger and Upper Senegal, is among the oldest grain cultivated in West Africa (Kanlindogbe et al. 2020). The biological diversity of FM encompasses different cultivated varieties worldwide: *D. exilis* and *D. iburua* are primarily cultivated in West Africa (Jideani 2012). *D. sanguinalis*, also referred to as European millet or red manna, is native to Eastern Europe, while *D. cruciata*, locally known as Raishan, is grown in regions of India and Vietnam (Kanlindogbe et al. 2020). Among these, *D. exilis* and *D. iburua* are the most frequently mentioned in various studies, with *D. exilis* being the predominant species cultivated. Outside of Africa, *D. exilis* is also cultivated in the Dominican Republic, primarily for food manufacturing (Morales-Payán et al. 2002).

The origins of FM date back to 5000 BCE, with domestication beginning in the western region of Sudan (Vodouhe et al. 2007). WFM originates from the Niger Delta region, with Togo as a secondary center of diversification. BFM is associated with Hausa culture and is believed to have originated in Nigeria, spreading subsequently to Zinder in Niger, Benin, and Togo. In West Africa, FM is cultivated across the region stretching between the 8th and 14th northern parallels, extending from Senegal to Lake Chad. *D. exilis* is predominantly grown in countries such as Guinea, Mali, Burkina Faso, Nigeria, Benin, Togo, Senegal, and Guinea-Bissau. In contrast, *D. iburua* has a more limited geographical distribution, mainly concentrated in parts of Central Nigeria (Hitu et al. 1997).

3.3 Cultivation

FMs, among the oldest cultivated cereals, have gained attention for their resilience and nutritional benefits (Burton et al. 2024). Cultivation of FM begins with selecting well-drained, sandy, or loamy soils as this drought-tolerant crop thrives in poor soils where other cereals might fail (Cruz et al. 2016). FM requires minimal external natural fertility of the soil and the rainfall patterns. *Striga hermonthica* is a major pest for FM, affecting crops like millet, dry-land rice, sorghum, and maize (Ballogou et al. 2015). FM is often considered an “easy” crop due to minimal maintenance, primarily involving weeding (Abrouk et al. 2020). During its growth, FM fields change color from dark green to soft green and eventually to red, brown, and yellow at maturity, based on the variety. The Berele oule variety, for example, turns a rusty color when mature (McIntosh and McIntosh 1988). While FM generally faces minimal pest issues, farmers might need to monitor for insects, birds, or mammals like wart hogs. Chemical inputs are rarely used; in Middle Guinea, some farmers apply nitrogen, phosphorous, and potassium (NPK) fertilizer occasionally, but it is more common to use manure or leguminous crops to enhance soil fertility (Gigou et al. 2009). Overall, FM is predominantly grown organically, offering an advantage in markets that value organic produce (Ezekiel et al. 2012). During planting, seeds are typically broadcast over prepared soil at a rate of 1–2 kg per acre, and they germinate within 3–4 days. To achieve optimal yields, manual weeding is essential, ideally conducted twice; once around the second week and again by the fourth week after sowing (Seetharam et al. 1990). The plants typically reach up to 75 cm of height, and each panicle branches into 2 –4 rachis, each about 15 cm long. FMs, which are approximately 1 mm in size, develop a yellow hue upon ripening and are prone to easy shattering. The crop matures quickly, becoming ready for harvest within 2–3 months (Mirza and Marla 2019). Harvesting involves cutting the entire straw at the base using a sickle, followed by threshing. Post-harvest, the grains are manually pounded to eliminate the chaff and then ground into smooth flour, which is typically used to make gruel. Alternatively, the whole grains can be steamed and mixed with ingredients such as palm oil, butter, or milk to prepare a form of couscous (Seetharam et al. 1990). Although the tiny grain size makes processing labor-intensive, FM is nutritionally rich and offers various advantages, such as excellent taste, a short growth cycle, and consistent yields even in drought-prone years when crops like sorghum or pearl millets have not grown completely (Singh 2023).

3.4 Storage and Processing

FMs, similar to rice, are typically preserved in their "paddy" or "dressed grain" form (Zargaran Khouzani 2023). Traditionally, they are stored in bulk within "banco" granaries, with a raw FM density ranging from 650 to 660 g/l (Zargaran Khouzani 2023). For commercial purposes, FM may be packed into sacks, either as raw or hulled grains. Modern storage methods are gradually being adopted, with some producers to store hulled grains using metal barrels and plastic containers (Cruz et al. 2016). Due to its limited size, FM usually experiences minimal attacks by the pest during harvesting and storage. When dried appropriately (moisture content <10–11%), FM can be stored for longer duration from several months to even years, even in traditional granaries (Cruz et al. 2016). When needed for consumption, FM is retrieved from storage, hulled, and whitened before cooking. Similar to rice, FM is primarily consumed after hulling and predominantly whitening (Blakeney et al. 1997). However, historical accounts suggest variations in traditional consumption (Zargaran Khouzani 2023). The traditional methods of storing FM, combined with its resistance to pests, make it a reliable staple in regions with limited resources. The transition to using more modern storage containers can further enhance its longevity and reduce spoilage. Processing FM involves hulling to remove the outer husk and whitening to improve texture and appearance. Innovations in processing technology are making these steps more efficient, ensuring that FM retains its nutritional value while becoming more accessible to wider markets (Cruz et al. 2016)

3.5 Appearance, Genetics, and Morphology

WFM and BFM are small-seeded grains that are prominently involved in the agricultural landscapes of West Africa (Burton et al. 2024). WFM, with its seeds measuring about 1.2 mm in length, presents a white to pale-yellow color, while BFM seeds are slightly larger and darker, ranging from black to dark brown. Both species share a similar grass-like appearance with slender, upright stems, reaching heights between 30 and 80 cm (Jideani 2012). Their shallow yet extensive root systems allow them to thrive in poor, sandy soils. The stems are thin, and flexible, while the leaves are linear and rough-textured. FM's inflorescences, or flower heads, differ slightly between species; WFM has loose, open panicles, whereas BF exhibits more compact panicles (Dayal et al. 2023).

Genetically, FMs possess relatively small genomes, with *D. exilis* having a genome size of approximately 574 megabase pairs (Mb). Both species are diploid, with 2n = 36 chromosomes (Arya et al. 2021). The genetic diversity within FM is substantial, particularly among the landraces cultivated across various regions, which is crucial for its adaptability and resistance to environmental stresses and diseases (Wang et al. 2021). Recent advancements in breeding techniques and genomic studies aim to enhance FM's yield, grain quality, and stress tolerance by utilizing several genetic tools and molecular markers (Govindaraj et al. 2015). The recent whole-genome sequencing of *D. exilis* by Abrouk et al. 2020 has provided deeper insights into the evolutionary history and domestication patterns of FM, revealing key genomic regions associated with seed size and shattering resistance. Moreover, transcriptomic approaches have begun to uncover stress-responsive genes for climate-resilient breeding (Hufnagel et al. 2021). Marker-assisted selection and the development of core germplasm collections are facilitating the identification of elite traits, while CRISPR/Cas-based genome editing is being explored as a frontier strategy for accelerating functional genomics research in minor millets, including FM (Santhoshini et al. 2025). Such molecular interventions, combined with participatory breeding programs, are expected to elevate FM's role in sustainable agriculture, particularly in marginal environments of sub-Saharan Africa.

3.6 Phytochemical Composition and the Health Attributes

Agriculturally, FMs are important because they possess a high nutritional content valued for their high nutritional content, especially their richness in essential amino acids like methionine and cysteine, which are often deficient in other cereals (Jacob et al. 2024). Methionine is crucial for various metabolic processes, including the synthesis of proteins and the detoxification of harmful substances, while cysteine is important for antioxidant defense and the production of collagen (Bello et al. 2022). The phytochemical profile of FM is extensive, featuring significant amounts of polyphenols, flavonoids, and tannins. These compounds possess potent antioxidant properties, which are vital in scavenging dreadful radicals in the body. The antioxidants contribute to suppressing the potential risk of harmful ailments, including cancer, CVS, and neurodegenerative diseases, by reducing oxidative stress (Zahra et al. 2024).

3.6.1 Nutrient Profile and Chemical Constituents of FM

3.6.1.1 Caloric Value

The calorific potential of FM has not been extensively documented in existing studies. However, a study conducted by Jideani and Akingbala in 1993 reported that the energy yield was approximately 19,400 KJ per KG for FM. This value is notably higher than those observed in other cereal grains (Saldivar 2016), which include rice (18,091 KJ/kg), maize (16,982 KJ/kg), and sorghum (16,245 KJ/kg).

3.6.1.2 Carbohydrates

Cereals are primarily composed of carbohydrates, which serve as the major provider of energy for the human body. Beyond their nutritional role, carbohydrates are essential to numerous sectors, especially within the food and feed industries. Additionally, they act as sustainable raw materials in a range of eco-friendly industrial processes, including the production of bio-based plastics, glues, and ethanol-derived fuels (Poutanen et al. 2022). In FM grains, carbohydrate levels have shown considerable variability, ranging from 67.1% to 91%, with an average concentration of 79.05% (Ballogou et al. 2013). Among the available data, the lowest carbohydrate content was documented by Saldivar in 2016, while the highest level was reported by Jideani and Akingbala in 1993 (Jideani and Akingbala 1993). Carbohydrates in foods appear in different molecular forms, such as monosaccharides, oligosaccharides, and polysaccharides—the latter including compounds like starch and cellulose, which are key structural components of plant cell walls (Lebrilla et al. 2022).

3.6.1.3 Starch and Soluble Sugars

Starch is the predominant carbohydrate in FM, as seen in other cereals, and has a promising role in energy requirement. However, the studies on the presence of starch in this grain are limited (Bojarczuk et al. 2022). A study reported a level of approximately 68%, which is somewhat lower than the average starch contents found in sorghum (73.8%) and rice (77.2%). Starch generally comprises amylose (≈25%) and amylopectin (≈75%), though these proportions vary across species (Cruz et al. 2016) . In a study by Jideani and Akingbala in 1993, FMs sourced from Nigerian markets were analyzed using the Robyt and Bemis method, revealing an amylose content of 28%, indicating a nonwaxy endosperm type. Variations in starch characteristics such as granule morphology, chain structure, lipid interaction, and crystallinity lead to differing functional properties across cereals. As for soluble sugars, FM contains small amounts, with an average content of 1%. Sugars such as sucrose, glucose, fructose, and raffinose have been identified, with sucrose, glucose, and fructose being the most prevalent (Cruz et al. 2016).

3.6.1.4 Dietary Fiber

Fiber in FM consists primarily of nonstarchy polysaccharides and lignin, which suppresses digestion in the gastrointestinal tract. Reported crude fiber content in FM ranges from 0.41% to 11.3%, with an average of 5.85%. The lowest value was noted by Jideani and colleagues (Jideani and Akingbala 1993), while Saldivar in 2016 recorded the highest. This wide variability may be linked to environmental conditions, cultivation practices, genetic diversity, and analytical methodologies. Additionally, compared to other cereals like sorghum (2.7%) and rice (0.9%), FM shows a higher average fiber content (Saldivar 2016). Specifically, its hemicellulose, cellulose, and lignin contents were reported as 3%, 4%, and 0.5%, respectively. Like other grains, FM is a notable source of insoluble fiber—mainly cellulose and hemicellulose. A fiber-rich diet is directly associated with low risks of cardiovascular diseases, colorectal cancers, and diabetes, highlighting its potential health benefits (Ballogou et al. 2013).

3.6.1.5 Proteins and Amino Acids

Cereal grains are a primary source of dietary protein globally, especially in countries with a low economy where they form the staple diet. FM grains have a protein content ranging between 5.1% and 11%, with an average of 8.05%, as determined primarily using the Association Official of Analytical Chemists methods (Temple and Bassa 1991; Chukwu and Abdul-kadir 2008). Compared to cereals like sorghum millet and rice, FM generally has a lower protein content (Zhu 2020). Limited studies are available on the protein fractions of FM; however, Jideani et al. 1994 reported its distribution across Osborne fractions, identifying four typical cereal protein types: albumin (3.5%), globulin (1.8%), prolamin (5.5%), and glutelin (14%), with glutelin being the predominant fraction. The prolamin content of FM (5.5%) is comparable to that of rice, though it is notably lower than levels found in maize, wheat, and barley. Amino acid profiling reveals that glutelin and prolamin serve as storage proteins, whereas albumin and globulin, located mainly in the germ, are rich in essential amino acids and serve metabolic functions (Sachdev et al. 2023). Various studies have assessed the amino acid composition of FM using different analytical methods and expression units (Temple and Bassa 1991; Chukwu and Abdul-kadir 2008). FM contains key essential amino acids such as valine, isoleucine, threonine, histidine, lysine, leucine, methionine, tryptophan, and phenylalanine. Like most cereals, it is low in lysine but stands out for its high methionine and cysteine content; amino acids typically limited in principle cereals like sorghum, rice, wheat, and barley (Vietmeyer et al. 1996).

3.6.1.6 Lipids

Although lipids are present in relatively small amounts in cereal grains, they play an essential role in human nutrition, serving as a vital energy source and providing essential fatty acids (Price and Parsons 1975). In FM, the lipid content has been reported to range from 1.3% to 5.2%, with an average of 3.25% . This average is quite similar to the lipid content found in sorghum (3.2%), slightly lower than that in millet (5.1%), and higher than in rice (2.5%) (Ballogou et al. 2013). The primary fatty acids typically found in cereal grains include linoleic, oleic, and palmitic acids. In FM, more than 75% of the lipid content comprises unsaturated fatty acids, predominantly linoleic acid (45.7%) and oleic acid (30.6%) (Ballogou et al. 2013). Additionally, minor quantities of other fatty acids such as palmitic, myristic, palmitoleic, stearic, linolenic, arachidic, and behenic acids have also been identified in FM lipids (Bora 2013).

3.6.1.7 Minerals

The ash content in FM has been reported to range from 1% to 6%, with an average of approximately 3.5% (Deriu et al. 2022). The lowest recorded value (1%) was observed by a few researchers (Jideani and Akingbala 1993). However, the highest ash content was documented by other scientists (Saldivar 2016). The primary minerals identified in FM include magnesium, phosphorus, and potassium (Sanusi et al. 2019). Additionally, the calcium content differs between 0.0067% and 0.03%, while the amount of phosphorus is between 0.09% and 0.25%. Potassium levels lie between 0.02% and 0.26%, and sodium is present in the range of 0.005–0.03%. Magnesium concentrations span from 0.07% to 0.85%, and sulfur is found at an average level of 0.16%. Iron content ranges between 36 and 133.6 ppm, copper from 1.5 to 15 ppm, manganese between 21.6 and 30 ppm, and zinc from 30 to 42.3 ppm (Ballogou et al. 2013).

3.6.1.8 Vitamins

Vitamins, though required in minute quantities, play a crucial role in cellular metabolism. Cereal grains are generally recognized as good sources of B vitamins, with the exception of vitamin B_{12} (cobalamin) (Hanna et al. 2022). However, scarce reports are available on the vitamin profile of FM. Among the B-complex vitamins, FM is particularly rich in thiamin, with concentrations ranging between 0.3 and

0.48 mg per 100 g. Riboflavin levels have been found to fall between 0.05 and 0.1 mg/100 g (Ballogou et al. 2013). All the nutrients present in FM are depicted in Table 3.1.

3.7 Modification/Isolation of Starch

Starch constitutes the principal component of FM, with the ratio of amylose to amylopectin playing a pivotal role in defining the functional attributes of starch-based food matrices and cereal grains (Kumar et al. 2024). Ballogou et al. 2013 quantified amylose levels in BFM at approximately 28%, whereas Mahajan et al. 2021 documented a significantly lower value of the amylose in WFM as 19%. Additionally, Zhu 2020 reported amylose concentrations of 23% and 26% for two distinct WFM cultivars, highlighting further varietal variation.

To enhance the physicochemical and functional attributes of FM starch, various modification techniques have been explored. Emerging literature focuses on the structural and functional transformation of FM starch for diversified industrial applications. The isolation procedure typically involves steeping FM flour in an alkaline medium (NaOH), followed by wet milling. The resultant slurry is sieved to eliminate fibrous matter and then subjected to centrifugation. The upper proteinaceous layer on the resultant starch sediment is carefully removed, and the starch is repeatedly washed to obtain high-purity material (Carcea and Acquistucci 1997).

In one such modification approach, WFM starch was chemically altered via succinylation, achieving a degree of substitution up to 0.2%. This modification contributes to enhanced water solubility and swelling capacity, while simultaneously reducing pasting viscosity, without significantly altering starch granule morphology (Arueya and Oyewale 2015). Furthermore, (Alimi and Workneh in 2018) investigated citric acid-induced modifications of starches derived from both BF and WFM. Their findings revealed that such treatment did not influence the crystalline polymorph or granule dimensions, although the starches exhibited notable solubilization upon pasting (Alimi & Workneh 2018).

Acid hydrolysis of WFM starch was carried out for the synthesis of starch nanocrystals (Akin-Ajani et al. 2014; Odeniyi et al. 2019). This hydrolytic process reduces both granular swelling power and pasting viscosity, while concurrently enhancing crystallinity and solubility. These effects are likely due to selective degradation of the amorphous domains within starch granules. To further improve solubility, the resulting starch nanocrystals were subjected to carboxymethylation (Odeniyi et al. 2019).

Collectively, the observed changes in physicochemical parameters of modified FM starch are consistent with findings from similar treatments applied to widely utilized starches such as those from maize and wheat (BeMiller and Whistler 2009; Xu et al. 2020). These findings showed that modified FM starch has potential as an alternative or complementary ingredient to conventionally modified cereal starches in specific functional and industrial contexts (Amarnath et al. 2024).

Table 3.1 Nutritional composition of fonio grains

Parameter	Reported range / average	Comparison with other cereals	Reference(s)
Caloric value	~19,400 KJ/kg	Higher than rice (18,091), maize (16,982), and sorghum (16,245) KJ/kg	Jideani and Akingbala (1993) and Saldivar (2016)
Carbohydrates	67.1–91%	Comparable to major cereals	Jideani and Akingbala (1993) and Saldivar (2016)
Starch content	~68%	Lower than sorghum (73.8%) and rice (77.2%)	Cruz et al. (2016)
Amylose content	22.1–28%	Indicates nonwaxy-type endosperm	Jideani and Akingbala (1993)
Soluble sugars	~1%; sucrose, glucose, and fructose most common	Similar profile to other cereals	Cruz et al.(2016)
Dietary fiber	0.41–11.3%	Higher than sorghum (2.7%), millet (2%), and rice (0.9%)	Jideani and Akingbala (1993), Saldivar (2016) and Ballogou et al. (2013)
Hemicellulose	3%	–	Ballogou et al. (2013)
Cellulose	4%	–	Ballogou et al. (2013)
Lignin	0.5%	–	Ballogou et al. (2013)
Protein content	5.1–11%	Lower than sorghum, millet, and rice	Temple and Bassa (1991) and Chukwu and Abdul-Kadir (2008)
Protein fractions	Albumin (3.5%), globulin (1.8%), prolamin (5.5%), and glutelin (14%)	Glutelin is predominant	Jideani et al. (1994)
Essential amino acids	High in methionine and cysteine; low in lysine	Better sulfur amino acid profile than other cereals	Sachdev et al. (2023) and Vietmeyer et al. (1996)
Lipid content	1.3–5.2%	Similar to sorghum (3.2%), > rice (2.5%), < millet (5.1%)	Ballogou et al. (2013)
Fatty acids	Linoleic (45.7%), oleic (30.6%), palmitic, myristic, etc.	>75% unsaturated fats	Bora (2013)
Ash content (minerals)	1–6%	Higher than many cereals	Deriu et al. (2022) and Jideani and Akingbala (1993)
Major minerals	Mg (0.07–0.85%), P (0.09–0.25%), K (0.02–0.26%), Ca (0.0067–0.03%), and S (~0.16%)	Notably higher in sulfur	Sanusi et al. (2019)

(continued)

Table 3.1 (continued)

Parameter	Reported range / average	Comparison with other cereals	Reference(s)
Trace elements	Fe (36–133.6 ppm), Cu (1.5–15 ppm), Mn (21.6–30 ppm), and Zn (30–42.3 ppm)	–	Ballogou et al. (2013)
Vitamin B1 (thiamin)	0.3–0.48 mg/100g	Good source	Ballogou et al. (2013)
Vitamin B2 (riboflavin)	0.05–0.1 mg/100g	Lower end of cereal range	Ballogou et al. (2013)

3.8 Food and Nonfood Applications

FMs, being gluten free, hold significant potential in both food and nonfood sectors due to their high nutritional profile. FM flour, with various other components, is used to enhance both the nutritional profile and texture of a wide range of food items, including malted products, drinks, sourdough preparations, breads, desserts like puddings, crackers, and cookies (Zhu 2020). FM starch has been utilized as a potential thickener and binder in formulations like soups, sauces, and infant foods. Additionally, its gluten-free property confers a reasonable substitute for gluten-sensitive or celiac individuals, particularly in gluten-free flour blends used for pasta, breakfast cereals, and snacks (Kumar et al. 2024). In a study conducted by Deriu and team, where isolated FM starch with an amylose content of 25% was used to make noodles (Deriu et al. 2022). These FM starch noodles exhibited low tensile strength and cooking loss compared to unspecified commercial noodles, which are typically prepared from starch with high amylose contents (Zhu 2020). However, the characteristics of FM starch noodles still need to be assessed with those of commercially significant starch noodles, such as those made from mung bean or sweet potato starches. Additionally, FM starch can serve as a complementary food item to meet balanced nutritional requirements (Deriu et al. 2022). It is blended with defatted moringa seed flour to develop a recombined food item with the desired nutritional quality. The FM starch applications primarily include pharmaceutical excipients (Raji et al. 2018). Odeku in 2013 reviewed the use of native and modified WFM and BFM starches as glidants in pharmaceutical granulation and as additive for solid dosage forms (Odeku 2013). Outcomes indicated that FM starch performed similarly to maize starch in pharmaceutical applications. Recent studies have further demonstrated that both native and modified FM starches have been used to manufacture tablets. FM starch nanocrystals and carboxymethylated nanocrystals have been utilized as carriers for the sustained release of drugs with low solubility, drug loading capacities, and efficiencies exceeding 50% and 75%, respectively (Odeniyi et al. 2019). Acid-hydrolyzed starch and related nanocrystals improved tablet compaction properties, resulting in tablets with higher tensile strength and increased disintegration efficiency compared to those made from native starch. Comparative studies indicated that acid-modified FM starch was suboptimal for tablet embedded with paracetamol compared to modified sweet potato starch (Akin-Ajani et al. 2016). In conclusion, the suitability of FM starch for pharmaceutical applications depends on the type of modification and the specific drugs involved (Builders and Arhewoh 2016). Native and modified starches from other plant sources have shown potential in various pharmaceutical and biomedical applications, suggesting that FM starch, in different forms, could be complementary to other starches in these fields (Compart et al. 2023). Furthermore, starch in various forms can encapsulate and deliver food ingredients, possessing similar encapsulation characteristics for food applications. The favorable protein content and rich mineral profile evaluate FM as a feed component for monogastric animals such as poultry and pigs (Clottey et al. 2006). In certain regions, FM starch is combined with clay to enhance the structural integrity of building materials. This mixture is traditionally used in constructing houses and walls, leveraging the fibrous nature of the straw to improve durability (Mabila et al. 2020). Among the important starches such as potato, cassava, and maize starches, FM starch is considered "novel," underutilized, and unconventional (Kumar et al. 2024). There are numerous modifications and applications for both native and modified starches, and it is anticipated that these approaches can be expanded to FM starch for enhanced quality.

3.9 Conclusion

FM, an indigenous grain, native to West Africa, holds considerable promise as a multifunctional crop capable of addressing modern nutritional and industrial demands. Its rich nutrient profile, including sulphur-containing amino acids such as methionine and cysteine, along with its antioxidant potential and low glycemic index, positions it as a valuable dietary component for the prevention and management of metabolic disorders. The gluten-free nature and resilience of FM to poor growing conditions further enhance its relevance in today's climate-challenged agricultural systems. The modification of FM starch has expanded its functional potential, enabling its use across a broad spectrum of industries. In the food sector, both native and processed FM are being incorporated into products like noodles, baked goods, porridges, and infant foods, contributing to global food security by offering an alternative, nutrient-dense grain. Its processed form demonstrates improved shelf stability, functional versatility, and wider consumer acceptability. Beyond food, FM starch has shown promising utility in pharmaceutical applications such as drug delivery systems, tablet formulation, and even biodegradable packaging, reflecting its significance beyond traditional uses. Despite being underutilized, advances in cultivation, breeding programs, and post-harvest processing are gradually overcoming existing challenges like low yield and limited market penetration. Continued scientific exploration and technological integration are vital to unlocking the full potential of FM. Its high translational value lies in its ability to bridge indigenous agricultural knowledge with modern food science and biotechnology, offering sustainable solutions for nutrition, health, and industrial innovation on a global scale.

Data Availability Statement Not applicable.

Ethics Approval and Informed Consent Not applicable.

Disclosure The authors declare no conflict of interest.

References

Abdul SD, Jideani AI (2019) Fonio millets (*Digitaria spp.*) breeding. In: Advances in plant breeding strategies: cereals. Springer, Switzerland

Abrouk M, Ahmed HI, Cubry P, Šimoníková D, Cauet S, Pailles Y, Bettgenhaeuser J, Gapa L, Scarcelli N, Couderc M, Zekraoui L (2020) Fonio millets genome unlocks African orphan crop diversity for agriculture in a changing climate. Nat Commun 11(1):4488–4501. https://doi.org/10.1038/s41467-020-18329-4

Aguyoh JN, Masiunas JB (2003) Interference of large crabgrass *(D. sanguinalis*) with snap beans. Weed Sci 51(2):171–176. https://doi.org/10.1614/0043-1745(2003)051[0171:IOLCDS]2.0.CO;2

Akin-Ajani OD, Itiola OA, Odeku OA (2014) Effect of acid modification on the material and compaction properties of fonio and sweet potato starches. Starch/Stärke 66(7–8):749–759. https://doi.org/10.1002/star.201300280

Akin-Ajani OD, Itiola OA, Odeku OA (2016) Evaluation of the disintegrant properties of native and modified forms of Fonio and sweet potato starches. Starch/Stärke 68(1–2):69–174. https://doi.org/10.1002/star.201500188

Alimi BA, Workneh TS (2018) Structural and physicochemical properties of heat moisture treated and citric acid modified acha and iburu starches. Food Hydrocoll 81:449–455. https://doi.org/10.1016/j.foodhyd.2018.03.027

Amarnath MS, Muhammed A, Antony AK, Yashini M, Sherin A, Sunil CK (2024) Millet starch modification: effect of thermal and non-thermal methods on its properties and their application-a review. Food and Hum 3:100356. https://doi.org/10.1016/j.foohum.2024.100356

Animasaun DA, Adedibu PA, Olawepo GK, Oyedeji S (2023) Fonio millets: an underutilized crop with potential as a future smart cereal. In: Neglected and underutilized crops. Academic. https://doi.org/10.1016/B978-0-323-90537-4.00028-4

Arueya GL, Oyewale TM (2015) Effect of varying degrees of succinylation on the functional and morphological properties of starch from acha (*Digitaria exilis Kippis Stapf*). Food Chem 177:258–266. https://doi.org/10.1016/j.foodchem.2015.01.019

Arya L, Singh M, Verma M, Singh M, Aglawe SB (2021) Molecular studies on millets and pseudocereals. In: Millets and pseudo cereals. Woodhead Publishing, India

Ballogou VY, Soumanou MM, Toukourou F, Hounhouigan JD (2013) Structure and nutritional composition of Fonio millets (*Digitaria exilis*) grains. Int Res J Biol Sci 2(1):73–79

Ballogou VY, Sagbo FS, Soumanou MM, Manful JT, Toukourou F, Hounhouigan JD (2015) Effect of processing method on physico-chemical and functional properties of two Fonio millets (*D. exilis*) landraces. J Food Sci Tech 52:1570–1577. https://doi.org/10.1007/s13197-013-1150-4

Barnaud A, Billot C, Abrouk M, Adjebeng-Danquah J, Akanvou L, Ibrahim Ahmed H, Adje CO, Bakasso Y, Barry MB, Achigan-Dako EG, Yerima AR (2022) From shade to light: Fonio millet, an African orphan crop, towards renewed challenges

Bello OM, Ogbesejana AB, Balkisu A, Osibemhe M, Musa B, Oguntoye SO (2022) Polyphenolic fractions from three millet types (Fonio, finger millet, and pearl millet): their characterization and biological importance. CCMP 2(1):100020–100030. https://doi.org/10.1016/j.ccmp.2022.100020

BeMiller JN, Whistler RL (2009) starch: chemistry and technology, 3rd edn. Academic, Cambridge, MA

Blakeney AB, Welsh LA, Martin M, Booth RI, Turner NE (1997) RVA viscograms of flours prepared from cooked rice. RVA World 10:2–4

Bojarczuk A, Skąpska S, Khaneghah AM, Marszałek K (2022) Health benefits of resistant starch: a review of the literature. J Funct Foods 93:105094–105105. https://doi.org/10.1016/j.jff.2022.105094

Bora P (2013) Nutritional properties of different millet types and their selected products. University of Guelph

Builders PF, Arhewoh MI (2016) Pharmaceutical applications of native starch in conventional drug delivery. Starch-Starke 68(9–10):864–873. https://doi.org/10.1002/star.201500337

Burton G, Gori B, Camara S, Ceci P, Conde N, Couch C, Magassouba S, Vorontsova MS, Ulian T, Ryan P (2024) Landrace diversity and heritage of the indigenous millet crop fonio (*Digitaria exilis*): socio-cultural and climatic drivers of change in the Fouta Djallon region of Guinea. PPP 7(3):704–718. https://doi.org/10.1002/ppp3.10490

Carcea M, Acquistucci R (1997) Isolation and physicochemical characterization of Fonio (*Digitaria exilis Stapf*) starch. Starch-stärke 49(4):131–135. https://doi.org/10.1002/star.19970490403

Chukwu O, Abdul-kadir AJ (2008) Proximate chemical composition of acha (*Digitaria exilis* and *Digitaria iburua*) grains. J Food Tech 6(5):214–216

Clottey VA, Avornyo F, Addo-Kwafo A, Agyare WA (2006) The potential of fonio (*Digitaria exilis, Stapf*) as feed for monogastrics. LRRD 18(7)

Compart J, Singh A, Fettke J, Apriyanto A (2023) Customizing starch properties: a review of starch modifications and their applications. Polymers 15(16):3491–3511. https://doi.org/10.3390/polym15163491

Cruz JF, Beavogui F (2016) Fonio millet, an African cereal. CIRAD 14(2):154. https://fonio.cirad.fr

Dayal P, Prajapati SK, Sow S, Ranjan S (2023) Millets: nutritional profile and health benefits as nutri cereals. Elite Publishing House, India

Deriu AG, Vela AJ, Ronda F (2022) Techno-functional and gelling properties of Acha (Fonio) (*Digitaria exilis stapf*) flour: a study of its potential as a new Gluten-free starch source in industrial applications. Foods 11(2):183–198. https://doi.org/10.3390/foods11020183

Ezekiel CN, Sulyok M, Warth B, Krska R (2012) Multi-microbial metabolites in fonio millet (acha) and sesame seeds in Plateau State, Nigeria. Eur Food Res Technol 235:285–293. https://doi.org/10.1007/s00217-012-1755-2

Gigou J, Stilmant D, Diallo TA, Cissé N, Sanogo MD, Vaksmann M, Dupuis B (2009) Fonio millet (*Digitaria exilis*) response to N, P and K fertilizers under varying climatic conditions in West Africa. Exp Agric 45(4):401–415. https://doi.org/10.1017/S0014479709990421

Govindaraj M, Vetriventhan M, Srinivasan M (2015) Importance of genetic diversity assessment in crop plants and its recent advances: an overview of its analytical perspectives. Genet Res Int 2015(1):431487–431501. https://doi.org/10.1155/2015/431487

Hanna M, Jaqua E, Nguyen V, Clay JB (2022) Vitamins: functions and uses in medicine. Perm J 26(2):89–97. https://doi.org/10.7812/TPP/21.204

Hitu KW, M'ribu K, Liang H, Mandelbaum C (1997) Fonio millets: ethnobotany, genetic diversity and evolution. S Afr J Bot 63(4):185–190. https://doi.org/10.1016/S0254-6299(15)30742-0

Hufnagel B, Soriano A, Taylor J, Divol F, Kroc M, Sanders H, Yeheyis L, Nelson M, Péret B (2021) Pangenome of white lupin provides insights into the diversity of the species. Plant Biotechnol J 19(12):2532–2543. https://doi.org/10.1111/pbi.13678

Jacob J, Krishnan V, Antony C, Bhavyasri M, Aruna C, Mishra K, Nepolean T, Satyavathi CT, Visarada KB (2024) The nutrition and therapeutic potential of millets: an updated narrative review. Front Nutr 11:1346869–1346883. https://doi.org/10.3389/fnut.2024.1346869

Jideani IA (2012) *Digitaria exilis* (acha/fonio), *Digitaria iburua* (iburu/fonio) and *Eluesine coracana* (tamba/finger millet) non-conventional cereal grains with potentials. SRE 7(45):3834–3843. https://doi.org/10.5897/SRE12.416

Jideani AI, Akingbala JO (1993) Some physicochemical properties of acha (*Digitaria exilis Stapf*) and iburu (*Digitaria iburua Stapf*) grains. J Sci Food Agric 63(3):369–374. https://doi.org/10.1002/jsfa.2740630317

Jideani IA, Owusu RK, Muller HG (1994) Proteins of acha (*Digitaria exilis Stapf*): solubility fractionation, gel filtration, and electrophoresis of protein fractions. Food Chem 51(1):51–59. https://doi.org/10.1016/0308-8146(94)90047-7

Kanlindogbe C, Sekloka E, Kwon-Ndung EH (2020) Genetic resources and varietal environment of grown fonio millets in West Africa: challenges and perspectives. Plant Breed Biotech 8(2):77–88. https://doi.org/10.9787/PBB.2020.8.2.77

Kheya SA, Talukder SK, Datta P, Yeasmin S, Rashid MH, Hasan AK, Anwar MP, Islam AA, Islam AM (2023) Millets: the future crops for the tropics-status, challenges and future prospects. Heliyon 9(11):1–23. https://doi.org/10.1016/j.heliyon.2023.e22123

Koreissi-Dembélé Y, Fanou-Fogny N, Hulshof PJ, Brouwer ID (2013) Fonio (*Digitaria exilis*) landraces in Mali: nutrient and phytate content, genetic diversity and effect of processing. J Food Compos Anal 29(2):134–143. https://doi.org/10.1016/j.jfca.2012.07.010

Kumar N, Singh R, Upadhyay A, Giri BS (2024) Application and functional properties of millet starch: wet milling extraction process and different modification approaches. Heliyon 10(3):1–20. https://doi.org/10.1016/j.heliyon.2024.e25330

Lebrilla CB, Liu J, Widmalm G, Prestegard JH (2022) Oligosaccharides and polysaccharides. In: Essentials of glycobiology, 4th edn. Cold Spring Harbor Laboratory Press, New York

Mabila E, Delvoie S, Toguyeni D, Attia S, Courard L (2020) An experimental study on the use of fonio straw and shea butter residue for improving the thermophysical and mechanical properties of compressed earth blocks. JMMCE 8(2):107–132. https://doi.org/10.4236/jmmce.2020.83008

Mahajan P, Bera MB, Panesar PS, Chauhan A (2021) Millet starch: a review. Int J Biol Macromol 180:61–79. https://doi.org/10.1016/j.ijbiomac.2021.03.063

McIntosh RJ, McIntosh SK (1988) From siècles obscurs to revolutionary centuries on the Middle Niger. World Archaeol 20(1):141–165. https://doi.org/10.1080/00438243.1988.9980062

Mirza N, Marla SS (2019) Finger millet (*Eleusine coracana L. Gartn.*) breeding. In: Advances in plant breeding strategies: cereals, pp 83–132. https://doi.org/10.1007/978-3-030-23108-8_3

Morales-Payán JP, Ortiz JR, Cicero J, Taveras F (2002) Digitaria exilis as a crop in The Dominican Republic. In: Supplement to: trends in new crops and new uses. ASHS Press, Alexandria, VA

Odeku OA (2013) Potentials of tropical starches as pharmaceutical excipients: a review. Starch-Starke 65(1–2):89–106. https://doi.org/10.1002/star.201200076

Odeniyi MA, Adepoju AO, Jaiyeoba KT (2019) Native and modified *D. exilis* starch nanoparticles as a carrier system for the controlled release of naproxen. Starch-Starke 71(9–10):1900067–1900077. https://doi.org/10.1002/star.201900067

Popkin BM, Ng SW (2022) The nutrition transition to a stage of high obesity and noncommunicable disease prevalence dominated by ultra-processed foods is not inevitable. Obes Rev 23(1):13366–13384. https://doi.org/10.1111/obr.13366

Porcuna-Ferrer A, Calvet-Mir L, Faye NF, Klappoth B, Reyes-García V, Labeyrie V (2024) Drought-tolerant indigenous crop decline in the face of climate change: a political agroecology account from South-Eastern Senegal. J Rural Stud 105:103163–103179. https://doi.org/10.1016/j.jrurstud.2023.103163

Poutanen KS, Kårlund AO, Gómez-Gallego C, Johansson DP, Scheers NM, Marklinder IM, Eriksen AK, Silventoinen PC, Nordlund E, Sozer N, Hanhineva KJ (2022) Grains–a major source of sustainable protein for health. Nutr Rev 80(6):1648–1663. https://doi.org/10.1093/nutrit/nuab084

Price PB, Parsons JG (1975) Lipids of seven cereal grains. J Am Oil Chem Soc 52(12):490–493. https://doi.org/10.1007/BF02640738

Raji AO, Nassam HO, Abolaji G, Ayorinde JO, Raji MO (2018) Nutritional properties of Fanio millets starch-defatted moringa seed flour blends prepared at different ratios. Croat J Food Sci Technol 10(2):145–156. https://doi.org/10.17508/CJFST.2018.10.2.01

Sachdev N, Goomer S, Singh LR, Pathak VM, Aggarwal D, Chowhan RK (2023) Current status of millet seed proteins and its applications: a comprehensive review. Appl Food Res 3(1):100288–110034. https://doi.org/10.1016/j.afres.2023.100288

Saldivar SS (2016) Cereals: dietary importance. Elsevier, Amsterdam

Santhoshini RT, Kumar R, Dubey N, Avinashe HA, Puri P, Singh S, Rai AK, Rustagi S, Shreaz S, Negi R, Yadav N (2025) Advancing in orphan crops through innovative breeding methods and genomic approaches: current research and future challenges. J Appl Biol Biotechnol 13(2):9–19. https://doi.org/10.7324/JABB.2025.183553

Sanusi SN, Sulaiman SA, Bako HK (2019) Comparative of proximate and mineral composition of commercially-available millet types in Katsina metropolis, Nigeria. WJFST 3(1):14–19. https://doi.org/10.11648/j.wjfst.20190301.13

Seetharam A, Riley KW, Harinarayana G (1990) Small millets in global agriculture. IBH Publishing Co. Pvt. Ltd, India

Singh AK (2023) Early presence/introduction of African and east Asian millets in India: integral to traditional agriculture. Nucleus 66(3):261–271. https://doi.org/10.1007/s13237-023-00435-5

Temple VJ, Bassa JD (1991) Proximate chemical composition of Acha (*Digitaria exilis*) grain. J Sci Food Agric 56(4):561–563. https://doi.org/10.1002/jsfa.2740560415

Touvier M, da Costa Louzada ML, Mozaffarian D, Baker P, Juul F, Srour B (2023) Ultra-processed foods and cardiometabolic health: public health policies to reduce consumption cannot wait. BMJ 383:e075294. https://doi.org/10.1136/bmj-2023-075294

Vodouhè RS, Achigan Dako, Dansi A, Adoukonou-Sagbadja H (2007) Fonio: A treasure for West Africa. In Plant genetic resources and food security in West and Central Africa. Regional Conference, Ibadan, Nigeria. pp. 219-222.

Vietmeyer ND, Borlaugh NE, Axtell J, Burton GW, Harlan JR, Rachie KO (1996) Fonio millets (*Acha)*. In: Lost crops of Africa, vol 1. National Academic Press, Washington D.C.

Wang X, Chen S, Ma X, Yssel AE, Chaluvadi SR, Johnson MS, Gangashetty P, Hamidou F, Sanogo MD, Zwaenepoel A, Wallace J (2021) Genome sequence and genetic diversity analysis of an under-domesticated orphan crop, white fonio (*Digitaria exilis*). Gigascience 10(3):1–12. https://doi.org/10.1093/gigascience/giab013

Xu J, Andrews TD, Shi YC (2020) Recent advances in the preparation and characterization of intermediately to highly esterified and etherified starches: a review. Starch-Starke 72:1900238–1900285. https://doi.org/10.1002/star.201900238

Yadav L, Upsana U (2024) Millets: rediscover ancient grains. Intech Open, London

Zahra M, Abrahamse H, George BP (2024) Flavonoids: antioxidant powerhouses and their role in nanomedicine. Antioxidants 13(8):922–948. https://doi.org/10.3390/antiox13080922

Zargaran Khouzani MR (2023) Evaluating the production, processing and storage stages of fonio, an African grain. CAJPSI 2(2):42–53. https://doi.org/10.22034/CAJPSI.2022.02.01

Zhu F (2020) Fonio grains: physicochemical properties, nutritional potential, and food applications. Compr Rev Food Sci Food Saf 19(6):3365–3389. https://doi.org/10.1111/1541-4337.12608

Chapter 4
Proso Millet: Chemical Composition, Starch Isolation, Modifications, Functionality, and Food/Nonfood Applications

Rohit Bhardwaj, Satyendra Kumar Rajput, Nrashant Singh, and Tarun Kumar

Abbreviations

PM	Proso millet
PMS	Proso millet starch
HCl	Hydrochloric acid
DS	Degree of substitution
HMT	Heat moisture treatment

4.1 Introduction

Cereals are a vital dietary staple, delivering critical energy and nutrients (Porwal et al. 2023). Their widespread availability and adaptability to various agricultural environments lead to global food safety and nutritional balance (Laskowski et al. 2019). The water shortage, pollution, climate change, population growth, and increasing food cost demands high agricultural stability, driving up food prices and exacerbating food security issues. Smallholder farmers face economic vulnerability and struggle with yield and profitability under these conditions (Gitz et al. 2016).

R. Bhardwaj (✉) · S. K. Rajput · T. Kumar
Department of Pharmaceutical Sciences, Faculty of Pharmaceutical Sciences, Gurukula Kangri (Deemed to be University), Haridwar, Haridwar, Uttarakhand, India
e-mail: rohit.bhardwaj@gkv.ac.in; rohitbhardwaj.aip@gmail.com; Satyendra.rajput@gkv.ac.in; satyendraj@gmail.com; t.drx1121@gmail.com

N. Singh
Department of Forensic Science, Amity University Dubai Campus, Dubai International Academic City, Dubai, United Arab Emirates
e-mail: nsingh@amityuniversity.ae

© The Author(s), under exclusive license to Springer Nature Switzerland AG 2026

J. Singh, M. Kumar Malik (eds.), *Millets and Millet Starch: Chemical Composition, Functional Modification, and Applications*,
https://doi.org/10.1007/978-3-032-09502-2_4

Consequently, there is an urgent need to explore alternative crops like millet (Gairola et al. 2022), which could offer a utritious and resilient solution to overcome these global challenges.

Proso millet (PM), also known as *Panicum miliaceum,* is a cereal crop predominantly grown for feed, forage, and food, especially in Asian countries (China and India) (Narciso and Nystrom 2023). It has been cultivated for a very long time in Russia, North America, the Middle East, and India, acknowledged by different names such as common millets, broomcorn, red, and hog millets (Bhat et al. 2019). This crop has traditionally been underutilized, due to limited awareness, inadequate infrastructure, traditional agricultural practices, and lower market demand compared to more mainstream crops (Kaur et al. 2006, 2025). However, with increasing concerns about water shortages globally, and the need for sustainable food sources, PMs are gaining considerable attention due to their ability to thrive in dry climates with minimal water (Singh and Chauhan 2023). Recent advancements in research and development have made efforts to enhance the value of PM. It contains 60–70% starch, up to 11% protein, 2–7% dietary fiber, and is enriched with various vitamins and minerals. Additionally, PM are the rich source of amino acid (methionine and cysteine), and vitamins (niacin and folic acid) (Balli et al. 2023; Dayakar et al. 2017). However, it is limited by its lysine content. According to the recent literature, the dietary benefits of PM are comparable to that of other cereals like rice, wheat, and oats (Kalinova and Moudry 2006). This chapter explores the historical significance, nutritional benefits, and potential of PM addressing contemporary agricultural and food security challenges.

4.2 Domestication and Expansion of PM

PM has long been cultivated in China for 10,000 years. While the wild species of these millets has not been definitively examined (Miller et al. 2016; Zohary et al. 2012). Phylogenetic and chromosomal data indicates that PM is an allotetraploid, derived from ancestors such as *Panicum capillare* or a close relative and *Panicum repens* (Hunt et al. 2014). Though the precise timing of millet domestication remains debated, it is generally accepted that domestication occurred independently in three primary regions: Inner Mongolia (Zhao 2005), Northwest China (Bettinger et al. 2007), and Central China (Lu et al. 2009). From the cultivated areas, PM later spread to high-altitude regions (Tibetan Plateau) and East Asia. By the end of the second millennium BCE, its cultivation had extended to Central Eurasia and Eastern Europe (Miller et al. 2016; Marcott et al. 2013; Chen et al. 2015; Guedes 2015). On the Tibetan Plateau, barley and wheat took over millet cultivation, though it remained a significant crop in northern China's low-lying plains (Boivin et al. 2012). The current trends in other regions (Himalayas) made millet cultivation possible there. By the fifth millennium BCE, PM cultivation had reached Kazakhstan (Frachetti et al. 2010) and Pakistan (Weber 2019), although it is uncertain whether it was grown in these regions before this period.

4.3 Cultivation and Environmental Requirements

PM is considered an annual grass that is cultivated as late summer grain and could fulfill its growth in approximately 100 days (Rao 1989; Baltensperger 2002). Its branches, resembling an old broom, give rise to its common name, "broom corn" (Changmei and Dorothy 2014). The millets are round, measuring almost 3 mm in length and 2 mm in width, and are encased in a smooth hull that can be white, creamy-white, yellow, red, gray, brown, or black (Mohanan et al. 2025). In the United States, white-seeded varieties are the most commonly cultivated, followed by red-seeded types (Baltensperger 2002; Williams et al. 2007; Changmei and Dorothy 2014). PM plants typically grow to a height of around 100 cm, feature some tillers, and possess a fibrous root system (Baltensperger 2002).

PM thrives further north than other millets, growing up to 54°N latitude, and is well-suited to high-altitude and plateau environments. For instance, it is cultivated up to 1200 meters in the mountains of the former Soviet Union and as high as 3500 meters in India (Baltensperger 2002). PM can be grown in arid regions with average annual precipitation ranging from 200 to 500 mm (Ceccarelli and Grando 1996) and can produce grain with as little as 330–350 mm of rainfall per year (Oelke et al. 1990). As a warm-season crop, it is frost-sensitive and requires warm temperatures for germination and growth, with optimal soil temperatures for seed germination between 20 and 30 °C (Baltensperger et al. 1995). Being a C4 plant with a low transpiration rate, PM effectively fixes carbon in conditions of drought, high temperatures, and limited nitrogen and CO_2. It mitigates drought sensitivity by maturing quickly (Baltensperger et al. 1995) and adapts to temperatures above 30 °C by halting vegetative growth, stopping flowering, and keeping its main stem shorter to better withstand drought (Changmei and Dorothy 2014).

PM can be cultivated on sandy loam, slightly acidic, saline, and low-fertility soils (Riley et al. 1989). However, it performs poorly on waterlogged soils (Seghatoleslami et al. 2008) and on coarse, sandy soils (Williams et al. 2007). It thrives in low pH conditions, with optimal seed germination occurring in soils with a pH between 5.5 and 6.5 (Lyon et al. 2008). In contrast, soils with a pH above 7.8 can cause iron chlorosis in PM plants (Habiyaremye et al. 2017; McFarland et al. 2015).

4.4 Morphology of Starch

The dimension of starch granules extracted from PM differs on the basis of plant species used. Typically, these granules are spherical or polygonal, with dimensions ranging from 0.3 to 17 μm (Wang et al. 2023a). Polygonal granules are generally bigger and exhibit more indentures compared to globular ones (Annor et al. 2014). The morphology of starch granules is significantly affected by their treatment or biomodification (Ashogbon and Akintayo 2014). Furthermore, the particle size of Proso millet starch (PMS) can vary based on the growing region due to local environmental factors. For instance, higher altitudes and lower mean temperatures can result in larger granules (Wen et al. 2014). The alignment of these starch granules within the grain's endosperm also affects their morphology (Zarnkow et al. 2007). Surface pores and protein bodies create random cavities throughout the outer layer of starch granules. These pores further coordinate with the granular central cavity and permit molecules to enter from the environment. This feature is advantageous for starch modification. The pores facilitate the infiltration of OH ions or water, which disrupts the amylose-containing amorphous regions, and due to this, amylose-restricted characteristics are suppressed, resulting in improved hydration and swelling of starch (Kumar et al. 2023).

4.5 Genetics

The genome of PM is relatively small, measuring 923 megabases (Zou et al. 2019). Despite its compact size, PM is an allotetraploid with a chromosome number of $2n = 4x = 36$ (Hamoud et al. 1994). The two subgenomes of PM are believed to have separated about 5.6 million years ago (Zou et al. 2019). There is notable genetic redundancy between these subgenomes. Since PM is an underutilized cereal, its economic resources are also scarce compared to the major cereals (Upadhyaya et al. 2014). In PM, diversity based on genes has been assessed using both phenotypic data (morphological, agronomic, and seed nutrient traits) and genotypic data (Johnson et al. 2019; Vetriventhan et al. 2019). Before the recent advances in genome sequencing (Shi et al. 2019; Zou et al. 2019), the genomic references of PM were restricted to nonsequenced-DNA markers, including AFLP, SSSR, RAPD, and sequence-based single-nucleotide polymorphisms (Al-Khayri et al. 2019; Johnson et al. 2019; Boukail et al. 2021; Li et al. 2021).

4.6 Phytochemical Composition and the Health Attributes

PM is rich in bioactive components, including caffeic acid, chlorogenic acid, ferulic acid, syringic acid, and p-coumaric acid (Zhang et al. 2014). These compounds indicate that PM may boost the antioxidant defense system and offer promising health benefits. Notably, incorporating PM into the diet could be a strategic approach to address prevalent health issues, including diabetes, obesity, and cardiovascular disease, which are exacerbated by sedentary lifestyles and poor dietary habits (Das et al. 2019). Therefore, understanding how PM contributes to health through its bioactive components is crucial, as detailed in Table 4.1, which outlines its preventive benefits against various diseases on the consumption of PM.

Table 4.1 In vitro and in vivo studies of PM in various diseases

Sl. no.	Diseases	Effect	References
1	Cancer	PM has antiproliferative capabilities against MDA PM has prebiotic potential It can prevent constipation and is recommended for intestinal cancers	Zhang et al. (2014) and Das et al. (2019)
2	Coronary heart disease (CAD)	PM dietary protein is important in cholesterol metabolism as it raises high-density lipoprotein (HDL) cholesterol levels, especially the HDL2 subfraction while decreasing low-density lipoprotein (LDL) cholesterol levels.	Shimanuki et al. (2006) and Turer and Scherer (2012)
3	Diabetes	PM has been shown to improve glycemic and insulin responses in genetically obese type 2 diabetic mice, fed on a high-fat diet Feeding high fat and PM diet boosted plasma levels of adiponectin and high-density lipoprotein (HDL) cholesterol in genetically obese type 2 diabetes mice Extract of PM shows α-glucosidase inhibitory activity of 8–55%	Park et al. (2008) and Kim et al. (2011)
4	Hepatitis or liver failure	Proso can be used to prevent liver damage such as hepatic encephalopathy in patients with chronic liver failure PM protein's inhibitory action on liver damage is attributed to the branched-chain amino acids to aromatic amino acids ratio PM used in the study has branched-chain amino acid to aromatic amino acid of protein in the ratio of 3:1, making it excellent for hepatitis treatment	Nishizawa et al. (2002)
5	Celiac	Celiac disease is an immune-mediated enteropathic illness caused by gluten ingestion in susceptible individuals	Chandrasekara et al. (2012)
6	Others	The fiber content of PM aids in the eradication of ailments such as constipation, excess gas, bloating, and cramping	Ambati and Sucharitha (2019)

4.6.1 Diabetes Mellitus

It is well documented that PM showed significant improvement in the blood sugar levels and glycemic responses (Park et al. 2008). Additionally, the protein concentrate from PM protects rats from liver damage induced by D-galactosamine, suggesting its potential as a therapeutic option for type 2 diabetes (Park et al. 2008). Furthermore, PM extract exhibits α-glucosidase inhibitory activity ranging from 8 to 55%, which, combined with α-amylase inhibition, may help manage diabetes by reducing glucose absorption (Kim et al. 2011).

4.6.2 Obesity and Cardiovascular Diseases

Obesity and cardiovascular diseases are largely driven by poor lifestyles and irregular dietary habits, demanding nutritional adjustments required for preventing and managing these metabolic disorders (Koliaki et al. 2019). Adiponectin is vital for enhancing insulin sensitivity and lipid metabolism (Das et al. 2019). Park et al. 2008 observed elevated adiponectin levels (10.5–28.7 μg/ml) in rats fed a high-fat diet with PM proteins (PMP). Additionally, PM's ethanol extract or butanol fraction has shown potential cytotoxic effects on 3 T3-L1 preadipocytes, suggesting its ability to reduce adipocyte numbers (Jun et al. 2014).

PMPs play a significant role in cholesterol metabolism by raising adiponectin levels and HDL cholesterol while lowering LDL cholesterol. Adiponectin enhances insulin sensitivity and lipid metabolism (Turer and Scherer 2012). Clinical studies have shown that normal individuals have higher adiponectin levels compared to obese individuals (Nigro et al. 2014). Lee et al. 2010 documented that this millet reduces triglyceride levels in hyperlipidemic rats, potentially lowering cardiovascular disease risk.

4.6.3 Cancer

Bioactive components in PM have shown potential in inhibiting cancer cell proliferation. Ramadoss and Sivalingam (2021) demonstrated that vanillin extracted from PM at a therapeutic concentration (250 and 1000 μg/ml) suppressed cell viability of breast cancer cells (MCF-7), to 25.60% and 8.13%, respectively. They also found that vanillin significantly promotes cell death through apoptosis in HT-29 cells (colon cancer cell lines). Cui et al. 2018 also reported that cationic peroxidase from PM protects against DNA damage and necroptosis in human colon cancer cells (HCT116 and HT29). For this reason, PM holds promise for dietary management in cancer patients.

4.6.4 Liver Diseases

The impact of PMP on hepatic injury caused by D-galactosamine and CCL_4 was investigated by Nishizawa et al. 2002. They prepared food regimen for rats that comprises 20% PMP, which showed significant reductions in serum enzyme activities, such as alanine aminotransferase, aspartate aminotransferase, and lactate dehydrogenase, compared to those fed casein protein (Nishizawa et al. 2002).

4.6.5 Renal Diseases

Renal diseases, characterized by reduced nephron counts and increased glomerular pressure, lead to proteinuria and substantial protein loss. Karthiga et al. 2019 developed nutri-cookies with a blend of Proso and other millets, featuring limited sodium, potassium, and phosphorus levels, for children with renal conditions. Polyphenols, tannins, and flavonoids in millets can inhibit xanthine oxidase activity, potentially preventing renal disorders (Chandrasekara et al. 2012).

4.7 Starch

Starch, a key component of millet, exists in two forms: amylose and amylopectin. Millets are categorized based on their amylose content into nonwaxy and waxy varieties. The former contains a high content of amylose, whereas the latter has a low content for the same (Yang et al. 2019). According to Yang et al. 2018, the starch content ranges from 59% to 77% in nonwaxy PM and from 55% to 69% in waxy millet. Starch is an essential energy source for humans and plays a significant role in the food industry. It is a natural, sustainable, ecological, and cost-effective material that helps in altering the texture of different foods. Starch is used as stabilizers, thickeners, and sweeteners (Zhang et al. 2021).

4.7.1 Isolation, Yield, and Composition of PMS

PMS granules have a strong affinity for binding with the surrounding protein matrix. To isolate the starch and dissolving the proteins in the grain, various methods and chemical reagents were employed (El Halal et al. 2019). Typically, starch extraction involves three main phases: fragmentation, cell disruption, and purification or separation. The wet milling is the common technique utilized currently to extract starch from millets. The process involves soaking the millets in water, alkali, and acidic solution for a specific duration depending upon its chemical characterization and architecture (Mahajan et al. 2021). The yield of the starch is significantly affected by the nature of solution used for extraction (Punia et al. 2021; Punia Bangar 2021). The process for isolation of starch from PMS is illustrated in Fig. 4.1.

Further, for the alkaline steeping method, millet grains are dissolved in an alkaline solution consisting of 0.3% NaOH, at 4 °C for the next 24 hours. Following this, the grains are pulverized into a wet slurry via mill and then filtered with the help of a sieve of 100 mesh size. The slurry is centrifuged at 3000 rpm for 15 minutes; the upper layer is discarded, whereas the remaining material is resuspended in

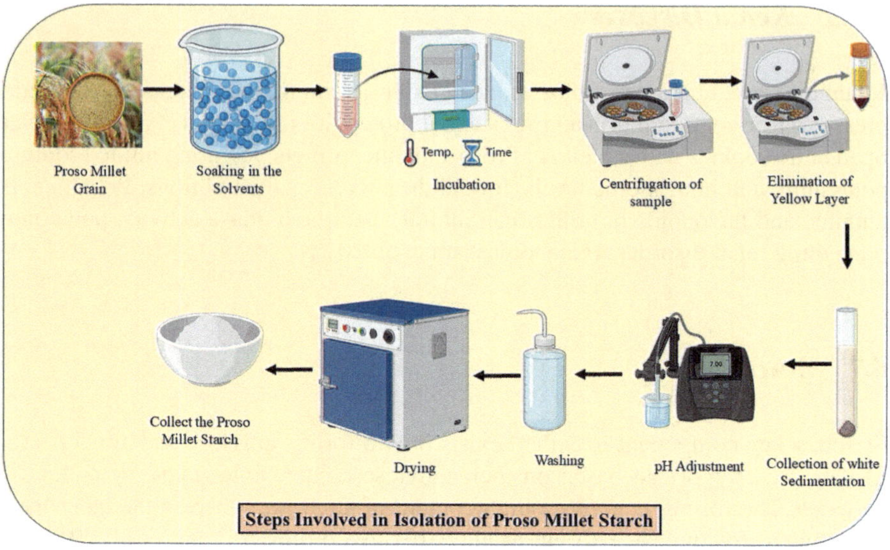

Fig. 4.1 The common process of isolating the starch from PM

water. This washing process is repeated three times, after which the slurry is neutralized with hydrochloric acid (HCl). The starch cake is then dehydrated at 40 °C for 48 hours (Sun et al. 2014).

In the acid steeping method, millets are impregnated in an acidic solution of 0.15% sulfur dioxide (SO_2) at 52 °C for 48 hours. The grains are then blended and filtered with the help of sieve of mesh size 40 and rinsed with water. Further, the mortar and pestle were utilized to crush the residue. The resulting solution was then filtered through Whatman filter paper, centrifuged, and air-dried at 45 °C. Notably, any starch extraction process requires a neutral solution of pH 6 along with a few amount of mercury chloride and 0.01% of sodium azide as it suppresses microbial growth and impede amylase activity. However, sodium azide and mercury (II) chloride can pose significant morbidity risks if ingested. For PM, small amounts of sulfur dioxide (0.5 g/L) and lactic acid (0.15 g/L) are added in the acid steeping procedure. Similarly, NaOH (0.1%) and a sodium borate buffer with SDS (0.5%) and $Na_2S_2O_5$ (0.5%) are used for starch isolation. The choice of isolating solutions can significantly influence the chemical composition and properties of the extracted starch. Comparative studies showed that acid steeping results in a higher residual protein content (4.3%) compared to alkaline steeping (0.7%) (Yanez and Walker 1986).

4.7.2 *Modification of Starch*

The structure and function of unmodified (native) starch is based on their botanical origin (Siroha and Bangar 2024). However, the starch is associated with certain drawbacks, including a high level of retrogradation, low thermal stability, and sheer strength, which prohibits its contribution in the food and nonfood industries (Singh and Adedeji 2017). The ease with which starch can be modified makes it a highly valuable ingredient in both sectors. Starch can be altered through various methods, including chemical, enzymatic, and physical modifications (Kumar et al. 2023). This section explores how these modifications impact the functionality and properties of PMS, with details provided in Table 4.2.

Table 4.2 Methods of starch modifications

Sl. no.	Modifications	Major findings and methodology used	References
Chemical modifications			
1	Acid modification	Water-holding capacity, amylose content, peak viscosity, setback viscosity, breakdown viscosity, holding strength, final viscosity, onset temperature, and peak temperature decreased while conclusion temperature, heat of enthalpy, swelling power, and solubility increased when PMS is modified with HCl (0.14 mol/L) at 6.5 pH	Singh and Adedeji (2017)
Physical modifications			
2	Heat moisture treatment (HMT)	Gelatinization parameters like onset temperature, peak temperature, and conclusion temperature increased except for the heat of enthalpy while swelling power increased at 70 to 90 °C when PMS (30 g/100 g moisture content) heated at 110 °C for 3 h	Singh and Adedeji (2017)
3	Ultra-high pressure (UHP)	Setback viscosity and solubility increased at 50 C to 60 C, and gelatinization parameters such as onset temperature; peak temperature; final viscosity, trough viscosity, peak time, and pasting temperature also increased when the PMS suspension (30% W/V) was treated with pressure level up to 600 MPa for 15 min	Kalinova and Moudry (2006)
4	Heat moisture treatment (HMT)	Water-holding capacity and solubility increased while pasting properties (breakdown viscosity, final viscosity, setback viscosity, peak viscosity), swelling power, and Mw significantly decreased when the PMS slurry (20 g starch with 80 ml water) was treated at 100 °C temperature for 30 min	Zheng et al. (2020)
5	Autoclave treatment (AT)	Water-holding capacity, amylose content, solubility increased while pasting properties (breakdown viscosity, final viscosity, setback viscosity, peak viscosity), and swelling power decreased when the PMS slurry (20 g starch with 80 ml water) was subjected to autoclave at 121 °C (0.1 MPa) for 30 min	Zheng et al. (2020)

4.7.2.1 Chemical Methods

Chemical modification of PMS includes the incorporation of functional group or new entities leading to altered starch's properties such as swelling, water absorption, gelling characteristics, and solubility as its entire structure gets modified. These modifications typically involve substitution and elongation reactions (Singh and Adedeji 2017).

4.7.2.1.1 Substitution Modifications

Starch, composed of glucose monomers, has free polyhydroxyl groups at the C-2, C-3, and C-6 positions that can participate in substitution reactions, including esterification and etherification (Egharevba, 2020). In the case of PMS, such chemical modifications have been explored to enhance its physicochemical properties, such as thermal stability and pasting behavior (Wang et al. 2023a, b) (Table 4.3).

Table 4.3 Substitution method of PMS: sources, techniques, and reagents used

Source	Modification type	Method	Reagents used	Inference
PMS	Acetylation	Aqueous acetylation at alkaline pH	Acetic anhydride, NaOH	Conventional method, low DS
PMS	Acetylation	Microwave-assisted acetylation	Acetic anhydride	Faster reaction, better uniformity
PMS	Acetylation	Anhydrous acetylation in organic solvent	Acetic anhydride, pyridine	Higher DS achievable
PMS	Acetylation	Enzyme-assisted acetylation	Acetic anhydride, α-amylase	Eco-friendly approach
PMS	Succinylation	Wet esterification	Succinic anhydride, NaOH	Enhances hydrophilicity and emulsification
PMS	Phosphorylation	Dry-heat esterification	Sodium tripolyphosphate (STPP), Na_2HPO_4	Increases paste clarity, stability
PMS	Citrate esterification	Aqueous medium, controlled pH	Citric acid, NaOH	Biodegradable and nontoxic modifier
PMS	Succinylation	Wet esterification	Succinic anhydride, NaOH	Enhances hydrophilicity and emulsification

4.7.2.1.1.1 Esterification

Esterification is a key method for modifying plant-based native starches to enhance their industrial applications. The process of formation of starch esters includes the substitution of hydroxyl groups with the acyl groups in glucose units. Esterification is a widely used technique for starch modification and employs various chemical agents, including inorganic acids (such as phosphates and sulfates) and organic acids (such as chlorides and anhydrides) (Compart et al. 2023).

4.7.2.1.1.2 Acetylation

Acetylation is a prevalent modification technique where the starch slurry is treated with acetic anhydride $(CH_3CO)_2O$. This process replaces the hydroxyl groups (OH–) in the starch chains with acetyl groups (CH_3CO–). The properties of the modified starch depend on the degree of substitution (DS) (Mehboob et al. 2020). The DS is influenced by factors such as the quality and quantity of the reagent, the reaction medium (anhydrous or aqueous), temperature, pH, treatment duration, and the use of catalysts or additional treatments like high pressure or microwaves (Golachowski et al. 2015). Acetylated starch produced in an anhydrous medium typically has a DS of around 2.9, while in an aqueous medium, the DS is generally less than 0.2. According to FDA regulations, the acetyl group content in modified starches should not exceed 2.5% when using acetic anhydride or vinyl acetate (Das et al. 2010). The acetylation process can also lead to morphological changes in starch granules, as observed by various researchers.

4.7.2.2 Degradation-Type Modification

4.7.2.2.1 Acid Hydrolysis

Acid modification represents a chemical strategy for altering starch, primarily involving the hydrolytic cleavage of α-glucan chains through the action of mineral acids. This process disrupts glycosidic linkages, resulting in significant structural and functional changes to the native starch (Compart et al. 2023). Typically, this modification is carried out using acids such as HCl, sulfuric acid (H_2SO_4), nitric acid (HNO_3), or phosphoric acid (H_3PO_4) (Hoover 2000). The hydrolysis leads to a higher proportion of short-chain fragments, particularly affecting amylose content. Although widely adopted, one major drawback of this technique is its dependence on strong chemical reagents, which raises concerns regarding environmental safety and sustainability (Mahajan et al. 2021). Beyond modifying starch's molecular structure, acid treatment also influences its physical and chemical behavior, making it useful in diverse applications such as the food industry and textile manufacturing. Products derived from acid-treated starch include gummy confections, specialty papers, and both cationic and amphoteric starch derivatives (Punia et al. 2021). When compared to unmodified PMS, the acid-treated form exhibits a reduced ability to retain water. This decrease is believed to be due to the contraction of the amorphous zones within the starch granules, limiting access to sites that typically interact with water.

4.7.3 Physical Modifications

Physical modifications of starch involve changes to its structure and morphology due to factors, including pressure, electromagnetic radiation, milling, moisture, and temperature. These modifications directly affect the starch's physicochemical and functional attributes (Zheng et al. 2020). Physical modification techniques are broadly categorized into thermal and nonthermal treatments. Thermal modifications include methods such as microwave treatment, dry heat treatment, heat moisture treatment (HMT), and autoclaving. Nonthermal modifications encompass techniques like ultra-high-pressure treatment. Since physical modifications do not use chemicals and face fewer regulatory restrictions in the food industry, physically modified starch is increasingly accepted (Amarnath et al. 2024).

4.7.3.1 Thermal Modifications

4.7.3.1.1 Heat and Moisture Treatment

Heat moisture treatment (HMT) significantly affect the physicochemical, structural, and digestibility properties of PM flour and starch. It involves applying controlled heat (above the glass transition temperature but below the gelatinization temperature) and moisture (less than 35% moisture w/w) for varying durations. In PM starch, HMT has been shown to alter functional and structural properties, including gelatinization temperature, swelling power, pasting behavior, retrogradation tendency, and amylose leaching. These modifications can enhance the starch's thermal stability and reduce its digestibility, making HMT a promising approach for tailoring the functional properties of PM starch for various food applications.

4.8 Food and Nonfood Applications of PM and Its Starch in Native and Modified Forms

PM, a versatile and nutrient-rich grain, is gaining recognition for its substantial role in both food and nonfood sectors (Khound et al. 2023). It is enriched with essential minerals like magnesium, phosphorus, and iron, along with B-complex vitamins and high-quality proteins, which make it a valuable grain for nutritional interventions (Sharma et al. 2021; Jha et al. 2022). The grain's low glycemic index and high dietary fiber content further enhance its appeal for managing chronic metabolic disorders such as diabetes and obesity (Jia et al. 2025). In culinary contexts, PM is processed into multiple forms—whole grains, flours, and flakes, making it adaptable for a range of food products. Its mild, nutty flavor and gluten-free nature have made it an ideal choice for baking and ready-to-eat (RTE) snacks (Porwal et al. 2023). PM flour is widely used to prepare cookies, muffins, flatbreads, and pancakes, especially for individuals suffering from gluten intolerance or celiac disease (Asrani et al. 2023). Studies have shown that replacing wheat flour with PM flour in baked goods can enhance their nutritional composition without compromising sensory qualities (Woomer et al. 2020). Additionally, its ability to retain moisture and improve texture in baked goods has made it a preferred ingredient in functional and health-promoting food formulations (Saleh et al. 2013). Traditional dishes such as fermented porridges, steamed cakes, and millet-based snacks are commonly prepared using PM in Asian and African cuisines. Furthermore, recent food product development has incorporated PM into breakfast cereals, extruded snacks, and nutritional bars, demonstrating its wide applicability in modern food systems (Ajagekar et al. 2023). Its high antioxidant capacity, due to the presence of polyphenols and flavonoids, adds an additional health-promoting aspect to its consumption (Sanjay Kumar et al. 2024). Beyond food applications, PM has also emerged as a promising crop in the nonfood sector. In livestock and poultry nutrition, it serves as a valuable component in feed formulations due to its high protein digestibility, essential amino acid content, and low antinutritional factors (Das et al. 2019). It supports better weight gain and health in animals, offering a sustainable alternative to conventional feed ingredients. PM's industrial potential is increasingly being recognized in the production of biodegradable polymers and biofuels. Its starch, being highly fermentable, can be converted into bioethanol, contributing to the development of renewable energy solutions (Rose and Santra 2013). Additionally, PMS has been investigated for use in the manufacturing of biodegradable films and packaging materials, which are critical to addressing the global plastic pollution crisis (Punia et al. 2021). Environmentally, PM contributes to sustainable agriculture by thriving in low-water and marginal conditions. Its short growth cycle and drought tolerance make it suitable for cultivation in arid and semi-arid regions, where it helps combat soil degradation, reduce erosion, and support agroecological balance (Kheya et al. 2023). The crop's adaptability to diverse agro-climatic conditions also positions it as a climate-resilient food source, aligning with global sustainability goals.

4.9 Conclusions

Recent shifts in dietary patterns emphasize nutrient intake and healthy eating habits, leading to increased attention in PM as it has high dietary value and gluten-free nature. The amylose content in PM makes it unique in terms of eating and cooking. PMS is categorized into waxy and nonwaxy types, with nonwaxy starch offering more diverse food applications due to its high-resistant and slowly digestible starch content. The pasting and thermal behavior of PMS, marked by A-type crystallinity, are affected by variations in amylose content and structural integrity. Although there is limited research on PM's use in food products such as pasta, bread, and porridge, its promising physicochemical properties, including cooking loss, water and oil holding capacity, and pasting properties, make it a viable alternative to commercially available starches. To optimize its functional properties and expand its applicability, modifications to PMS are being explored.

Data Availability Statement Not applicable.

Ethics Approval and Informed Consent Not applicable.

References

Ajagekar AA, Sali SD, Borse OD et al (2023) Millets based fermented products: a review. ASNH 7(6):2582–1423. https://doi.org/10.31080/ASNH.2023.07.1251

Al-Khayri JM, Jain SM, Johnson DV (2019) Advances in plant breeding strategies: cereals. Springer, Berlin

Amarnath MS, Muhammed A, Antony AK et al (2024) Millet starch modification: effect of thermal and non-thermal methods on its properties and their application-a review. Food Humanit 3:100356. https://doi.org/10.1016/j.foohum.2024.100356

Ambati K, Sucharitha KV (2019) Millets-review on nutritional profiles and health benefits. Int J Recent Sci Res 10(7):33943–33948. https://doi.org/10.24327/ijrsr.2019.1007.3786

Annor GA, Marcone M, Bertoft E et al (2014) Physical and molecular characterization of millet starches. Cereal Chem 91(3):286–292. https://doi.org/10.1094/CCHEM-08-13-0155-R

Ashogbon AO, Akintayo E (2014) Recent trend in the physical and chemical modification of starches from different botanical sources: a review. Starch Starke 66(1–2):41–57. https://doi.org/10.1002/star.201300106

Asrani P, Ali A, Tiwari K (2023) Millets as an alternative diet for gluten-sensitive individuals: a critical review on nutritional components, sensitivities and popularity of wheat and millets among consumers. Food Rev Int 39(6):3370–3399. https://doi.org/10.1080/87559129.2021.2012790

Balli D, Bellumori M, Masoni A et al (2023) Proso millet (*Panicum miliaceum L.*) as alternative source of starch and phenolic compounds: a study on twenty-five worldwide accessions. Molecules 28(17):6339–6354. https://doi.org/10.3390/molecules28176339

Baltensperger DD (2002) Progress with proso, pearl and other millets. In: Janick J, Whipkey A (eds) Trends in new crops and new uses. Alexandria: ASHS Press, pp 100–103

Baltensperger DD, Nelson LA, Frickel GE (1995) Registration of 'Earlybird' proso millet. Crop Sci 35:1204–1205

Bettinger RL, Barton L, Richerson PJ et al (2007) The transition to agriculture in northwestern China. Dev Quat Sci 9:83–101. https://doi.org/10.1016/S1571-0866(07)09008-2

Bhat S, Nandini C, Srinathareddy S et al (2019) Proso millet (*Panicum miliaceum L.*)-a climate resilient crop for food and nutritional security: a review. Environ Conserv J 20(3):113–124. https://doi.org/10.36953/ECJ.2019.20315

Boivin N, Fuller DQ, Crowther A (2012) Old World globalization and the Columbian exchange: comparison and contrast. World Archaeol 44(3):452–469. https://doi.org/10.1080/00438243.2012.729404

Boukail S, Macharia M, Miculan M et al (2021) Genome wide association study of agronomic and seed traits in a world collection of Proso millet (*Panicum miliaceum L.*). BMC Plant Biol 21:1–12. https://doi.org/10.1186/s12870-021-03111-5

Ceccarelli S, Grando S (1996) Drought as a challenge for the plant breeder. Plant Growth Regul 20:149–155. https://doi.org/10.1007/BF00024011

Chandrasekara A, Naczk M, Shahidi F (2012) Effect of processing on the antioxidant activity of millet grains. Food Chem 133(1):1–9. https://doi.org/10.1016/j.foodchem.2011.09.043

Changmei S, Dorothy J (2014) Millet-the frugal gain. Int J Sci Res and Rev 3(4):75–90

Chen FH, Dong GH, Zhang DJ et al (2015) Agriculture facilitated permanent human occupation of the Tibetan plateau after 3600 BP. Science 347(6219):248–250. https://doi.org/10.1126/science.1259172

Compart J, Singh A, Fettke J et al (2023) Customizing starch properties: a review of starch modifications and their applications. Polymers 15(16):3491–3511. https://doi.org/10.3390/polym15163491

Cui X, Wang R, & Wang Z (2018) Cationic peroxidase from proso millet induces human colon cancer cell necroptosis by regulating autocrine TNF-α and RIPK3 demethylation. Food Funct 9(3):1878–1888

Das AB, Singh G, Singh S et al (2010) Effect of acetylation and dual modification on physicochemical, rheological and morphological characteristics of sweet potato (Ipomoea batatas) starch. Carbohydr Polym 80(3):725–732. https://doi.org/10.1016/j.carbpol.2009.12.018

Das S, Khound R, Santra M et al (2019) Beyond bird feed: Proso millet for human health and environment. Agriculture 9(3):64–83. https://doi.org/10.3390/agriculture9030064

Dayakar Rao B, Bhaskarachary K, Arlene Christina GD et al (2017) Nutritional and health benefits of millets. ICAR_Indian Institute of Millets Research (IIMR) Rajendranagar, Hyderabad, 2

Egharevba HO (2020) Chemical properties of starch and its application in the food. In: Chemical properties of starch, vol 63. https://doi.org/10.5772/intechopen.87777

El Halal SLM, Kringel DH, Zavareze EDR et al (2019) Methods for extracting cereal starches from different sources: a review. Starch-Stärke 71(11–12):1900128. https://doi.org/10.1002/star.201900128

Frachetti MD, Spengler RN, Fritz GJ et al (2010) Earliest direct evidence for broomcorn millet and wheat in the central Eurasian steppe region. Antiquity 84(326):993–1010. https://doi.org/10.1017/S0003598X0006703X

Gairola S, Sinha S, Singh I (2022) Novel millet husk crop-residue based thermoplastic composites: waste to value creation. Ind Crop Prod 182:114891–114904. https://doi.org/10.1016/j.indcrop.2022.114891

Gitz V, Meybeck A, Lipper L et al (2016) Climate change and food security: risks and responses. Food and Agriculture Organization of the United Nations (FAO) Report 110(2):3–36

Golachowski A, Zieba T, Kapelko-Zeberska M et al (2015) Current research addressing starch acetylation. Food Chem 176:350–356. https://doi.org/10.1016/j.foodchem.2014.12.060

Guedes JDA (2015) Rethinking the spread of agriculture to the Tibetan plateau. Holocene 25:1498–1510. https://doi.org/10.1177/0959683615585835

Habiyaremye C, Matanguihan JB, D'Alpoim Guedes J et al (2017) Proso millet (*Panicum miliaceum L.*) and its potential for cultivation in the Pacific northwest, US: a review. Front Plant Sci 7:1961–1978. https://doi.org/10.3389/fpls.2016.01961

Hamoud MA, Haroun SA, Macleod RD et al (1994) Cytological relationships of selected species of *Panicum L.* Biol Plant 36:37–45. https://doi.org/10.1007/BF02921265

Hoover R (2000) Acid-treated starches. Food Rev Int 16(3):369–392. https://doi.org/10.1081/FRI-100100292

Hunt HV, Badakshi F, Romanova O et al (2014) Reticulate evolution in Panicum (Poaceae): the origin of tetraploid broomcorn millet, P miliaceum. J Exp Bot 65(12):3165–3175. https://doi.org/10.1093/jxb/eru161

Jha UC, Nayyar H, Parida SK et al (2022) Ensuring global food security by improving protein content in major grain legumes using breeding and 'omics' tools. Int J Mol Sci 23(14):7710–7737. https://doi.org/10.3390/ijms23147710

Jia H, Ren F, Liu H (2025) Development of low glycemic index food products with wheat resistant starch: a review. Carbohydr Polym 361:123637–123664. https://doi.org/10.1016/j.carbpol.2025.123637

Johnson M, Deshpande S, Vetriventhan M et al (2019) Genome-wide population structure analyses of three minor millets: Kodo millet, little millet, and Proso millet. Plant Genome 12(3):190021–190030. https://doi.org/10.3835/plantgenome2019.03.0021

Jun DY, Lee JY, Han CR, Kim K, Seo MC, Nam MH & Kim YH (2014) Pro-apoptotic and Anti-adipogenic Effects of Proso Millet (Panicum miliaceum) Grains on 3T3-L1 Preadipocytes. J Life Sci 24(5):505–514

Kalinova J, Moudry J (2006) Content and quality of protein in Proso millet (*Panicum miliaceum L.*) varieties. Plant Foods Hum Nutr 61(1):43–47. https://doi.org/10.1007/s11130-006-0013-9

Karthiga A, Hemamalini AJ, Prabha S (2019) Mep 3s - the Nutri cookie for children with renal diseases. Curr Res Nutr Food Sci 7(3). https://doi.org/10.12944/CRNFSJ.7.3.24

Kaur L, Singh J, Singh N (2006) Effect of cross-linking on some properties of potato (*Solanum tuberosum L.*) starches. J Sci Food Agric 86(12):1945–1954. https://doi.org/10.1002/jsfa.2568

Kaur S, Kaur G, Kumari A et al (2025) Resurrecting forgotten crops: food-based products from potential underutilized crops–a path to nutritional security and diversity. Future Foods 11:100585–100603. https://doi.org/10.1016/j.fufo.2025.100585

Kheya SA, Talukder SK, Datta P et al (2023) Millets: the future crops for the tropics-status, challenges and future prospects. Heliyon 9(11). https://doi.org/10.1016/j.heliyon.2023.e22123

Khound R, Mathivanan RK, Santra DK (2023) Proso millet nutraceutomics for human health and nutritional security. In: Compendium of crop genome designing for nutraceuticals, pp 1–31. https://doi.org/10.1016/j.foodhyd.2019.105497

Kim JS, Hyun TK, Kim MJ (2011) The inhibitory effects of ethanol extracts from sorghum, foxtail millet and Proso millet on α-glucosidase and α-amylase activities. Food Chem 124(4):1647–1651. https://doi.org/10.1016/j.foodchem.2010.08.020

Koliaki C, Liatis S, Kokkinos A (2019) Obesity and cardiovascular disease: revisiting an old relationship. Metabolism 92:98–107. https://doi.org/10.1016/j.metabol.2018.10.011

Kumar SR, Tangsrianugul N, Suphantharika M (2023) A review on isolation, characterization, modification, and applications of Proso millet starch. Foods 12(12):2413–2429. https://doi.org/10.3390/foods12122413

Laskowski W, Górska-Warsewicz H, Rejman K et al (2019) How important are cereals and cereal products in the average polish diet? Nutrients 11(3):679–700. https://doi.org/10.3390/nu11030679

Lee SH, Chung IM, Cha YS et al (2010) Millet consumption decreased serum concentration of triglyceride and C-reactive protein but not oxidative status in hyperlipidemic rats. Nutr Res 30(4):290–206. https://doi.org/10.1016/j.nutres.2010.04.007

Li C, Liu M, Sun F, Zhao X, He M, Li T, Lu P, Xu Y (2021) Genetic divergence and population structure in weedy and cultivated broomcorn millets (*Panicum miliaceum L.*) revealed by specific-locus amplified fragment sequencing (SLAF-Seq). Front Plant Sci 12:688444–688458. https://doi.org/10.3389/fpls.2021.688444

Lu H, Zhang J, Liu KB et al (2009) Earliest domestication of common millet (*Panicum miliaceum*) in East Asia extended to 10,000 years ago. Proc Natl Acad Sci U S A 106(18):7367–7372. https://doi.org/10.1073/pnas.0900158106

Lyon DJ, Burgener PA, DeBoer KL et al (2008) Producing and marketing Proso millet in the Great Plains. Univ. of Nebraska, Lincoln

Mahajan P, Bera MB, Panesar PS et al (2021) Millet starch: a review. Int J Biol 180:61–79. https://doi.org/10.1016/j.ijbiomac.2021.03.063

Marcott SA, Shakun JD, Clark PU et al (2013) A reconstruction of regional and global temperature for the past 11,300 years. Science 339:1198–1201. https://doi.org/10.1126/science.1228026

McFarland CR, Huggins DR, Koenig RT (2015) Soil pH and implications for management: an introduction

Mehboob S, Ali TM, Sheikh M et al (2020) Effects of cross linking and/or acetylation on sorghum starch and film characteristics. Int J Biol Macromol 155:786–794. https://doi.org/10.1016/j.ijbiomac.2020.03.144

Miller NF, Spengler RN, Frachetti M (2016) Millet cultivation across Eurasia: origins, spread, and the influence of seasonal climate. Holocene 26(10):1566–1575. https://doi.org/10.1177/0959683616641742

Mohanan MM, Vijayakumar A, Bang-Berthelsen CH et al (2025) Millets: journey from an ancient crop to sustainable and healthy food. Foods 14(10):1733–1755. https://doi.org/10.3390/foods14101733

Narciso JO, Nystrom L (2023) The genetic diversity and nutritional quality of Proso millet (Panicum miliaceum) and its Philippine ecotype, the ancient grain "kabog millet": a review. J Agr Food Res 11:100499–100511. https://doi.org/10.1016/j.jafr.2023.100499

Nigro E, Scudiero O, Monaco ML et al (2014) New insight into adiponectin role in obesity and obesity-related diseases. Biomed Res 2014:658913–658927. https://doi.org/10.1155/2014/658913

Nishizawa N, Sato D, Ito Y et al (2002) Effects of dietary protein of Proso millet on liver injury induced by D-galactosamine in rats. Biosci Biotechnol Biochem 66(1):92–96. https://doi.org/10.1271/bbb.66.92

Oelke EA, Oplinger ES, Putnam DH et al (1990) Millets: alternative field crop manual. University of Wisconsin Extension, Cooperative Extension, University of Minnesota Center for Alternative Plant and Animal Products CAPAP and the Minnesota Extension Service. https://hort.purdue.edu/newcrop/afcm/millet.html

Park KO, Ito Y, Nagasawa T et al (2008) Effects of dietary Korean Proso-millet protein on plasma adiponectin, HDL cholesterol, insulin levels, and gene expression in obese type 2 diabetic mice. Biosci Biotechnol Biochem 72(11):2918–2925. https://doi.org/10.1271/bbb.66.92

Porwal A, Bhagwat G, Sawarkar J et al (2023) An overview of millets-the nutri-cereals: its nutritional profile, potential health benefits and sustainable cultivation approach. IJSRA 10(1):841–859. https://doi.org/10.30574/ijsra.2023.10.1.0828

Punia S, Kumar M, Siroha AK et al (2021) Pearl millet grain as an emerging source of starch: a review on its structure, physicochemical properties, functionalization, and industrial applications. Carbohydr Polym 260:117776–117792. https://doi.org/10.1016/j.carbpol.2021.117776

Punia Bangar S, Sandhu KS, Rusu AV et al (2021) Proso-millet-starch-based edible films: an innovative approach for food industries. Coatings 11(10):1167–1181. https://doi.org/10.3390/coatings11101167

Ramadoss DP, Sivalingam N (2021) Vanillin extracted from Proso and barnyard millets induces cell cycle inhibition and apoptotic cell death in MCF-7 cell line. J Cancer Res Ther 17(6):1425–1433. https://doi.org/10.4103/jcrt.JCRT_1128_19

Riley KW, Setharam A, Harinarayana G (1989) Recommendations on breeding and varietal improvement. In: Small millets in global agriculture. Oxford and IBH Publishing, New Delhi

Rose DJ, Santra DK (2013) Proso millet (*Panicum miliaceum L.*) fermentation for fuel ethanol production. Ind Crop Prod 43:602–605. https://doi.org/10.1016/j.indcrop.2012.08.010

Saleh AS, Zhang Q, Chen J (2013) Millet grains: nutritional quality, processing, and potential health benefits. Compr Rev Food Sci Food Saf 12(3):281–295. https://doi.org/10.1111/1541-4337.12012

Sanjay Kumar T, Nageswari R, Somasundaram S et al (2024) Significance of millets for food and nutritional security—an overview. Discov Food 4(1):73–89. https://doi.org/10.1007/s44187-024-00149-w

Seghatoleslami MJ, Kafi M, Majidi E (2008) Effect of drought stress at different growth stages on yield and water use efficiency of five Proso millet (*Panicum miliaceum L.*) genotypes. Pak J Bot 40(4):1427–1432

Sharma R, Sharma S, Dar BN et al (2021) Millets as potential nutri-cereals: a review of nutrient composition, phytochemical profile and techno-functionality. Int J Food Sci Technol 56(8):3703–3718

Shi J, Ma X, Zhang J et al (2019) Chromosome conformation capture resolved near complete genome assembly of broomcorn millet. Nat Commun 10(1):464–473. https://doi.org/10.1038/s41467-018-07876-6

Shimanuki S, Nagasawa T, Nishizawa N (2006) Plasma HDL subfraction levels increase in rats fed Proso-millet protein concentrate. Int J Exp Clin Res 12(7):221–2266

Singh M, Adedeji AA (2017) Characterization of hydrothermal and acid modified Proso millet starch. LWT-Food Sci Technol 79:21–26. https://doi.org/10.1016/j.lwt.2017.01.008

Singh S, Chauhan AK (2023) Proso millet (*Panicum miliaceum*): bioactive composition, pharmacological impact and techno-functional attributes. In: Nutri-Cereals. CRC Press, United States

Siroha AK, Bangar SP (2024) Millet-based food products: an overview. Curr Food Sci Technol Rep:1–8. https://doi.org/10.1007/s43555-024-00031-x

Sun Q, Gong M, Li Y et al (2014) Effect of dry heat treatment on the physicochemical properties and structure of Proso millet flour and starch. Carbohydr Polym 110:128–134. https://doi.org/10.1016/j.carbpol.2014.03.090

Turer AT, Scherer PE (2012) Adiponectin: mechanistic insights and clinical implications. Diabetologia 55:2319–2326. https://doi.org/10.1007/s00125-012-2598-x

Upadhyaya HD, Dwivedi SL, Singh SK et al (2014) Forming core collections in barnyard, kodo, and little millets using morphoagronomic descriptors. Crop Sci 54(6):2673–2682. https://doi.org/10.2135/cropsci2014.03.0221

Vetriventhan M, Azevedo VC, Upadhyaya HD et al (2019) Variability in the global Proso millet (*Panicum miliaceum L.*) germplasm collection conserved at the ICRISAT genebank. Agriculture (Basel) 9(5):112–128. https://doi.org/10.3390/agriculture9050112

Wang L, Ming H, Chen Q et al (2023a) Analysis of starch structure and pasting characteristics of millet thick wine during fermentation. Foods 12(9):1840–1856. https://doi.org/10.3390/foods12091840

Wang J, Ye Y, Chai Z et al (2023b) Physicochemical properties and in vitro digestibility of Proso millet starch modified by heat-moisture treatment and annealing processing. Int J Biol Macromol 235:123829. https://doi.org/10.1016/j.ijbiomac.2023.123829

Weber SA (2019) Plants and Harappan subsistence: an example of stability and change from Rojdi. Taylor and Franscis, Routledge

Wen Y, Liu J, Meng X et al (2014) Characterization of Proso millet starches from different geographical origins of China. Food Sci Biotechnol 23:1371–1377. https://doi.org/10.1007/s10068-014-0188-z

Williams MM, Boydston RA, Davis AS (2007) Wild Proso millet (*Panicum miliaceum*) suppressive ability among three sweet corn hybrids. Weed Sci 55(3):245–251. https://doi.org/10.1614/WS-06-123.1

Woomer J, Singh M, Vijayakumar PP et al (2020) Physical properties and organoleptic evaluation of gluten-free bread from Proso millet. Br Food J 122(2):547–560. https://doi.org/10.1108/BFJ-07-2019-0555

Yanez GA, Walker CE (1986) Effect of tempering parameters on extraction and ash of Proso millet flours, and partial characterization of Proso starch. Cereal Chem 63(2):164–167

Yang Q, Zhang P, Qu Y et al (2018) Comparison of physicochemical properties and cooking edibility of waxy and non-waxy Proso millet (*Panicum miliaceum L.*). Food Chem 257:271–278. https://doi.org/10.1016/j.foodchem.2018.03.009

Yang Q, Zhang W, Li J et al (2019) Physicochemical properties of starches in Proso (non-waxy and waxy) and foxtail millets (non-waxy and waxy). Molecules 24(9):1743–1760. https://doi.org/10.3390/molecules24091743

Zarnkow M, Mauch A, Back W et al (2007) Proso millet (*Panicum miliaceum L.*): an evaluation of the microstructural changes in the endosperm during the malting process by using scanning-electron and confocal laser microscopy. J Inst Brew 113(4):355–364. https://doi.org/10.1002/j.2050-0416.2007.tb00762.x

Zhang L, Liu R, Niu W (2014) Phytochemical and antiproliferative activity of Proso millet. PLoS One 9(8):104058–104068. https://doi.org/10.1371/journal.pone.0104058

Zhang T, Li K, Ding X et al (2021) Starch properties of high and low amylose Proso millet (*Panicum miliaceum L.*) genotypes are differentially affected by varying salt and pH. Food Chem 337:127784. https://doi.org/10.1016/j.foodchem.2020.127784

Zheng MZ, Xiao Y, Yang S et al (2020) Effects of heat–moisture, autoclaving, and microwave treatments on physicochemical properties of Proso millet starch. Food Sci Nutr 8(2):735–743. https://doi.org/10.1002/fsn3.1295

Zohary D, Hopf M, Weiss E (2012) Domestication of plants in the old world: the origin and spread of domesticated plants in Southwest Asia, Europe, and the Mediterranean Basin. Oxford University Press, New York

Zou C, Li L, Miki D et al (2019) The genome of broomcorn millet. Nat Commun 10(1):436–447. https://doi.org/10.1038/s41467-019-08409-5

Zhao Z (2005) Palaeoethnobotany and its new achievements in China. Kaogu (Archaeology) 7(42):49

Chapter 5
Kodo Millet: Chemical Composition, Starch Isolation, Modifications, Functionality, and Food/Nonfood Applications

Tarun Kumar, Vikas, Rekha Rikhari, and Tanvi Singh

Abbreviations

GI glycemic index
ATPS Aqueous two-phase systems
HMT Heat moisture treatment
PEF Pulsed electric field
UHP Ultra-high pressure

5.1 Introduction

Kodo millet (*Paspalum scrobiculatum*), part of the *Poaceae* family, is considered one of the oldest cultivated grains in regions like India and Africa, with its roots tracing back to India (Bunkar et al. 2021). This millet is often referred to as "Varagu" in India, where it has been a staple diet for generations (Paschapur et al. 2021). In the last couple of years, it has also become popular for its nutritional richness and because it is a very resilient grain for very difficult growing conditions. Kodo millet

T. Kumar
Department of Pharmaceutical Sciences, Gurukula Kangri (Deemed to be University), Haridwar, Uttarakhand, India
e-mail: t.drx1121@gmail.com

Vikas · R. Rikhari
Department of Chemistry, Motherhood University, Roorkee, Uttarakhand, India

T. Singh (✉)
Department of Chemistry, Delhi Technological University, Delhi, India

© The Author(s), under exclusive license to Springer Nature Switzerland AG 2026
J. Singh, M. Kumar Malik (eds.), *Millets and Millet Starch: Chemical Composition, Functional Modification, and Applications*,
https://doi.org/10.1007/978-3-032-09502-2_5

is one of those notable crops, and the interesting feature is that it thrives well in arid, drought-stress-prone areas, and thus can be a strong candidate for sustainable agriculture in places that are prone to climate change consequences. Even though it has been a mainstay of traditional diets and is highly beneficial, it is not as popular as cereals like rice, wheat, and maize (Saini et al. 2021).

Kodo millet is rich in protein, dietary fiber, and essential minerals like iron, calcium, and magnesium. The major carbohydrate in it is starch, accounting for about 60–65% of the grain, and is an effective source of energy (Kumari et al. 2024). Kodo millet is thus well placed because of its nutrient and starch rich composition within the growing market for functional foods, especially gluten-free or associated with metabolic health conditions like diabetes. In addition, it has a naturally low glycemic index, making it a wise choice for people who wish to maintain steady blood sugar levels (Anitha et al. 2021).

Researchers have shown interest in starch derived from Kodo millet on account of its unique properties. Starch plays an important role in shaping the texture, consistency, and shelf life of the products in the food industry (Wang et al. 2019). Kodo millet starch has specific gelling and thickening characteristics; therefore, it is used in many food products such as porridge, noodles, and baked foods (Bangar et al. 2021). Several modification techniques have been studied to enhance functionality of kodo millet starch for food as well as nonfood utilization, including physical and chemical treatment, and an enzymatic process (Kaimal et al. 2021; Amarnath et al. 2024).

Somewhat interestingly, the nutrition that Kodo millet provides is not the only thing it provides. Now, its starch is being looked at for nonfood applications such as biodegradable films and bio-based plastics. Finally, these eco-friendly materials could serve as an alternative to ordinary nonbiodegradable polymers based products (Punia Bangar et al. 2023). Excluding starch, Kodo millet has antioxidants such as phenolic acids and flavonoids that are known to be good for health. These bioactives are being explored for their potential roles in pharmaceutical and cosmetic products, particularly for their anti-inflammatory effects and skin-protective benefits (Dey et al. 2022).

In general, Kodo millet offers many opportunities, from being a nutritious grain to being used as a raw material in eco-conscious industries. Due to its unique composition and functional versatility, it can be used to address some important global challenges such as health, nutrition, and sustainability (Patil et al. 2023). This chapter attempts to bring together what is known, or at least described, about the millet's chemical composition, how its starch can be extracted and modified, and what has been done with the starch in all of its myriad forms. The aim was to establish a base for additional research and demonstrate why this frequently overlooked grain should be accorded more attention.

5.2 Appearance, Morphology, and Genetics

Kodo millet (*Paspalum scrobiculatum*) is a small hardy cereal crop with elliptical and grayish grains. Seeds are hard and fibrous, are covered in a husk, and must be removed before they are consumed. Kodo millet is an erect annual grass, branching from the base and stem, becoming slender, tufted stems up to 1 m high. The leaves are narrow and linear, the inflorescence is racemose, with several spikelets each having numerous seeds (Hariprasanna 2017).

Genetically, Kodo millet is a diploid (2n = 40) species with a lot of genetic diversity in the stimulation of drought and pest resistance genes. The diversity of the species ensures that it can be highly adaptable to adverse environmental conditions. Recent genomic studies have identified several gene loci responsible for stress tolerance and grain yield, making Kodo millet a promising crop for breeding programs to enhance food security in marginal environments (Sao et al. 2024).

5.3 Geographical Distribution

Kodo millet is primarily produced in India. Kodo millet is mainly grown in the Indian states of Madhya Pradesh, Maharashtra, Chhattisgarh, Karnataka, Odisha, and Tamil Nadu (Yadav et al. 2024).

Madhya Pradesh is considered the "millet bowl" of India, with Kodo millet being a significant part of the agricultural landscape, accounting for approximately 60% of the country's cultivation area for this crop. Dindori, Mandla, Umaria, Shahdol, Betul, and Chhindwara are prominent regions for Kodo millet cultivation. It also thrives in the Bundelkhand region, where farmers grow it in sandy and rocky soils (Yadav et al. 2024).

Chhattisgarh is enhancing its Kodo millet production through initiatives like "Mission Millet" and collaborations with the Indian Institute of Millet Research (IIMR). Kanker, Kondagaon, Bastar, Dantewada, Bijapur, Sukma, Narayanpur, Rajnandgaon, Kawardha, Gaurela-Pendra-Marwahi, Balrampur, Koriya, Surajpur, and Jashpur have signed MoUs to boost Kodo millet cultivation (Babu et al. 2024).

Kodo millet is grown in hilly and drought-prone areas of the southern states of Karnataka and Maharashtra, and is an important food for the local communities. Traditionally cultivated in Karnataka, Kodo millet is called "haraka." Famous places of Karnataka for Kodo millet farming are Koppal, Raichur, and Ballari (Veerabadhran et al. 2023). Kodo millet is grown in the regions of Sangli, Yavatmal, and Jalgaon in Maharashtra.

Some of the millets cultivated in Tamil Nadu include Kodo millet. The cultivation of this millet is also popular in Villupuram, Cuddalore, Thoothukudi, and Tiruvannamalai (Maheswari 2024).

In Andhra Pradesh, districts such as Prakasam, Kurnool, Ananthapuram, Nandyala, and Nellore are known for millet farming, including Kodo millet (Ramakrishna 2024).

Apart from India, Kodo millet is also cultivated in other Asian countries. These include Nepal and Bangladesh, where Kodo millet is grown in hilly regions and is a staple food in many communities. Its cultivation helps sustain local diets and agriculture (Kheya et al. 2023).

It is also cultivated in countries such as Nigeria, Mali, Ethiopia, South Africa, and Zimbabwe (Ravikesavan et al. 2023). Nigeria is considered to be one of those countries where Kodo millet is commonly grown, used in traditional dishes across the country, and serves as a staple food for most communities. They are, however, significant producers of the northern regions like Kano and Jigawa. The grain is grown in Mali's semi-arid regions, where it is an important food source that allows farmers to deal with climatic variability. Kodo millet is grown in a variety of regions in Ethiopia, and especially in the highlands. Kodo millet is cultivated in marginal lands in South Africa and Zimbabwe (de Wet et al. 1983).

5.4 Cultivation

Kodo millet is extensively cultivated in rainfed, arid, and semi-arid regions. Although the Kharif or rainy season used to be during the months of June–September, it is currently cultivated throughout the year due to the presence of the dreaded locust. It is usually grown in these regions on sandy or rocky soils with very little inputs, in smallholdings, which makes it a sustainable crop for smallholder farmers (Gangaiah and Yadav 2024).

The fact that the crop needs very little water and can tolerate long dry periods makes it very appropriate for regions where rainfall is erratic and water is scarce. Line sowing with 10 cm spacing between plants and 22.5 cm spacing between lines, seed rate of 10kg/h is ideal for achieving the highest yield (Kalaisekar et al. 2016). The crop is often intercropped with legumes for local soil improvement. Kodo millet takes 90–120 days to grow, but requires comparatively less pest resistance intervention, which reduces the need for chemical interventions. Kodo millet is an adaptable crop that can be used for food security because of its ability to grow successfully in harsh conditions. It can be harvested when the crop dries and is processed manually or mechanically (Jeeva et al. 2024).

5.5 Storage and Processing

Kodo millet is wrapped up and stored to preserve its nutritional value, extending its shelf life for further processing. Kodo millet seeds are then typically sun-dried to reduce moisture content prior to storage in order to prevent mold growth. In rural areas, traditional storage methods are practiced, such as storing in earthen or metal containers (Sruthi and Rao 2021).

Steeping and dehusking the grain of Kodo millet to create flour is the processing step involved. Labor intensity in dehusking varies, depending on whether it is done mechanically or manually. Food products manufactured from Kodo millet flour are porridge, breads, snacks, etc. It improves the grain's digestibility and extends potential food applications (Bisht 2022).

5.6 Chemical Composition

The chemical composition of Kodo millet is predominantly characterized by its carbohydrate, protein, fat, and mineral content, making it a rich source of energy and essential nutrients (Krishna Kumari and Thayumanavan 1998). Table 5.1 briefly explains the chemical composition of Kodo millet.

5.6.1 Carbohydrates

About 60–65 percent of the grain of Kodo millet are carbohydrates. Kodo millet carbohydrate content is mostly as starch (carbohydrate plus fiber), with some simple sugars and lesser amounts of dietary fiber. But, in fact, Kodo millet starch consists mainly of amylose and amylopectin, and it is important for the digestibility and glycemic index of the grain (Ranjan et al. 2023; Bhatt et al. 2022). Kodo millet is also known to have a high resistant starch content, and this has also been studied for its possible roles in reducing glucose absorption and assisting in the management of blood sugar in diabetic diets (Yadav et al. 2024b).

5.6.1.1 Starch

Kodo millet contains around 50–55% starch. The starch comprises two main polysaccharides: amylose and amylopectin (Louis et al. 2023).

- *Amylose*: This is a molecule of a glucose unit connected by alpha 1,4 glycosidic bonds and is a straight chain that forms a helical structure. Around 20–28% of the total starch in Kodo millet is the pattern of this linear glucose polymer, slightly higher than other cereal grains such as rice (15–20% amylose). Starches with different amounts of amylose affect starch's gelatinization and its ability to retrograde, hence its digestibility and food textural properties (Bhatt et al. 2022).
- *Amylopectin*: The major glucan of Kodo millet starch, consisting of 72–80% of the starch. The side chain branches of amylopectin are made up of about 30 glucose units linked with $1\alpha \rightarrow 6$ linkages nearly every 20 to 30 glucose units along the chain. Cooked millet has a viscous and sticky texture due to the contribution of amylopectin, which is appropriate for porridges and soft food products. The high amylopectin content in Kodo millet starch makes it easily digestible (Kumari et al. 2024).

Table 5.1 Chemical composition of Kodo millet (Krishna Kumari and Thayumanavan, 1998)

Constituent	Sub-constituents	Content (%)	Details	References
Carbohydrates (60–65%)	Starch	50–55%	Made up of amylose (20–28%) and amylopectin (72–80%).	Rawat et al. (2021)
	Resistant starch	8–10%	It acts as dietary fiber, lowers glycemic index, and promotes gut health	Arya and Shakya (2021)
	Sugars	1–2%	Small amounts of glucose, fructose, and sucrose	Malathi et al. (2024)
Protein		8–12%	It contains essential amino acids that are beneficial for plant-based diets	Sachdev et al. (2021) and Ranjan et al. (2023)
Fat		2–3%	Predominantly unsaturated fats	Ranjan et al. (2023)
Dietary fiber		8–10%	It includes both soluble and insoluble fibers and improves digestion	Shankar Bunkar et al. (2021)
Nonstarch polysaccharides (8–10%)	Hemicellulose	3–4%	Provides structure to plant cells	Suri et al. (2024) and Ranjan et al. (2023)
	Cellulose	2–3%	Makes the cell wall	
	Pectin	1–2%	Water-holding capacity	
Mineral (2.5–3.5%)	Calcium	0.01–0.02%	Contributes to bone health and muscle function	Arya and Bisht (2022b), Rani et al. (2024) and Ranjan et al. (2023)
	Iron	0.0017–0.004%	It is important for oxygen transport in the blood and combats anemia	
	Magnesium	0.11–0.13%	Supports enzyme function and muscle health	
	Phosphorus	0.15–0.18%	Key role in energy metabolism and bone strength	

5.6.1.2 Resistant Starch

Kodo millet is known for its relatively high resistant starch content, around 8–10% of the total carbohydrate content. Resistant starch is a type of starch that resists digestion in the small intestine, acting like dietary fiber. It contributes to lower glycemic index values and provides health benefits such as improved gut health by promoting the growth of beneficial gut bacteria (Vijayakumar et al. 2024).

5.6.1.3 Simple Sugars

In Kodo millet, the simple sugars constitute only a small fraction of the total carbohydrates, typically around 1–2%. The major simple sugars present include glucose, fructose, and small amounts of sucrose. These sugars contribute to the mild sweetness of the grain and are readily fermentable, which is beneficial in products like fermented millet foods or beverages (Paschapur et al. 2021).

5.6.1.4 Nonstarch Polysaccharides

The dietary fiber component is that of hemicellulose, cellulose, and pectin in Kodo millet. Grain has around 8–10% of nonstarch polysaccharides (Dewan et al. 2022).

- *Hemicellulose*: Comprising about 3–4% of the grain, hemicellulose is a heteropolysaccharide that consists of xylose, arabinose, and other sugar residues. It contributes to the structural integrity of the plant cell walls (Serna Saldívar Sergio and Hernández 2020).
- *Cellulose*: Kodo millet is a high-fiber food; it contains around 2–3% cellulose, which is a linear polymer of β (1 $\rightarrow$ 4)-linked glucose units. It is an insoluble dietary fiber that forms the rigid part of the cell wall, and the cell walls act as a rigid component (Anjali and Vijayaraj 2020).
- *Pectin*: Pectin, which has a crystalline structure, is a gel-forming polysaccharide, comprising about 1–2% in the grain and takes part in maintaining the cell wall structure, water-holding capacity, and texture properties of cooked millet (Amadou 2022).

5.6.2 Proteins

Kodo millet contains about 8–12% protein, comparable to other cereals like rice and wheat. The protein quality of Kodo millet is high, containing essential amino acids such as lysine, methionine, and tryptophan (Dey et al. 2024b). Particularly regarding plant-based diets where protein variety can be limited, these amino acids are highly important. Kodo millet protein has recently been shown to have good digestibility, and it can be an adequate source of dietary protein (Sudharshana et al. 1988).

5.6.3 *Fats*

Although the fat content in Kodo millet does not exceed much, around 2–3%, the fats present are particularly unsaturated, which is good for heart health. It has 35–40% of total fatty acids in linoleic acid, which is a necessary fatty acid for keeping healthy cell membranes and fighting inflammation (Sridhar and Lakshminarayana 1992).

5.6.4 *Minerals*

Kodo millet is rich in minerals such as calcium, magnesium, iron, and phosphorus. Subsequently, these minerals play a vital role in humans and other animals in a number of physiological functions, including bone health, enzyme activity, and oxygen transport. Kodo millet is noted for its iron content because Fe level is much higher compared to other cereals like rice. Kodo millet, therefore, becomes a valuable food to combat iron deficiency anemia, especially where there is no alternative source of meat and other animal products (Ranjan et al. 2023; Morya et al. 2017).

5.6.5 *Secondary Metabolites of Kodo Millet*

Kodo millet is also a good source of bioactive compounds, phytochemicals, and plant-based chemicals that are referred to as health-promoting. The antioxidant activity, as well as disease prevention properties, is a known property of these compounds (Samtiya et al. 2023). Table 5.2 briefly explains the phytochemical composition of Kodo millet.

5.6.5.1 Phenolic Compounds

Ferulic and caffeic acids are the phenolic acids present in high concentration in Kodo millet (Malathi et al. 2024). These phenolic compounds are antioxidants and hence serve to neutralize free radicals in the body (Martemucci et al. 2022). These antioxidants may help reduce oxidative stress and, in turn, lower the risk of chronic diseases like cardiovascular disease and cancer. It is also in the case of Kodo millet its phenolic content also supports its anti-inflammatory properties, which helps lower inflammation in the body (Dey et al. 2022).

Table 5.2 Phytochemical composition of Kodo millet (Shah et al. 2021)

Constituent	Sub-constituents	Percentage/content (mg/g)	Details	References
Phytochemicals	Phenolic compounds	1–2 mg/g	It includes ferulic acid and p-coumaric acid, which contribute to antioxidant activity	Dubey and Tripathy (2024)
	Flavonoids	0.5–1 mg/g	Quercetin, catechins, is known for its anti-inflammatory and antioxidant effects	Shikha et al. (2024)
	Phytates	1–2%	It acts as an antinutrient by binding minerals, but it can also have health benefits	Shukla et al. (2024)
	Tannins	Trace amounts	It may affect protein digestibility but exhibit antioxidant properties	Bheemaiah Balyatanda et al. (2024)
	Saponins	Trace amounts	They are known for their antimicrobial and immune-boosting properties	Tripathi et al. (2021)

5.6.5.2 Flavonoids

Another group of phytochemicals present in Kodo millet are flavonoids. Nonetheless, these compounds have been intensively studied for their antioxidant, anti-inflammatory, and anticancer properties. Kodo millet contains queretin, a known flavonoid, which helps support heart health and reduce the risk of cancer by protecting cells from oxidative damage (Upadhyay et al. 2015).

5.6.5.3 Phytates

Like other grains, Kodo millet has phytates, which are antinutritional factors that reduce the bioavailability of minerals such as calcium and iron. Owing to the high levels of phytate, it is processed, but different processing methods caused reduction in the phytate levels and increase the absorption of minerals. In spite of the presence of phytate, the mineral content of Kodo millet is still high and therefore makes a healthy food supplement to the diet, especially in areas with nutritional deficiencies (Soni et al. 2022).

5.6.6 Nutritional Composition

However, the nutritional profile of all Kodo millet in general is very good, and in being a combination of nutrients, Kodo millet is very highly nutritious and has several health benefits (Malleshi et al. 2021). Table 5.3 briefly explains the nutritional composition of Kodo millet.

Table 5.3 Nutritional composition of Kodo millet (Malleshi et al. 2021)

Constituent	Sub-constituents	Percentage/content (mg/g)	Details	References
Vitamins	Vitamin B1 (thiamine)	0.2–0.4 mg/100 g	Helps in carbohydrate metabolism and nerve function	Raza et al. (2023)
	Vitamin B2 (riboflavin)	0.1–0.2 mg/100 g	Supports energy production and antioxidant function	
	Vitamin B3 (niacin)	1.2–1.5 mg/100 g	Important for skin health and converting food into energy	
Minerals	Calcium	10–12 mg/100 g	Essential for bone health, muscle function, and nerve transmission	Arya and Bisht (2022a, b) and Rani et al. (2024)
	Iron	1.7–4 mg/100 g	Crucial for blood production and oxygen transport	
	Magnesium	110–130 mg/100 g	Supports enzyme function, heart health, and energy production	
	Phosphorus	150–180 mg/100 g	Important for bone structure and energy metabolism	
Dietary fiber	Soluble fiber	8–10%	Helps in lowering cholesterol and maintaining stable blood glucose levels	Arya and Bisht (2022a)
Antioxidants	Carotenoids	Trace amounts	May support eye health and act as antioxidants	Awasthi et al. (2024)
	Polyphenols	1–2 mg/g	Protects against oxidative damage and inflammation	

5.6.7 *Dietary Fiber*

Kodo millet is rich in dietary fiber, one of the most important nutritional aspects. Maintaining digestive health is important, and keeping your system regular and avoiding constipation is crucial, which is why dietary fiber is important. Total dietary fiber content (6.39 g) is higher in Kodo millet than rice (2.81 g) causing low glycemic index (Deshpande et al. 2015). Kodo millet is, hence, the perfect go-to and an excellent choice for diabetics or those who would like to remain away from it (Sharma et al. 2017).

5.6.8 *Low Glycemic Index*

Kodo millet has a lower glycemic index (GI) than refined grains such as white rice. Glycemic index of Kodo millet starch (47.81) is lower than rice flour (53.25) (Annor et al. 2013). Food with a low glycemic index is digested and absorbed slowly, and by slow absorption, it gradually increases blood sugar (Arya and Shakya 2021). Kodo millet is therefore a good food for persons with diabetes or who would like to regulate their blood sugar level through diet. Studies have proven that, apart from providing the body with energy, regular Kodo meal intake can reduce the chances of your body developing type 2 diabetes because it enhances insulin sensitivity (Yadav et al. 2024b).

5.6.9 *Micronutrients*

Kodo millet is also an excellent source of B vitamins as well as niacin (B3), thiamine (B1), and riboflavin (B2). This is because these vitamins are necessary for energy metabolism and nerve function, and healthy skin. They add to the nutritional value of Kodo millet, thus helping us as a whole to be a bit helpful in our overall nutrition (Sharma et al. 2017; Singh et al. 2024b, c).

5.7 Starch Isolation Methods

Kodo millet starch isolation is done by extracting pure starch granules from other components like proteins, lipids, and fibers. The goal is to get high-purity starch for food, industrial, and improvement purposes. Optimized methods have been developed and used for starch extraction from Kodo millet, falling under major mechanical, chemical, and enzymatic categories (Mahajan et al. 2021). The methods used for the isolation of starch from Kodo millet are different in complexity, efficiency, and cost. Among the methods most often used in the large-scale production of starch is a wet steeping method, whereas alkali and enzymatic methods are based on higher purity and functional properties of starch. Aqueous two-phase systems (ATPS) is emerging technique based on innovative solutions, which is, however, not yet fully adopted by the industry.

5.7.1 *Wet/Water Steeping Method*

Kodo millet is one of the cereals most commonly used for the isolation of starch using the wet steeping method. There are several key steps in extracting starch from Kodo millet. Thus, the soaking in water or a dilute sodium bisulfite solutthick paste.

The starch slurry is passed through a sieve to remove the larger fiber and protein particles from the fine starch granules, and the starch is left in suspension in water with nonstarch components deposited on the sieve. Starch is centrifuged and settled in the bottom as a pellet, leaving the supernatant, which contains soluble proteins and other compounds. The final step is to dry the wet starch with air or in an oven, or by freeze-drying it to yield a fine starch powder. However, high-purity starch is produced with little time spent and minimal use of water using this method, but it is expensive and used for large-scale production (Heena et al. 2024).

5.7.2 *Alkali Extraction Method*

In this method, sodium chloride is used in combination with an alkaline solution (commonly sodium hydroxide) to assist in breaking down the Kodo millet non-starch components (proteins and lipids) for further isolation of starch. The first step in the alkali extraction method of isolating starch from Kodo millet is the soaking of grains for 12–24 hours in a dilute acid solution (e.g., 0.1–0.5 per cent NaOH). It helps break the protein matrix that surrounds the starch granules. The grain is soaked, then washed with water so as to remove residual alkali, and milled or ground to release the starch (Kurichh et al. 2025; Sonkar et al. 2023b; Singh et al. 2024a, b, c). Then, the starch slurry is subjected to centrifugation in order to get the starch separated out from its nonstarchy fraction. However, the starch is then neutralized—that is, the starch is treated with a dilute acid, such as hydrochloric acid, and it is washed again with water once separated. Finally, the pure starch is dried into a fine powder. It has the advantage of efficient removal of proteins, which gives high-purity starch, and it is simple, so it is scalable for industrial use. The process is, however, fraught with handling alkali solutions with great caution, and starch degradation may be inevitable if not employed with great caution (Kaur et al. 2023).

5.7.3 *Enzymatic Method*

Enzymatic methods of starch isolation use enzymes that only enzymatically digest proteins, lipids, or fibers to leave the starch granules intact. The process begins with soaking and grinding Kodo millet grains in water to release the starch, and the same amount is credited to the wet steeping process (Annor et al. 2013). To degrade any protein present in the starch slurry, proteolytic enzymes such as papain or protease can be added to the starch slurry; alternatively, amylolytic enzymes such as alpha-amylase may also be included to partially hydrolyze the soluble starch fractions for the isolation of pure starch. After the starch slurry has been filtered and centrifuged, the purified starch granules are separated from the enzymatically degraded materials. The starch product is finally dried with traditional means, such as an oven or air drying. The main advantage of this method consists precisely of a selective removal of a chosen set of proteins or lipids and reduced dependence on the use of harsh chemicals. This is more expensive, since the enzymes are costly, but limited to controlling the activity of enzymes to prevent degradation of starch (Pragyani Bora and Marcone 2019).

5.7.4 Aqueous Two-Phase System (ATPS)

A relatively advanced technique for isolating starch consists of the aqueous two-phase system (ATPS), which makes use of the partitioning of components into two aqueous phases formed by the use of polymers like polyethylene glycol and salts such as potassium phosphate. This is the method where a starch suspension is mixed into an aqueous solution of the polymer and salt (Lamsal and Wanasundara 2024). As a result, two different phases are formed: a salt-rich phase and a polymer-rich phase. Typically, starch granules partition into the salt-rich phase, and proteins and other impurities are partitioned into the polymer-rich phase. The starch is then recovered from the salt-rich phase after the separation of the phases by centrifugation. The starch is washed to remove residual salts and dried afterward. It provides high selectivity for starch isolation, but the degradation of starch properties is minimal. However, it is expensive, and the process involves the use of polymers and the complex phase separation (Zhang et al. 2022).

Another method, such as dry steeping, when water usage is a concern, is also possible for starch isolation, although it is less common. The method involves grinding Kodo millet grains into a fine powder without soaking. Then, the ground floor is sieved to separate starch-rich fraction from the protein and fiber fractions on the ground floor. The lighter, fibrous components of the cake can be further separated from the starch granules by air classification. The starch-rich fraction may be purified to final purity either by dry sieving or minimal wet washing to extract residual impurities. The implementation of this method is simple and low-cost, especially when used on small-scale applications where no water is needed. It, however, has a lower starch purity than wet methods, and its application for large-scale production is limited (Kumar et al. 2024).

5.8 Modifications in Kodo Millet Starch

Starch modification is essential in improving the functional properties of Kodo millet starch toward some industrial and food applications. The modification process changes the physicochemical properties of starch, such as insolubility, diffusion, solubility, gelatinization, retrogradation, and digestibility. Its functionality is improved by a variety of techniques such as physical, chemical, or enzymatic modifications (Zhu 2014).

5.8.1 Physical Modifications

Physical modifications of starch are generally nonchemical processes that involve changes in temperature, pressure, or mechanical treatments. These methods are considered more environmentally friendly as they do not involve chemical reagents (BeMiller and Huber 2015).

5.8.1.1 Heat Moisture Treatment (HMT)

Heat moisture treatment (HMT) involves heating the starch at less than its gelatinization temperature (80–120°C) for a long time, in the presence of moisture, below 20–30% moisture. It is found that the HMT of Kodo millet starch improved the thermal stability and altered starch crystalline structure, hence, it improved resistance to enzymatic digestion and retrogradation properties. HMT provides an advantage for the resistance of starch to dietary applications, such as managing the blood sugar level (Kurichh et al. 2025; Schafranski et al. 2021). HMT raises the gelatinization temperature of millet starch, indicating increased thermal stability, which is desirable in food products that undergo heat processing (Wang et al. 2021). Millet starch treated with HMT exhibits slower digestibility and a reduced glycemic response, making it ideal for use in functional foods for diabetic patients (Kumar et al. 2023).

5.8.1.2 Annealing

Annealing is therefore the second physical modification technique whereby starch is treated with excess water at a temperature near but below the gelatinization temperature. In this process, the granular structure of the starch is changed but not destroyed. To increase the viscosity of the Kodo millet starch and decrease the gelatinization time, the process of annealing was used (Krishna Kumari and Thayumanavan 1998; Schmiele et al. 2019). The annealed starch from the sorghum had better thermal stability with less disruption to the granules. Annealing of Kodo millet starch, which could be possible in certain conditions, could be needed for modifying Kodo millet starch for thermal food applications like ready-to-eat meals (Nanje Gowda et al. 2025).

5.8.1.3 Pre-Gelatinization (Drum Drying and Extrusion)

Starches are obtained by cooking starch with water in the presence of water, pregelling the starch granules, and afterward drying with drum drying or extrusion. The starch finally becomes cold water soluble (Chen et al. 2024). They can be used for making instant mixes, bakery, or for weaning due to the good functional characteristics of pre-gelatinized starch, such as instant solubility, increased water absorption, and increased viscosity (Hong and Liu 2023). For example, obtaining the form of pre-gelatinized barnyard millet starch under drum drying showed significantly improved solubility and rapid hydration, which are highly desired in the field of convenience food (Idrishi et al. 2025). Similar to other minor millets, Kodo millet starch was likely to show similar functional improvement when subjected to similar processing, thus promising an ingredient for the development of ready-to-use food products (Dekka et al. 2023).

5.8.1.4 Microwave and Ultrasound Assisted Modification

In recent years, the physical modification usage of microwave treatment has developed as an emerging technique through the use of electromagnetic radiation to induce rapid, uniform heating within starch granules. Intense change both in the crystalline and amorphous regions of the starch matrix occurs in this process due to the disruption of internal hydrogen bonding (Yi et al. 2024). To achieve that, some modifications of the pasting profile (i.e., reducing peak viscosity and shifting gelatinization parameters) are observed for microwave-treated starches. Furthermore, microwave irradiation may improve the resistant starch fraction formation under certain processing conditions as it has dietary as well as functional health benefits (Farooq and Yu 2024). Particularly, the research on millets such as proso and foxtail has shown that microwave exposure not only lowers the viscosity but also enhances resistance to enzymatic hydrolysis to make the starch suitable for controlled release and low glycemic food applications (Olawoye et al. 2023). Microwave-assisted modification of Kodo millet starch is extrapolated for the development of value-added products in the health food and nutraceutical sectors. Vijayakumar et al. 2024 explored the impact of dual modification ie, esterification followed by ultrasound modification, on Kodo millet starch characteristics. It was reported that dual modification showed change in the surface of the starches, improved functional and structural properties of the starches, and increased relative crystallinity, nutrient bioavailability. Adhikary et al. 2025 investigated the impact of ultrasonication on Kodo millet starch properties and reported that modified Kodo millet starch have lower syneresis (4.90%), solubility (8.97%), swelling power (9.83%), higher elasticity, firmer gel structure, enhanced digestion, and superior pasting properties (Adhikary et al. 2025).

5.8.1.5 Pulsed Electric Field (PEF) Treatment

Starch modification with the emerging techniques, such as nonthermal energy, bearing cold plasma and pulsed electric field (PEF) treatments, has been studied to tailor the functional characteristics. Kumar et al. 2025 investigated the impact of cold plasma treatment on powder flow characteristics and rheological properties of Kodo millet starch. It was concluded that clod plasma treatment caused significant reduction in size of the granules (8.79 to 7.77 μm), the surface of starch granule became more rough and irregular in shape (Kumar et al. 2025). Kodo millet starch was treated in a plasma reactor at different voltages of 30 kV, 20 kV, and 10 kV for a duration of 30 min, 20 min, and 10 min and the impact of cold plasma treatment on the starch hydration, structural, and physicochemical properties were explored. It has been concluded in the study that water absorption and oil absorption capacities enhanced from 1.43 to 1.55 g/g and 1.01–1.34 g/g respectively. While, significant decrease in the amylose content (31.06 to 25.94 g/hg) and pH of starch (7.5 to 6.9) was observed (Sonkar et al. 2023b). It has been reported that cold plasma treatment can improve the surface hydrophilicity, yet the granule morphology, and the water absorption behavior in starches derived from rice and potato (Sun et al. 2022). Such modifications improve emulsifying properties and decrease the retrogradation tendencies desirable for consumption in emulsion, gel, or consumption in other texturally sensitive food systems (Kaavya et al. 2023). With extrapolation of these findings to Kodo millet starch, it is possible to develop advanced functional ingredients.

5.8.2 *Chemical Modifications*

Chemical modifications involve introducing new functional groups into the starch molecule, altering its properties. These processes are widely used to improve starch performance in food processing, adhesives, and other industrial applications (Vijayakumar et al. 2024).

5.8.2.1 Acid Hydrolysis

Starch is treated by dilute acids, for example, hydrochloric acid, in the process called acid hydrolysis, which partially hydrolyzes the glycosidic bonds in the starch granules. It therefore leads to a reduced molecular weight and changes structural properties (Olawoye et al. 2023). Kodo millet starch is reduced to a smaller size and subsequently increased in solubility by acid hydrolysis. This modification improves the application of the starch in product required low-viscosity, such as some beverages or sauces. Furthermore, it enhances film-forming ability of Kodo millet starch to serve as a biodegradable packaging material (Majumder et al. 2024).

5.8.2.2 Esterification

Acetylation is the most frequently applied esterification technique, is defined as the acetylation of starch, that is, starch with acetyl groups introduced into the starch structure primarily by the reaction of starch with acetic anhydride. The functional properties of starch are thereby enhanced, and its hydrophobicity is improved by this modification (Subroto et al. 2023). Kodo millet starch esterified using octenyl succinic anhydride has a reduced water absorption (1.85 to 1.32 g/g), swelling, and solubility and is hence suitable for frozen and refrigerated foods (Bist et al. 2026).

5.8.2.3 Oxidation

Oxidation of starch includes the use of oxidizing agents, such as sodium hypochlorite, which needs to have carbonyl and carboxyl groups coupled into the starch chains. Such a modification is ordinarily used to reduce the viscosity of starch pastes and make them more stable. Oxidation of Kodo millet starch improves the clarity and stability, which makes it suitable for use in food products like sauces, where a clear, stable gel is needed. Oxidized starch also improves the freeze–thaw stability of products stored at low temperatures (Singh et al. 2024a, b, c).

5.8.2.4 Cross-Linking

One of the most commonly used method for starch functional performance enhancement in industrial applications is through the cross-linking (Chen et al. 2015). The formation of covalent bonds between starch molecules using bi- or multifunctional reagents like phosphorus oxychloride or epichlorohydrin originate cross-linking (Amaraweera et al. 2021). These cross-links strengthen the internal structure of the starch granules and improve the resistance to shear, heat, and acidic conditions during the processing (Kumari and Sit 2023; Dey et al. 2024a, b). Cross-linking could greatly enhance paste clarity and stability in the context of Kodo millet starch, making Kodo millet starch a more suitable candidate for applications such as sauces, fillings, and thermally processed foods. Such cross-linked starches are less reported in studies on similar minor millets, which have revealed higher gelatinization temperatures, reduced swelling and solubility, and better viscosity retention after cooking (Lv et al. 2023). Controlling degree of cross-linking by reagent concentration and reaction conditions plays a key role in tailoring these properties according to the purpose of end use, which provides great potential for functional diversification of Kodo millet starch (Kaur et al. 2024).

5.8.3 Enzymatic Modifications

The enzymatic modification makes use of enzymes like amylases or glucosidases to do selective hydrolysis of starch. It is a more controlled and greener method than the chemical one (Punia Bangar et al. 2022). However, Kodo millet starch is still unexplored for enzymatic modification.

Starch is partially hydrolyzed using enzymes like alpha-amylase or glucoamylase, giving maltodextrins or glucose syrups. On hydrolysis, the starch structure is changed to make it more soluble, and the molecular weight is reduced. Kodo millet starch is hydrolyzed enzymatically for increased solubility and decreased viscosity, which can be fitted for use in food products requiring low-viscosity starches. This modification is commonly used to tailor the starch functionality for application in syrup, beverage, and infant food production (Bunkar et al. 2021).

5.9 Food and Nonfood Applications of Kodo Millet

Kodo millet (*Paspalum scrobiculatum*) is an ancient grain that is increasingly being noticed for its nutritional benefits and scope of applications. Kodo millet is a rich source of nutrients and has functional properties with a lot of environmental sustainability, and it is increasingly being incorporated into various food and nonfood products.

5.9.1 Food Applications

Kodo millet is utilized in various food products owing to its health benefits and functional characteristics.

5.9.1.1 Traditional and Ethnic Foods

Traditional dishes often make up much of the use of this food, in regions such as India and Africa. In some cultures, Kodo millet is used to prepare porridge, which is the main breakfast item. However, it serves as a nutritious alternative to rice and wheat (Bajpai and Ravichandran 2024). In many regions, Kodo millet flour is used in making unleavened flatbreads, locally called "roti" or "chapati." They are gluten-free and suitable for people with gluten intolerance (Mishra and Satheesh 2024; Siroha and Bangar 2024).

5.9.1.2 Processed Foods

Kodo millet starch has a unique property that makes it suitable for many food processing applications. It may be worked up to make puffed snacks or chips. Effective puffing of the starch properties leads to a crunchy texture (Hema et al. 2022). As more bakery products, like cookies, cakes, and bread, are made using Kodo millet flour, the trend is gaining popularity in the market. The high fiber and nutrient content give baked goods a much-improved quality. Ultrasound modified Kodo millet starch was utilized to develop gummies. It was reported that gummies exhibited desirable lower water activity (0.544), textural properties, improved gel strength, and chewiness compared to control gummies (Adhikary et al. 2025).

5.9.1.3 Health Foods

Kodo millet is low in glycemic index and high in resistant starches. Kodo millet is helpful in controlling blood sugar levels and hence is commonly used in diabetic foodstuffs. Health bars, cereals, and functional foods that constitute health-conscious consumers contain Kodo millet (Saleem et al. 2023).

5.9.1.4 Fermented Products

Kodo millet can also be used in the preparation of fermented foods (Gupta and Gaur 2025). Some regions have a tradition of brewing traditional beverages from Kodo millet that has been fermented. The fact that the fermentation process enhances the bioavailability of nutrients and improves digestibility is a key factor. Kodo millet is evaluated as a substrate for producing probiotic foods, and due to its prebiotic potential, it is thought to be able to contribute to gut health (Amadou 2019).

5.9.2 Nonfood Applications

Aside from its culinary uses, Kodo millet has several nonfood applications based on its functional properties.

5.9.2.1 Biodegradable Packaging

Its potential use as a biodegradable film and packaging material for the starch produced from Kodo millet is being investigated. Due to the good mechanical properties and moisture barrier capabilities, it was found that starch-based films with Kodo millet starch can be successfully produced for food packaging applications. It contributes to reducing plastic waste and facilitating sustainability (Majumder et al. 2024). Suryavanshi et al. 2025 developed antimicrobial biodegradable packaging film using Kodo millet starch and gum tragacanth. It has been reported that the developed film showed strong UV barrier properties.

5.9.2.2 Animal Feed

Kodo millet can be used in the formulation of animal feed. Due to a high fiber and energy content, the grain is a valuable ingredient in livestock and poultry feed. Basically, it supports the nutritional requirements of various animals while replacing the former feed grains.

5.9.2.3 Pharmaceutical and Cosmetic Industries

Kodo millet starch is biocompatible, which means it can be used in the pharmaceutical and cosmetic sectors. Kodo millet starch is a candidate for use as the carrier of the active pharmaceutical ingredients for drug delivery systems (Sachdev et al. 2023). Kodo millet phytochemicals possess antioxidant properties, which can be utilized in cosmetics and skin care products for their antiaging and skin protective effects (Sonkar et al. 2023a).

5.9.2.4 Biotechnology

Kodo millet may be used in biotechnological processes. Kodo millet can be utilise to produce bioethanol, which is a renewable energy source and a sustainable practice (Nanda et al. 2024).

5.10 Conclusion

This chapter provides the unique chemical composition of Kodo millet, which makes it clinically valuable in terms of its high nutritional value. The methods of isolation of the starches prove varying degrees of efficiency and purity. Further, Kodo millet starch in modified form can be used for different food and nonfood applications, like pharmaceuticals, eco-friendly, and biodegradable materials. Apart from nutritional benefits, Kodo millet is a versatile crop and can be explored as a sustainable replacement for conventional crops in several industries. At the same time, it emphasizes the importance of further research for the development of innovative applications and processing methods of Kodo millet, which can ensure food security and sustainable agriculture practices. With growing health and environment concerns, Kodo millet presents itself as a promising candidate for future foray into the field due to the global extension toward healthier and environmentally friendly foods. The potential benefits of Kodo millet that can be unlocked by advancing our understanding of its properties and applications will be realized, and Kodo millet can be promoted in food and nonfood sectors through its future adoption.

Conflict of Interest There is no conflict of interest regarding the publication of this chapter.

References

Adhikary J, Dutta D, Kumari B, et al (2025). Effect of Ultrasonication Parameters on Properties of Kodo Millet Starch and Development of Gummies Using the Modified Starches. Starch-Stärke, 77(9),e70078.

Amadou I (2019) Millet based fermented beverages processing. Sci Beverages 5:433–472. https://doi.org/10.1016/b978-0-12-815271-3.00011-7

Amadou I (2022) Millet based functional foods: bio-chemical and bio-functional properties. Funct Foods 303–329. https://doi.org/10.1002/9781119776345.ch9

Amaraweera SM, Gunathilake C, Gunawardene OHP et al (2021) Development of starch-based materials using current modification techniques and their applications: a review. Molecules 26(22):6880. https://doi.org/10.3390/molecules26226880

Amarnath MS, Muhammed A, Antony AK et al (2024) Millet starch modification: Effect of thermal and non-thermal methods on its properties and their application – a review. Food Humanit 3:100356. https://doi.org/10.1016/j.foohum.2024.100356

Anitha S, Kane-Potaka J, Tsusaka TW et al (2021) A systematic review and meta-analysis of the potential of millets for managing and reducing the risk of developing diabetes mellitus. Front Nutr 8:687428. https://doi.org/10.3389/fnut.2021.687428

Anjali P, Vijayaraj P (2020) Functional food ingredients from old age cereal grains. In: Phytoc-FP, pp 47–92. https://doi.org/10.1016/B978-0-12-818593-3.00002-6

Annor GA, Marcone M, Bertoft E, Seetharaman K (2013) In Vitro starch digestibility and expected glycemic index of kodo millet (*Paspalum scrobiculatum*) as affected by starch–protein–lipid interactions. Cereal Chem 90(3):211–217. https://doi.org/10.1094/cchem-06-12-0074-r

Arya C, Bisht A (2022a) Nutritional profile of small millets. In: Srivastava S (ed) Small millet grains. cereals, pulses and oilseeds. Springer, Singapore, pp 15–47. https://doi.org/10.1007/978-981-16-9306-9_2

Arya C, Bisht A (2022b) Small millets: path to food and nutrition security. In: Srivastava S (ed) Small millet grains. cereals, pulses and oilseeds. Springer, Singapore, pp 161–190. https://doi.org/10.1007/978-981-16-9306-9_8

Arya SS, Shakya NK (2021) High fibre, low glycaemic index (GI) prebiotic multigrain functional beverage from barnyard, foxtail and kodo millet. LWT 135:109991. https://doi.org/10.1016/J.LWT.2020.109991

Awasthi J, Mishra A, Rathore S et al (2024) A review of the nutraceutical composition of millets and their health benefits. Curr Funct Foods 3(2):e160424228953. https://doi.org/10.2174/0126668629298743240320035203

Babu S, Tandon S, Pandey SN (2024) Magic of millets in Chhattisgarh: status, opportunities and challenges. Indian Farming 73(12):3–7. https://epubs.icar.org.in/index.php/IndFarm/article/view/147473

Bajpai P, Ravichandran R (2024) The potential of millet grains: a comprehensive review of nutritional value, processing technologies, and future prospects for food security and health promotion. J Cereal Res 15(2):157–169. https://epubs.icar.org.in/index.php/jwr/article/view/136711

Bangar SP, Ashogbon AO, Dhull SB et al (2021) Proso-millet starch: Properties, functionality, and applications. Int J Bio Macro 190:960–968. https://doi.org/10.1016/j.ijbiomac.2021.09.064

BeMiller JN, Huber KC (2015) Physical modification of food starch functionalities. Annu. Rev. Food Sci. Technol. 6:19–69. https://doi.org/10.1146/annurev-food-022814-015552

Bhatt D, Fairos M, Mazumdar A (2022) Millets: nutritional composition, production and significance: a review. Pharma Innov J 7:1577–1582

Bheemaiah Balyatanda S, Gowda NN, Subbiah J et al (2024) Physiochemical, bio, thermal, and non-thermal processing of major and minor millets: a comprehensive review on antinutritional and antioxidant properties. Foods 13(22):3684. https://doi.org/10.3390/foods13223684

Bisht A (2022) Processing of small millets. In: Srivastava S (ed) Small millet grains. cereals, pulses and oilseeds. Springer, Singapore, pp 49–77. https://doi.org/10.1007/978-981-16-9306-93

Bist Y, Nagar M, Kumar Y et al (2026) Cold plasma pretreatment enhances OSA esterification efficiency in Kodo millet starch: Morphological, structural, rheological, and functional insights. Innov Food Sci Emerg Technol 107:104358.

Bunkar DS et al (2021) Nutritional, functional role of kodo millet and its processing: a review. Int J Curr Microbiol App Sci 10(1):1972–1985. https://doi.org/10.20546/ijcmas.2021.1001.229

Chen Q, Yu H, Wang L et al (2015) Recent progress in chemical modification of starch and its applications. RSC Adv 5(83):67459–67474. https://doi.org/10.1039/c5ra10849g

Chen K, Wei P, Jia M et al (2024) Research progress in modifications, bioactivities, and applications of medicine and food homologous plant starch. Foods 13(4):558. https://doi.org/10.3390/foods13040558

de Wet JMJ, Brink DE, Rao KP, Mengesha MH (1983) Diversity in kodo millet, *Paspalum scrobiculatum*. Eco Bot 37(2):159–163. https://doi.org/10.1007/bf02858779

Dekka S, Paul A, Vidyalakshmi R, Mahendran R (2023) Potential processing technologies for utilization of millets: an updated comprehensive review. J Food Process Eng 46(10):e14279. https://doi.org/10.1111/JFPE.14279

Deshpande SS, Mohapatra D, Tripathi MK et al (2015) Kodo millet: nutritional value and utilization in Indian foods. J Grain Process Storage 2(2):16–23.

Dewan A, Tiwari M, Chhikara N, Khatkar BS (2022) Millet based functional food. In: Functional Foods, pp 91–16. https://doi.org/10.1002/9781119776345.ch4

Dey S, Singh G, Ishwarya P, Mateen A (2022) Understanding the antinutritional factors and bioactive compounds of kodo millet (*Paspalum scrobiculatum*) and little millet (*Panicum sumatrense*). J Food Qual 2022(1):1578448. https://doi.org/10.1155/2022/1578448

Dey D, Singh G, Ishwarya P, Mateen A (2024a) Functionality and extrusion processing of millets - a review. Food Reviews Int 40(10):3605–3633. https://doi.org/10.1080/87559129.2024.2367643

Dey S, Kumar Y, Das A (2024b) Kodo millet: biology, functional potential and sustainable utilization. In: Thakur M (ed) Millets: the multi-cereal paradigm for food sustainability. Springer, Switzerland, pp 123–134. https://doi.org/10.1007/978-3-031-64237-1_8

Dubey A, Tripathy PP (2024) Evaluation and multivariate analysis of antinutrients, bioactive and techno-functional attributes of finger millet and kodo millet as influenced by various germination regimes. J Food Meas Charact 18(7):6035–6049. https://doi.org/10.1007/s11694-024-02627-2/figures/5

Farooq MA, Yu J (2024) Recent advances in physical processing techniques to enhance the resistant starch content in foods: a review. Foods 13(17):2770. https://doi.org/10.3390/foods13172770

Gangaiah B, Yadav AK (2024) Modern crop management practices for pearl millet cultivation in Asia. In: Tonapi VA, Thirunavukkarasu N, Gupta S, Gangashetty PI, Yadav O (eds) Pearl millet in the 21st century: food-nutrition-climate resilience-improved livelihoods. Springer, Singapore, pp 479–511. https://doi.org/10.1007/978-981-99-5890-0_18

Gupta R, & Gaur S (2025) Deciphering the Effect of Millet Type on the Structure of the Bacterial Community in an Indian Fermented Beverage. Food Safety and Health 3:3(455–465)

Hariprasanna K (2017) Kodo millet, *Paspalum scrobiculatum* l. In: Patil JV (ed) Millets and Sorghum: biology and genetic improvement. Wiley, Hoboken, pp 199–225. https://doi.org/10.1002/9781119130765.ch8

Heena, SR, Singh N, Kumar BP, Kaur A, Upadhyay (2024) Effects of Wet-Milling Extraction Methods on Nutritional Functional and Structural Properties of Barnyard Millet Starch Abstract Starch - Stärke 76(11–12). https://doi.org/10.1002/star.v76.11-12, https://doi.org/10.1002/star.202300136

Hema V, Hema V, Ramaprabha M, Saraswathi R et al (2022) Millet food products. In: Anandharamkrishnan C, Rawson A, Sunil CK (eds) Handbook of millets - processing, quality, and nutrition status. Springer, Singapore, pp 265–299. https://doi.org/10.1007/978-981-16-7224-8_12

Hong Y, Liu X (2023) Pre-gelatinized modification of starch. In: Sui Z, Kong X (eds) Physical modifications of starch. Springer, Singapore, pp 91–102. https://doi.org/10.1007/978-981-99-5390-5_6

Idrishi R, Singha S, Rangan L (2025) Recent advances in flaking, puffing/parching, popping, and instant technologies for production of ready-to-eat cereals. In: Idrishi R, Singha S, Rangan L (eds) Ready-to-eat snacks. Taylor & Francis, New York, pp 131–157. https://doi.org/10.1201/9781003570127-5

Jeeva G, Suhasini B, Pramitha L et al (2024) Unlocking the potential of kodo millet: Reviving an indigenous super grain for tomorrow's nutrition. Planta 259(6):140. https://doi.org/10.1007/s00425-024-04414-1

Kaavya R, Pandiselvam R, Gavahian M et al (2023) Cold plasma: a promising technology for improving the rheological characteristics of food. Crit Rev Food Sci Nutr 63(32):11370–11384. https://doi.org/10.1080/10408398.2022.2090494

Kaimal AM, Mujumdar AS, Thorat BN (2021) Resistant starch from millets: Recent developments and applications in food industries. Trends Food Sci Techno 111:563–580. https://doi.org/10.1016/J.tifs.2021.02.074

Kalaisekar A, Padmaja PG, Bhagwat VR et al. (2016). Insect pests of millets: systematics, bionomics, and management. Academic Press. p 1–25

Kaur B, Singh A, Suri S et al (2023) Minor millets: a review on nutritional composition, starch extraction/modification, product formulation, and health benefits. J Sci Food Agri 103(10):4742–4754. https://doi.org/10.1002/jsfa.12493

Kaur G, Ahmadzadeh-Hashemi S, Amir S et al (2024) Exploring sustainable novel millet protein: a look at the future foods through innovative processing. Future Foods 9:100367. https://doi.org/10.1016/j.fufo.2024.100367

Kheya SA, Talukder SK, Datta P et al (2023) Millets: the future crops for the tropics - status, challenges and future prospects. Heliyon 9(11):e22123. https://doi.org/10.1016/j.heliyon.2023.e22123

Krishna Kumari S, Thayumanavan B (1998) Characterization of starches of proso, foxtail, barnyard, kodo, and little millets. Plant Foods Human Nutr 53(1):47–56. https://doi.org/10.1023/A:1008083020810

Kumar SR, Tangsrianugul N, Sriprablom J et al (2023) Effect of heat-moisture treatment on the physicochemical properties and digestibility of proso millet flour and starch. Carbo Poly 307:120630. https://doi.org/10.1016/J.carbpol.2023.120630

Kumar HN, Singh R, Upadhyay A, Giri BS (2024) Application and functional properties of millet starch: wet milling extraction process and different modification approaches. Heliyon 10(3):e25330. https://doi.org/10.1016/j.heliyon.2024.e25330

Kumar Y, Bist Y, Nagar M et al (2025) Cold plasma induced morphological, structural, powder flow and rheological properties of Kodo millet starch. Innovative Food Science & Emerging Technologies 100:103908. https://doi.org/10.1016/j.ifset.2024.103908

Kumari B, Sit N (2023) Comprehensive review on single and dual modification of starch: methods, properties and applications. Int J Bio Macro 253(3):126952. https://doi.org/10.1016/j.ijbiomac.2023.126952

Kumari P, Pareek V, Kajla P, Khurana S (2024) Composition, structure and functionality of starch isolated from Kodo millet. In: Lorenzo JM, Bangar SP (eds) Non-conventional starch sources: properties, functionality, and applications. Academic, New York, pp 253–278. https://doi.org/10.1016/B978-0-443-18981-4.00009-4

Kurichh R, Kumar Y, Bist Y et al (2025) Enhancing Kodo millet starch-based hybrid gel for fat replacement in cookies: Dual modification (heat treatment-octenyl succinic anhydride) for improved properties Food Chemistry 478143709.-https://doi.org/10.1016/j.foodchem.2025.143709

Lamsal BP, Wanasundara JPD (2024) Processing of protein ingredients from plant sources. In: Wanasundara JPD, Schmitt C, Lamsal BP (eds) Functionality of plant proteins: properties, methods of assessment, modifications and applications. Academia, New York, pp 23–45. https://doi.org/10.1016/b978-0-323-91721-6.00018-9

Louis IR, Balamurugan M, Periasamy M et al (2023) Revealing the biochemical profile and antioxidant activity of kodo millet (*Paspalum scrobiculatum* L.). J Phyto 15(S1):S8–S13. https://doi.org/10.25081/jp.2023.v15.8583

Lv P, Liu J, Wang Q et al (2023) Influence of accelerating storage of foxtail millet on the edible and cooking quality of its porridge: an insight into the structural alteration of the *in-situ* protein and starch and physicochemical properties. Int J Biol Macro 240:124375. https://doi.org/10.1016/j.ijbiomac.2023.124375

Mahajan P, Bera MB, Panesar PS et al (2021) Millet starch: a review. Int J Bio Macromo 180:61–79. https://doi.org/10.1016/j.ijbiomac.2021.03.063

Maheswari B (2024) Economic significance of millet production in Tamil Nadu. ACTA Scientiae 7(1):10.17648/acta.scientiae.6838

Majumder R, Miatur S, Roy S et al (2024) Millet starch-based film: a review. Trends Food Sci Techno 151:104618. https://doi.org/10.1016/J.TIFS.2024.104618

Malathi VM, Jacob J, Venkateswarlu R et al (2024) Nutritional aspects, phytochemical composition and potential health benefits of small millets. In: Mishra S, Kumar S, Srivastava RC (eds) Genetic improvement of small millets. Springer, Singapore, pp 129–152. https://doi.org/10.1007/978-981-99-7232-6_7

Malleshi NG, Agarwal A, Tiwari A, Sood S (2021) Nutritional quality and health benefits. In: Singh M, Sood S (eds) Millets and pseudo cereals: genetic resources and breeding advancements. Woodhead Publishing, Cambridge, pp 159–168. https://doi.org/10.1016/B978-0-12-820089-6.00009-4

Martemucci G, Costagliola C, Mariano M et al (2022) Free radical properties, source and targets, antioxidant consumption and health. Oxygen 2(2):48–78. https://doi.org/10.3390/oxygen2020006

Mishra R, & Satheesh N (2024) Kodo Millet (*Paspalum scrobiculatum*)-A Nutri-rich Millet Crop: Nutritional Composition, Health Benefits and Food Product Developments. In Millets and Other Potential Crops (pp. 157–179). CRC Press.

Morya GC, Vinita V, Mishra HS et al (2017) Millets: the indigenous food grains. Int J Adv Ayurveda, Yoga, Unani, Siddha and Homeopathy 6(1):447–452. https://doi.org/10.23953/cloud.ijaayush.328

Nanda S, Mishra A, Priyadarsini A et al (2024) A comparative thermo-chemical characterisation of oilseed, millet, and pulse stem biomass for bioethanol production. Heliyon 10(17):e36946. https://doi.org/10.1016/j.heliyon.2024.e36946

Nanje Gowda NA, Kambhampati V, Pulivarthi MK et al (2025) Thermal and non-thermal bioprocessing: a comprehensive review on millet starch properties and digestibility. J Food Meas Charact 19(2):806–832. https://doi.org/10.1007/s11694-024-02998-6

Olawoye B, Jolayemi OS, Akinyemi TY et al (2023) Modification of starch. In: Sharanagat VS, Saxena DC, Kumar K, Kumar Y (eds) Starch: advances in modifications, technologies and applications. Springer, Cham, pp 11–54. https://doi.org/10.1007/978-3-031-35843-2_2

Paschapur AU, Joshi D, Mishra KK et al (2021) Millets for life: a brief introduction. In: Kumar A, Tripathi MK, Joshi D, Kumar V (eds) Millets and millet technology. Springer, Singapore, pp 1–32. https://doi.org/10.1007/978-981-16-0676-2_1

Patil PB, Goudar G, Preethi K et al (2023) Millets: empowering the society with nutrient-rich superfoods to achieve sustainable development goals. J Drug Res Ayurvedic Sci 8(Suppl 1):S100–S114. https://doi.org/10.4103/jdras.jdras_207_23

Pragyani Bora SR, Marcone M (2019) Characterisation of several types of millets as functional food ingredients. Int J Food Sci Nutr 70(6):714–724. https://doi.org/10.1080/09637486.2019.1570086

Pu H, Huang J, Guo S (2023) Ultrahigh pressure treatment. In: Sui Z, Kong X (eds) Physical modifications of starch. Springer, Singapore, pp 169–188. https://doi.org/10.1007/978-981-99-5390-5_9

Punia Bangar S, Ashogbon AO, Singh A et al (2022) Enzymatic modification of starch: a green approach for starch applications. Carbo Poly 287:119265. https://doi.org/10.1016/j.carbpol.2022.119265

Punia Bangar S, Ilyas RA, Chaudhary N et al (2023) Plant-based natural fibers for food packaging: a green approach to the reinforcement of biopolymers. J Poly Env 31(12):5029–5049. https://doi.org/10.1007/s10924-023-02849-3

Ramakrishna P (2024) A study on market opportunities for value added millet products in urban areas of Prakasam district of Andhra Pradesh. Doctoral dissertation, Acharya NG Ranga Agricultural University https://krishikosh.egranth.ac.in/handle/1/5810206773

Rani R, Pandey C, Kapoor N (2024) Nurturing health through millet derived nutraceuticals. Food Humanit 2:100277. https://doi.org/10.1016/j.foohum.2024.100277

Ranjan R, Singh S, Dhua S et al (2023) Kodo millet (*Paspalum scrobiculatum*): bioactive profile, health benefits and techno-functionality. In: Ranjan R, Nanda V, Sharma S (eds) Nutri-cereals: nutra techno-functional potential. Taylor & Francis, Boca Raton, pp 193–211. https://doi.org/10.1201/9781003251279-8

Ravikesavan R, Jeeva G, Jency JP et al (2023) Kodo millet (*Paspalum scorbiculatum* L.). In: Farooq M, Siddique KHM (eds) Neglected and underutilized crops: future smart food. Academic, New York, pp 279–304. https://doi.org/10.1016/B978-0-323-90537-4.00019-3

Rawat L, Karnatak AK, Bisht TS, Kukreti A (2021) Minor millets: profile and ethnobotanical scenario. In: Kumar A, Tripathi MK, Joshi D, Kumar V (eds) Millets and millet technology. Springer, Singapore, pp 51–80. https://doi.org/10.1007/978-981-16-0676-2_3

Raza AM, Khaitan J, Ghosh J, Kar S et al (2023) Indian millets (finger millet, kodo, sorghum and pearl millet): potent functional foods and processing scopes. J Adv Zool 44(S6):2012. 10.17762/jaz.v44is6.2688

Sachdev N, Goomer S, Singh LR (2021) Foxtail millet: a potential crop to meet future demand scenario for alternative sustainable protein. J Sci Food Agric 101(3):831–842. https://doi.org/10.1002/jsfa.10716

Sachdev N, Goomer S, Singh LR et al (2023) Current status of millet seed proteins and its applications: a comprehensive review. App Food Res 3(1):100288. https://doi.org/10.1016/j.afres.2023.100288

Saini S, Saxena S, Samtiya M et al (2021) Potential of underutilized millets as nutri-cereal: an overview. J Food Sci Techno 58(12):4465–4477. https://doi.org/10.1007/s13197-021-04985-x

Saleem S, Mushtaq NU, Shah WH et al (2023) Millets as smart future food with essential phytonutrients for promoting health. J Food Compost Anal 124:105669. https://doi.org/10.1016/j.jfca.2023.105669

Samtiya ME, Aluko RE, Dhaka N et al (2023) Nutritional and health-promoting attributes of millet: Current and future perspectives. Nutr Rev 81(6):684–704. https://doi.org/10.1093/nutrit/nuac081

Sao A, Pali V, Patil HE et al (2024) Genetic improvement for yield, quality, biotic, and abiotic stresses in little millet (*Panicum sumatrense* roth. ex roem. and schult.). In: Mishra S, Kumar S, Srivastava RC (eds) Genetic improvement of small millets. Springer, Singapore, pp 571–599. https://doi.org/10.1007/978-981-99-7232-6_28

Schafranski K, Ito VC, Lacerda LG (2021) Impacts and potential applications: A review of the modification of starches by heat-moisture treatment (HMT). Food Hydrocol 117:106690. https://doi.org/10.1016/J.foodhyd.2021.106690

Schmiele M, Sampaio UM, Gomes PT et al (2019) Physical modifications of starch. In: Clerici MTPS, Schmiele M (eds) Starches for food application: chemical, technological and health properties. Academic, New York, pp 223–269. https://doi.org/10.1016/b978-0-12-809440-2.00006-X

Serna Saldívar Sergio O, Hernández DS (2020) Dietary fiber in cereals, legumes, pseudocereals and other seeds. In: Welti-Chanes J, Serna-Saldívar S, Campanella O, Tejada-Ortigoza V (eds) Science and technology of fibers in food systems, Food engineering series. Springer, Cham, pp 87–122. https://doi.org/10.1007/978-3-030-38654-2_5

Shah P, Kumar A, Kumar V, Tripathi MK (2021) Millets, phytochemicals, and their health attributes. In: Kumar A, Tripathi MK, Joshi D, Kumar V (eds) Millets and millet technology. Springer, Singapore, pp 191–218. https://doi.org/10.1007/978-981-16-0676-2_9

Sharma S, Sharma N, Handa S, Pathania S (2017) Evaluation of health potential of nutritionally enriched Kodo millet (*Eleusine coracana*) grown in Himachal Pradesh, India. Food Chem 214:162–168. https://doi.org/10.1016/j.foodchem.2016.07.086

Shikha D, Kumar Y, Sharanagat VS (2024) Kodo millet: technological impact and nutritional benefits for value-addition in food products. J Food Process Eng 47(6):e14655. https://doi.org/10.1111/jfpe.14655

Shukla V, Srivastava S, Singh S et al (2024) Unveiling the intricacies of phytate antinutrients in millets and their therapeutic implications in breast cancer. Intell Pharma 2(4):516–527. https://doi.org/10.1016/j.ipha.2023.12.005

Singh DK, Kurichh R, Bist Y et al (2024a) Effect of microwave roasting on the physicochemical, functional, rheological, and antioxidant properties of kodo millet flour. J Food Process Preserv 2024(1):6861190. https://doi.org/10.1155/2024/6861190

Singh P, Singh R, Bhadauria V, Singh H (2024b) The functionality and extraction of protein from sorghum, finger millet, and kodo millet: a review. Int J Food Sci Techno 59(1):512–521. https://doi.org/10.1111/ijfs.16747

Singh S, Sharma H, Ramankutty R, Ramaswamy S (2024c) Review on nutritional potential of underutilized millets as a miracle Grain. Curr Pharma Biotech 25(9):1082–1098. https://doi.org/10.2174/0113892010248721230921093208

Siroha AK, Bangar SP (2024) Millet-based food products: an overview. Curr Food Sci Technol Rep 2(2):213–220. https://doi.org/10.1007/s43555-024-00031-x

Soni K, Samtiya M, Krishnan V, Dhewa T (2022) Antinutritional factors: Nutrient bioavailability and health beneficial effects. In: Ramesh SSV, Praveen S (eds) Conceptualizing plant-based nutrition: bioresources, nutrients repertoire and bioavailability. Springer, Singapore, pp 157–179. https://doi.org/10.1007/978-981-19-4590-8_8

Sonkar S, Jaddu S, Dwivedi M, Pradhan RC (2023a) Impact of multi-pin atmospheric cold plasma on dynamic rheological characteristics of kodo millet starch. J Food Process Eng 46(12):e14485. https://doi.org/10.1111/jfpe.14485

Sonkar S, Jaddu S, Pradhan RC et al (2023b) Effect of atmospheric cold plasma (pin type) on hydration and structure properties of kodo-millet starch. Lwt, 182, p.114889.

Sridhar R, Lakshminarayana G (1992) Lipid class contents and fatty acid composition of small millets: little (*Panicum sumatrense*), kodo (*Paspalum scrobiculatum*), and barnyard (*Echinochloa colona*). J Agric Food Chem 40(11):2131–2134. https://doi.org/10.1021/jf00023a018

Sruthi NU, Rao PS (2021) Effect of processing on storage stability of millet flour: a review. Trends Food Sci Technol 112:58–74. https://doi.org/10.1016/j.tifs.2021.03.043

Subroto E, Cahyana Y, Indiarto R, Rahmah TA (2023) Modification of starches and flours by acetylation and its dual modifications: a review of impact on physicochemical properties and their applications. Polymers 15(14):2990. https://doi.org/10.3390/polym15142990

Sudharshana L, Monteiro PV, Ramachandra G (1988) Studies on the proteins of kodo millet (*Paspalum scrobiculatum*). J Sci Food Agric 42(4):315–323. https://doi.org/10.1002/jsfa.2740420405

Sun X, Saleh AS, Sun Z (2022) Modification of multi-scale structure, physicochemical properties, and digestibility of rice starch via microwave and cold plasma treatments. LWT 153:112483. https://doi.org/10.1016/J.LWT.2021.112483

Suri S, Grover A, Nagpal K, Patil AG (2024) Millet processing by-products: sustainable applications in food and non-food sectors. Nutrire 49(2):1–16. https://doi.org/10.1186/s41110-024-00274-7/tables/2

Suryavanshi VR, Santhosh R, Singhi H et al (2025) Preparation and characterization of kodo millet starch/gum tragacanth/copper oxide nanoparticles-based antimicrobial food packaging films. Materials Today Communications 42:111443

Tripathi MK, Mohapatra D, Jadam RS et al (2021) Nutritional composition of millets. In: Kumar A, Tripathi MK, Joshi D, Kumar V (eds) Millets and millet technology. Springer, Singapore, pp 101–119. https://doi.org/10.1007/978-981-16-0676-2_5

Upadhyay R, Jha A, Singh SP et al (2015) Appropriate solvents for extracting total phenolics, flavonoids and ascorbic acid from different kinds of millets. J Food Sci Technol 52(1):472–478. https://doi.org/10.1007/s13197-013-0976-0

Veerabadhran V, Palakuru HR, Pratap GP et al (2023) A critical review of the history and practices of millets in India. J Drug Res Ayurvedic Sci 8(1):S18–S25. https://doi.org/10.4103/jdras.jdras_113_23

Vijayakumar P, Mounika A, Shanmugam A, Farzana W (2024) A comparative study on effect of succinylation on kodo millet and chickpea starch with ultrasound treatment. Food Biosci 62:105104. https://doi.org/10.1016/j.fbio.2024.105104

Wang X, Reddy CK, Xu B (2019) Toward functional starch foods with longer shelf life: effects of inulin and konjac glucomannan on freeze-thaw stability, retrogradation, and pasting characteristics of selected starches. Starch - Stärke 71(5–6):1800285. https://doi.org/10.1002/star.201800285

Wang Q, Li L, Zheng X (2021) Recent advances in heat-moisture modified cereal starch: structure, functionality and its applications in starchy food systems. Food Chem 344:128700. https://doi.org/10.1016/j.foodchem.2020.128700

Yadav OP, Singh DV, Kumari V et al (2024) Production and cultivation dynamics of millets in India. Crop Sci 64(5):2459–2484. https://doi.org/10.1002/csc2.21207

Yadav M, Bansal C, Shukla U, Zainab S (2024b) A critical review on the efficacy of kodo, sorghum, and foxtail millets in the management of type 2 diabetes mellitus. Ind J Health Sci Biomed Res KLEU 17(3):200–208. https://doi.org/10.4103/kleuhsj.kleuhsj_243_24

Yi M, Tang X, Liang S et al (2024) Effect of microwave alone and microwave-assisted modification on the physicochemical properties of starch and its application in food. Food Chem 446:138841. https://doi.org/10.1016/J.foodchem.2024.138841

Zhang H, Zhao W, Bai T et al (2022) Sustainable extraction of polyphenols from millet using switchable deep eutectic solvents. LWT 170:114082. https://doi.org/10.1016/j.lwt.2022.114082

Zhu F (2014) Structure, physicochemical properties, and uses of millet starch. Food Res Int 64:200–211. https://doi.org/10.1016/j.foodres.2014.06.026

Chapter 6
Little Millet: Chemical Composition, Starch Isolation, Modification, Functionality, and Food/Nonfood Applications

Payal Sharma, Manila, and Bharat Bhushan

Abbreviations

HMT	heat moisture treatment
LMS	little millet starch
WBC	water-binding capacity
LM	little millet
XRD	X-ray diffraction
FTIR	Fourier-transform infrared spectroscopy
RS	Resistant Starch
RDS	Rapidly Digestible Starch
SDS	Slowly Digestible Starch

6.1 Introduction

Little millet (LM) is associated with the *Poaceae* family, which is native to the Indian peninsula (Indirani 2021). It is known as the king of cereals because of its superior nutritional and therapeutic properties. "Cool food" is a different name for LM due to the cooling effect on the body during the summer (Pradeep and Sreerama 2018). It is a species of plant within the order Poales. Their subfamily is Panicoideae,

P. Sharma
Department of Chemistry, Gurukula Kangri (Deemed to be University), Haridwar, UK, India

Manila · B. Bhushan (✉)
Department of Chemistry, Gurukula Kangri (Deemed to be University), Haridwar, UK, India

Department of Botany, Cluster University of Jammu, Jammu, J&KUT, India

© The Author(s), under exclusive license to Springer Nature Switzerland AG 2026

J. Singh, M. Kumar Malik (eds.), *Millets and Millet Starch: Chemical Composition, Functional Modification, and Applications*,
https://doi.org/10.1007/978-3-032-09502-2_6

Fig. 6.1 Little millet plant and grains with husk

and the tribe is Paniceae. Its botanical genus is *Panicum* (Ladumore et al. 2021; Shweta et al. 2018). In India, LM is referred to as Samai (Tamil), Sama (Telugu), and Kutki (Hindi) (Singh et al. 2023). LM is categorized as a minor millet. It contributes to a nutritious, eco-friendly food system (Paschapur et al. 2021). Researchers currently focus on collecting, conserving, and utilizing millet germplasm in breeding programs to improve their qualities and promote wider adoption in agricultural systems. Governmental bodies are also crucial in shaping policies encouraging millet cultivation and consumption (Jadhav and Londhe 2023). These millets contribute to diversifying food baskets and achieving food security. This chapter reviews the diverse and modified approaches adapted for starch isolation from LM and their impact on the quality and functionality of starch. Additionally, the food and non-food application of little millet starch (LMS) will also be discussed in this chapter. Hence, there is a captivating need nowadays to provide a detailed focus on LMS composition, the isolation method of starch, functionality, and its application (Fig. 6.1).

6.2 Geographical Distribution

In India, it is primarily grown in several states, including Tamil Nadu, Karnataka, Andhra Pradesh, Odisha, and other regions in the central and southern parts of the country (Panda et al., 2024). It is cultivated in more than 100 nations globally which includes Sudan, Ethiopia, Nepal, Mali, Bangladesh, Srilanka, Western Myanmar and Nigeria (Chapke et al. 2020; Taylor and Duodu 2015). India, China, and Nigeria are the largest manufacturers, accounting for over 55% of the world's millet production (Meena et al. 2021). In China, LM is the second most extensively produced cereal crop. Table 6.1 shows the geographical distribution of LM.

Table 6.1 Geographical distribution of little millet

Region	Distribution
India	Bihar, Karnataka, Madhya Pradesh, Tamil Nadu, Odisha, Andhra Pradesh, and Maharashtra
Global	Main production areas include Japan, China, India, and Eastern Europe

Table 6.2 Different varieties of LM, maturity period, cultivation state and their yield

Varieties of little millets	Days of maturity	Yields (quintal/hectare)	State
BL 6 Chhattisgarh-Kutki 2, (BL-4)	90–95	8–14	Chhattisgarh
JK 36, Jawahar-Kutki 4 (JK 4),	75–80	11–20	Madhya Pradesh
Paiyur 2	95–100	8–16	Tamilnadu
OLM 217	105–110	10–16	Orissa
Tarini (OLM)203, JK 8	100–108	11–17	Karnataka
OLM 203	90–95	9–15	Gujarat

6.3 Cultivation of Little Millet

LM is mainly cultivated in the kharif season. The best time for sowing rainfed kharif crops is June–July (Sao et al. 2024). The history of LM cultivation is extensive and spans over to thousands of years. It is considered the oldest crop, and archaeological evidence suggests that it was cultivated by early human societies in the Indian subcontinent (Tripathi and Vyas 2023). The accurate origin of LM is not closely documented; however, it is to be accepted as indigenous to India, as proved by its worldwide vernacular name around the nation (Swamy 2023). It was first cultivated in the Eastern Ghats of India around 4000 years ago when it became a significant part of the diet of tribal communities. From there, it spread across to Nepal, Myanmar, Sri Lanka, and other countries (Panda et al. 2024). It is mainly a crop that self-pollinates, and it is cultivated all through India at altitudes up to 2150 m (Nirmalakumari et al. 2010). It is widely cultivated in the tribal areas of Madhya Pradesh, Chhattisgarh, and Andhra Pradesh. Cultivation requires adequate spacing between plants to ensure optimal growth and yield. Additionally, sequential cropping of LM with niger, lentil, or guar can be considered. In the conditions of south Bihar, LM is generally succeeded by niger (Chapke et al. 2018). In Gujarat state, it is cultivated across 10,634 hectares, yielding 9526 tons with an output of 896 kg/ha (Rajput et al. 2022). LM crop is primarily grown in the districts of Navsari (South Gujarat), Dangs, Panchmahal (Middle Gujarat), and Valsad. This is adapted to both climates: tropical and temperate; it can withstand drought and waterlogging (Padulosi et al. 2015). It is a staple food for the rural population, particularly in areas with challenging climatic conditions. LM is well-suited for fragile and exposed agroecosystems due to its eco-friendly nature (Patil et al. 2015). Table 6.2 shows the different varieties, yields, maturity period and state of cultivation of LM.

6.4 Storage and Processing of Little Millet

LM grains need primary and secondary processing to obtain the food and goodness value. The primary processing initiates with threshing the grain, followed by dehusking because LM consists of a hard seed coat. Further sorting, grading, and polishing/pearling are carried out to improve grain quality and transform the grain into a more useful form (Birania et al. 2020; Rao et al. 2022). After primary processing, the grains goes through unit operation that convert into useful products generally for direct consumption. They are usually done away from farms either in unorganized or organized sectors (Kate and Singh 2021). In the end, polishing is optional and it does not affect the nutritional value as it helps improve the appearance and removes any remaining husk fragments (Chapke et al. 2020). Figure 6.2 shows the flowchart of processing after harvesting of LM.

Once cleaned, it is suggested to store LM in a moisture content of approximately 10–12% to prevent insect infestation and spoilage. Grains should be stored in a dry, cool, well-ventilated space to prevent them from rancidity. Sealed airtight containers and food-grade plastic bags are good options for storage. It is essential to store the seeds in the proper pest control region and also avoid temperature fluctuations for long-term storage (Gupta et al. 2018; Mobolade et al. 2019). Table 6.3 shows the methods of storage of LM.

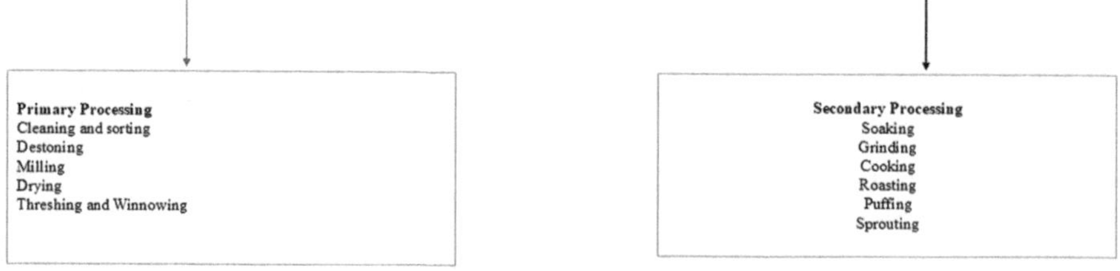

Fig. 6.2 Flowchart of processing after harvesting of little millet. (Source image 2: WikimediaCommons Retrievedch 26, 2025. Source image 1:www.needpix.com)

Table 6.3 Storage methods of little millet

Storage method	Description	Reference
Aerial storage	The temporary method exposes grains to air	Nimbkar et al. (2022)
Open platform	Temporary, exposes grains to air and sunlight	Ghode (2023)
Earthenware pots	Protects from pests and moisture	Gupta et al. (2018)
Hermetic storage	For bulk scale storage, better quality grain retention, no visible mold/pest infestation	Dip and Sharma (2023)

Table 6.4 General morphology of little millet grain and harvesting time

Millet type	Little millet
Color	Gray to white
Length of grain	1.7 to 1.8 mm
Shape of grain	Elliptical to oval
Size	1.0–8.0 μm
Kernel weight	Approximately 1.8 gram
Origin	Asia region
Crop	Kharif
Harvesting time	September to October

6.5 Little Millet Plant Morphology

LM is classified into two subspecies according to inflorescence morphology: *sumatrense and psilopodium.* The races "robusta" and "nana" are subspecies of *sumatrense* (Dayakar et al. 2017). The "nana" variety typically rises to a height of 65–75 cm, with an inflorescence of 15–16 cm long, upright, open, and extensively branched. The branches may be droop as they mature. In contrast, the "robusta" variety is taller, reaching a height of 120–190 cm (Upadhyaya et al. 2008; Arya et al. 2021). Its inflorescence is around 21–48 cm long, compact, and extensively branched. The inflorescence of LM is known as a panicle (Natesan et al. 2020). LM grains are gray to white in color (Kundgol et al. 2014). The shape of the LM grains is elliptical to oval, and its length is around 1.7–1.9 mm long (Ganapathy 2017; Patil et al. 2024). The leaves of LM are linear and arranged alternately along the stem. Table 6.4 shows the general morphology and harvesting season of LM (Kumar et al. 2024).

6.6 Economic and Nutritional Importance of Little Millet

LM is an impeccable source of powerful antioxidants such as flavonoids, polyphenols, tannins, and phenolic compounds. These antioxidants are crucial for improving health by preventing diseases such as inflammation, diabetes, cardiovascular disease, gastrointestinal problems, and cancer (Singh and Chaudhry 2023). LM can help in tackling lifestyle diseases because of its slow digestibility and high fibre content. According to a 2017 study published in the Nutrition Society proceedings, millet grains are healthier and have higher levels of antioxidants and polyphenols than other millet flour or flakes (Singh and Chaudhry 2023). Their high fiber content helps in digestion and prevents constipation. They are also rich in anti-inflammatory compounds (Yadav et al. 2023).

6.6.1 Health Benefits of LM

LM is more nutritious than other cereals and are currently regarded as "nutri-cereals." It is a useful cereal crop that boasts several nutraceutical components. Its richness in healthy polyunsaturated fatty acids makes it a nutrient-dense grain. A study on schoolgoing girls was performed (Bhat et al. 2018; Chandrakar and Shahi 2023). They were given millet-based food like upama, khichdi, and paddu. The study demonstrated significant enhancements in hemoglobin levels, height, and weight among participants following the LM feeding trial compared to the control group (Anitha et al. 2019). Further, LM demonstrated the highest calorie content, with a value of 322.5 kcal (Karuna and Nazni 2018). The approximate composition of the grains is depicted in Table 6.5 shows the health benefits of LM.

- *Diabetes management*: LM has a low glycemic index, which helps in maintaining blood sugar level. The study done through the National Library of Medicine found that LM helps stabilize blood sugar levels, which is beneficial for diabetics who need to manage glucose fluctuations (Jacob et al. 2024; Agrawal et al. 2023).

Table 6.5 Nutritional value of LM per 100 g (Kumari et al. 2022; Indirani 2021)

Nutritional content	Amount per 100 g
Calcium	20 mg
Fats	5.4 g
Proteins	8.45g
Isoleucine	580 mg
Carbohydrate	67 g
Vitamin B2	0.09 mg
Ferrous	9.4 mg
Vitamin B1	0.29 mg
vitamin B3	3.5 mg
Zinc	1.7 mg
Sodium	5 mg
Potassium	115 mg
L-tryptophan	180 mg
Crude fiber	8.0 g

- *Weight loss and detoxification*: It offers several health benefits related to weight management, tissue repair, and post-workout energy production. Its high water-soluble fiber content prolongs gastric emptying and supports weight management (Nithiyanantham et al. 2019).
- *Provides high-quality protein*: This millet is like a powerhouse of tiny protein, providing 7.7 grams per 100 grams comparable to eggs, wheat, rice, and chicken. Moreover, it contains all the essential amino acids, the fundamental components of protein, essential for our well-being, as our body cannot produce them alone (Sachdev et al. 2023).
- *Antioxidant properties*: It is rich in antioxidants that neutralize harmful free radicals in the body, potentially reducing the risk of chronic disorders (Chandrasekara and Shahidi 2011; Yadav et al. 2023; Pradeep and Guha 2011).
- *Gluten-free option*: A gluten-free diet is essential for people who have a gluten-related medical condition, such as celiac disease or gluten ataxia (Jacob et al. 2024). If you are following a gluten-free diet or have celiac disease, then LM is an excellent alternative. Its nutritional value makes it a valuable addition to balanced diets, especially for those who engage in yoga, workouts, or cardio (Sachdev et al. 2023).
- *Good source of potent antioxidants*: LM contains compounds like phenols, flavonoids, and tannins, which provide antioxidant properties (Mahajan et al. 2021a). While these antioxidants do not directly provide nourishment, they play a crucial role in enhancing health by assisting in managing conditions, including heart disease, inflammation, cataracts, cancer, and digestive disorders (Shweta et al. 2018).
- *Helps in reducing cholesterol*: LM is abundant in magnesium, which provides benefits to cardiovascular disorders. Additionally, it has a significant amount of niacin, which helps lower cholesterol levels (Yadav et al. 2023; Kaimal et al. 2021). Phenolic extracts from the kernel of LM, at a concentration of 0.49 mg/mL, reduced LDL cholesterol oxidation by 1–40% (Dey et al. 2022).

6.7 Little Millet Starch: Isolation and Functional Properties

Millet starch can be isolated using three different methods. The commonly applied three consecutive phases involved in these methods include anatomic fragmentation of starch, cell disruption, and finally purification process of starch (Mahajan et al. 2021a, b). It is necessary to choose a method that gives high starch yields with lower protein residues (Mahajan et al. 2022). The process of starch isolation (e.g., alkaline, enzymatic, or acidic) has an important effect on the yield. In comparative research between alkaline and acid steeping methods, the acid steeping method results in higher residual protein content, around 4.2%, compared to the alkaline steeping method around 0.8% (Yanez and Walker 1986). The alkaline isolation approach is widely favored due to its effectiveness in removing associated proteins (Kumar et al. 2024).

6.7.1 Methods for Starch Isolation of Little Millet

Several researchers reported that kernels should be processed into flour before processing. The LM flour is ground in a fine powder and steeped in a 0.1% $Na_2S_2O_5$ solution for 18–20 h. The slurry was allowed to settle for 4–5 h before the upper layer was carefully removed. The settled starch layer was then centrifuged at 3000 rpm for 6 to 7 min. The upper off white layer was scraped off to discard the nonstarch portion. To eliminate any contaminants, the white layer was resuspended in water and centrifuged. The process of steeping and centrifugation is carried out multiple times to eliminate impurities. The starch was collected and dried (Singh and Singh 2005).

In another method, LM grain flour was steeped in a 0.20% NaOH solution for 24 h at 4 degrees Celsius. The soaking flour was then passed through British standard sieves. The resulting slurry underwent cleaning and centrifugation. For further purification, the extracted starch was suspended in a 1:1 solution of 0.1 N NaCl and toluene, stirred, centrifuged, and washed with distilled water. The white starch of LM was air-dried and stored in sealed containers (Thabet et al. 2016; Suma and Urooj 2015; Mahajan et al. 2022). Figure 6.3 represents methods used for starch isolation from the LM kernel. The impact of the isolation method on starch yield and its components is presented in Table 6.6.

6.7.2 Chemical Composition of Little Millet Starch

The starch content in LM kernels falls within the range of approximately 42–64% on a dry weight basis (w/w) (Kumari and Thayumanavan 1997). It has been documented that the amylose and amylopectin content in the LM kernel was around 28.5% and 50–60%, respectively (Nazni and Bhuvaneswari 2016; Kaimal et al. 2021). Further studies reported that the amylose content of LM genotypes (improved variety sukshema and local genotype) was around 26.48 and 23.32 g/100 g, respectively. The results also show that the amylopectin content observed in the improved variety sukshema contained 33.75 g/100 g and the local variety comprised 32.81g/100 g (Roopa et al. 2013). Kulkarni and Naik (2009) reported that the value of moisture content of LM flour is 10.8%, which was the second highest value, succeeding proso millet (10.59%). The protein, ash, and fat content values were also recorded at 8.70, 3.00, and 4.93 g/100 g, respectively. The glycemic index of LMS (65.23) was found to be lower than the Kodo starch (68.47) (Annor et al. 2013). Table 6.7 represents starch fractions of LM.

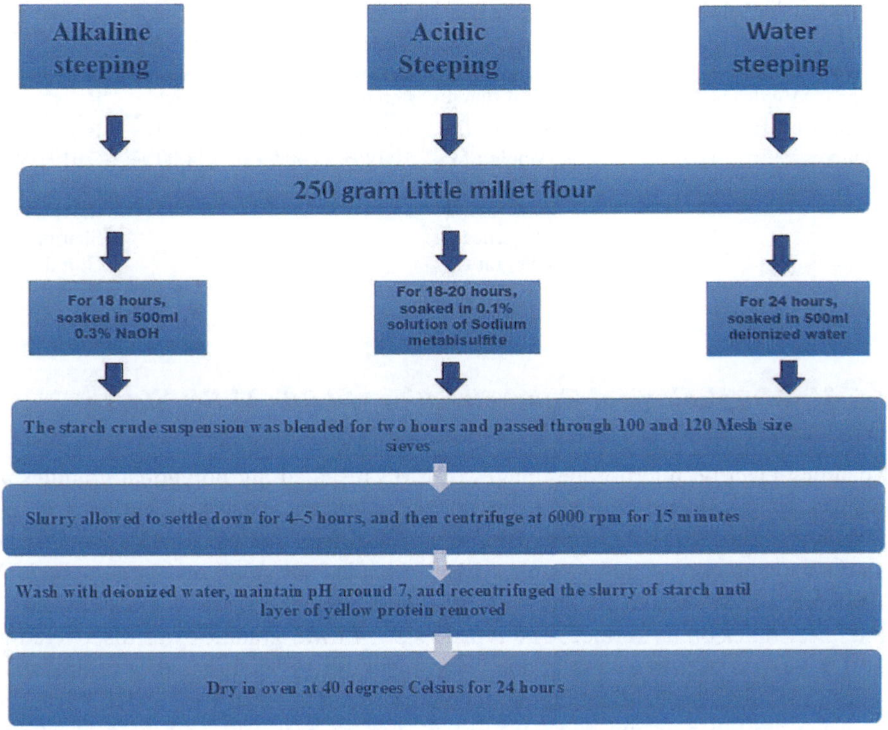

Fig. 6.3 Process for starch isolation from little millet kernel. (Kumar et al. 2024; Adkins and Greenwood 1966; Krishna Kumari and Thayumanavan 1998; Sandhu and Singh 2005)

Table 6.6 Impact of isolation method on starch yield and their components

S. no	Method of starch isolation	Reagent/ chemical	Protein content	Starch content	Amylose content	Reference
1.	Alkaline steeping method	NaOH	7.7%	60 to 70%	32.06%	Mahajan et al. (2022) and Kumar (2023)
2.	Alkaline steeping method	NaOH	6.36%	52–80%	33.02%	Mahajan et al. (2021a, b) and Nambi et al. (2012)
3.	Alkaline steeping method	–	7.96%	53.12%	32.09%	Mahajan et al. (2024), Bora (2020) and Akshitha et al. (2022)
4.	Acidic steeping method	Acetate buffer	8.49%	–	18.81%	Krishna Kumari and Thayumanavan (1998) and Roopa et al. (2013)
5.	Acidic steeping method	Acetate buffer	–	44.3%	18.8%	Nazni and Bhuvaneswari (2016)

Table 6.7 Starch fractions in LM kernels (g/100 g starch)

Millet type	Amylose	Amylopectin	Total starch	Slowly digestible starch	Resistant starch	Rapid digestible starch	Reference
Little millet	12–18%	83.10–90.00%	57% to 60% (whole LM grains) 77% to 87% (decorticated LM grains)	20.00–31.12%	47.54–48.99%	21.19–34.00%	Bora et al. (2019), Vali Pasha et al. (2018) and Sharma and Gujral (2020)

6.7.3 *Morphological Characteristics of Little Millet Starch*

Scanning electron microscope images of LMS revealed big polygonal granules as well as spherical ones similar to proso, pearl, finger, and foxtail millet starches (Mahajan et al. 2021b). A comparative study found that the LMS granules are smoother than corn starch granules. LMS granules are small and large spherical. LMS has a lower water-binding capacity (WBC) than corn starch granules due to the fewer holes on its surface. The typical size of LMS granules was discovered to be 1.0–9.0 μm (Krishna Kumari and Thayumanavan 1998). In another comparative study, the area mean diameter and volume mean diameter of starches ranged between 4.69 to 5.88 μm and 6.90 to 9.50 μm, respectively, and the highest and smallest values were identified for barnyard and LMS, respectively (Bangar et al. 2021).

6.7.4 *Functionality and Crystalline Structure*

In a comparative study of some millet starches, XRD results proved that LMS has an A-type crystalline structure, which is similar to cereal starches (Bangar et al. 2021). LMS has a type A crystalline structure, with a distinct peak at 2θ around 15°, single peak at 22°, and doublet peaks at 18° and 19° (Mahajan et al. 2021a, b). The Debye–Scherrer formula was used to find the crystallite size (6–10 nm) of LMS. LMS has a less distinct peak. The reduced crystallinity of LM is attributed to its greater amorphous region, which hinders enzymatic degradation caused by the dense packing of amylose chains (Mahajan et al. 2022). FT-IR spectroscopy of LMS exhibits distinct broadbands at 929.3–2930.2 cm^{-1} and 2983.2–2987.1 cm^{-1} assigned to C–H deformation in the stretching vibrations of the CH2 bond and the broad band of OH stretching vibrations at around 3383–3377 cm^{-1} due to hydroxyl groups in the unit of glucose, which shows the characteristic carbohydrate peaks (Mahajan et al. 2022).

6.8 Physicochemical Properties of Little Millet Starch

The functional attributes of LMS, including solubility, gelatinization, swelling, viscosity, textural qualities, and so on, make it suitable for use in both food and non-food applications. The functional qualities of different LM cultivars can vary significantly depending on the distribution of amylopectin chain lengths (Mahajan et al. 2022). The potential uses of millet starch in food product production are largely determined by its functional properties, including solubility, thermal behavior, gelatinization, swelling, viscosity, and digestibility as detailed in the subsequent section (Kumar et al. 2024).

6.8.1 Swelling Power and Solubility

Swelling power, which indicates the water-holding ability of starch, is commonly used to distinguish different types of starches. The swelling power of LMS is less than that of rye, wheat, and potato starches, suggesting that LM starch granules are more resistant to swelling due to robust intra- and intermolecular bonding. The swelling power of LMS at 49°C, 59 °C, 69 °C, 79 °C, and 89°C was found to be 3.23,3.85, 5.79, 8.30, and 9.39 (g/g), respectively (Mahajan et al. 2022; Mahajan et al. 2024). In a comparative study, the solubility and swelling power of LMS were observed to range between 14.9% to 25.8% (Yang et al. 2020). Phattanakulkaewmorie (2011) showed that the presence of elevated levels of protein, lipid, fiber, and a greater amount of amylose-lipid complex in starch might restrict starch swelling. The water solubility of starch is influenced by amylose content and its crystalline structure. For LMS, solubility typically ranges from 15.2% to 25.9% (Bangar et al. 2021; Mahajan et al. 2021a, b; Krishna Kumari and Thayumanavan 1998). The high solubility power of LMS allow it to absorb a significant amount of water, beneficial for baked products like bread and pasta (Mahajan et al. 2021a, b).

6.8.2 Water Absorption Capacity and Oil Binding Capacity (OBC)

A study reported that the WBC and OBC of LMS are approximately 0.96 g/g and 1.09 g/g, respectively (Butt and Batool 2010). Another study concluded that the WBC was 1.38 g/g, significantly lower than corn starch (Mahajan et al. 2022). Another study found that the OBC of native LMS was found to be 84.37 ± 3.30 ml/100 g, and modified millet flour had the highest OBC, which is 86.91 ± 2.13 ml/100 g (Thilagavathi et al. 2015). The WBC and OBC of the native LMS were obtained as 1.50 g/g and 0.69 ml/100 g, respectively, which exhibited the lowest value compared to the modified LMS.

6.8.3 Pasting Property

The pasting properties of food are influenced by the type of starch, its composition, concentration, cooking, cooling temperatures, and the presence of solutes like sugar, lipids, and pH levels. The higher pasting properties of LMS suggest its usage in products that demand great gel power (Mahajan et al. 2022; Mahajan and Bera 2023). A study reported that the pasting temperature of LMS is 95°C (Mahajan et al. 2022). In a comparative study, it was reported that LMS had the lowest breakdown viscosity, indicating greater stability during shearing and cooking. The setback viscosity of LMS ranged from 810 to 1906 cP, with LMS having the lowest value compared with other millet starch. A comparative study with different millet starch revealed that the pasting temperature ranged from 79.1 to 84.1°C. The lowest values were recorded for LMS (Bangar et al. 2021). It has also been previously reported that the pasting viscosity of LMS was 450 BU (Kumari and Thayumanavan 1997).

6.8.4 Gelatinization Properties

Starch gelatinization determines its suitability for thickening, gelling, and extrusion (Sharma and Niranjan 2018). When heated with moisture, it gelatinizes and shows specific behaviors. The thermal characteristics of starch, such as peak temperature (Tp), onset temperature (To), and enthalpy of fusion (the energy required to change starch from a solid to a liquid), are commonly measured using a differential scanning calorimeter (DSC). Starch gelatinization involves the disruption of intermolecular bonds of starch through the application of heat. For LMS, the gelatinization temperature typically falls from 64°C to 74°C (Mahajan et al. 2022; Mahajan and Bera 2023). The proportion of amylopectin crystals in millet starch affects the gelatinization temperature. A high gelatinization temperature for starch shows that more energy is required to start the gelatinization process (Bangar et al. 2021). In a comparative study between proso millet starch and LMS, the onset temperatures were found to be between 63.57–81.08°C and 72.27–83.29°C, respectively (Kumar et al. 2020).

6.8.5 Starch Digestibility

The starch is tested for in vitro enzyme digestibility to determine digestibility-related characteristics. There are three types of digestible starch: slowly digestible starch (SDS), resistant starch (RS), and rapidly digestible starch (RDS) (Sajilata et al. 2006). According to earlier research, the amylose content, along with the RS, RDS, and SDS contents of LM, varies by approximately 28.6%, 40.1%, and 10..7%, and 21.8% respectively (Bora et al. 2019; Sharma and Gujral 2020; Krishna Kumari and Thayumanavan 1998). Another study reported that LM contains 30.11% resistant starch (Mahajan et al. 2022). It has been documented that in vitro starch digestibility of LM decreases by 29% due to cooking (Kumar et al. 2020). Sharma and Gujral (2020) concluded in their study that when LM was soaked and then sprouted, the content of RS (12.01–27.22%) and SDS (9.00–26.05%) significantly decreased while the content of RDS (16.16–27.21%) increased in comparison to native starch.

6.8.6 Analysis of Texture

The texture properties of starch were evaluated through the analysis of starch gels. The hardness of LMS gels varies approximately 53.14 g to 55.15 g. Cohesiveness is regarded as the ideal measure for identifying a quality set of gels, and the value of cohesiveness of LMS was found to be around 0.40–0.34 g. The chewiness ranges from 44.30 g to 45.67 g. The adhesiveness of the LMS was comparatively lower (Mahajan et al. 2022). In an evaluative study, the textural properties of LM flour showed maximum and minimum hardness values around 27.85 N and 7.84 N, respectively (Saini and Yadav 2018). The viscosity indices of LMS are low, ranging from 2.16cP to 2.39cP, which indicates that their viscosities undergo rapid changes with temperature (Dey and Sit 2017). A recent study observed the increment in the hardness of bread due to the incorporation of LMS from 1.77 to 5.72 N (Mannuramath et al. 2015).

6.9 Modification of Little Millet Starch

Native starch exhibits drawbacks such as thermal instability, elevated hot-paste viscosity, and susceptibility to shear stress. It degrades rapidly during cold storage and has poor resilience to acidic conditions and high temperatures, which impacts product quality and limits its suitability for widespread use in food processing. Modified

Table 6.8 Different methods of modification of LMS and their impact on various properties

Methods of modification		
Physical modifications	Altered properties	References
Microwave treatment	Increase swelling ability, viscosity, crystalline structure, property of gelatinization, and oxidation resistance	Zhi et al. (2022)
Ultrasonic treatment	Enhanced peak time, viscosity, and temperature of pasting	Kumar et al. (2024)
Chemical modifications		
Cross-linking	Low digestibility, showing viscosity, and degradation	Mahajan et al. (2021a)
Enzymatic method (lysine, threonine, and aspartic acid)	Enhanced swelling power and improved properties of textural	Mahajan et al. (2021b)
Enzymatic method (lysine)	Viscosity and swelling power	Mahajan et al. (2024)
Enzymatic method (lysine)	Gelling power and viscosity	Mahajan and Bera (2023)

native starch addresses its inherent limitations, allowing for broader industrial applications (Kumar et al. 2024; Mahajan et al. 2024). These modifications enhance functional properties, extend shelf life, and facilitate use in diverse sectors such as medicine, pharmaceuticals, papermaking, and food processing. Native starch can be modified using chemical, enzymatic, physical, and combined methods (Adewale et al. 2022; Guida et al. 2021). Physical modifications of starch involve techniques such as hydrothermal treatment HMT (including heat) (Zia-ud-Din et al. 2017), and annealing treatments, and other physical treatments such as HMT, irradiation, ultra-high pressure, and microwave treatment) (Thilagavathi et al. 2015). Chemical-based modification of starch methods includes esterification, cross-linking, etherification, acid hydrolysis, and oxidation (Thakur et al. 2023). Physical modification (nonthermal) of starch affects its functional, structural, and physical characteristics, including alteration in crystallinity, swelling ability, solubility, heat stability, and viscosity (Krishna Kumari and Thayumanavan 1998; Kumar et al. 2024). Table 6.8 shows various methods of modification of LMS.

6.9.1 Physical Modification of Little Millet Starch

Physical modification is a common technique for treating native starch under varying conditions of pressure, moisture, irradiation conditions, and temperature. These treatments are considered cost-effective and safer because they do not involve any chemicals or biological agents. As a result, physical modification methods reduce granule size and improve water solubility (Kushwaha and Kaur 2018). Annealing, HMT, and pre-gelatinization are some of the most common physical modifications. Additionally, advanced techniques such as ball milling (Han et al. 2022), pulsed electric field (Chen et al. 2021), ultrasonication (Rahaman et al. 2021), and microwave (Babu and Parimalavalli 2018) were used for starch modification.

- *Properties of Physically Modified LMS*

Physical modification of LMS by the autoclaving cooling cycle method found that the OBC of LMS was higher in modified flour compared to native flour. The findings also conclude that modified LMS had higher soluble, insoluble, and total dietary fiber content than native flour (Thilagavathi et al. 2015) Roopa et al. (2012). LMS has a soluble dietary fiber, insoluble dietary fiber, and total dietary fiber content ranging from 2.99 to 3.30 g/100 g, 5.20 to 5.30 g/100 g, and from 8.25 to 8.60 g/100 g, respectively. Further, it has been reported that the swelling index of modified LMS is 60 ± 1.15g/g (Tamaka 2019). It has also been reported in previous studies that microwave treatment applied to LMS considerably increases the rapidly digestible and slowly digestible starch content ($p < 0.05$) while lowering the resistant starch (Zhi et al. 2022; Babu et al. 2019). The results also show that the peak viscosity of microwave-treated LMS decreased with increasing microwave time than the native starch. Furthermore, physical modification reduces the relative crystallinity and short-range structure of LMS (Zhi et al. 2022). In a previous study, it was found that pan-cooked LMS and microwave-cooked LMS had a deviation in the region of carbohydrate (1200-901 cm^{-1}) compared to raw LMS but did not show any changes in protein region (1700–1601 and 1571–1535 cm^{-1}) (Kumar et al. 2020).

6.9.2 Chemical Modification of LMS

Chemical modification is a common approach to enhance the properties of starch that involves the use of chemical reagents to introduce native functional groups into starch granules (Jiranuntakul et al. 2014). However, limited studies have been conducted on modifying LMS by utilizing chemicals (Mahajan et al. 2021a, b; Mahajan et al. 2022).

- *Properties of Chemically Modified LMS*

Modified starches exhibited a significant increase in short-chain fragments, varying from 20% to 55% depending on the type of modification method of starch. This impact is amplified in acidic treatment due to the breakage of long-chain amylopectin and amylose branches. Similar findings were observed in starch samples treated with succinic anhydrate, which was attributed to the breakdown of amylose chains and disruption of amylopectin branches at C–O–C connections (Babu et al. 2019; Kumar et al. 2024). The modified LMS can be used in capsule and tablet formation for the pharmaceutical industry.

6.9.3 Enzymatically Modified Starch

Over the past few years, enzymatic modification have gained significant attention as a replacement for physical and chemical starch modification methods because of their safety, environmental friendliness, and ability to produce safe by-products (Uthumporn et al. 2010). Enzymatic modification changes the solubility, viscosity, and gelation properties of native starch, making it valuable in both food and nonfood applications (You and Izydorczyk 2007). Additionally, such modifications lead to granular disorganization, molecular rearrangements within starch granules and disruption of polymer networks, changes in amylose and amylopectin ratios, crystallinity, particle size, morphological changes, and so on (Lin et al. 2016). Enzymatic modification of starches, especially LMS, is rare in the literature. According to the literature survey, fewer studies have been conducted on the structural characteristics of enzymatically modified millet starches (Kumar et al. 2024).

- *Properties of Enzymatically Modified LMS*

The modification of LMS with amino acids as a cross-linker influenced a variety of characteristics. The swelling power of lysine-modified LMS has a higher value, and the amylose content varies significantly, influencing their hardness, gel setting, and adhesiveness (Mahajan et al. 2024). It was observed that charged amino acids had a greater influence on starch properties, whereas neutral amino acids had a moderate effect on LMS. In a study, the modification of LMS with aspartic acid, threonine, and lysine was carried out for improved function, and the result indicated that lysine-modified LMS had better gelling and textural properties (Mahajan et al. 2021b). Amino acids like threonine (neutral), lysine (positively charged), and aspartic acid (negatively charged) can serve as cross-linkers at various pH levels. They facilitate the modification of LMS by adding functional groups between its chains, leading to the formation of hydrogels. The results also prove that lysine amino acids may be a more effective cross-linker for the modification of LMS, which is also used in the development of gelling agents to deliver nutraceuticals (Siroha et al. 2021; Wang et al. 2022). The amino acid-modified LMS gels showed the lowest hardness (54.123) and were lower than the native. Lysine-modified LMS gels show the maximum cohesiveness range (0.43–0.85) over the different pH (Mahajan et al. 2021a, b).

6.10 Food and Nonfood Applications

There are diverse applications of LM and LMS in both food and nonfood contexts (Zhu 2014). LM contains abundant magnesium, which contributes to heart health improvement (Kaimal et al. 2021). It is also rich in phosphorus, which supports weight management, tissue repair, and energy production after intense workouts (Singh and Chaudhry 2023).

6.10.1 Food Applications of LM

LM has become a popular choice in the food industry for developing value-added products (Kaimal et al. 2021). It serves as an excellent gluten-free alternative. LMS can be incorporated in various baking recipes, including bread, cookies, biscuits, and flakes (Lydia et al. 2023). Value-added products from LM are given next:

- *Weaning mix from LM*: A study was conducted on a weaning mix for youngsters that included Little millet, kodo millet, foxtail millet, and finger millet in various concentrations. All millets were germinated before use. In comparison to the wheat-based market sample, millet-based weaning mixtures had more protein (14.3 to 23.17%), ash (2.8 to 7.6%), and crude fiber (2.5 to 6.0%) (Prasanna et al. 2020).
- *Flakes from LM*: LM flakes are convenient for quick breakfasts or snacks. Takhellambam et al. (2016) evaluate the production and quality evaluation of LM chocoflakes and their nutritional values. Further research revealed that flakes from LM also unveil a low glycemic load and glycemic index in normoglycemic subjects (Patil et al. 2015).
- *Biscuits and cookies from LM*: Incorporating LM flour into baked goods enhances their nutritional value. Wheat flour biscuits were standardized with the incorporation of LM flour. The developed biscuits were evaluated for sensory attribute and overall consumer appeal (Hiregoudar and Mamatha 2022). Biradar et al. (2021) made cookies using LMS as a substitute for refined wheat flour. LMS-based biscuit has high ash, iron, calcium, riboflavin, fiber, and thiamine content (Hemalatha et al. 2006).
- *Pasta from LM:* Devi et al. (2014) conducted a research experiment to develop LM-based ready-to-cook pasta mixed with wheat grain flour at various ratios. This study also estimates storage stability of the developed pasta food product by sensory, biochemical, and physicochemical changes when stored under selected packaging materials.

- *Rice analogs*: Innovations in rice analogs using LM are gaining popularity. It has been reported that anaras were prepared with various types of LM flour and rice to evaluate the suitability of LM for the preparation of traditional rice-based fried snack anaras (Ninganagoudar et al. 2012).
- *Bread*: LM flour was substituted for wheat flour in various quantities in the bread-making process to produce a fiber-rich functional loaf (Mannuramath et al. 2015).

6.10.2 Nonfood Applications

LMS has many nonfood applications and benefits. Some are discussed next:

- *Hydrogels*: In the research by Mahajan et al. (2021a, b), modified starch of LM was used to fabricate hydrogels for the delivery of various drugs.
- *Edible film*: LMS was used to make edible film (Bangar et al. 2021).
- *Nanoparticle*: The synthesis of silver nanoparticles from LMS with different varieties of millet has been reported. The developed silver nanoparticles were found to have effective antimicrobial activity against specific microorganisms (Shivakumar and Tejaswini 2023)
- *Fat replacer and gelling agent*: It has been reported that LMS is utilized as a thickening and stabilizing agent in many food products and also as a gelling agent (Mahajan et al. 2021a, b).

6.11 Conclusion

This chapter focuses on the chemical composition of LM, isolation of starch, modification, functionality, and food/nonfood applications. With the above knowledge, it is understandable that LM is highly nutritious and offers various health benefits. It also concludes that LMS has various significant functionalities, for example, swelling power and digestibility, which are beneficial for food/nonfood applications. This millet is a nutritious grain that can positively impact children's health. It offers a sustainable choice for modern agriculture, contributing to both food security and environmental adaptability. The significant functional properties of LMS make it suitable for pharmaceutical applications. LMS-based foods support proper growth and development, boost immunity, provide sustained energy, and contribute to cognitive function. However, due to its significance in food processing, further research is essential to explore its applications, high-yield method of starch isolation, nutraceutical value, and safety for incorporation into food products.

Conflict of Interest The authors have declared that there is no conflict of interest.

References

Adewale P, Yancheshmeh MS, Lam E (2022) Starch modification for non-food, industrial applications: market intelligence and critical review. Carbohydr Polym 291:119–590. https://doi.org/10.1016/j.carbpol.2022.119590

Adkins GK, Greenwood CT (1966) Studies on starches of high amylose-content. Part VII. Observations on the potentiometric iodine-titration of amylomaize starch. Carbohydr Res 3:81–88. https://doi.org/10.1016/S0008-6215(00)82299-5

Agrawal P, Singh BR, Gajbe U et al (2023) Managing diabetes mellitus with millets: a new solution. Cureus 15(9). https://doi.org/10.7759/cureus.44908

Akshitha R, Puskuri JS, Devi TS et al (2022) Comparative study of starch content in Foxtail, Little, Kodo, Proso, Barnyard and Brown Top Millets processed by traditional and modern methods. Biol Forum 14(3):1265–1270

Anitha S, Kane-Potaka J, Tsusaka TW et al (2019) Acceptance and impact of millet-based mid-day meal on the nutritional status of adolescent school going children in a peri-urban region of Karnataka State in India. Nutrients 11(9):2077. https://doi.org/10.3390/nu11092077

Annor GA, Marcone M, Bertoft E et al (2013) In vitro starch digestibility and expected glycemic index of kodo millet (Paspalum scrobiculatum) as affected by starch–protein–lipid interactions. Cereal Chem 90(3):211–217. https://doi.org/10.1094/CCHEM-06-12-0074-R

Arya L, Singh M, Verma M et al (2021) Molecular studies on millets and pseudocereals. In: Millets and Pseudo Cereals. Woodhead Publishing, pp 65–117

Babu AS, Parimalavalli R (2018) Effect of pullulanase debranching and storage temperatures on structural characteristics and digestibility of sweet potato starch. J Saudi Soc Agric Sci 17(2):208216. https://doi.org/10.1016/j.jssas.2016.04.005

Babu AS, Mohan RJ, Parimalavalli R (2019) Effect of single and dual-modifications on stability and structural characteristics of foxtail millet starch. Food Chem 271:457–465. https://doi.org/10.1016/j.foodchem.2018.07.197

Bangar SP, Siroha AK, Nehra M et al (2021) Structural and film-forming properties of millet starches. a comparative study. Coatings 11(8):954. https://doi.org/10.3390/coatings11080954

Bhat S, Nandini C, Tippeswamy V (2018) Significance of small millets in nutrition and health – a review. Asian J Dairy Food Res 37(1):35–40. https://doi.org/10.18805/ajdfr.DR-1306

Biradar SD, Kotecha PM, Godase SN et al (2021) Studies on nutritional quality of cookies prepared from wheat flour and little millet. Int J Chem Stud 9(1):1675–1680. https://doi.org/10.22271/chemi.2021.v9.i1x.11466

Birania S, Rohilla P, Kumar R et al (2020) Post harvest processing of millets: a review on value added products. Int J Chem Res 8(1):1824–1829

Bora P (2020) Nutritional potentiality of millets. In: Shukla R (ed) Food technology and nutrition, 14th volume. AkiNik Publications, New delhi

Bora P, Ragaee S, Marcone M (2019) Characterisation of several types of millets as functional food ingredients. Int J Food Sci Nutr 70(6):714–724. https://doi.org/10.1080/09637486.2019.1570086

Butt MS, Batool R (2010) Nutritional and functional properties of some promising legumes protein isolates. Pak J Nutr 9:373–379. https://doi.org/10.3923/pjn.2010.373.379

Chandrakar L, Shahi S (2023) Millets and their nutritional value: a review. J Adv Zoo 44:1690–1697. https://doi.org/10.17762/jaz.v44iS3.2387

Chandrasekara A, Shahidi F (2011) Determination of antioxidant activity in free and hydrolyzed fractions of millet grains and characterization of their phenolic profiles by HPLC-DAD-ESI-MSn. J Funct Foods 3(3):144–158. https://doi.org/10.1016/j.jff.2011.03.007

Chapke RR, Prabhakar SG, DasIK DIK, Tonapi VA (2018) Improved millet production technologies and their impact, Technology Bulletin. ICAR-Indian Institute of Millets Research, Hyderabad

Chapke RR, Shyam Prasad G, Das IK et al (2020) Latest millet production and processing technologies, Booklet. ICAR–Indian Institute of Millets Research, Hyderabad

Chen BR, Wen QH, Zeng XA et al (2021) Pulsed electric field assisted modification of octenyl succinylated potato starch and its influence on pasting properties. Carbohydr Polym 254:117294. https://doi.org/10.1016/j.carbpol.2020.117294

Dayakar RB, Bhaskarachary K, Arlene GD et al (2017) Nutritional and health benefits of millets, vol 2. ICAR_Indian Institute of Millets Research (IIMR) Rajendranagar, Hyderabad

Devi GS, Kumar PA, Palanimuthu V (2014) Studies on development and storage of little millet-based pasta. Madras Agric J 101(1–3):96–101. https://doi.org/10.29321/MAJ.10.001159

Dey A, Sit N (2017) Modification of foxtail millet starch by combining physical, chemical and enzymatic methods. Int J Biol Macromol 95:314–320. https://doi.org/10.1016/j.ijbiomac.2016.11.067

Dey S, Saxena A, Kumar Y et al (2022) Understanding the antinutritional factors and bioactive compounds of kodo millet (Paspalum scrobiculatum) and little millet (Panicum sumatrense). J Food Qual 2022(1):1578448. https://doi.org/10.1155/2022/1578448

Dip G, Sharma N (2023) Physiochemical study on storage of little millet (Panicum miliare) in hermeticstorage. Int J Adv Biochem Res 7(2):353–357. https://doi.org/10.33545/26174693.2023.v7.i2Se.234

File: A closeup fo Samai millet with husk. JPG. (2025, February 9). Wikimedia Commons. Retrieved 07:21, March 26, 2025 from https://commons.wikimedia.org/w/index.php?title=File:A_closeup_fo_Samai_millet_with_husk.JPG&oldid=996384949

Ganapathy KN (2017) Genetic improvement in little millet. In: Patil JV (ed) Millets and sorghum: biology and genetic improvement, 1st edn. Wiley, New York. https://doi.org/10.1002/9781119130765.ch6

Ghode N (2023) Tribal methods for millets preservation and storage in Bastar District of Chhattisgarh. Pharma Innov 12(2):3749–3753

Guida C, Aguiar AC, Cunha RL (2021) Green techniques for starch modification to stabilize Pickering emulsions: a current review and future perspectives. Curr Opin Food Sci 38:52–61. https://doi.org/10.1016/j.cofs.2020.10.017

Gupta AK, Nayak P, Patil M (2018) Processing of minor millets: a review. Int J Farm Sci 8(1):50–58. https://doi.org/10.5958/2250-0499.2018.00013.7

Han N, Fan JL, Chen N et al (2022) Effect of ball milling treatment on the structural, physicochemical, and digestive properties of wheat starch, A- and B-type starch granules. J Cereal Sci 104:103439. https://doi.org/10.1016/j.jcs.2022.103439

Hemalatha G, Amutha S, Malathi D et al (2006) Development of little millet (Panicum sumatrense) substituted biscuits and characterization of packaging requirements. J Trop Agric 18. https://dl.nsf.gov.lk/handle/1/12151

Hiregoudar S, Mamatha HS (2022) Development and consumer acceptability of little millet (Panicum sumatrense) based biscuits. Pharma Innov 11(2):1658–1661

https://www.needpix.com/photo/112580/millet-cultivation-cereals-agriculture-spike-free-pictures-free-photos-free-images-royalty-free

Indirani K (2021) Review on nutritional profiles and health benefits of little millets–India. Int J Res Sci Eng 9(11):07–11. http://ir.psgcas.ac.in/id/eprint/1372

Jacob J, Krishnan V, Antony C et al (2024) The nutrition and therapeutic potential of millets: an updated narrative review. Front Nutr 11:1346869. https://doi.org/10.3389/fnut.2024.1346869

Jadhav N, Londhe DJ (2023) Policy support for the promotion of millets: current status and its impact. J Drug Res Ayurvedic Sci 8(1):148–151. https://doi.org/10.4103/jdras.jdras_181_23

Jiranuntakul W, Puncha-Arnon S, Uttapap D (2014) Enhancement of octenyl succinylation of cassava starch by prior modification with heat moisture treatment. Starch-Starke 66(11–12):10711078. https://doi.org/10.1002/star.201400115

Kaimal AM, Mujumdar AS, Thorat BN (2021) Resistant starch from millets: recent developments and applications in food industries. Trends Food Sci Technol 111:563–580. https://doi.org/10.1016/j.tifs.2021.02.074

Karuna TD, Nazni P (2018) Formulation, texture and sensory characteristics of little millet-based RTC (ready-to-cook) soup mix. Int J Food Sci Nutr 3(2):187–191

Kate A, Singh A (2021) Processing technology for value addition in millets. In: Millets and millet technology. Springer, Singapore, pp 239–254. https://doi.org/10.1007/978-981-16-0676-2_11

Krishna Kumari S, Thayumanavan B (1998) Characterization of starches of proso, foxtail, barnyard, kodo, and little millets. Plant Foods Hum Nutr 53:47–56. https://doi.org/10.1023/a:1008083020810

Kulkarni LR, Naik RK (2009) Composition of minor millets. J Agric Sci 8(1):88–92

Kumar SR, Sadiq MB, Anal AK (2020) Comparative study of physicochemical and functional properties of pan and microwave cooked underutilized millets (proso and little). Food Sci Technol 128:109465. https://doi.org/10.1016/j.lwt.2020.109465

Kumar SR, Tangsrianugul N, Suphantharika M (2023) A review on isolation, characterization, modification, and applications of proso millet starch. Foods 12(12):2413. https://doi.org/10.3390/foods12122413

Kumar N, Singh R, Upadhyay A et al (2024) Application and functional properties of millet starch: wet milling extraction process and different modification approaches. Heliyon 10(3):25330. https://doi.org/10.1016/j.heliyon.2024.e25330

Kumari SK, Thayumanavan B (1997) Comparative study of resistant starch from minor millets on intestinal responses, blood glucose, serum cholesterol and triglycerides in rats. J Sci Food Agric 75(3):296–302. https://doi.org/10.1002/(SICI)1097-0010(199711)75:3<296::AID-JSFA877>3.0.CO;2-X

Kumari P, Singh YV, Yadav R et al (2022) Analytical study of little millets and therapeutic effect of little millets based on its components in type 2 diabetes mellitus. Int J Herb Med 10(5):11–14. https://doi.org/10.22271/flora.2022.v10.i5a.832

Kundgol NG, Kasturiba B, Math KK et al (2014) Screening of little millet landraces for chemical composition. Int J Farm Sci 4(2):33–38

Kushwaha R, Kaur D (2018) Recent techniques used in modification of starches: a review. In: Food Technology From Health to Wealth & Future Challenges. Bharti Publication, Delhi, p 115

Ladumore V, Jumabhai DG, Patel SN (2021) Genetic diversity studies in little millet (Panicum sumatrense Roth. Ex. Roemer and Schultes) genotypes: a multivariate PCA analysis. Pharma Innov 10(9):1067–1071

Lin L, Cai C, Gilbert RG et al (2016) Relationships between amylopectin molecular structures and functional properties of different-sized fractions of normal and high-amylose maize starches. Food Hydrocoll 52:359–368. https://doi.org/10.1016/j.foodhyd.2015.07.019

Lydia PJ, Ganesan J, Francis N et al (2023) Revitalization of small millets for nutritional and food security by advanced genetics and genomics approaches. Front Genet 13:1007552. https://doi.org/10.3389/fgene.2022.1007552

Mahajan P, Bera MB (2023) Application of free volume concept and manipulation of network structure parameters for optimum loading of gallic acid in the modified kutki (*Panicum sumatrense*) millet starch hydrogel matrix. Food Hydrocoll 135:108218. https://doi.org/10.1016/j.ijbiomac.2021.09.072

Mahajan P, Bera MB, Panesar PS et al (2021a) Millet starch: a review. Int J Biol Macromol 180:61–79. https://doi.org/10.1016/j.ijbiomac.2021.03.063

Mahajan P, Bera MB, Panesar PS (2021b) Modification of Kutki millet (Panicum sumatrense) starch properties by the addition of amino acids for the preparation of hydrogels and its characterization. Int J Biol Macromol 191:9–18. https://doi.org/10.1016/j.ijbiomac.2021.09.072

Mahajan P, Bera MB, Panesar PS (2022) Structural, functional, textural characterization and in vitro digestibility of underutilized Kutki millet (*Panicum sumatrense*) starch. Food Sci Technol 154:112831. https://doi.org/10.1016/j.lwt.2021.112831

Mahajan P, Bera MB, Panesar PS et al (2024) Thermodynamic and network characteristics of optimized lysine-modified kutki (*Panicum sumatrense*) millet starch hydrogels. J Food Sci Technol 61(9):1743–1757. https://doi.org/10.1007/s13197-024-05953-x

Mannuramath M, Yenagi N, Orsat V (2015) Quality evaluation of little millet (Panicum miliare) incorporated functional bread. J Food Sci Technol 52(12):8357–8363. https://doi.org/10.1007/s13197-015-1932-y

Meena RP, Joshi, D, Bisht JK et al (2021) Global scenario of millets cultivation. In: Kumar et al (eds) Millets and millet technology. Springer, Singapore. https://doi.org/10.1007/978-981-16-0676-2_2

Mobolade AJ, Bunindro N, Sahoo D et al (2019) Traditional methods of food grains preservation and storage in Nigeria and India. Ann Agric Sci 64(2):196–205. https://doi.org/10.1016/j.aoas.2019.12.003

Nambi VA, Eganathan P, Philip PIM (2012) Proximate and mineral composition analysis of little millet collected from some millet growing areas in Tamil Nadu. Indian J Plant Genet Res 25(2):189–191

Natesan S, Venugopalan G, Selvamani SB et al (2020) Characterization of little millet (*Panicum sumatrense*) varieties using morphological descriptors and SSR based DNA fingerprinting. J Phytol 12:29–34. https://doi.org/10.25081/jp.2020.v12.6317

Nazni P, Bhuvaneswari J (2016) Functional, pasting and thermal characteristics of finger millet and little starch. Int J Food Nutr Sci 5:191–198

Nimbkar S, Raja V, Shanmugasundaram S et al (2022) Post-harvest treatments and storage of millets. In: Anandharamakrishnan C, Rawson A, Sunil CK (eds) Handbook of millets-processing, quality, and nutrition status, 1st edn. Springer, Singapore. https://doi.org/10.1007/978-981-16-7224-8_8

Ninganagoudar GS, Yenagi NB, Prashanti KS (2012) Evaluation of Anaras prepared from different types of rice and little millet (Panicum sumatrense) flours. Karnataka J Agric Sci 25(4):503–505

Nirmalakumari A, Salini K, Veerabadhiran P (2010) Morphological characterization and evaluation of little millet (*Panicum sumatrense* Roth. ex. Roem and Schultz) germplasm. Electron J Plant Breed 1(2):148–155

Nithiyanantham S, Kalaiselvi P, Mahomoodally MF et al (2019) Nutritional and functional roles of millets—a review. J Food Biochem 43(7):12859. https://doi.org/10.1111/jfbc.12859

Padulosi S, Mal B, King OI et al (2015) Minor millets as a central element for sustainably enhanced incomes, empowerment, and nutrition in rural India. Sustainability 7(7):8904–8933. https://doi.org/10.3390/su7078904

Panda D, Muni P, Panda A et al (2024) Nutritional and nutraceutical richness of neglected little millet genotypes from Eastern Ghats of India: implications for breeding and food value. Planta 259(2):37. https://doi.org/10.1007/s00425-023-04314-w

Paschapur AU, Joshi D, Mishra KK, et al (2021) Millets for life: a brief introduction. A Kumar et al (ed) Millets and millet technology Springer Singapore. https://doi.org/10.1007/978-981-16-0676-2_11

Patil KB, Chimmad BV, Itagi S (2015) Glycemic index and quality evaluation of little millet (Panicum miliare) fakes with enhanced shelf life. J Food Sci Technol 52(9):6078–6082. https://doi.org/10.1007/s13197-014-1663-5

Patil HE, Pali V, Sao A et al (2024) Floral biology, pollination, genetics, origin and diversity in little millet (*Panicum sumatrense* L. Roth ex. Roem. and Schultz). In: Mishra S, Kumar S, Srivastava RC (eds) Genetic improvement of small millets, 1st edn. Springer, Singapore. https://doi.org/10.1007/978-981-99-7232-6_27

Phattanakulkaewmorie N, Paseephol T, Moongngarm A (2011) Chemical compositions and physico-chemical properties of malted sorghum flour and characteristics of gluten free bread. Int J Food Eng 5(9):532–538

Pradeep SR, Guha M (2011) Effect of processing methods on the nutraceutical and antioxidant properties of little millet (*Panicum sumatrense*) extracts. Food Chem 126(4):1643–1647. https://doi.org/10.1016/j.foodchem.2010.12.047

Pradeep PM, Sreerama YN (2018) Phenolic antioxidants of foxtail and little millet cultivars and their inhibitory effects on α-amylase and α-glucosidase activities. Food Chem 247:46–55. https://doi.org/10.1016/j.foodchem.2017.11.103

Prasanna MS, Sowjanya VS, Jaya E et al (2020) Development of millet based instant weaning mix. J Pharmacogn Phytochem 9(4):1908–1913

Rahaman A, Kumari A, Zeng XA et al (2021) Ultrasound-based modification and structural-functional analysis of corn and cassava starch. Ultrason Sonochem 80:105795. https://doi.org/10.1016/j.ultsonch.2021.105795

Rajput JS, Thakur AK, Nag NK et al (2022) Effect of nano fertilizer in relation to growth, yield and economics of little millet (*Panicum sumatrense*roth) under rainfed conditions. Pharma Innov 11(7):153–156. https://doi.org/10.59797/ija.v66i4.2883

Rao VV, BVS P, Vardhan SV et al (2022) Performance evaluation of different Millet Dehullers. Andhra Agric J 69:580–584

Roopa UB, Naik KR, Malagi U et al (2012) Comparison of physico-chemical and functional properties of little millet genotypes. Int J Food Sci Nutr 1(1):30–36

Roopa U, Kasturiba B, Rama Naik RN et al (2013) Physico-chemical and functional properties of little millet genotypes. Karnataka J Agric Sci 26(4):539–542. http://14.139.155.167/test5/index.php/kjas/article/viewFile/7023/7240

Sachdev N, Goomer S, Singh LR et al (2023) Current status of millet seed proteins and its applications: a comprehensive review. Appl Food Res 3(1):100288. https://doi.org/10.1016/j.afres.2023.100288

Saini R, Yadav KC (2018) Development and quality evaluation of little millet (*Panicum sumatrense*) based extruded product. J Pharmacogn Phytochem 7(3):3457–3463

Sajilata MG, Singhal RS, Kulkarni PR (2006) Resistant starch–a review. Compr Rev Food Sci Food Saf 5(1):1–17. https://doi.org/10.1111/j.1541-4337.2006.tb00076.x

Sao A, Pali V, Patil HE (2024) Genetic improvement for yield, quality, biotic, and abiotic stresses in little millet (Panicum sumatrense Roth. ex Roem. and Schult.). In: Mishra, S., Kumar S, Srivastava RC (ed) Genetic improvement of small millets. Springer, Singapore. https://doi.org/10.1007/978-981-99-7232-6_28

Sharma B, Gujral HS (2020) Modifying the dough mixing behavior, protein & starch digestibility and antinutritional profile of minor millets by sprouting. Int J Biol Macromol 153:962–970. https://doi.org/10.1016/j.ijbiomac.2019.10.225

Sharma N, Niranjan K (2018) Foxtail millet: properties, processing, health benefits, and uses. Food Rev Int 34(4):329–363. https://doi.org/10.1080/87559129.2017.1290103

Shivakumar N, Tejaswini KL (2023) Green synthesis of silver nanoparticles using millets, its characterization and evaluation of antimicrobial activity. Int J Pharm Sci Res 14(9):4645–4657. https://doi.org/10.13040/IJPSR.0975-8232.14(9).4645-57

Shweta S, Sindhu S, Sushma K et al (2018) Little millets: properties, functions and future prospects. Int Agric Eng J 11(Special Issue):179–181. https://doi.org/10.15740/HAS/IJAE/11.Sp.Issue/179-181

Singh N, Chaudhry A (2023) Little Millets –Review on nutritional profile and health benefits. J Emerg Technol Innov Res 10(6):2349–5168. http://www.jetir.org/papers/JETIR2306A89.pdf

Singh SK, Singh N (2005) Relationships between selected properties of starches from different corn lines. Int J Food Prop 8(3):481–491. https://doi.org/10.1080/10942910500267711

Singh P, Tiwari A, Vimal SC, Meena NR (2023) Millets: climate resilient, nutritious, marginal crop in todays era. Int J Stat Appl Math 8(6):82–87

Siroha AK, Bangar SP, Sandhu KS et al (2021) Effect of cross-linking modification on structural and film-forming characteristics of pearl millet (*Pennisetum glaucum L.*) starch. Coatings 11(10):1163. https://doi.org/10.3390/coatings11101163

Suma PF, Urooj A (2015) Isolation and characterization of starch from pearl millet (Pennisetum typhoidium) flours. Int J Food Prop 18(12):2675–2687. https://doi.org/10.1080/10942912.2014.981640

Swamy KRM (2023) Origin, domestication, taxonomy, botanical description, genetics and cytogenetics, genetic diversity and breeding of little millet (Panicum sumatrense Roth. ex. Roem. and Schult.). Int J Curr Res 15(10):26170–26192. https://doi.org/10.24941/ijcr.46069.10.2023

Takhellambam RD, Chimmad BV, Prkasam JN (2016) Ready-to-cook millet flakes based on minor millets for modern consumer. J Food Sci Technol 53:1312–1318. https://doi.org/10.1007/s13197-015-2072-0

Tamaka K (2019) Physico-chemical properties of different millets and relationship with cooking quality of millets. Int J Multidiscip Res Dev 6(4):97–101

Taylor JR, Duodu KG (2015) Effects of processing methods on the physicochemical and nutritional properties of millet grain. Compr Rev Food Sci Food Saf 14(1):1–18. https://doi.org/10.1002/jsfa.6713

Thabet M, Boukid F, Mejri M (2016) Comparative study of physical, chemical, structural and functional properties of starches isolated from wild and improved Tunisian millet. Int J Adv Sci Eng Inf Technol 3(1):256–261

Thakur K, Sharma S, Sharma R (2023) Morphological and functional properties of millet starches as influenced by different modification techniques: a review. Starch-Stärke 75(3–4):2200184. https://doi.org/10.1002/star.202200184

Thilagavathi T, Banumathi P, Kanchana S et al (2015) Effect of heat moisture treatment on functional and phytochemical properties of native and modified millet flours. Plant Archive 15(1):15–21

Tripathi T, Vyas S (2023) From ancient grains to modern solutions: a history of millets and their significance in agriculture and food security. Int J Home Sci 9:72–78

Upadhyaya HD, Gowda CLL, Reddy VG et al (2008) Diversity of small millets germplasm in genebank at ICRISAT. Paper presented at the 5th International Symposium on New Crops and Uses: their role in a rapidly changing world, University of Southampton, Southampton, UK 3–4 September, 2007. http://oar.icrisat.org/id/eprint/6828

Uthumporn U, Zaidul IS, Karim AA (2010) Hydrolysis of granular starch at sub-gelatinization temperature using a mixture of amylolytic enzymes. Food Bioprod Process 88(1):4754. https://doi.org/10.1016/j.fbp.2009.10.001

Vali Pasha K, Ratnavathi CV, Ajani J et al (2018) Proximate, mineral composition and antioxidant activity of traditional small millets cultivated and consumed in Rayalaseema region of South India. J Sci Food Agric 98(2):652–660. https://doi.org/10.1002/jsfa.8510

Wang L, Wang M, Zhou Y et al (2022) Influence of ultrasound and microwave treatments on the structural and thermal properties of normal maize starch and potato starch: a comparative study. Food Chem 377:131990. https://doi.org/10.1016/j.foodchem.2021.131990

Yadav A, Chaudhary S, and Shah S (2023) Little millet (Kutki/Shavan). In Singh S,Pandey M, Chaudhary S Millets: forgotten grains regaining prominence, 1 P.K. Publishers & Distributors, New Delhi

Yanez GA, Walker CE (1986) Effect of tempering parameters on extraction and ash of proso millet flours. Cereal Chem 63(2):164–167

Yang Y, Li T, Li Y et al (2020) Understanding the molecular weight distribution, in vitro digestibility, and rheological properties of the deep-fried wheat starch. Food Chem 331:127315. https://doi.org/10.1016/j.foodchem.2020.127315

You S, Izydorczyk MS (2007) Comparison of the physicochemical properties of barley starches after partial α-amylolysis and acid/alcohol hydrolysis. Carbohydr Polym 69(3):489502. https://doi.org/10.1016/j.carbpol.2007.01.002
Zhi W, Zhou Y, Wang R et al (2022) Effect of microwave treatment on the properties of starch in millet kernels. Starch-Stärke 74(9–10):2200063. https://doi.org/10.1002/star.202200063
Zhu F (2014) Structure, physicochemical properties, and uses of millet starch. Food Res Int 64:200–211. https://doi.org/10.1016/j.foodres.2014.06.026
Zia-ud-Din, Xiong H, Fei P (2017) Physical and chemical modification of starches: a review. Crit Rev Food Sci Nutr 57(12):2691–2705. https://doi.org/10.1080/10408398.2015.1087379

Chapter 7
Pearl Millet: Chemical Composition, Starch Isolation, Modifications, Functionality, and Food/Non-food Applications

Meenu Teotia, Neeraj Ravi, Shifa Altaf, and Rakesh Kumar Soni

Abbreviation

DMSO	Dimethyl sulfoxide
DS	Degree of substitution
DSC	Differential scanning calorimetry
ESA	East/Southern Africa
EPI	Epichlorohydrin
FAO	Food and Agriculture Organization
FDA	Food and Drug Administration
FTIR	Fourier transform infrared
GI	Glycemic Index
HMT	Heat moisture treatment
HPP	High-Pressure Processing
JECFA	Joint Expert Commitee on Food Additives
KHSM	Kofler hot stage microscopy
NMR	Nuclear magnetic resonance
OSA	Octenyl succinic anhydride
PEF	Pulse Electric Field
PMS	Pearl millet starch
RDS	Rapid digestible starch
RS	Resistant starch
SDS	Slow digestible starch
STMP	Sodium trimetaphosphate
STPP	Sodium tripolyphosphate
T_c	Conclusion temperature

M. Teotia (✉) · N. Ravi · S. Altaf · R. K. Soni
Department of Chemistry, Chaudhary Charan Singh University, Meerut, UP, India
e-mail: neerajravi3791@gmail.com; shifa0530@gmail.com; rksoni_rks@yahoo.com

© The Author(s), under exclusive license to Springer Nature Switzerland AG 2026

J. Singh, M. Kumar Malik (eds.), *Millets and Millet Starch: Chemical Composition, Functional Modification, and Applications*,
https://doi.org/10.1007/978-3-032-09502-2_7

T_o	Onset temperature
T_p	Peak temperature
WCA	West/Central Africa
WHC	Water holding capacity

7.1 Introduction

The term "millet" refers to a variety of grass crops that produce seeds suitable for human and animal consumption. Pearl millet (*Pennisetum glaucum*) holds significant importance among millet crops, serving as a nutritious cereal grain. It belongs to the Poaceae family under the Genus *Pennisetum* and species *P. glaucum*. This versatile crop goes by various vernacular names across different countries: in India, it's referred to as "bajra" (in Hindi); in Nigeria, as "Gero" (in Hausa); in Niger, as "hegni" (in Djerma); in Mali, as "sanyo"; in Sudan, as "dukhon" (in Arabic); and in Namibia, as "mahangu" (Shaikh et al. 2017a, b; Taylor 2016; De Wet 1986). This millet has recently gained attention for its potential starch content, which can reach up to 70% depending on the cultivar. It is mainly cultivated for food and fodder purposes in arid and semi-arid tropical areas. This millet is also called cattail millet or bulrush millet (Yadav and Rai 2013). Pearl millet is a robust and cost-effective crop that flourishes in harsh environments such as drought and salt-affected areas. Its distinct characteristics, including tolerance to high temperatures and barren soils, distinguish it from other cereals. This ecologically beneficial C_4 cereal (plants that utilize the C4 pathway of photosynthesis, a specialized adaptation for environments with high temperatures, low CO_2 levels, and high light intensity) not only absorbs more CO_2 but also assists in dealing with climate uncertainty and lowering carbon emissions, making it an ideal answer for sustainable agriculture (Kumar et al. 2018). The cultivation, production, and consumption of millets exhibit regional variations influenced by geographic factors. China predominantly cultivates foxtail millet, whereas pearl millet finds its primary cultivation regions in India, Africa, and Nepal. Proso millet, on the other hand, is particularly emphasized in North America. Pearl millet stands out as a superior alternative to mainstream starch sources like corn, wheat, rice, and potatoes due to its higher nutritional content and gluten-free nature. In India, these millets are often preferred over wheat, rice, or maize, especially during Hindu fasting rituals (Obilana 2003; Punia et al. 2021).

7.2 Origin and Distribution

Millet originated in tropical West Africa about 4000 years ago. The largest numbers of both wild and cultivated forms of this species are found in this region. It was then divided into the *globsum* race, which moved toward the west, and the *typhoides* race, which invaded Eastern Africa and expanded to Southern Africa and India around 2000–3000 years ago. Pearl millet evolved to be able to withstand the arid, nutrient-depleted soils and high temperatures of the Indian and African hot deserts better than other grains, such as wheat and rice. It may be found in regions in which yearly rainfall varies between 125 and 900 mm. The optimal growing temperature varies from 21 to 35 °C. Pearl millet is a versatile crop that can thrive in both acidic sandy and saline soils (Sharma et al. 2021). It is now grown on more than 30 million hectares globally, with Africa (>18 million ha) and Asia (>10 million ha) accounting for the majority of the crop (Raheem et al. 2021). Pearl millet grain is a staple meal for around 90 million people in Africa's Sahel area and northern India (Srivastava et al. 2020). Pearl millet ranks as the sixth largest cereal crop globally, trailing behind maize, rice, wheat, barley, and sorghum. It spans approximately 30 million hectares across the arid and semi-arid tropical regions of Asia and Africa. This crop constitutes around 50% of the total millet production worldwide, with Africa contributing approximately 60% of the cultivated area while Asia contributing about 35% (Jukanti et al. 2016) (Fig. 7.1).

India stands out as the leading producer of pearl millet, both in terms of area under cultivation and production. Between 2010 and 2012, the average millet cultivation area in India spanned an impressive 8.5 million hectares, yielding around 9.4 million tons annually. Furthermore, the average production was 8.61 million tons, with an area of 6.93 million hectares, boasting a productivity rate of 1243 kg per hectare according to the Directorate of Millets Development (2020). In Africa, pearl millet cultivation is primarily concentrated in two key regions: West/Central Africa (WCA) and East/Southern Africa (ESA). The significance of pearl millet as the third major crop in sub-Saharan Africa cannot be overstated, particularly in countries like Niger, Nigeria, Chad, Senegal, Mali, and Burkina Faso. The cultivation of pearl millet in WCA holds immense socio-economic, nutritional, and environmental impact, especially for the marginalized communities of the region. With WCA accounting for a staggering 95% of the global pearl millet cultivation area, it emerges as the epicenter of production dominance (Jukanti et al. 2016). Additionally, West Africa, spearheaded by Nigeria (41%), Niger (16%), Burkina Faso (7%), Mali (6.4%), and Senegal and Sudan (4.8%), emerges as the largest producer within the region. Across Africa, pearl millet cultivation spans approximately 18.50 million

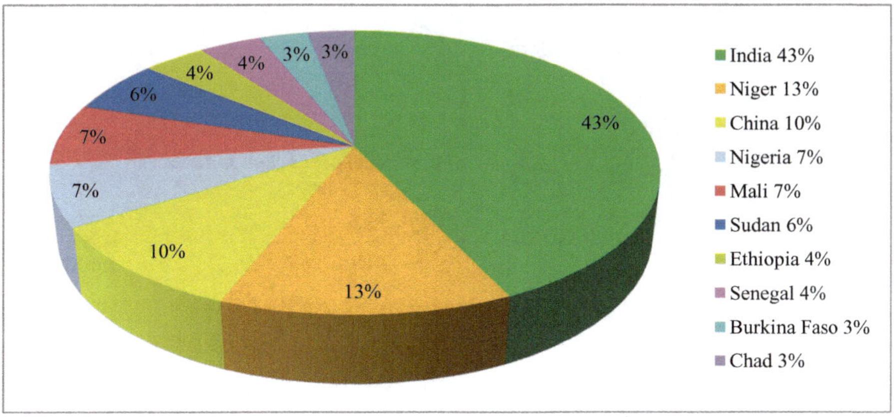

Fig. 7.1 World millet production. (FAO 2019)

hectares, involving 28 countries and yielding an impressive 11.36 million tons annually, covering 30% of the continent's diverse agroecologies. With its significant share of 49% in the global millet cultivation area, pearl millet holds substantial significance in sustaining agricultural economies and food security initiatives across Africa, as recognized by the Food and Agriculture Organization (FAO 2019).

Pearl millet holds a significant position in India's agricultural landscape, emerging as the fourth most widely cultivated cereal crop, trailing only behind rice, wheat, and maize. Uttar Pradesh, Rajasthan, Gujarat, Haryana, Madhya Pradesh, Maharashtra, Karnataka, and Tamil Nadu stand out as the major pearl millet-producing states, collectively contributing a whopping 90% of the nation's

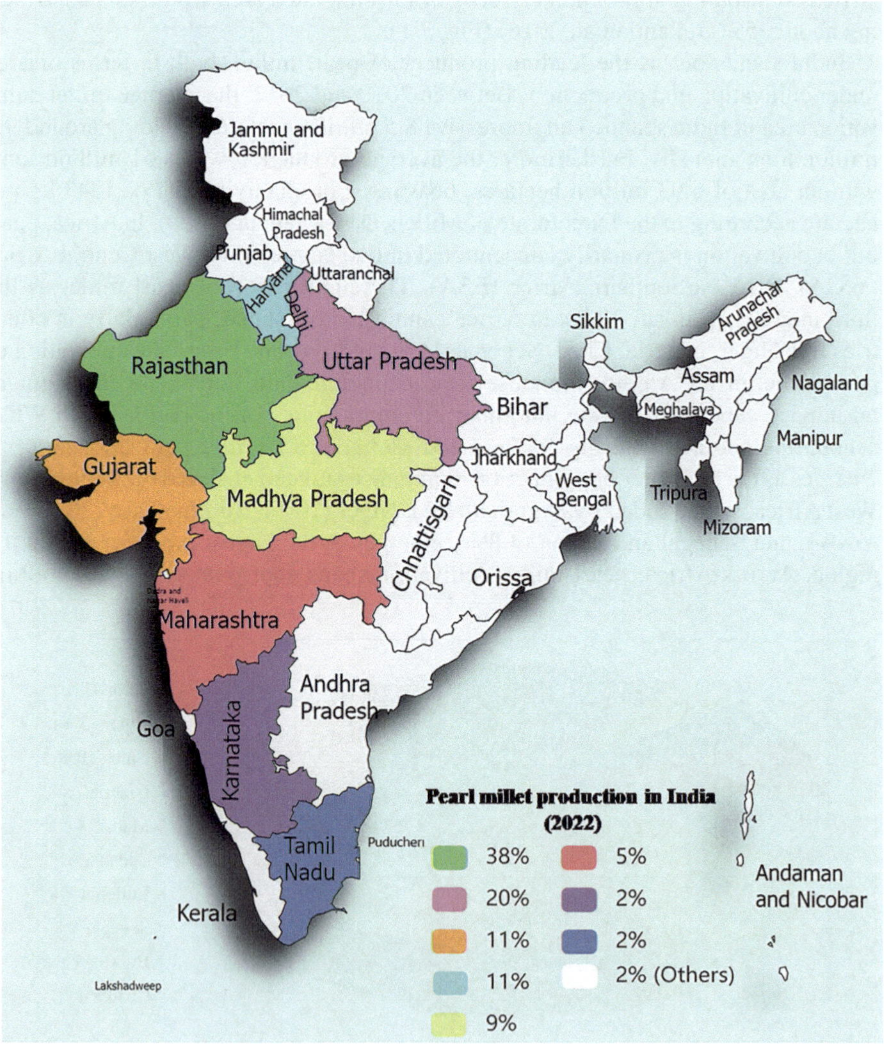

Fig. 7.2 Pearl millet production in India. (FAO 2022)

total production (Fig. 7.2). Among these, Rajasthan leads the charge, contributing around 4.283 million 5 tons, followed closely by Uttar Pradesh (1.949 million tons), Haryana (1.12 million tons), Gujarat (1.056 million tons), Madhya Pradesh (0.869 million tons), Maharashtra (0.475 million tons), and Tamil Nadu (0.157 million tons) (FAO, 2022). Cultivation primarily occurs during the rainy (kharif) season, spanning from June to October, although it also extends to some regions of Gujarat, Rajasthan, and Uttar Pradesh during the summer season (February–May). Additionally, smaller-scale cultivation takes place in Maharashtra and Gujarat during the post-rainy (rabi) season from November to February (Satyavathi et al. 2021). Recognized for their climate resilience and nutritional value, millets, including pearl millet, are rightfully referred to as "nutricereals." Emphasizing their importance, the Government of India declared the year 2018 as the "Year of Millets," while the FAO Committee on Agriculture (COAG) designated 2023 as the "International Year of Millets," aiming to promote their cultivation and harness their superior nutritional qualities on a global scale (Chellapilla et al. 2022).

7.3 Plant Description

Pearl millet (*Pennisetum glaucum*) is a warm-season annual grass characterized by its robust and upright growth habit (Fig. 7.3). The plant typically reaches a height of 2 to 4 feet, with slender stems that may vary in color from green to purple. The leaves are long, narrow, and finely serrated, contributing to the overall feathery appearance of the plant (Brunken et al. 1977; Newman et al. 2010).

Pearl millet produces distinctive inflorescences in the form of dense, cylindrical spikes at the tips of the stems (Sanjana Reddy 2017). The seeds, which are the valuable part of the plant, are tightly packed on the seed head, forming a

Fig. 7.3 Different stages and forms of *Pennisetum glaucum:* (**a**) standing field crop of *Pennisetum glaucum* with mature grain-bearing panicles; (**b**) close-up view of a single, mature panicle showing compact grain arrangement; (**c**) harvested and cleaned pearl millet grains

characteristic pearl-like arrangement from which the name "pearl millet" is derived. The grain size of pearl millet ranges from 3 to 4 mm, exceeding that of other millets, and the weight of 1000-kernel ranges from 2.5 to 14 g with an average of 8 g (Singh et al. 1997; Serna-Saldivar and Espinosa-Ramírez 2019). The seeds are small, round, and typically ivory or light brown in color, and the kernel (the edible part of the seed) color may range from white, pale yellow, brown, gray, slate blue, to purple, reflecting a pale color with variations depending on the specific variety and growing conditions (Pattanashetti et al. 2016).

7.4 Chemical Composition

Pearl millet exhibits a diverse chemical composition, positioning it as a valuable and nutritious cereal grain. Its carbohydrate-rich profile, primarily composed of complex starches, makes it an excellent source of energy. Pearl millet surpasses other cereals like rice and maize in protein content, offering essential amino acids crucial for human nutrition. With a substantial presence of dietary fiber, it supports digestive health and aids in preventing constipation. The grain contains moderate amounts of healthy unsaturated fats while maintaining low levels of saturated fats. Pearl millet contributes significantly to vitamin intake, particularly B vitamins such as niacin, thiamine, and folate, essential for energy metabolism. Abundant in minerals like magnesium, phosphorus, potassium, and zinc, it promotes bone health, energy production, and immune function (refer to Tables 7.1 and 7.2). The antioxidant content in pearl millet aids in neutralizing free radicals, contributing to overall health. Noteworthy is its gluten-free nature, making it a suitable option for individuals with gluten sensitivity or celiac disease (Asrani et al. 2023).

In terms of composition, pearl millet stands out with its richness in resistant starch (RS), soluble and insoluble dietary fibers, minerals, and antioxidants. With approximately 92.5% dry matter, the grain comprises 2.1% ash, 2.8% crude fiber, 7.8% crude fat, 13.6% crude protein, and 63.2% starch, attesting to its nutritional significance (Ali et al. 2003).

The starch in pearl millet, a significant component, comprises two main constituents: amylose and amylopectin. Amylose consists of amorphous sections, while amylopectin contains crystalline portions. Amylose is a linear polymer composed of glucose monomers linked together through glycosidic linkage [α-D-(1,4)], making up approximately 20–30% of the total starch. Conversely, amylopectin is a branched polymer with both α-D-(1,4) and (1,6) linkages in the linear chain and at branch points, respectively. Amylose content of PMS has been reported as 28.8–31.9% (Hoover et al. 1996), 15.64–19.46% (Bhupender et al. 2013), 3.96–4.96% (Suma and Urooj 2015), 28.14% (Wu et al. 2014), 32.5% (Annor et al. 2014), 13.6–18.1% (Sandhu and Siroha 2017), 11.57–21.93% (Siroha et al. 2020a), and 14.8% (Siroha et al. 2020b). These variations in amylose content underscore its pivotal role as a significant component, conferring potential applications to starch. The change in amylose quantity may be attributed to differences in climate, soil type during grain

Table 7.1 Nutrient content of pearl millet (per 100 g)

Constituent	Value
Energy	378 kcal
Carbohydrates	72.8 g
Protein	11.0 g
Dietary fiber	8.5 g
Fat	4.2 g
Polyunsaturated fatty acid	2.1 g
Monounsaturated fatty acid	0.8 g
Saturated fatty acid	0.7 g
Omega (ω)-6 fatty acid	2.0 g
Omega (ω)-3 fatty acid	0.1 g
Minerals	
Phosphorous (P)	285 mg
Potassium (K)	195 mg
Magnesium (Mg)	114 mg
Calcium (Ca)	8.0 mg
Sodium (Na)	5.0 mg
Iron (Fe)	3.0 mg
Zinc (Zn)	1.7 mg
Manganese (Mn)	1.6 mg
Copper (Cu)	0.8 mg
Selenium (S)	2.7 μg
Vitamins	
Vitamin B_1 (Thiamine)	0.42 mg
Vitamin B_2 (Riboflavin)	0.29 mg
Vitamin B_3 (Niacin)	4.72 mg
Vitamin B_5 (Pantothenic acid)	0.85 mg
Vitamin B_6 (Pyridoxine)	0.38 mg
Vitamin B_9 (Folate)	85 μg
Vitamin K (Phylloquinone)	0.9 μg

Adapted from Kapri et al., Taylor & Francis (2017). Used with permission of the author

growth, and source of starch. Starches rich in amylose are ideal for making RS food items like noodles, snacks, and bakery goods, which exhibit a low glycemic index (GI) (Zhong et al. 2023). Furthermore, the advancement of pearl millet cultivars with elevated amylose levels holds promise for expanding the applications of PMS in various food formulations.

Protein is the second main ingredient in millet. It is estimated that pearl millet has around 11.6% protein, which is more than the protein content of 7.2% in rice, 11.5% in barley, 11.1% in maize, and 10.4% in sorghum (Jha et al. 2013). Moreover, Adeola and Orban (1995) found that pearl millet boasts a richer amino acid profile compared to sorghum and corn proteins. It contains higher levels of lysine, threonine, methionine, and cysteine, placing it on par with the protein profiles of wheat,

Table 7.2 Nutrient content of various cereal crops (per 100 g)

Constituent	Rice[a]	Wheat[b]	Sweet corn[c]
Energy (kcal)	365	327	86
Carbohydrates (g)	79	71.2	19
Protein (g)	7	12.6	3
Dietary fiber (g)	1	12.2	3
Fat (g)	1	1.5	1
Phosphorous (P) (mg)	115	288	89
Potassium (K) (mg)	115	363	270
Magnesium (Mg) (mg)	25	126	37
Calcium (Ca) (mg)	28	29	2
Iron (Fe) (mg)	0.8	3.2	0.5
Zinc (Zn) (mg)	1.1	2.6	0.5
Manganese (Mn) (mg)	1.1	3.9	0.2
Vitamin B_1 (Thiamine) (mg)	0.1	0.38	0.2
Vitamin B_2 (Riboflavin) (mg)	>0.1	0.1	0.1
Vitamin B_3 (Niacin) (mg)	1.6	5.5	1.8
Vitamin B_5 (Pantothenic acid) (mg)	1.0	0.9	0.7
Vitamin B_6 (Pyridoxine) (mg)	0.2	0.3	0.1
Vitamin B_9 (Folate) (μg)	8	38	42

Adapted from Kapri et al., Taylor & Francis (2017). Used with permission of the author
[a]White, long grain, raw
[b]Hard red winter
[c]Sweet, yellow, raw

barley, and rice (Rooney et al. 1981; Hoseney 1994). Interestingly, pearl millet aligns closely with maize in terms of protein distribution, particularly with true prolamins, which are believed to be soluble in alcohol (Ejeta et al. 1987).

Moreover, the abundant fiber content in pearl millet (8–9%) plays a crucial role in regulating the pace at which food moves through the digestive system, creating prolonged intervals between meals. This extended duration aids in preventing obesity by mitigating excessive food intake (Mustafa et al. 2008). Furthermore, owing to its characteristic of slow digestion, pearl millet contributes to an extended transit time, resulting in a gradual release of glucose into the bloodstream. This attribute proves beneficial for individuals with non-insulin-dependent diabetes, as evidenced by a lower incidence of diabetes in millet consumers (Shobana et al. 2009). The fiber content in millet also contributes to a multifaceted impact on cholesterol levels. It not only works to decrease harmful cholesterol but also promotes the elevation of beneficial cholesterol, thus fostering cardiovascular health. Additionally, the preventive effect on the secretion of bile acids by millet's fiber content serves as a safeguard against the formation of gallstones within the body (Malik 2015).

The lipid content of pearl millet, estimated as 5–7%, exceeds that of maize (3.21–7.71%) (Ullah et al. 2010). Chauhan et al. (2018) reported 5% of lipid in pearl millet. Furthermore, pearl millet exhibits high concentrations of palmitic, stearic, and linoleic acids, with a lower presence of oleic acid compared to corn (Adeola and

Orban 1995). The overall lipid content ranges from 1.5 to 6.8%, making it one of the lipid-rich millet varieties (Taylor 2004). Free and bound lipid components in pearl millet range from 5.6 to 6.1% and 0.6 to 0.9%, respectively, with identified triglycerides, diglycerides, and monoglycerides (Hoseney 1986).

Mineral composition of pearl millet has been tabulated in Table 7.1. According to Martínez-Ballesta et al. (2011), the mineral composition of food is impacted by environmental stresses such as high salt content, limited water accessibility, and extreme temperatures. The composition of pearl millet is based on the total mineral content and trace elements found in the soil. Ash varies in concentration from 0.86 to 1.35% in maize and from 1.6 to 3.6% in pearl millet. Pearl millet has higher quantities of minerals than maize, including calcium, phosphorus, magnesium, manganese, zinc, iron, and copper (Adeola and Orban 1995).

7.5 Isolation of Starch from Alkaline, Acidic, and Neutral Method

In the starch isolation procedure, freshly harvested pearl millet is used, and the grains are thoroughly washed with water to remove all impurities from the pearl millet grains. Process the purified pearl millet grains into a fine powder using a mixture grinder, and excess bran is removed with the help of a sieve shaker. To prevent microbiological growth, the semi-refined flour of pearl millet is soaked in distilled water containing antimicrobial agents such as sodium azide and mercury chloride (Beleia et al.1980; Krishna Kumari and Thayumanavan 1998) for a whole night to help separate the starch from other ingredients. The soaked flour is filtered using a 60 or 150 mesh standard sieve, and the process is repeated until no more starch can be separated. After being repeatedly cleaned with distilled water, the resulting slurry is subjected to centrifugation to recover the separated starch before further processing (Krishna Kumari and Thayumanavan 1998). Using a spatula, the top layer (protein) of the residue was scraped off and disposed away. After stirring the slurry for 15 min and adding diluted NaOH to reduce its pH to approximately neutral, the excess alkali was removed by repeatedly washing it with distilled water. After being suspended in distilled water and agitated for 5 h, the bottom layer (starch) is centrifuged and cleaned until the pH approaches neutral. This procedure, which is carried out twice, facilitates the complete elimination of proteins. Centrifugation and several washes with distilled water are performed after the crude starch was suspended in NaCl (0.1 N):Toluene (1:1) and stirred for 3 h. Centrifugation was performed after adding the alcohol and stirring the mixture for 3 h. Acetone is used to repeat this procedure (Sowbhagya and Bhattacharya 1971). The resulting white prime starch is air-dried and kept for later use in airtight polythene bags. These conditions may be neutral or nearly neutral, alkaline, or acidic. Thus, while comparing results from various studies, the starch isolation process should be carefully taken into consideration. It should be mentioned that an alkaline, neutral, or

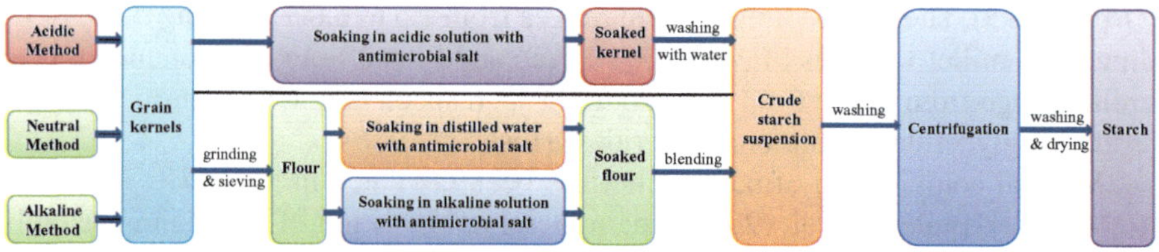

Fig. 7.4 General method used for isolation of starch from pearl millet

acidic solution may degrade starch or result in a low amount of total starch being recovered from the flour. These problems may be overcome by developing extraction methods based on ethanol/methanol precipitation (Carpita and Kanabus 1987) and dimethyl sulfoxide (DMSO) dissolution (Klucinec and Thompson 1998) for laboratory analysis (Fig. 7.4).

7.5.1 *Isolation of Starch Through DMSO Method*

Isolating starch from pearl millet using DMSO involves several steps aimed at extracting and purifying the starch. Initially, the pearl millet grains are cleaned and ground into a fine powder to increase the surface area for extraction. The grounded pearl millet is then mixed with DMSO in a suitable container and heated gently to facilitate dissolution, with continuous stirring to ensure thorough mixing. Once the starch is dissolved in the DMSO solution, it is filtered to separate insoluble impurities such as fibers, proteins, and lipids from the soluble starch-DMSO solution (Klucinec and Thompson 1998; Whistler et al. 2012). Filtration is typically achieved using filter paper or a fine mesh sieve to achieve effective separation. Subsequently, the starch is precipitated from the solution by adding a non-solvent, usually a large volume of a polar solvent like ethanol or methanol (Carpita and Kanabus 1987). This addition causes the starch to re-precipitate due to its reduced solubility in the mixed solvent system. The resulting starch precipitate is then subjected to multiple washing steps with a suitable solvent to remove any residual DMSO or impurities, thereby purifying the starch. After washing, the isolated starch is dried to remove any remaining solvent and moisture, resulting in the production of pure starch. Various drying methods, such as air-drying, vacuum drying, or using a desiccator, can be employed based on the desired moisture content and intended application of the starch (Singla et al. 2020).

7.6 Modification of Starch

Depending on the plant source, the native starches have different structures and functions. Due to their high degree of retrogradation and low shear and thermal durability, native starches have limited uses in food and non-food applications (Jyothi et al. 2006). Furthermore, native starches are prone to syneresis and show gelling properties in their pastes. Therefore, to enhance and optimize specific functional attributes for various applications in the food, paper, and textile industries, starches are frequently subjected to modification through chemical, physical, and enzymatic methods. Numerous modifications have been made to PMS to provide a variety of functional characteristics. These modifications induce structural changes in starch, influenced by various variables, including environmental conditions and processing methods. Chemical modification of starch entails using hydroxyl groups to achieve desired properties to fulfill specific applications. Meanwhile, physical modifications of starches, using methods like moisture, heat, shear, or radiation, are gaining wider acceptance due to their chemical-free nature. The continuous process of starch modification aims to develop a substantial marketplace for several value-added products with novel functional qualities (Kaur et al. 2012). The structural and functional properties of starch molecules are influenced by modifications, which can have favorable or bad outcomes. The following subsections provide further details on the modifications that have been made to PMS in recent years to produce a wide range of functionalities.

7.6.1 Chemical Modifications

Modifying starch through chemical processes involves the introduction of novel functional groups or chemical entities, thereby altering its structure and physicochemical attributes for various uses in food and non-food applications. These modifications primarily target the hydroxyl groups in starch through processes such as etherification, esterification, cross-linking, or degradation (e.g., hydrolysis and oxidation).

7.6.1.1 Modifications Through Substitution of Hydroxyl Group

The reactivity of starch, influenced by the poly-hydroxyl functional groups within glucose monomers, plays an important role in determining the chemical properties of starches. Specifically, at the C-2, C-3, and C-6 positions, poly-hydroxyl groups remain unbound to glycosidic bonds, making them available for substitution reactions. These reactions can involve either the complete hydroxyl groups or the attached hydrogen (Egharevba 2019). Modifications through the substitution of hydroxyl groups primarily target the hydroxyl groups in starch through reactions such as esterification and etherification.

7.6.1.1.1 Esterification

Modifying the physicochemical and structural properties of starch through esterification is a fundamental practice essential for its improved industrial application. Esterified starches can be synthesized through the application of inorganic acids, such as sulfate and phosphate, as well as organic acids, including anhydrides and chlorides. Because of these qualities, starch phosphates can be used as bulking agents, adhesives, thickeners, and stabilizers in medication delivery systems. The esterification process of PMS demonstrates a high degree of substitution (DS) of the hydroxyl functional group, making it well-suited for applications in the food industry. Esterification of PMS has the potential to improve flexibility, enhance crystallinity, facilitate processability, and impart hydrophobic properties. Two notable types of esterification reactions involved in this context are acetylation and succinylation.

7.6.1.1.1.1 Acetylation

Acetylation is defined as the esterification of starch according to criteria issued by the Joint Expert Committee on Food Additives (JECFA), FAO, and WHO. This is achieved by introducing acetate groups through the substitution of hydroxyl groups within the starch structure. Typically, acetic anhydride and vinyl acetate are used as key agents in the preparation of acetylated starch, conducted by an addition of an alkaline catalyst such as KOH, NaOH, or Na_2CO_3. Acetylation replaces hydroxyl groups in starch's anhydroglucose units with acetyl groups (Wani et al. 2012). The amount of acetyl group that is incorporated into starch depends on the catalyst, reactant concentration, starch supply, and reaction duration (Singh et al. 2004). The Food and Drug Administration (FDA) states that there should be no more than 2.5% acetyl group in starch (JECFA 2001). The utilization of acetylated starches is determined by the DS. Starches with low DS (ranging from 0.01 to 0.2) can serve various purposes such as binding agents, stabilizers, thickeners, and texturizers. Additionally, starches with low DS are used in salad dressings, sauces, baked goods, and snacks for young children. Starches with moderate (0.2 to 1.5) and high (1.5 to 3) DS are preferred for their excellent solubility in solvents like chloroform and acetone, making them ideal for thermoplastic applications (Luo and Shi 2012). These starches contribute to the production of acetylated PMS with DS and acetyl levels ranging from 0.040 to 0.081 and 1.07% to 2.15%, respectively. The introduction of acetyl groups disrupts the ordered structure of native starch, hindering the re-association of amylose and amylopectin in gelatinized starch. This disruption results in reduced gelatinization temperatures (Onset temperature (T_o): 69.6 to 48.58 °C, Peak temperature (T_p): 73.67 to 56.38 °C, and Conclusion teperature (T_c): 87.23 to 67.38 °C), increased swelling power (14 to 17.6 g/g), solubility (10.4% to 19.2%), peak viscosities (1069 to 1772 mPa.s), setback viscosities (352 to 851 mPa.s), and final viscosities (913 to 2120 mPa.s) of PMS, alongside enhanced storage stability (Siroha et al. 2019). According to Siroha et al. (2019), there was a decrease in rapid digestible starch (RDS) (49.9–45.4) and slow digestible starch (SDS) (38.1–27.3) along with an increase in RS (12–27.3) when the concentration of acetic anhydride in PMS increased. In another research, Shaikh et al. (2019) used acetylated 4% and 8% PMS to create edible films. They observed that acetylation increased the flexibility and moisture barrier qualities of the films but reduced their transparency. Acetylated starches also serve as effective thickening agents in the production of instant soups, dehydrated soups, and sauces, enhancing their texture and consistency.

7.6.1.1.2 Succinylation

In the presence of an alkaline catalyst, native starches are esterified with succinic anhydride to produce starch succinate. In food systems, starch esterification with octenyl succinic anhydride (OSA) is allowed up to 3% (Code of Federal Regulation 2001). Functional characteristics, including consistency, swelling power, thickening, paste clarity, resilience to pH and pressure, heat and shear stability, and cold storage stability, are all improved by succinylation (Shaikh et al. 2017a). Shaikh et al. (2017a, b) have conducted a study on the preparation, structure, and functional characteristics of succinylated PMS. While enhancing PMS solubility power and improving specific power, succinylation substantially lowered enthalpy (ΔH) and transition temperatures. PMS was found to have lower retrogradation enthalpies (ΔH_{ret}) than gelatinization enthalpies (ΔH_{gel}). Both food and non-food products used starches that have been modified with OSA as an emulsifier. Modified starches with OSA have a variety of applications, including use in salad dressings, beverages, fragrances, flavors, clouding agents, and encapsulating agents (Jeon et al. 1999; Park et al. 2004). In order to successfully prevent the separation of the water and oil phases, the glucose component of starch is bound with water, and the oil component is bound with octenyl in the OSA modification process (Mitchell and Hill 2021). The DS for OSA-modified starches is capped at 3 (CFR 2001). Through OSA modification, properties such as paste clarity (2.5% to 5.4%), swelling power (13.1 g/g to 31.6 g/g) (Shaikh et al. 2015), solubility (0.14 to 0.15), and viscosity (peak viscosity: 300 BU to 394 BU; hot paste viscosity: 162.5 BU to 262 BU; setback viscosity: 124.5 BU to 187 BU) are all enhanced. Additionally, OSA modification leads to a decrease in gelatinization temperature (To: 69.6 °C to 65.5 °C, Tp: 73.7 °C to 69.4 °C, and Tc: 87.23 °C to 74.3 °C) (Shaikh et al. 2017a). There's a notable ($p < 0.05$) increase in RS content during octenyl succinate derivatization, rising from 2.4% in its native state to 13.7% (OSA-5H, DS: 0.022) for PMS (Sharma et al. 2016). The influence of OSA was investigated within ice-cream formulations, uncovering its ability to emulate fat-like characteristics owing to the hydrophobic bulky group within starch (Sharma et al. 2015). Following this discovery, OSA-modified PMS was successfully employed as a substitute for fats (at concentrations ranging from 0.5% to 2.0%) in low-fat yogurt recipes containing 1.5% fats (Sharma et al. 2018). Through analysis using the Herschel–Bulkley model, the pseudoplastic behavior was evaluated, indicating that OSA-modified PMS contributed to enhanced firmness and increased storage modulus (ranging from 58.9 to 72.7 Pa) and loss modulus (from 22.2 to 32.3 Pa) in proportion to the concentration of OSA (Sharma et al. 2018). Starches modified with OSA have the potential to form a uniform blend with soft-leveling properties, making them a suitable choice as substitutes for fats. Because of their improved viscosity and structural integrity, OSA-modified starch pastes are a good substitute for fat and may find use in the manufacturing of dairy and low-fat food products.

7.6.1.1.3 Etherification

In the starch etherification process, epoxide reagents such as propylene oxide and ethylene oxide are commonly utilized. Initially, these epoxides undergo reduction to diols via nucleophilic ring opening, followed by the eventual condensation of one hydroxyl group with starch (Carey 2003). This approach facilitates the incorporation of lipophilic alkyl groups into the starch chains, thereby reducing hydrophilicity and the degree of both intermolecular and intramolecular hydrogen bonding (Neelam et al. 2012). Etherification contributes to enhancing the stability and viscosity of food products. Specifically, in PMS etherification, the process involves forming an ether bond within the chain by replacing the -OH groups typically present on glucose units with carboxymethyl, hydroxypropyl, or other groups.

7.6.1.1.3.1 Hydroxypropylation

In this process, an alkaline medium is used to combine with propylene oxide, an etherifying reagent, to etherify PMS. This process results in the replacement of hydroxyl groups on the starch chains with hydrophilic hydroxypropyl groups. (Shaikh et al. 2017a; Shaikh et al. 2019)) used different concentrations of propylene oxide (10% and 30%) to synthesize PMS with a high degree of hydroxypropylation. It was found that hydroxypropylation led to an increase in swelling power (9.49–38.32 g/g) and solubility (0.14–0.34 g/g) of PMS although enthalpy (ΔH) and gelatinization temperatures (T_o: 69.6–33.32 °C; T_p: 73.67 to 57.6 °C; T_c: 87.23–63.08 °C) were decreased (Code of Federal Regulation 2001). Intermolecular and intramolecular hydroxypropyl groups break hydrogen bonds, weakening starch granules and increasing water absorption, which is the cause of the increase in swelling power. Custards can be stabilized at low temperatures by adding hydroxypropylated starch, which reduces syneresis (Code of Federal Regulation 2001). Modifying starch retrogradation in hydroxypropylated PMS can notably influence the reassociation of starch chains and alter water activity within starch-water systems. Hydroxypropylated PMS finds application in producing edible films and exhibits enhanced retrogradation capabilities compared to fermented foods. The incorporation of hydroxypropyl groups disrupts the internal bond structure of starch granules, which significantly changes the physicochemical characteristics and uses of the material. Hydroxypropylation lowers the temperature at which gelatinization occurs while improving solubility, water holding capacity (WHC), and swelling power. This modified starch variety seems to be a good option for a stabilizing and thickening agent (Shaikh et al. 2017b).

7.6.1.2 Elongation Type Modifications

Elongation modification increases the strength and stability of the polymer by introducing intra- and intermolecular bonding at different places inside the starch granule. For the purpose of producing the desired product, these processes frequently entail oxidation, hydrolysis, esterification, or etherification, either alone or together. The objective is to achieve certain physicochemical qualities that are appropriate for a range of food applications, including viscosity, gelatinization, retrogradation, and textural properties.

7.6.1.2.1 Cross-Linking

Cross-linking modifications of pearl millet involve chemically or physically linking polymer chains within the starch granules, enhancing its properties for various applications. Chemical methods utilize agents like sodium trimetaphosphate (STMP) or oxidizing agents to form covalent bonds between starch molecules, while physical methods involve high-pressure processing (HPP) or controlled heat and moisture treatment to induce cross-links. These modifications improve the thermal stability, viscosity, gel strength, and freeze-thaw stability of PMS (Chung et al. 2004). Cross-linked starches have a short paste texture, high paste consistency, strong shear stability, resistance to gelatinization, and increased stability against retrogradation (Egharevba 2019). Improved phosphorus levels, reduced swelling power, higher temperatures, and lowered ΔH are characteristics of cross-linked RS (Woo and Seib 2002). Sodium tripolyphosphate (STPP), STMP, epichlorohydrin (EPI), and phosphoryl chloride ($POCl_3$) are the most widely used cross-linking agents for PMS (Siroha and Sandhu 2018). EPI readily hydrolyzes into the carcinogen 3-chloropropane-1,2-diol when it comes into contact with water. Hence, the necessity for readily available and environmentally friendly cross-linking reagents becomes evident. Starch has primary (6-OH) and secondary (2-OH and 3-OH) hydroxyl groups (Shah et al. 2016). When these groups interact with cross-linkers, they reduce the swelling and breakage of granules, thereby strengthening the starch granules against heat, acidic conditions, and shear forces (Ayoub and Rizvi 2009; Eliasson 2006). The FDA has specified that $POCl_3$, STMP, and adipic-acetic mixed anhydride have maximum limits of 0.1, 1, and 12% with respect to potential effects on food. It has been found that $POCl_3$ cross-linking produces a higher viscosity than modified starches made using STMP and EPI (Hirsch and Kokini 2002). EPI (0.5%)-modified PMS was examined for its pasting, functional properties, and in vitro digestibility (Siroha and Sandhu 2018). Cross-linking treatment led to significant changes in the properties of starch (Chung et al. 2004). It resulted in decreased amylose content (13.6–18.1%–6.2–8.5%), pasting viscosity (1860–3071 mPa.s to 299–1164 mPa.s), final viscosity (1430–2961 mPa.s to 354–906 mPa.s), solubility (10.4–16.2%–3.4–9%), and swelling power (14.1–17.9 g/g to 6.72–7.64 g/g), while increasing the pasting temperature (70.4–71 °C to 87.4–90.7 °C). Additionally, cross-linking reduced the levels of RDS (46.3–51.6%–46.1–50.6%) and SDS (37.2–38.7% to 34.5–36.4%), while increasing the content of RS (9.7–16.5%–13.6–19.4%). Cross-linking primarily takes place within the structure of amylopectin, establishing covalent bonds that reinforce the starch granules and increase their resistance to chemical or mechanical disruption. Consequently, they exhibit reduced vulnerability to acid attack, heat, or shear.

The prolonged and consistent viscosity of cross-linked starches makes them highly suitable for serving as thickening agents in a range of food applications. In the case of cross-linked PMS with high RS, it proves to be an efficient substitute for fat in bakery goods, providing enhanced baking properties.

7.6.1.3 Modifications Through Degradation

Enzymes and acids facilitate degradation modifications, increasing the production of short linear chains of amylose. This favors the retrogradation process, serving as a mechanism for generating RS. Starch may be treated with acids, enzymes, and an alcoholic-alkaline solution to break down PMS and produce compounds with diverse functions (Bhardwaj and Saxena 2017).

7.6.1.3.1 Acid Hydrolysis

Acid hydrolysis involves the breakdown of starch using mineral acids like sulfuric acid (H_2SO_4), hydrochloric acid (HCl), nitric acid (HNO_3), and phosphoric acid (H_3PO_4). Before reaching the inside part of the starch granules, mineral acids act on their surface. Below its gelatinization temperature, acid alters the structure of starch without impacting its granular morphology or other structural and functional characteristics (Wang and Copeland 2015). The solubility, pasting properties, WHC, and swelling power are all decreased by this modification (Kaur et al. 2011). The acid hydrolysis with HCl improved the color values (L,[1] a,[2] and b[3]), pasting qualities, paste clarity, and freeze-thaw stability (Balasubramanian et al. 2014). Bhardwaj and Saxena (2017) used acid hydrolysis combined with ultra-sonication and H_2SO_4 hydrolysis (3.2 M) to prepare starch nanoparticles from pearl millet. According to the findings, PMS can be utilized to produce bio-nanocomposites. Guar gum was hydrolyzed by Dangi et al. (2019), and the resultant hydrolysates were blended with PMS. The WHC, swelling power, and solubility of PMS were all increased by acid hydrolysis. Acid-hydrolyzed gums, when combined with PMS, lead to increased pasting viscosities, accompanied by a decrease in pasting temperatures. These acid-hydrolyzed starches are utilized as gelling agents in gum candies, substitutes for fat, and serve as RS in cookies.

7.6.1.3.2 Oxidation

The physicochemical (solubility, paste stability, and thermo-mechanical shearing) and structural characteristics of starch are modified by the process of adding carbonyl and carboxyl groups, which enhances its applications (El Halal et al. 2015).

---

[1] L represents lightness axis.

[2] +/− a value represents redness/greenness axis.

[3] +/− b represents yellowness/blueness axis.

The extent of physicochemical and structural modification is greatly influenced by the source of starch, oxidizing agents, and the surrounding environment for oxidation reactions. Typically, oxidants such as hydrogen peroxide (H_2O_2), potassium permanganate ($KMnO_4$), and sodium hypochlorite (NaOCl) are used to modify the natural structure of starch. Oxidized starch has distinct qualities such as low viscosity, reduced molecular size, high stability, enhanced clarity, and film-forming abilities that make it appropriate for use in food applications (Shaikh et al. 2019). Examined the effects of oxidation using NaOCl on the functional characteristics of PMS. They found that whereas starch oxidation increased the surface pores, it did not destroy the starch granules. Oxidation reactions cause depolymerization in addition to the incorporation of carboxyl and carbonyl groups on starch. This ultimately leads to intragranular modifications, the enlargement of surface pores, and an overall alteration in the microstructure of starch. In comparison to native starch, oxidized starch showed improved cold-storage stability in terms of springiness (0.6–1.0), cohesiveness (0.2–1.8), chewiness (29.4–19.8 g), and hardness (109 to 43.8 g). Oxidation modifications reduced ΔH_{gel} from 9.86 to 2.5 J/g, suggesting a lower energy requirement for starch granule gelatinization (Siroha et al. 2019).

7.6.1.3.3 Enzymatic Modifications

The effect of starch type and structure on starch breakdown has been investigated using a variety of enzymatic modifications (Hoover et al. 1996; Wankhede et al. 1990). PMS was enzymatically hydrolyzed using human saliva (Wankhede et al. 1990) and porcine pancreatic α-amylase (Hoover et al. 1996). When compared to corn and wheat starches, these starches are said to digest more swiftly (Hoover and Vasanthan 1994). For the hydrolysis of PMS, fungal amylase (0.1%) that was isolated from *Aspergillus oryzae* and had an enzyme activity of 2000 U/kg was used. Better transmittance or paste clarity was obtained by enzymatic treatment over acid-modified and heat-treated PMS. To achieve the necessary consistency in meals, such as fruit pastes and jellies, improved paste clarity is crucial. Because of the decrease in syneresis caused by PMS enzymatic alteration, PMS starches have better freeze-thaw stability (Balasubramanian et al. 2014).

7.6.2 Physical Modifications

Physical modifications of PMS involve altering its structure or properties without the use of chemical reactions and can be divided into thermal and nonthermal methods. Thermal modification includes pre-gelatinization and hydrothermal techniques like heat moisture treatment (HMT), whereas nonthermal methods include ultrasonication, pulse electric field (PEF), irradiation, and HPP. These approaches induce changes in the three-dimensional structure and morphological characteristics of starches by manipulating factors such as moisture, temperature, pressure, ultrasonic waves, radiation, and PEF. Consequently, they alter surface properties, particle size, functional attributes, solubility index, and gelation and pasting behavior of starch molecules. The physical modification of starch is often perceived as a straightforward and natural process, facilitating its versatile application in food without being restricted by regulatory limitations (Punia et al. 2021).

7.6.2.1 Thermal Modifications

7.6.2.1.1 Heat and Moisture Treatment

Heat and moisture treatment (HMT) typically involves heating the starch slurry to specific temperatures and maintaining it at those temperatures for a set duration under controlled moisture conditions. This treatment can lead to changes in the molecular structure of starch, including gelatinization and retrogradation, as well as modifications in its functional properties such as viscosity, swelling power, WHC, and thermal stability (Balasubramanian et al. 2014; Hoover 2010).

HMT induces significant structural modifications in both the crystalline and amorphous regions of starch by enhancing helical structures and starch chain mobility (Hoover 2010). The effects of under different moisture levels (20–30%) on the thermal properties of PMS were examined. The findings indicated that higher moisture content led to an elevation in gelatinization temperatures (T_o: 59.9–78.12 °C, T_p: 62.59–84.05 °C, and T_c: 73.55–89.04 °C). Furthermore, HMT led to softer gels with enhanced spreadability, attributed to decreased flow points and yield of PMS gels. This softening effect was accompanied by a reduction in peak viscosity, indicative of enhanced chain linkage post-HMT. The change from the amorphous phase to a more crystalline state during HMT contributed to this phenomenon. Additionally, HMT resulted in an increase in RS content (2.41–7.07%) as moisture levels increased. These findings suggest potential applications of HMT in heat-processed and frozen food products. Further studies suggested a decrease in RDS content and an increase in SDS and RS content following HMT treatment of pearl millet (Siroha et al. 2020a, b).

7.6.2.2 Nonthermal Modifications

Nonthermal modifications of starch involve altering its properties without the application of heat, employing techniques like enzymatic treatment, acid hydrolysis, oxidation, and cross-linking. Among these methods, gamma irradiation emerges as a nonthermal approach to modifying PMS. In gamma radiation treatment, starch samples are exposed to ionizing radiation, inducing structural changes in the starch molecules. This process can lead to modifications in crystallinity, molecular weight, and functional properties. In γ-irradiation, high-energy emissions from cobalt-60 (^{60}Co) or cesium-137 (^{137}Cs) are used (Ocloo et al. 2014). According to US FDA guidelines and international regulations, foods can be irradiated with either ^{60}Co or

^{137}Cs. Since ^{137}Cs can be difficult to handle, ^{60}Co is often used. Pearl millet was exposed to two doses of 1.0 and 2.5 kGy of radiation using ^{60}Co γ sources, and the results showed that the irradiation procedure increased the PMS in vitro digestibility (71.70–77.89%) (Sujatha et al. 2018). They also concluded that, although PMS digestibility was enhanced at low doses of 0–2.5 kGy, starch digestibility was reduced by irradiation at doses of 5–10 kGy. Higher concentrations cause the production of RS or amylolytic enzyme inhibitors, which reduce the digestibility of starch (Kong 2019). The modification and application of PMS have come a long way over the last 10 years, although certain nonthermal approaches (including HPP, ultrasonication, and PEF) have not yet been thoroughly investigated or used for PMS modification. Furthermore, the effects of nonthermal modifications on PMS will provide fresh perspectives on starch modifications.

7.6.3 Dual Modifications

A growing topic in nutrition science is the use of combination modification to improve the functionalities of starch in the food system. Before succinylation, PMS was exposed to acid hydrolysis at 0.1 M and 1 M HCl by Shaikh et al. (2015). When starches are acid-thinned, they form a stiff gel and have a high rate of retrogradation, which causes syneresis when stored for an extended period of time in a refrigerator. These disadvantages can be avoided by succinylation following acid hydrolysis. Every textural characteristic, including gumminess (40 to 15.7 g), chewiness (26.7–15.2 g), hardness (127.3–21.5 g), and springiness (0.6 to 0.8), was enhanced.

The hydrophilic succinyl groups of these modified starches provide them with certain significant qualities, such as high viscosity, low gelatinization temperature, and improved thickening and film-forming abilities. Foods that must be stored at a low temperature thus need succinylated-acid-thinned starches.

7.7 Functionality of PMS

The functionality of PMS, including its water absorption capacity, oil absorption capacity, swelling power, solubility, gelatinization temperature, digestibility, viscosity, and textural qualities, significantly impacts the potential uses of the starch in the advancement of food items.

7.7.1 Absorption Capacity of Water and Oil

The absorption capacity of water and oil in PMS is crucial for its functionality in various food and non-food applications. PMS exhibits significant water absorption capacity, which is vital for moisture retention, texture improvement, and shelf life enhancement in baked goods (Suma and Urooj 2015). Additionally, its oil absorption capacity is significant in applications like batter and breading systems, where starches act as binding agents and contribute to the desired crispy texture upon frying. The absorption capacities for water and oil vary depending on factors such as granule size, amylose content, and processing conditions. Experimental methods such as centrifugation or gravimetric techniques can be employed to determine these capacities accurately. Understanding the absorption properties of PMS informs its optimization in food products like baked goods, fried snacks, sauces, and dressings, as well as its potential applications in non-food industries such as pharmaceuticals, textiles, and cosmetics, where starches are valued for their functional attributes (Abdalla et al. 2009).

$$\text{Water / Oil absorption capacity} = \frac{\text{Weight of the sample} - \text{Weight of the sample after centrifuge}}{\text{Weight of the sample}}$$

The water absorption capacity and oil absorption capacity of pearl millet flour and isolated starch samples were investigated using standardized methods, revealing insights into their functional properties (Suma and Urooj 2015). The study revealed that flour samples exhibited higher water absorption capacity compared to their corresponding isolated starch due to differences in protein and fiber content (Taki 1962; Kinsella and Melachouris 1976). On the other hand, the oil absorption capacity of flour samples surpassed that of isolated starches, indicating potential differences in particle characteristics. Despite having a lower oil absorption capacity compared to corn starch, PMS still demonstrated significant potential (Adebowale et al. 2005). Furthermore, HMT was found to enhance both water and oil absorption capacities, with treated samples showing a notable increase in absorption capacity (Aruna and Parimalavalli 2023). These findings underscore the importance of understanding water absorption capacity and oil absorption capacity in pearl millet-based food formulations, offering valuable insights for enhancing product quality and shelf life.

7.7.2 Swelling Power and Solubility

The structural behavior of starches in food is marked by a progression through various stages, starting with water binding, followed by swelling, and culminating in the disintegration of starch granules (Punia 2020). The essential properties of swelling power and solubility provide strong evidence of the degree of interaction between starch chains in both amorphous and crystalline states. The standard temperature range used to determine the swelling power and solubility of PMS is 60–90 °C. Notably, these properties exhibit temperature dependence, with increased values as the temperature rises. This observation suggests a strong correlation between temperature fluctuations and the water absorption rate as well as the swelling capacity of PMS. When starch is heated in excess water, it induces the breakdown of the crystalline structure of starch. This breakdown occurs through the disruption of hydrogen bonds, consequently facilitating the formation of new bonds between the -OH group of water and the -OH group of starch. Consequently, the swelling power and solubility of starch increase, showcasing the dynamic interplay between temperature, water absorption, and structural changes in starch (Wani et al. 2012).In general, solubility and swelling power are integral properties that unveil a substance's intricate interactions with solvents, offering insights into its behavior, particularly exemplified in the transformation of starch under varying temperature conditions and excess water exposure. The swelling power and solubility are calculated as (Aruna and Parimalavalli 2023):

$$\text{Swelling Power} = W_{sg} / W_s \times 100$$

$$\%\text{Solubility} = W_{ds} / W_s \times 100$$

W_{sg} represents the weight of swollen granules, W_s represents the weight of the sample, and W_{ds} represents the weight of dried starch in the petri dish.

The swelling power is positively correlated to the amylopectin content, while the presence of amylose leads to dilution, acting as a swelling inhibitor (Singh et al. 2003). When PMS has low amylose content, the swollen starch granules break down and disintegrate at temperatures exceeding 65 °C, indicating shear-thinning behavior in PMS (Suma and Urooj 2015). Both the swelling power and the solubility of starch are influenced by lipid–amylose complexes, the ratio of starch chains in amylopectin, molecular weight, and the relative distribution of amylose and amylopectin (Wani et al. 2010). Hydrogels and lubricants are two products made from shear-thinned starches. The solubility patterns and swelling power of millet starches, which range from 60 to 90 °C, are less than those of other commercial starch sources like potato and wheat. This implies that millet starches are more resistant to swelling, which is a sign of a strong bonding force inside the starch granules (Wankhede et al. 1979). OSA modification can be used to improve the functional properties of PMS, such as swelling power, peak viscosity, emulsion activity, and lowered setback viscosity. In many food applications, higher starch solubility and swelling power levels are beneficial, especially when used as thickeners or binders.

Beleia et al. (1980) found that starches belonging to the RMP categories showed significantly reduced swelling properties at 65 °C and 70 °C. However, at temperatures ranging from 80 °C to 90 °C, they exhibit the highest swelling behavior but insignificant values. The swelling behavior and amylose leaching from starches isolated from pearl millet were studied by Suma and Urooj (2015). They found that swelling factor and amylose leaching increased proportionately with temperature, with the largest effects occurring between 60 and 70 °C. Additionally, the swelling factor of PMS surpassed that of wheat and rice, indicating a higher swelling capacity. The study concluded that the swelling power depends on the amount of amylopectin as well as chemical factors such as dispersion across the granule and the distribution of amylose and amylopectin. The swelling power of PMS was found to range from 2.03 to 14.5 g/g with a temperature range from 60 °C to 90 °C.

Studies indicate that a decrease in swelling power results from the formation of an intermolecular network through the presence of phosphorous residues following cross-linking reactions. Hoover and Manuel (1996) suggested that the starch modification may have increased the starch crystallinity, which would have limited the amount of water that could enter the starch matrix and resulted in limited swelling behavior. However, earlier reports suggested that a higher degree of cross-linking could reduce swelling power (Jyothi et al. 2006). The reduced swelling behavior of cross-linked starch varied from 1.81 to 8.62 g/g as the temperature changed from 50 to 90 °C. In contrast, oxidized starch exhibited an increase in swelling behavior, ranging from 2.14 to 11.4 g/g with a temperature increase from 50 to 90 °C. This increase in swelling power is due to the gelatinization process, which opens up the structure of the starch granule and introduces functional groups (Lawal 2004; Betancur et al. 1997).

HMP550 and HMP1700 cultivars exhibited low solubility at 75 °C (Beleia et al. 1980), although starches extracted from RMP cultivars exhibited the opposite behaviour. Starch from the RMP78 cultivar was noticeably more soluble than other starches, and it was most soluble at 95 °C. The authors also observed that starch exhibits differential swelling and solubility behavior depending on temperature; this phenomenon can be attributed to the two-stage relaxation of bonding forces inside the starch granules during swelling. It was also observed that millet starches had a two-stage solubility pattern.

Hoover and Vasanthan (1994) investigated the leaching of PMS from 50 to 95 °C and observed that amylose leaching was most common between 60 and 70 °C as millet starch temperatures increased. There is variation in the solubility pattern with different temperatures, which has been attributed to the occurrence of variable chain lengths inside the starch molecule. The amount of amylose that leaks out of the starch granule during the swelling process is known as the solubility behavior of the starch at that specific temperature; as a result, raised amylose leaching will cause the swelling behavior to increase (Larsson et al. 1998).

Oxidized starch exhibited lower solubility compared to native starch, attributed to the formation of aldehyde groups that minimized the creation of oxidant hemiacetal cross-links. These cross-links stabilized the swelling behavior of starch granules but hindered amylose molecules from leaching out, thus decreasing solubility (Femi-Oyewo et al. 2015). Similarly, cross-linked and dual starches demonstrated lower solubility than native starch, likely due to the strengthening of granule structure during the cross-linking process, which prevented starch molecules from being leached out and minimized granule disintegration during gelatinization (Wurzburg 1986). Moreover, oxidized cross-linked starch exhibited a lower solubility pattern than native starch. In contrast, Beleia et al. (1980) found that HMP550 and HMP1700 displayed low solubility at 75 °C, while RMP78 cv. was more soluble, with the highest solubility observed at 95 °C. This suggests that starches exhibit low swelling and solubility behavior at 70 °C compared to higher swelling and solubility behavior at 80–95 °C. These properties play an important role in various applications, influencing characteristics such as texture, viscosity, and overall functionality in fields like food science, pharmaceuticals, and materials engineering.

7.7.3 Gelatinization

Granular starch experiences a phase change known as gelatinization when heated in the presence of sufficient water throughout a range of temperatures. This process includes the amorphous region absorbing water, the granules expanding radially, the crystalline region disintegrating and breaking double helices, and the starch molecules leaking. These are the main features that characterize the gelatinization phenomena (Hoover 2010). Many techniques such as a hot stage microscope equipped polarized filter, Kofler hot stage microscopy (KHSM), differential scanning calorimetry (DSC), thermal analytical techniques, NMR spectroscopy, and FTIR are frequently used to determine the degree of gelatinization in millet starch. During analysis, parameters including enthalpy change (ΔH), T_c, T_p, and T_o are regularly recorded (Table 7.3). Significant differences exist between KHSM and DSC in both experimental and instrumental settings, which can cause changes in the findings of gelatinization temperature for the same sample. DSC has been the most widely used technique to measure starch gelatinization compared to KHSM in recent years. The first technique evaluates the enthalpy change (ΔH) as well as the temperatures (T_o, T_p, and T_c), whereas the second technique just analyzes the temperatures. The gelatinization parameters determined by DSC are highly influenced by variables like the starch:water ratio and the temperature scanning rate (Annor et al. 2014).

Table 7.3 Thermal properties of starch derived from pearl millet

	Gelatinization parameters				
Starch: water ratio (w/w)	T_o (°C)	T_p(°C)	T_c (°C)	ΔH_{gel} (J/g)	References
1:3	60.9–64.5	67.5–70.0	74.0–78.0	2.5–3.5	Hoover et al. (1996)
1:2	66.2–67.2	69.7–71.4	86.3–91.0	14.3–14.7	Wurzburg (1986)
1:2	60.8–64.8	67.1–69.9	74.1–78.0	5.6–7.7	Bhupender et al. (2013)
1:3	62.8	67.9	72	12.3	Annor et al. (2014)
1:2	64.7	69.7	74.6	3.5	Wu et al. (2014)
1:3	69.6	73.7	87.2	9.8	Kong (2019)
1:4	59.9	62.6	73.6	11.9	Gaffa et al. (2004)
1:2	63.4–67.7	69.3–71.6	74.5–76.3	10.6–12.4	Sandhu and Siroha (2017)
1:3	67.29	71.83	77.81	1.21	Balasubramanian et al. (2014)
1:3	69.60	73.67	87.23	9.86	Shaikh et al. (2019)
1:3	62.92–76.16	67.95–81.05	73.78–84.50	7.9–10.8	Sharma et al. (2016)
1:2	62.5	68.5	74.2	10.3	Punia et al. (2021)

7.7.4 Digestibility

The metabolic process of starch digestibility is important because it affects the rate and degree of intestinal digestion, which in turn affects blood glucose levels (Magallanes-Cruz et al. 2017). There are three types of digestible starch, namely SDS, RS, and RDS. Unlike the other types, RS travels to the large intestine unhydrolyzed because enzyme activity is absent in the small intestine (Leong et al. 2019). RDS breaks down quickly during digestion and is quickly absorbed in the small intestine. This quick absorption raises blood glucose levels and increases the risk of obesity, type-2 diabetes, and cardiovascular disease (Sajilata et al. 2006). Meals with a high amount of SDS breakdown and absorb more slowly in the small intestine than meals with high levels of RDS. As a consequence, there is a modest glycemic response and a gradual and continuous release of glucose into the circulation. The favorable effects of SDS, attributed to its low GI, help to reduce the risks related to obesity, overweight, and type 2 diabetes. RS is protective against several diseases, such as gallstones, cancer, high cholesterol, decreased fat buildup, and enhanced absorption of minerals. The GI and enzymatic starch hydrolysis are predicted to be lower in millet starches that have a high concentration of long amylose chains and prolonged chain segments between branch terminals (Annor et al. 2013). The content of RDS, SDS, and RS in PMS varies between studies. RDS, SDS, and RS were observed ranging from 11.3% to 12.2%, 7.9% to 9.0%, and 1.4% to 2.2%, respectively. In contrast, another study reported higher values, with RDS, SDS, and RS between 46.3% and 51.6%, 37.2% and 38.7%, and 9.7% and 16.5%, respectively. These variations in starch digestibility are attributed to differences in the crystalline structure of the starches. Starches with A-type crystallinity are more sensitive to enzymatic hydrolysis, resulting in higher RDS and SDS content and lower RS content than those with B-type crystallinity. The decreased RS level in A-type starches is due to structural changes that occur during starch separation, removing barriers to amylase hydrolysis (Perera et al. 2010). The effect of a diet on blood glucose levels is evaluated by the GI. Diet is very important in managing diabetes, and low-GI foods are becoming more and more popular. These meals are preferred because they help to slow down the bloodstream's absorption of glucose (Geetha et al. 2020). A lower glycemic and insulin response is associated with the structural properties of millet starch and the interactions that occur between starch, protein, and fat. Millet starch's structural properties, as well as interactions between starch, protein, and lipid components, help to explain its low glycemic and insulin response. The lower ratio of amylose to amylopectin in PMS can be attributed to its low GI, since the lower amylopectin content results in a lower postprandial blood glucose concentration when compared to other starch sources. As a result, PMS presents itself as a feasible alternative to high-GI cereals in daily nutritional intake, especially for carbohydrate-sensitive persons. Due to its low GI, incorporating PMS in dietary treatment can help to reduce the risk of diabetes in developed as well as developing countries.

7.7.5 Textural Quality

The textural qualities of a starch gel play a pivotal role in determining the overall appeal of a food product to consumers. Various factors, including starch concentration, types of starch modification, cross-linking agents, and time-temperature conditions, influence the formation of a starch gel. Lower cohesiveness and springiness values indicate that the gel may have a tendency to break apart easily during chewing. Typically, starch gels derived from natural sources exhibit suboptimal textures, but desired textures can be achieved through the incorporation of external gelling agents, cross-linkers, or modification of native starches. For instance, the addition of guar gum and its hydrolysate notably affected the textural properties of PMS gels. The inclusion of guar gum hydrolysate led to an increase in gel hardness, likely due to enhanced water immobilization facilitated by the hydrolysate. Conversely, increasing the concentration of native guar gum resulted in decreased gel hardness, possibly because of increased interaction between amylose and guar gum molecules, hindering starch retrogradation. Additionally, there was a slight reduction in springiness with the addition of guar gum/hydrolysate, resulting in gels that were easier to chew due to their less rubbery texture (Lau et al. 2000). Comparative analysis between native starch gels and modified starch gels revealed slight declines in cohesiveness in the latter, indicating increased susceptibility to disintegration during mastication. Noteworthy alterations in textural characteristics were observed in gel formation processes involving modified pearl millet starches, including acetylated, succinylated, octenylsuccinylated, and oxidized variants (Shaikh et al. 2017a, b). Esterification of pearl millet starches disrupts the network development of amylose gel by introducing exogenous groups such as OSA, resulting in gels that are less rigid than those made from native starches. While increased amylose concentration typically correlates with increased hardness, the addition of starch increases water absorption, yielding a softer gel. Hence, the selection of suitable starch and modification techniques is crucial and should be tailored according to the desired outcome.

7.7.6 Transmittance

The way that light travels through starch paste may be understood by its light transmittance. Paste clarity is impacted by the chain lengths of amylose and amylopectin, granular swelling, molecular structure, and leached amylase (Jan et al. 2017; Suriya et al. 2016). Wu et al. (2014) found a transmittance of 6% for PMS paste, and the observed light transmittance in the case of PMS paste varied from 1.80% to 2.60% (Bhupender et al. 2013). For food items like fruit pastes and jellies to have the proper consistency, starch pastes must be clear, as demonstrated by their light transmittance.

7.7.7 *Bulk Density*

The bulk density of pearl millet can vary depending on factors such as moisture content and processing methods. Generally, the bulk density of pearl millet flour ranges from about 0.80 to 1.00 g/cm^3 (Pawase et al. 2019; Akinola et al. 2017). It is calculated by dividing the mass of the sample (M) by its total volume (V), expressed in g/cm^3(Varnamkhasti et al. 2008).

$$BD = (M / V)$$

To determine the bulk density of pearl millet, the method involves weighing a known quantity of pearl millet grains using a digital balance and pouring them into a graduated cylinder or any container with a known volume, such as a 500-mL cylinder. Subsequently, the grains are gently tapped to settle them and remove any air voids. The volume of the settled grains in the cylinder is then recorded accurately. It is essential to repeat the process with multiple samples to ensure accuracy and take the average of the calculated bulk densities for reliable results (Mariotti et al. 2006). This method provides a quantitative measure of the compactness of pearl millet grains.

7.8 Food/Non-Food Applications

7.8.1 *Food Applications*

Pearl millet stands out as a highly nutritious food option, boasting a wealth of essential nutrients. Its nutritional richness makes it a versatile ingredient suitable for use in both modern and traditional cooking. From traditional to baked products, extruded products, and health-enhancing snacks, pearl millet continues to inspire culinary innovation and contribute to global food diversity (Fig. 7.5). Pearl millet grains are used to make a broad range of dishes in Africa and India, including porridge, couscous, flatbreads, desserts, and alcoholic (opaque beer, chibuku shake, merissa, mbeg) and non-alcoholic beverages (boza, bushera, kunun Zaki, mahewu, marewa, pito, pombe, oskikundu) (Adebiyi et al. 2018).

7.8.1.1 Traditional Products

Pearl millet is a staple in various traditional dishes worldwide, serving as unleavened bread, porridge, and dessert. Its versatility allows for the creation of culturally significant foods like khichri, roti, and couscous, which are prevalent in India, Africa, and the Middle East. West African countries have signature dishes like "Tuwo" or "Tô" porridges, while steam-cooked couscous is prevalent in Francophone

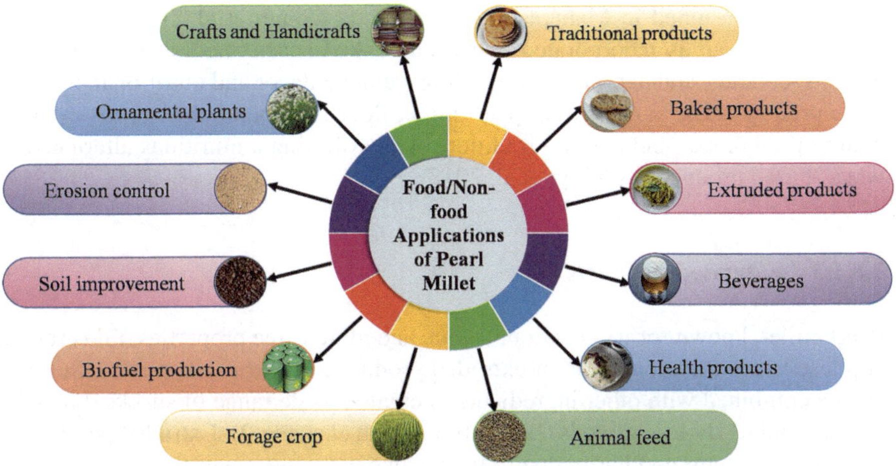

Fig. 7.5 Food/non-food applications of pearl millet

nations like Senegal and Mali. Nigeria and Niger favor thin porridges such as "Fourra" and the fried cake "Masa." Porridges made from pearl millet flour, whether fermented or unfermented, exhibit a range of consistencies and flavors, while flatbreads resembling pancakes are popular in India and Saudi Arabia (Kaur et al. 2014). Additionally, couscous, a staple in North African cuisines, is commonly consumed with vegetables or yogurt, adding to the rich culinary tradition. Sweets like Ladoo and Dakuwa, prepared from roasted pearl millet grain flours, offer a delightful treat, with potential ingredients like popped pearl millet grain and malted flour enhancing their nutritional value (Ajeigbe et al. 2020; Singh and Sehgal 2008).

7.8.1.2 Baked Products

Pearl millet flour serves as a versatile substitute for wheat flour, offering nutritional diversity in baked goods such as "whole-grain" breads, pretzels, crackers, tortillas, and cereals. Additionally, pancake-like flatbreads made with unfermented or fermented pearl millet flour are popular in various regions, often accompanied by complementary condiments and sauces (Dahlberg et al. 2003). In India, pearl millet is extensively used in baked products, particularly in regions where it is a staple grain. Pearl millet flour serves as a key ingredient in various traditional Indian baked goods, offering a gluten-free alternative to wheat-based products. One of the most common uses of pearl millet in baked products in India is in the preparation of unleavened bread, such as roti or chapatti. These flatbreads are a staple in many Indian households and are made by mixing bajra flour with water and salt, then rolling out the dough and cooking it on a hot griddle or tawa (Siroha et al. 2016). Additionally, pearl millet flour is used to make traditional snacks like bhakri, a type of crispy flatbread popular in Maharashtra, and thalipeeth, a multigrain pancake commonly consumed in Maharashtra and Gujarat (Singh et al. 2013). Moreover, pearl millet flour is incorporated in various baked snacks and sweets, including biscuits, cookies, and nan-khatai, providing a unique flavor and nutritional profile to these products. The versatility of pearl millet flour in Indian baked goods caters to dietary preferences and cultural traditions while offering a nutritious alternative to wheat-based products (Adebiyi et al. 2018).

7.8.1.3 Extruded Products

Pearl millet, known for its nutritional value and gluten-free properties, finds several applications in the production of extruded products. In extrusion, pearl millet flour can be combined with other ingredients to create a wide range of snacks, breakfast cereals, and ready-to-eat foods (Balasubramanian et al. 2012). Extruded pearl millet snacks are gaining popularity due to their crunchy texture, savory or sweet flavors, and enhanced shelf life. These snacks can be seasoned with various spices, herbs, or sweeteners to cater to different taste preferences (Selladurai et al. 2023). Additionally, pearl millet can be extruded into breakfast cereals like flakes, puffs, or clusters, offering a nutritious and convenient breakfast option. Extruded pearl millet products are not only tasty but also offer health benefits, as pearl millet is rich in essential nutrients such as protein, fiber, vitamins, and minerals (Dias-Martins et al. 2018).

7.8.1.4 Beverages

Pearl millet finds extensive applications in the beverage industry, contributing to both alcoholic and non-alcoholic drinks. Traditional alcoholic beverages like opaque beer and Chibuku Shake are crafted through the fermentation of pearl millet grains, offering distinctive flavors and textures enjoyed across Africa (Singh et al. 2023; Embashu and Nantanga et al. 2019). Additionally, merissa, another fermented beverage, is notable for its sour undertones and is often savored during social gatherings. Pearl millet also serves as a key ingredient in various non-alcoholic drinks such as boza, bushera, kunun Zaki, mahewu, oskikundu, marewa, pito, and pombe, providing refreshing options particularly in warmer climates. Furthermore, malted pearl millet is utilized in the production of malted drinks, prized for their nutritional content and energy-boosting properties (Modha and Pal 2011; Embashu and Nantanga 2019).

7.8.1.5 Health Products

Pearl millet is increasingly recognized for its valuable contributions to health products, offering nutritional richness and diverse functional benefits. Its high protein content, along with essential minerals such as calcium, phosphorus, potassium, iron, and zinc, makes it a desirable ingredient for fortified foods and supplements (Rai et al. 2008). Pearl millet flour is incorporated into various health-focused products like fortified bread, biscuits, pasta, and weaning foods, catering to diverse dietary needs and preferences. Additionally, the gluten-free nature of pearl millet makes it suitable for individuals with gluten sensitivities or celiac disease, expanding the accessibility of nutritious options. Moreover, pearl millet's potential in non-dairy fermented beverages contributes to probiotic-rich drink options, promoting gut health and digestion (Dahlberg et al. 2003). With its ability to enhance the nutritional profile of products and cater to specialized dietary requirements, pearl millet plays a pivotal role in the development of a wide range of health products, aligning with the growing demand for functional foods and dietary supplements (Nambiar et al. 2011).

7.8.2 Non-food Applications

While pearl millet (*Pennisetum glaucum*) is primarily recognized for its importance as a staple food crop, its utility extends far beyond culinary purposes. This hardy and versatile grain has found diverse applications in various sectors, ranging from agriculture and renewable energy to environmental conservation and cultural heritage. In this comprehensive exploration, we review the diverse non-food uses of pearl millet, elucidating its significant contributions across different domains (Fig. 7.5).

7.8.2.1 Animal Feed

Pearl millet serves as a crucial component in animal feed formulations due to its high nutritional content and adaptability to different climates (Reddy et al. 2018). The grains are rich in protein, carbohydrates, vitamins, and minerals, making them an excellent feed option for poultry, livestock, and even fish. Its resilience to drought conditions and efficient water use make it particularly suitable for regions with limited water availability (Bastos et al. 2002). As a fodder crop, pearl millet provides valuable nutrition for grazing animals, contributing to their growth, health, and productivity (Rostagno et al. 2011; Manwar and Mandal 2009).

7.8.2.2 Forage Crop

In addition to grain production, pearl millet is cultivated as a forage crop for its nutritious foliage. The green leaves and stems of pearl millet are rich in protein and digestible carbohydrates, making them an ideal feed source for ruminant animals such as cattle, sheep, and goats. Pearl millet forage is especially valuable in arid and semi-arid regions where other forage options may be scarce. Its deep root system enhances soil structure and fertility, contributing to sustainable agriculture practices and soil conservation efforts (Harinarayana et al. 2005; Sedivec and Schatz 1991).

7.8.2.3 Biofuel Production

Pearl millet biomass holds promise as a renewable energy source for biofuel production. The crop's high biomass yield, rapid growth rate, and efficient nutrient utilization make it an attractive feedstock for bioenergy applications (Umakanth et al. 2024). Pearl millet residues, including stalks, leaves, and husks, can be processed through biochemical or thermochemical conversion methods to produce biofuels such as ethanol, biogas, or biochar. By utilizing pearl millet biomass for bioenergy production, farmers can diversify their income streams while contributing to climate change mitigation efforts and reducing dependence on fossil fuels (Packiam et al. 2018; Blessie et al. 2021).

7.8.2.4 Soil Improvement

Pearl millet plays a vital role in soil improvement and conservation through its agronomic characteristics and cultivation practices. As a C_4 crop, pearl millet exhibits efficient photosynthesis and water use efficiency, enabling it to thrive in hot and dry conditions. Its deep root system penetrates the soil profile, improving soil structure, enhancing water infiltration, and reducing erosion risk. Pearl millet residues left after harvest contribute organic matter to the soil, enhancing nutrient cycling and soil fertility. Moreover, the crop's ability to fix atmospheric nitrogen can reduce the need for synthetic fertilizers, promoting sustainable agriculture practices and soil health (Bidinger and Parthasarathy Rao 1990).

7.8.2.5 Erosion Control

Pearl millet is effective in controlling soil erosion and mitigating land degradation, particularly in vulnerable landscapes prone to erosion. Its extensive root system stabilizes soil structure, binds soil particles together, and reduces surface runoff, thus preventing soil erosion on slopes and barren lands. By establishing pearl millet cover crops or buffer strips in erosion-prone areas, farmers can protect valuable topsoil, preserve soil fertility, and safeguard water resources. The crop's resilience to adverse environmental conditions makes it a valuable tool in erosion control strategies, contributing to sustainable land management practices and ecosystem resilience (Michels and Bielders 2006).

7.8.2.6 Ornamental Plant

Certain varieties of pearl millet are cultivated for ornamental purposes, valued for their striking appearance and architectural form. These ornamental millet varieties, characterized by their colorful foliage and distinctive seed heads, are cultivated in gardens, parks, and landscaping projects for their aesthetic appeal. Their tall, upright growth habit and unique textures add visual interest and structural diversity to outdoor spaces. Ornamental millets are appreciated not only for their ornamental value but also for their drought tolerance and low maintenance requirements, making them popular choices for landscape designers and garden enthusiasts (Rai et al. 2014).

7.8.2.7 Crafts and Handicrafts

In some cultures, pearl millet stalks are utilized in crafts and handicrafts, showcasing the crop's cultural significance and practical utility beyond agriculture. The sturdy stalks of pearl millet can be woven, braided, or twisted to create various decorative items such as baskets, mats, hats, and traditional artwork. These handcrafted products not only showcase the artisanal skills of local communities but also serve functional purposes in daily life, such as storage, shelter, or ceremonial use. By incorporating pearl millet stalks into crafts and handicrafts, artisans preserve traditional knowledge and cultural heritage while promoting sustainable livelihoods and rural entrepreneurship (Bharati et al. 2019).

7.9 Conclusion

Pearl millet stands out as a tough and nutritionally rich grain, offering significant promise across various sectors. Its chemical composition, featuring essential nutrients and functional components, underscores its value for addressing food security challenges. Through detailed exploration of starch isolation, modifications, and diverse applications, pearl millet emerges as a sustainable and versatile resource with untapped potential. Pearl millet holds immense promise not only as a food source but also for various non-food applications. Its remarkable nutritional profile, boasting high protein content, dietary fiber, and essential minerals, all at a low production cost, makes it a valuable resource for both culinary and non-food purposes. Moreover, its agronomic resilience, including tolerance to high temperatures and minimal water requirements, coupled with its low susceptibility to mycotoxins and non-transgenic nature, positions it as a key contributor to food security and environmental sustainability. Therefore, there is a great need to pay more attention to genetic enhancement for developing high-yielding varieties and in-depth exploration of the potential of pearl millet in different regions. By promoting advances in millet cultivation and use, food security can be bolstered, environmental sustainability can be promoted, and viable solutions can be offered to meet the diverse needs of societies around the world.

Conflict of Interest The authors declare no conflicts of interest.

References

Abdalla AA, Ahmed UM, Ahmed AR et al (2009) Physicochemical characterization of traditionally extracted pearl millet starch (Jir). J Appl Sci Res 5:2016–2027

Adebiyi JA, Obadina AO, Adebo OA et al (2018) Fermented and malted millet products in Africa: expedition from traditional/ethnic foods to industrial value-added products. Crit Rev Food Sci Nutr 58:463–474. https://doi.org/10.1080/10408398.2016.1188056

Adebowale KO, Afolabi TA, Olu-Owolabi BI (2005) Hydrothermal treatments of finger millet (Eleusinecoracana) starch. Food Hydrocoll 19:974–983. https://doi.org/10.1016/j.foodhyd.2004.12.007

Adeola O, Orban JI (1995) Chemical composition and nutrient digestibility of pearl millet (Pennisetumglaucum) fed to growing pigs. J Cereal Sci 22(2):177–184. https://doi.org/10.1016/0733-5210(95)90048-9

Ajeigbe HA, Angarawai II, Inuwa AH, et al (2020) Hand book on improved pearl millet production practices in north eastern Nigeria. http://oar.icrisat.org/id/eprint/11530

Akinola SA, Badejo AA, Osundahunsi OF et al (2017) Effect of preprocessing techniques on pearl millet flour and changes in technological properties. Int J Food Sci Technol 52(4):992–999. https://doi.org/10.1111/ijfs.13363

Ali MAM, Tinay AH, Abdalla AH (2003) Effect of fermentation on the in vitro protein digestibility of pearl millet. Food Chem 80(1):51–54. https://doi.org/10.1016/S0308-8146(02)00234-0

Annor GA, Marcone M, Bertoft E et al (2013) In vitro starch digestibility and expected glycemic index of kodo millet (Paspalum scrobiculatum) as affected by starch–protein–lipid interactions. Cereal Chem 90:211–217. https://doi.org/10.1094/CCHEM-06-12-0074-R

Annor GA, Marcone M, Bertoft E et al (2014) Physical and molecular characterization of millet starches. Cereal Chem 91(3):286–292. https://doi.org/10.1094/CCHEM-08-13-0155-R

Aruna M, Parimalavalli R (2023) Effect of dual modification on the properties of native pearl millet starch. Asian J Dairy Res 42(3):355–360. https://doi.org/10.18805/ajdfr.DR-2089

Asrani P, Ali A, Tiwari K (2023) Millets as an alternative diet for gluten-sensitive individuals: a critical review on nutritional components, sensitivities and popularity of wheat and millets among consumers. Food Rev Int 39(6):3370–3399. https://doi.org/10.1080/87559129.2021.2012790

Ayoub AS, Rizvi SSH (2009) An overview on the technology of crosslinking of starch for non-food applications. J Plast Film Sheet 25(1):25–45. https://doi.org/10.1177/8756087909336493

Balasubramanian S, Singh KK, Patil RT et al (2012) Quality evaluation of millet-soy blended extrudates formulated through linear programming. J Food Sci Technol 49:450–458. https://doi.org/10.1007/s13197-011-0297-0

Balasubramanian S, Sharma R, Kaur J et al (2014) Characterization of modified pearl millet (Pennisetum typhoides) starch. J Food Sci Technol 51:294–300. https://doi.org/10.1007/s13197-011-0490-1

Bastos AO, Landell Filho LDC, Passipieri M et al (2002) Different levels of pearl millet (Pennisetum glaucum (L.) R. Brown) in the pigs feeding. Rev Bras Zootec 31:1753–1760. https://doi.org/10.1590/S1516-35982002000700018

Beleia A, Varriano Marston E, Hoseney RC (1980) Characterization of starch from pearl millets. Cereal Chem 57(5):300–303

Betancur AD, Chel GL, Cañizares HE (1997) Acetylation and characterization of Canavaliaensiformis starch. J Agric Food Chem 45(2):378–382. https://doi.org/10.1021/jf960272e

Bharati K, Dhiman SR, Puja S et al (2019) Utilization of agri-horticultural crops/wastes for making dry and value-added products to generate employment. Indian J Ecol 46:325–329

Bhardwaj M, Saxena DC (2017) Synthesis of starch nanoparticles by acid hydrolysis alone and coupled with other technique and their subsequent characterization and utilization. J Nanomed Nanotechnol. https://doi.org/10.4172/2157-7439-C1-053

Bhupender SK, Rajneesh B, Baljeet SY (2013) Physicochemical, functional, thermal and pasting properties of starches isolated from pearl millet cultivars. Int Food Res J 20(4):1555–1561

Bidinger FR, Parthasarathy Rao P (1990) Genetic and cultural improvement in the production of pearl millet. A comparative assessment for India and Sahelian/ Sudanian Africa. In: Proceedings of the International Congress of Plant Physiology, Vol. 1, June 1988, New Delhi, India. Society for Plant Physiology and Biochemistry, New Delhi, India, pp 194–206

Blessie JP, Kennady ZJ, Karthikeyan S et al (2021) Pearl millet (Pennisetumglaucum) as a sustainable feedstock for bioethanol production by catalytic down flow liquid contact reactor. Crop Res 56:270–275. https://doi.org/10.31830/2454-1761.2021.043

Brunken J, De Wet JMJ, Harlan JR (1977) The morphology and domestication of pearl millet. Econ Bot 31(2):163–174

Carey FA (2003) Organic chemistry, 5th edn. Mcgraw-Hill, New York

Carpita NC, Kanabus J (1987) Extraction of starch by dimethyl sulfoxide and quantitation by enzymatic assay. Anal Biochem 161(1):132–139. https://doi.org/10.1016/0003-2697(87)90662-2

CFR (Code of Federal Regulation) (2001) Food starch modified. Food additives permitted in food for human consumption. Govt. Printing office, Washington, USA. 21/1/172/ 172.892

Chauhan M, Sonawane SK, Arya SS (2018) Nutritional and nutraceutical properties of millets: a review. Clin J Nutr Diet 1(1):1–10

Chellapilla TS, Ambawat S, Gurjar NR (2022) Millets: role and responses under abiotic stresses. In: sustainable remedies for abiotic stress in cereals. Springer, Singapore, pp 171–207. https://doi.org/10.1007/978-981-19-5121-3_8

Chung HJ, Woo KS, Lim ST (2004) Glass transition and enthalpy relaxation of cross-linked corn starches. Carbohydr Polym 55(1):9–15. https://doi.org/10.1016/j.carbpol.2003.04.002

Dahlberg JA, Wilson JP, Snyder T (2003) Sorghum and pearl millet: health foods and industrial products in developed countries. Alternative uses of sorghum and pearl millet in Asia 12(3):42–54

Dangi N, Yadav BS, Yadav RB (2019) Pasting, rheological, thermal and gel textural properties of pearl millet starch as modified by guar gum and its acid hydrolysate. Int J Biol Macromol 139:387–396. https://doi.org/10.1016/j.ijbiomac.2019.08.012

De Wet JMJ (1986) Pearl millet (Pennisetumglaucum) in Africa and India. In: International pearl millet workshop 3–4

Dias-Martins AM, KLF P, Pacheco S et al (2018) Potential use of pearl millet (Pennisetumglaucum (L.) R. Br.) in Brazil: food security, processing, health benefits and nutritional products. Food Res Int 109:175–186. https://doi.org/10.1016/j.foodres.2018.04.023

Directorate of Millets Development (2020) Department of Agriculture, Cooperation & Farmers Welfare, Ministry of Agriculture & Farmers Welfare, Government of India

Egharevba HO (2019) Chemical properties of starch and its application in the food industry. In: Chemical properties of starch, vol 9. IntechOpen, London

Ejeta G, Hassen MM, Mertz ET (1987) In vitro digestibility and amino acid composition of pearl millet (Pennisetumtyphoides) and other cereals. Proc Natl Acad Sci U S A 84(17):6016–6019. https://doi.org/10.1073/pnas.84.17.6016

El Halal SLM, Colussi R, Pinto VZ et al (2015) Structure, morphology and functionality of acetylated and oxidised barley starches. Food Chem 168:247–256. https://doi.org/10.1016/j.foodchem.2014.07.046

Eliasson AC (2006) Carbohydrates in food.CRC Press.https://doi.org/10.1201/9781420015058

Embashu W, Nantanga KKM (2019) Pearl millet grain: a mini-review of the milling, fermentation and brewing of ontaku, a non-alcoholic traditional beverage in Namibia. Trans R Soc S Afr 74(3):276–282

Embashu W, Iileka O, Nantanga KK (2019) Namibian opaque beer: a review. J Inst Brew 125(1):4–9. https://doi.org/10.1002/jib.533

FAO (2019) Food and agriculture Organization of the United Nations. Rome

FAO (2022) Food and agriculture Organization of the United Nations. Rome

Femi-Oyewo MN, Ajala TO, Babs-Awolowo D (2015) The compaction, mechanical and disintegration properties of modified Pennisetumglaucum (Poaceae) starch in directly compressed chloroquine tablet formulations. J Appl Pharm Sci 5(2):043–050. https://doi.org/10.7324/JAPS.2015.50207

Gaffa T, Yoshimoto Y, Hanashiro I et al (2004) Physicochemical properties and molecular structures of starches from millet (Pennisetumtyphoides) and sorghum (Sorghum bicolor L. Moench) cultivars in Nigeria. Cereal Chem 81:255–260. https://doi.org/10.1094/CCHEM.2004.81.2.255

Geetha K, Yankanchi GM, Hulamani S et al (2020) Glycemic index of millet based food mix and its effect on pre diabetic subjects. J Food Sci Technol 57:2732–2738. https://doi.org/10.1007/s13197-020-04309-5

Harinarayana G, Melkania NP, Reddy BV, et al (2005) Forage potential of sorghum and pearl millet Forage potential of sorghum and pearl millet 292–321

Hirsch JB, Kokini JE (2002) Understanding the mechanism of cross-linking agents (POCl3, STMP, and EPI) through swelling behaviour and pasting properties of cross-linked waxy maize starches. Cereal Chem 79:102–107. https://doi.org/10.1094/CCHEM.2002.79.1.102

Hoover R (2010) The impact of heat-moisture treatment on molecular structures and properties of starches isolated from different botanical sources. Crit Rev Food Sci Nutr 50:835–847. https://doi.org/10.1080/10408390903001735

Hoover R, Manuel H (1996) The effect of heat–moisture treatment on the structure and physicochemical properties of normal maize, waxy maize, dull waxy maize and amylomaize V starches. J Cereal Sci 23:153–162. https://doi.org/10.1006/jcrs.1996.0015

Hoover R, Vasanthan T (1994) Effect of heat-moisture treatment on the structure and physicochemical properties of cereal, legume, and tuber starches. Carbohydr Res 252:33–53. https://doi.org/10.1016/0008-6215(94)90004-3

Hoover R, Swamidas G, Kok LS et al (1996) Composition and physicochemical properties of starch from pearl millet grains. Food Chem 56(4):355–367

Hoseney RC (1986) Principles of cereal science and technology, A general reference on cereal foods, p 327

Hoseney RC (1994) Principles of cereal science and technology (No. Ed. 2). American Association of Cereal Chemists (AACC)

Jan KN, Panesar PS, Rana JC et al (2017) Structural, thermal and rheological properties of starches isolated from Indian quinoa varieties. Int J Biol Macromol 102:315–322. https://doi.org/10.1016/j.ijbiomac.2017.04.027

Jeon YS, Lowell AV, Gross RA (1999) Studies of starch esterification: reactions with alkenylsuccinates in aqueous slurry systems. Starch-Stärke 51(2–3):90–93. https://doi.org/10.1002/(SICI)1521-379X(199903)51:2<90::AID-STAR90>3.0.CO;2-M

Jha A, Tripathi AD, Alam T et al (2013) Process optimization for manufacture of pearl millet-based dairy dessert by using response surface methodology (RSM). J Food Sci Technol, J Food Sci Technol 50:367–373. https://doi.org/10.1007/s13197-011-0347-7

Jukanti AK, CLL G, Rai KN et al (2016) Crops that feed the world 11. Pearl millet (Pennisetumglaucum L.): an important source of food security, nutrition and health in the arid and semi-arid tropics. Food Secur 8:307–329. https://doi.org/10.1007/s12571-016-0557-y

Jyothi AN, Moorthy SN, Rajasekharan KN (2006) Effect of cross-linking with epichlorohydrin on the properties of cassava (ManihotesculentaCrantz) starch. Starch-Stärke 58(6):292–299. https://doi.org/10.1002/star.200500468

JECFA (2001) Starches; Summary of evaluations performed by the Joint FAO/WHO expert committee on food additives.

Kapri M, Verma DK, Kumar A et al (2017) Modified pearl millet starch: a review on chemical modification, characterization and functional properties. In: Engineering interventions in agricultural processing, pp 191–226

Kaur M, Oberoi DPS, Sogi DS et al (2011) Physicochemical, morphological and pasting properties of acid treated starches from different botanical sources. J Food Sci Technol 48:460–465. https://doi.org/10.1007/s13197-010-0126-x

Kaur B, Ariffin F, Bhat R et al (2012) Progress in starch modification in the last decade. Food Hydrocoll 26(2):398–404. https://doi.org/10.1016/j.foodhyd.2011.02.016

Kaur KD, Jha A, Sabikhi L et al (2014) Significance of coarse cereals in health and nutrition: a review. Crit Rev Food Sci Nutr 51:1429–1441. https://doi.org/10.1007/s13197-011-0612-9

Kinsella JE, Melachouris N (1976) Functional properties of proteins in foods: a survey. Crit Rev Food Sci Nutr 7:219–280. https://doi.org/10.1080/10408397609527208

Klucinec JD, Thompson DB (1998) Fractionation of high-amylose maize starches by differential alcohol precipitation and chromatography of the fractions. Cereal Chem 75(6):887–896. https://doi.org/10.1094/CCHEM.1998.75.6.887

Kong X (2019) Starches modified by nonconventional techniques and food applications. In: Starches for food application. Academic. https://doi.org/10.1016/B978-0-12-809440-2.00007-1

Krishna Kumari S, Thayumanavan B (1998) Characterization of starches of proso, foxtail, barnyard, kodo, and little millets. Plant Foods Hum Nutr 53:47–56. https://doi.org/10.1023/A:1008083020810

Kumar A, Tomer V, Kaur A et al (2018) Millets: a solution to agrarian and nutritional challenges. Agric Food Secur 7(1):1–15

Larsson CT, Khoshnoodi J, Ek B et al (1998) Molecular cloning and characterization of starch-branching enzyme II from potato. Plant Mol Biol 37:505–511. https://doi.org/10.1023/A:1005908305456

Lau MH, Tang J, Paulson AT (2000) Texture profile and turbidity of gellan/gelatinmixed gels. Food Res Int 33(8):665–671. https://doi.org/10.1016/S0963-9969(00)00111-3

Lawal OS (2004) Composition, physicochemical properties and retrogradation characteristics of native, oxidised, acetylated and acid-thinned new cocoyam (Xanthosomasagittifolium) starch. Food Chem 87:205–218. https://doi.org/10.1016/j.foodchem.2003.11.013

Leong SY, Duque SM, Abduh SBM et al (2019) Carbohydrates. In: Innovative Thermal and Non-Thermal Processing, Bioaccessibility and Bioavailability of Nutrients and Bioactive Compounds. Woodhead Publishing, United Kingdom, p 171–206

Luo ZG, Shi YC (2012) Preparation of acetylated waxy, normal, and high-amylose maize starches with intermediate degrees of substitution in aqueous solution and their properties. J Agric Food Chem 60(37):9468–9475. https://doi.org/10.1021/jf301178c

Malik S (2015) Pearl millet-nutritional value and medicinal uses. Int J Adv Res Innov Ideas Educ 1(3):414–418.

Magallanes-Cruz PA, Flores-Silva PC, Bello-Perez LA (2017) Starch structure influences its digestibility: a review. J Food Sci 82:2016–2023. https://doi.org/10.1111/1750-3841.13809

Manwar SJ, Mandal AB (2009) Effect of high moisture storage of pearl millet (Pennisetum typhoides) with or without feed enzymes on growth and nutrient utilization in broiler chickens. J Anim Sci 80:438–445. https://doi.org/10.1111/j.1740-0929.2009.00656.x

Mariotti M, Alamprese C, Pagani MA et al (2006) Effect of puffing on ultrastructure and physical characteristics of cereal grains and flours. J Cereal Sci 43(1):47–56. https://doi.org/10.1016/j.jcs.2005.06.007

Martínez-Ballesta MC, Dominguez-Perles R, Moreno DA et al (2011) Minerals in plant food: effect of agricultural practices and role in human health. In: Lichtfouse E (ed) Sustainable Agriculture, vol 2. Springer, Dordrecht, pp 111–128. https://doi.org/10.1007/978-94-007-0394-0_8

Michels K, Bielders CL (2006) Pearl millet growth on an erosion-affected soil in the Sahel. Exp Agric 42:1–17. https://doi.org/10.1017/S0014479705002899

Mitchell JR, Hill SE (2021) Starch. In: Phillips GO, Williams PA (eds) Handbook of hydrocolloids, Woodhead publishing series in food science, technology and nutrition, 3rd edn. Woodhead Publishing, pp 239–271. https://doi.org/10.1016/B978-0-12-820104-6.00027-9

Modha H, Pal D (2011) Optimization of Rabadi-like fermented milk beverage using pearl millet. J Food Sci Technol 48:190–196. https://doi.org/10.1007/s13197-010-0146-6

Mustafa AF, Seguin P, Belair G et al (2008) Chemical composition and ruminal degradability of grain pearl millet grown in southwestern Quebec. Can J Anim Sci 88(1):71–77. https://doi.org/10.4141/CJAS07075

Nambiar VS, Dhaduk JJ, Sareen N et al (2011) Potential functional implications of pearl millet (Pennisetum glaucum) in health and disease. J Appl Pharm Sci 10:62–67

Neelam K, Cheung CMG, Ohno-Matsui K et al (2012) Choroidal neovascularization in pathological myopia. Prog Retin Eye Res 31(5):495–525. https://doi.org/10.1016/j.preteyeres.2012.04.001

Newman Y, Jennings E, Vendramini J et al (2010) Pearl millet (*Pennisetum glaucum*): overview and management. In: SSAGR-337. Florida cooperative extension service, Institute of Food and Agricultural Sciences. University of Florida, Gainesville, FL, pp 1–6. https://doi.org/10.32473/edis-ag347-2010

Obilana AB (2003) Overview: importance of millets in Africa. World (all cultivated millet species) 38(2):28

Ocloo FC, Minnaar A, Emmambux NM (2014) Effects of gamma irradiation and stearic acid, alone and in combination, on functional, structural, and molecular characteristics of high amylose maize starch. Starch-Stärke 66:624–635. https://doi.org/10.1002/star.201300221

Packiam M, Subburamu K, Desikan R et al (2018) Suitability of pearl millet as an alternate lignocellulosic feedstock for biofuel production in India. J Appl Environ Microbiol 6:51–58. https://doi.org/10.12691/jaem-6-2-4

Park S, Chung MG, Yoo B (2004) Effect of octenylsuccinylation on rheological properties of corn starch pastes. Starch-Stärke 56(9):399–406. https://doi.org/10.1002/star.200300274

Pattanashetti SK, Upadhyaya HD, Dwivedi SL et al (2016) Pearl millet. In: Singh M, Upadhyaya HD (eds) Genetic and genomic resources for grain cereals improvement. Academic, pp 253–289. https://doi.org/10.1016/B978-0-12-802000-5.00006-X

Pawase PA, Shingote A, Chavan UD (2019) Studies on evaluation and determination of physical and functional properties of millets (Ragi and Pearl Millet). AJDFR 38(3):203–212. https://doi.org/10.18805/ajdfr:DR-1407

Perera A, Meda V, Tyler RT (2010) Resistant starch: a review of analytical protocols for determining resistant starch and of factors affecting the resistant starch content of foods. Food Res Int 43:1959–1974. https://doi.org/10.1016/j.foodres.2010.06.003

Punia S (2020) Barley starch: structure, properties and in vitro digestibility-a review. Int J Biol Macromol 155:868–875. https://doi.org/10.1016/j.ijbiomac.2019.11.219

Punia S, Kumar M, Siroha AK et al (2021) Pearl millet grain as an emerging source of starch: a review on its structure, physicochemical properties, functionalization, and industrial applications. Carbohydr Polym 260:117776. https://doi.org/10.1016/j.carbpol.2021.117776

Raheem D, Dayoub M, Birech R et al (2021) The contribution of cereal grains to food security and sustainability in Africa: potential application of UAV in Ghana, Nigeria, Uganda, and Namibia. Urban Sci 5(1):8. https://doi.org/10.3390/urbansci5010008

Rai KN, Gowda CLL, Reddy BVS et al (2008) Adaptation and potential uses of sorghum and pearl millet in alternative and health foods. Compr Rev Food Sci Food Saf 7(4):320–396. http://oar.icrisat.org/id/eprint/7913

Rai KN, Patil HT, Yadav OP et al (2014) Dhanashakti: a high-iron pearl millet variety. Indian Farming 64(7):32–34. http://oar.icrisat.org/id/eprint/8602

Reddy AA, Raju SS, Suresh A et al (2018) Analysis of pearl millet market structure and value chain in India. J Agribus Dev Emerg Econ 8(2):406–424. https://doi.org/10.1108/JADEE-02-2016-0007

Rooney LW, Miller FR, Mertin JV (1981) Variation in the structure and kernel characteristics of sorghum. In: Proceedings of the international symposium on sorghum grain quality. ICRISAT, Patancheru, India 28, pp 143–162

Rostagno HS, Albino LFT, Donzele JL et al (2011) Brazilian tables for poultry and swine: composition of feedstuffs and nutritional requirements, 3rd edn. UFV, Viçosa

Sajilata MG, Singhal RS, Kulkarni PR (2006) Resistant starch–a review. Compr Rev Food Sci Food Saf 5(1):1–17. https://doi.org/10.1111/j.1541-4337.2006.tb00076.x

Sandhu KS, Siroha AK (2017) Relationships between physicochemical, thermal, rheological and in vitro digestibility properties of starches from pearl millet cultivars. LWT-Food Sci Technol 83:213–224. https://doi.org/10.1016/j.lwt.2017.05.015

Sanjana Reddy P (2017) Pearl millet, Pennisetum glaucum (L.) R. Br. In: Patil JV (ed) Millets and sorghum: biology and genetic improvement. Wiley-Blackwell, pp 49–86. https://doi.org/10.1002/9781119130765.ch2

Satyavathi CT, Ambawat S, Khandelwal V et al (2021) Pearl millet: a climate-resilient nutricereal for mitigating hidden hunger and provide nutritional security. Front Plant Sci 12:659938. https://doi.org/10.3389/fpls.2021.659938

Sedivec KK, Schatz BG (1991) Pearl millet: forage production in North Dakota. North Dakota State University Extension Service, Fargo, ND

Selladurai M, Pulivarthi MK, Raj AS et al (2023) Considerations for gluten free foods-pearl and finger millet processing and market demand. Grain Oil Sci Technol 6(2):59–70. https://doi.org/10.1016/j.gaost.2022.11.003

Serna-Saldivar SO, Espinosa-Ramírez J (2019) Grain structure and grain chemical composition. In: Serna-Saldivar SO (ed) Sorghum and millets. AACC International Press, pp 85–129. https://doi.org/10.1016/B978-0-12-811527-5.00005-8

Shah N, Mewada RK, Mehta T (2016) Crosslinking of starch and its effect on viscosity behaviour. Rev Chem Eng 32(2):265–270. https://doi.org/10.1515/revce-2015-0047

Shaikh M, Ali TM, Hasnain A (2015) Post succinylation effects on morphological, functional and textural characteristics of acid-thinned pearl millet starches. J Cereal Sci 63:57–63. https://doi.org/10.1016/j.jcs.2015.02.006

Shaikh M, Ali TM, Hasnain A (2017a) Effects of different modification reagents on functional properties of pearl millet starches. Starch-Stärke 69(3–4):1600142. https://doi.org/10.1002/star.201600142

Shaikh M, Ali TM, Hasnain A (2017b) Utilization of chemically modified pearl millet starches in preparation of custards with improved cold storage stability. Int J Biol Macromol 104:360–366. https://doi.org/10.1016/j.ijbiomac.2017.05.183

Shaikh M, Haider S, Ali TM et al (2019) Physical, thermal, mechanical and barrier properties of pearl millet starch films as affected by levels of acetylation and hydroxypropylation. Int J Biol Macromol 124:209–219. https://doi.org/10.1016/j.ijbiomac.2018.11.135

Sharma M, Yadav DN, Singh AK et al (2015) Rheological and functional properties of heat moisture treated pearl millet starch. J Food Sci Technol 52:502–6510. https://doi.org/10.1007/s13197-015-1735-1

Sharma M, Singh AK, Yadav DN et al (2016) Impact of octenylsuccinylation on rheological, pasting, thermal and physicochemical properties of pearl millet (Pennisetumtyphoides) starch. LWT 73:52–59. https://doi.org/10.1016/j.lwt.2016.05.034

Sharma M, Yadav DN, Singh AK et al (2018) Impact of octenyl succinylated pearl millet (Pennisetum typhoides) starch addition as fat replacer on the rheological, textural and sensory characteristics of reduced-fat yoghurt. Int J Dairy Technol 71(3):723–733. https://doi.org/10.1111/1471-0307.12461

Sharma S, Sharma R, Govindaraj M et al (2021) Harnessing wild relatives of pearl millet for germplasm enhancement: challenges and opportunities. Crop Sci 61(1):177–200. https://doi.org/10.1002/csc2.20343

Shobana S, Sreerama YN, Malleshi NG (2009) Composition and enzyme inhibitory properties of finger millet (Eleusine coracana L.) seed coat phenolics: mode of inhibition of α-glucosidase and pancreatic amylase. Food Chem 115(4):1268–1273. https://doi.org/10.1016/j.foodchem.2009.01.042

Singh G, Sehgal S (2008) Nutritional evaluation of ladoo prepared from popped pearl millet. Nutr Food Sci 38(4):310–315. https://doi.org/10.1108/00346650810891360

Singh F, Rai KN, Reddy BV, et al (1997) Development of cultivars and seed production techniques in sorghum and pearl millet. https://oar.icrisat.org/2842/

Singh N, Singh J, Kaur L et al (2003) Morphological, thermal and rheological properties of starches from different botanical sources. Food Chem 81:219–231. https://doi.org/10.1016/S0308-8146(02)00416-8

Singh N, Chawla D, Singh J (2004) Influence of acetic anhydride on physicochemical, morphological and thermal properties of corn and potato starch. Food Chem 86:601–608. https://doi.org/10.1016/j.foodchem.2003.10.008

Singh BP, Jha A, Sharma N et al (2013) Optimization of a process and development of a shelf life prediction model for instant multigrain dalia mix. J Food Process Eng 36:811–823. https://doi.org/10.1111/jfpe.12050

Singh P, Gupta N, Hameed F et al (2023) Pearl millet as an underutilized source of nutrition: a review on its nutritional composition, value addition and industrial applications. Indian J Agric Biochem 36(1):26–30. https://doi.org/10.5958/0974-4479.2023.00003.5

Singla D, Singh A, Dhull SB et al (2020) Taro starch: isolation, morphology, modification and novel applications concern-a review. Int J Biol Macromol 163:1283–1290. https://doi.org/10.1016/j.ijbiomac.2020.07.093

Siroha AK, Sandhu KS (2018) Physicochemical, rheological, morphological, and in vitro digestibility properties of cross-linked starch from pearl millet cultivars. Int J Food Prop 21(1):1371–1385. https://doi.org/10.1080/10942912.2018.1489841

Siroha AK, Sandhu KS, Kaur M (2016) Physicochemical, functional and antioxidant properties of flour from pearl millet varieties grown in India. J Food Meas Charact 10:311–318. https://doi.org/10.1007/s11694-016-9308-1

Siroha AK, Sandhu KS, Kaur M et al (2019) Physicochemical, rheological, morphological and in vitro digestibility properties of pearl millet starch modified at varying levels of acetylation. Int J Biol Macromol 131:1077–1083. https://doi.org/10.1016/j.ijbiomac.2019.03.179

Siroha AK, Punia S, Kaur M et al (2020a) A novel starch from Pongamia pinnata seeds: comparison of its thermal, morphological and rheological behaviour with starches from other botanical sources. Int J Biol Macromol 143:984–990. https://doi.org/10.1016/j.ijbiomac.2019.10.033

Siroha AK, Punía S, Sandhu KS et al (2020b) Physicochemical, pasting, and rheological properties of pearl millet starches from different cultivars and their relations. Acta Aliment 49(1):49–59. https://doi.org/10.1556/066.2020.49.1.7

Sowbhagya CM, Bhattacharya KR (1971) A simplified colorimetric method for determination of amylose content in rice. Starch-Stärke 23(2):53–56. https://doi.org/10.1002/star.19710230206

Srivastava RK, Singh RB, Pujarula VL et al (2020) Genome-wide association studies and genomic selection in pearl millet: advances and prospects. Front Genet 10:1389. https://doi.org/10.3389/fgene.2019.01389

Sujatha M, Hymavathi TV, Devi KU et al (2018) Effect of heat treatment and gamma irradiation on in vitro starch digestibility of selected millet grains. Pharma Innov 7:409–413. https://doi.org/10.20546/ijcmas.2017.611.118

Suma PF, Urooj A (2015) Isolation and characterization of starch from pearl millet (Pennisetum typhoidium) flours. Int J Food Prop 18(12):2675–2687. https://doi.org/10.1080/10942912.2014.981640

Suriya M, Baranwal G, Bashir M et al (2016) Influence of blanching and drying methods on molecular structure and functional properties of elephant foot yam (Amorphophallus paeoniifolius) flour. LWT 68:235–243. https://doi.org/10.1016/j.lwt.2015.11.060

Taki M (1962) Fractionation of starch components by paper chromatography. Agric Biol Chem 26:1–9. https://doi.org/10.1080/00021369.1962.10857924

Taylor JRN (2004) Millet Pearl encyclopedia of grain sciences, vol 2, pp 253–261. https://doi.org/10.1016/B0-12-765490-9/00097-5

Taylor JRN (2016) Grain production and consumption: Africa. In: Encyclopedia of Food Grains, vol 4, pp 359–366

Ullah I, Ali M, Farooqi A (2010) Chemical and nutritional properties of some maize (Zea mays L.) varieties grown in NWFP, Pakistan. Pak J Nutr 9(11):1113–1117. https://doi.org/10.3923/pjn.2010.1113.1117

Umakanth AV, Reddy PS, Karthikeyan S et al (2024) Biofuel opportunities in pearl millet. In: Pearl millet in the 21st century: food-nutrition-climate resilience-improved livelihoods, pp 567–588. https://doi.org/10.1007/978-981-99-5890-0_22

Varnamkhasti MG, Mobli H, Jafari A et al (2008) Some physical properties of rough rice (Oryza Sativa L.) grain. J Cereal Sci 47(3):496–501. https://doi.org/10.1016/j.jcs.2007.05.014

Wang S, Copeland L (2015) Effect of acid hydrolysis on starch structure and functionality: a review. Crit Rev Food Sci Nutr 55:1081–1097. https://doi.org/10.1080/10408398.2012.684551

Wani IA, Sogi DS, Wani AA et al (2010) Physico-chemical properties of starches from Indian kidney bean (Phaseolus vulgaris) cultivars. Int J Food Sci Technol 45:2176–2185. https://doi.org/10.1111/j.1365-2621.2010.02379.x

Wani IA, Sogi DS, Gill BS (2012) Physicochemical properties of acetylated starches from some Indian kidney bean (Phaseolus vulgaris L.) cultivars. Int J Food Sci Technol 47(9):1993–1999. https://doi.org/10.1111/j.1365-2621.2012.03062.x

Wankhede DB, Shehnaz A, Raghavendra Rao MR (1979) Preparation and physicochemical properties of starches and their fractions from finger millet (Eleusinecoracana) and foxtail millet (Setariaitalica). Starch-Stärke 31:153–159. https://doi.org/10.1002/star.19790310505

Wankhede DB, Rathi SS, Gunjal BB et al (1990) Studies on isolation and characterization of starch from pearl millet (Pennisetum americanum (L.) Leeke) grains. Carbohydr Polym 13:17–28. https://doi.org/10.1016/0144-8617(90)90048-W

Whistler RL, BeMiller JN, Paschall EF (2012) Starch: chemistry and technology. Academic, London

Woo KS, Seib PA (2002) Cross-linked resistant starch: preparation and properties. Cereal Chem 79(6):819–825. https://doi.org/10.1094/CCHEM.2002.79.6.819

Wu Y, Lin Q, Cui T et al (2014) Structural and physical properties of starches isolated from six varieties of millet grown in China. Int J Food Prop 17(10):2344–2360. https://doi.org/10.1080/10942912.2013.803119

Wurzburg OB (1986) Nutritional aspects and safety of modified food starches. Nutrition Reviews, 44(2):74–79. https://doi.org/10.1111/j.1753-4887.1986.tb07590.x

Yadav OP, Rai KN (2013) Genetic improvement of pearl millet in India. Agric Res 2:275–292. https://doi.org/10.1007/s40003-013-0089-z

Zhong Y, Tai L, Blennow A et al (2023) High-amylose starch: structure, functionality and applications. Crit Rev Food Sci Nutr 63(27):8568–8590. https://doi.org/10.1080/10408398.2022.2056871

Chapter 8
Barnyard Millet: Chemical Composition, Starch Isolation, Modification, and Industrial Applications

Tarun Kumar, Mohit Kumar, Rohit Bhardwaj, and Satyendra Kumar Rajput

8.1 Introduction

Millets are an outstanding source of essential nutrients and are often referred to as "Nutri-Cereals". With over 38.4% of the world's total millet production, India is the world's leading producer of millet. Compared to wheat and rice, millets offer high nutritional value, being particularly rich in vitamins, proteins, and minerals (FAO 2022). Barnyard millet (BM) is a rich source of carbohydrates, protein and is particularly known for its higher micronutrient content, especially Fe and Zn, compared to other major cereals. However, despite its impressive nutritional value and agronomic advantages, it continues to be an underutilized crop (Renganathan et al. 2020). Among the cultivated species of BM, *Echinochloa esculenta* (Japanese barnyard millet) and *Echinochloa frumentacea* (Indian barnyard millet) are the most prominent, both cultivated thousands of years ago and valued for their adaptability to marginal soils and ability to withstand both biotic and abiotic stresses (Sood et al. 2015). The Indian BM is recognized as a primitive domesticated cereal, with archaeological evidence indicating its cultivation far back in 5000 B.C. and 3000 B.C. in India and China, respectively (Bhinda et al. 2023a, b). Ancient Indian literature also mentioned this millet, highlighting its historical significance. Some studies suggest its domestication began in India, while some researchers suggest

T. Kumar · R. Bhardwaj · S. K. Rajput (✉)
Department of Pharmaceutical Sciences, Faculty of Pharmaceutical Sciences, Gurukula Kangri (Deemed to be University), Haridwar, Haridwar, Uttarakhand, India
e-mail: t.drx1121@gmail.com; tarun.kumar@gkv.ac.in; rohit.bhardwaj@gkv.ac.in; rohitbhardwaj.aip@gmail.com; satyendra.rajput@gkv.ac.in; satyendraj@gmail.com

M. Kumar
School of Pharmaceutical Sciences and technology, Sardar Bhagwan Singh University, Dehradun, Uttarakhand, India

© The Author(s), under exclusive license to Springer Nature Switzerland AG 2026
J. Singh, M. Kumar Malik (eds.), *Millets and Millet Starch: Chemical Composition, Functional Modification, and Applications*,
https://doi.org/10.1007/978-3-032-09502-2_8

that BM originated in Central Asia before spreading to Europe and America. It is widely regarded as indigenous to India, where it has been cultivated for more than 4000 years, particularly in the central and eastern parts of the country (Nugroho et al. 2024). Figure 8.1 represents agriculture land with young BM crop.

The Japanese BM holds a significant place in Japan's Historical and cultural heritage, it is the only crop mentioned in the country's two oldest historical texts, the Nihon Shoki and Kojiki, compiled in 720 and 712 AD, respectively. These ancient records describe the millet as having emerged from the body of a deceased deity, highlighting its symbolic and mythical value. Archaeological findings and historical accounts further emphasize the crop's longstanding role in Japanese society. A prominent example of BM's use is from the twelfth century, when three members of the influential Fujiwara family were mummified and placed in coffins. In Hidehira's coffin, a member of the Fujiwara family who passed away in 1187, approximately 1.8 L of unhulled BM seeds were discovered scattered at the bottom, likely used as pillow stuffing. This finding offers clear evidence of the grain's utilization during that era. Additionally, historical records from the Nanbu feudal lords document the existence of 102 varieties (out of which 29 of early-maturing, 29 of medium-maturing, and 44 of late-maturing varieties) of Japanese BM, showcasing the crop's agricultural diversity and significance (Yabuno 1987).

Japanese BM is known as Xiao Su in China, Ya-plonglaman/Ya-khaonok in Thailand, Rajamaratu or Tavali in Srilanka, Smao bek kbol/bayokibok in Cambodia, Pi/Jo in Korea, Kibi/Hie in Japan, bayokibok (Tagalog)/lagtom (Bikol)/marapagay

Fig. 8.1 Cultivation of Barnyard millet in agriculture land (Credit: Mr. Balkrishna, Resident of Almora, Uttarakhand)

(Ilokano) in Philippines, Padi burong, Rumput sambau/kekusabasar/jawan in Malaysia, Padi burung/beras burung in Indonesia, Shogai/Baryu in Mangolia, Long vuc me or co gạo in Vietnam, Taiwan oil millet in Taiwan, Swank/sawa in Pakistan, shyama/Shyama chal in Bangladesh, Mannik in Russia, and Gewohnliche Huhnerhirse in Germany (Swamy 2023). BM is also commonly known by several other names in India as shown in Table 8.1.

BM grains are tiny, concave or oval-rounded in shape, and generally display a pale yellow or dull white color. It is a self-pollinated crop and is the hardest millet (Kaur and Sharma 2020). It is classified under the Kingdom: Plantae, Phylum: Tracheophyta, Class: Liliopsida, Order: Poales, Family: Poaceae, Sub-family: Panicoideae, Genus: *Echinochloa*, and Tribe: Paniceae. BM is an underutilized, climate-resilient cereal crop that holds significant potential in contributing to food and nutritional security, especially in regions prone to harsh environmental conditions. It is oftenly cultivated in regions where conventional crops like rice or wheat struggle due to poor soil fertility or erratic rainfall. Its resilience, rapid growth cycle, and minimal input requirements make it a crucial resource for smallholder farmers, especially in marginal lands (Joshi et al. 2021; Sood et al. 2015). Beyond its agronomic advantages, BM has gained attention for its rich nutritional profile, being an abundant source of dietary fiber, vital minerals such as Fe and Ca, and bioactive compounds like polyphenols and antioxidants (Singh et al. 2022; Bhinda et al. 2023a, b). Notably, its starch content, which represents a major fraction of the grain, holds significant promise for both food and non-food industries. Recent advancements in starch isolation and modification techniques, including physical, chemical, and enzymatic approaches, have expanded the potential uses of BM starch, enhancing properties like gelatinization, viscosity, and resistance to enzymatic breakdown (Gomez and Gupta 2003; Tripathi and Vyas 2023). Such modifications have

Table 8.1 Common names of barnyard millet in India

Languages	Names
English	Indian Millet, Japanese Millet, Billion Dollar Grass, Poor man's millet, Sawa Millet, Siberian millet, Marsh millet, White millet
Hindi	Jhangora, Samvat, Sanwa, and Samwa
Sanskrit	Shyamaka, Nivara, Hastishyamaka, Ambhushyamaka
Kannada	Oodalu
Tamil	Kuthiravalli, Koodai pul, Kuthiraivolly, Kuthiraivaal, Pullu Rail Pullu, Mondi Chamai
Malayalam	Kavadapullu, Varak
Telugu	Kodisama and Udalu
Punjabi	Swank
Marathi	Bhagar, Vari, and Shamul
Gujarati	Sama
Bengali	Shyama and Shamula
Oriya	Khira, Bila Suan
Chhattisgarhi	Bhagar or Varai, Jhangon

practical implications for producing functional foods, gluten-free products, packaging materials, and other value-added items. These techniques also expand their industrial relevance, supporting product development with health benefits, improved texture, and stability. The unique physicochemical properties of BM starch, such as small granule size, high digestibility, and structural flexibility, have positioned it as a promising raw material for biodegradable, pharmaceutical, and food industries (Krishijagran 2023). Its naturally gluten-free nature makes it an ideal choice for those with celiac disease or gluten intolerance, further increasing its relevance in modern dietary patterns. A detailed overview is provided in this chapter regarding the chemical composition of BM, methods of starch extraction, and strategies for starch modification. Furthermore, the potential applications of native and modified BM starch in various industrial sectors are discussed, highlighting the crop's role as an underutilized, sustainable resource in the context of food security and emerging raw material for industrial needs.

8.2 Geographical Distribution

"Barnyard millet" is the common name that refers to several species within the *Echinochloa* genus, many of which originate from Southern or Eastern Asia. This millet is particularly valued for its rapid growth and ability to adapt to diverse environmental conditions. The genus *Echinochloa* comprises approximately 40 to 50 species, including both annual and perennial grasses, distributed predominantly across temperate and tropical regions worldwide, spanning latitudes from 50°North to 40°South (Hoste and Verloove 2022; Nilavarasi et al. 2020; Bajwa et al. 2015; Clayton and Renvoize 2006). It is recognized as one of the oldest millet crops domesticated in the semi-arid regions of Asia (especially in India, China, Korea, and Japan) and Africa (Sood et al. 2015). BM ranks as the fourth most commonly cultivated minor millet and plays a vital role in promoting food security among economically disadvantaged populations worldwide (Renganathan et al. 2020) (Fig. 8.2).

Echinochloa crus-galli (Barnyard grass) is a widespread wild species and is considered one of the ancestors or progenitors of some cultivated BM. It is native to temperate and tropical regions of the world, including Asia, North America, and Europe. It is characterized by the presence of awns, native to temperate regions of Eurasia, and has been cultivated in Japan for approximately 4000 years (Wallace et al. 2015). It is believed that this millet species first emerged in Japan, derived from its wild ancestor *E. crus-galli*, around 4000 years ago. From there, it spread as a cultivated crop to China, Korea, and neighboring parts of Russia. Originally native to the temperate regions of Europe and Asia, wild *E. crus-galli* is now widely distributed worldwide, particularly prevalent in both tropical and temperate regions such as Southeast Asia (Partohardjono and Jansen 2023).

Echinochloa utilis (Japanese Barnyard Millet) is thought to have been domesticated from *E. crus-galli* in East Asia, particularly Japan, Korea, and parts of China.

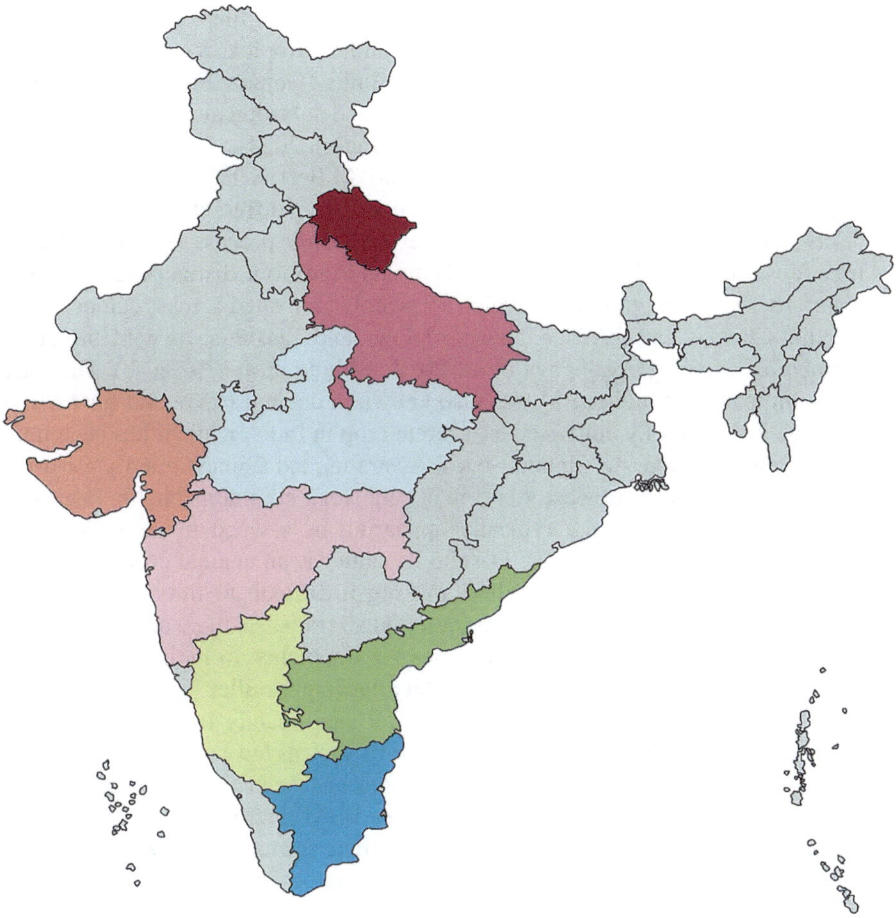

Fig. 8.2 Geographical distribution of barnyard millet in India

This species is cultivated mainly in Japan, Korea, and northeastern China, where it is grown as a cereal crop. It has been selectively bred for larger seeds, improved yield, and better adaptability (Yabuno 1987).

Echinochloa esculenta (Japanese Barnyard Millet) is an annual species primarily cultivated in the temperate regions of Japan, China, Russia, Korea, and Germany. This crop originated from the wild grass *E. crus-galli* (L.) from which it was domesticated. It is adapted for cultivation in temperate climates. It has characteristics typical of domesticated crops, such as larger seeds and reduced shattering. Archaeological findings indicate its cultivation in Japan during the Yayoi period, around 4000 to 5000 years ago. Other research suggests domestication occurred during the Jomon period, around 2000 BC (Nozawa et al. 2004). It is often cultivated as an alternative to rice in regions where paddy cultivation is not feasible. Additionally, this crop has attracted interest for its use as animal fodder in countries like the United States and Japan (Joshi et al. 2021).

Echinochloa colona, commonly referred to as Sawa millet or Indian barnyard millet, is widespread throughout the tropical and subtropical regions of the world and is believed to have been domesticated in India (Gomez and Gupta 2003). Its wild species is now widespread across tropical and subtropical areas globally, particularly in Southeast Asia (Partohardjono and Jansen 2023).

Echinochloa frumentacea (Indian Barnyard Millet) is believed to have been domesticated from *Echinochloa colona* (Jungle Rice), a different wild ancestor than *E. crus-galli.* It has undergone a parallel evolutionary process in both India and Africa (Joshi et al. 2021). This species is divided into four distinct races: *robusta, stolonifera, laxa,* and *intermedia*, which are widely cultivated across regions including India, China, Central Africa, Nepal, Malawi, and Tanzania. Its wild ancestor is *Echinochloa colona* (Upadhyaya et al. 2014; Bhinda et al. 2023a, b). Historical records indicate that Indian BM was also known in ancient Egypt and East Africa. Currently, it is primarily cultivated as a cereal crop in India, while it has been introduced to countries like the United States, Australia, and Canada mainly for use as forage. In mainland Southeast Asia, it is relatively common, but in Peninsular Malaysia, it is rarely found, typically appearing as a weed in cultivated fields (Partohardjono and Jansen 2023). Despite its status as an ancient crop, the precise genetic origin of BM remains unclear, making it difficult to fully understand its evolutionary development based on morphological traits. As there are very few visible differences or phenotypic mutations seen in the plants. In India, BM is considered the second most important minor millet after finger millet (Singh et al. 2022).

Both *E. esculenta* and *E. frumentacea* differ significantly from their wild relatives, *E. crus-galli* and *E. colona*, exhibiting traits such as fewer vegetative branches, a more compact plant structure, larger inflorescences, increased grain size, and reduced seed shattering. The ability of cultivated BM to cross with its wild ancestors, along with the presence of natural intermediate forms, offers valuable insights into the domestication process and the genetic factors involved in shaping these traits (Sood et al. 2015).

8.3 Morphology, Appearance, and Genetics of Barnyard Millet

BM can grow at altitudes of up to 2000 meters above sea level during the summer. It shows a lot of variation in traits such as flowering time, inflorescence shape, plant structure, and spikelet color (Upadhyay et al. 2014; Nozawa et al. 2006). This crop is a tall and strong annual plant that can reach a height of about 220 cm. It has a short life cycle and is the fastest growing among the small millets, usually taking 45 to 60 days to grow from seed to seed, though it may take longer in hilly regions. It can grow in different seasons, but at higher altitudes, it may need 3 to 4 months to fully mature (Nugroho et al. 2024). The characteristics of various BM species is discussed in Fig. 8.3.

It also exhibits distinct traits like panicle structure, glume texture, and features of lower lemma, distinguishing the two primary cultivated species, i.e., Indian BM, characterized by tiny, awnless spikelets having membranous glumes, and Japanese BM, which has awned spikelets with papery upper glumes and lower lemma (Sood et al. 2015; Yabuno 1987). Japanese BM is commonly cultivated as a late-season

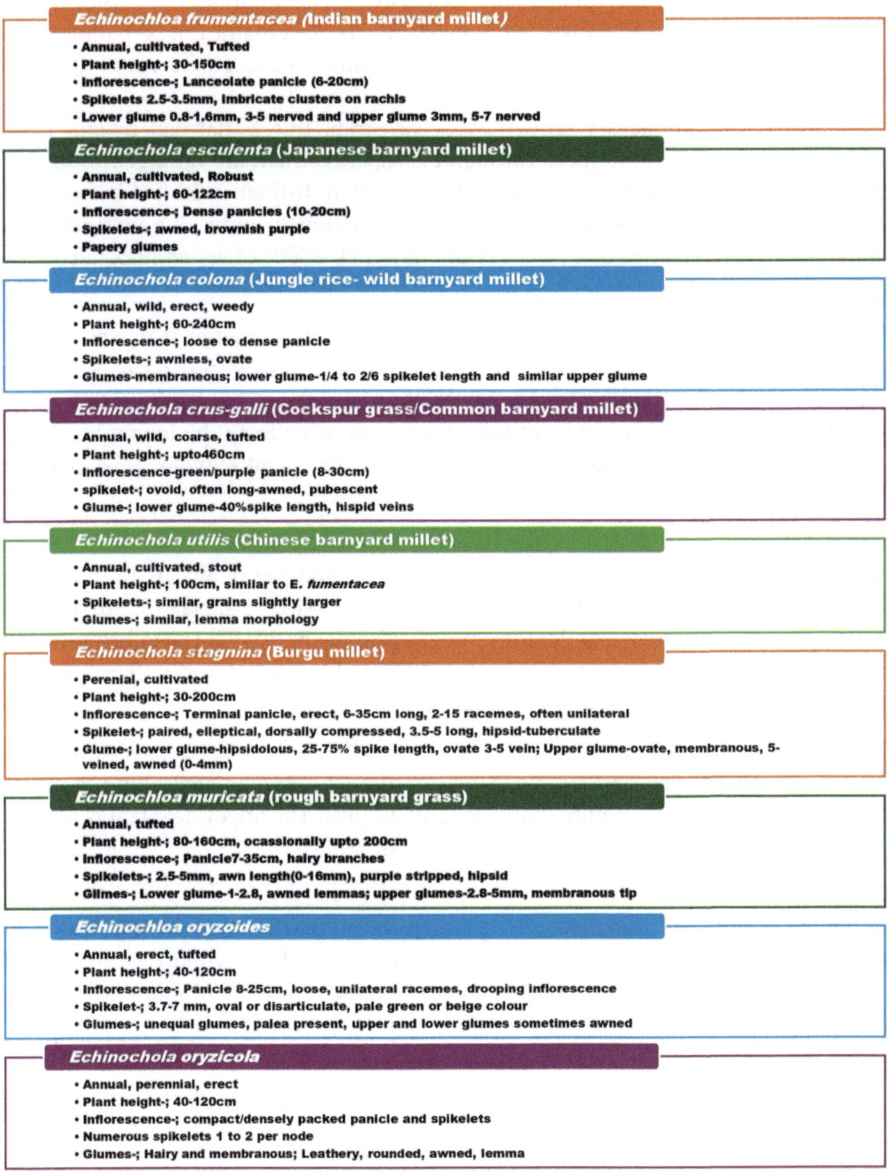

Fig. 8.3 Characteristics of various barnyard millet species

forage crop in temperate climates with humid or sub-humid conditions. Under favorable weather conditions, it grows rapidly and can sometimes produce mature grains after 45 days of seeding. The annual grass is an erect plant reaching a height of 2 to 4 ft. Its panicle inflorescence consists of 5 to 15 erect, sessile branches, each bearing brownish to purplish spikelets aligned on one side of each branch. Seeds are slightly longer and larger than those of barnyard grass. Japanese BM makes its best growth on good soils (soil type: medium to heavy). The duration of planting is from April to July. Although it shows good resistance to most major fungal pathogens, it remains vulnerable to various species of head smut (Vázquez García et al. 2021; Gomashe 2017).

E. crus-galli produces clumps having prostrate tillers that can extend up to 150 cm in height. It reproduces through caryopses, which are arranged in upright panicles. The plant features dull green leaves, often distinctly marked by anthocyanin pigmentation. Its leaf sheaths are glabrous, flattened, and lack both ligules and auricles (Vázquez García et al. 2021; Damalas et al. 2008; Maun and Barrett 1986).

Roots

BM develops two types of roots: seminal roots, which emerge from the seed to establish the seedling, and adventitious roots, which originate from basal nodes near the soil surface. Seminal roots stay active but make up only a small portion of the mature root system. While adventitious roots that grow from the lower part of the stem (near the soil surface) and later form the plant's main root network (Gomashe 2017).

Clums/Stem

The stem of BM is typically cylindrical, hollow, and segmented with nodes and internodes. It can be smooth or slightly hairy, and side branches or tillers often develop from buds located at the lower nodes (Gomashe 2017). The culms typically range from 20 to 60 cm in length or even more. Their color varies from green to purple. The nodes are dark and smooth (Bohara et al. 2024).

Leaves

Leaves are broad, flat, and simple, with an alternate arrangement along the stem. They may be smooth or slightly hairy, lacking ligules. The uppermost leaf is referred to as the flag leaf (Gomashe 2017; Bohara et al. 2024).

Panicle

The plant develops terminal panicles ranging from 10 to 25 cm in length, which are typically upright and display a variety of shapes, including cylindrical, pyramidal, globular, or elliptical forms. These panicles bear several to many compact racemes containing spikelets measuring about 4 to 6 mm, positioned in four uneven rows along the three-angled/triangular rachis (Joel et al. 2006; Sood et al. 2015; Napper 1965). The spikelets exhibit different colors such as green, brown, or purple, and are commonly clustered along one side of the rachis, though some forms have spikelets distributed all around it. Depending on the species, the spikelets may be either awned or awnless. The pedicellate spikelets are enclosed by unequal glumes, with the upper glume extending to the length of the spikelet and displaying multiple veins (Gupta et al. 2010; Sundararaj and Thulasidas 1976; Mitich 1990).

Racemes
The plant produces racemes that can be few or numerous, with spikelets densely packed in four irregular rows along the triangular-shaped rachis. These spikelets, typically found on one side of the rachis, display colors such as green, brown, or purple. Each spikelet contains two florets and may either lack awns or possess red or green awns. The spikelets are positioned on short, rough stalks, each supported by a pair of glumes (Nugroho et al. 2024).

Leaf Blades
The leaf blades are wide, glabrous, and flat, and they lack ligules. It typically measures between 5 and10 cm long, while the width is between 3 and 10 mm. It may occasionally display purple blotches or horizontal bands. The leaf sheaths are flattened, open, and keeled (Bohara et al. 2024).

Flower
Flowering in BM begins at the apex of the inflorescence and grows downward, usually completing within 10 to 15 days. The flowers typically open between 5 a.m. to 10 a.m. with peak blooming activity observed between 6 a.m. to 7 a.m. Within each raceme, flowering starts at the periphery and gradually moves toward the center. The flowers are hermaphroditic. The stigma branches spread and the flowers open just before the anthers release pollen (Fig. 8.4). Late-season florets often remain closed (cleistogamous). The crop predominantly undergoes self-pollination and exhibits self-compatibility, though some level of cross-pollination occurs, facilitated by wind. Under hill regions, treating the inflorescence with hot water for 4 to 5 minutes at 48 °C has proven effective in inducing male sterility in both cultivated species (Sood et al. 2015).

The lower floret is sterile and consists of a lemma and a small palea. The lemma is sterile and displays five distinct veins, while the fertile floret is bisexual, featuring a plano-convex, elliptical, shiny, smooth, pointed, or sharp tip. The margins of the fertile lemma roll inward beneath the palea, leaving the tip of the palea exposed. The palea is flat and has a surface texture similar to the fertile lemma. There are three stamens present, while the superior ovary bears two separate styles with feathery stigma (Swamy 2023).

Seed/Caryopsis
The seeds are small, smooth, and rounded at the base with a pointed tip, measuring about 2–3 mm in length and width up to 1 to 2 mm. They are encased in hard, shiny, white lemma and palea layers (Gomashe 2017). Figure 8.5 represents the caryopsis of BM seed.

Genetics
The genus *Echinochloa* consists of approximately 50 species (Michael 2003). These species include only a limited number of cultivated types and are primarily grown as minor millets by marginal farmers. The main cultivated species include *E.*

Fig. 8.4 Close view of barnyard millet panicle. (Credit:Balkrishna, Resident of Pauri Garhwal, Uttarakhand)

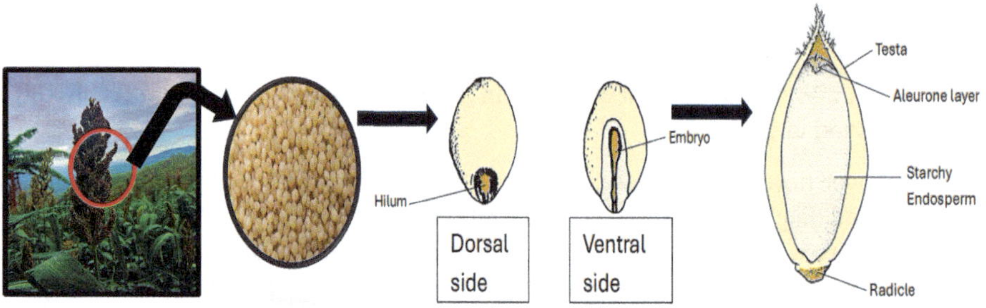

Fig. 8.5 Sectional view of barnyard millet seed

frumentacea, *E. esculenta*, and *E. utilis* (Upadhyaya et al. 2014; De Wet et al. 1983; Yabuno 1987). Species such as *E. crus-galli*, *E. colona*, and *E. oryzicola* are recognized as invasive weeds in key agricultural systems (Kraehmer et al. 2016; Yamaguchi et al. 2005; Yuichiro et al. 1999; Wanous 1990; Yabuno 1987). Among them, *E. crus-galli* stands out as a dominant rice field weed in over 60 countries, primarily due to its ability to germinate under low-oxygen conditions (up to 100 mm depth), its fast growth rate, rice-like appearance, adaptability to various environments, and high seed output (Barrett 1983; Renganathan et al. 2020).

However, it is difficult to clearly identify *Echinochloa* species on the basis of morphological traits, owing to minimal interspecific and intraspecific variation and their high phenotypic plasticity (Chauhan and Johnson 2011; Renganathan et al. 2020). Nevertheless, till now around 35 species within this genus have been classified based on morphological traits, cytological analysis, and molecular marker studies to understand their taxonomy and phylogenetic relationships (Yamaguchi et al. 2005; Yabuno 1987; Yuichiro et al. 1999).

E. crus-galli (L.) Beauv., with chromosome numbers 2n = 42, 48, and 54, is the most widespread (Yabuno 1983). Another common species, *E. colona* (L.) ranks alongside *E. crus-galli* as one of the world's most problematic weeds, classified as the third and fourth worst weed globally, respectively (Ye et al. 2014; Holm et al. 1991; Rao 2021). The genus *Echinochola* has a basic chromosome number of x = 9 and exhibits a wide range of polyploidy. Yabuno (1962) documented interspecific polyploidy across various cytotypes, including 2n = 36 (tetraploid = 4x), 54 (hexaploidy = 6x), 63 (heptaploid = 7x), 108 (dodecaploid = 12), and 126 (tetradecaploid = 14) (Gomashe 2017) (Table 8.2).

8.4 Cultivation, and Processing

Millets are considered among the earliest domesticated cereal grains and have served as staple foods for humans since ancient times. Millets serve as a staple food for about 90 million people in Asia and Africa, while over 500 million people in more than 30 nations rely on sorghum as a dietary staple (ICRISAT). Despite their nutritional value and historical significance, the cultivation of these traditional crops has declined over the past five decades due to the increasing focus on major crops like rice, wheat, maize, and soybeans (FAO 2018). In 2019, millet was grown on 71.8 million hectares globally, yielding 86.3 million tons. Africa led with 48.9 million hectares and 42.3 million tons, followed by other continents (FAO 2022; Ministry of Agriculture & Farmers Welfare 2022; Bhatt et al. 2023a, b).

BM is considered among the most climate-resilient, summer crop, fast-growing, and nutrient-rich grains (Dhanalakshmi et al. 2019). India is the leading producer of BM, covering an area of approximately 146,000 hectares and producing about 147,000 tones, with an average yield of 1034 kg per hectare over the past three years (Renganathan et al. 2020; IIMR 2018; Madhusudhana et al. 2018). Barnyard millet cultivation in a natural field in India is shown in Fig. 8.6. Its exceptional drought tolerance, rapid growth, and ability to thrive in marginal soils makes it a valuable food and fodder source, especially in regions vulnerable to famine and erratic weather conditions (Sharma et al. 2021a, b). It is also flood-resistant and utilizes nitrogen more efficiently than traditional cereal crops (Goron and Raizada 2015). Appreciated for its high nutritional value in terms of genetic resources and enhancement, BM remains one of the least explored crops (Manimekalai et al. 2018).

Improving grain yield remains a key focus in BM breeding programs. However, a significant gap persists between the potential and actual yields obtained by

Table 8.2 Ploidy levels and Karyotype of various barnyard millet species

Echinochloa species	Ploidy levels	Karyotype	References
Echinochloa frumentacea	Hexaploid	2n = 6x = 54	Renganathan et al. (2020)
Echinochola esculenta	Hexaploid	2n = 6x = 54	Gomashe (2017)
Echinochola Colona	Hexaploid	2n = 6x = 54	Rao (2021) and Renganathan et al. (2020)
Echinochola crus-galli	Hexaploid	$2n = 6x = 54$	Ye et al. (2014), Rao (2021) and Renganathan et al. (2020)
Echinochola utilis	Hexaploid	$2n = 6x = 54$	Gomashe (2017)
Echinochola oryzicola	Tetraploid	2n = 4x = 36	Kraehmer et al. (2016), Renganathan et al. (2020), Yamaguchi et al. (2005) and Gomashe (2017)
Echinochola obtusiflora	Diploid	$2n = 2x = 18$	Gomashe (2017)

farmers, largely due to the continued use of traditional cultivation methods. In many regions where BM is grown, local landraces and traditional varieties dominate. However, the crop has the capacity to yield over 2.0 tons/hectare, and certain *E. esculenta* lines have recorded yields as high as 3.0 tons/hectare (Sood et al. 2015). Using improved BM varieties along with modern farming practices can greatly increase crop yield and help reduce the gap between actual and potential production. BM straw is also preferred more than rice straw. So, choosing a variety that gives both good grain yield and plenty of straw for fodder is the best option (Swamy 2023; Sood et al. 2015).

In Japan, agricultural regions were divided into two regions, i.e., region A and region B. Region A, located at a higher elevation, was historically known for upland farming due to its vulnerability to cold weather, making rice cultivation difficult. As a result, residents traditionally consumed Japanese BM either as a substitute for rice or mixed with it in equal proportions. This practice continued until the 1940s. In contrast, region B had more favorable conditions for paddy farming, allowing for double-cropping systems such as rice followed by wheat or barley. Over time, advancements such as improved irrigation, development of rice varieties tolerant to cooler climates, and better farming techniques led to a significant increase in rice cultivation from 40,000 hectares during 1881–1917 to 90,000 hectares by the late 1960s. Consequently, BM cultivation declined steadily. In 1879, millet was grown on 22,700 hectares, but by 1984, this had dropped dramatically to only 285 hectares, mostly in the northern part of Region A. Another factor contributing to this decline was the reduced use of horses, which previously required BM as fodder. With the introduction of power cultivators and vehicles in the 1950s, the need for horse feed diminished, leading to a further reduction in millet cultivation (Yabuno 1987).

Over time, the area under BM and other minor millets has declined. This decrease is due to several factors, including changes in farming practices, economic priorities, and social perceptions.

Fig. 8.6 Barnyard millet cultivation in a natural field during 2025 at Almora district of Uttarakhand (Credit: Balkrishna, Resident of Pauri Garhwal, Uttarakhand)

The Green Revolution of the 1970s brought significant changes to Indian agriculture by encouraging the widespread production of major cereals like rice, wheat, and maize. High-yielding varieties such as IR8 rice helped prevent food shortages, especially during famines, but this shift led to small millets being pushed into less favorable regions and labeled as “poor man’s crops”. Government policies also focused more on mainstream cereals, neglecting traditional millets. For example, in Kenya, maize replaced finger millet (Dida et al. 2008), and in Japan, cold-tolerant rice varieties were preferred over BM (Yabuno 1987). This consistent neglect has not only reduced cultivation but also led to the loss of genetic diversity and traditional knowledge, making BM production less efficient today (Rengalakshmi 2005; Swamy 2023).

Most BM varieties produce non-glutinous (non-sticky) seeds, which lead to a less desirable texture in food products. In many countries, people prefer grains with a soft, sticky texture like waxy rice. To make BM more appealing for consumption, improving its texture through selective breeding is necessary (Sood et al. 2020).

Genetic research and available resources for BM are still very limited. The unavailability of good-yielding varieties, poor-quality seeds, traditional farming methods, and limited use of modern technology make it difficult for this crop to gain wider acceptance globally. Additionally, there is little market integration, which further reduces its popularity and profitability. Several social and economic challenges also contribute to the reduced interest in millet farming. These crops require substantial manual labor, and in many areas, especially in India, the post-harvest processing is still done manually by local women (Padulosi et al. 2015; Rengalakshmi 2005).

Moreover, some yield loss is caused by sheath blight disease, banding leaf disease, and grain smut (Kumar 2013; Kumar and Prasad 2009; Pall et al. 1980). Although many of these diseases result in only minor losses, a significant biotic challenge for BM in countries like India, China, and Japan is grain smut caused by *Ustilago panici-frumentacei*. However, studies have found that *E. esculenta* is generally more resistant to this disease than *E. frumentacea* (Sood et al. 2020).

Another major issue with millet farming is its low yield compared to other staple crops (Plaza-Wüthrich and Tadele 2012). Investments from public and private sectors, along with agricultural research initiatives, have mostly focused on major cereals like rice, wheat, and maize, leaving millets underfunded and overlooked (Pingali 2015). Farmers also earn less from millets than from high-value cash crops, making millet farming less attractive financially. The impact of various processing techniques on the characteristics of BM is shown in Fig. 8.7.

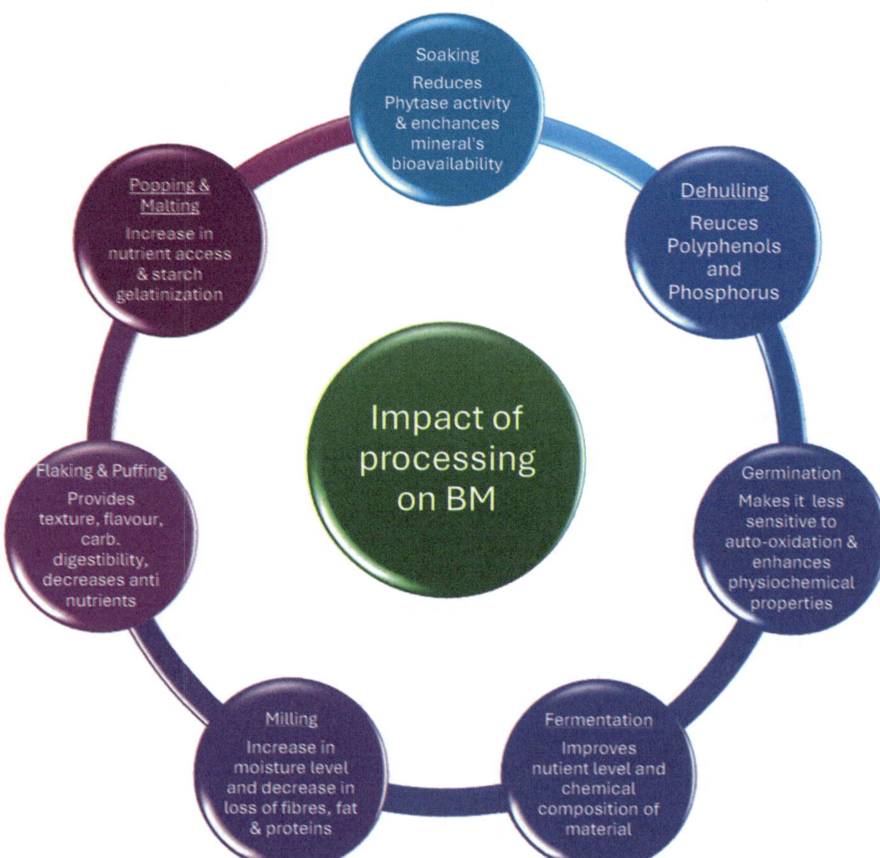

Fig. 8.7 Processing and its impact on barnyard millet characteristics

8.5 Isolation and Modification of Barnyard Millet Starch

Various approaches were explored by various researchers to isolate starch from BM with the highest yield and purity. Bhatt et al. (2023a) utilize an alkali isolation approach for the extraction of starch from BM and reported that on using 3-4 times consecutive treatment with alkali followed by treatment with water yields 14.68 ± 0.52% g/100g starch. In another study, Choudhary et al. (2025) also used the alkali isolation approach for the isolation of starch. However, in this approach researcher explored alkali treatment followed by successive sieving and centrifugation. It was concluded in the study that the starch yield was 62.12% (Choudhary et al. 2025). In another study, it was reported that sodium hydrogen sulphite was used to isolate starch from BM (Sandhu and Singh 2007). BM starch was also isolated by acid steeping approach (Bangar et al. 2021). Verma et al. (2018a) used alkaline treatment for the isolation of BM starch and the yield was about 58%. The modification of starch can be performed by numerous techniques, including physical and chemical approaches or by the combination of both. The chemical modification of starch potentially alters the functional and physicochemical attributes via the replacement of the hydroxyl group and incorporation of other functional groups. Physical modification includes dry heat treatment, ultrasonication, microwave treatment, gamma irradiation, and cold plasma treatment. However, physical tailoring usually alters only the packing arrangement of the starch chain (Fan and Picchioni 2020; Rostamabadi et al. 2023). Despite its surge in consumption interest, BM starch in native and modified form is still underutilized and unexplored in comparison to other millets.

8.5.1 *Physical Modification of Barnyard Millet Starch*

Ultrasound-assisted modification, a key physical approach, is utilized to tailor the functionality of starch by using high-intensity ultrasound (>1 W/cm^2). Dimri and Singh (2024) investigated the impact of ultrasonication and gamma irradiation on the characteristics of Indian BM starch. BM starch subjected to ultrasonication for 30 and 60 min at 20 KHz showed a decrease in total protein content and peak viscosity, along with an increase in amylose content (21.42% from 20.56%), swelling power (14.21% from 13.57%), and solubility (11.23% from 10.45%). Ultrasonication resulted in a change in water absorption capacity (1.51 to 1.22 g/g), foaming capacity (18.0% to 22.0%), total phenolic (167.82 from 148.88 mg GAE/100 g), and antioxidant activity (21.92 to 25.31% RSA) of BM flour (Bhandari et al. 2025). Dry heat treatment is another way to enhance the content of resistant starch in BM. In an investigation, a significant rise in the resistant starch in BM starch after microwave modification (dry heat treatment) at 110°C and 120°C for 10 and 20 min was observed. It has been claimed that the resistant starch content was increased to

15.08/100 g from 0.61/100 g when treated with microwave processing for 20 min at 170 °C (Kanagaraj et al. 2019). Wani et al. (2021a, b) investigated the impact of gamma irradiation on the functionality of BM. It has been reported that gamma irradiation significantly lowered the swelling behavior, synersis, and pasting profile and enhanced the water absorption, solubility, and oil absorption characteristics. However, augmentation in characteristics after gamma irradiation is dose dependent.

8.5.2 Chemical Modification of Barnyard Millet Starch

Crosslinking of starch is considered as one of the promising techniques to tailor the starch functionality by introducing new chemical bonds (Malik et al. 2022). BM starch was subjected to crosslinking using sodium trimetaphosphate (1%, 3% and 5%) and assessed for change in functional characteristics. It has been reported that upon crosslinking with sodium trimetaphosphate, amylose content decreased from 15% to 7.86%. Furthermore, light transmittance, peak viscosity, breakdown viscosity, swelling, and solubility also decreased, but gelatinization temperature increased after crosslinking (Sharma et al. 2021a, b). Bhatt et al. (2023a, b) investigated the impact of crosslinking on functional and tableting characteristics of BM starch. It has been concluded in the study that crosslinking improves the tableting characteristics of alkali-isolated BM starch, such as the Hausner ratio, moisture content, and flow properties. Furthermore, decrement in the swelling power, amylose content (9.65 from 19.96), and digestibility in simulated fluids was observed. Esterification or acetylation using acetic anhydride or other esterification reagents is another technique to insert the acetyl group in starch. In another study, the same author investigated the effect of varying concentrations of crosslinking agent (sodium trimetaphosphate) on physiological properties and stated that the starch crosslinked with sodium trimetaphosphate is safe for consumption when crosslinked with the specified amount of sodium trimetaphosphate (Bhatt et al. 2023a). Enzymatic hydrolysis is another method of modification applied to BM starch. During enzymatic hydrolysis, starch and other complex carbohydrates are hydrolyzed and changed into simpler sugars. However, the amount of enzyme and its activity play a key role in the hydrolysis of starch. Kumar et al. (2016) investigated the effect of enzymatic hydrolysis (bacterial alpha amylase) on BM starch pasting, viscoelastic, and rheological characteristics. The study concluded that complete hydrolysis of BM starch requires an enzyme-to-substrate ratio of 50:1 (w/v) and an incubation period of 24 h. Furthermore, a significant change was observed in the typical pasting characteristics of native BM starch (Kumar et al. 2016).

8.6 Physicochemical Properties

8.6.1 Color

The native BM seed flour showed an "L" value of 71.91, which slightly decreased at a 2.5-kGy radiation dose, though the change was statistically significant. However, when exposed to 5 kGy, the reduction in the "L" value became statistically significant. This is due to Maillard reactions forming dark-colored melanoidins. The "a" value for the untreated flour remained unchanged after irradiation, indicating no significant variation. The "b" value for the control sample showed a slight, nonsignificant increase after irradiation. These changes, including a decline in the "L" value and a rise in the "b" value, suggested that the irradiated samples developed a darker, more yellowish hue (Bhat et al. 2016; Wani et al. 2021a, b). In another finding, the color profile of BM starches revealed significant differences in (L*), (a*), and (b*) values among various BM starches. The two varieties, BMV2 and BMV1, exhibited the highest lightness values, indicating better whiteness. Slightly lower brightness in other variants was linked to the presence of pigments like carotenoids and polyphenols. All starch samples showed low chroma and high hue angles, and the L* values were consistently above 90, reflecting good whiteness across cultivars (Dimri and Singh 2022).

8.6.2 Amylose Content

The amylose content of BM starch, isolated using the alkaline steeping method, was found to be 19.97 ± 3.57% w/w. After crosslinking, this value significantly declined to 9.65 ± 2.14% w/w. This reduction is attributed to the intermolecular interactions between amylose and amylopectin in the crosslinked BM starch, which lead to structural modifications within the starch matrix and apparently result in a lower amylose content (Bhatt et al. 2025). In another study, the amylose content in BM starches shows significant variation, ranging from 21.27 ± 0.0896% in the BMV1 variety to 23.3 ± 0.27% in BMV2. A comparable value of 20.02% amylose content was previously reported by Krishna Kumari and Thayumanavan (1998), whereas Verma et al. (2018) documented a considerably higher level of amylose content approximately 33%. This variation in amylose levels could be attributed to several influencing factors, such as the plant's genetic background, environmental conditions, harvesting time, and soil characteristics. These variations can directly impact the starch's functional attributes, including its gelatinization behavior, pasting properties, and retrogradation tendencies (Krishna Kumari and Thayumanavan 1998; Verma et al. 2018b; Dimri and Singh 2022).

8.6.3 Swelling Power and Solubility

It was reported that BM starch exhibited a swelling capacity of 194.4 ± 0.0065% w/w. However, after crosslinking, the starch showed a notably lower swelling ability, recorded as 180.2 ± 0.2860% w/w. This decline indicates that the crosslinking process alters internal starch chain interactions, leading to reduced water uptake and expansion (Bhatt et al. 2025).

An increase in temperature from 50°C to 90°C, along with gamma irradiation doses ranging from 0 to 5 kGy, led to a significant rise in the solubility index of BM flours. A sharp elevation in solubility was particularly noticeable at higher temperatures (80–90°C) (Dar et al. 2017). The enhanced solubility in irradiated samples may be attributed to radiation-induced starch fragmentation, which weakens the hydrogen bonds between starch molecules and increases molecular polarity, thereby improving water solubility (Liu et al. 2012; Bhatt et al. 2025).

8.6.4 Syneresis

A notable rise in syneresis was observed during the storage period from 0 to 120 hours. Initially, syneresis in both the native and irradiated flour gels ranged from 56.09 to 57.0 g per 100 g, which increased from 63.69 to 65.24 g per 100 g after 120 hours. Compared to the raw BM starch, gamma irradiation caused a significant reduction in syneresis (Wani et al. 2021a, b).

8.6.5 Pasting Property

The transparency of modified BM starch paste was assessed by measuring light transmission at 650 nm, which yielded a value of 3.41 ± 0.3209% indicating low clarity at 37 °C. After 24 hours of storage at 40 °C, the transmittance (%T) of the alkali isolated BM starch paste slightly declined to 2.4%. In comparison, the cross-linked starch paste initially exhibited a paste clarity of 2.5%, which further dropped to 2.2% following storage at 4 °C. This reduced clarity in crosslinked starches can be attributed to their enhanced structural integrity. Unlike native starch granules, which completely disintegrate during gelatinization, crosslinked starches tend to retain their granular structure, thereby reducing light transmission and overall paste transparency (Bhatt et al. 2025)

In another study, it has been noted that following gelatinization, crosslinked starches retain their structure, whereas native starch granules completely dissolve, leading to a decrease in clarity (Bhatt et al. 2025; Bhatt et al. 2023b).

8.6.6 Oil and Water Binding Capacity

The water binding capacity (WBC) is particularly important, as excessive water retention can negatively affect the product's shelf life. In this context, native alkali-isolated BM starch demonstrated a WBC of 0.868 ± 0.0430% g/g, whereas the crosslinked variant exhibited a significantly higher WBC of 1.655 ± 0.1520% g/g. This increase in WBC can be attributed to the presence of negatively charged phosphate groups introduced during crosslinking. These groups cause electrostatic repulsion, which reduces intermolecular forces and enhances the starch's capacity to absorb water (Bhatt et al. 2025). Oil absorption capacity (OAC) is an important functional property that contributes to flavor retention and extends shelf life in food products (Bhat et al. 2008). It reflects the efficiency of proteins to bind with lipids within food systems. The native flour sample exhibited an OAC of 4.90 g/g. Upon gamma irradiation, a significant enhancement in OAC was observed, increasing from 5.02 to 5.15 g/g with higher treatment levels. This increase may be attributed to irradiation-induced denaturation, which exposes apolar peptide residues, enhancing their interaction with fats. Additionally, structural alterations in the native starch matrix likely facilitated greater oil entrapment (Abu et al. 2006).

8.6.7 Emulsion Capacity

The native flour showed an emulsion capacity (EC) of 1.31%, whereas irradiated samples showed values ranging from 4.61% to 4.66%. The increase in EC for flours treated with 2.5 and 5 kGy doses did not show a significant difference between the untreated and treated samples. Gamma irradiation likely disrupts the native protein structure through deamination, enhancing protein solubility and promoting their adsorption at oil–water interfaces (Malik and Saini 2017). Regarding emulsion stability, the untreated flour exhibited 100% stability, whereas treated flours showed reduced values ranging from 29.16% to 79.62%. This decline in stability may result from the breakdown, reassociation, and crosslinking of proteins triggered by irradiation (Lee and Song 2002). Moreover, the availability of soluble proteins typically enhances emulsion stability. The emulsifying properties of protein-enriched flour are affected by both protein and non-protein constituents (Njintang et al. 2001).

8.6.8 Moisture Content

The moisture level in BM flour was recorded as 11.93% (Verma et al. 2015). Moisture content plays a critical role in determining the stability of excipients. In a study, the moisture content of alkali-isolated BM starch was found to be 15.33 ± 1.01% w/w. However, after crosslinking, the moisture content of the modified BM starch decreased by approximately 20%. This reduction is attributed to the chemical interaction between the hydroxyl groups of the glucose units in the native starch and the bi- or poly-functional reagents used in the crosslinking process, which likely reduced the starch's ability to retain water (Bhatt et al. 2023b).

8.7 Phytochemical Properties

A significant portion of India's population lives in poverty, making it difficult for them to access even the most basic nutritious diets. Due to its wide-ranging benefits, BM has the potential to improve food security and nutrition, particularly in hilly areas where nutrient deficiencies are more prevalent. However, further research is needed on the metabolic impacts, bioavailability, and functional food development using BM, particularly with improved processing, texture, and sensory quality. From a nutritional standpoint, BM is highly valuable due to its richness in proteins, lipids, B-complex vit B1, vit B2, and nicotinic acid, as well as its abundance of amino acids and antioxidant polyphenolic compounds as compared to common cereal grains (Watanabe 1999; Maithani et al. 2023). It contains phytochemicals like flavonoids, phenolic acids, and tannins that act as natural antioxidants (Panwar et al. 2016). It is particularly abundant in bioactive constituents, including both free and bound phenolic acids (e.g., protocatechuic and ferulic acid, respectively), chlorogenic acid, lignans, β-glucan, phytates, sterols, tocopherols, carotenoids, inulin, and resistant starch (Sarita and Singh 2016). Additionally, BM includes various polyphenols, such as coumarins, stilbenes, and tannins, though phenolic acids and tannins are the dominant types, while flavonoids are present in smaller quantities (Singh et al. 2022; Chandrasekara and Shahidi 2010).

BM contains a total phenolic content of approximately 45 ± 3.8 mg FAE/g with phenolic acids as a major compound (Anis and Sreerama 2020). The flavonoid profile mainly comprises flavonols, flavanones, flavones, and anthocyanins, with a flavonoid concentration of 8.7 ± 0.17 mg CE/g. These flavonoids are known to contribute significantly to human health by offering antioxidant, anti-allergic, anti-carcinogenic, anti-inflammatory, and gastro-protective effects (Ofosu et al. 2020).

In one analysis, three antioxidant phenolic compounds were isolated from an ethanolic extract of Japanese BM, including a serotonin derivative and two flavonoids (Kumar and Prasad 2009; Watanabe 1999).

BM contains 7.1 ± 0.13 mg CE/g condensed tannins and consist largely of flavan-3-ol structures, exhibiting properties like antioxidant, anti-inflammatory, antibacterial, and antiviral activity (Anis and Sreerama 2020; Ofosu et al. 2020). The tannin content significantly decreases after processing, with raw BM flour having higher levels than processed forms (Sarita and Singh 2016).

Another important bioactive, phytic acid (inositol hexaphosphate), is mainly found in the bran or aleurone layer of whole grains and acts as a powerful antioxidant (Bhatt et al. 2023a, b). Compared to other cereals, BM has lower levels of phytic acid and reduced trypsin inhibition activity. It shows greater effectiveness in inhibiting enzymes like α-galactosidase, acid phosphatase, and α-amylase, with phytic acid levels ranging between 3.30 and 3.70 mg/100 g (Panwar et al. 2016).

Phytates combat oxidative stress by binding iron involved in Fenton's reaction (Yabuno 2001). Generally, phytates, alkyl resorcinols, tocols, carotenoids, and lignans are richer in antioxidants compared to their processed counterparts (Anis and Sreerama 2020). However, processing methods like germination, soaking, fermentation, heating, and hydrothermal treatment can significantly reduce or even eliminate phytate levels (Gupta et al. 2015a, b).

It also contains important bioactive compounds like gamma-aminobutyric acid (GABA) and β-glucan, known for their roles in antioxidation and blood lipid regulation (Sharma et al. 2016; Kofuji et al. 2012; Sampath et al. 1989).

Minor millets show lower insulin, blood glucose, adiponectin, and tumor necrosis factor-α (TNF-α) (Nishizawa et al. 2009). Japanese BM has nearly double the protein content as compared to rice (Yabuno 1987). Studies suggest that whole grains, including BM, help lower the risk of serious diseases such as cancer. In fact, the phytochemical content in whole grains may surpass that found in fruits and vegetables combined (Liu 2007).

BM contains all key macronutrients like carbohydrates, proteins, fats, fiber and is rich in antioxidants, Zn, Fe, and Ca (Manimekalai et al. 2018) (Table 8.3). The Fe and fiber content of BM exceeds that of many major cereals and other millet varieties (Devi et al. 2014; Saleh et al. 2013). BM contains Ca (4.5 g/kg), P (2.2 g/kg), K (27.4 g/kg), Na (2.2 g/kg), Mg (203 mg/kg), Zn (39 mg/kg), and Cu (8 mg/kg) The dry matter content is about 26%, while ash makes up 14.5% of the total composition (Borkar et al. 2016). BM grains contain minerals ranging from 1.5% to 4%, and their proteins are easily digestible (Ugare et al. 2014; Veena et al. 2005). Due to its

Table 8.3 Chemical constituents in barnyard millet

Chemical constituents	Amount	References
Starch (g/100 g)	51.5–62	Dimri and Singh (2022), Saleh et al. (2013), and Veena et al. (2005)
Protein (%)	6–15%	Renganathan et al. (2020), Ugare et al. (2014), and Veena et al. (2005)
Fat (%)	2.02–4.5	Wani et al. (2021a, b), Ugare et al. (2014), and Veena et al. (2005)
Fibre (g)	8.5–14.4	Rajeswari and Priyadharshini (2021) and Verma et al. (2015
Ca (mg 100 g^{-1})	17–32	Veena et al. (2005)
Na (mg 100 g^{-1})	13–20	Bhatt et al. (2023a, b)
K (mg 100 g^{-1})	298	Renganathan et al. (2020)
Fe (mg 100 g^{-1})	1.5.6–1.8.6	Verma et al. (2015), Yadav et al. (2007), and Veena et al. (2005)
P (mg 100 g^{-1})	298	Nazni and Devi (2016)
Zn (mg 100 g^{-1})	4.9	Bhatt et al. (2023a, b)
Cu (mg 100 g^{-1})	0.6-	Renganathan et al. (2020)

high content of micronutrients and proteins, it can help combat malnutrition, particularly in children and women in developing regions. Therefore, it is often suggested as a staple food for pregnant women and young children (Underwood 2000).

8.8 Health Benefits

BM is low in calories and serves as a rich source of easily digestible protein. Its consumption makes one feel light and energized. BM is not just a grain reserved for fasting days; it is nutritious enough to be included in daily meals for people of all age groups. For those aiming to follow a nutritious diet, BM is a great option (Ugare et al. 2014; Maurya et al. 2023).

BM is a highly nutritious grain that has been cultivated for centuries. Known for its resilience and short growing season, this underutilized crop is gaining recognition for its exceptional health benefits. Packed with essential nutrients like iron, calcium, fiber, and antioxidants, BM plays a valuable role in managing lifestyle-related disorders, such as diabetes, obesity, and cardiovascular diseases, owing to its anti-inflammatory properties (Selvan et al. 2022). Its low glucose value, high dietary fiber, and antioxidant-rich profile make it an ideal choice for modern diets focused on wellness and disease prevention. Regular consumption of BM contributes to better overall health by lowering the risk of cardiovascular diseases, type II diabetes, constipation, and certain types of cancer (Bhatt et al. 2023a, b).

Studies have investigated the plant's potential in cancer prevention (Bohara et al. 2024). Peerzada et al. (2016) observed cytotoxic effects on cancer cells of *E. colona* L. extracts. These findings indicate a possible application in cancer treatment (Peerzada et al. 2016).

E. colona is traditionally used in folk medicine across several Asian regions. Different parts of the plant or the whole plant is applied in treating a variety of ailments, such as wounds, diabetes, ulcers, and for antiseptic purposes. A study investigated the wound-healing properties of different extracts of *E. colona* (Borkar et al. 2016).

8.8.1 Cardiovascular Diseases

Risk factors such as obesity, smoking, unhealthy eating habits, and lack of physical activity significantly raise the risk of developing heart disease and stroke. A growing number of countries worldwide are witnessing high and continuously increasing rates of cardiovascular-related illnesses. In studies comparing rice and other minor millets, it has been observed that rodents fed on diets containing both native and processed starch derived from BM exhibited the lowest concentrations of blood glucose, serum cholesterol, and triglycerides (Oswal 2023; Anitha et al. 2021; Maurya et al. 2023).

8.8.2 Diabetes Mellitus

BM is enriched in dietary fiber and contains a relatively moderate amount of carbohydrates, with slow digestibility, which can help in managing blood glucose levels (Veena et al. 2005). BM is rich in starch but contains very low sugar, making it a suitable option for people suffering from diabetes. Replacing white rice with BM in the diet can help manage high blood sugar levels. It is especially recommended for those with diabetes and heart-related conditions, as it supports the reduction of blood glucose and cholesterol levels (Kaur and Sharma 2020). Compared to other millets and rice, BM has shown greater effectiveness in reducing blood sugar and cholesterol (Bhatt et al. 2023a, b). Studies have shown that the prevalence of diabetes is lower in communities where BM is a staple food. Phenolic compounds present in millet inhibit enzymes such as α-glucosidase and pancreatic amylase, resulting in a decrease in post-meal blood sugar levels. This occurs through the partial inhibition of the enzymatic breakdown of complex carbohydrates. Additionally, enzyme inhibitors like aldose reductase help prevent the buildup of sorbitol, thereby reducing the risk of diabetes and associated complications. According to Ugare et al. (2014), the glycemic index of dehulled BM was recorded as 50.04, while that of heat-treated millet was 41.72. These findings suggest that both dehulled and thermally processed BM are effective in managing type II diabetes (Kaur and Sharma 2020; Maurya 2019; Ugare et al. 2014; Anitha et al. 2021).

8.8.3 Gastrointestinal Disorder

Fermented BM-based foods function as natural probiotics, supporting healthy digestion and offering protection against diarrhea in young children. These products help improve gut health due to their probiotic and prebiotic properties (Sarita and Singh 2016). It also maintains stomach pH by neutralizing acidity.

E. colona L. has traditionally been used in folk medicine to treat digestive disorders. Uremis et al. (2000) examined the possible gastroprotective activity of this plant by suggesting its potential contribution to support gastrointestinal health. This effect may be due to its anti-inflammatory and antioxidant characteristics, which could aid in preserving the integrity of the gastrointestimucosa (Swain et al. 2017; Uremis et al. 2000).

8.8.4 Celiac Disease

BM is a naturally gluten-free grain, making it an excellent substitute for wheat and other common gluten-containing cereals, which can pose health risks for patients with celiac disease or gluten sensitivity (Kaur and Sharma 2020; Saleh et al. 2013).

According to Thompson (2009), gluten-free grains such as BM, corn, oats, buckwheat, quinoa, sorghum, little, kodo, and pearl millet have increasingly been used in place of wheat, rye, and barley, offering safer dietary choices for those following gluten-free diets. As a result, incorporating BM into meals can help reduce the adverse effects associated with celiac disease (Dey et al. 2022; Bunkar et al. 2021).

8.8.5 Hypolipidemic

Its consumption has shown positive effects in individuals by increasing levels of High-Density Lipoprotein (HDL) and decreasing levels of Low-Density Lipoprotein (LDL). Additionally, supplementation with BM has been associated with a modest reduction in triglyceride levels and a rise in HDL cholesterol (Ugare et al. 2014; Bhatt et al. 2023a, b).

8.8.6 Fights Against Anemia and detoxification

The high iron content in BM makes it an excellent choice to fight anemia. Including controlled amounts of this grain in the diet provides essential nutrients important for supporting a healthy metabolic system. For individuals who frequently suffer from tiredness and low energy levels as a result of anemia, incorporating BM into their meals may help reduce these symptoms. Studies have revealed that specific types of BM contain notably high amounts of Fe (18.6 mg/ 100 g). This makes BM the leading millet in terms of iron content. Research indicates that incorporating BM into the diet can enhance blood hemoglobin and serum ferritin levels, making it a valuable iron source, particularly for individuals following a vegetarian diet (Maurya et al. 2023). The liver is essential for detoxification processes, and hepatoprotective substances aid in protecting it from harm (Pallerla et al. 2019). Research conducted by the Southern Weed Science Society explored the liver-protective potential of E. colona, suggesting possible advantages for liver function. These benefits may be linked to its antioxidantAntioxidants capabilities and its role in reducing oxidative stress within liver tissues (Southern Weed Science Society 1998). Antioxidants (Curcumin, Quercetin, Ellagic acid) neutralize free radicals, prevent cell disruption, and exhibit strong radical cation scavenging activity (Dykes and Rooney 2006; Choi et al. 2005; Eduru et al. 2021).

8.8.7 Diuretic Properties

In traditional medicinal practices, *E. colona* L. has been commonly used as a diuretic, aiding in the removal of excess fluids from the body. Diuretic properties can be helpful in treating health conditions like edema and high blood pressure. However, to fully understand the diuretic potential and safety of *E. colona* L., more comprehensive and scientifically controlled studies are required (Quattrocchi 2006; Bohara et al. 2024).

8.8.8 Phytoremediation

It is a developing, environmentally friendly, and cost-effective technique for removing metals from contaminated soil. In 2014, a study was published in which *E. colona* was used for the phytoextraction of soils polluted with heavy metals such as cadmium, lead, and chromium. Additionally, the bio-concentration factor (BCF) was determined. The results of the experiment indicated that *E. colona* was an effective accumulator of Cd and Cr (Subhashini and Swamy 2014). Another study was conducted in 2015 to validate the effectiveness of *E. colona* in the phytoremediation of Pb, Cd, and Cr. Based on the total metal accumulation and the calculated BCF values, *E. colona* absorbed the metals in the following decreasing order: Cr > Cd > Pb. Therefore, *E. colona* can be considered a highly efficient plant for phytoextraction of heavy metals (Subhashini and Swamy 2015). Nutritionally, it is a highly versatile grain, offering a rich supply of macronutrient content and dietary fiber (Ugare et al. 2014). Its grains are packed with essential amino acids, minerals, and vitamins (Vinoth and Ravindhran 2017).

8.9 Applications of Barnyard Millet

BM is mainly grown as a staple food source for human consumption. BM has recently gained attention due to its potential in various industrial sectors. Owing to its fast growth, climate resilience, and high starch and nutrient content, this millet is finding applications beyond food, particularly in sectors like pharmaceuticals, packaging, textiles, and cosmetics, owing to its functional characteristics such as swelling, pasting behavior, and gelatinization.

8.9.1 Food Industry

BM has found significant use in the food industry due to its nutritional composition. Its low glycemic value and rich dietary fiber content make it a preferred ingredient in functional foods aimed at managing diabetes and obesity (Renganathan et al.

2020). It is used in the formulation of health-oriented products such as whole grain flours, breakfast cereals, baby food, ready-to-eat snacks, and bakery items (Manimekalai et al. 2018). BM holds great potential as a fundamental ingredient in various regional food preparations. It can be infused into a wide range of products, including bread (both fermented and unfermented), porridge, millet-based beverages such as wine, and nutritional powders (Nisha and Saravanakumar 2019; Chandrasekara and Shahidi 2012). It can be used in proportions that lie between 30% and 50% for the development of bakery items such as bread, cookies, cakes, soup sticks, and Khari. Additionally, it can be a suitable replacement for rice and wheat in several traditional and value-added food preparations like breakfast dishes, sweets, and snacks.

Popular recipes include idli, dosa, upma, adai, vermicelli, pittu, idiyappam, rotti, chapathi, porridge, and khakra (Salunke et al. 2019). One strategy to promote its consumption is blending BM flour with other cereal flours then followed by proper processing (Singh et al. 2005). Researchers have created composite flour by replacing a portion of traditional cereal flour with millet for making both traditional and instant foods (Jaybhaye et al. 2014). Nazni and Karuna (2016) prepared rusks and muffins using blends of BM bran and wheat flour. For snack production, thin, rectangular cold extrudates were created and then puffed using high-temperature short-time technology to make BM (*E. frumentacea*)-based ready-to-eat products (Nazni and Karuna 2016).

Furthermore, BM is utilized in producing value-added items like sweets, vermicelli, biscuits, multi-grain flour, and ready mixes (Sood et al. 2015).

8.9.2 Fodder Industry and Livestock Feed

The straw of this crop also serves as high-quality cattle fodder due to its high digestibility and nutritional value and is often used to prepare hay and silage, particularly useful in hilly regions where fodder is scarce during winter. BM straw is often preferred over rice straw by livestock farmers for its palatability and fiber content (Sood et al. 2015). It is also fed to birds (Maithani et al. 2023).

8.9.3 Pharmaceuticals

Due to its rich polyphenolic content and antioxidant properties, BM holds potential as a functional ingredient in nutraceuticals and dietary supplements. Its starch can also be utilized as a binder and disintegrant in tablet formulations, offering a natural alternative to synthetic excipients (Bhatt et al. 2023a; Sood et al. 2020). Fermented BM-based products also have natural pre- and probiotic properties (Sarita and Singh 2016).

8.9.4 Starch-Based Applications

Starch is among the most abundant and essential polysaccharides found in food and plays a vital role in the digestive process (Bangar et al. 2024). One of the most promising components of BM is its starch, which constitutes a significant portion of the grain. The starch can be isolated and further modified through physical, chemical, or enzymatic methods to enhance its functional properties.

Modified BM starch demonstrates favorable quality and functional traits that make it suitable for various industrial applications. Its high pasting temperature (typically above 80 °C) and low breakdown viscosity indicate strong thermal and shear stability, which is desirable for food processing. Additionally, its high swelling capacity and greater amylose leaching enhance its performance as a thickener, stabilizer, texturizer, gelling agent, and emulsifier in food processing. Additionally, BM starch in antive and modified form have applications in biodegradable film production, owing to their film-forming ability (Dhanalakshmi et al. 2019; Padulosi et al. 2015). Heat-treated resistant starch of BM shows a favorable impact on serum lipid and blood glucose in individuals (Bangar et al. 2024). BM starch based tablets were explored for the delivery of a drug (Bhatt et al. 2023a, b).

8.9.5 Biofilms and Coating

BM starch (both in its native and modified forms) also presents potential for use in producing biodegradable and environmentally friendly films and coatings. Due to its high amylose content, BM starch serves as a suitable base for bio-composite films that exhibit low oxygen permeability. This characteristic makes it ideal for creating films with effective barrier properties.

Cao et al. (2017) investigated that incorporating borage seed oil into BM starch-based films can significantly increase their hydrophobic nature. This addition improved the film's resistance to water vapor and reduced its solubility. Furthermore, the films demonstrated antioxidant properties, making them promising candidates for use as antioxidant packaging or coatings in food applications. Enhancing the functional attributes of BM starch through modification could further expand its utility in the development of biodegradable films and coatings (Cao et al. 2017; Bangar et al. 2024).

8.9.6 Textile Industry

BM starch has also been explored as a natural sizing agent in textile manufacturing. Its biodegradable nature and ease of processing make it a suitable alternative to synthetic chemicals used in fabric finishing (IIMR 2018).

8.9.7 Biofuels

The high carbohydrate content of BM makes it a promising feedstock for bioethanol production. Recent studies have explored the feasibility of utilizing BM starch in the development of biodegradable plastics, contributing to sustainable industrial practices (Padulosi et al. 2015).

8.9.8 Cosmetics and Personal Care

BM provides significant benefits for both skin and hair health. Being high in phenolic compounds and flavonoids, it serves as a strong source of antioxidants that help support healthy and vibrant skin. Moreover, the presence of iron and zinc plays a vital role in promoting strong and healthy hair growth. The grain's finely milled flour and starch are being incorporated into skincare products such as exfoliating scrubs, facial masks, and lotions. Their thickening and emollient properties support their inclusion in natural cosmetics (Renganathan et al. 2020). In conditions like atopic dermatitis, BM is sometimes used as a substitute for wheat and rice (Watanabe 1999; Maithani et al. 2023).

8.10 Conclusion

BM stands out as a nutritionally rich and functionally versatile grain with promising applications across food, pharmaceutical, and industrial sectors. Its chemical composition, particularly the presence of resistant starch, dietary fiber, and essential micronutrients, makes it a valuable alternative to conventional cereals for health-conscious people. The process of starch isolation from BM has revealed distinctive physicochemical and functional properties, influenced by factors like variety, environmental conditions, and post-harvest handling. Furthermore, chemical and physical modifications such as crosslinking and irradiation enhance the starch's performance characteristics—improving parameters like water and oil absorption, swelling capacity, paste clarity, and emulsion stability. These enhancements broaden its applicability in diverse formulations, including thickeners, binders, controlled-release systems, and biodegradable films. Overall, BM starch, with its modifiability and favorable properties, holds substantial potential for sustainable and innovative industrial uses.

References

Abu JO, Muller K, Duodu KG, Minnaar A (2006) Gamma irradiation of cowpea (Vigna unguiculata L. Walp) flours and pastes: effects on functional, thermal and molecular properties of isolated proteins. Food Chem 95:138–147

Anis MA, Sreerama YN (2020) Inhibition of protein glycoxidation and advanced glycation end-product formation by barnyard millet (Echinochloa frumentacea) phenolics. Food Chem 315:126265

Anitha S, Botha R, Kane-Potaka J et al (2021) Can millet consumption help manage hyperlipidemia and obesity?: a systematic review and meta-analysis. Front Nutr 8:700778. https://doi.org/10.3389/fnut.2021.700778

Bajwa A, Jabran K, Shahid M et al (2015) Eco-biology and management of Echinochloa crus-galli. Crop Protect 75:151e162. https://doi.org/10.1016/j.cropro.2015.06.001

Bangar SP, Siroha AK, Nehra M et al (2021) Structural and film-forming properties of millet starches: A comparative study. Coatings 11(8):954. https://doi.org/10.3390/coatings11080954

Bangar SP, Balakrishnan G, Navaf M, Sunooj KV (2024) Recent advancements on barnyard millet starch: a sustainable alternative to conventional starch. Starch-Stärke 76:9–10. https://doi.org/10.1002/star.202300232

Barrett SH (1983) Crop mimicry in weeds. Econ Bot 37:255–282. https://doi.org/10.1007/bf02858881

Bhandari M, Sharma R, Bobade H et al (2025) Ultrasonication-induced modification of finger millet and barnyard millet flours: phytoconstituents and antioxidant capacity, functional properties, and microstructural arrangement. J Food Process Preserv 2025(1):6640001. https://doi.org/10.1155/jfpp/6640001

Bhat R, Sridhar RK, Young CC et al (2008) Composition and functional properties of raw and electron beam–irradiated Mucuna pruriens seeds. Int J Food Sci Technol 43:1338–1351. https://doi.org/10.1111/j.1365-2621.2007.01617.x

Bhat NA, Wani IA, Hamdani AM et al (2016) Physicochemical properties of whole wheat flour as affected by gamma irradiation. LWT Food Sci Technol (Lebensmittel-Wissenschaft -Technol) 71:175–183. https://doi.org/10.1016/j.lwt.2016.03.024

Bhatt P, Kumar V, Rastogi H et al (2023a) Functional and tableting properties of alkali-isolated and phosphorylated barnyard millet (Echinochloa esculenta) starch. ACS Omega 8(33):30294–30305. https://doi.org/10.1021/acsomega.3c03158

Bhatt D, Rasane P, Singh J et al (2023b) Nutritional advantages of barnyard millet and opportunities for its processing as value-added foods. J Food Sci Technol 60(11):2748–2760. https://doi.org/10.1007/s13197-022-05602-1

Bhatt P, Singh S, Garg S et al (2025) Impact of cross-linking on the physicochemical and physiological characteristics of barnyard millet (Echinochloa frumentacea) grains starch. Starch-Starke 77(1):2300285. https://doi.org/10.1002/star.202300285

Bhinda MS, Hasan N, Joshi DC (2023a) Barnyard millet improvement: from pre-genomics to post-genomics era. In: Smart plant breeding for field crops in post-genomics era. Springer, Singapore. https://doi.org/10.1007/978-981-19-8218-7_8

Bhinda MS, Joshi DC, Parihar M et al (2023b) Genetics, breeding, and genomics of Indian barnyard millet (Echinochloa frumentacea): status and perspectives. In: Neglected and underutilized crops. https://doi.org/10.1016/B978-0-323-90537-4.00017-X

Bohara B, Oli S, Kafle B et al (2024) A review on biology, status, and medicinal importance of Echinochloa colona L. Matrix Sci Pharma 8(1):10–17. https://doi.org/10.4103/mtsp.mtsp_3_24

Borkar VS et al (2016) Ethnomedical properties of Echinochloa colona and Hydrolea zeylanica: a review. World J Pharm Res 5(3):354–360

Bunkar DS, Goyal SK, Meena KK, Kamalvanshi V (2021) Nutritional, functional role of kodo millet and its processing: a review. Int J Curr Microbiol App Sci 10(01):1972–1985

Cao TL, Yang SY, Song KB (2017) Characterization of barnyard millet starch films containing borage seed oil. Coatings 7(11):183

Chandrasekara A, Shahidi F (2010) Content of insoluble bound phenolics in millets and their contribution to antioxidant capacity. J Agric Food Chem 58(11):6706–6714

Chandrasekara A, Shahidi F (2012) Bioaccessibility and antioxidant potential of millet grain phenolics as affected by simulated in vitro digestion and microbial fermentation. J Funct Foods 4(1):226–237

Chauhan B, Johnson DE (2011) Ecological studies on Echinochloa crus-galli and the implications for weed management in direct-seeded rice. Crop Prot 30:1385–1391. https://doi.org/10.1016/j.cropro.2011.07.013

Choi YY, Osada K, Ito Y et al (2005) Effect of dietary protein of Korean foxtail millet on plasma adiponectin, HDL-cholesterol, and insulin levels in genetically type 2 diabetic mice. Biosci Biotechnol Biochem 69(1):31–37

Choudhary A, Kumar M, Chaudhary V (2025) Isolation and characterization of starch from underutilized barnyard millet (*Echinochloa frumentacea*) and its utilization in the development of biopolymer film. J Food Meas Charact 1-5. https://doi.org/10.1007/s11694-025-03455-8

Clayton WD, Renvoize SA (2006) Genera Graminum: grasses of the world in Kew bulletin Royal Botanical Gardens Kew, U.K

Damalas CA, Dhima KV, Eleftherohorinos IG (2008) Morphological and physiological variation among species of the genus Echinochloa in Northern Greece. Weed Sci 56(3):416–423

Dar MZ, Deepika K, Jan K et al (2017) Modification of structure and physicochemical properties of buckwheat and oat starch by irradiation. Int J Biol Macromol. https://doi.org/10.1016/j.ijbiomac.2017.11.067

De Wet JMJ, KEP Rao, MH Mengesha, DE Brink (1983) Domestication of Sawa millet (Echinochloa colona). Econ Bot 37:283–291. https://doi.org/10.1007/BF02858883

Devi PB, Vijayabharathi R, Sathyabama S et al (2014) Health benefits of finger millet (Eleusine coracana L.) polyphenols and dietary fiber: a review. J Food Sci Technol 51(6):1021–1040. https://doi.org/

Dey S, Saxena A, Kumar Y et al (2022, 2022) Understanding of the antinutritional factors and bioactive compounds kodo millet (Paspalum scrobiculatum) and little millet (Panicum sumatrense). J Food Qual (1):1578448. https://doi.org/10.1155/2022/1578448

Dhanalakshmi R, Subramanian A, Thirumurugan T et al (2019) Genetic variability and association studies in barnyard mil let Echinochloa frumentacea Roxb. Link germplasm under sodic soil condition. Electron J Plant Breed 10(2):430–439

Dida MM, Wanyera N, MLH D et al (2008) Population structure and diversity in finger millet (Eleusine coracana) germplasm. Trop Plant Biol 1:131–141. https://doi.org/10.1007/s12042-008-9012-3

Dimri S, Singh S (2022) Structural, thermal, and rheological properties of starches isolated from four different varieties of Indian barnyard millet (Echinochloa frumentacea). J Food Process Preserv 46(7):e16718. https://doi.org/10.1111/jfpp.16718

Dimri S, Singh S (2024) Transformations in properties of Indian barnyard millet starch as induced by ultra-sonication and gamma irradiation. Starch-Stärke 76(9–10):2300218. https://doi.org/10.1002/star.202300218

Dykes L, Rooney LW (2006) Sorghum and millet phenols and antioxidants. J Cereal Sci 4(3):236–251. https://doi.org/10.1016/j.jcs.2006.06.007

Eduru A, Kamboj A, Reddy PM, Pal B (2021) Nutritional and health benefits of millets, present status and future prospects: A review. J Pharm Innov 10(5):859–868

Fan Y, Picchioni F (2020) Modification of starch: a review on the application of "green" solvents and controlled functionalization. Carbohydr Polym 241:116350. https://doi.org/10.1016/j.carbpol.2020.116350

FAO (2018) The future of food and agriculture—alternative pathways to 2050. Food and Agriculture Organization of the United Nations, Rome. https://www.fao.org/3/CA1553EN/ca1553en.pdf

FAO (2022) International year of millets 2023. Food and Agriculture Organization of the United Nations. https://www.fao.org/millets2023

Gomashe SS (2017) Barnyard millet: present status and future thrust areas. Millets and sorghum: biology and genetic improvement https://doi.org/10.1002/9781119130765.ch7, Wiley

Gomez MI, Gupta SC (2003) Millets in: encyclopedia of food sciences and nutrition 2

Goron T, Raizada M (2015) Genetic diversity and genomic resources available for the small millet crops to accelerate a new green revolution. Front Plant Sci 1:157. https://doi.org/10.3389/fpls.2015.00157

Gupta A, Mahajan V, Gupta HS (2010) Genetic resources and varietal improvement of small millets for Indian Himalaya. In: Tewari LM, Pangtey YPS, Tewari G (eds) Biodiversity potentials of the Himalaya. Gyanodaya Prakashan, Nainital, India

Gupta RK, Gangoliya SS, Singh NK (2015a) Reduction of phytic acid and enhancement of bioavailable micronutrients in food grains. J Food Sci Technol 52(2):676–684. https://doi.org/10.1007/s13197-013-0978-y

Gupta A, Sood S, Agrawal PK, Bhatt JC (2015b) B 29: an easy dehulling barnyard millet (Echinochloa frumentacea link) genotype. Natl Acad Sci Lett 38(1):21–24. https://doi.org/10.1007/s40009-014-0295-7

Holm LG, Plucknett DL, Pancho JV, Herberger JP (1991) The world's worst weeds: distribution and biology. The University Press of Hawaii, Malabar, Florida

Hoste I, Verloove F (2022) taxonomy of the weed species of the genus Echinochloa (Poaceae, Paniceae) in southwestern Europe: exploring the confused current state of affairs. PhytoKeys 197:1. https://doi.org/10.3897/phytokeys.197.79499c

IIMR (2018) Annual report 2017–18. Indian Institute of Millets Research, Hyderabad

Jaybhaye RV, Pardeshi IL, Vengaiah PC, Srivastav PP (2014) Processing and technology for millet based food products: a review. J ready eat food 1(2):32–48

Joel JA, Raveedran TS, Vanniarajan C, Gopalan A (2006) Fundamentals of crop botany. Directorate of open and distance learning, Tamil Nadu Agricultural University (TNAU), Coimbatore

Joshi RP, Jain AK, Malhotra N, Kumari M (2021) Origin, domestication, and spread. In: Millets and pseudo cereals. Woodhead Publishing

Kanagaraj SP, Ponnambalam D, Antony U (2019) Effect of dry heat treatment on the development of resistant starch in rice (Oryza sativa) and barnyard millet (Echinochloa furmantacea). J Food Process Preserv 43(7):e13965

Kaur JH, Sharma S (2020) An overview of barnyard millet (Echinochloa frumentacea). J Pharmacogn Phytochem 9(4):819–822

Kofuji K, Aoki A, Tsubaki K et al (2012) Antioxidant activity of β-glucan. ISRN Pharm:1–5. https://doi.org/10.5402/2012/125864

Kraehmer H, Jabran K, Mennan H, Chauhan BS (2016) Global distribution of rice weeds–a review. Crop Prot 80:73–86. https://doi.org/10.1016/j.cropro.2015.10.027

Krishijagran (2023) Barnyard Millets. https://krishijagran.com/cultivation-of-barnyard-millets

Krishna Kumari S, Thayumanavan B (1998) Characterization of starches of proso, foxtail, barnyard, kodo, and little millets. Plant Foods Hum Nutr 53(1):47–56

Kumar B (2013) Management of grain smut disease of barnyard millet (Echinochloa frumentacea). Indian Phytopathol 66:403–405

Kumar B, Prasad D (2009) First record of banded sheath blight disease of barnyard millet caused by Rhizoctonia solani. J Mycol Plant Pathol 39:352–354

Kumar PA, Pushpadass HA, Franklin M et al (2016) Effect of enzymatic hydrolysis of starch on pasting, rheological and viscoelastic properties of milk-barnyard millet (Echinochloa frumentacea) blends meant for spray drying. Int J Biol Macromol 91:838–845. https://doi.org/10.1016/j.ijbiomac.2016.06.027

Lee Y, Song KB (2002) Effect of gamma-irradiation on the molecular properties of myoglobin. J Biochem Mol Biol 35(6):590–594

Liu RH (2007) Whole grain phytochemicals and health. J Cereal Sci 46:207–219. https://doi.org/10.1016/j.jcs.2007.06.010

Liu T, Ma Y, Xue S, Shi J (2012) Modifications of structure and physicochemical properties of maize starch by γ-irradiation treatments. LWT-Food Sci Technol 46(1):156–163

Madhusudhana R, Padmaja PG, Cheruku D (2018) ICAR-IIMR millets annual report 2017–18. https://millets.res.in/annual_report/ar18-19.pdf
Maithani D, Sharma A, Gangola S et al (2023) Barnyard millet (Echinochloa spp.): a climate resilient multipurpose crop. Vegetos 36(2):294–308. https://doi.org/10.1007/s42535-022-00420-4
Malik MA, Saini CS (2017) Gamma irradiation of alkali extracted protein isolate from dephenolized sunflower meal. LWT-Food Sci Technol 84:204–211
Malik MK, Kumar V, Sharma PP et al (2022) Improvement in digestion resistibility of mandua starch (Eleusine coracana) after cross-linking with epichlorohydrin. ACS Omega 7(31):27334-27346.
Manimekalai M, Dhasarathan M, Karthikeyan A et al (2018) Genetic diversity in the barnyard millet (Echinochola frumentacea) germplasms revealed by morphological traits and simple sequence repeat markers. Curr Plant Biol 14:71–78. https://doi.org/10.1016/j.cpb.2018.09.006
Maun MA, SCH B (1986) The biology of Canadian weeds.: 77. Echinochloa crus-galli (L.) Beauv. Can J Plant Sci 66(3):739–759
Maurya NK (2019) Therapeutic effects of soya bean chunks supplementation during hemodialysis. Plant Arch 19(1):728–732
Maurya NK, Singh J, Gupta E et al (2023) Barnyard millet's health and nutritional significance. Int J Agric Innov 8(2):156–160. https://doi.org/10.46492/IJAI/2023.8.2.2
Michael PW (2003) Echinochloa P. Beauv. In: Barkworth (ed) Flora of North America. Oxford University Press, New York
Ministry of Agriculture & Farmers Welfare (2022) International Year of Millets (IYoM)-2023. National Conference on Kharif Campaign2022 Ministry of Agriculture & Farmers Welfare:1–45
Mitich LW (1990) Intriguing world of weeds. Barnyard grass. Weed Technol 4:918–920
Napper DM (1965) Grasses of Tanganyaika: with keys for identification. Ministry of Agriculture Forest and Wildlife, Tanzania
Nazni P, Devi SR (2016) Effect of processing on the characteristics changes in barnyard and foxtail millet. J Food Process Technol 7(3):1–8
Nazni P, Karuna TD (2016) Development and quality evaluation of barnyard millet bran incorporated rusk and muffin. J Food Ind Microbiol 2(116):2
Nilavarasi R, Sujatha K, Geetha R, Vanniarajan C (2020) Morphological characterization and evaluation of barnyard millet varieties, cultivars and mutants for varietal identification. J Pharmacogn Phytochem 9(1):1600–1604. https://doi.org/10.22271/phyto.2020.v9.i1aa.10691
Nisha N, Saravanakumar M (2019) Development of reciprocating cutter bar test rig for measurement of cutting force of finger millets. Int J Curr Microbiol App Sci 8(04):2441–2447
Nishizawa N, Togawa T, Park KO et al (2009) Dietary Japanese millet protein ameliorates plasma levels of adiponectin, glucose, and lipids in type 2 diabetic mice. Biosci Biotechnol Biochem 73:351–360. https://doi.org/. https://doi.org/10.1271/bbb.80589
Njintang NY, Mbofung CMF, Waldron KW (2001) In vitro protein digestibility and physicochemical properties of dry red bean flour (Phaseolus vulgaris) flour: effect of processing and incorporation of soybean and cowpea flour. J Agric Food Chem 49:2465–2471
Nozawa S, Nakai H, Sato YI (2004) Characterization of microsatellite and ISSR polymorphisms among Echinochloa (L.) Beauv. spp. in Japan. Breed Res 6:87–93
Nugroho RB, Chusnah U, Suwarno WB, Ardie SW (2024) Flowering phenology and SiDREB2-based SNAP marker-assisted hybridity confirmation for artificial hybridization of Indonesian foxtail millet (Setaria italica L. Beauv) genotypes. J Biosci 31(6):1082–1094
Ofosu FK, Elahi F, Daliri EBM et al (2020) Phenolic profile, antioxidant, and antidiabetic potential exerted by millet grain varieties. Antioxidants 9(3):254
Oswal MS (2023) Millet: the old food revolution to our new life. Agric Eng Today 47(1):27–28
Padulosi S, Mal B, King OI, Gotor E (2015) Minor millets as a central element for sustainably enhanced incomes, empowerment, and nutrition in rural India. Sustain 7:8904–8933. https://doi.org/10.3390/su7078904
Pall BS, Jain AC, Singh SP (1980) Diseases of lesser millet
Pallerla P, Vanapatla SR, Vatsavaya SR et al (2019) Hepatoprotective studies on methanolic extract of caryopses of Echinochloa colona link. Natl Lib Med 32:1949–1956

Panwar P, Dubey A, Verma AK (2016) Evaluation of nutraceutical and antinutritional properties in barnyard and finger millet varieties grown in Himalayan region. J Food Sci Technol 53(6):2779–2787. https://doi.org/10.1007/s13197-016-2250-8

Partohardjono S Jansen PCM (2023) Echinochloa (PROSEA). Plant Resources of South-East Asia. https://uses.plantnet-project.org/en/Echinochloa_(PROSEA)

Peerzada AM, Bajwa AA, Ali HH, Chauhan BS (2016) Biology, impact, and management of Echinochloa colona (L.) link. Crop Prot 83:56–66. https://doi.org/10.1016/j.cropro.2016.01.011

Pingali P (2015) Agricultural policy and nutrition outcomes–getting beyond the preoccupation with staple grains. Food Secur 7:583–591. https://doi.org/10.1007/s12571-015-0461-x

Plaza-Wüthrich S, Tadele Z (2012) Millet improvement through regeneration and transformation. Biotechnol Mol Biol Rev 7:48–61. https://doi.org/10.5897/BMBR12.001

Quattrocchi U (2006) World dictionary of grasses: common names, scientific names, eponyms, synonyms and etymology. CRC Press, Taylor and Francis Group, Boca Raton, USA. https://doi.org/10.4103/mtsp.mtsp_3_24

Rajeswari N, Priyadharshini VP (2021) Evaluation of nutritional and nutraceutical content of polished and unpolished barnyard millet - an analytical study. Curr Res Nutr Food Sci 9(3). https://doi.org/10.12944/CRNFSJ.9.3.31

Rao AN (2021) Echinochloa colona and Echinochloa crus-galli. In: Biology and management of problematic crop weed species. https://doi.org/10.1016/B978-0-12-822917-0.00013-6

Rengalakshmi R (2005) Folk biological classification of minor millet species in Kolli Hills, India. J Ethnobiol 25:59–70. https://doi.org/10.2993/0278-0771_2005_25_59_fbcomm_2.0.co_2

Renganathan VG, Vanniarajan C, Karthikeyan A, Ramalingam J (2020) Barnyard millet for food and nutritional security: current status and future research direction. Front Genet 11:500. https://doi.org/10.3389/fgene.2020.00500

Rostamabadi H, Bajer D, Demirkesen I et al (2023) Starch modification through its combination with other molecules: gums, mucilages, polyphenols and salts. Carbohydr Polym 314:120905. https://doi.org/10.1016/j.carbpol.2023.120905

Saleh AS, Zhang Q, Chen J, Shen Q (2013) Millet grains: nutritional quality, processing, and potential health benefits. Compr Rev Food Sci Food Saf 12:281–295. https://doi.org/10.1111/1541-4337.12012

Salunke P, Chavan U, Kotecha PM, Lande SB (2019) Studies on nutritional quality of barnyard millet cookies. Int J Chem Stud 7:651–657

Sampath TV, Razvi SM, Singh D, Bondale KV (1989) Small millets in Indian agriculture. In: Seetharam A, Riley KW, Harinarayana G (eds) Small millets in global agriculture. Oxford & IBH Publishing Co. Pvt. Ltd., India

Sandhu KS, Singh N (2007) Some properties of corn starches II: Physicochemical, gelatinization, retrogradation, pasting and gel textural properties. Food Chem 101(4):1499-1507

Sarita ES, Singh E (2016) Potential of millets: nutrients composition and health benefits. J Sci Innov Res 5(2):46–50

Selvan JT, Kalaiselvi C, Jothimanivannan C et al (2022) Pharmacological activities of Barnyard millets-A https://doi.org/10.20959/wjpr20233-26169

Sharma S, Saxena DC, Riar CS (2016) Isolation of functional components β-Glucan and γ-amino butyric acid from raw and germinated barnyard millet (Echinochloa frumentaceae) and their characterization. Plant Foods Hum Nutr 71:231–238. https://doi.org/10.1007/s11130-016-0545-6

Sharma VK, Bhatt P, Malik M (2021a) Chemical composition of Jhingora (Indian barnyard millet) varieties cultivated in Uttarakhand and their uses. In climate Change, Water Availability and Food Security: Impacts & Sustainable Management. ISBN No: 9789354269981

Sharma V, Kaur M, Sandhu KS et al (2021b) Barnyard millet starch cross-linked at varying levels by sodium trimetaphosphate (STMP): film forming, physico-chemical, pasting and thermal properties. Carbohydr Polym Technol Appl 2:100161. https://doi.org/10.1016/j.carpta.2021.100161

Singh Sandhu K, Singh N (2005) Relationships between selected properties of starches from different corn lines. Int J Food Prop 8(3):481–91

Singh P, Gurmukh Singh GS, Srivastava S, Agarwal P (2005) Physico-chemical characteristics of wheat flour and millet flour blends. J Food Sci Technol (Mysore) 42(4):340–343

Singh A, Bharath M, Kotiyal A et al (2022) Barnyard millet: the underutilized nutraceutical minor millet crop. J Pharm Innov 11(6):115–128

Sood S, Khulbe RK, Gupta AK et al (2015) Barnyard millet–a potential food and feed crop of future. Plant Breed 134(2):135–147. https://doi.org/10.1111/pbr.12243

Sood S, Joshi DC, Pattanayak A (2020) Breeding advancements in barnyard millet. In: Gosal S, Wani S (eds) Accelerated plant breeding. Springer, Cham. https://doi.org/10.1007/978-3-030-41866-3_15

Southern Weed Science Society (Illinois) (1998) Weeds of the United States and Canada. Southern Weed Science Society

Sowmyalatha BS, Shubhashree KS, Sahana SR (2023) Response of barnyard millet (Echinochloa esculenta L.) to liquid biofertilizer consortium and its methods of application. Environment and Ecology 41(1A):302—307

Subhashini V, Swamy AVVS (2014) Phytoremediation of cadmium and chromium contaminated soils by Cyperus rotundus L. Am Int J Res Sci Technol Eng Math 6:97–101

Subhashini V, Swamy AVVS (2015) Efficiency of Echinochloa colona on phyto remediation of lead, cadmium and chromium. Int J Curr Sci 15:E71–76

Sundararaj DP, Thulasidas G (1976) Botany of field crops. Macmillan Publisher, India

Swain D, Paroha S, Singh M, Subudhi HN (2017) Evaluations of allelopathic effect of Echinochloa colona weed. J Environ Biol 33(5):881–889

Swamy KRM (2023) Origin, domestication, taxonomy, botanical description, genetics and cytogenetics, genetic diversity and breeding of barnyard millet (Echinochloa frumentacea Link.). Int J Curr Res 15(9):25839–25864. https://doi.org/10.24941/ijcr.45982.09.2023

Thompson T (2009) The nutritional quality of gluten-free foods. In: Gallagher E, editor. Gluten-free food science and technology. Oxford, UK: John Wiley and Sons, Ltd. 42–51

Tripathi T, Vyas S (2023) From ancient grains to modern solutions: a history of millets and their significance in agriculture and food security. Int J Home Sci 9(2):72–78

Ugare R, Chimmad B, Naik R et al (2014) Glycemic index and significance of barnyard millet (Echinochloa frumentacae) in type II diabetics, J Food Sci Technol 51:392–395.https://doi.org/https://doi.org/10.1007/s13197-011-0516-8

Underwood MP (2000) Genetic enhancement of sorghum and millet residues fed to ruminants: farmers' perceptions of fodder quality in livelihood systems: summary report: PRA case studies in Andhra Pradesh, Gujarat, Maharashtra, Karnataka and Rajasthan States

Upadhyaya H, Dwivedi SL, Singh SK et al (2014) Forming core collections in Barnyard, Kodo, and little millets using morpho-agronomic descriptors. Crop Sci 54:2673e2682. https://doi.org/10.2135/cropsci2014.03.0221

Uremis I, Ulger AC, Uludag A et al (2000) Weed control in corn in the Cukurova Region of Turkey. Abstracts, Third International Weed Science Congress

Vázquez García JG, Rojano Delgado AM, de la Cruz A et al (2021) Distribution of glyphosate-resistance in Echinochloa crus-galli across agriculture areas in the Iberian Peninsula. Front Plant Sci 12:617040. https://doi.org/10.3389/fpls.2021.617040

Veena B, Chimmad BV, Naik RK, Shantakumar G (2005) Physicochemical and nutritional studies in barnyard millet. Karnataka J Agric Sci 18:101–105

Verma S, Srivastava S, Tiwari N (2015) Comparative study on nutritional and sensory quality of barnyard and foxtail millet food products with traditional rice products. J Food Sci Technol 52:5147–5155. https://doi.org/10.1007/s13197-014-1617-y

Verma VC, Kumar A, Zaidi MG et al (2018a) Starch isolation from different cereals with variable amylose/amylopectin ratio and its morphological study using SEM and FT-IR. Int J Curr Microbiol App Sci 7(10):211–28

Verma VC, Verma VC, Singh A, Agrawal S (2018b) Ethno-botanical study of small millets from India: prodigious grain for nutritional and industrial aspects. Int J Chem Stud 6:2155–2162

Vinoth A, Ravindhran R (2017) Biofortification in millets: a sustainable approach for nutritional security. Front Plant Sci 8:29

Wallace JG, Upadhyaya HD, Vetriventhan M et al (2015) The genetic makeup of global barnyard millet germplasm collection. Plant Genome 8(1):1e7. https://doi.org/10.3835/plantgenome2014.10.0067

Wani HM, Sharma P, Wani IA, Kothari SL, Wani AA (2021a) Influence of γ-irradiation on antioxidant, thermal and rheological properties of native and irradiated whole grain millet flours. Int J Food Sci Technol 56(8):3752–3762. https://doi.org/10.1111/ijfs.15062

Wani IA, Sultan N, Qadir N, Ahmad T (2021b) Physico-chemical, functional and antioxidant properties of wild barnyard grass (Echinochloa crusgalli L.) seed flour as affected by gamma-irradiation. Radiat Phys Chem 183:109409. https://doi.org/10.1016/j.radphyschem.2021.109409

Wanous MK (1990) Origin, taxonomy and ploidy of the millets and minor cereals. Plant Var Seeds 3:99–112

Watanabe M (1999) Antioxidative phenolic compounds from Japanese barnyard millet (Echinochloa utilis) grains. J Agric Food Chem 47:4500–4505. https://doi.org/10.1021/jf990498s

Yabuno T (1962) Cytotaxonomic studies on the two cultivated species and the wild relatives in the genus Echinochloa. Cytologia, 27(3):296–305.

Yabuno T (1983) Biology of Echinochloa species. In Proceeding of the Conference on Weed Control in Rice 31:307–318.

Yabuno T (1987) Japanese barnyard millet (Echinochloa utilis, Poaceae) in Japan. Econ Bot 41(4):484–493. https://doi.org/10.1007/BF02908141

Yabuno T (2001) Taxonomy and phylogeny of the genus Echinochloa. In: Natural history of genus Echinochloa. Zennoukyo Shuppan, Tokyo, Japan

Yadav RB, Khatkar BS, Yadav BS (2007) Morphological, physicochemical and cooking properties of some Indian rice (Oryza sativa L.) cultivars. J Agric Technol 3(2):203–210

Yamaguchi H, Utano A, Yasuda K et al (2005) A molecular phylogeny of wild and cultivated Echinochloa in East Asia inferred from non-coding region sequences of trnT-L-F. Weed Biol Manag 5:210–218. https://doi.org/10.1111/j.1445-6664.2005.00185.x

Ye CY, Lin Z, Li G et al (2014) Echinochloa chloroplast genomes: insights into the evolution and taxonomic identification of two weedy species. PLoS One 9:e113657. https://doi.org/10.1371/journal.pone.0113657

Yuichiro M, Shinya U, Yamaguchi H (1999) Identification of polyploid groups in the genus Echinochloa by isozyme analysis. J Weed Sci Technol 44:205–217. https://doi.org/10.3719/weed.44.205

Chapter 9
Sorghum Millet: Chemical Composition, Starch Isolation, Modifications, Functionality, and Food/Non-Food Applications

Monika Rani, Hemendra Tripathi, and Anjali Goel

Abbreviations

ADH	adhesiveness
C_3H_6O	propylene oxide
$C_4H_6O_3$	acetic anhydride
CFTRI	Central Food Technological Research Institute
CHW	chewiness
COH	cohesiveness
DSC	differential scanning calorimetry
Fe	iron
FV	final viscosity
GBSSI	granule-bound starch synthase I
GT	gelatinization temperature
GUM	gumminess
H_2O_2	hydrogen peroxide
HD	hardness
HHP	high-hydrostatic pressure
HMT	heat-moisture treatment

M. Rani (✉)
GRD-Institute of Management and Technology, Dehradun, India
e-mail: monika.rani@grdedu.in

H. Tripathi
Shivalik College of Engineering, Dehradun, India
e-mail: hemendra.tripathi@sce.org.in

A. Goel
Department of Chemistry, Kanya Gurukul Campus, Gurukula Kangri
(Deemed to be University), Haridwar, Uttarakhand, India
e-mail: anjali.goel@gkv.ac.in

© The Author(s), under exclusive license to Springer Nature Switzerland AG 2026
J. Singh, M. Kumar Malik (eds.), *Millets and Millet Starch: Chemical Composition, Functional Modification, and Applications*,
https://doi.org/10.1007/978-3-032-09502-2_9

MM	molecular mass
MW	molecular weight
NaOCl	sodium hypochlorite
NaOH	sodium hydroxide
ORAC	oxygen radical absorbance capacity
PEG	polyethylene glycol
PT	peak temp.
PUL	pullulanase
PV	peak viscosity
RDS	rapid digestible starch
RS	raw sorghum
RTE	ready-to-eat
RVA	rapid visco-analyzer
SB	set back value
SBEIIb	starch branching enzyme
SDS	slowly digestible starch
SP	swelling power
SPR	springiness or elasticity
ß-galactosidase	beta-galactosidase
SSSIIa	soluble starch synthase
STMP	sodium trimetaphosphate
STPP	sodium tripolyphosphate
Zn	zinc
α-amylase	alpha-amylase
β-amylase	beta-amylase
ΔH	enthalpy change
(v/v)	volume/volume
(w/v)	weight/volume

9.1 Introduction

Superfoods like millets are currently quite popular in the realm of healthy eating. These tiny gluten-free grains are like health superheroes and support weight loss while maintaining overall health. Incorporating these organic millets into the diet may revolutionize eating habits if current consumption is limited to wheat and rice (Krishnan 2024). It is difficult to choose one millet, due to the abundance of options, relative caloric and health benefits of each.

Sorghum, scientifically known as *Sorghum bicolor* (L.) Moench., is a warm-season cereal that is valued for its potential applications such as feed, food, and fodder. After barley, rice, corn, and wheat, sorghum millet is one of the most significant cereal in the world. It is classified as a C4 plant with low water requirements (about 400–600 mm annually) and is resistant to drought and other climatic stress. It has a better potential for agricultural production and a faster rate of photosynthesis than other crops. It has a reduced photorespiration rate and is more resistant to oxygen-induced inhibition (Serna-Saldivar 2010; Abah et al. 2020). Sorghum serves as a major base for traditional meals and drinks throughout Africa and Asia, such as traditional beers, wine, and non-alcoholic beverages, through widely industrialized techniques. Reports indicate that sorghum millet is a rich source of energy, minerals, vitamins, carbohydrates, proteins, beneficial flavonoid and phenolic components for human consumption, potentially offering health benefits such as antioxidant, antibacterial, and anti-carcinogenic properties. Various methods, such as those related to milling, fermentation, malting, and distillation, have been studied and applied in the production of bread, tortillas, cookies, and noodles, among other foods. Those with allergies or celiac disease to wheat-based products can take also benefit from this millet because it does not contain gluten proteins (Aruna and Visarada 2019; Smolensky et al. 2018).

9.2 Geographical Distribution

Sorghum is cultivated over 42 million hectares across the globe, with Africa contributing the largest share (24.5 million ha), followed by Asia with around 10.6 million hectares. A considerable amount of this crop is also grown in the Americas (Argentina, Brazil (0.7 million ha), Bolivia) (6.6 million ha) and Australia (0.7 million ha). It is also commonly distributed in Mali (1.2 million ha), Burkina Faso (1.7 million ha), Niger (3.1 million ha), Nigeria (6.1 million ha), Sudan (6.8 million ha), Cameroon, Ethiopia (1.6 million ha), Tanzania, India, China, Mexico (1.8 million ha) and USA (2.5 million ha) (Taylor 2025). India has the greatest sorghum area, with 9.1 million acres, ranking sixth (Anugrau 2022). Sorghum is cultivated in 31 countries in sub-Saharan Africa, with an average harvested area of 24.5 million hectares from 2001 to 2020.

9.3 Cultivation

Sorghum is grown all over the world. The fact that millets and sorghum often require less water is a crucial agronomic feature. Sorghum need 400 mm of water at minimum, most millets require 300 mm, and maize requires 500–600 mm. Further, Sorghum is also renowned for its capacity to survive severe droughts. For crop development, sorghum typically requires soil temperatures above 15°C. Temperatures between 20 to 39°C are necessary for flowering and seed set; pollen generation is an especially critical process, with higher or lower temperatures resulting in sterility (Cruickshank 2016). Nitrogen, phosphorus, and potassium are the main nutrients needed for the growth of sorghum. A soil pH of 6.5 to 7.5 is ideal for maximum productivity. Farmers may need to raise seeding rates by 15–20% if the crop is planted late. Sorghum is typically grown in two seasons in India: Rabi (the post-rainy season) and kharif (April to October), with the kharif season producing the majority of the crop (Taylor 2025). Sorghum millet is commonly grown as a rain-fed plant with fertilizer quantities of very low, 60 kg/ha in dry areas with low soil fertility. Under the All India Coordinate Sorghum Improvement Project (AICSIP), 18 hybrids (CSH 1-CSH 18R) and 15 sorghum cultivars (CSV 1–CSV 15) were created. The finest for the kharif season are CSH 1, 6, and 9, and for the Rabi season crop, CSH 15R and 18R work best. Several rainfall-prone areas of India are recommended for the hybrids and cultivars (Kannababu et al. 2020). Before the monsoon arrives, seeds are sown at a depth of 5 cm. Pre-monsoon sowing, also known as dry seeding, is planted 1 or 2 weeks prior to the monsoon. The majority of hybrids and high-yielding varieties reach maturity in 100–115 days. When the grain becomes hard and contains less than 25% moisture, it is prepared for harvesting. Since hybrid sorghum does not allow the straw and plants to dry, it remains green for prolonged period. Harvesting is feasible at physiological maturity. After a drying period of 2–5 days, the entire plant is harvested, or the upper portion is removed first, followed by the chopping of the plants. The ear heads are threshed using a mechanical thresher. The threshed grains are sun-dried for a week to cut its moisture content by roughly 10–12% in order to guarantee safe storage. In rainy seasons, the yield of grain ranges from 2 to 3 t/ha, whereas the dried Stover yield ranges from 8 to 10 t/ha.

9.4 Storage and Processing

The goal of storage is to maximize the grain's worth for its planned use in the future. Either the grain's food value is maintained or a significant percentage of viable seeds are preserved for the subsequent season of cultivation. The grain is linked to the glumes, where the black spots are found. This shows that the sorghum millet has reached physiological maturity. According to Ballagh et al. (2021), grain typically contains 25% moisture when it achieves physiological maturity. The crop is highly vulnerable to outside influences after it reaches maturity; therefore, it's critical to harvest it as soon as possible to preserve grain quality. Bird damage, pre-harvest sprouting from precipitation, and lodging from dry conditions are all threats to field-grown crops. Sorghum is mechanically harvested in America and Australia using combine harvester; with this technique, the seeds are extracted straight from the plant without the need for additional shelling (Prasad and Staggenborg 2009). Hand harvesting is the method used in Asia and Africa. Inflorescences (The heads of plants) are collected, bundled, and threshed manually when the moisture content reaches 12% (Dharmaputra 2013).

In manual threshing, Sorghum heads are spread out on a clean floor and are regularly turned to dry in the sun. To separate the grains, the heads are either trampled beneath the feet of animals, stone rollers, or tractor wheels, or they are manually threshed with wooden sticks to separate the grains. Following threshing, the grains are cleaned by wind-driven winnowing to remove rachis, glumes, and chaff before being stored. Grain that has been threshed may be further dried in some areas using forced air at a temperature of no more than 40°C (Arama et al. 2014). Depending on the weather, low- or high-temperature drying systems are used in the United States to dry grain. For the grains to yield the highest market price and earnings, the right temperature and moisture content must be present. After harvesting and drying, grain must be stored under optimal conditions to prolong its shelf life and maintain seed quality, particularly when the moisture level exceeds 10% (Hussen 2021). The environment's temperature and humidity levels should be continuously monitored and controlled for prolonged storage. Granaries developed utilizing reeds, sorghum stems, bamboo and other elastic, bendable long sticks are also used for storage of sorghum especially in rural areas. Besides these, small containers, plastic bags, sealed gourds or small containers such as sealed gourds are also utilized for sorghum grain storage. For efficient and bulk storage of sorghum, however, concrete bins, mud bins with plant linings, iron or other metal bins have been proposed more recently (Beta and Ndolo 2025). Processing is regarded as a necessity for the manufacturing and/or consumption of food products made from sorghum. Several processing techniques such as soaking, cooking, fermentation, germination, roasting, steaming, and other treatments are extensively employed as traditional, easy, and effective technological procedures to boost the nutritional value of sorghum grains. Germination is a commonly utilized procedure in sorghum and other legumes grains that alters their biochemical, nutritional, and sensory properties, resulting in more appealing flavor and taste. Afify et al. reported that germination of sorghum millet increases protein digestibility, solubility, nitrogen solubility index but reduced protein content, fat content, phytic acid, phosphorus, potassium, magnesium, sodium, calcium, iron, zinc, and manganese content (Afify et al. 2012a, b, c). Fermentation of sorghum is another way to tailor the nutritive value. It was reported that fermentation of sorghum significantly increased the available lysine (33.2 mg/g N when fermented at 25 and 34.5 mg/g N when fermented at 35°C) and methionine (25.68 mg/g N at 25°C and 26.79 mg/g N at 35°C) (AU and Fields 1981). Roasting is a classical dry heat treatment approach used in snack manufacture that has been shown to enhance the edibility, digestibility, and organoleptic qualities of a variety of food grains. However, microwave roasting is now gaining substantial importance due to its cost-effective, quick, and simple handling. Sharanagat et al. investigated the impact of microwave roasting on functional, antioxidant and pasting characteristics of sorghum and reported that roasting had a substantial effect on the antioxidant activities (92.50 to 83.95% DPPH free radical-scavenging activity), total phenolics (2.64 to 4.67 mg GAE/100 g) and total flavonoid (3.89 to 2.73 mg QE/100 g) of sorghum flour (Sharanagat et al. 2019). During the popping process, the grains are exposed to high temperatures for small periods of time, which caused breaking of pericarp and quick expansion of endosperm. Popped foods such as snacks are gaining popularity due to low-calorie, and nutrient-dense characteristics. Saravanabavan et al. 2013 investigated the impact of popping on the quality characteristics of sorghum carbohydrate and confirmed that popping significantly alters the starch digestibility, resistant starch, amylose content and protein digestibility with significant decrease in phytic acid (20%–25% reduction).

Fig. 9.1 Sorghum young plant in agriculture field

9.5 Appearance, Morphology, and Genetics

Sorghum belongs to the Gramineae family of grasses. Farmers need to be well-informed about the crop they are cultivating to develop the most effective production strategy (Fig. 9.1).

The typical characteristics of sorghum leaves are glassy, flat, green, and less broad than those of maize (Fig 9.1). The leaf area of sorghum plant is less than that of maize plant. The leaf blade is sharp, long, and narrow. When leaves are young, their blades are straight, but as the leaves get older, they begin to sag. The leaf has stomata on both surfaces. The characteristic of sorghum leaves is the rows of motor cells on the upper surface that span the length of the midrib. These cells have the ability to rapidly coil up the leaves when they are under moisture stress (Muimba-Kankolongo 2018). The leaves are wrapped in a thin layer of wax and emerge in opposite direction (down the stem). The quantity of leaves fluctuates based on environmental conditions, ranging from 8 to 22 leaves per plant. The stem's diameter ranges from 5 to 30 mm. The internodes are blue-white due to the heavy layer of wax that coats them. The waxy layer reduces transpiration, making the plants more drought-tolerant (Chemelewski et al. 2023). Primary roots come from the root band of nodes that are either below or slightly above the soil's surface. As the bulb grows, the secondary roots appear. Water and nutrients from the soil are supplied to the seedling by the main roots. The growth of primary roots is restricted and secondary roots quickly take over early in a plant's life. Roots can spread 1 m laterally and descend up to 2 m if there are no obstacles in the soil. Compared to maize plant

Table 9.1 Different varieties, their grains and fodder yield, and maturing duration (Kumar 2021)

Name of variety	Number of days (Duration)	Yield of grain (t/ha)	Yield of fodder (t/ha)
K10	110–115	1.6	16.00
CO 25	115–120	3.68	13.25
APK 1	105–110	2.60	8.00
CO 26	105–110	4.5	14.00
BSR 1	105–110	3.00	7.20
K tall	90	3.75	11.25
K 8	85	2.4	7.30

roots, these roots branch about twice as often and are finer (Muimba-Kankolongo 2018). Below the soil's surface, nodes give rise to secondary roots. The permanent root system spreads out freely into the soil and laterally as well. The inflorescences of sorghum are compact panicles. The panicle varies in color and shape throughout cultivars. The peduncle, or main stalk, on which the florets are carried, also holds the heads. The usual straight length of the peduncle is between 75 and 500 mm. Each panicle has 800–3000 kernels, the majority of which are partially covered in glumes (Kazungu et al. 2023). The hue of the glumes may be black, red, brown, or tan. Sorghum blooms from midnight until 10:00 in the morning, with the best blooming taking place between 6:00 and 8:00 AM. The uppermost flowers open first, and the entire panicle takes 6–9 days to bloom. Due to the flower's structure, self-pollination is the primary process. Natural cross-pollination occurs in very small amount (2–10%) (Chiluwal et al. 2020; Sobetski 2015). There are a number of varieties of sorghum millet on basis of maturation and yield as listed in Table 9.1.

9.6 Phytochemical Composition and Their Health Attributes

Macronutrient content in sorghum is 72.6% carbohydrates, 1.6% crude fiber, 1.9% fat, and 10.4% crude protein. However, it depends on the genotypes used. It was reported that different genotypes of sorghum contains: fat (2.98–5.14%), protein (6.24–9.34%), fiber (1.58–2.49%), and starch (62.27–69.43%) (Htet et al. 2022). A sample of sorghum millet grain was taken from Kansas and it has been reported that sorghum contained 11% protein (Malleshi and Klopfenstein 1998).

Sorghum millet has a superior amino acid profile compared to rice, wheat, and barley (Ejeta et al. 1987; Abdalla et al. 1998; Hadimani et al. 1995). Sorghum millet generally has a better protein efficiency ratio than wheat (Rao et al. 1964; Pushpamma et al. 1972). The sorghum carbohydrate consists of 16–20% non-starchy polysaccharides (NSPs) (Adebo 2020).

Dietary fiber is composed primarily of NSPs, which is available in the cell walls of endosperm and bran. A comparison of 16 different sorghum varieties' malts was made to determine their amylase activity. The results showed that some sorghum varieties malt had α-amylase 178–183 μg, which was comparable to commercially barley malt (189 μg), and that β-amylase was lower (37–41 μg) than barley malt (52 per μg) (Beta et al. 1995). According to recent reports, micronutrient malnutrition, particularly linked to vitamin A, iron, and zinc, is a global health concern related to food, especially for those living in developing nations where outreach to fruits, vegetables, and animal products is limited (Mason and Garcia 1993). To address this issue sustainably and economically, the International Crops Research Institute for the Semi-Arid Tropics (ICRISAT) has been working on genetically enhancing the zinc and iron content in sorghum by biofortification (Bouis 1996). Genetically modified sorghum millet shows greater iron and zinc levels (Gaddameedi et al. 2022). ICRISAT has found that sorghum breeding lines and germplasm had more than 50 ppm Zn (equivalent to maize and wheat) and more than 75 ppm Fe (20% more than wheat and maize). Some sorghum germplasm accessions have been reported to have up to 91 ppm Zn (zinc) and 93 ppm Fe (iron) (Rai et al. 2008). Phytates and polyphenols are examples of anti-nutritional factors that can decrease the bioavailability of nutrients. By increasing their bioavailability through processing or other means, these substances can be more readily absorbed and have a higher overall nutritional value. Polyphenols, which are mostly found in the seed's periphery, inhibit the hydrolytic enzymes, including ß-galactosidase, cellulases, trypsin, amylases, and chymotrypsin (Singh 1984). As a result, less protein and starch are utilized (Pawar and Parlikar 1990; Thompson and Yoon 1984). Additionally, they have a lesser availability of vitamins and minerals. p-coumaric acid, gallic acid, vanillic acid, ferulic acid, caffeic acid, and cinnamic acid are the major phenolic acids available in sorghum millet (Li et al. 2021). Sorghum also contains naringenin, luteolin, eriodictyol, and apigenin as its main flavonoids (Apea-Bah et al. 2021). Polycosanol have numerous positive effects on health such as decrease in cholesterol levels and improvement in cardiovascular health. It was reported that polished and unpolished sorghum grain contains 74.5 mg/100 g and 9.8 mg/100 g of total polycosanol in the dry kernel (Tanwar et al. 2023). Sorghum grains have high concentrations of 3-deoxyanthocyanidins and proanthocyanidins. These compounds have been found to be useful for human health as these provide protection against inflammation, oxidative stress, and diabetes. Further, 3-deoxyanthocyanidins inhibit the proliferation of gastrointestinal cancer cell lines. Sorghum flavonoids have been found to induce proinflammatory immune responses, which can be beneficial in certain health conditions (Rhodes et al. 2014; Smolensky et al. 2018).

9.7 Sorghum Starch Isolation

9.7.1 Sorghum Starch Isolation by Alkaline Extraction Process

Sorghum has a wide range of grain colors, including white, cream, yellow, brown, red, and black, as well as varying preferences of starch extraction method for those color. One popular technique for extracting starch is by sodium hydroxide (NaOH) extraction method (in which 450 g of sorghum flour is mixed with 2.7 L of 0.15 N NaOH to improve starch yield). Using NaOH solution, the protein and lipid complexes in the grain are broken down and starch is separated. The alkali extraction approach is widely employed due to its ease of use, affordability, and high starch production (Zhu 2017). A laboratory scale method for isolation of starch from sorghum was developed by Sira and Amaiz (2004). For this purpose, 0.25% NaOH solution was used to dip the sorghum grains overnight. Following distilled water wash until pH reaches neutral. Finally, wet milling extraction was used for starch isolation.

9.7.2 Sorghum Starch Isolation by Wet Milling Process

While sorghum kernel starch extraction is similar to maize starch extraction (Caransa and Bakker 1987), there are additional challenges, as the starch and protein in the kernels have a strong interaction. The industrial production of sorghum starch using a wet milling technique has been reported in earlier studies (Eckhoff and Watson 2009). Briefly, the grains are cleansed before being steeped and ground. In the wet-milling process (Xie and Seib 2002), 1.5 parts of fresh water were utilized for each part of the grain to mitigate drying and transfer losses. A model study was conducted using 100 g of commercially No. 2 grain of sorghum (dry solids), varying steep time (1–3 h), steep temperature (25–60 °C), and coarse-grinding speed (7500–12,500 rpm); the responses measured were starch recovery, starch lightness (L), and damaged starch content. Grain sorghum was soaked in twice its weight of process water containing 0.2% sulfur dioxide, and the steeped kernels were combined with an equivalent volume of process water and pulverized for 6 min in a Waring blender equipped with dull blades ($d = 3.2$ cm). The coarsely grounded material was sieved (mesh size 1190 μm) to separate the bran and germ, while the other particles were permitted to settle. After passing through a plate mill once, the sedimented part was finely ground, and the fine fiber was eliminated using sieving (opening 73 μm). The starch was separated from the protein fraction on a starch table after the slurry was adjusted to a specific gravity of 1.04. The protein fraction was mixed with the bran/fiber and fine-fiber fractions, along with the steep-liquor concentrate adjusted to 54% solids, resulting in a co-product that had 27% protein (dry basis/db) and 70% moisture (wet basis/wb). In the surface response study, L (Lightness) varied from 90 to 93, the protein content of the starch concentrate ranged from 0.4% to 0.5%, and starch recovery varied from 57% to 89%. The amount of damaged starch remained consistent at 0.4%.

9.8 Sorghum Starch Functionality

9.8.1 *Swelling Power and Solubility Properties*

Swelling and solubilization of sorghum starches occur in two stages. This pattern is characterized by an initial rapid phase, a slight reduction, and then another rapid phase. This pattern explains that starch granules contain two independent sets of intragranular bonding forces that relax at different temperatures (Udachan et al. 2012). The starch becomes soluble as a result of amylose leaching out during the swelling process. Since the starch granules swell when heated in water, the swelling power (SP) of starch is a measure of both intragranular and intergranular water absorption. Instead, solubility generally relates to how well starch components and amylose dissolve in water (Semwal and Meera 2023). The assessment of swelling power and water solubility index was conducted using the methodology (Leach et al. 1959) with certain changes. The starch suspension in distilled water was heated at 50 °C, 60 °C, and 70 °C in a water bath for 30 min. The suspension was thereafter cooled to ambient temperature and centrifuged for 20 min at 5000 revolutions per minute. Subsequently, supernatant was transferred to a weighing plate and was dried at 120 °C for 2 h. The residual supernatant was carefully collected, and weighed to determine the water solubility index of the starch granules. The swelling power of starch isolated from different Sorghum cultivars at diverse temperature is shown in Table 9.2.

9.8.2 *Pasting Properties*

The crystalline structure of starch granules is broken during gelatinization in excess water, allowing for water absorption and a notable enlargement in size (Bao and Bergman 2018; Khoddami et al. 2023). A rapid visco-analyzer (RVA) and a rheometer are used to closely monitor changes in viscosity. Starch has been subjected to regulated heating and cooling cycles under constant shearing force and water (Chen et al. 2023). The pasting characteristics of the flour samples and porridges were analyzed using an Anton-Paar Rheometer (MCR 72, Graz, Austria) by a modified approach (Adeboye and Emmambux 2017). 3.4 gram of dry sample and distilled

Table 9.2 Swelling power variation by temperature difference (in g/g) (Udachan et al. 2012)

	Swelling Power (g/g)		
Variety of Sorghum	50 °C	60 °C	70 °C
CSH-5	11.60	13.40	17.60
CSH-9	12.20	14.10	16.60
Dadar	13.00	15.10	17.20
Parbhani	13.20	15.80	17.20

water was used to make final slurry. The slurry was stirred for 10 min at 50 °C and 960 rpm to start the pasting process. The stirring speed was then reduced to 160 rpm for the remainder of the cycle. The suspension was heated from 50 to 95 °C at a rate of 5.5 °C per minute, held for 15 min, and then cooled to 50 °C at the same rate and held there for another 15 min. The following parameters were noted: breakdown viscosity (BV), setback viscosity (SV), final viscosity (FV), peak viscosity (PV), and pasting temperature. The PV results show that the pasting properties of sorghum flour, sorghum okara composite flour, and their respective porridges differ significantly ($p < .05$). The maximum PV (953.10 cP) was shown by 100% raw sorghum flour, indicating its high starch content and notable swelling potential (Adeyanju et al. 2025).

9.8.3 Thermal Properties

Gelatinization is an irreversible phenomenon that occurs when starch granules are heated in the presence of water and break their molecular structure (Olayinka et al. 2015). Differential scanning calorimetry (DSC) is used to measure important thermal parameters, including enthalpy change (ΔH) and gelatinization temperature (GT), to examine these changes. The thermograms of freeze-dried porridge powder, sorghum flour, and okara flour were analyzed using DSC. The first point of a phase change, melting, is represented by the onset temperature (To). The raw sorghum (100%) had the highest onset temperature, and okara (100%) had the lowest in the raw flours, with onset temperatures ranging from 76.33 to 96.17 °C. This suggests that sorghum flour contains starch content and requires the highest temperature to start gelatinizing (Adeyanju et al. 2025).

9.8.4 Retrogradation Properties

When gelatinized starch cools below its crystalline melting point, the amylose and amylopectin polymers realign and re-associate into more ordered molecular configurations over time. This process is known as retrogradation (Bao and Bergman 2018). Physical changes, including increased PV, turbidity, gel formation, and improved crystallinity, are characteristics of this process. According to Shi and Seib (1992), DSC was used to assess the gelatinization and retrogradation of starches. Water and starch in a 1:2 (w/w) ratio were heated from 25 to 130°C at a rate of 10°C per minute. After 7 days of storage at 4°C, gelatinized starch was rescanned. By using DSC, the retrogradation of amylopectin in these sorghum starches was examined following their heating in two parts of water to 130°C and 7 days of storage at 4°C. Heterowaxy and waxy sorghum starch had a greater degree of retrogradation than normal sorghum starch based on the total weight of starch.

9.8.5 Textural Properties of Starch Gels

According to Boudries et al. (2009), a texture profile analyzer can be used to efficiently examine the textural characteristics of sorghum starch gels, particularly when performing RVA analysis. The textural parameters assessed for various sorghum starches include, gumminess (GUM), cohesiveness (COH), springiness (SPR) or elasticity, chewiness (CHW), and adhesiveness (ADH). The amylose matrix and the interactions between the continuous and dispersed phases of the starch gel, which affect the mechanical properties of sorghum starch, have a major impact on these characteristics (Singh et al. 2010). Depending on the type and amount of starch, the batters used to make the loaves had varying consistencies. Compared to batters made with pregelatinized starch, sorghum-based batters using native starch had a lower viscosity (Onyango et al. 2011). Batters made from sorghum and pregelatinized starch, on the other hand, were inelastic, solid, and resembled wheat dough. Global research on sorghum-wheat composites indicated that up to 30% sorghum flour can be blended with a wheat flour of "reasonable strength" to produce high-quality bread (Frederick 2007).

9.8.6 Digestion Properties

The sorghum starch digestion property differs depending on whether it is cooked or raw. After 20 min of alpha-amylase incubation at 37 °C, raw sorghum starch contains 35% RS, 25% slowly digestible starch (SDS), and 40% rapid digestible starch (RDS) (Liang et al. 2024). The RS content of a white, uncooked sorghum starch was 43.7% (Giuberti et al. 2019). In contrast to raw sorghum starch, cooked starch has a significantly lower RS due to the gelatinization that occurs during cooking, which facilitates easier digestion by α-amylase (Hamad et al. 2019). Consuming sorghum is known to have positive health impacts, like lowering blood glucose and cholesterol levels. The study indicates that adding sorghum extracts to food products can reduce the Estimated Glycemic Index (EGI) and the starch digestibility while increasing the amount of RS in corn porridge (Lemlioglu-Austin et al. 2012). Additionally, their study showed that these crop groats with higher RS content and lower EGI value were comparable to or greater than whole grain pasta, legumes, or cereals (Magaletta et al. 2010).

9.8.7 Fermentation Properties

One of the most prevalent types of fermentation in sorghum is lactic acid fermentation. This fermentation, by LAB (lactic acid bacteria), restricts all other microbes by utilizing present carbohydrates and accumulating organic acids like lactic acid, along with other metabolites present in the food matrix, lipolysis, proteolysis, and glycolysis. LAB strains enhance nutrient levels, including B-complex vitamins, and improve texture and flavor through the formation of proteins, exopolysaccharides (EPS), and lipids, while also decreasing antinutritional factors (Adebo 2020). In sorghum, the levels of phenolic compounds, including gallic acid, catechin, and quercetin, rise during fermentation. On the other hand, flavonoids, total phenolics, and total tannin compounds become less abundant. Reports indicate that LAB fermentation resulted in a conversion of phenolic compounds into structurally similar molecules. These modifications were all related to fermentation-induced hydrolysis, esterification, and decarboxylation (Adebo et al. 2018). Large biomolecules are broken down in this process by the action of several enzymes (such as proteases, amylases, xylanases, and hemicellulases), which enables the synthesis and release of a variety of bioactive compounds with immunomodulatory, antimicrobial, antioxidant, and other beneficial attributes (Rodríguez-España et al. 2022). Strong and light soy sauce, sweet honey, and other sorts are the five main categories of Chinese liquor, a well-known traditional beverage (Fan and Qian 2006). Depending on the sorghum genotype, the local environment, and specific brewing methods used, the taste of sorghum-based liquor differs highly (Wu et al. 2007).

9.9 Sorghum Starch Modification

Although natural sorghum starch can be used in food processing, it lacks the necessary physical and chemical modifications due to its high degree of retrogradation, poor shear stability, and poor heat stability (Han et al. 2020). Therefore, to make native starches suitable for some commercial uses, changes are required.

9.9.1 Chemical Modification

By adding functional groups to the polymer molecules, native sorghum starch can be chemically modified to alter its intra- and intermolecular structure and physicochemical properties, such as pasting, gelatinization, and retrogradation (Ogundiran et al. 2022). The functional characteristics of chemically modified starches are described by the type, degree, concentration, reaction conditions, and source of the starch (Han et al. 2020).

9.9.1.1 Phosphorylation

According to Han et al. (2020), the functionality of chemically modified starches is shown by the starch source, concentration, reaction conditions, degree of substitution. Sodium trimetaphosphate (STMP) and sodium tripolyphosphate (STPP) can change the structure of the starch by creating phosphate diester bridges between the glucose units at the C6 position between the starch molecules (Bangar et al. 2024). According to García-Gurrola et al. (2019), phosphorylated sorghum starch with 4% STPP demonstrated improved thermal stability. In comparison to native starch, sorghum starch phosphorylation (with a 99:1 ratio of 3% STMP and STPP) exhibited greater solubility and a thicker film of starch (Mehboob et al. 2020).

9.9.1.2 Oxidation

The sorghum starch is oxidized, with oxidizing chemicals such as hydrogen peroxide (H_2O_2) or sodium hypochlorite (NaOCl) in alkaline medium. The physicochemical properties of starch molecules are impacted by this oxidative cleavage, which takes place through carboxyl and carbonyl functional groups to glucose units (at OH groups of C2, C3, and C6) (Dimri et al. 2023). According to Ali and Hasnain (2014), oxidizing sorghum starch using a 1:10 NaOCl to starch ratio results in a decrease in SP and retrogradation, indicating a reduction in starch's ability to recrystallize following gelatinization.

9.9.1.3 Ozonation

Ozonation alters shear resistance, temperature stability, low syneresis, pH stability, and gelling qualities by oxidizing the hydroxyl groups on C2 or C6 of glucose units and C–H bonds at glycosidic links (Maniglia et al. 2021). Ozone treatment of sorghum starch results in a significant decrease in SP. On the other hand, PV and FV are found to be greater than BD and SB. This is attributed to the disruption of the amorphous regions of the starch granules, which affects their ability to reassociate following gelatinization (Thakur et al. 2023).

9.9.2 Physical Modifications

Physical treatments such as heat, shear, radiation, or moisture to modify sorghum starch achieve favorable changes. This method requires no chemical reagents (Anggreini et al. 2021).

9.9.2.1 Heat-Moisture Treatment

Heat-moisture treatment (HMT) exposes starch to temperatures higher than its GT for a defined period while keeping proper moisture conditions (Schafranski et al. 2021). During HMT, temperature, moisture content, and pH all affect swelling power and solubility (Sun et al. 2014). After 10 h of HMT at 100 °C, the amount of insoluble amylose in sorghum starch rises, most likely as a result of improved interactions between lipids and starch polymers (Sun et al. 2014). Additionally, white sorghum starch treated at 110 °C with different moisture contents exhibits notable decreases in PV (Peak viscosity), FV, and SB while polymer crosslinking causes a modest rise in PT (Olayinka et al. 2008).

9.9.2.2 Ultrasonication

The physical process of ultrasonication produces three main mechanical forces: high-frequency vibrations, high-velocity liquid jets, and localized high pressure. Together, these forces break the polymer chains in starch by producing shearing effects, free radicals, and high temperatures (Li et al. 2023). SP, solubility, and oil absorption capacity all significantly increase when sorghum starch is treated with ultrasonics at a frequency of 20 kHz (Sharma et al. 2023).

9.9.3 *Enzymatic Modification*

The physicochemical and structural characteristics of sorghum starch change by enzymatic treatment. Treatment of sorghum starch with protease (Neutrase, 0.625 AU/mL activity) degrades protein fractions, increasing peak viscosity and decreasing peak temperature (Mezo-Villanueva and Serna-Saldivar 2004). Similarly, the starch hydrolysis rate was significantly increased when a-amylase-treated sorghum starch was combined with protease treatment (Pérez-Carrillo and Serna-Saldivar 2007). Rearranging the intramolecular bonds of sorghum starch with pullulanase and infrared radiation treatment caused increased relative crystallinity and decreased digestibility of the starch (Semwal and Meera 2023). Waxy sorghum starch was treated with the starch branching enzyme IIb, which showed an increase in the number of A-chains. However, when waxy starch was treated with granule-bound starch synthase I, the Bl-chain concentration of amylopectin increased considerably (Kang et al. 2023). But when waxy starch was treated with the soluble starch synthase enzyme, the number of A-chains was significantly reduced, leading to the polymerization of A-chains into B-chains (Kang et al. 2023). During the malting process, endogenous enzymes and physical factors (such as temperature and steeping duration) can change granule surface pores, lowering the PV of sorghum starch and the branching density of amylopectin and amylose (Oseguera-Toledo et al. 2020).

9.9.4 Food and Non-Food Applications

9.9.4.1 Traditional Food Products

The most well-liked and simplest traditional foods prepared from pearl and sorghum millet are thick and thin porridge (fermented and non-fermented). Foods such as unleavened fermented breads like dosa and injera and unfermented breads like roti have a smooth texture. In addition, they can be used to prepare products such as couscous, a steamed granular dish, or foods that resemble boiled rice (Young et al. 1993). The process of making fermented bread involves combining flour, water, and a starter; the mixture is then allowed to ferment for 12–24 h before being cooked (Poonam 2002). Sorghum is also used in India to make a variety of snacks. According to Rekha (1997), these items were very well received and demonstrated a prolonged shelf life of up to 3 months. According to a previous study by Chaudhary (1993), traditional items made from sorghum millet, such as khichri, chapati, popped grain, bhakri, dalia, and shakkarpara, were not only acceptable but also had improved starch and protein digestibility. Nutritious ladoo and puttu can be made using sorghum flour by adding 30% soy flour. It has been demonstrated that sorghum-based chapatis are organoleptically acceptable. The manufacture of bhakri with sorghum flour has proven effective, and extensive research is being conducted in India to extend the durability of sorghum bhakri that is deemed acceptable (Unhale et al. 2012).

Since baked foods made from sorghum millet flour lack gluten and create dough with an uneven consistency, they are not suitable as raw materials for the baking industry. Baked foods like cookies, muffins, yeast-leavened bread, cakes, and biscuits have been made with blends of sorghum and wheat flour (Torres et al. 1993; Suhendro et al. 1998; Rooney et al. 1980; Morad et al. 1984). Typically, wheat flour is replaced with 5–50% by sorghum flour. The composite flour of sorghum used to make bread with 30% boiling malt flour had better taste and texture, as well as improved water retention, staling resistance, and crumb structure compared to bread made with the same percentage of unmalted sorghum flour composited (Hugo et al. 2000). It has been confirmed that adding 15% of sorghum flour results in bread that is acceptable without changing the crumb texture, color, or volume of the loaf (Iwuoha et al. 1997). Bread and cakes that were deemed acceptable were made with a mixture of 30% detoxified cassava starch and 70% sorghum flour (Olatunji et al. 1990). Organoleptically acceptable biscuits were produced using a 1:1 ratio composite flour containing sorghum and wheat (Orewa and Iloh 1989; Priyolkar 1989).

9.9.4.2 Extruded Products

The production of ready-to-eat (RTE) foods is increasingly prepared by extrusion. Cereals are roasted at a high temperature during extrusion procedures. Their digestibility is increased by denatured protein and gelatinized starch. It is possible that anti-nutritional factors are inactive. Research has shown that high-pressure, high-temperature expansion and compression can modify the molecular mass of tannin sorghum, commonly referred to as brown sorghum (Awika and Rooney 2004). During this process, the degradation of proanthocyanidin occurs, and a higher molecular weight polymer disintegrates into its monomers. These products boost antioxidant capacity and are more easily absorbed. The elimination of most bacteria extends the product's shelf life. Additives serve as an effective means to enhance product strength. Friction extruders represent a cost-effective technique for the direct expansion of sorghums, whether in whole or milled form. It found that the extrudates from sorghum compare closely to those from corn and rice, depending on their moisture content, particle size, degree of decortications, etc. (Awika and Rooney 2004). Sorghum is used in twin and single-screw extruders to create light-colored, bland-tasting, inexpensive extrudates that are seasoned and flavored similar to rice. Sorghum, for example, is sometimes less expensive than rice and can expand correctly without additions in cases where rice cannot. Rice, on the other hand, occasionally requires proper expansion by adding potato starch or other expensive ingredients. Flour and sorghum millet grits can be utilized to make RTE products.

Nutritiously balanced supplemental foods can be produced by extruding sorghum millet in combination with soy and other protein-rich components (peanut or legume cake) (Malleshi et al. 1996). Millet flour can be used to make noodles, macaroni, and pasta-like extruded items (Desikachar 1975). Although sorghum millet has a distinct flavor and aroma that offers dietary fiber, it is extremely uncommon to use it to make soft-cooked goods like vermicelli noodles. However, a method for making noodles has been given by researchers of the Central Food Technological Research Institute in Mysore, India (Sowbhaghya and Ali 2001a). After boiling in water, the noodles maintained their strand texture and firmness without crumbling, and less than 6% of their solid mass was lost (Sowbhaghya and Ali 2001b). Both the savory and sweet recipes worked well with the sorghum millet noodles. Research was conducted on sorghum flaking, leading to the standardization of certain process parameters such as wet or dry heat treatment conditions, temperature, and soaking duration. After that grain has been covered with water to its equilibrium moisture content, it is thoroughly gelatinized by steaming or roasting. The moisture contents were then dried to 18%, conditioned, and decorticated before being flaked with two heavy-duty rollers. After boiling and seasoning, the flakes can also be used to make traditional snacks "uppitu." According to the CFTRI procedure, the ideal conditions for grain popping involve bringing sorghum millet to a moisture equilibrium of approximately 16% and then treating the grains in a popcorn maker developed by the institute, at a high temperature of around 230 °C for a short duration. The popped millet is utilized in supplemental diets for mothers and children due to its nutritional value (Bhaskaran et al. 1999). Sorghum millet can be utilized to make a variety of nutritious dishes and food ingredients. Sorghum contains a slightly higher proportion of insoluble dietary fiber. Sorghum foods are beneficial to people with diabetes. The release of sugar from sorghum food is very slow, which lowers the risk of diabetes. Specifically, in comparison to products derived from dehulled sorghum and wheat, sorghum (whole grain) snacks like upama, dhokla, and chapatti are prepared by both. Sorghum dishes exhibit reduced peak spikes, decreased glucose levels, and a smaller area under the curve in diabetes (Lakshmi and Vimala 1996). For diabetics, a variety of cookies and biscuits were made with both malted and unmalted sorghum flour. The fiber content in cookies composed of 40% wheat and 60% malted sorghum flour was elevated. It is widely acknowledged that applying moist heat to sorghum reduces its digestibility. Due to this unique quality, sorghum flakes could be sold as diabetic flakes. In the flaking process, starch has been retrograded, which either increases the amount of dietary fiber or forms resistant starch (Mangala et al. 1999). Tannin-rich sorghums reduce indigestion. Some African societies prefer tannic sorghums because they provide a longer feeling of fullness than other grains. Sorghum tannins may therefore be useful in diabetic-friendly diets (Awika and Rooney 2004). A physiological condition known as gluten intolerance, which results in protein allergies (specifically, gliadin allergy), affects around 0.5 million people in the United States alone (Dahlberg et al. 2003). Since sorghum millet contains no gluten, it has a good chance of being promoted as a dietary remedy for this condition.

Sorghum is a source of nutraceuticals that lower cholesterol and antioxidant phenolics (Taylor et al. 2006). Given this, sorghum flakes with controlled fiber content may be a perfect snack for those who are overweight or watch their calorie intake (Hadimani and Malleshi 1993). It has been discovered that brown sorghum has an oxygen radical absorbance capacity that is more than three times greater than that of blueberries (Zheng and Wang 2003). The pericarp and testa of these sorghums contain the majority of the flavonoids and tannins, which can be isolated and subjected to abrasive milling to be used as phytonutrients in food. Sorghum bran tannin was used to prepare bread that has a natural dark color, perfect flavor, higher phenol content, antioxidant activity, and nutritional fiber (Tanwar et al. 2023). In the United States, barley flour, flax seed, and tannin-rich sorghum bran have all been included in health bread mixes (Rudiger 2003). Research indicates that the use of wheat flour with 15% brown sorghum bran produces bread loaves characterized by a distinct dark brown hue and enhanced levels of phenols and dietary fiber. This suggests that sorghum bran can improve the color and nutritional quality of bread products (Xu et al. 2021).

9.9.5 Non-Food Products of Sorghum Millet

9.9.5.1 Bioethanol and Films

Bioethanol manufacturing is one of the bio-industrial products in which sorghum can be used for the production of bioethanol. Thus, sorghum grain would be suitable for fermentation processes that employ maize to produce bioethanol. Its application would be especially advantageous in nations where maize growth is stunted by insufficient rainfall. Reports show that in 2008, 248 million bushels of grain, including 128 million bushels of sorghum, were used to produce 644 million gallons of ethanol (Voegele 2009). Comparatively less research has been reported on the process of converting sorghum grain into bioethanol. Several reports have been published on the utilization of sweet-stemmed sorghum for the production of bioethanol (House et al. 2002; Schaffert 1992). A method (simultaneous saccharification and fermentation) was investigated (Suresh and Rao 1999; Suresh et al. 1999) to produce ethanol from rice grains, damaged sorghum, and premium sorghum. Their sample of damaged grains consisted of half sound grains and half damaged grains. The damaged portion contained unclean, discolored, insect-attacked, cracked, and shattered kernels. The superior sorghum flour came from a nearby source. They discovered that the damaged grain sample produced 3.5% (v/v) ethanol when 25% (w/v) of the substrate was used. On the other hand, the premium sorghum flour produced 5.0% (v/v) of ethanol. Even though yields were lower, the damaged grain sample was reportedly ten times less expensive than high-quality grain, suggesting that it would be a cost-effective method of producing ethanol. The authors also emphasized how using raw starch would result in energy savings.

The nisin-containing films are made from sorghum flour and starch (Rojas et al. 2002). The growth of *Lactobacillus debrueckii*, a Gram-positive bacterium, was hindered by these "active" films. Kafirin, the prolamin storage protein of sorghum, exhibits hydrophobic properties in all prolamins (Belton et al. 2006), and its digestibility decreases with heat-moisture treatment (Duodu et al. 2003). It is excellent for bioplastic production. Researchers discovered that kafirin, a by-product of wet milling, isolated from sorghum gluten, may be used to manufacture plastic films (Buffo et al. 1997). The tensile and water vapor barrier qualities of kafirin films plasticized with glycerol and polyethylene glycol (PEG) 400 were comparable to those of films prepared from commercial maize zein plasticized in the same manner.

Films made with kafirin from red pericarp sorghum were darker than those made with kafirin from white pericarp sorghum, while films made with kafirin from bran were darker than those made with flour. Phenolics, which are abundant in the bran and co-extracted with the kafirin, were blamed for the films' color. The kafirin films were found to have greater water vapor permeability, lower strain, and higher tensile strength when plasticized with a combination of glycerol, PEG 400, and lactic acid, as compared to films made from commercial zein plasticized in the same manner (Da Silva and Taylor 2005).

9.9.5.2 Malt and Beer

Specialized traits found in GM cultivars are used to manufacture commercial malt. GM and GH cultivars are used to make industrial malt. GH cultivars' condensed tannins are neutralized before malting starts. Sorghum beer is manufactured industrially using malt (Taylor and Duodu 2019).

Beer preparation is a laborious and time-consuming three-day procedure. Malt, meal, and yeast are ingredients used to make beer (Lyumugabe et al. 2012).

9.9.5.3 Livestock Feed and Other Animal Products

The primary market for surplus sorghum is livestock feed since it effectively competes with other grain products on both quality and price. Sorghum plays a significant role in poultry feed, and significant advancements have been achieved in the production of food for dogs, pigeons, and ostrich feed (Kanthaswamy et al. 2025).

9.10 Conclusion

The high nutritional content, resilience to biotic and abiotic challenges, and ability to thrive in marginal areas with low fertility make sorghum and millets crucial crops for food security in semi-arid and arid countries. These crops have a variety of uses, including food, brewing, feed, fodder, forage, biofuel, and building materials. Sorghum starch is used as gelling and thickening agent in noodles, sauces and gravies etc. With some modifications Sorghum starch can also be utilized for commercial purposes as it increases the durability of food products. Grain sorghum is mostly used as food in Africa and Asia, but it's also fed to cattle in the United States and Australia. It is currently being used as chicken feed in Africa and India. Additionally, gluten-free sorghum millets can be significant dietary sources for millions of people worldwide who suffer from gluten intolerance (celiac disease), especially those with diabetes. This applies to both developed and developing nations. Sorghum's high biomass yield, photoperiod sensitivity, and sweet stalks make it a promising crop for bioenergy. Millets work effectively in a variety of cropping systems due to their vast range of crop cycles, some of which are even quite short.

Sorghum millet is produced under subsistence farming, using few resources. However, waxy and low waxy sorghum often react favorably to fertilizer inputs and can offer intriguing advantages in terms of economic production. Sorghum millet has a high potential yield, but its productivity in farmers' fields is low, primarily because they are being pushed into more marginal farmlands. In Asia and Africa's subsistence cropping systems, sorghum millet will remain essential for ensuring food security and increasing the financial returns on germplasm. Wider applications for millets and sorghum are highly promising. In terms of sorghum malt brewing, it appears that genetic modification of the grain is necessary to lower the temperature at which starch gelatinization occurs and raise the activity of malt beta-amylase without the use of external enzymes. Technically, sorghum can be used in bio-industrial settings. Commercialization, however, most likely requires cost reductions in the manufacture of bioethanol, including the use of co-products such as bioplastics and nutraceuticals, as well as proactive government initiatives to support a more "green economy."

References

Abah CR, Ishiwu CN, Obiegbuna JE, Oladejo AA (2020) Sorghum grains: nutritional composition, functional properties and its food applications. Eur J Nutr Food Saf 12(5):101–111. https://doi.org/10.9734/ejnfs/2020/v12i530232

Abdalla AA, El Tinay AH, Mohamed BE, Abdalla AH (1998) Proximate composition, starch, phytate and mineral contents of 10 pearl millet genotypes. Food Chem 63(2):243–246

Adebo OA (2020) African sorghum-based fermented foods: past, current and future prospects. Forum Nutr 12(4):1111. https://doi.org/10.3390/nu12041111

Adebo OA, Njobeh PB, Kayitesi E (2018) Fermentation by *Lactobacillus fermentum* strains (singly and in combination) enhances the properties of ting from two whole grain sorghum types. J Cereal Sci 82:49–56. https://doi.org/10.1016/j.jcs.2018.05.008

Adeboye AS, Emmambux NM (2017) Physicochemical, morphological, thermal and pasting properties of marama (*Tylosema esculentum*) storage root starch. Starch-Stärke 69(3–4):1600084. https://doi.org/10.1002/star.201600084

Adeyanju AA, Emmanuel PO, Adetunji AI, Adebo OA (2025) Nutritional, pasting, rheological, and thermal properties of sorghum–okara composite flours and porridges. IJFST 60(1):vvae021. https://doi.org/10.1093/ijfood/vvae021

Afify AE, El-Beltagi HS, Abd El-Salam SM, Omran AA (2012a) Protein solubility, digestibility and fractionation after germination of sorghum varieties. Plos one 7(2):e31154

Afify AE, El-Beltagi HS, Abd El-Salam SM, Omran AA (2012b) Oil and fatty acid contents of white sorghum varieties under soaking, cooking, germination and fermentation processing for improving cereal quality. Notulae Botanicae Horti Agrobotanici Cluj-Napoca 40(1):86–92

Afify AE, El beltagi HS, Abd El-Salam SM, Omran AA (2012c) Effect of soaking, cooking, germination and fermentation processing on proximate analysis and mineral content of three white sorghum varieties (Sorghum bicolor L. Moench). Notulae Botanicae Horti Agrobotanici Cluj-Napoca 40(2):92–8

Ali TM, Hasnain A (2014) Morphological, physicochemical, and pasting properties of modified white sorghum (*Sorghum bicolor*) starch. Int J Food Prop 17(3):523–535. https://doi.org/10.1080/10942912.2012.654558

Anggreini RA, Choiriyah NA, Athennia A (2021) Modification of *Sorghum bicolor* (L) moench starch: review of HMT (heat moisture treatment), autoclaving cooling, and annealing methods. Int J Adv Trop Food 3(2):57–66. https://doi.org/10.26877/ijatf.v3i2.9927

Angrau (Acharya N.G. Ranga Agricultural University) (2022) Sorghum outlook report—January to December 2022. https://angrau.ac.in/downloads/AMIC/OutlookReports/2022/2-Sorghum%20outlook%20report%20-%20January%20to%20Decemeber%202022.pdf

Apea-Bah FB, Li X, Beta T (2021) Phenolic composition and antioxidant properties of cooked rice dyed with sorghum-leaf bio colorants. Foods 10(9):2058

Arama P, Ochola SO, Kange AM et al (2014) Pre-and post harvest factors affecting sorghum production (*Sorghum bicolor* L. Moench) among smallholder farming communities. IJAAR 5(4):40–47

Aruna C, Visarada KBRS (2019) Sorghum grain in food and brewing industry. In: Aruna C, Visarada KBRS, Bhat BV, Tonapi VA (eds) Breeding sorghum for diverse end uses, 1st edn. Woodhead Publishing. https://doi.org/10.1016/B978-0-08-101879-8.00013-9

AU PM, Fields ML (1981) Nutritive quality of fermented sorghum. Journal of Food Science 46(2):652–4

Awika JM, Rooney LW (2004) Sorghum phytochemicals and their potential impact on human health. Phytochemistry 65(9):1199–1221. https://doi.org/10.1016/j.phytochem.2004.04.001

Ballagh A, Kezar S, Harris SC, Lofton J (2021) Grain sorghum desiccation: impacts on plant and grain dry-down. Agrosyst Geosci Environ 4(4):e20222. https://doi.org/10.1002/agg2.20222

Bangar S, Sunooj K, Navaf M et al (2024) Recent advancements in cross-linked starches for food applications—a review. Int J Food Prop 27(1):411–430. https://doi.org/10.1080/10942912.2024.2318427

Bao J, Bergman CJ (2018) Rice flour and starch functionality. In: Sjöö M (ed) Starch in food: structure, function and applications, 2nd edn. Woodhead Publishing/Elsevier. https://doi.org/10.1016/B978-0-08-100868-3.00010-X

Baskaran V, Mahadevamma, Malleshi NG et al (1999) Acceptability of supplementary foods based on popped cereals and legumes suitable for rural mothers and children. Plant Foods Hum Nutr 53:237–247. https://doi.org/10.1023/A:1008016828185

Belton PS, Delgadillo I, Halford NG, Shewry PR (2006) Kafirin structure and functionality. J Cereal Sci 44(3):272–286. https://doi.org/10.1016/j.jcs.2006.05.004

Beta TR, Rooney LW, Waniska RD (1995) Malting characteristics of sorghum cultivars. Cereal Chem 72(6):533–538

Beta T, Ndolo VU (2025) Postharvest technologies for sorghum and millets. In:Taylor JR, Bean RS and Duodu KG (Eds) Sorghum and Millets 3rd Edn. Elsevier. https://doi.org/10.1016/B978-0-443-23954-0.00005-7

Boudries N, Belhaneche N, Nadjemi B et al (2009) Physicochemical and functional properties of starches from sorghum cultivated in the Sahara of Algeria. Carbohydr Polym 78(3):475–480. https://doi.org/10.1016/j.carbpol.2009.05.010

Bouis H (1996) Enrichment of food staples through plant breeding: a new strategy for fighting micronutrient malnutrition. Nutr Rev 54(5):131–137. https://doi.org/10.1111/j.1753-4887.1996.tb03915.x

Buffo RA, Weller CL, Gennadios A (1997) Films from laboratory-extracted sorghum kafirin. Cereal Chem 74(4):473–475. https://doi.org/10.1094/CCHEM.1997.74.4.473

Caransa A, Bakker WG (1987) Modern process for the production of sorghum starch. Starch-Stärke 39(11):381–385. https://doi.org/10.1002/star.19870391104

Chaudhary S (1993) Preparation of nutritional evaluation of some traditional foods from pearl millet. Dissertation, CCS Haryana Agricultural University

Chemelewski R, McKinley BA, Finlayson S, Mullet JE (2023) Epicuticular wax accumulation and regulation of wax pathway gene expression during bioenergy sorghum stem development. Front Plant Sci 14:1227859. https://doi.org/10.3389/fpls.2023.1227859

Chen Z, Yao Y, Pu H et al (2023) The effects of high-pressure treatment on starch pasting, retrogradation, thermal, crystalline, rheological and digestive PROPERTIES—A review. IJFST 58(5):2189–2198. https://doi.org/10.1111/ijfs.16342

Chiluwal A, Bheemanahalli R, Kanaganahalli V et al (2020) Deterioration of ovary plays a key role in heat stress-induced spikelet sterility in sorghum. Plant Cell Environ 43:448–462

Cruickshank A (2016) Encyclopedia of Food Grains Sorghum Grain Its Production and Uses: Overview Elsevier 153–158

da Silva LS, Taylor JR (2005) Physical, mechanical, and barrier properties of kafirin films from red and white sorghum milling fractions. Cereal Chem 82(1):9–14. https://doi.org/10.1094/CC-82-0009

Dahlberg JA, Wilson JP, Snyder T (2003) Sorghum and pearl millet: health foods and industrial products in developed countries. In: Alternative uses of sorghum and pearl millet in Asia, Proceedings of the Expert Meeting ICRISAT, India

Desikachar HS (1975) Processing of maize, sorghum and millets for food uses. JSIR 34:231–237. http://ir.cftri.res.in/id/eprint/5046

Dharmaputra OS (2013) Postharvest quality improvement of sorghum (*Sorghum bicolor* (L.) Moench) grains. Biotropia 19(2) https://journal.biotrop.org/index.php/biotropia/article/view/255

Dimri S, Aditi Bist Y, Singh S (2023) Oxidation of starch. In: Sharanagat VS, Saxena DC, Kumar K, Kumar Y (eds) Starch: advances in modifications, technologies and applications. Springer, Cham. https://doi.org/10.1007/978-3-031-35843-2_3

Duodu KG, Taylor JR, Belton PS, Hamaker BR (2003) Factors affecting sorghum protein digestibility. J Cereal Sci 38(2):117–131. https://doi.org/10.1016/S0733-5210(03)00016-X

Eckhoff SR, Watson SA (2009) Corn and sorghum starches: production. In: Whistler RL, BeMiller JN, Paschall EF (eds) Starch, 3rd edn. Academic, Cambridge, MA. https://doi.org/10.1016/B978-0-12-746275-2.00009-4

Ejeta G, Hassen MM, Mertz ET (1987) In vitro digestibility and amino acid composition of pearl millet (*Pennisetum typhoides*) and other cereals. Proc Natl Acad Sci USA 84(17):6016–6019. https://doi.org/10.1073/pnas.84.17.6016

Fan W, Qian MC (2006) Characterization of aroma compounds of Chinese "Wuliangye" and "Jiannanchun" liquors by aroma extract dilution analysis. J Agric Food Chem 54(7):2695–2704. https://doi.org/10.1021/jf052635t

Frederick EJ (2007) Effect of sorghum flour composition and particle size on quality of gluten-free bread BS. Kansas State University

Gaddameedi A, Sheraz S, Kumar A (2022) The location of iron and zinc in grain of conventional and biofortified lines of sorghum. J Cereal Sci 107:103531

García-Gurrola A, Rincón S, Escobar-Puentes AA et al (2019) Microencapsulation of red sorghum phenolic compounds with esterified sorghum starch as encapsulant materials by spray drying. Food Technol Biotechnol 57(3):341. https://doi.org/10.17113/ftb.57.03.19.6146

Giuberti G, Marti A, Gallo A et al (2019) Resistant starch from isolated white sorghum starch: functional and physicochemical properties and resistant starch retention after cooking. A comparative study. Starch-Stärke 71(7–8):1800194. https://doi.org/10.1002/star.201800194

Hadimani NA, Malleshi NG (1993) Studies on milling, physico-chemical properties, nutrient composition and dietary fibre content of millets. J Food Sci Technol 30(1):17–20

Hadimani NA, Ali SZ, Malleshi NG (1995) Physico-chemical composition and processing characteristics of pearl millet varieties. J Food Sci Technol 32(3):193–198

Hamad SAA, Mustafa AI, Magboul BI et al (2019) Nutritional quality of raw and cooked flours of a high β-glucan sorghum inbred line. J Cereal Sci 90:102857. https://doi.org/10.1016/j.jcs.2019.102857

Han Z, Shi R, Sun DW (2020) Effects of novel physical processing techniques on the multi-structures of starch. Trends Food Sci Technol 97:126–135. https://doi.org/10.1016/j.tifs.2020.01.006

House LR, Murty DS, Gomez M et al (2002) The development of agricultural industries in several African and Asian countries. In: Smith CW, Frederiksen RA (eds) Sorghum: origin, history, technology, and production. Wiley, New York. https://doi.org/10.1016/j.foodchem.2013.08.129

Htet MNS, Feng B, Wang H et al (2022) Comparative assessment of nutritional and functional properties of different sorghum genotypes for ensuring nutritional security in dryland agro-ecosystem. Front Nutr 9:1048789. https://doi.org/10.3389/fnut.2022.1048789

Hugo LF, Rooney LW, Taylor JR (2000) Malted sorghum as a functional ingredient in composite bread. Cereal Chem 77(4):428–432. https://doi.org/10.1094/CCHEM.2000.77.4.428

Hussen A (2021) Impact of grain storage method on sorghum grain quality in Ethiopia: a review. J Curr Res Food Sci 2(1):40–45

Iwuoha CI, Anyadike AC, Eke OS (1997) The effect of flour-blending on the physico-chemical and sensory qualities of bread. J Food Sci Technol 34(4):311–315

Kang X, Gao W, Cui B, Abd El-Aty AM (2023) Structure and genetic regulation of starch formation in sorghum (*Sorghum bicolor* (L.) Moench) endosperm: a review. Int J Biol Macromol 239:124315. https://doi.org/10.1016/j.ijbiomac.2023.124315

Kannababu N, Dinni S, Talwar HS et al (2020) Sorghum Hybrid Seed Production and Quality Management: Important Considerations. In: Tonapi VA, Talwar HS, Are AK, Bhat BV, Reddy CR, Dalton TJ. (eds) Sorghum in the 21st Century: Food – Fodder – Feed – Fuel for a Rapidly Changing World. Springer, Singapore. https://doi.org/10.1007/978-981-15-8249-3_12

Kanthaswamy H, Das A, Venkatachalam S et al (2025) Strengthening livestock systems through sorghum-based fodder value chain innovations in southern India. Agric Syst 225:104273. https://doi.org/10.1016/j.agsy.2025.104273

Kazungu FK, Muindi EM, Mulinge JM (2023) Overview of sorghum (*Sorghum bicolor* L.), its economic importance, ecological requirements and production constraints in Kenya. Int J Plant Soil Sci 35(1):62–71. https://doi.org/10.9734/IJPSS/2023/v35i12744

Khoddami A, Messina V, Vadabalija Venkata K et al (2023) Sorghum in foods: functionality and potential in innovative products. Crit Rev Food Sci Nutr 63(9):1170–1186. https://doi.org/10.1080/10408398.2021.1960793

Krishnan V (2024) Editorial: millets: superfood of the century—challenges and opportunities. Front Nutr 11:1462567. https://doi.org/10.3389/fnut.2024.1462567

Kumar R (2021) Cultivation of sorghum and bajra. Dissertation, Monad University

Lakshmi KB, Vimala V (1996) Hypoglycemic effect of selected sorghum recipes. Nutr Res 16(10):1651–1658. https://doi.org/10.1016/0271-5317(96)00184-4

Leach HW, McCowen LD, Schoch TJ (1959) Structure of the starch granule. I. Swelling and solubility patterns ofvarious starches. Cereal Chem 36:534–44

Lemlioglu-Austin D, Turner ND, McDonough CM, Rooney LW (2012) Effects of sorghum [*Sorghum bicolor* (L.) Moench] crude extracts on starch digestibility, estimated glycemic index (EGI), and resistant starch (RS) contents of porridges. Molecules 17:11124–11138. https://doi.org/10.3390/molecules170911124

Li Z, Zhao X, Zhang X, Liu H (2021) Bioactive compounds and biological activities of sorghum grains. Foods 10(11):2868

Li M, Daygon VD, Solah V, Dhital S (2023) Starch granule size: does it matter? Crit Rev Food Sci Nutr 63(19):3683–3703. https://doi.org/10.1080/10408398.2021.1992607

Liang W, Lin Q, Zeng J et al (2024) Understanding the improvement of sorghum starch acid hydrolysis modification by E-beam irradiation: a supramolecular structure perspective. Food Chem 437:137820. https://doi.org/10.1016/j.foodchem.2009.05.086

Lyumugabe F, Gros J, Nzungize J et al (2012) Characteristics of African traditional beers brewed with sorghum malt: a review. Base 16(4):509–530

Magaletta RL, DiCataldo SN, Dong L et al (2010) In vitro method for predicting glycemic index of foods using simulated digestion and an artificial neural network. Cereal Chem 87:363–369. https://doi.org/10.1094/CCHEM-87-4-0363

Malleshi NG, Klopfenstein CF (1998) Nutrient composition and amino acid contents of malted sorghum, pearl millet and finger millet and their milling fractions. J Food Sci Technol 35(3):247–249

Malleshi NG, Hadimani NA, Chinnaswamy R, Klopfenstein CF (1996) Physical and nutritional qualities of extruded weaning foods containing sorghum, pearl millet, or finger millet blended with mung beans and nonfat dried milk. Plant Foods Hum Nutr 49:181–189. https://doi.org/10.1007/BF01093214

Mangala SL, Malleshi NG, Tharanathan RN, Mahadevamma (1999) Resistant starch from differently processed rice and ragi (finger millet). Eur Food Res Technol 209:32–37. https://doi.org/10.1007/s002170050452

Maniglia BC, Castanha N, Le-Bail P et al (2021) Starch modification through environmentally friendly alternatives: a review. Crit Rev Food Sci Nutr 61(15):2482–2505. https://doi.org/10.1080/10408398.2020.1778633

Mason JB, Garcia M (1993) Micronutrient deficiency–the global situation. SCN News 9:11–16

Mehboob S, Ali TM, Sheikh M, Hasnain A (2020) Effects of cross linking and/or acetylation on sorghum starch and film characteristics. Int J Biol Macromol 155:786–794. https://doi.org/10.1016/j.ijbiomac.2020.03.144

Mezo-Villanueva M, Serna-Saldívar SO (2004) Effect of protease addition on starch recovery from steeped sorghum and maize. Starch-Stärke 56(8):371–378. https://doi.org/10.1002/star.200300264

Morad MM, Doherty CA, Rooney LW (1984) Utilization of dried distillers grain from sorghum in baked food systems. Cereal Chem 61(5):409–414

Muimba-Kankolongo A (2018) Food crop production by smallholder farmers in Southern Africa. Elsevier. https://doi.org/10.1016/B978-0-12-814383-4.00008-6

Ogundiran OO, Ogundiran AA, Afolabi TA (2022) Physicochemical properties of cross-linked and hydroxypropyl white sorghum (*Sorghum bicolor*) starch. NJPAS 35:4375–4384. https://doi.org/10.48198/NJPAS/22.A15

Olatunji O, Adesina AA, Koleoso OA (1990) Use of sorghum as composite flour in baking. In: Symposium on the current status and potential of industrial uses of sorghum in Nigeria, Kano, Nigeria, 4–6 Dec 1989

Olayinka OO, Adebowale KO, Olu-Owolabi BI (2008) Effect of heat-moisture treatment on physicochemical properties of white sorghum starch. Food Hydrocoll 22(2):225–230. https://doi.org/10.1016/j.foodhyd.2006.11.004

Olayinka FS, Olayinka OO, Olu-Owolabi BI, Adebowale KO (2015) Effect of chemical modifications on thermal, rheological and morphological properties of yellow sorghum starch. J Food Sci Technol 52(12):8364–8370. https://doi.org/10.1007/s13197-015-1891-3

Onyango C, Mutungi C, Unbehend G, Lindhauer MG (2011) Rheological and textural properties of sorghum-based formulations modified with variable amounts of native or pregelatinised cassava starch. LWT 44(3):687–693. https://doi.org/10.1016/j.lwt.2010.08.019

Orewa GO, Iloh A (1989) Current status of composite flour technology and prospects with particular reference to the production of biscuits in Nigeria. In: Symposium on the current status and potential of industrial uses of sorghum in Nigeria, Kano, Nigeria, 4–6 Dec 1989

Oseguera-Toledo ME, Contreras-Jiménez B, Hernández-Becerra E, Rodriguez-Garcia ME (2020) Physicochemical changes of starch during malting process of sorghum grain. J Cereal Sci 95:103069. https://doi.org/10.1016/j.jcs.2020.103069

Pawar VD, Parlikar GS (1990) Reducing the polyphenols and phytate and improving the protein quality of pearl millet by dehulling and soaking. J Food Sci Technol 27(3):140–143

Pérez-Carrillo E, Serna-Saldívar SO (2007) Effect of protease treatment before hydrolysis with α-amylase on the rate of starch and protein hydrolysis of maize, whole sorghum, and decorticated sorghum. Cereal Chem 84(6):607–613. https://doi.org/10.1094/CCHEM-84-6-0607

Poonam (2002) Effect of acid and heat treatment on nutrient composition and shelf life of pearl millet (*Pennisetum glaucum*) flour. Dissertation, Chaudhary Charan Singh Haryana Agricultural University

Prasad PV, Staggenborg SA (2009) Growth and production of sorghum and millets. In: Verheye WH (ed) Soils, plant growth and crop production EOLSS, developed under the auspices of the UNESCO. Eolss Publishers, Oxford

Priyolkar VS (1989) Use of sorghum flour in biscuit and wafer production: the Nasco [Foods] experience. In: Symposium on the current status and potential of industrial uses of sorghum in Nigeria, Kano, Nigeria, 4–6 Dec 1989

Pushpamma S, Parrish DB, Deyoe CW (1972) Improving protein quality of millet, sorghum, and maize diets by supplementation. Nutr Rep Int 5(2):93–100

Rai KN, Gowda CL, Reddy BV, Sehgal S (2008) Adaptation and potential uses of sorghum and pearl millet in alternative and health foods. Comprehensive Reviews in Food Science and Food Safety 7(4):320–96

Rao SV, Swaminathan M, Parpia HA (1964) Mutual supplementation of dietary proteins for meeting protein needs and overcoming protein shortage in developing countries. Indian J Nutr Diet 1:128–138

Rekha (1997) Efficacy of processing techniques in the utilization of pearl millet for value-added products. Dissertation, Chaudhary Charan Singh Haryana Agricultural University

Rhodes DH, Hoffmann Jr L, Rooney WL (2014) Genome-wide association study of grain polyphenol concentrations in global sorghum [Sorghum bicolor (L.) Moench] germplasm. Journal of agricultural and food chemistry 62(45):10916–27

Rodríguez-España M, Figueroa-Hernández CY, Figueroa-Cárdenas JDD et al (2022) Effects of germination and lactic acid fermentation on nutritional and rheological properties of sorghum: a graphical review. Curr Res Food Sci 5:807–812. https://doi.org/10.1016/j.crfs.2022.04.014

Rojas C, Schause A, Brenes CH (2002) Sorghum edible films active against *Lactobacillus delbrueckii* ATCC 11842. In: Institute of Food Technologists Annual Meeting, Anaheim, CA, 100B-39. https://repository.up.ac.za/server/api/core/bitstreams/e0b395c5-919d-4ff8-b1bf-0fc08fbdea79/content

Rooney LW, Khan MN, Earp CF (1980) The technology of sorghum products. In: Inglett GE, Munck L (eds) Cereals for food and beverages. Academic/Elsevier

Rudiger CR (2003) The formulation of a nutraceutical bread mix using sorghum, barley, and flaxseed. Dissertation, Texas A&M University

Saravanabavan NS, Shivanna MM, Bhattacharya S (2013) Effect of popping on sorghum starch digestibility and predicted glycemic index. Journal of food science and technology 50(2):387–92

Schaffert RE (1992) Sweet sorghum substrate for industrial alcohol. In: Gomez MI, House LR, Rooney LW, Dendy DAV (eds) Utilization of sorghum and millets. Proceedings of a workshop held at Bulawayo, Zimbabwe, 8–12 February, 1992

Schafranski K, Ito VC, Lacerda LG (2021) Impacts and potential applications: a review of the modification of starches by heat-moisture treatment (HMT). Food Hydrocoll 117:106690. https://doi.org/10.1016/j.foodhyd.2021.106690

Semwal J, Meera MS (2023) Modification of sorghum starch as a function of pullulanase hydrolysis and infrared treatment. Food Chem 416:135815. https://doi.org/10.1016/j.foodchem.2023.135815

Serna-Saldivar SO (2010) Cereal grains: properties, processing, and nutritional attributes. CRC Press. Taylor & Francis Group, Florida. https://doi.org/10.1201/9781439882092

Sharma S, Thakur K, Sharma R, Bobade H (2023) Molecular morphology & interactions, functional properties, rheology and in vitro digestibility of ultrasonically modified pearl millet and sorghum starches. Int J Biol Macromol 253:127476. https://doi.org/10.1016/j.ijbiomac.2023.127476

Sharanagat VS, Suhag R, Anand P et al (2019) Physico-functional, thermo-pasting and antioxidant properties of microwave roasted sorghum [Sorghum bicolor (L.) Moench]. Journal of Cereal Science 85:111–9

Shi YC, Seib PA (1992) The structure of four waxy starches related to gelatinization and retrogradation. Carbohydr Res 227:131–145. https://doi.org/10.1016/0008-6215(92)85066-9

Singh U (1984) The inhibition of digestive enzymes by polyphenols of chickpea (*Cicer arietinum* L.) and pigeonpea (*Cajanus cajan* (L.) Millsp.). Nutr Rep Int 29(3):745–753. http://pascal--francis.inist.fr/vibad/index.php?action=getRecordDetail&idt=9622660

Singh H, Sodhi NS, Singh N (2010) Characterisation of starches separated from sorghum cultivars grown in India. Food Chem 119(1):95–100. https://doi.org/10.1016/j.foodchem.2009.05.086

Sira EE, Amaiz ML (2004) A laboratory scale method for isolation of starch from pigmented sorghum. Journal of food engineering 64(4):515–9

Smolensky D, Rhodes D, McVey DS et al (2018) High-polyphenol sorghum bran extract inhibits cancer cell growth through ROS induction, cell cycle arrest, and apoptosis. J Med Food 21(10):990–998. https://doi.org/10.1089/jmf.2018.0008

Sobetski H (2015) Sorghum pollination. ECHO development notes no. 126. http://edn.link/9w43fa

Sowbhagya CM, Ali SZ (2001a) A process for manufacturing high fibre, low fat vermicelli noodles from millets. Indian Patent 1128, 2001

Sowbhagya CM, Ali SZ (2001b) Vermicelli noodles and their quality assessment. J Food Sci Technol 38(5):423–432

Suhendro EL, McDonough CM, Rooney LW et al (1998) Effects of processing conditions and sorghum cultivar on alkaline-processed snacks. Cereal Chem 75(2):187–193. https://doi.org/10.1094/CCHEM.1998.75.2.187

Sun Q, Han Z, Wang L, Xiong L (2014) Physicochemical differences between sorghum starch and sorghum flour modified by heat-moisture treatment. Food Chem 145:756–764

Suresh K, Rao LV (1999) Utilization of damaged sorghum and rice grains for ethanol production by simultaneous saccharification and fermentation. Bioresour Technol 68(3):301–304. https://doi.org/10.1016/S0960-8524(98)00135-7

Suresh K, Kiransree N, Venkateswar Rao L (1999) Production of ethanol by raw starch hydrolysis and fermentation of damaged grains of wheat and sorghum. Bioprocess Eng 21:165–168. https://doi.org/10.1007/s004490050657

Tanwar R, Panghal A, Chaudhary G et al (2023) Nutritional, phytochemical and functional potential of sorghum: a review. Food Chem Adv 3:100501. https://doi.org/10.1016/j.focha.2023.100501

Taylor JRN, Duodu KG (2019) Traditional sorghum and millet food and beverage products and their technologies. In: Taylor JRN, Duodu KG (eds) Sorghum and millets, 2nd edn. AACC International Press/Elsevier. https://doi.org/10.1016/B978-0-12-811527-5.00009-5

Taylor JR, Schober TJ, Bean SR (2006) Novel food and non-food uses for sorghum and millets. J Cereal Sci 44(3):252–271. https://doi.org/10.1016/j.jcs.2006.06.009

Taylor JR. (2025) Sorghum and millets: Taxonomy, history, distribution, production, and end-uses. In:Taylor JR, Bean RS and Duodu KG (Eds) Sorghum and Millets 3rd Edn. Elsevier. https://doi.org/10.1016/B978-0-443-23954-0.00009-4

Thakur K, Sharma S, Sharma R et al (2023) Characterization of ozonated pearl millet and sorghum starches: pasting profile, functional properties, in vitro-digestibility, molecular interactions, and morphology. Starch-Stärke 75(11–12):2300020. https://doi.org/10.1002/star.202300020

Thompson LU, Yoon JH (1984) Starch digestibility as affected by polyphenols and phytic acid. J Food Sci 49(4):1228–1229. https://doi.org/10.1111/j.1365-2621.1984.tb10443.x

Torres PI, Ramirez-Wong B, Serna-Saldivar SO, Rooney LW (1993) Effect of sorghum flour addition on the characteristics of wheat flour tortillas. Cereal Chem 70(1):8–13

Udachan IS, Sahoo AK, Hend GM (2012) Extraction and characterization of sorghum (*Sorghum bicolor* L. Moench) starch. Int Food Res J 19(1):315–319

Unhale DS, Sakhale BK, Ranveer RC, Pawar VD (2012) Studies on shelf life extension of sorghum roti. Int Food Res J 19(2):733–736

Voegele E (2009) Ethanol plants utilize of U.S. grain sorghum production. Report posted 2009. https://ethanolproducer.com/articles/ethanol-plants-utilize-29-percent-of-u.s.-grain-sorghum-production-6165

Wu X, Zhao R, Bean SR et al (2007) Factors impacting ethanol production from grain sorghum in the dry-grind process. Cereal Chem 84(2):130–136. https://doi.org/10.1094/CCHEM-84-2-0130

Xie XJ, Seib PA (2002) Laboratory wet-milling of grain sorghum with abbreviated steeping to give two products. Starch-Stärke 54(5):169–178. https://doi.org/10.1002/1521-379X(200205)54:5<169::AID-STAR169>3.0.CO;2-7

Xu J, Wang W, Zhao Y (2021) Phenolic compounds in whole grain sorghum and their health benefits. Foods 10(8):1921–1934. https://doi.org/10.3390/foods10081921

Young R, Gomez MH, McDonough CM et al (1993) Changes in sorghum starch during parboiling. Cereal Chem 70(2):179–183

Zheng W, Wang SY (2003) Oxygen radical absorbing capacity of phenolics in blueberries, cranberries, chokeberries, and lingonberries. J Agric Food Chem 51(2):502–509. https://doi.org/10.1021/jf020728u

Zhu F (2017) Structures, properties, modifications, and uses of oat starch. Food Chem 229:329–340. https://doi.org/10.1016/j.foodchem.2017.02.064

Chapter 10
Nutritional Composition and Ethnobotanical Insights of Browntop Millet (*Urochloa ramosa*): A Traditional Grain for Modern Health

Neeraj Pandey, Amrit Kumar, and Nitin Kamboj

10.1 Introduction

On a global scale, India is recognized as the largest producer and consumer of millets, whereas the United States of America is the leading exporter of millets. Millet farming is mainly indigenous to the African continent; however, overall global production has increased by 36% from 1961 (575 kg/ha) to 2018 (900 kg/ha). Meanwhile, the area of cultivable land dedicated to millets has been gradually declining. Niger holds the second position in terms of both production and consumption, with China, Nigeria, and Mali following closely behind (Meena et al. 2021). Malnutrition among adolescents and childern is major public health problem. The globe is currently facing malnutrition crisis that includes both overnutrition (obesity, underweight conditions, metabolic syndrome, and diseases related to poor lifestyle habits) and undernutrition deficiencies in vitamins, minerals, and proteins) (Jose et al. 2020). This has led to a growing recognition of the need to include millets as part of a healthier, more affordable, and easily accessible diet. Not only are millets naturally high in nutrients, free of gluten, but they are also a very good source of protein, dietary fiber, vital fatty acids, and vitamins (Asrani et al. 2023). Among the first foods humans ate were millets, which are tiny, seeded turfs that grow in four seasons and are commonly grown as grain crops (Abbasi et al. 2018). Millets are often referred to as "the nutri-cereals of the present, and the coarse grain of yesterday".

N. Pandey
Wildlife Institute of India, Dehradun, Uttarakhand, India
e-mail: neerajk847@gmail.com

A. Kumar · N. Kamboj (✉)
Department of Zoology and Environmental Science, Gurukula Kangri (Deemed to University), Haridwar, Uttarakhand, India

© The Author(s), under exclusive license to Springer Nature Switzerland AG 2026

J. Singh, M. Kumar Malik (eds.), *Millets and Millet Starch: Chemical Composition, Functional Modification, and Applications*,
https://doi.org/10.1007/978-3-032-09502-2_10

Millets are considered the grain of the future because they can be grown in harder and sometimes unfavorable conditions, such as arid and semi-arid regions across much of Asia-particularly in southern areas-and Africa. Moreover, millets are resilient to most pests and insects (Paschapur et al. 2021). Millet is the staple food source for millions of people in the arid and steppe climatic regions. Typically, millets are planted in regions where major crops like rice (*Oryza sativa*), maize (*Zea mays*), and wheat (*Triticum aestivum*) cannot sustainably support agricultural output (Ahmed et al. 2017; Alam et al. 2016). The millet crop has numerous beneficial qualities, like quick maturation, and the capability to be stored for a sufficient duration of time without being harmed by any variety of insects (Amadou et al. 2013). Since the era of the Green Revolution, crops such as millet have received very little attention, as the majority of people on the planet eat mostly wheat, rice, and maize (Ayele and Tarekegn 2020). Millets are also known as a "crop" because these crop grains are important for animal feed and food for humans as well. Due to a switch in dietary preferences and status symbols, the crop was formerly disregarded; nevertheless, the nutria crop is nowadays seeing a significant upturn in the world-scale crop production industries (Braicu et al. 2013; Bora et al. 2019; Babele et al. 2022). The main aims of this chapter are to shed light on the nutritional value and historical uses of browntop millet, exploring its integral role in traditional cuisines and its association with cultural practices. By unraveling the culinary traditions surrounding this ancient grain, we can gain insights into the rich tapestry of human dietary history and the sustainable practices that have sustained communities for generations.

10.2 Classification of Millets

Millets are divided into major millets and minor millets according to their grain size. The pseudo millets are so named as they are not a member of the *Poaceae* family, to which "perfect or original" grains belong. However, they are nutritionally similar to the *Poaceae* family grains and are consumed in almost similar ways to other cereals grains. Figure 10.1 depicts the division of millet into different groups.

10.2.1 Browntop Millet

Browntop millet is an annual spikelet grass primarily grown in various regions, including Africa, Australia, Western Asia, China, and the Arab countries (Sravani et al. 2020). During the starting of the 20th century, its introduction to the US was governed from India (Maitra et al. 2023). Browntop millet (*Urochloa ramosa*), also known as Korra in India, represents an ancient grain with a rich history of cultivation. It is also known as signal grass, was initially domesticated in South India from where it spread to other region of the country. In India, it is cultivated in Andhra Pradesh specifically in Ananthpur district and Dharwar, Chitradurga,

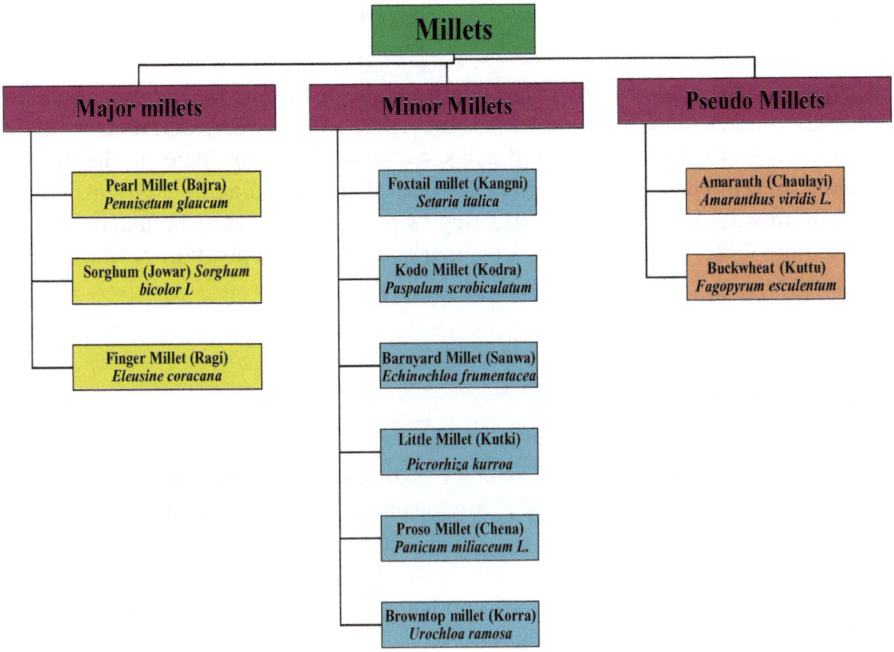

Fig. 10.1 Classification of millets

Tumkuru, Gulbarga, and Chikkaballapura districts. Further, it is also cultivated in the Bundelkhand region of Central India (Maitra et al. 2023; Maitra, 2020). It is a member of the Poaceae family and the Paniceae tribe, grown for both grain and forage. Morphologically, this plant attains a height ranging from approximately 0.3 to 1 m, characterized by smooth stems, lanceolate leaves measuring between 2 and 18 cm in length and 6 to 18 mm in width, along with a spreading open inflorescence that produces ellipsoid seeds. Further, roots of Browntop millet is fibrous and can reach up to 60 cm inside soil. It bears indeterminate white flowers. Panicles are non-bristly and looser. The inflorescences is 1–8 cm long, originated from a central axis, open and spreading. Panicles are looser and nonbristlyIt undergoes maturation swiftly, typically within a span of 60 to 90 days. Seeds are similar to foxtail millet, with ellipsoid and tan colored (Kingwell-Banham and Fuller 2014; Sheahan 2014). In the face of evolving dietary patterns and the rising prevalence of diseases related to lifestyle, there is a growing interest in exploring traditional and lesser-known food sources to enhance nutritional diversity and overall well-being. Browntop millet (*Urochloa ramosa*), a prehistoric cereal grain that had been cultivated for centuries, is one such underappreciated yet highly promising crop. With its rich cultural history and potential health benefits, browntop millet deserves closer attention as an important component of the human diet (Laryea et al. 2024). Indigenous communities have relied on browntop millet for its versatility in culinary applications, using it as a staple in a huge array of traditional food dishes. From flatbreads to porridges, its presence in regional cuisines has stood the test of time, underscoring its cultural significance.

10.3 Nutritional Composition of Browntop Millet

In addition to its historical and cultural importance, browntop millet holds immense potential as a nutritionally dense food source. As a gluten-free grain, it caters to the dietary needs of individuals with gluten sensitivities and contributes to the diversification of grain-based diets (Singh et al. 2022). Understanding the nutrient composition of browntop is crucial for evaluating its potential impact on human health and well-being. Table 10.1 represents a comparison of the composition of macronutrients in browntop millet and other millets.

10.3.1 Macronutrients

Carbohydrates, protein and fats are the building blocks of the body, collectively referred to as macronutrients as they provide energy via oxidation (Costa-Pinto and Gantner 2020). Browntop millet is a good source of macronutrients. It has a balanced ratio of proteins, fats, and carbohydrates, making it a wholesome dietary choice (Fig. 10.2). The following is a summary of the macronutrients that browntop millets normally contain in a 100 g serving:

(a) *Carbohydrates:* Browntop millet is a great energy source due to rich content of carbs. Browntop millet normally provide 60–70 g of carbohydrates per 100 g of food (Nevedhitha and Priya 2022). These carbs comprise dietary fiber, which facilitates digestion and increases satiety, and starch, which acts as the body's main energy source.

(b) *Protein:* Unlike other grains, sprouted browntop millet have a much higher protein content. For every 100 g, they usually contain 10–12 g of protein. For vegetarians and vegans, browntop millet is a magnificent source of plant-based protein since they have all of the amino acids that are essential for the body and have a high grade of protein (Langyan et al. 2022; Akhtar et al. 2022).

(c) *Fat:* Browntop millet contains low-fat, including approximately 3–4 g per 100 g of meal (Dayakar Rao et al. 2017). The omega-3 fatty acids and omega-6 fatty

Table 10.1 Composition of macronutrients in browntop millets and other millets (per 100 g) (Maitra 2020; Singh et al. 2022)

S. No.		Proteins (g)	Carbohydrates (g)	Fats (g)	Reference
1.	Browntop millet	8.98-11.5	71.3	1.8	Roopa (2015)
2.	Pearl Millet	10.96 ± 0.2	61.78 ± 0.8	5.43 ± 0.6	Longvah et al. (2017)
3.	Finger Millet	7.16 ± 0.63	66.82 ± 0.7	1.92 ± 0.1	
4.	Foxtail Millet	12.3	60.9	4.3	
5.	Proso Millet	8.92 ± 1.0	66.19 ± 1.1	2.55 ± 0.1	Longvah et al. (2017)
6.	Little Millet	10.13 ± 0.4	65.55 ± 1.2	3.89 ± 0.3	
7.	Sorghum Millet	9.97 ± 0.4	67.68 ± 1.0	1.73 ± 0.6	

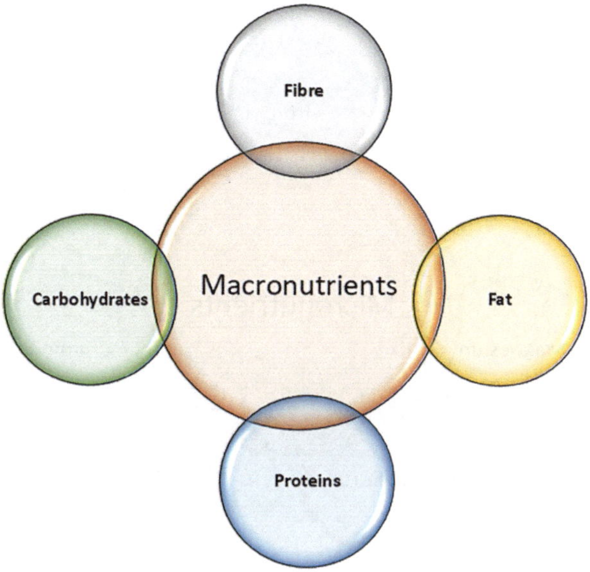

Fig. 10.2 Macronutrients present in browntop millets

acids are among the useful unsaturated fats that make up the majority of the fat content. These fats are essential for many biological processes, such as the synthesis of steroid hormones and the integrity of cell membranes (Khairi et al. 2014).

(d) *Fiber:* With about 3–4 grams of dietary fibers per 100 g serving, browntop millets are an excellent source of dietary fiber (Harish et al. 2024; Singh et al. 2022). Dietary fibers are necessary for regular bowel movements, constipation prevention, and digestive health. It also aids in controlling cholesterol and blood sugar levels (Gill et al. 2021).

10.3.2 Micronutrients

Like other whole grains, browntop millet contains vital elements crucial for general health and well-being. Most cereal grains are rich in carbohydrates but deficient in micronutrients, but browntop millet is a excellent reservoir of micronutrients, making it an ideal food source for a balanced diet. Among the important micronutrients present in browntop millet are the following (Tiwari et al. 2024) (Fig. 10.3):

(a) *Iron:* Iron is an essential element for the synthesis of hemoglobin and red blood cells, and is present adequately in browntop millet. Adequate iron consumption promotes general health and aids in the prevention of anemia. For every 100 g, browntop millet normally provide 3–5% of the recommended daily intake dose

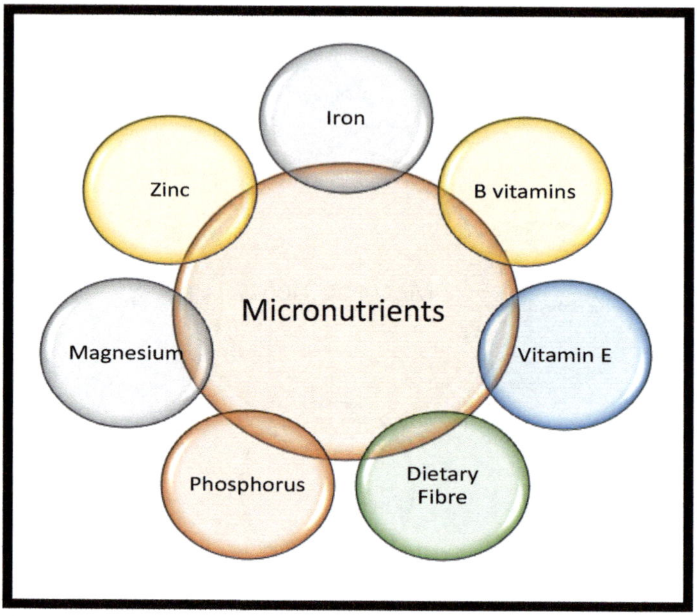

Fig. 10.3 Micronutrients present in browntop millets

(RDI) of iron (Tiwari et al. 2024). However, soil's composition and growing environment are two examples of variables that can affect iron concentration in browntop millet.

(b) *Zinc:* Zinc deficiency is widespread in developing nations and affects nearly two billion people, leading to numerous diseases. Zinc is adequately available in browntop millet, which helps meet the daily need for this important mineral (Jagati et al. 2021). It is an essential element for DNA synthesis, wound healing, immune system activation, and activity. Kishor et al. 2021 and Niharika et al. 2020 reported 2.5 and 2.11 mg zinc in browntop millet respectively. For every 100 g of browntop millet, there are approximately 1–2 mg of zinc, or 7–14% of the adult RDI (Habwe 2005).

(c) *Magnesium:* Magnesium is essential for biochemical processes, such as the health of bones, tissue, blood sugar management, and cardiac and peritoneal muscle and neuron function (Fiorentini et al. 2021). The magnesium content of browntop millet is comparatively high; each 100 g serving provides 10–15% of the RDI (Babu 2016).

(d) *Phosphorus:* Phosphorus is necessary for the metabolism of energy, cell signaling, and bone health (Bird and Eskin 2021). Phosphorus from browntop millet aids in these processes. For every 100 g, browntop millet supply 8–10% of the RDI in phosphorus.

(e) *B-complex vitamins:* Vitamin B3, vitamin B1, and vitamin B2 are among the B vitamins that can be found in browntop millet (Ramashia 2018). Among their many functions, these vitamins are involved in energy metabolism, neuromuscular function, and skin health maintenance. The following represents an approximation of the percentage of important B vitamins in 100 g of browntop millet:

- *Niacin (Vitamin B3):* Browntop millet are especially high in niacin, offering 10–15% of the RDI per 100 g.
- *Thiamine (Vitamin B1):* A single serving of 100 g of browntop millet provides around 10–15% of the RDI of thiamine.
- *Pyridoxine (Vitamin B6):* Browntop millet typically contain around 10–15% of the RDI for vitamin B6 per 100 g serving. However, these may change depending on growing conditions, processing, and variety. Furthermore, methods of farming, cooking techniques, and soil quality can all have an impact on the actual B-vitamin content.

(f) *Vitamin E:* Vitamin E is one of the fat-soluble antioxidant vitamin found in browntop millet that aids in shielding cells from harm caused due to free radicals. It also affects skin health and immunological function. A 100g meal of browntop millet provides about 4–6% of the RDI in vitamin E.

(g) *Dietary fiber:* Although it isn't classified as a micronutrient, it is crucial for gut health, and this may help control the blood sugar and cholesterol levels. It is also helpful in digestive health and in weight management. Dietary fiber from browntop millet contributes to these advantages. For every 100 grams of grain, browntop millets provide about 3–4 grams of dietary fiber on average (Dayakar Rao et al. 2017). According to dietary standards, this amount of dietary fiber corresponds to around 12–16% of the RDI for adults, which is about 25 g for women and 38 g for men. Browntop millet offers several micronutrients critical to health and well-being. It can support overall dietary goals and assist in diversifying nutrient intake when incorporated into a balanced diet.

10.4 Uses of Browntop Millet

There are multiple uses for browntop millet due to its rich nutritional constituents. Browntop millet has several culinary uses. It is used as livestock feed and forage, catch crop, cover crop, nurse crop. It is also used as green manure crop and for the maintenance of wildlife (Maurya et al. 2023).

10.4.1 Browntop Millet as livestock feed, forage and cover crop

Poultry: A common feed item for chickens, ducks, and turkeys is browntop millet. Because of its higher protein content, vital amino acids, and grains that are high in energy, it is frequently used in mixes for chicken feeding. It can be given as cracked grains, whole grains, or as a component of a diet that is well-balanced diet.

As a source of energy, protein, and vital nutrients, browntop millet can be fed to livestock as a supplement. As whole grains or as a part of haylage or silage, it can be fed to cattle in feed regimens. When provided, browntop millet is usually appetizing to cattle and is quickly consumed. Goats and sheep can be given browntop millet as part of a well-balanced diet. Feeding options for browntop millet include whole grains, chopped fodder, and mixed feed rations. It can be added to pig feed rations in the form of whole grains, crushed grains, or pelleted feed. Pigs may effectively use browntop millet for their growth and development, and usually find it to be appetizing. Although forage-based diets are the main requirement for horses, browntop millet can be added to horse feed to provide additional nutrition and energy. For horses, it should be given in moderation and ideally, as a component of a regular feed ration to fulfill their unique nutritional needs (Reddy et al. 2025; Jindal and Nikhanj 2023). It can be a useful addition to livestock feed to help meet their nutritional demands. The browntop millet stover is rich in calcium (9.0 g/kg), magnesium (5.9 g/kg), crude protein (56.7 g/kg), phosphorus (1.5 g/kg) and total digestible nutrients (594.2 g/kg) confirming it can be an ideal fodder in the semiarid and drylands of the world (Kering and Broderick 2018).

Palatability: Browntop millet is generally palatable to livestock, which makes it simpler to feed them (Myers 2018). To make sure the animals are eating browntop millet well and enduring it without experiencing any negative side effects, it is crucial to introduce it gradually, just like with any other feed ingredient.

Cover Crops: Browntop millet is frequently utilized in agriculture as a cover crop. It can be planted in between major crops to stop erosion, reduce weed growth, and enhance soil health. Its thick growth keeps weeds at bay and shields the soil from wind and water erosion (Myers 2018).

Environmental Restoration: To repair damaged areas and revitalize ecosystems, browntop millet can be employed in environmental restoration initiatives with disturbed areas. It can aid with soil stabilization, erosion management, and biodiversity enhancement (Brown et al. 2011*). Soil Stabilization: Browntop millet's large root system aids in stabilizing the soil by binding soil particles together (Cheraghi et al.* 2023*). The roots of millet help stabilize the soil, limiting erosion from wind and water when planted in regions that are prone to soil erosion, such as slopes or degraded lands. Browntop millet works well to control erosion because of its wide root system and rapid development (Maitra* 2020). To strengthen the soil and stop topsoil loss, it can be grown on slopes, banks, and other locations that are vulnerable to soil erosion.

Browntop millet's deep root system binds soil particles together and aids in keeping the soil in place. Because of the robust soil matrix this root arrangement produces, there is less chance of erosion from wind and water. Root reinforcement: Browntop millet's firmly embedded roots strengthen the soil's structure and make it more resistant to erosion (Sharma 2018). This root strengthening is beneficial in preserving the integrity of the soil, particularly in eroding locations like slopes, embankments, and disturbed lands. Addition of organic matter: By shedding leaves and turning over roots, browntop millet enriches the soil with organic matter as it develops. Because of the way that this organic matter promotes soil structure, water infiltration, and soil moisture retention, erosion is minimized.

Weed Suppression: Browntop millet suppresses weeds by competing with them for nutrients, water, and sunlight (Giri et al. 2024). It helps to limit weed development by outcompeting weeds and forming a dense stand, which lessens competition for resources and preserves the undisturbed soil. Weeds and browntop millet fight over resources, including sunlight, water, and nutrients. Browntop millet has the potential to inhibit weed development as other millets showed, when utilized as a cover crop or in restoration projects (Samarajeewa et al. 2006). Habitat restoration: Browntop millet can be utilized to restore habitat in degraded habitats (Volis 2019). It can aid in the restoration of ecosystem function and biodiversity by enhancing soil fertility, stabilizing the soil, and offering food and habitat to wildlife.

Ornamental Grass: Browntop millet can occasionally be cultivated as a decorative grass in landscaping projects. Gardens, parks, and other outdoor areas are visually interesting due to their tall, slender stalks and beautiful seed heads.

Aesthetic Appeal: Browntop millet is visually appealing because of its tall, slender stalks and beautiful seed heads, which give gardens and landscapes visual interest. A gentle and realistic atmosphere can be produced by the delicate appearance of the seed heads blowing in the air.

Crop for Climate Change Rought Areas: Browntop millet thrives in environments with harsh growing conditions because of its tenacity and drought tolerance (Pattanashetti et al. 2024). Once planted, it needs little care and can flourish in arid, hot settings with poor soil. Wildlife Habitat: Birds, insects, and other animals can find food and shelter in the dense growth of browntop millet. The seeds might attract birds, and the stalks may provide cover for insects, increasing the diversity of life in the garden.

Nutrient Cycling: Browntop millet is involved in the cycling of nutrients in ecosystems, just like other plants. Browntop millet takes up nitrogen, phosphate, and potassium from the soil as it grows (Anitha et al. 2020). These nutrients go back to the soil during the senesce and decomposition of millet plants, improving it and promoting the growth of other flora.

Some of the important uses of browntop millet in ceremonial offerings are illustrated below:

Religious Ceremonies: Browntop millet can be used in religious ceremonies and festivals to represent abundance, prosperity, and fertility (Ankita and Seth 2025). It is frequently used as an offering to deities or spirits to obtain blessings for a plentiful harvest or good agricultural endeavors (Kumar 2023). Harvest festivals: As a representation of the richness of agriculture and the outcomes of labor, browntop millet may be a main attraction during these events. It can be utilized to honor the harvest and express gratitude for nature's benefits in ceremonial rites, decorative displays, and group dinners. Rites of passage: In ceremonies related to marriage, childbirth, and adulthood, browntop millet may be used. It can be served as a part of customary rituals and celebrations commemorating these important life milestones. It can also be used symbolically to symbolize fertility, prosperity, and the life cycle.

Sacred Offerings: Browntop millet is regarded as a sacred grain with spiritual significance in certain civilizations. It can be offered as a sign of respect, thanks, and supplication during rituals honoring ancestors, spirits, or other divine entities. Rituals within the community: Browntop millet can be employed in rituals and ceremonies that unite people within a community to commemorate common customs, values, and beliefs. It could be given as a part of ceremonies, prayers, or offerings made for the prosperity, well-being, and harmony of the community.

10.4.2 Traditional Uses

For centuries, browntop millet has been a mainstay in many cultures, with a diverse range of traditional applications that have supported societies and enhanced culinary practices. Browntop millet is primarily grown for its nutritional content and adaptability, and it has long been a mainstay in traditional diets in Asia and Africa (Ambati et al. 2024). It is believed to help support general wellness and aid in the removal of toxins from the body.

Its grains provide a substantial source of carbs and important nutrients when crushed into flour and used to make classic flatbreads, porridges, and savory foods. In addition, its straw is prized for feeding cattle, giving farm animals in agrarian civilizations something to eat. In addition to being a good food source, browntop has cultural importance and is frequently used extensively in religious festivals and rituals. Due to its adaptability to a variety of climates and low water needs, it has become important for farmers in arid regions, ensuring their livelihoods and access to food. The traditional applications of browntop millet are still honored in modern countries, serving as a constant reminder of the grain's lasting significance in both our culinary and common histories (Shanker 2024; Ankita and Seth 2025).

10.4.3 Medicinal Applications

Browntop milletBrowntop millets is employed in traditional medical procedures in certain cultures. It is thought to offer several health advantages, such as enhancing general well-being, lowering inflammation, and enhancing digestion (Kumari et al. 2025; Singh et al. 2022). Browntop millet has therapeutic qualities that have been acknowledged by traditional medical systems (Table 10.2). In addition to its

Table 10.2 Disease-controlling properties of millets (Malik 2015)

S. No.	Illness	Benefits of millets	Positive factors
1.	Diabetes	Curing	Due low glycemic index of millet
2.	Obesity	Helpful in reducing weight	Due to the high content of fibers
3.	Heart health	Blood pressure is controlled by a high amount of magnesium Tannins are a potent antioxidant that helps in heart health	Due to the presence of antioxidants
4.	Bone	Phosphorus is the essential component of ATP, which is present in high amounts in the millet; phosphorus is also essential for bone formation	High phosphorus content
5.	Anti-allergic	Highly digestible	It has hypoallergenic properties
6.	Ulcer	Prevent the ulcers of the stomach	Control the acidity of the stomach
7.	Constipation	The presence of fiber helps in treating constipation	High fiber content
8.	Anemia	Increase the hemoglobin level	The concentration of Zn and Fe is very high

culinary benefits, browntop millet has several medical uses that have been recognized by traditional medical systems. Enriched with minerals including iron (Fe), magnesium, and Vitamins B-complex, it promotes overall health and wellness. Because of its higher fiber content, it also helps with digestion and gut health, and there may be a decreased incidence of gastrointestinal diseases. Furthermore, browntop millet has been utilized in traditional and Ayurvedic medicine to treat illnesses like diabetes and cardiovascular disease (Singh et al. 2022, 2024; Keshwa et al. 2023). Its low glycemic index makes it a good nutritional grain for people who are controlling their blood sugar, and its ability to decrease cholesterol is good for heart health. Additionally, due to presence of phenolics (Gallic acid, ferulic acid and caffeic acid) and flavonoids (myricetin and kaempferol) it showed antioxidant qualities, which aid in the fight against inflammation and oxidative stress, it may be beneficial for diseases like inflammatory disorders and arthritis. Because browntop millet has good content of vitamins and minerals, it's also said to be good for skin and hair health. Browntop millet is a vital ingredient in both ancient medicines and modern health practices, as research continues to find its medicinal potential (Bhavani et al. 2024). It embodies the relationship between food and medicine. In conventional medicine, browntop millet is sometimes employed to boost immunity. It can be taken regularly to boost the body defense system and fight against infectious agents and diseases.

10.4.4 Health Benefits and Uses (Ashoka and Sunitha 2020; Maitra et al. 2023)

Gluten-Free Alternative: Browntop millet is a naturally gluten-free cereal, making it a suitable grain for those with gluten sensitivity or celiac disease.

Weight Management: The higher fiber content in browntop millet can aid in weight management by promoting a feeling of fullness stomach and supporting digestive health.

Blood Sugar Regulation: The complex carbohydrates, minerals, and fiber in browntop millet may contribute to better blood sugar regulation, making it a good choice for people with diabetes or those at risk of developing diabetes or onset of diabetes.

Heart Health: The presence of nutrients like magnesium (Mg), potassium (K), and fiber supports heart health by regulating blood pressure and cholesterol levels.

Nutrient-Dense Alternative: Millet can be used as a nutrient-dense alternative to refined grains, contributing essential nutrients and amino acids to the diet.

10.4.5 Culinary Uses

Whole Grain: Browntop millet can be cooked and consumed as a whole grain, similar to rice or quinoa.

Flour: Millet flour can be used to make flatbreads and pancakes, or added to other flours for baking.

Porridge: Browntop millet can be used to make porridge, which can be flavored and customized with various toppings for a nutritious breakfast.

Side Dish: It can be used as a side dish in salads or mixed with vegetables.

However, before incorporating browntop millet into your diet, it's necessary to consult with a healthcare professional or a nutritionist, especially if you have any specific dietary concerns or health conditions or are allergic to any kind of cereals (Kaushik et al. 2024).

10.5 Conclusion

Browntop millet stands as a versatile grain, celebrated not only for its nutritional richness but also for its deep-rooted cultural relevance. As global interest in diverse and sustainable food sources grows, understanding the nutrient composition and traditional use of browntop can pave the way for its revival as a cherished and nourishing grain in contemporary diets. Through a comprehensive examination of both the historical and nutritional aspects of browntop millet, it is clear that processed browntop millet can be used to create a variety of value-added products. In this sense, adding brown top millet to a regular diet is good for human health. Additionally, browntop millet has a higher nutritional content than other livestock rations when used as animal feed or fodder. Unfortunately, such a promising crop has remained underutilized; therefore, adequate research emphasis, as well as marketing efforts, may be coordinated to promote the browntop millet among farmers and consumers in order to achieve agricultural sustainability in drylands and ensure food and nutritional security in the future.

References

Abbasi YF, See OG, Ping NY et al (2018) Diabetes knowledge, attitude, and practice among type 2 diabetes mellitus patients in Kuala Muda District, Malaysia—a cross-sectional study. Diabetes Metab Syndr Clin Res Rev 12(6):1057–1063. https://doi.org/10.1016/j.dsx.2018.06.025

Ahmed SM, Luo L, Namani A et al (2017) Nrf2 signaling pathway: pivotal roles in inflammation. Biochim Biophys Acta 1863(2):585–597. https://doi.org/10.1016/j.bbadis.2016.11.005

Akhtar S, Ismail T, Layla A et al (2022) An overview of plant-based protein-rich products. Plant Protein Foods 29:27–60. https://doi.org/10.1007/978-3-030-91206-2_2

Alam MA, Subhan N, Hossain H (2016) Hydroxycinnamic acid derivatives: a potential class of natural compounds for the management of lipid metabolism and obesity. Nutr Metab 13(1):27. https://doi.org/10.1186/s12986-016-0080-3

Amadou I, Gounga ME, Le GW (2013) Millets, nutritional composition, some health benefits, and processing. Emir J Food Agric 25(7):501–508. https://doi.org/10.9755/ejfa.v25i7.12045

Ambati S, Anuragi H, Prasad KR, Vidhyadhar B, Marathi B (2024) Botanical description, a brief history of Browntop millet and its spectacular adaptations as a hardy food and feed crop. In Mishra S, Kumar S, Srivastava RC (eds) Genetic Improvement of Small Millets 637–644. Springer Nature Singapore. https://doi.org/10.1007/978-981-99-7232-6_31

Anitha KV, Chikkaramappa T, Basavaraja PK et al (2020) Effect of different levels of fertilizers on productivity and nutrient dynamics of browntop millet. Int Res J Pure Appl Chem 21(24):308–320. https://doi.org/10.9734/IRJPAC/2020/v21i2430368

Ankita, Seth U (2025) Millets in India: exploring historical significance, cultural heritage, and ethnic foods. J Ethn Food 12:2. https://doi.org/10.1186/s42779-024-00262-2

Asrani P, Ali A, Tiwari K (2023) Millets as an alternative diet for gluten-sensitive individuals: a critical review on nutritional components, sensitivities, and popularity of wheat and millets among consumers. Food Rev Int 39(6):3370–3399. https://doi.org/10.1080/87559129.2021.2012790

Ashoka P, Sunitha NH (2020) Review on Browntop Millet- A Forgotten Crop. Journal of Experimental Agriculture International 54-60. https://doi.org/10.9734/jeai/2020/v42i730553

Ayele A, Tarekegn K (2020) The impact of urbanization expansion on agricultural land in Ethiopia: a review. Environ Socio-Econ Stud 8(4):73–80. https://doi.org/10.2478/environ-2020-0024

Babele PK, Kudapa H, Singh Y et al (2022) Mainstreaming orphan millets for advancing climate-smart agriculture to secure nutrition and health. Front Plant Sci 13:902536. https://doi.org/10.3389/fpls.2022.902536

Babu VR (2016) Biofortification: improvement zinc contents in rice grains through conventional and molecular approaches. In: Reshaping agriculture and nutrition linkages for food and nutrition security. ICAR—Central Research Institute for Dryland Agriculture, Hyderabad, 326 p

Bhavani P, Nandini C, Maharajan T et al (2024) Brown-top millet: an overview of breeding, genetic, and genomic resources development for crop improvement. Planta 260(1):10. https://doi.org/10.1007/s00425-024-04446-7

Bird RP, Eskin NM (2021) The emerging role of phosphorus in human health. In: Advances in food and nutrition research, vol 96. Academic, pp 27–88. https://doi.org/10.1016/bs.afnr.2021.02.001

Bora P, Ragaee S, Marcone M (2019) Characterisation of several types of millets as functional food ingredients. Int J Food Sci Nutr 70(6):714–724. https://doi.org/10.1080/09637486.2019.1570086

Braicu C, Ladomery MR, Chedea VS (2013) The relationship between the structure and biological actions of green tea catechins. Food Chem 141(3):3282–3289. https://doi.org/10.1016/j.foodchem.2013.05.122

Brown DR, Dettmann P, Rinaudo T et al (2011) Poverty alleviation and environmental restoration using the clean development mechanism: a case study from Humbo, Ethiopia. Environ Manag 48(2):322–333. https://doi.org/10.1007/s00267-010-9590-3

Cheraghi M, Mousavi SM, Zarebanadkouki M (2023) Functions of rhizosheath on facilitating the uptake of water and nutrients under drought stress: a review. Plant Soil 491(1):239–263. https://doi.org/10.1007/s11104-023-06126-z

Costa-Pinto R, Gantner D (2020) Macronutrients, minerals, vitamins and energy. Anaesthesia & Intensive Care Medicine 21(3):157–161

Dayakar Rao B, Bhaskarachary K, Arlene Christina GD et al (2017) Nutritional and health benefits of millets. ICAR_Indian Institute of Millets Research (IIMR) Rajendranagar, Hyderabad, vol 2, pp 38–57

Fiorentini D, Cappadone C, Farruggia G, Prata C (2021) Magnesium: biochemistry, nutrition, detection, and social impact of diseases linked to its deficiency. Forum Nutr 13(4):1136. https://doi.org/10.3390/nu13041136

Gill SK, Rossi M, Bajka B, Whelan K (2021) Dietary fibre in gastrointestinal health and disease. Nat Rev Gastroenterol Hepatol 18(2):101–116. https://doi.org/10.1038/s41575-020-00375-4

Giri S, Das NK, Kundu AK (2024) Weed management in millets. In: Millets: the miracle grains of the 21st century, 1st edn. Kripa-Drishti Publications, pp 62–72

Habwe FO (2005) Evaluation of iron and zinc levels, microbiological load and acceptability of crackers, noodles and cookies modified using selected underutilized crops. Doctoral dissertation, Maseno University

Harish MS, Bhuker A, Chauhan BS (2024) Millet production, challenges, and opportunities in the Asia-pacific region: a comprehensive review. Front Sustain Food Syst 8:1386469. https://doi.org/10.3389/fsufs.2024.1386469

Jagati P, Mahapatra I, Dash D (2021) *Finger millet* (Ragi) as an essential dietary supplement with key health benefits: a review. Int J Home Sci 7(2):94–100. https://doi.org/10.22271/23957476.2021.v7.i2b.1152

Jindal P, Nikhanj P (2023) A review on processing technologies for value-added millet products. J Food Process Eng 46(10):e14419. https://doi.org/10.1111/jfpe.14419

Jose S, Gulati A, Khurana K (2020) Achieving nutritional security in India: vision 2030. Indian Council for Research on International Economic Relations, p 24

Kaushik A, Saxena DC, Singh S (2024) Functional, Thermal, Pasting, and Antioxidant Properties of Flour from Indian Browntop Millet (Brachiaria ramosa) Cultivars. Food Biophysics 19, 637–652. https://doi.org/10.1007/s11483-024-09858-2

Kavi Kishor PB, Anil Kumar S, Naravula J et al (2021) Improvement of small seed for big nutritional feed. Physiol Mol Biol Plants 27(10):2433–2446. https://doi.org/10.1007/s12298-021-01071-6

Keshwa K, Singh V, Itani N, Parveen T (2023) Millets: ayurvedic and modern perspectives of utilization in the current era. Int J Ayurveda Pharma Res 12:39–43. https://doi.org/10.47070/ijapr.v11iSuppl4.2922

Kering MK, and Broderick C (2018) Potassium and Manganese Fertilization and the Effects on Millet Seed Yield, Seed Quality and Forage Potential of Residual Stalks Agricultural Sciences 9: 888–900.

Khairi NA, Gobouri AA, Dhahawi HO (2014) A review on omega-3 and omega-6 essential fatty acids: uses, benefits, and their availability in pumpkins (*Cucurbita maxima*) seed and desert dates (*Balanites aegyptiaca*) seed kernel oils. Pak J Biol Sci 17(12):1195–1208. https://doi.org/10.3923/pjbs.2014.1195.1208

Kingwell-Banham E, Fuller DQ (2014) Brown Top Millet: Origins and Development. In: Smith C. (eds) Encyclopedia of Global Archaeology. Springer, New York, NY. https://doi.org/10.1007/978-1-4419-0465-2_2318

Kishore AS, Rekha KB, Hussain HS et al (2021) Quality Enhancement of Nutri-Cereal BTM Through Agronomic Practices. Current Science 120:468–470

Kumar S (2023) Millets an ingredient of celebrating harvest festivals and religion in Bharat. Akshara Multidiscip Res J 4(8):13–17

Kumari A, Sadh PK, Kamboj A, Yadav B et al (2025) Exploring the benefits of nutrition of little millet: unveiling the effect of processing methods on bioactive properties. J Food Biochem 2025(1):2488816. https://doi.org/10.1155/jfbc/2488816

Langyan S, Yadava P, Khan FN et al (2022) Sustaining protein nutrition through plant-based foods. Front Nutr 18(8):772573. https://doi.org/10.3389/fnut.2021.772573

Laryea D, Mills ST, Dapuliga CC, Zaukuu JL (2024) New horizons for millets: food and nutrition security. In: Sustainable and functional foods from plants. Apple Academic Press, pp 25–54

Longvah T, Anantan I, Bhaskarachary K et al (2017) Indian food composition tables. National Institute of Nutrition, Indian Council of Medical Research, Hyderabad

Maitra S (2020) Potential horizon of brown-top millet cultivation in drylands: a review. Crop Res 55(1 and 2):57–63. https://doi.org/10.31830/2454-1761.2020.012

Maitra S, Patro TSSK, Reddy A et. al (2023) in Neglected and Underutilized Crops. Brown top millet (Brachiaria ramosa L. Stapf; Panicum ramosum L.)2014a neglected and smart crop in fighting against hunger and malnutrition, Academic Press, 2212013;245. https://doi.org/10.1016/B978-0-323-90537-4.00012-0

Malik S (2015) Pearl millet-nutritional value and medicinal uses. Int J Adv Res Innov Ideas Educ 1(3):414–418

Maurya R, Boini T, Misro L et al (2023) Comprehensive review on millets: nutritional values, effect of food processing and dietary aspects. J Drug Res 8(1):S82–S98. https://doi.org/10.4103/jdras.jdras_123_23

Meena RP, Joshi D, Bisht JK, Kant L (2021) Global Scenario of Millets Cultivation. In: Kumar A, Tripathi MK, Joshi D, Kumar, V. Millets and Millet Technology. Springer, Singapore, 33–50.

Myers R (2018) Growing millets for grain, forage, or cover crop use. Division of Plant Sciences. Extension University of Missouri

Nevedhitha KG, Priya KL (2022) Effect of pre-treatments on the quality of browntop millet flour and development of cookies. J Trop Agric, 60(1). https://doi.org/10.5555/20230065919

Niharika V, Rao GB, Tushara M, Rao VS (2020) Studies on performance of browntop millet indigenous collections for grain yield and nutritional traits. J Pharmacogn Phytochem 9(5):2636-2638

Paschapur AU, Joshi D, Mishra KK et al (2021) Millets for life: a brief introduction. In: Millets and millet technology. Springer Singapore, Singapore, pp 1–32. https://doi.org/10.1007/978-981-16-0676-2_1

Pattanashetti BM, Supritha Raj DS, Ragi S et al (2024) Breeding browntop millet (*Brachiaria ramose*) for biotic and abiotic stress resistance. In: Genetic improvement of small millets. Springer, Singapore, pp 645–654. https://doi.org/10.1007/978-981-99-7232-6_32

Ramashia SE (2018) Physical, functional and nutritional properties of flours from finger millet (*Eleusine coracana*) varieties fortified with vitamin B_2 and zinc oxide. Doctoral dissertation

Reddy MP, Sen A, Swamy CT et al (2025) Chemistry of macro-and micronutrients. In: Colored cereals. CRC Press, pp 96–118

Roopa OM (2015) Nutritional analysis and development of value-added products from browntop millet. Doctoral dissertation, University of Agricultural Sciences, GKVK

Samarajeewa KB, Horiuchi T, Oba S (2006) Finger millet (Eleusine coracana L. Gaertn.) as a cover crop on weed control, growth and yield of soybean under different tillage systems. Soil & Tillage Research, 90:93–99. https://doi.org/10.1016/j.still.2005.07.003

Shanker N (2024) Millets: the ancient grain for modern health. In: Millets: the multi-cereal paradigm for food sustainability. Springer, Cham, pp 31–42. https://doi.org/10.1007/978-3-031-64237-1_2

Sharma RA (2018) Land management practices for in-situ rainwater conservation and soil erosion control. Any error in this proceeding is silent testimony of the fact that it was a human effort vol 15, pp 88–96

Sheahan CM (2014) Plant guide for browntop millet (Urochloa ramosa). USDA-Natural Resources Conservation Service, Cape May Plant Materials Center, Cape May.

Singh S, Suri S, Singh R (2022) Potential and unrealized future possibilities of browntop millet in the food sector. Front Sustain Food Syst 6:974126. https://doi.org/10.3389/fsufs.2022.974126

Singh H, Aalam Z, Baheti SR et al (2024) Exploring scientific validation of millets in contemporary healthcare: a traditional food supplement. Recent Adv Food Nutr Agric 15(1):2–12. https://doi.org/10.2174/012772574X265711231109195603

Sravani M, Kuna A, Devi SS, Rao KS, Gayatri B (2020) Effect of processing on the Physico-chemical properties of Browntop millet (Brachiaria ramosa). J. Pharmacogn. Phytochem. 9(5): 1480–1483

Tiwari M, Kaur D, Puranik V (2024) Comprehensive analysis of proximate, antioxidant, functional, and anti-nutritional properties of millet flour: A comparative study of Brown Top Millet, Little Millet, and Foxtail Millet. J Food Agric Res, 4(2):169–180. https://doi.org/10.47509/JFAR.2024.v04i02.04

Volis S (2019) Conservation-oriented restoration—a two-for-one method to restore both threatened species and their habitats. Plant Divers 41(2):50–58. https://doi.org/10.1016/j.pld.2019.01.002

Chapter 11
Teff Millet: Chemical Composition, Starch Isolation, Modification, Functionality, and Food/Non-Food Applications

Ratnakar Garg, Mohan Lal Jat, Preeti Badoni, Harish Nath, and Vidhi Upadhyaya

11.1 Introduction

Ethiopia is the birthplace of teff (*Eragrostis tef(Zucc.) Trotter*), a member of the Poaceae family, an ancient cereal grain that has been a staple food source for millennia (Lemecha et al. 2018; Woldemariam et al. 2019). Teff millet (*Eragrostis tef (Zucc.) Trotter*) is grown primarily for its small grain, which is used to make injera (Amharic, Tigrinya), a staple dish that resembles pancakes or bread (Arendt and Zannini 2013). There are over 350 different species in the genus *Eragrostis*. However, according to Habtegebrial et al. (2007), teff (Eragrostis tef (Zucc.) Trotter) is the only species that is cultivated. It is tolerant to stress. From lowlands to mountains, it may grow in a variety of environmental settings. Additionally, teff millet is resistant to storage pests (Girma et al. 2014). Because it can be grown and thrive in a variety of ecological circumstances, even harsher ones where many other cereals fail, teff can therefore be regarded as a low-risk crop. Because of its agronomic qualities and applications, Teff is preferred by farmers and consumers above many other typical Eritrean and Ethiopian grain crops, such as wheat. However, the scientific and academic sectors outside of Ethiopia and Eritrea know less about the crop. For a variety of reasons, researchers should focus more on this underutilized

R. Garg
Research and Development (R&D) Organisation, Akums Drugs and Pharmaceutical Pvt Ltd., Sidcul, Haridwar, Uttarakhand, India

M. L. Jat
Department of Horticulture, Govt of Haryana, Hisar, Haryana, India

P. Badoni · H. Nath
College of Pharmacy, COER University, Roorkee, Uttarakhand, India

V. Upadhyaya (✉)
Faculty of Pharmaceutical Sciences, Motherhood University, Roorkee, Uttarakhand, India

© The Author(s), under exclusive license to Springer Nature Switzerland AG 2026
J. Singh, M. Kumar Malik (eds.), *Millets and Millet Starch: Chemical Composition, Functional Modification, and Applications*,
https://doi.org/10.1007/978-3-032-09502-2_11

and undiscovered grain crop. With the world's population growing and plant resources being depleted, focusing more on underutilized and poorly studied food crops like teff millet helps to ensure food and nutrition security by improving access to nutrient-dense food, increasing revenue generation for small-scale farmers' livelihoods, and providing environmental benefits (Zhu 2018).

The crop has recently attracted more attention due to its grain's exceptional nutritional qualities (Gebremariam et al. 2014). Because of its alleged nutritional advantages, teff grains have gained popularity recently and are now grown in Canada, China, North America, India, the United Kingdom, Australia, Switzerland, and other African nations, including Uganda and Cameroon (Abraham 2015; Shumoy and Raes 2016). In Ethiopia alone, there are already over 30 million people who regularly consume teff products. Its gluten-free status is its most alluring feature, which offers it great potential for use in a variety of food products to benefit those who suffer from celiac disease (Do Nascimento et al. 2018).When planted during warmer seasons, the crop is proven to withstand drought (Miller 2014). Ethiopia produces the majority of the world's teff (around 90%). It constitutes 20% of Ethiopia's cereal grain output and is regarded as the most significant crop in the nation in terms of both yield (1.664 tons ha^{-1}) and acreage (~3 million ha). Teff also yields higher profits in Ethiopia, where it commands a premium price over other cereal crops. In other nations, however, consumers are willing to pay more for teff-based goods due to the desire for healthier options (Lee 2018).

11.2 Geographical Distribution

Teff millet is primarily grown in the Horn of Africa, especially Ethiopia and Eritrea, where it is a major staple crop and deeply integrated into local agriculture and cuisine (NRC 1996). Ethiopia accounts for the vast majority of global teff production due to its favorable highland climate, which provides the cool temperatures and well-drained soils that the crop requires. In 2016, Ethiopia grew more than 90% of the world's teff (O'Connor 2016). In recent years, teff millet cultivation has expanded to other regions with similar agro-ecological conditions, including parts of the United States (particularly in states like Idaho, Nevada, and Oregon), Australia, India, Germany, the Netherlands, Spain, and South Africa. Although its global distribution is still limited compared to other cereals, teff's adaptability to drought and marginal soils is encouraging its gradual spread into new agricultural zones worldwide (Abraham 2015; Ketema 1997; Roseberg et al. 2005; Haileselassie et al. 2011). The crop has been grown as the main cereal in 10 of Ethiopia's 18 broad Agro-Ecological Zones (AEZs) (MoA 1998).

11.3 Cultivation

Teff is typically grown in marginal production regions that are unable to meet the crop water requirements necessary to complete the life-cycle of the majority of other cereal crops, as well as high potential production areas that receive enough rain to support growth (Harrouni 1989; Ketema 1997). Teff has been cultivated as a low-risk crop in several nations where drought is frequent and cultivation of other cereals is not feasible due to moisture stress (Ketema 1997; Admas and Belay 2011). Teff millet has been grown extensively in a variety of soil types, climates, and altitudes (300–2800 m a.s.l.). According to Ketema (1983) and van Delden (2011), teff millet is a short-day plant that is sensitive to day length. Although the number of days needed to reach maturity varies throughout crop cultivars, the majority of cultivars complete their lifecycle in a duration of 60–180 days, contingent upon the specific agro-ecological conditions. However, 90–130 days is the ideal duration for maturity (Deckers et al. 2001). According to van Delden (2011), the crop demands 10°C as a minimum temperature to initiate germination. For optimal crop growth and development of Teff Millet, 750–850 mm of rainfall per year and 450–550 mm of rainfall per season and temperatures between 10 and 27°C is required (Sharma and Adera 2016). Teff millet is primarily cultivated at moderate elevations between 1800 and 2100 m above sea level (Habtegebrial et al. 2007; Ketema 1997). While the crop flourishes in a range of soils from heavy black to light red and in both acidic and alkaline conditions, optimal yield is achieved when cultivated in light sandy to heavy clay soils and in environments that are moderately acidic to neutral (Ketema 1997; Takele et al. 2001). According to Ketema (1993), there is a range of 4–12 days between the sowing and germination periods, with an average of 5 days. However, some of the researchers reported the time to emergence ranged from 6 to 14 days (Araya et al. 2011; Evert et al. 2009). The mean number of days till heading was 37, with a range of 26–54 days. The time to heading is greatly extended when the plant is subjected to extended photoperiods (Ketema 1983; van Delden 2011). According to Ketema (1997), the crop performs exceptionally well when the day lasts roughly 12 h. According to van Delden (2011), Ethiopia's main growing season day length ranges from 13 h during planting to 10 h at harvest. Additionally, cultivar-to-cultivar variations in time to maturity have been documented (Ketema 1993; Teklu and Tefera 2005; van Delden 2011). According to several studies (Ketema 1993; Teklu and Tefera 2005; Paff and Asseng 2018; Assefa et al. 2011), the average time to maturity for various tef cultivars was 93 days, with a range of 62–123 days.

11.4 Handling, Storage, and Processing

The harvested teff panicles are conically heaped up on sack sheets or pavement for a brief time before being subjected to human stick pounding or ox foot stomping for the threshing process. Mechanized threshing with a thresher or combination harvester is practiced in some places. It is possible to store teff for up to 5 years in a traditional storage setting by keeping it in gotera or gota. Harvested grains are packed into synthetic polymer bags in the absence of these features (IFPRI 2018). It retains the viability of its seeds for up to 3 years in storage. Since teff grains are not eaten by fungi or grain weevils like other frequently afflicted cereal grains, post-harvest maintenance expenses, such as the application of insecticides, are decreased (Mekonen et al. 2014; Kassier 2002). Prior research has documented that teff grains contain chemical residues, such as organochlorines and pyrethroids, even in the absence of pesticide treatment however, this was solely attributable to contamination in the storage and field conditions (Mekonen et al. 2014). The traditional method for storing teff in the United States involves storing the grains in towers or silos (U.S. Patent No. 565,375, 2006).

11.5 Appearance, Morphology, and Genetics

Tall and fine-stemmed, teff plants have drooping panicles with little grains. There is significant phenotypic diversity among the various tef accessions, and the Ethiopian Biodiversity Institute possesses around 5000 tef accessions (Tesema et al. 2013). Their shape, developmental timing, and capacity to withstand environmental stress vary greatly (Ebba 1975; Assefa et al. 2015). According to Adnew et al. (2005), cultivar variations are typically greater within a region than between regions.

Although they can have up to 13 leaves per culm, teff plants typically produce 2–5 leaves per culm (Ebba 1975; van Delden 2011). The leaves have an extended, sharp apex and are long, flat, and narrow (Ebba 1975). Different cultivars differ greatly in the area of their flag leaves (Assefa et al. 2015). Teff's leaves have Kranz-type anatomy because it is a C4 plant. According to Kebede et al. (1989), this indicates that they have bundle sheath cells surrounding vascular tissues, which concentrate CO_2 around Rubisco and boost photosynthetic efficiency. C4 plants can have reduced stomatal conductance during droughts, saving water and enabling them to develop in drier regions thanks to their efficient utilization of CO_2 during photosynthesis (Taylor et al. 2014).

According to Ebba (1975), the fibrous roots of the teff plant often grow 2–8 cm deep and near the soil floor. However, some times roots can grow deep as 1 m (van Delden 2011). Rarely, internodes above the plant's base may give rise to adventitious roots (Ebba 1975). The crown roots of teff frequently appear above the soil surface because the tiny seeds of teff can only be placed on the soil surface or at shallow depths. These develop into lignified bundles that descend into the soil vertically. Root bundles start to extend at an angle from the vertical at a depth of around 1 cm below the soil surface (van Delden et al. 2010).

Although only a small portion of the genetic variety between various teff types has been examined thus far, recent research has started to clarify this diversity (Assefa et al. 2015). Teff exhibits significant genotype environment interaction variations in addition to genetic heterogeneity among cultivars. This implies that distinct cultivars might react to particular environmental conditions with phenotypic variations that are noticeably different (Martinelli et al. 2018).

11.6 Phytochemical Composition and Their Health Attributes

11.6.1 Physical Properties of Teff Millet Seed

According to Morris (2016) and Abewa et al. (2019), grain quality is frequently assessed using its nutrient composition and physical characteristics. Teff grains are divided into four color categories (ESA 2012): brown (key), mixed (sergegna), white (nech), and very white (magna). For food production, the color classifications of white and very white are more favored and typically command the highest rates. In contrast to the darkest classes, which include frequently marketed to low-income consumers, these classes need a better growing environment.

With an oval form, a diameter of 0.71–0.87 mm, and a length of 1.01–1.27 mm, teff grains are among the smallest cereals (Zewdu and Solomon 2007; Akansha and Chauhan 2018; Serna-Saldivar and Espinosa-Ramirez 2019). The grains have a volumetric weight of 850–870 kg m^{-3} and a kernel weight of 0.2–0.4 g (Belay et al. 2009; Bultosa 2007). The layers of endosperm, pericarp (mesocarp, endocarp, and cuticle), and germ make up the teff grain, just like other cereals. A monolayer aleurone that is high in fatty acids and proteins is also present in the kernels (Serna-Saldivar and Espinosa-Ramirez 2019). The endosperm layer of teff makes up a sizable section of the kernel, which has a corneous endosperm area encircling a starchy floury zone (Serna-Saldivar and Espinosa-Ramirez 2019). Teff grains are rich in lipids and protein components, and their germ layer makes up a greater percentage of the grains than that of other cereals (Adebowale et al. 2011; Bultosa 2007). Rich in polyphenols and tannins, the pigmentations found in colored teff classes may offer antioxidant advantages (Abebe et al. 2025).

Table 11.1 Nutritional composition of teff millet (Akansha and Chauhan 2020; Gebru et al. 2020; Shumoy et al. 2017)

Nutrient	Amount per 100 g	Unit	Remarks
Energy	367	kcal	High energy content
Carbohydrates	73.13	g	Rich source of complex carbohydrates
Dietary fiber	8.0	g	Excellent fiber content
Protein	13.3	g	Higher than most cereals
Fat	2.4	g	Low fat content
Calcium	180	mg	Very high—beneficial for bone health
Iron	7.6	mg	Excellent source, especially for vegetarians
Magnesium	184	mg	Helps in enzymatic activities
Phosphorus	429	mg	Vital for bones and energy metabolism
Potassium	427	mg	Supports heart and muscle function
Zinc	4.4	mg	Important for immune function
Thiamin (Vitamin B1)	0.39	mg	Important for energy metabolism
Riboflavin (B2)	0.27	mg	Helps in body growth and red cell production
Niacin (B3)	3.36	mg	Supports skin health and digestion
Folate (B9)	87	µg	Essential during pregnancy
Moisture	11	g	Typical for most dry grains

11.6.2 Nutritional Composition of Teff Millet

11.6.2.1 Proteins

In order to overcome protein deficits and substitute the pricey animal proteins, cereals offer an alternate source of protein. It is commonly known that the physiological processes and food processing of cereals' seed are mainly responsible for the seed storage proteins (SSPs) for their nutritional value and food product attributes (Gorinstein et al. 2002). The protein content in teff millet was found to be between 12.8% and 20.9% (Zhu 2018; El-Alfy et al. 2012). Further, the essential amino acid content (40.9%) and total protein content (9.37%) (Table 11.1). Lysine had the greatest concentration of all the amino acids, at 12.8% of dry weight per gram flour in white teff millet and Leucine was highest in brown teff millet (17.51) (Gebru et al. 2019). Glutamins (46.62%) made up the majority of the proteins in the SSP fractions of teff, followed by prolamins (29.39%), globulins (21.1%), and albumins (39.1%) (Gebru et al. 2019). Teff proteins' rheological characteristics, including water hydration, foaming, and gelation, were investigated in various solutions. When the concentration of NaCl increased up to 1 M, the teff protein's least gelatin concentration decreased from 6% to 2% (Arogundade 2006).

11.6.2.2 Carbohydrates: Starch and Dietary Fiber

Teff grain's small size increases the amount of insoluble polysaccharides in the grain's whole flour (Marti et al. 2017). The starch content of teff grain ranged from 74% to 75.5%, whereas the total carbohydrate content was 85.6% (Abebe and Ronda 2014). Its amylose and amylopectin had molecular weights ranging from 1.0 × 106 g/mol to 1.3 × 106 g/mol and 10.1 × 107 g/mol to 16.5 × 107 g/mol, respectively, while its amylose content varied from 20% to 30% (Bultosa et al. 2002, 2008). Compared to maize starch, teff starch paste was less viscous (Bultosa et al. 2002). In comparison to maize starch, the water absorption index (WAI) was higher (108%) and the water solubility index (WSI) was lower (0.34%) for teff starch. Whole grain teff has a total dietary fiber level of 9.8% (dry basis) significantly higher than wheat (3.4%), oats (4.05%), rice (0.43%), quinoa (7.14%), sorghum (4.51%), buckwheat (2.18%), and maize (2.62%) (Hager et al. 2012a, b). These suggest that consuming more teff has significant health advantages.

11.6.2.3 Lipids

Teff grain flour has a lipid concentration of (4.4%) higher than wheat (3.6%), sorghum (3.5%), rice (0.9%), and maize (2.5%) grain flours, whereas lower than quinoa (8.6%) grain and oat (6.7%) flours (Hager et al. 2013). Eighty percent of the lipid in teff is in the free state significantly greater than the lipid's free content of buckwheat, rye, and wheat. In contrast to wheat, quinoa, buckwheat, and rye flours, acid hydrolysed found comparatively less lipid attached to starch (Collar and Angioloni 2014). Oleic and linoleic acids made up 29% and 50% of the teff lipids, respectively, indicating that the bulk of the lipids were unsaturated (Hager et al. 2013). Teff millet, therefore, has the potential to be just as beneficial as other common grains in terms of providing healthy lipids.

11.6.2.4 Bioactive Phytochemicals

In addition to being high in macronutrients and micronutrients, teff is also a good source of bioactive compounds, including saponins and phenolic compounds. According to in vitro investigation, the chemical components of teff millet grain include anti-inflammatory and antioxidant qualities as well as other health-promoting benefits. However, the constituents of these phenolic compounds are exceedingly complex and can be in bound, conjugated, and free forms, it is very difficult to estimate their precise quantities (Valdes et al. 2015; Gomez-Caravaca et al. 2011). The total phenolic content (TPC) present in teff grain flour was found to be 123.6 GAE (gallic acid equivalent)/100 g dry weight. Compared to other Ethiopian main grains like maize and wheat, this value is significantly greater (Forsido et al. 2013). Total flavonoid contents (TFCs) and TPCs of five varieties of teff millet were examined in another study (Kotaskova et al. 2016). Compared to the white teff variety, the brown teff millet had greater TFCs and TPCs. The TFCs and TPCs content of teff millet was found to be 2.19–1.41 mg GAE/g and 1.16–0.62 mg RE (rutin equivalent)/g, respectively (Gebru et al. 2020).

Ferulic acid is a key phenolic acid present in bound fractions. However, another study reported that in the soluble fraction concentration of total flavones in white and brown teff ranged from 1398–2049 μg/g, and 1720–1847 μg/g, db, respectively (Ravisankar et al. 2018). The findings indicates a marked variation in the phenolic profiles of brown and white teff millet varieties, with the white teff millet containing apigenin glycosides, while the brown teff millet is mostly composed of luteolin glycosides. The elevated levels of flavones in teff relative to other cereals provide a basis for proposing that unusual dietary phenolic chemicals can be found in considerable amounts in the grain (Ravisankar et al. 2018).

11.6.2.5 Minerals

A comparative analysis of the mineral content in teff grain revealed that phosphorus, magnesium, manganese, and copper are present in greater concentrations than in other common cereals such as wheat and maize (Forsido et al. 2013). Information from the Agricultural Research Service Database of the United States Department of Agriculture indicated that the mineral contents of the grain on a wet basis are calcium (180 mg/100 g), iron (7.63 mg/100 g), magnesium (184 mg/100 g), phosphorus (429 mg/100 g), potassium (427 mg/100 g), sodium (12 mg/100 g), and zinc (3.63 mg/100 g) (USDA 2017). Differences in mineral content have been observed between the grain types of brown and white teff millet. The types of brown teff exhibited an increased level of iron and calcium compared to the white varieties. Conversely, the white teff varieties were found to contain greater amounts of copper than their brown counterparts (Gujral et al. 2012). Nevertheless, the majority of minerals' bioavailability is affected by chelating agents like phytic acid. This poses a nutritional challenge for the grain, given that its whole grain phytic acid level is predicted to be as high as 1544 mg/100 g dry weight (Baye et al. 2014). Therefore, processing techniques like adding fungi and lactic acid bacteria to teff millet batter increase the minerals' bioavailability in teff.

11.6.2.6 Vitamins

Further research using experimental techniques is still needed to fully understand the vitamin levels and composition of teff. The USDA Food Composition Databases (USDA 2017) have included data on certain vitamins. Raw teff contains niacin (3.363 mg/100 g), vitamin B6 (0.482 mg/100 g), thiamin (0.39 mg/100 g), riboflavin (0.27 mg/100 g), vitamin K (phylloquinone) (1.9 μg/100 g), vitamin A (9 IU), and α-tocopherol (0.08 mg/100 g).

11.6.3 Teff Millet Health Attributes

1. *Celiac Disease and Immune System Support*

   Gluten intolerance, a protein present in barley, rye, and wheat, causes aberrant nutritional absorption in the small intestine's inner lining, which is the symptom of celiac disease (Green and Cellier 2007). It is estimated that the disease affects approximately 1% of the global population (Baye et al. 2014). The incidence of celiac disease is more prevalent in individuals who suffer from symptoms of iron deficiency anemia and type 1 diabetes. The condition affects 10–15% of people with symptomatic iron deficiency anemia and 3–6% of those with type 1 diabetes (Phamm-Short et al. 2015). A rigorous, lifelong gluten-free diet is the only safe food for those with celiac disease (Bascunan et al. 2017). Even without fortification, teff, a nutrient-dense gluten-free grain, is becoming more widely acknowledged as the best option for those with celiac disease in nations like the Netherlands (Bascunan et al. 2017). Furthermore, teff grain is adaptable, boasting an appealing taste and texture, making it suitable for a wide range of food items, including non-sourdough bread, sourdough bread, cookies, weaning foods, pasta, and extruded products, as well as in addition to drinks like beer and lactic acid drinks (Mancebo et al. 2015; Gebremariam et al. 2014; Ronda et al. 2015; Hager et al. 2012a, b). Furthermore, a high level of Zn in teff millets supports the immune system as zinc plays an important role in making, maturing, and signaling immune cells. Besides, iron (196 mg/kg in red teff millet and 115 mg/kg in white teff millet) is considered for its metabolism because it carries oxygen and makes energy. As, teff millet contains higher iron content than corn and wheat, it will be beneficial for pregnant women (Dame 2020).

2. *Diabetes, Heart and Bone Health*

   Diabetes, often known as diabetes mellitus, is a metabolic condition marked by elevated blood glucose levels and irregular insulin regulation. Contemporary trends toward high-fat diets, rising obesity rates, and a more sedentary lifestyle are contributing to a worldwide surge in diabetes cases (Guariguata et al. 2014). This has emerged as a significant global health issue, with an estimated 415 million individuals affected by diabetes as of 2015 (Ogurtsova et al. 2017). The primary determinants of insulin responses and glucose levels in the human body are the amount and type of carbohydrates consumed (Wolever 2000).

   The glycemic index depends on how much fiber the food has. Fiber helps to slow down the digestion and absorption of carbohydrates (Gutschall et al. 2009; Wolever 1990). It is well proven that eating low glycemic foods may help prevent type 2 diabetes (Augustin et al. 2015). Teff millet has more fiber than most other grains thus helpful in lowering blood sugar (Wolter et al. 2013). Teff foods are expected to play a major role in preventing and managing diabetes. However, due to a lack of enough proof to see how eating teff can help people with diabetes, there is a need for further research. It was reported that Ca intake may reduce the risk of cardiovascular diseases by decreasing intestinal lipid absorption, promoting lipid excretion, and reducing blood cholesterol. It has been reported that teff millet contains 78.8–147 mg/100 g Ca, supporting bone health and cardiovascular health. Further, teff millet also contains healthy unsaturated lipids, including omega-3 and omega-6 polyunsaturated fatty acids, which can help improve lipid profiles supporting the heart health. It is also rich in magnesium, which helps regulate blood pressure and ensures the heart functions properly, thereby reducing the risk of heart disease (Puranik et al. 2017).

3. *Anemia*

   Anemia is a condition characterized by a reduction in the number of red blood cells, which indicates a decrease in hemoglobin content, thereby impairing the blood's ability to transport oxygen throughout the body. Anemia affects over 25% of the global population, with a higher prevalence in developing nations (Kassebaum et al. 2014). The primary cause of anemia worldwide is a deficiency of dietary iron. In both low-income developed nations, iron deficiency anemia mostly impacts the lives of young children and premenopausal women (McLean et al. 2009). Although food fortifications and iron supplements can increase dietary iron consumption, these strategies have gastrointestinal adverse effects, such as diarrhea, nausea, and constipation (Kassebaum et al. 2014). Eating plants or seeds that have a lot of iron, like teff millet, is better and safer alternative than taking iron pills. Wheat bread made with 30% teff flour had twice as much iron as bread made with regular flour. Teff is a type of grain that can help pregnant women keep their iron levels normal (Alaunyte et al. 2012; Kassebaum et al. 2014).

11.7 Starch Isolation from Teff Millet

Starch isolation from Teff millet involves extracting pure starch granules while removing other components like proteins, lipids, and fibers. The primary objective is to obtain high-purity starch suitable for food, industrial, or research applications. Various methods have been developed and optimized for starch extraction from Teff millet, and they generally fall into mechanical, chemical, and enzymatic categories (Mahajan et al. 2021). The wet milling method is most commonly used in large-scale starch production, while alkali and enzymatic methods offer higher-purity starch with specific functional properties.

Sidhu et al. (2024) extracted starch from teff millet using alkali isolation approach. For this purpose, the Teff millet grains were soaked in an aqueous sodium hydroxide solution (NaOH 0.5%) for 12 h. Following this, the grains were subjected to wet milling using a grinder and subsequently filtered three times using a muslin cloth. The resulting slurry was then processed through 100 and 150 mesh sieves. After several steps of grinding and filtering, the slurry was then centrifuged for 15 min at 5000 rpm, the residue or solid layer was collected, and the supernatant was discarded. The collected starch was then dried at 45°C for 24 h. Upon completion of the drying process, it was ground and stored in an air-tight container (Sidhu et al. 2025).

Starch was extracted from teff millet using the method established by Bultosa et al. (2002), with some adjustments made. Thoroughly cleaned teff grains were ground with a grinder. The resulting powder was then sieved through a 600-mesh sieve. A 1000 ml beaker was filled with 100 g of teff grain powder and 500 ml of deionized water. The mixture was thoroughly mixed at 30°C for an hour. The resultant slurry was thoroughly pulverized using wet milling processes. Furthermore, it was filtered twice using a 300 mesh sieve. The filtrate was then centrifuged at 9000 rpm for 6 min at 30°C. After removing the supernatant, the gray layer was carefully separated with a spatula. The resultant residue was further diluted with distilled water. These steps were continued until the gray coating was no longer noticeable. Deionized water was used to disseminate the final product, which was then dried for 30 h at 55°C. The isolated starch was ground in a mortar and then sieved through a 300-mesh sieve (Prabhu et al. 2021).

Further, starch from teff millet was extracted in accordance with the method described by Gebremariam and Schimit (1996), with minor modifications. Teff flour was mixed in a 0.075% sodium metabisulfite solution for 2 h, stirring periodically during the first hour. The supernatant containing soluble materials was decanted, and the suspension was treated repeatedly with sodium metabisulfite solution. Insoluble materials such as fibers, sand, lipids, etc., were eliminated from the suspension by continuous filtration using sieves and nylon cloth. The suspension was allowed to sit overnight to facilitate the separation of starch, after which the supernatant was carefully decanted. The starch was then dried at 50°C. The dried starch was ground using a mortar and pestle, sieved through a 224 μm sieve, and stored in a air tight glass jar (Assefa et al. 2020).

11.8 Modification of Teff Millet Starch

Citric acid was used to crosslink teff starch in accordance with the methodology described by Kim et al. (2017). After dissolving 10% and 20% w/w of citric acid based on starch in distilled water, 10 M NaOH was added to bring the pH of the mixture down to 3.5. This solution was then mixed with 100 g of starch and left to stand at room temperature at 30°C for 16 h. To remove any remaining citric acid, the resultant suspension was centrifuged for 5 min at 5000 rpm and then rinsed thrice. After being cleaned, the starch was pulverized with a pestle and mortar, dried for 24 h at 40°C in an oven, and kept at 4°C until it could be examined further. The result shows that after modification of starch with citric acid (10% and 20%), the resistant starch content increases from 21.28% to 23.43% and 24.11 % respectively. While the swelling power of citrate starch decreases from 10.41% to 10.07% and 8.90%, respectively (Sidhu et al. 2025).

The oxidation of teff starch was carried out using NaOCl by Zhou et al. (2016). This slurry (40% starch w/v) was kept at 35°C, and 2 N NaOH was used to bring its pH down to 9.5. While keeping the pH at 9.5, sodium hypochlorite (NaOCl) was added gradually and stirred constantly for 30 min to reach final concentrations of

0.8 and 1.2 g of active chlorine per 100 g of starch. After the addition of NaOCl, the mixture was continuously stirred for an additional 50 min using 1 N NaOH to maintain the pH 9.5. Following a pH 7.0 neutralization with 1 N H_2SO_4, the slurry were filtered, cleaned with distilled water, dried for 24 h at 40°C in an oven, pulverized using a pestle and mortar, and then kept for further examination. The result shows that after oxidation of starch the resistant starch content increases from 21.28% to 26.48 %and 31.42% respectively. While the swelling power of oxidised starch decreases from 10.41% to 9.3% and 8.93% respectively (Sidhu et al. 2025).

The enzymatic modification of teff millet starch was conducted following the method described by Dura et al. (2014). Specifically, 50 g of teff millet starch was suspended in 25 ml of pH 4.7 sodium acetate buffer for the enzymatic modification. In this starch suspension, 6.66 mg of amyloglucosidase (200 U/g) and 125 mg of amylase (250 U/g) were added and the suspension was incubated for 90 min at 37°C in a shaking water bath. Than the suspension was heated at 70°C for 10 min and brought to room temperature, to stop the reaction. After that, the solution was centrifuged for 10 min at 3000 rpm, dried for 24 h at 40 °C, and cleaned with distilled water. After enzymatic modification of starch, the swelling power of starch decreases from 10.41% to 5.55% and resistant starch content increases (Sidhu et al. 2025).

11.9 Food and Non-Food Applications

Teff is primarily transformed into an array of foods and drinks, including sweets, breads, porridges, biscuits, pancakes, cookies, stir-fry dishes, cakes, casseroles, stews, puddings, and soups (Alaunyte et al. 2012; Arendt and Zannini 2013). Breads made from white teff millet show increased dietary fiber and protein content (from 10.7 up to 13.1%) (Hofmanova et al. 2014).

Numerous studies have looked at teff flour as a key component for improving the quality of gluten-free goods, mostly because of its remarkable nutritional profile. The technological characteristics of teff suggest a substantial potential for its application in diverse food products, especially in baked goods (Campo et al. 2016). Teff grain flour is extensively utilized in Ethiopia for preparing injera (a staple for most Ethiopians, this is a fermented, pancake-like, soft, sour, circular flatbread), local spirit, sweet unleavened bread, soups, and porridges (Fischer et al. 2014; Bultosa 2007). Teff flour is an excellent substitute for gluten in gluten-free goods since it is not only gluten-free but also naturally enhanced with increased quantities of calcium, fiber, and iron (Kenney et al. 2011).

Lately, the incorporation of teff in to food systems has become increasingly popular—both in the infant food industry as a nutrient-dense ingredient and as a naturally gluten-free alternative to wheat products. Teff flour incorporated wheat flour bread has been developed and investigated for nutritional value. It was reported that incorporation of teff flour significantly increased the in vitro antioxidant activity while decreasing the sensory and textural characteristics (Alaunyte et al. 2012; Ronda et al. 2015). Breads made from white teff millet shows increased dietary fiber and protein content (from 10.7 up to 13.1%) (Hofmanova et al. 2014). Even though teff flour is gluten-free, it has been shown to produce excellent leavened flatbread that remains fresh longer than bread prepared from other cereals, especially sorghum, which is frequently used in both conventional flatbreads and gluten-free baked products (Adebowale et al. 2011). Teff flour was explored for the production of gluten free egg spaghetti and the developed spaghetti was compared with that of oat and wheat based spaghetti. It was reported that the texture of oat and teff millet flour spaghetti were similar to that of wheat spaghetti. However, the cooking time for teff spaghetti (12 min) was found to be slightly higher than the time required for oat and wheat ones (10 min) (Hager et al. 2012a).

In Ethiopia, the cereal is similarly employed in the fabrication of indigenous alcoholic beverages that are characteristic of the region, including a non-transparent beer recognized as tella, a distilled beverage designated as shamit, and katikala/arake at the domestic level (Tekle et al. 2019).

Tella possesses a smoky flavor derived from the incorporation of bread that has been darkened through specific baking techniques and fermentation processes that utilize a vessel smoked by inversion over smoldering weyra wood. Besides the cereal, the most crucial ingredient in producing tella is the leaves of Gesho (*Rhamnus prinoides*), which contribute to the distinctive bitterness of the drink. Research findings have indicated that Gesho influences the microflora responsible for the fermentation process (Kleyn and Hough 1971). Furthermore, it has been observed that the bitterness of the brew is directly correlated with the quantity of Gesho added. Tella, a locally produced beverage, is not governed by stringent government regulations, with its alcohol content varying between 2% and 4% vol, whereas filtered tella exhibits a higher alcohol concentration ranging from 5% to 6% vol (Selinus 1971).

Arake is a distilled beverage made from ground Gesho leaves and water, which are left to steep for a duration of 3–4 days. After this period, unleavened bread known as kita, made from teff or other grains, along with germinated wheat and barley, is introduced. This amalgamation is then permitted to ferment for a period of 5–6 days before being distilled. In rural areas, distillation performed by distillation system developed from gourds and wood, typically undergoing redistillation, resulting in an alcoholic concentration that falls within the range of 45–50% vol (Selinus 1971).

Shamit is a locally brewed beer produced by the Gurage ethnic group. Kita, or unleavened teff bread, and bekel, or germinated barley, are ground and mixed with water. After 3–4 days of fermentation, the combination is filtered. To prepare the beverage for consumption the following day, dehusked barley is toasted on a metal plate (known as milted), pulverized, and added to the mixture (Selinus 1971).

Teff has been thoroughly examined regarding its potential as a base material for biofilms. In this research, a biofilm was created using the starch of teff millet grains. Additionally, the mechanical, optical, and water-resistance properties of starch-based films of teff millet grain were analyzed. The findings indicated that edible film made from teff starch is highly suitable for food packaging applications. Furthermore, when compared to the standard cassava-starch-based biofilm, the optical and mechanical properties demonstrated that films of teff starch are more fitting as packaging materials (Prabhu et al. 2021).

The research conducted by Teshome et al. in 2024 assesses resistant starch derived from teff millet grain as a film coating material to deliver metronidazole to the colon, which serves as a model drug. Starch was extracted from teff, and resistant starch was isolated from the total starch. Metronidazole core tablets were produced through wet granulation, compressed, and subsequently coated with a resistant starch-based film. The physicochemical properties of the tablets were evaluated in vitro. The targeted drug release from the film material is attributed to the breakdown of the component of resistant starch by bacterial enzymes within the colon, and the bacterial microflora flourishes it. Based on these findings, resistant starch from teff millet grain appears to hold promise as a colon-targeting excipient.

11.10 Conclusions

Teff millet stands out as a nutritionally superior and functionally versatile grain with significant potential in both food and non-food industries. Its rich chemical composition particularly its high content of essential minerals, amino acids, and dietary fiber makes it a valuable ingredient for developing health-focused and gluten-free food products. The unique physicochemical characteristics of teff starch, including its small granule size and high amylopectin content, offer numerous advantages in food processing and product formulation. However, to expand its applicability, especially in industrial contexts, modification of native starch is often essential. This chapter has detailed various starch isolation techniques and explored physical, chemical, and enzymatic modifications that improve teff starch's functional properties, such as thermal stability, solubility, pasting behavior, and digestion resistance. These modifications not only enhance its performance in food systems but also open avenues for non-food uses such as biodegradable films, controlled drug delivery systems, and other biopolymer-based applications. The integration of teff into diverse sectors contributes to agricultural sustainability, value addition, and economic opportunities, particularly in regions where the crop is traditionally cultivated. As global demand for functional and clean-label ingredients grows, teff millet emerges as a promising, underutilized grain deserving of further research and commercialization.

References

Abebe W, Ronda F (2014) Rheological and textural properties of tef [*Eragrostis tef* (Zucc.) Trotter] grain flour gels. J Cereal Sci 60(1):122–130. https://doi.org/10.1016/j.jcs.2014.02.001

Abebe BS, Aprodu I, Istrati DI, Vizireanu C (2025) Nutritional characteristics, health-related properties, and food application of teff (Eragrostis tef): An overview. Int J Mol Sci 26(19):9293

Abewa A, Adgo E, Yitaferu B et al (2019) Teff grain physical and chemical quality responses to soil physicochemical properties and the environment. Agronomy 9(6):283. https://doi.org/10.3390/agronomy9060283

Abraham, R (2015) Achieving Food Security in Ethiopia by Promoting Productivity of Future World Food Tef: A Review. Adv Plants Agric Res 2;86–95. https://doi.org/10.15406/apar.2015.02.00045

Adebowale AR, Emmambux MN, Beukes M, Taylor JR (2011) Fractionation and characterization of teff proteins. J Cereal Sci 54(3):380–386. https://doi.org/10.1016/j.jcs.2011.08.002

Admas S, Belay G (2011) Drought-resistance traits variability in *Eragrostis tef* × *Eragrostis pilosa* recombinant inbred lines. Afr J Agric Res 6(16):3755–3761. https://doi.org/10.5897/AJAR10.612

Adnew T, Ketema S, Tefera H, Sridhara H (2005) Genetic diversity in tef [*Eragrostis tef* (Zucc.) Trotter] germplasm. Genet Resour Crop Evol 52:891–902. https://doi.org/10.1007/s10722-003-6099-0

Alaunyte I, Stojceska V, Plunkett A et al (2012) Improving the quality of nutrient-rich Teff (Eragrostis tef) breads by combination of enzymes in straight dough and sourdough breadmaking. J Cereal Sci 55(1):22–30

Akansha K, Chauhan E (2018) Nutritional composition, physical characteristics, and health benefits of teff grain for human consumption: a review. J Pharm Innov 7:03–07

Akansha CS, Chauhan ES (2020) Teff millet: nutritional, phytochemical and antioxidant potential. Int J Pharmaceut Sci Res 11:6005-9. https://doi.org/10.13040/IJPSR.0975-8232.11(12).6005-09

Araya A, Stroosnijder L, Girmay G, Keestra SD (2011) Crop coefficient, yield response to water stress and water productivity of teff (*Eragrostis tef* (Zucc.)). Agric Water Manag 98:775–783. https://doi.org/10.1016/j.agwat.2010.12.001

Arendt EK, Zannini E (2013) Cereal grains for the food and beverage industries. Elsevier.

Arogundade LA (2006) Functional characterization of tef (*Eragostics tef*) protein concentrate: influence of altered chemical environment on its gelation, foaming, and water hydration properties. Food Hydrocoll 20(6):831–838. https://doi.org/10.1016/j.foodhyd.2005.07.012

Assefa K, Yu JK, Zeid M et al (2011) Breeding tef [*Eragrostis tef* (Zucc.) trotter]: conventional and molecular approaches. Plant Breed 130(1):1–9. https://doi.org/10.1111/j.1439-0523.2010.01782.x

Assefa K, Cannarozzi G, Girma D et al (2015) Genetic diversity in tef [*Eragrostis tef* (Zucc.) Trotter]. Front Plant Sci 6:177. https://doi.org/10.3389/fpls.2015.00177

Assefa F, Dilebo J, Gabriel T et al (2020) Characterization and tablet property evaluation of pregelatinized starch of teff (*Eragrostis tef*). Asian J Pharm Res Dev 8(5):18–23. https://doi.org/10.22270/ajprd.v8i5.819

Augustin LSA, Kendall CWC, Jenkins DJA et al (2015) Glycemic index, glycemic load and glycemic response: an international scientific consensus summit from the International Carbohydrate Quality Consortium (ICQC). Nutr Metab Cardiovasc Dis 25(9):795–815. https://doi.org/10.1016/j.numecd.2015.05.005

Bascunan KA, Vespa MC, Araya M (2017) Celiac disease: understanding the gluten-free diet. Eur J Nutr 56(2):449–459. https://doi.org/10.1007/s00394-016-1238-5

Baye K, Mouquet-Rivier C, Icard-Verniere C et al (2014) Changes in mineral absorption inhibitors consequent to fermentation of Ethiopianinjera: implications for predicted iron bioavailability and bioaccessibility. Int J Food Sci Technol 49(1):174–180. https://doi.org/10.1111/ijfs.12295

Belay G, Zemede A, Assefa K, Metaferia G, Tefera H (2009) Seed size effect on grain weight and agronomic performance of tef [*Eragrostis tef* (Zucc.) Trotter]. Afr J Agric Res 4(9):836–839

Bultosa G (2007) Physicochemical characteristics of grain and flour in 13 tef [*Eragrostis tef* (Zucc.) Trotter] grain varieties. Res J Appl Sci 3(12):2042–2051

Bultosa G, Hall AN, Taylor JR (2002) Physico-chemical characterization of grain tef [*Eragrostis tef* (Zucc.) Trotter] starch. Starch-Starke 54(10):461–468. https://doi.org/10.1002/1521-379X(200210)54:10<461::AID-STAR461>3.0.CO;2-U

Bultosa G, Hamaker BR, Bemiller JN (2008) An SEC−MALLS study of molecular features of water-soluble amylopectin and amylose of tef [*Eragrostis tef* (Zucc.) Trotter] starches. Starch-Starke 60(1):8–22. https://doi.org/10.1002/star.200700642

Campo E, del Arco L, Urtasun L et al (2016) Impact of sourdough on sensory properties and consumers' preference of gluten-free breads enriched with teff flour. J Cereal Sci 67:75–82. https://doi.org/10.1016/j.jcs.2015.09.010

Collar C, Angioloni A (2014) Pseudocereals and teff in complex breadmaking matrices: impact on lipid dynamics. J Cereal Sci 59(2):145–154. https://doi.org/10.1016/j.jcs.2013.12.008

Dame ZT (2020) Analysis of major and trace elements in teff (Eragrostis tef). J King Saud Univ Sci 32(1):145–148

Deckers J, Yizengaw T, Negeri A, Ketema S (2001) Tef [*Eragrostis tef* (Zucc.) Trotter]. In: Raemaekers RH (ed) Crop production in tropical Africa. DGIC, Brussels

Do Nascimento KD, Paes SD, de Oliveira IR et al (2018) Teff: suitability for different food applications and as a raw material of gluten-free, a literature review. J Food Nutr Res 6(2):74–81. https://doi.org/10.12691/jfnr-6-2-2

Dura A, Błaszczak W, Rosell CM (2014) Functionality of porous starch obtained by amylase or amyloglucosidase treatments. Carbohydr Polym 101:837–845. https://doi.org/10.1016/j.carbpol.2013.10.013

Ebba T (1975) Tef (*Eragrostis tef*) cultivars: morphology and classification. Part II, experiment station bulletin 66, Addis Ababa University, College of Agriculture, Dire Dawa

El-Alfy TS, Ezzat SM, Sleem AA (2012) Chemical and biological study of the seeds of *Eragrostis tef* (Zucc.) Trotter. Nat Prod Res 26(7):619–629. https://doi.org/10.1080/14786419.2010.538924

Ethiopian Standard Agency (2012) Teff seed specification. ES 416: 2000. Ethiopia

Evert S, Staggenborg S, Olson B (2009) Soil temperature and planting depth effects on tef emergence. J Agron Crop Sci 195:232–236. https://doi.org/10.1111/j.1439-037X.2008.00343.x

Fischer MM, Egli IM, Aeberli I et al (2014) Phytic acid degrading lactic acid bacteria in tef-injera fermentation. International Journal of Food Microbiology 190:54–60

Forsido SF, Rupasinghe HPV, Astatkie T (2013) Antioxidant capacity, total phenolics and nutritional content in selected ethiopian staple food ingredients. Int J Food Sci Nutr 64(8):915–920. https://doi.org/10.3109/09637486.2013.806448

Gebremariam T, Schmidt PC (1996) Isolation and physicochemical properties of Enset starch. Starch-Starke 48(6):208–214. https://doi.org/10.1002/star.19960480603

Gebremariam MM, Zarnkow M, Becker T (2014) Teff (*Eragrostis tef*) as a raw material for malting, brewing and manufacturing of gluten-free foods and beverages: a review. J Food Sci Technol 51(11):2881–2895. https://doi.org/10.1007/s13197-012-0745-5

Gebru YA, Hyun-Il J, Young-Soo K et al (2019) Variations in amino acid and protein profiles in white versus brown teff (*Eragrostis tef*) seeds, and effect of extraction methods on protein yields. Foods 8(6):202. https://doi.org/10.3390/foods8060202

Gebru YA, Sbhatu DB, Kim KP (2020) Nutritional composition and health benefits of teff (*Eragrostis tef* (Zucc.) Trotter). J Food Qual 2020(1):9595086. https://doi.org/10.1155/2020/9595086

Girma D, Assefa K, Chanyalew S et al (2014) The origins and progress of genomics research on Tef (*Eragrostis tef*). Plant Biotechnol J 12(5):534–540. https://doi.org/10.1111/pbi.12199

Gomez-Caravaca AM, Segura-Carretero A, Fernandez-Gutierrez A, Caboni MF (2011) Simultaneous determination of phenolic compounds and saponins in quinoa (*Chenopodium quinoa* Willd) by a liquid chromatography-diode array detection-electrospray ionization-Time-of-flight mass spectrometry methodology. J Agric Food Chem 59(20):10815–10825. https://doi.org/10.1021/jf202224j

Gorinstein S, Pawelzik E, Delgado-Licon E et al (2002) Characterisation of pseudocereal and cereal proteins by protein and amino acid analyses. J Sci Food Agric 82(8):886–891. https://doi.org/10.1002/jsfa.1120

Green PHR, Cellier C (2007) Celiac disease. N Engl J Med 357(17):1731–1743. https://doi.org/10.1056/nejmra071600

Guariguata L, Whiting DR, Hambleton I et al (2014) Global estimates of diabetes prevalence for 2013 and projections for 2035. Diabetes Res Clin Pract 103(2):137–149. https://doi.org/10.1016/j.diabres.2013.11.002

Gujral N, Freeman HJ, Thomson ABR (2012) Celiac disease: prevalence, diagnosis, pathogenesis and treatment. World J Gastroenterol 18(42):6036–6059. https://doi.org/10.3748/wjg.v18.i42.6036

Gutschall MD, Miller CK, Mitchell DC, Lawrence FR (2009) A randomized behavioural trial targeting glycaemic index improves dietary, weight and metabolic outcomes in patients with type 2 diabetes. Public Health Nutr 12(10):1846–1854. https://doi.org/10.1017/s1368980008004680

Habtegebrial K, Singh BR, Haile M (2007) Impact of tillage and nitrogen fertilization on yield, nitrogen use efficiency of tef (*Eragrostis tef* (Zucc.) Trotter) and soil properties. Soil Tillage Res 94(1):55–63. https://doi.org/10.1016/j.still.2006.07.002

Hager AS, Lauck F, Zannini E, Arendt EK (2012a) Development of gluten-free fresh egg pasta based on oat and teff flour. Eur Food Res Technol 235(5):861–871. https://doi.org/10.1007/s00217-012-1813-9

Hager AS, Wolter A, Jacob F et al (2012b) Nutritional properties and ultra-structure of commercial gluten free flours from different botanical sources compared to wheat flours. J Cereal Sci 56(2):239–247. https://doi.org/10.1016/j.jcs.2012.06.005

Hager AS, Czerny M, Bez J et al (2013) Starch properties, in vitro digestibility and sensory evaluation of fresh egg pasta produced from oat, teff and wheat flour. J Cereal Sci 58(1):156–163. https://doi.org/10.1016/j.jcs.2013.03.004

Haileselassie B, Stomph T, Hoffland E (2011) Teff (*Eragrostis tef*) production constraints on Vertisols in Ethiopia: farmers' perceptions and evaluation of low soil zinc as yield limiting factor. Soil Sci Plant Nutr 57(4):587–596. https://doi.org/10.1080/00380768.2011.593482

Harrouni MC (1989) The effect of water stress on growth and development of barley and tef. PhD thesis, University College of North Wales, Bangor, UK

Hofmanová T, Hrušková M, Švec I (2014) Evaluation of wheat/non-traditional flour composite. Czech Journal of Food Sciences 32(3):288

International Food Policy Research Institute (IFPRI) (2018) The economics of teff: exploring Ethiopia's biggest cash crop. https://doi.org/10.2499/9780896292833

Kassebaum NJ, Jasrasaria R, Naghavi M et al (2014) A systematic analysis of global anemia burden from 1990 to 2010. Blood 123(5):615–624. https://doi.org/10.1182/blood-2013-06-508325

Kassier SB (2002) Comparative responses of fodder and grain teff (*Eragrostis tef* (Zucc.) Trotter) cultivars to spatial, temporal and nutritional management. Thesis (M.Sc. Agric.), University of Natal, Pietermaritzburg

Kebede H, Johnson R, Ferris D (1989) Photosynthetic response of *Eragrostis tef* to temperature. Physiol Plant 77(2):262–266. https://doi.org/10.1111/j.1399-3054.1989.tb04979.x

Kenney ES, Butler C, Moore C et al (2011) The effect of substituting teff flour in gluten-free sugar cookies and peanut butter cookies. J Am Diet Assoc 111(9):A63. https://doi.org/10.1016/J.JADA.2011.06.230

Ketema S (1983) Studies on lodging, floral biology and breeding techniques in tef [*Eragrostis tef* (Zucc.) Trotter]. PhD thesis, University of London, UK

Ketema S (1993) Phenotypic variations in tef (*Eragrostis tef*) germplasm-morphological and agronomic traits, a catalog. Technical manual no. 6. Institute of Agricultural Research, Addis Ababa, Ethiopia

Ketema S (1997) Tef-*Eragrostis tef* (Zucc.). Bioversity International, Rome

Kim JY, Lee YK, Chang YH (2017) Structure and digestibility properties of resistant rice starch cross-linked with citric acid. Int J Food Prop 20(2):2166–2177. https://doi.org/10.1080/10942912.2017.1368551

Kleyn JO, Hough JA (1971) The microbiology of brewing. Ann Rev Microbiol 25(1):583–608

Kotaskova E, Sumczynski D, Mlcek J, Valasek P (2016) Determination of free and bound phenolics using HPLC-DAD, antioxidant activity and in vitro digestibility of *Eragrostis tef*. J Food Compos Anal 46:15–21. https://doi.org/10.1016/j.jfca.2015.11.001

Lee H (2018) Teff, a rising global crop: current status of teff production and value chain. Open Agric J 12(1):185–193. https://doi.org/10.2174/1874331501812010185

Lemecha M, Morino K, Seifu D et al (2018) Improved glucose metabolism by *Eragrostis tef* potentially through beige adipocyte formation and attenuating adipose tissue inflammation. PLoS One 13(8):e0201661. https://doi.org/10.1371/journal.pone.0201661

Mahajan P, Bera MB, Panesar PS, Chauhan A (2021) Millet starch: a review. Int J Biol Macromol 180:61–79. https://doi.org/10.1016/j.ijbiomac.2021.03.063

Mancebo CM, Picon J, Gomez M (2015) Effect of flour properties on the quality characteristics of gluten free sugar-snap cookies. LWT—Food Sci Technol 64(1):264–269. https://doi.org/10.1016/j.lwt.2015.05.057

Marti A, Marengo M, Bonomi F et al (2017) Molecular features of fermented teff flour relate to its suitability for the production of enriched gluten-free bread. LWT 78:296–302. https://doi.org/10.1016/j.lwt.2016.12.042

Martinelli F, Cannarozzi G, Balan B et al (2018) Identification of miRNAs linked with the drought response of tef [*Eragrostis tef* (Zucc.) Trotter]. J Plant Physiol 224:163–172. https://doi.org/10.1016/j.jplph.2018.02.011

McLean E, Cogswell M, Egli I et al (2009) Worldwide prevalence of anaemia, WHO vitamin and mineral nutrition information system, 1993–2005. Public Health Nutr 12(4):444–454. https://doi.org/10.1017/s1368980008002401

Mekonen S, Ambelu A, Spanoghe P (2014) Pesticide residue evaluation in major staple food items of Ethiopia using the Quechers method: a case study from the Jimma Zone. Environ Toxicol Chem 33:1294–1302. https://doi.org/10.1002/etc.2554

Miller D (2014) Teff grass—crop overview and forage production guide, 2nd edn. Cal/West Seeds. https://kingsagriseeds.com/wp-content/uploads/2014/12/Teff-Grass-Management-Guide.pdf. Accessed 20 Apr 2020

Ministry of Agriculture MoA (1998) Agro-ecological zonations of Ethiopia. Addis Ababa, Ethiopia

Morris C (2016) Grain quality attributes for cereals other than wheat. Encycl Food Grains 3:257–261. https://doi.org/10.1016/b978-0-12-394437-5.00248-5

National Research Council, NRC (1996). Global Affairs, Office of International Affairs, Board on Science, Technology for International Development. Lost crops of Africa: volume I: grains. National Academies Press.

O'Connor A (2016) Is teff the new super grain. The New York Times 8:16.

Ogurtsova K, da Rocha Fernandes JD, Huang Y et al (2017) IDF Diabetes Atlas: global estimates for the prevalence of diabetes for 2015 and 2040. Diabetes Res Clin Pract 128:40–50. https://doi.org/10.1016/j.diabres.2017.03.024

Paff K, Asseng S (2018) A review of tef physiology for developing a tef crop model. Eur J Agron 94:54–66. https://doi.org/10.1016/j.eja.2018.01.008

Pham-Short A, Donaghue KC, Ambler G (2015) Screening for celiac disease in type 1 diabetes: a systematic review. Pediatr 136(1):e170–e176. https://doi.org/10.1542/peds.2014-2883

Prabhu SV, Hemalatha P, Tizazu BZ, Gosu A (2021) Development of teff-starch based edible film: mechanical and optical properties. IOP Conf Ser Mater Sci Eng 1091(1):012050. https://doi.org/10.1088/1757-899X/1091/1/012050

Puranik S, Kam J, Sahu PP et al (2017) Harnessing finger millet to combat calcium deficiency in humans: challenges and prospects. Frontiers in Plant Science 8:1311

Ravisankar S, Abegaz K, Awika JM (2018) Structural profile of soluble and bound phenolic compounds in teff (*Eragrostis tef*) reveals abundance of distinctly different flavones in white and brown varieties. Food Chem 263:265–274. https://doi.org/10.1016/j.foodchem.2018.05.002

Ronda F, Abebe W, Perez-Quirce S, Collar C (2015) Suitability of tef varieties in mixed wheat flour bread matrices: a physico-chemical and nutritional approach. J Cereal Sci 64:139–146. https://doi.org/10.1016/j.jcs.2015.05.009

Roseberg RJ, Norberg S, Smith J et al (2005) Yield and quality of teff forage as a function of varying rates of applied irrigation and nitrogen. Klamath Exp Stat 1069:119–136. https://agsci.oregonstate.edu/sites/agscid7/files/kbrec/documents/hort/teff.pdf

Selinus R (1971) The traditional foods of the central Ethiopian highlands, research report no. 7. The Scandinavian Institute of African Studies. http://ethnomed.org/clinical/nutrition/the-traditional-foods-of-the-central-ethiopian. Accessed 02 Oct 2010

Serna-Saldivar SO, Espinosa-Ramirez J (2019) Grain structure and grain chemical composition. In: Sorghum and millets, 2nd edn. AACC International Press. https://doi.org/10.1016/B978-0-12-811527-5.00005-8

Sharma JJ, Adera S (2016) Response of Tef [Eragrostis tef (Zucc.) Trotter] to Different Blended Fertilizer Rates on Vertisols in Jama District. Northeastern Ethiopia (Doctoral dissertation, Haramaya University).

Shumoy H, Raes K (2016) Antioxidant potentials and phenolic composition of tef varieties: an indigenous Ethiopian cereal. Cereal Chem 93(5):465–470. https://doi.org/10.1094/CCHEM-10-15-0210-R

Shumoy H, Lauwens S, Gabaza M et al (2017) Traditional fermentation of tef injera: impact on in vitro iron and zinc dialysability. Food Res Int 102:93–100. https://doi.org/10.1016/j.foodres.2017.09.092

Sidhu RK, Riar CS, Singh S (2024) Influence of heat moisture, ultrasonication, and gamma irradiation on pasting, thermal, morphological, and physicochemical properties of Indian Teff (*Eragrostis tef*) starch. Starch-Starke 76(7–8):2300241. https://doi.org/10.1002/star.202300241

Sidhu RK, Riar CS, Singh S (2025) In vitro starch digestibility and physicochemical properties of chemical and enzymatic modified Indian Teff (*Eragrostis tef*) starch for industrial applications. Int J Biol Macromol 307(1):141910. https://doi.org/10.1016/j.ijbiomac.2025.141910

Takele A, Kebede H, Simane B (2001) Physiological research in tef. In: Tefera H, Belay G, Sorerels M (eds) Narrowing the rift: tef research and development, proceedings of the international workshop on tef genetics and improvement, 16–19 October 2000. Addis Ababa, Ethiopia

Taylor SH, Ripley BS, Martin T et al (2014) Physiological advantages of C4 grasses in the field: a comparative experiment demonstrating the importance of drought. Glob Chang Biol 20(6):1992–2003. https://doi.org/10.1111/gcb.12498

Tekle B, Jabasingh SA, Fantaw D et al (2019) An insight into the Ethiopian traditional alcoholic beverage: Tella processing, fermentation kinetics, microbial profiling and nutrient analysis. Lwt 107:9–15

Teklu Y, Tefera H (2005) Genetic improvement in grain yield potential and associated agronomic traits of tef (Eragrostis tef). Euphytica 141(3):247–54. https://doi.org/10.1007/s10681-005-7094-7

Tesema A, Assefa K, Chanyalew S, Tadele Z (2013) Genetic resources of tef in Ethiopia. Tef improvement. In: Proceedings of the second international workshop, University of Bern, Switzerland, 7–9 November, 2011, Debre Zeit, Ethiopia, EIAR, Addis Ababa, Ethiopia; Institute of Plant Sciences. Printed at Stampfli AG, 3001 Bern, Switzerland 15–20

Teshome Y, Belete A, Gebre-Mariam T (2024) Evaluation of resistant starch from teff (*Eragrostis tef*) grain as a film coating material for colon-targeted drug delivery. bioRxiv. https://doi.org/10.1101/2024.08.06.606940

USDA Food Composition Databases, Food Data Central (2017) USDA. Washington, DC, USA

Valdes A, Vidal L, Beltran A et al (2015) Microwave-assisted extraction of phenolic compounds from almond skin byproducts (*Prunus amygdalus*): a multivariate analysis approach. J Agric Food Chem 63(22):5395–5402. https://doi.org/10.1021/acs.jafc.5b01011

Van Delden SH (2011) On seed physiology, biomechanics and plant phenology in *Eragrostis tef*. Wageningen University, Wageningen

Van Delden S, Ennos A, Stomph T (2010) Analysing of lodging of the panicle bearing cereal teff (*Eragrostis tef*). New Phytol 186(3):696–707

Woldemariam F, Mohammed A, Fikre Teferra T, Gebremedhin H (2019) Optimization of amaranths–teff–barley flour blending ratios for better nutritional and sensory acceptability of injera. Cogent Food Agric 5(1):1565079. https://doi.org/10.1080/23311932.2019.1565079

Wolever TM (1990) Relationship between dietary fiber content and composition in foods and the glycemic index. Am J Clin Nutr 51(1):72–75. https://doi.org/10.1093/ajcn/51.1.72

Wolever TMS (2000) Dietary carbohydrates and insulin action in humans. Br J Nutr 83(S1):S97–S102. https://doi.org/10.1017/s0007114500001021

Wolter A, Hager AS, Zannini E, Arendt EK (2013) In vitro starch digestibility and predicted glycaemic indexes of buckwheat, oat, quinoa, sorghum, teff and commercial gluten-free bread. J Cereal Sci 58(3):431–436. https://doi.org/10.1016/j.jcs.2013.09.003

Zhou F, Liu Q, Zhang H, Chen Q, Kong B (2016) Potato starch oxidation induced by sodium hypochlorite and its effect on functional properties and digestibility. Int J Biol Macromol 84:410–417. https://doi.org/10.1016/j.ijbiomac.2015.12.050

Zewdu AD, Solomon WK (2007) Moisture-dependent physical properties of tef seed. Biosystems engineering 96(1):57–63

Zhu F (2018) Chemical composition and food uses of teff (*Eragrostis tef*). Food Chem 239:402–415. https://doi.org/10.1016/j.foodchem.2017.06.101

Chapter 12
Job's Tears Millet: Chemical Composition, Starch Isolation, Modification, and Industrial Applications

Sushmita Paramanick, Mayank Kumar Malik, and Satya Prakash

12.1 Introduction

12.1.1 History of Job's Tears Millet

The origin of Job's tears (JT) as a crop and cultigen in India is associated with the Aryan invaders who grew it on the Himalayan slopes or the Mongolian conquests that spread the crop from the eastern Himalayan region to lower-humid subtropical regions (Arora 1977). Prior to the widespread popularity of maize in South Asia, JT was considered an important minor cereal and served as a secondary food source as well as a crop with economic value and fodder, particularly in northern India (Fu et al. 2019).

There are regional variations in the names given to JT throughout India. In Sanskrit, it is known as Gavedhuka, Varshani, Garala, Ashrubeeja, Kuruvinda, or Todaparni. In Hindi, it is called Kasi, Kasai, Saṃkrū, or Vaijanti; in Bengali, Gurgur; in Marathi, Ran Maka; in Kannada, Ashrubeeja; in Tamil, Kattukandumani; in Telugu, Adaviguruginja; in Malayalam, Kattugotampu; in Manipuri, Ngum or Changing; in Mizoram, Mim; and in Meghalaya, Sohriew. JT is also known by other common names such as *Coix* seed, Gromwell Reed, Tear grass, Adlay Millet, Pearl barley, Indian Beads, Mayuen, or Hatomugi (Biswas and Das 2022). JT is commonly called Medicinal Corn, *Coix* Pearl, etc. (Feng et al. 2020). Apart from its regional names in India, JT is recognized by various names across the globe, as outlined in Table 12.1.

S. Paramanick (✉) · M. K. Malik
Department of Chemistry, Gurukula Kangri (Deemed to be University), Haridwar, Uttarakhand, India
e-mail: sushmitaparamanick@gmail.com; comdt.malik@gmail.com

S. Prakash
Department of Life Sciences, Imperial College, London, UK

© The Author(s), under exclusive license to Springer Nature Switzerland AG 2026

J. Singh, M. Kumar Malik (eds.), *Millets and Millet Starch: Chemical Composition, Functional Modification, and Applications*,
https://doi.org/10.1007/978-3-032-09502-2_12

Table 12.1 Common names of Job's tears millet around the world (Lim 2012)

Countries	Common names of Job's tears
Arab	Badrang, Damu Ayub, Amadrayan, Damudad
Brazil	Capim-De-Nossa-Senhora, Lagrima-De-Nossa-senhora, Capim-De-Contas, Capim-Rosario
China	Mayuen, San Yi Mai, Chuan Gu, Ma Yuen, Yi Yi Yan
Columbia	Lagrimas De San Pedro
Czech Republic	Slzovka Obecna, Jobovy Slzy, Slzovka Porcelanova
Denmark	Jobstare
Netherland	Jobstranen
France	Larmilles, Herbe A Chapelets
Germany	Tranengras, Hiobstrane
India	Baru, Samkru, Sankhlu
Indonesia	Jali Watu, Senjeali, Hanjere
Italy	Erba Da Corone
Japan	Juzudama, Hatomugi
Korea	Kusulyulmu, Yulmu
Malaysia	Senjelai, Jelai, Batak, Buah Jali, Biji Bali
Mauritius	Herbe Jobe and Collier Cipaye
Nepal	Jabe and Bhirkoulo
Norway	Jobstare
Philippines	Bitongan, Paias, Tigbi, Atakai
Portugal	Lagrimas De Nossa Senhora, Erva Dos Rosaries, Lagrimas De Job
Spain	Capin De Nossa, Lagrima De San Pedro, Santa Maria
Sweden	Jobs Tarar
Thailand	Maduai, Duai
Turkey	Yashe Out and Gozyasi Otu
Vietnam	Hot Bo Bo, Y Di, Curom Gao, Bo Bo

The name Job's tears (*Coix lacryma-jobi*) comes from the hard, tear-like seed pods that have a bluish-gray, porcelain-like glossy appearance as shown in Fig. 12.1 (Biswas and Das 2022). JT belongs to the Kingdom:- Plantae, Phylum:- Streptophyta, Class:- Equisetopsida, Order:- Poales, Family:- Poaceae, Sub-family:- Panicoideae, Genus:- *Coix*, and Tribe:- Maydeae (formerly known as Tripsaceae) (Cai et al. 2013). It is characterized as a grain-producing tropical plant (Zhou et al. 2010). Originally, it is indigenous to Asia and has now been widely naturalized across the tropical and subtropical regions, extending approximately between 22°N and 22°S (Fu et al. 2019). It is an annual or perennial crop but the main growth period is during the rainy season in marshy and low-lying areas. Hot and humid conditions with sufficient moisture in the soil are the optimal conditions for its growth and development (Cai et al. 2013; Hedayetullah and Zaman 2022). This crop is introduced to different regions; hence, it can be cultivated in a wide range of environments from very dry to wet zones, temperate to tropical regions, swamps, forest margins, and other cool and moist areas (Corke et al. 2016). The species exists in

Fig. 12.1 Close view of Job's tears millet seeds

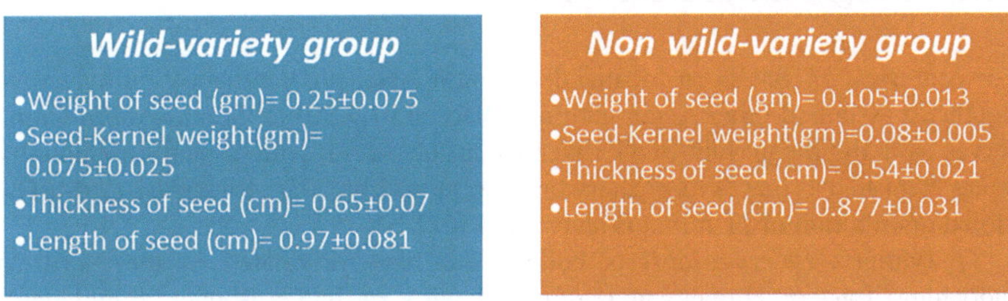

Fig. 12.2 Seeds characteristics of wild and non-wild varieties of Job's tears millet (Xi et al. 2016)

two primary varieties: a wild form and a cultivated (non-wild) form. The wild variety of JT is characterized by its pearly-white, hard-shelled, oval form and is used as beads for making necklaces, rosaries, and other objects (Biswas and Das 2022). The non-wild (soft-shelled) variants are edible, grown as a cereal, and are also used for medicinal purposes in some parts of Asia (Corke et al. 2016). The characteristics of both varieties of JT seeds are given in Fig. 12.2.

It is extremely beneficial and productive grass that is being considered as a possible source of energy. This non-traditional food crop has higher nutritional benefits than major crops (wheat, maize, rice, etc.). It is an alternative to rice. Additionally, the crop has a high degree of genetic variability that can be used for sustainable and productive uses (Kumar 2021). In Asian countries, the roots, leaves, and kernels of JT are commonly utilized as a traditional medicine (Lim 2012).

Considering the potential of this unexplored and underutilized crop, this chapter provides a comprehensive insight to the potential of this ancient grain in food and non-food industries. Understanding its chemical composition is crucial as it reveals valuable information about its nutritional profile, including carbohydrates, proteins, fats, vitamins, and minerals, as well as its antioxidant and bioactive compounds. This knowledge is important for evaluating its use in food products, health supplements, and pharmaceuticals.

Considering all these, this chapter will provide a detailed discussion on the cultivation, processing, and storage of the JT millet. Furthermore, isolation of JT starch from various methods and its modification is discussed in this chapter. Modification of the JT starch through physical, chemical, or enzymatic methods enhances its properties, making it suitable for use as a thickening agent, stabilizer, or emulsifier in food products, or even in the creation of bio-based materials and biodegradable plastics. These modifications open the opportunities for using JT in diverse industrial sectors, including food, pharmaceuticals, cosmetics, and environmental sustainability. By exploring these topics, the chapter provides a deep understanding of JT versatility and potential for various applications.

12.2 Geographical Distribution

For centuries, JT has been localized to South Asia, predominantly cultivated in Southeast Asia and other Asian nations (Diao and Jia 2017). Southwestern China serves as a significant hub of JT diversity in the world, with Guangxi potentially being identified as one of its primary origin region (Lim 2012; Lu and Li 1996). The global distribution of JT millet is shown in Fig. 12.3.

JT (variety: *lacryma-jobi*) is believed to have originated in the Northern Indochina, Eastern Himalayan region and parts of India, Bhutan, Bangladesh, Nepal, Myanmar, Thailand, Laos, China, and Vietnam (Venkateswarlu and Chaganti 1973; Arora 1977). Later, this plant was introduced to Europe, South America, Africa, Australia, West Indies, and various islands in the Indian and Pacific Oceans. Among its varieties, *lacryma-jobi* is the most widespread, found across many countries in Africa, Oceania, and America. The *stenocarpa* is primarily distributed in Northeast India, the Philippines, Myanmar, Indonesia, Vietnam, Southern China, and Papua New Guinea. The *ma-yuen* variety is native to regions such as India, Bhutan, Malaysia, Thailand, Myanmar, Laos, the Philippines, Indonesia, Korea, Japan, Vietnam, and China and has also been introduced to countries like Ecuador and Honduras (Zuloaga et al. 2003; Shouliang and Phillips 2006).

In India, the mountainous regions of the Northeast and the tribal areas of Jharkhand are the main locations for its cultivation (Yadav 2021). In the southern part of India, JT has been cultivated for over 4000 years. These seeds are commonly grown near rice fields. However, in commercial rice cultivation, this grass is often regarded as a weed (Patel et al. 2017). Table 12.2 discusses the regional distribution of JT in Asian countries.

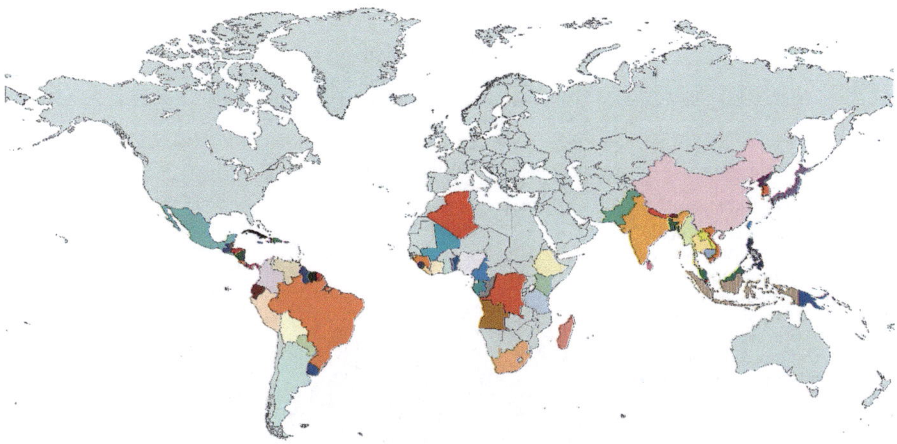

Fig. 12.3 Global distribution of Job's tears millet

Table 12.2 Regional distribution of Job's tears millet in Asia

Region	Distribution	References
China	Guangdong, Yunnan,Guizhou,Guangxi, Sichuan, Shandong, Fujian, Shanxi, Taiwan, Hainan, Henan, Liaoning, Hebei	Feng et al. (2020)
India	Mizoram, Manipur, Meghalaya, Nagaland, Arunachal Pradesh, Andaman and Nicobar Island	Soni et al. (2023) and Arora (1977)
Malaysia	Nyeli, Sarawak, Borneo Island	Harrison (1965)
Philippines	Bohol, Batuan, Carmen, Catigbian, Sagbayan, Mindanao Island, Luzon, Visaya Island	Quion et al. (2024) and Bon et al. (2023)
Thailand	Nong Khai, Chiang Rai, Chiang Mai, Loei, Sa Kaeo, Mae Hong Son	Sukkho et al. (2022) and Rupitak et al. (2024)
Myanmar (Burma)	Irrawaddy, Tenasserim, Karenni, MoBye, Shan State	Ochiai (2007)

12.3 Morphology, Appearance, and Genetics of Job's Tears Millet

The morphology and appearance of JT depend on the type of variety. Figure 12.4 represents the morphology and appearance of the different varieties of JT seeds. JT has prop/brace roots growing from the lower nodes. It is a robust and strong-tillering tall grass, characterized by its thick, upright stems (Acharya et al. 2024). JT plant typically grow to a maximum height of 1–3 m with knobby, spongy, and bamboo-like stems. The new tillers arise from the base of the stems and branching stems in the internodal region (Shahrajabian et al. 2022; Corke et al. 2016). Leaves are fairly elongated, flat, and broad enough. This grass has distinct male and female flowers. Inflorescences emerge from the axils of higher leaves on small peduncles that are

Fig. 12.4 Different varieties of Job's tears and their morphology and appearance

3–6 mm in length (Hedayetullah and Zaman 2022; Poddar et al. 2022). The JT plant is monoecious, having unisexual fertile spikelets devoid of hermaphrodite florets. On the same plant are the spikelets of the two sexually distinct forms, i.e., either only male or only female are found, and each one possesses multiple nerves. The fertile spikelets' gumes are multiple-nerved, and they are eventually enclosed in a hard involucre and contain two sterile and one tiny fertile flower (Ramadhan et al. 2023a). The male, pollen-producing flowers grow on slender stalks that emerge from the bead through a tiny orifice. The fertile female flower has two feathery stigmas that protrude from the same pore, ready to collect pollen from male flowers (Zheng 2014; Poddar et al. 2022). The male flower in the same inflorescence grows first, and its maturation period differs from the female flowers. At the end of the peduncle, a hollow, round, bread-like structure develops, which encapsulates pistillate spikelets (Corke et al. 2016). This herb is well known worldwide for its pods

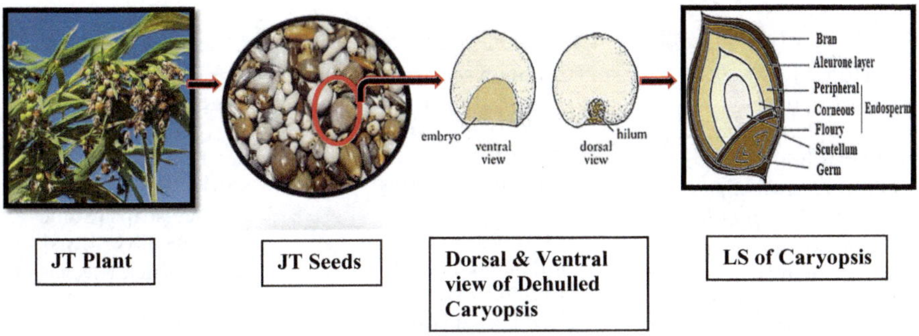

Fig. 12.5 Longitudinal section (LS) of the Job's tears caryopsis

(pericarps), which naturally have a hole at each end and used to make string of beads. The stringing seeds are referred to as "Vaijanti Mala" (Patel et al. 2017). It produces big and shiny pear/drop-shaped fruits that have a resemblance to tears, and it contains rice-like grains in it that are white, light brown, and purple. The seeds are fairly stony. The seeds are combination of yellow, brown, purple, and white. After dehulling, the seeds have an exterior surface that ranges from milky white to black. The seed weighs 16–17 grams per 100 grams. The seeds are shattered apart easily, and the flowering is asynchronous. The plant matures in 150–160 days. The longitudinal section of the JT caryopsis is represented in Fig. 12.5.

The oval-shaped seeds of JT have 5 mm diameter (Feng et al. 2020). When classifying JT species and variants, the form of texture of the seed, color, and size are important factors, as shown in Fig. 12.6.

12.3.1 Genetics of Job's Tears Millet

The crossbreed varieties of seeds are maintained by breeders with the help of genetic diversity that helps to sustain desirable traits. The genetic diversity can be obtained from local varieties, introductory breeding strains, wild varieties, and from high-yielding varieties of plants that are gathered in the germplasm collection (Dwipa et al. 2022; Ramadhan et al. 2023a).

Understanding genotypic relationships plays a crucial role in developing new crop varieties, as it gives information on genetic diversity, which is essential for stratified sampling in breeding programs (Laxmisha et al. 2023). Despite the high economic value of JT, there has been limited progress in its genetic improvement (Ma et al. 2006). A study was conducted to characterize the agro-morphological traits of JT germplasm to determine the extent of genetic diversity to use it in crop improvement (Laxmisha et al. 2023). Dwipa et al. found that JT samples from four regencies/cities in West Sumatra province can be grouped into two main groups based on a 67% genetic similarity. This suggests that the two populations are genetically quite different from each other (Dwipa et al. 2022).

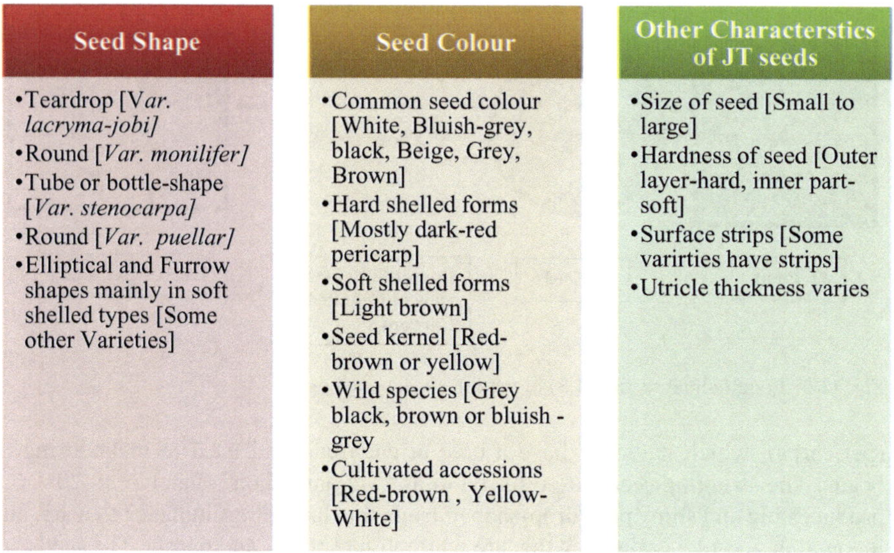

Fig. 12.6 Characteristics of Job's tears seeds (Jain and Banerjee 1974)

Table 12.3 Karyotypic formula of different Job's tears varieties (Huang et al. 2003)

Job's tears varieties	Karyotype formulas
Coix puellarum	$2n = 20 = 20m$ (Standard)
Coix stenocarpa	$2n = 20 = 18m + 2sm$
Coix lacryma-jobi var. lacryma-jobi	$2n = 20 = 18m + 2sm$ and $2n = 20 = 16m + 4sm$
Coix lacryma-jobi var. monilifer	$2n = 20 = 12m + 2m$ (sat) $+ 6sm$
Coix lacryma-jobi var. formosana	$2n = 20 = 14m + 6sm$ (+ 1–2B)
Coix lacryma-jobi var. mayuen (Roman) Stapf	$2n = 20 = 18m$ (2 sats) $+ 2sm$

n = chromosome number, m = metacentric chromosome number, sm = submetacentric chromosome number

There are 9–11 species in the *Coix* genus that exhibit various ploidy levels. The cultivated JT is a tetraploid, with a basic chromosome number of $x = 5$, which means it has 20 chromosomes in total ($2n = 20$) (Clayton 1980; Rao and Nirmala 2010; Christopher and Jacob 1990). The standard (usual) karyotype formula for *Coix puellarum* is $2n = 20 = 20\ m$. However, variations in the standard have been observed. Other research studies have reported chromosome complements of $2n = 10$ and $2n = 20$ in different JT varieties originating from India, as shown in Table 12.3 (Huang et al. 2003).

12.4 Cultivation

The cultivation area of JT in China is estimated at approximately 73,000 hectares, producing an annual grain yield of about 220,000 tons (Shahrajabian et al. 2022). JT has been reported to produce 45–53 t/ha of fresh fodder in Mali, with seed yields

Fig. 12.7 Close view of a mature Job's Tears plant with fruit

reaching 3.5 tons per hectare in the Philippines and 2.1 tons per hectare in Sri Lanka (Poddar et al. 2022). In Thailand, JT is primarily cultivated in the northern and northeastern regions. Each year, approximately 4000–6000 tons are exported to Japan and Taiwan from these areas (FAO; Rick and Ellen 2012). According to the FAO, forage from JT in India yields around 13.9 tons/hect of green material (equivalent to 6.1 non-metric tons per acre). The species *Coix lacryma-jobi* L. is considered well-suited for cultivation in regions such as Northern Bihar, West Bengal, Assam, and Odisha, with average yields of 34.6 tons per hectare (15.22 non-metric tons per acre) of green fodder and yields about 6.9 tons per hectare (3.04 non-metric tons per acre) of dry matter. Multiple harvests of JT fodder can be obtained annually. JT produces 2–4 tons per hectare (0.9–1.8 non-metric tons per acre) of husked grain, according to reports from the Plant Resource of Southeast Asia and the Food and Agriculture Organization of the United Nations. The hulling percentage (defined as the proportion of grain retained after husking) ranges from 30% to 50% (FAO 1995; Bergh and Iamsupasit 1996). The close view of the mature JT plant with fruit is shown in Fig. 12.7.

12.4.1 Soil, Climate Requirements, and Ecological Adaptability

JT is a light-sensitive, short-day, and frost-free plant, so it can tolerate annual precipitation between 6.1 and 42.9 decimeters. It is typically grown during the Kharif season, either as a single crop or in mixed cropping systems. The ideal average temperature for cultivating JT ranges between 9.6°C and 27.8°C, and the pH value should be between 4.5 and 8, providing an optimal condition for healthy growth and yield. Although the plant has a slow initial growth rate, its full growth cycle ranges from 125 to 180 days. On average, JT yields about 10–12 q/hec, and the market price for farmers ranges from ₹50 to ₹60 per kg (Corke et al. 2016).

Varieties having a maturing period of more than 160 days in South China, roughly 150 days in North China, and 125 days in Northeast China are appropriate for cultivation (Zhao and Liu 1992). Early March is the time to plant early maturing varieties, late March to early April is the time to plant middle maturing varieties, and late April to early May is the time to plant late maturing varieties. Early sowing is preferable to late sowing. Late seeding will reduce production, and if the seeds are sown too late, they won't even be able to mature by autumn (Li and Liu 2012).

For the cultivation of JT, early maturing varieties are considered the most suitable. JT should be cultivated during the summer-autumn period, with seeds sown in late spring or early summer and the crop typically harvested by early October. This season has a lot of rainfall and even some waterlogging in the area, but it only fits JT's hygroscopicity requirements, which can help the plant grow and develop and increase the yield (Feng et al. 2020). JT can be alternated with overwintering crops (such as garlic, onions, and spinach). In the first 10 days of June, overwintering crops are typically harvested. After harvesting these crops, the field may undergo minimal or no preparation before JT is sown at the appropriate time (Li and Jin 2010). In addition to being easily damaged by dry and hot weather in the early stages, spring sowing makes it vulnerable to drought. This results in hollow seeds and decreased yield due to the pollen losing its vitality and being unable to be pollinated (Yang 2010).

It is suitable for cultivation on sloped terrains with higher rainfall and is mostly grown in mountainous tracts of the northeastern region of India and its surrounding areas. Tribes from several communities primarily inhabit this area where they practice shifting cultivation, commonly known as the "Jhum Cultivation," to grow a variety of crops. The tribes use these sloping or undulating terrains (between 300 and 1500 m) to grow a variety of crops, including corn, chillis, brinjal, rainfed paddy, finger millet, foxtail millet, and others. Here, JT is planted as a side crop. After the hand hoeing and ploughing, the jhum farming is prepared for seeding by April, and by May the JT seeds are sown. It is frequently grown with close spacing (30 cm) along the field boundaries (Laxmisha et al. 2023). Like other rain-fed crops such as paddy, it relies on monsoon rainfall for its growth and development. According to reports, there could be up to 50 tillers (Laxmisha et al. 2023). All tiller bears inflorescences that produce fruit between September and October. When the tillers are loaded with ripe fruits, they bend downward. As the plant matures, it gradually loses its vibrant green color and the leaves turn to a grayish appearance, distinguishing it from its younger stage. This stage usually marks the appropriate time for harvesting (Arora 1977). The plants are cut off at the root, packed in bundles, and transported to the tribe homes, which are usually located distant from the jhum lands for threshing. The grain product is threshed, laid out on mats or the floor, and properly sun-dried outside of tribal homes before being kept for domestic use (Arora 1977). Generally, rice and soybeans are considered the most suitable preceding crops for JT cultivation. JT is a gramineous crop with lower autotoxicity (Qian 2016).

Being a hygrophilic plant, it prefers moist soil. As a result, low-lying and wet areas of land can be chosen for cultivation (Li and Liu 2012). However, an effective irrigation system is needed to guarantee that there is enough soil moisture during its early growth and flowering stages. If there is a drought or insufficient water in the soil during the early growth and flowering stages of JT, the yield will be reduced (Zheng 2014; Salami et al. 2025). Density has a significant impact on JT yield. To achieve high yield, an appropriate planting density must be maintained, which depends on the cultivar type, soil fertility, and the season of cultivation (Hu and Jin 2007). The planting density typically entails row spacing of 25–30 cm, plant spacing of 10–13 cm, and 20,000–26,000 seedlings per 667 m^2 (Wu et al. 2014). The key to achieving a consistent and high production of JT is water management (Song et al. 2015).

In recent years, it has been demonstrated that JT has strong moisture tolerance rather than a dry crop due to advancements in cultivation technology and growers' practices. To achieve a consistent, high-quality yield, it is important to rectify the misconception that it is a dryland crop and instead cultivate it using wet farming techniques similar to those used for rice (Pang et al. 1996). As suggested by Ferreira et al., the "two wet, two dry, and two water" technique should be adopted for effective water management. This practice involves keeping the soil moist during the early growth phase to support seedling development, applying dry farming practices at the jointing stage, ensuring water availability in the field during the booting and heading stages, keeping the soil moist during the grain filling stage, and allowing the soil to dry out during the harvesting period (Ferreira et al. 2020).

12.5 Job's Tears Seed Starch Isolation

The isolation of starch from JT millet focuses on separating high-purity starch granules by eliminating non-starch components such as proteins, lipids, and fibers. Various methods have been developed for extracting starch from JT seeds, suggesting that the JT seed kernels should first be milled into flour prior to starch extraction.

The most widely used approach is the wet milling technique, where the seeds are first cleaned, then soaked in water or in a mild alkaline solution to soften the structure, and then mechanically ground. The resulting slurry is sieved and washed repeatedly, and the starch is separated by sedimentation or centrifugation to enhance its purity (Liu et al. 2016). Starch can also be extracted using the dry milling method, which involves mechanical grinding without the use of solvents (Puchongkavarin et al. 2005; Chaisiricharoenkul et al. 2011).

Another method is the alkaline steeping method, which involves soaking of seeds in an alkaline solution, such as sodium hydroxide (Chao et al. 2014; Subroto et al. 2022; Igbokwe et al. 2021; Yang et al. 2021; Kim et al. 2016b; Capule and Trinidad 2016; Kang and Song 2019; Kim et al. 2008). This process facilitates the solubilization of proteins and hemicelluloses, enhancing the separation of starch. The alkaline steeping method is particularly effective in increasing the purity of the extracted starch. Additionally, enzymatic extraction techniques have been explored by using enzymes like proteases and cellulases to selectively degrade proteins and

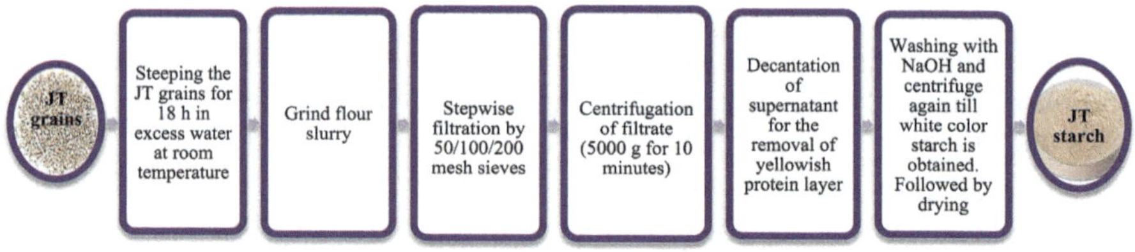

Fig. 12.8 Isolation process of Job's tears seed starch by alkaline steeping method

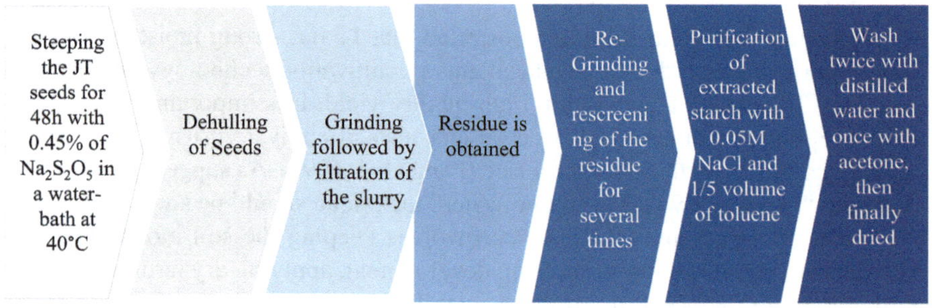

Fig. 12.9 Isolation process of Job's tears seed starch by sodium metabisulfite steeping

cell wall materials. This method offers a more environment friendly alternative, as it reduces the need for toxic chemical reagents and preserves the structural integrity of the starch granules. Enzymatic extraction has shown promise in improving starch recovery with minimal structural damage (Luithui and Meera 2019; Kim et al. 2009). Another study reported that starch was extracted using the glutamic acid–gluten method (Sira and Amaiz 2004).

These methods differ in complexity, efficiency, and impact on starch properties, but all aim to obtain high-purity starch that is suitable for food, pharmaceutical, and industrial applications. Figure 12.8 illustrates the alkaline steeping method, while Fig. 12.9 demonstrates sodium metabisulfite steeping method.

Studies have shown that the starch content in JT varies between 50% and 80% (Bhandari et al. 2012). Variation in starch content is influenced by factors such as the isolation method, crop variety, temperature, environmental conditions, and enzymatic activity. Table 12.4 represents the starch content percentages reported in different studies by various authors.

12.6 Modification of Job's Tears Seed Starch

Raw starch has several limitations for its use in industries. To overcome the drawbacks and to enhance the industrial sustainability, JT millet starch has been explored by numerous researchers for modification. There are several methods of modification applied to increase the industrial suitability of JT starch; some are as follows:

Table 12.4 Job's tears varieties and their starch content

Job's tears varieties	Starch content (%)	References
Coix lacryma-jobi	52.70	Luo et al. (2024)
Coix lacryma-jobi	58.00	Zhang et al. (2021) and Igbokwe et al. (2021)
Coix lacryma-jobi	62.00	Laxmisha et al. (2022)
Coix lacryma-jobi	67.20	Huang et al. (2003)
Coix-mayuen	70.00	Subroto et al. (2022)
Coix lacryma-jobi	61–83	Devaraj et al. (2020)
Coix lacryma-jobi	79.71–80.57	Yang et al. (2021)

12.6.1 Physical Modifications

JT starch modification was done via Ozonation and Freeze Moisture Treatment (FMT). Modification of JT starch through FMT and ozonation increases its functional properties and produces more pores and amorphous structures without introducing new functional groups. The modified JT starch also exhibited higher hot-paste viscosity, along with reductions in gelatinization temperature, hold, peak, setback, breakdown, and viscosities (Subroto et al. 2022). JT starch was modified using heat-moisture treatment (HMT) process. It was found that the HMT method significantly reduced the moisture content of JT starch while inducing notable changes in its physical and functional properties. As a result, HMT-modified JT starch demonstrates promising potential for use in food applications that require high water absorption capacity, low swelling and solubility, enhanced viscosity, improved gel strength with minimal retrogradation, and high thermal stability (Liu et al. 2016; Capule and Trinidad 2016). Resistant starch (RS) of JT was obtained using an autoclaving–cooling method to analyze physiochemical properties (Yang et al. 2021). In another approach, HMT was applied to JT starch, resulting in an increase in amylose content from 6.32% to 8.10%, thereby improving its structural and functional characteristics (Wang et al. 2020). Modification via ultra-high-pressure treatment was used to tailor its physicochemical properties. Key conditions were determined by single-factor experiments followed by orthogonal optimization to enhance RS yield. This treatment significantly increased RS content by 90% and improved properties such as freeze-thaw stability, gel formation, and solubility. However, slight reductions in swelling power and light transmittance were observed (Zhiyong 2018).

12.6.2 Chemical Modifications

The octenyl succinic anhydride-modified JT seed starch has been prepared to improve its emulsification efficiency, and the octenyl-modified starch was applied to improve the behavior of astaxanthin-loaded emulsion (Xu et al. 2024). To enhance the emulsification efficiency of JT seed starch through esterification, it was coupled with ultrasonication and enzymatic treatment. The impact of these treatments on the structure and properties of JT starch was assessed to clarify their effects in Pickering emulsion (Xu et al. 2023). Kim et al. (2016a, b) demonstrated that modification of JT starch with amylosucrase improves its structural properties and digestibility. The enzyme facilitated the addition of glucosyl units to the non-reducing ends of the branch chains, leading to a reduction in short chains and an increase in longer chain fractions, hence exhibiting notable structural changes (Kim et al. 2016a, b). Liu et al. (2025) employed a sequential enzymatic process using α-amylase, followed by β-amylase, transglucosidase, and pullulanase to alter the starch structure, digestibility and solubility (Liu et al. 2025). Furthermore, in another approach, the reaction conditions were optimized to enhance the resistant starch (RS) content in JT starch by esterifying it with glutaric acid, and it has been found that modification caused a 66% increase in RS (Kim et al. 2008). Acid modification led to significant changes in physicochemical properties of JT starch, such as degree of hydrolysis, swelling power, pasting properties, and water solubility index, which varies with the degree of acid hydrolysis. In this study, it is evaluated that there is a reduction in SP and paste viscosity of modified JT starch (at all temperatures), with an increase in hydrolysis time. Moreover, the solubility of modified starch increases with an increase in hydrolysis time (Kim et al. 2009). Modification of JT through *Lactobacillus* (LAB) fermentation resulted in changes in functional properties. This research was conducted to evaluate physicochemical and functional characteristics. A significant decrease in moisture content was observed in the modified one as compared to the native (control) one. Additionally, there is a significant decrease in pasting viscosity, solubility, WAC, and swelling power of modified JT. The oil-binding capacity varies depending upon the type of bacterial strain used (Kristanti and Ratnawati 2022). These modification strategies contribute to the development of JT starch with improved functional and health-promoting properties.

12.7 Physiochemical Compositions of Job's Tear Grain and Starch

12.7.1 Moisture Content

The moisture content of JT seed starch highlights variations influenced by cultivar type and processing methods. A study comparing the moisture content in big coix seed (BCS), translucent coix seed (TCS), and small coix seed (SCS) starches reported moisture contents of 7.96%, 5.96% and 5.85%, respectively. These differences are attributed to inherent varietal characteristics and environmental factors during cultivation and storage (Liu et al. 2017). Further investigations of the hygroscopic properties of JT seeds revealed that moisture adsorption and desorption behaviors vary with temperature and relative humidity. The study determined that the safe moisture content for storage at 25 °C ranges between 12.45% and 13.69%, depending on the relative humidity levels (Luo et al. 2024). The moisture content also varies according to the process (10.4460% by wet milling of JT flour and 8.1868% by dry milling of JT flour (Mulyono et al. 2019).

12.7.2 Swelling Power and Solubility

As temperature increases, starches exhibit greater water solubility and swelling power due to the breakdown of granules and exposure of hydrophilic groups (Khawas and Deka 2017). RSs initially exhibited lower solubility than native starches, but exceeded them at 90°C. Additionally, RSs exhibited greater swelling power than native starches within the temperature range of 50–90°C (Li et al. 2018). The hot water solubility (HWS) of both raw and control starches at 80 °C was notably greater than their cold-water solubility (CWS). However, glutarate-modified starch exhibited only a minimal rise in HWS compared to CWS. Hence, due to cross-linking in glutarate starch, the granules swelled less and decreased solubility in hot water. This decreased swelling may hinder digestive enzyme access and lower enzymatic hydrolysis (Kim et al. 2008).

Research indicates that the solubility of JT seed starch differs among its cultivars. For instance, the BCS starch demonstrates a solubility of approximately 49.62%, the SCS is around 41.91%, and TCS is about 10.06%. These variations are primarily attributed to differences in amylose content, with TCS having a higher amylose content, which is known to inhibit starch swelling and solubility (Liu et al. 2017).

Further, it has been reported by researchers that, amylose content can reduce starch swelling (Li et al. 2018). The white varieties (whole grain and degermed) showed greater swelling than black varieties due to their lower protein, lipid, and amylose levels. High protein content can limit water absorption by strengthening molecular interactions, and amylose can bind with lipids to form complexes that restrict starch granule expansion (Chaisiricharoenkul et al. 2011). Research indicates that the swelling power of JT seed starch varies significantly between waxy and non-waxy varieties. Waxy JT seed starch, characterized by low amylose content (0.7–1.1%), demonstrates higher swelling power, ranging from 28.6 to 41.0 g/g (Kim et al. 2009). In contrast, non-waxy varieties with higher amylose content (15.9–25.8%) exhibit lower swelling power. This difference is attributed to amylose's tendency to form complexes with lipids, which can inhibit water absorption and swelling (Li and Corke 1999). Table 12.5 displays the physiological properties of JT seed starch.

Table 12.5 Physiochemical properties of raw Job's tears seed starch

Physiochemical properties	Raw starch	References
Moisture content (%)	6–10	(Wu et al. 2007; Liu et al. 2015; Ibañez 2024)
Swelling volume (g/g)	18.120–41.0	(Subroto et al. 2022; Corke et al. 2016)
Ash content (%)	0.9–1.83	(Ramadhan et al. 2023a, b; Liu et al. 2015; Ibañez 2024; Wu et al. 2007)
Solubility (%)	4–6	(Chaisiricharoenkul et al. 2011)

Table 12.6 RS, RDS and SDS content of modified Job's tears seed starch

Modification	Variety	Rapidly digestible starch (RDS%) (Mean ± SD)	Slow digestible starch (SDS%)	Resistant starch (RS%)	References
Amylosucrase enzyme modification	(*Coix lachryma-jobi* var. *mayuen*)	24.5 ± 1.2	24.0 ± 0.1	51.5 ± 1.4	Kim et al. (2016b)
HMT modified	(*Coix lachryma-jobi*)	25.36 ± .1.0	24.41 ± .1	50.23 ± .2	Liu et al. (2016)
Sequential enzyme modification	(*Coix lachryma-jobi*)	11.24 ± .006	31.21 ± .010	57.55 ± .006	Liu et al. (2025)

12.7.3 Digestibility

The in vitro digestion profile of raw starch was found to be typical for granular starch, while the control starch exhibited a high percentage of rapid digestibile starch (RDS). The RDS concentration of amylosucrase (AS)-modified starch was half that of the control starch's value, while the levels of slowly digestible starch (SDS) and RS showed an increase. Both in vitro and in vivo investigation confirmed the decreased digestibility of AS-modified starch, indicating its potential applicability in food applications with a low glycemic response (Kim et al. 2016a, b). Table 12.6 displays the RS, RDS and SDS content of modified JT starch.

12.7.4 Water Absorption Capacity

Water absorption capacity (WAC) plays an important role in determining functional characteristics like viscosity, which is crucial for the bulk and texture of products. It is related to the physical characteristics and granular composition of the JT starch. It is a key factor that influences the application of starch in various products. The WAC of JT modified starch that has undergone ozonation and FMT showed an increase in WAC. The ozonation procedures considerably improved the solubility of JT starch (Subroto et al. 2022). In another study, when JT starch underwent HMT, its WAC increased notably from 209.50% to 265.60% (Capule and Trinidad 2016). The WAC ranges between 103% and 108% (Corke et al. 2016). The native starches of JT show WAC ranging from 97.5% to 114.6%, which significantly increased to 323.1% and 369.9% after autoclaving and cooling treatment, respectively (Yang et al. 2021).

12.7.5 Protein Content

Protein is the second most abundant nutrient in JT after carbohydrates. The seed kernels of JT is composed of approximately 20% protein, while the seed shells have 2.17–2.80% protein, and the wild variety has approximately 31.72% protein (Lee et al. 2011a, b; Corke et al. 2016; Zhu 2017; Ottoboni et al. 1993). In six tested varieties, protein content ranged from 10.53% to 13.82%. The lower protein levels observed in this study may be due to environmental factors and this value is below the 14–20% range reported by the ICAPOSTRD (Indonesian Center for Agriculture Postharvest Research and Development) (Pertanian 2020). The protein content was 14.0050% by wet milling and 15.2750% by dry milling (Mulyono et al. 2019). Further, Liu et al. (2016) reported that JT seeds contain approximately 14% protein (Liu et al. 2016).

12.7.6 Fat Content

Fat provides more energy than carbohydrates and proteins, and it also plays a vital role in maintaining the human body healthy. Among the six accessions analyzed, five showed similar fat content percentages, ranging from 7.98% to 8.44%, whereas one accession exhibited a lower fat content of 6.83% (Ramadhan et al. 2023b). According to Nurmala (2011), JT seeds contain up to 7.9% fat, which is higher than that found in several other crops, such as rice (2.1%), barley (2.4%), sorghum (4.2%), and corn (4.9%) (Nurmala 2011).

12.8 Phytochemical Properties

JT is abundant in nutrients and possesses a diverse group of phytochemicals and bioactive substances (Table 12.7). It contains tannins, terpenoids, flavonoids, alkaloids, phenols, phytosterols, and saponins (Mandal and Mandal 2015). JT also provides essential nutrients such as niacin, vitamin B1, B2, E, dietary fiber and other beneficial compounds (Zhang et al. 2002).

Table 12.7 Phytoconstituents of Job's tears seeds

Class of phytoconstituents	Compound names	Plant part	References
Vitamins	Vitamin C, B1, B2, B6, B12	Seeds, root tissues, and leaf	Yin et al. (2018) and Weng et al. (2022)
Phytosterol	Beta-sitosterol	Hull	Manosroi et al. (2016) and Ahmad and Ghosh (2022)
Coix oils	Coixenolide	Seeds	Lin et al. (2022) and Lakkham et al. (2009)
Polyphenols	Caffeicacid, ferulic acid, beta-hydroxy benzaldehyde	Seed and hull	Huang et al. (2021)
Lignans	N-Neolignans	Hull, bran, and seed	Kim et al. (2016a, b)
Flavonoids	Tangeretin, naringenin, rutin, quercetin, and kaempferol	Hull, seed kernel	Kuo et al. (2012) and Wang et al. (2016)
Carotenoids	Beta-carotene	Hull and endosperm	Huda et al. (2023)
Vitamin E	Tocopherol, tocotrienols	Seeds	Huda et al. (2023)

12.8.1 Phenolic Compounds

JT is rich in phenolic acids like p-coumaric, ferulic, vanillic, chlorogenic, caffeic, and tannic acid, along with catechol, which have been identified and extracted from JT's husk (Chang et al. 2018). JT seeds have both types of polyphenols, i.e., conjugated and free polyphenols (Wang et al. 2012). These phenolic compounds have strong antioxidant properties, which help neutralize free radicals in the body. By reducing oxidative stress, these antioxidants may help lower the risk of chronic diseases such as cardiovascular disease and cancer (Wang et al. 2016; Chen et al. 2012).

12.8.2 Terpenoids and Steroids

JT seeds are a rich source of various sterols, including feruloyl-stigmasterol, brassicol, and ferulyl-brassinosterol, campesterol, phytosterols, as well as α, β, and γ sito and stigmasterol (Hou et al. 2018). Chang et al. successfully isolated two triterpenoids (iso-arborinol and friedelin) from the kernels of JT (Chang et al. 2020).

12.8.3 Flavonoids

Flavonoids represent a key group of bioactive compounds found in JT seeds, with isoquercetin, rutin, eriodictyol, and luteolin being the predominant constituents. These compounds comprise approximately 45.35%, 32.77%, 6.04%, and 5.37% of the total flavonoid content in JT, respectively (Huang et al. 2019). These are associated with a range of physiological functions, including metabolic regulation, antioxidants, antibacterial, anti-inflammatory, anti-tumor, antiviral activities, and neuroprotective effects. Quercetin, alongside other significant compounds such as Rutin and kaempferol, has been extensively studied for its role in supporting cardiovascular health and its potential to reduce cancer risk by mitigating oxidative stress (Li et al. 2023)

12.8.4 Alkaloids and Saponins

The term "saponin" originates from the Latin word *sapo*, meaning soap, due to its characteristic of producing a soap-like foam when plant extracts containing these compounds are shaken in water (Timilsena et al. 2023). In JT seeds, saponins are recognized for their therapeutic roles, including anticancer and immunomodulatory effects. Studies also indicates that they may support liver function, help regulate blood glucose levels, and influence gut microbiota. Although present in relatively small quantities, saponins contribute significantly to the seeds' antimicrobial and anti-inflammatory properties (Diningrat et al. 2021). Moreover, tannins identified in JT seeds exhibit antimicrobial action by mechanisms such as iron chelation, hydrogen bonding, and interaction with essential microbial proteins and enzymes (Devi et al. 2015; Wang et al. 2016).

12.8.5 Phytosterols and Squalene

JT seeds contain squalene, a hydrocarbon involved in cholesterol biosynthesis (Moreda et al. 2001), which supports serum cholesterol reduction and may influence tumor proliferation (Khor and Chieng 1997). Table 12.7 represents various phytoconstituents found in JT seeds.

According to Kumar et al., JT seeds have the following nutritional values. It contains 380 calories, 65.3 g total carbohydrate, 1.5% fiber, 18.2% albuminoids, 6.2 g fat, 4.3 mg niacin, 25 mg Ca, 0.28 mg thiamine, 435 mg P, 0.19 mg riboflavin, and 5.0 mg Fe. Additionally, the seeds have 29.9% dry matter (Kumar et al. 2014). The lipid content of JT seeds (4.65%) is also higher than that of most of the cereal crops. Recently, Li et al. (2022) reported 11 triglycerides in JT seed oil (Li et al. 2022). Compared to other cereals, they are higher in protein and fat but lower in minerals.

12.8.6 Amino Acids and Proteins

The protein content of JT surpasses that of other cereals, making it a valuable nutritional source for both humans and animals (Jain and Banerjee 1974; Pestana 1946). The predominant protein component, coixin, constitutes around 79% of the total seed protein and is classified as a prolamin-type storage protein (Lee et al. 2011a, b; Corke et al. 2016; Ottoboni et al. 1993; Zhu 2017). These prolamins can be classified into α-, β-, and γ-types based on their solubility. The α-prolamins are particularly rich in amino acids such as proline, alanine, glutamine, and leucine and have low lysine and tryptophan because of high prolamins in the seed (Djaja et al. 2019). The primary protein fractions of JT include albumin (1.43%), globulin (6.20%), gluten (37.38%), and gliadin (44.74%) (Liu et al. 2015). Notably, glutamic acid comprises 25.85%, while a total of 23.5% is contributed by branched-chain amino acids such as leucine and valine. Antioxidant amino acids including methionine, tyrosine, tryptophan, and cysteine make up 10.8%, and the levels of tryptophan and lysine surpass those found in wheat, maize, and rice (Djaja et al. 2019). Additionally, the non-seed parts of the plant, such as the endothelium, roots, and husk, contain up to 18 amino acids, including seven essential ones, with the endothelium exhibiting the highest concentration (Liu et al. 2015).The JT seeds also contain an average of 20.8 mg per 100 grams of gamma-aminobutyric acid (also known as GABA). GABA is a non-protein amino acid known for its health benefits such as lowering blood pressure, enhancing cognitive function, exhibiting anticonvulsant properties, managing epilepsy, supporting kidney and liver activity, and improving sperm profile (Li et al. 2017).

12.8.7 Lipids

JT seeds are notably rich in fats, particularly functional lipids. The concentration of fatty acids is highest, i.e., up to 36% in the bran, while it is approximately 7% in seed kernels, nearly 6 times more than that of rice (Liu et al. 2015). Seed kernel shells are another source of fatty acids. The fat is primarily composed of more than 83% unsaturated fatty acids, mainly linoleic and oleic acids, which are recognized for their health-promoting benefits. Additionally, it contains other important lipids such as coix acid, palmitic acid, coix ester, myristic acid, stearic acid, and fat-soluble hormones (Huang et al. 2009b). Coicis oil, a medicinal extract derived from JT seeds and clinically known as Kanglaite, includes components like coixenolide, myristic acid, octadecenoic acid, and various esters (Bai et al. 2018; Chang et al. 2020). At low amounts, coicis oil has been shown to enhance respiratory function and stimulate both cardiac and smooth muscle activity, contributing to improved pulmonary blood flow (Chen et al. 2018; Liu et al. 2019).

12.8.8 Carbohydrates

Coixan, a water-soluble polysaccharide with a pH of approximately 6.8 is considered a key bioactive compound contributing to the medicinal properties of JT seeds. Coixan exhibits strong thermal stability. JT seeds are rich in various biologically active polysaccharides, comprising A, B, and C types along with neutral dextrans and acidic polysaccharides (Li et al. 2019; Seo 2019). These components are primarily composed of sugars like rhamnose, arabinose, mannose, galactose, and dextran (Hong et al. 2016; Hu et al. 2022). The polysaccharide content is relatively consistent across different plant parts, with concentrations recorded as 54.374 mg/g in roots, 56.9322 mg/g in stems, and 52.7549 mg/g in leaves. However, changes in seasonal variation significantly impact the content of polysaccharides, especially in the stem (increased from June to November). Additionally, the content of these polysaccharides can differ significantly depending on the region where the JT is cultivated (Liu et al. 2018). JT is rich in dietary fiber. The fiber content supports regular bowel movements and helps to prevent constipation by promoting efficient digestion. Furthermore, it plays a role in stabilizing blood glucose levels. The dietary fiber in JT seeds also contributes to a balanced gut microbiome, supporting overall gastrointestinal function (Zhou et al. 2024; Jinnouchi et al. 2021).

12.8.9 Minerals and Vitamins

JT seeds contain high levels of vitamin E and niacin. Studies have shown that the tocopherol (vitamin E) content in JT seed oil ranges from 0.64 to 1.57 mg per gram, surpassing that of olive oil (Chang et al. 2020). The seed coat holds greater concentrations of niacin, vitamins (B1, B2, C, B6, B12, E), and folic acid than the seed kernel, and brown variety JT is considered more nutritious than the white variety (Huang et al. 2019). Additionally, vit B1, vit B2, and vit B12 levels are highest in the roots, while leaves are richer in vitamin B6 and folic acid (Zhu 2017). JT is also a good source of minerals such as calcium, iron, phosphorus, zinc, magnesium, and manganese. Table 12.8 shows the nutritional composition of JT seeds.

Table 12.8 Nutritional composition of Job's tears seeds

Minerals	Content (per 100 g)	Part of plant	References
Carbohydrate (g)	65–68	Seed kernel	(Huang et al. 2003; Igbokwe et al. 2021; Tensiska et al. 2021)
Protein (%)	9–15	Seed kernel	(Laxmisha et al. 2022; Igbokwe et al. 2021; Liu et al. 2015; Ibañez 2024)
Fat (%)	4–9	Seed kernel	(Yang et al. 2008; Wu et al. 2007)
Fibre (%)	0.8–8	Seeds	(Ahmad et al. 2024; Feng et al. 2020; Huang et al. 2003)
P(mg)	298–435	Seed	(Tajalifar 2018; Liu et al. 2015; Huda et al. 2023)
Mg (mg)	125–130	Seed kernel	(Huang et al. 2003; Liu et al. 2015)
Ca (mg)	25–49	Seed	(Tajalifar 2018; Shahrajabian et al. 2022)
Zn (mg)	1.88–1.90	Seed	(Liu et al. 2015; Huda et al. 2023)
Mn (mg)	1.88–1.90	Seeds	(Shahrajabian et al. 2022; Huda et al. 2023)
Fe (mg)	2.90–13.6	Seeds, roots	(Tajalifar 2018; Liu et al. 2015; Shahrajabian et al. 2022)

12.9 Health Attributes of Job's Tears Millet

JT has both culinary and therapeutic properties, which have led to its increasing demand and popularity as a healthy food choice (Lim 2012). The medicinal benefits of JT have been demonstrated by recent pharmacological studies. These include antioxidant (Kuo et al. 2002; Huang et al. 2009a), anti-inflammatory (Huang et al. 2009a; Otsuka et al. 1988), anti-diabetic (Kim et al. 2000), anti-allergic (Hsu et al. 2003), anti-obesity (Huang et al. 2005; Kim et al. 2007), and anticancer properties (Chang et al. 2003; Lee et al. 2008; Chen et al. 2012), as well as digestive tonic, analgesic, and antispasmodic effects (Xi et al. 2016). The hypoglycemic, antioxidant, and hypolipidemic properties of JT seeds have all been described by Chung et al. 2011; Huang et al. 2005; Chiang and Chiang 2006 respectively. According to studies conducted by several researchers, JT seed oil has unusual carbon chain fatty acids with potent anticancer action that significantly inhibits the growth of tumor cells. The unique active ingredients could be an important source of JT's anti-cancer properties. A clinically tested anticancer drug called Kanglaite extracted from JT seed oil has been used successfully to treat various types of cancer (Rao and Nirmala 2010).

JT has a wide range of therapeutic properties for stopping diarrhea, discharging pus, diuretics, spleen, lung, dispelling moisture and heat, eliminating edema, etc. Moreover, JT seed has sedative, analgesic, and antipyretic properties. It also had an influence on uterus, heart, gut, and other organs (Yao 2005).

According to Zhang et al., JT roots, stems, leaves, and seeds can all be utilized as medicine (Zhang et al. 2002; Xie et al. 2016). JT was found to have many chemically active components, including lignans, lipids, and adenosine (Zhao 2002). JT helps maintain delicate and glossy skin. It can get rid of pregnancy spots, freckles, senile plaque, acne, and butterfly spots. It also works well to treat rough skin and hemorrhoids (Yao 2005). Clinical trials on humans and animals have shown that consuming dehulled flour of JT seeds can enhance lipid metabolism and hence lower the risk of heart disease. Additionally, it may lessen the accumulation of fat in the liver (Yu et al. 2005). JT contains a few bioactive compounds that protect against viral infection, including coixenolide (Chang et al. 2003; Shih et al. 2004). Those with celiac disease benefit from its gluten-free grains (Soni et al. 2023). So, in-depth research and development in the fields of food, medicine, and health related to JT is necessary to conduct (Feng et al. 2020). Table 12.9 represents the major chemical compounds found in JT seeds and their therapeutic benefits.

12.10 Industrial Applications of Job's Tears Millet

Polished grains of JT are used in food applications. JT transforms into flour, so that it can be used as a primary ingredient in a variety of products, leading to a rise in demand for JT. Whole grain intake has gained popularity recently since it retains its functional components. To make JT more suited for various applications, defatted JT flour is utilized (Mayasti et al. 2021).

Table 12.9 Chemical compounds found in Job's tears seeds, their properties, and structures

Compounds	Properties	Structures	References
Oleic acid Linoleic acid	Anti-inflammation, wound healing, blood clotting	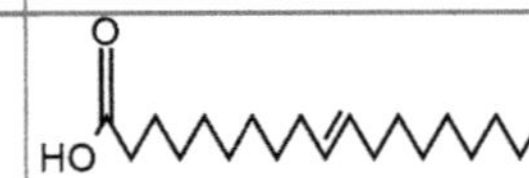 Oleic acid 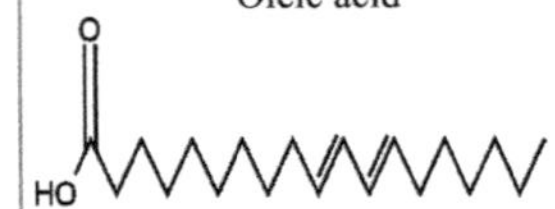 Linoleic Acid	Yu et al. (2015), Xi et al. (2016), Xuan et al. (2020), Igbokwe et al. (2021) and Ramadhan et al. (2023b)
Coixenolide	Anti-cancer and anti-tumor, rejuvenates skin, anti-wrinkle, regulates cell growth	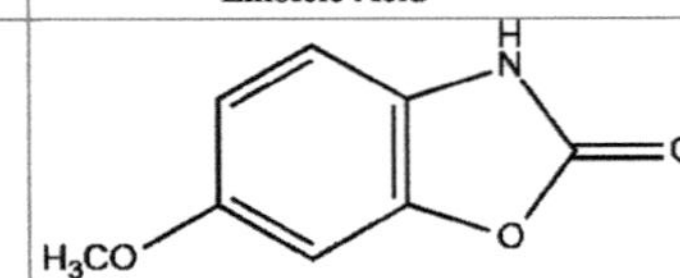 **Coixol**	Qian et al. (2019), Weng et al. (2022) and Guo et al. (2019)
Coixol	Sedation, anti-inflammation, anti-pyretic, anti-thrombotic, and muscle relaxation	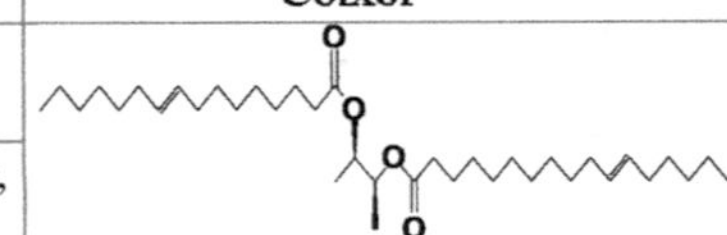 Coixenolide	Meng et al. (2010) and Zhang et al. (2019)
Coixan	Anti-diabetic, improves immune function, anti-cancer, stimulates the immune system, reduces drug toxicity, reduces body weight		Huang et al. (2009b), Luo et al. (2017), Yin et al. (2019), Weng et al. (2022) and Kang and Song (2019)
Tocopherol Tocotrienols	Antioxidant, anti-aging, anti-hyperlipidemia, anti-cancer	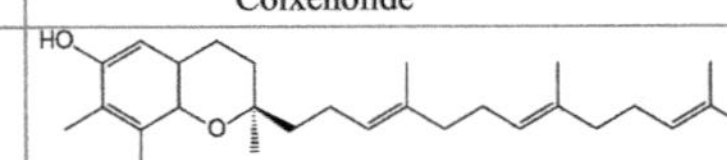 Gamma-Tocotrienol 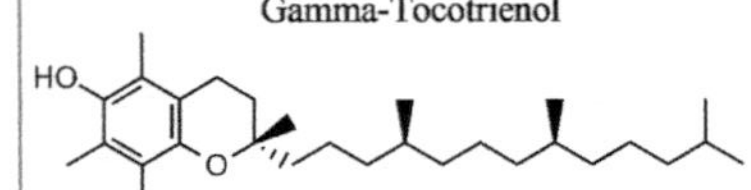 a-Tocopherol	Luithui et al. (2021)

Compounds	Properties	Structures	References
Beta-carotene	Antioxidant, anti-cancer	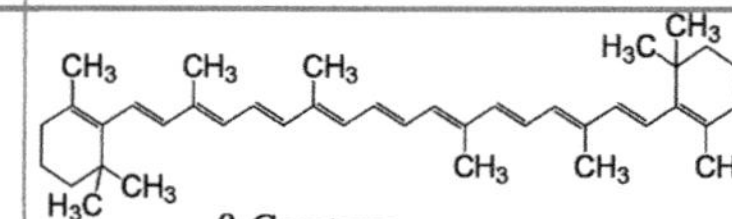 ß-Carotene	Bai et al. (2018) and Huda et al. (2023)
Beta-sitosterol and stigmasterol	Anti-hypoglycemic, anti-obesity	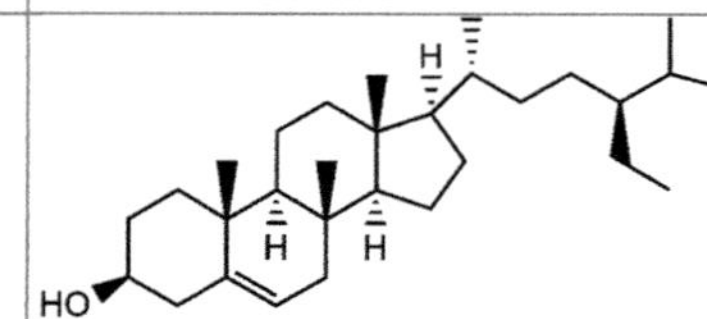 ß-Sitosterol 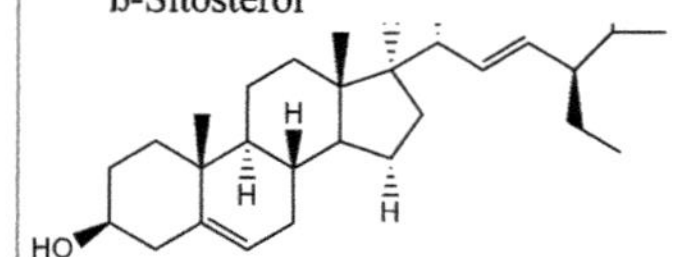 Stigmasterol	Chen et al. (2012) and Ragasa et al. (2014)

Compounds	Properties	Structures	References
Ferulic acid Caffeic acid	Anti-inflammatory, anti-hyperurecemic Anti-ulcer activity	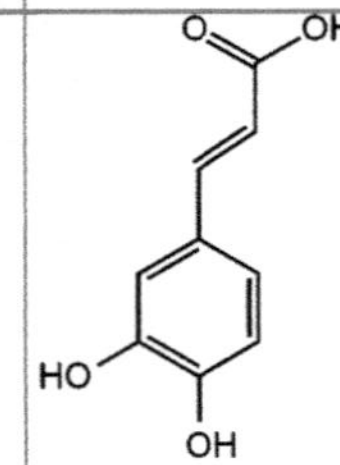Caffeic Acid 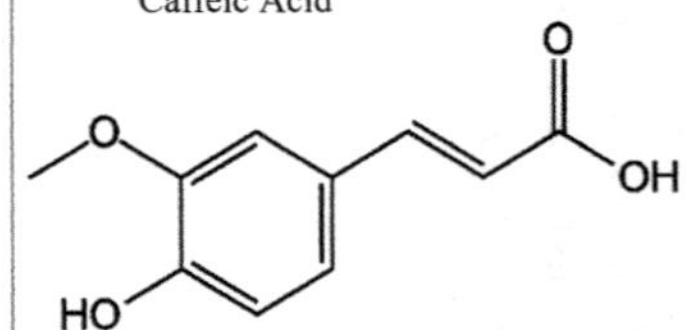Ferulic Acid	Sui et al. (2023) and Weng et al. (2022)

12.10.1 Beverage Industry

Chinese people take JT as soup, whereas Japanese people prepare a flavored beverage called "Dzu" that tastes like buttermilk and cider (Schaaffhausen 1952). JT is added as whole grain or flour to soups and broths. A non-dairy drink made from JT is sold as a substitute for healthy food products in Thailand and Japan (Chaisiricharoenkul et al. 2011). In order to prepare wine, JT seeds are fermented either by themselves or in conjunction with Poria fungus (Wu et al. 2017). Coixenolides is an immunoregulatory compound. JT can be processed into wine at elevated temperatures because of its high boiling point (Liu et al. 2016; Zheng et al. 2016). JT seeds are utilized to make half-solid fermented glutinous wine with a sour-sweet flavor (Wu 2010). It is also used to prepare teas and other beverages. In Korea and Thailand, a popular beverage is made from roasted JT seeds and later grind them into a fine powder. In China, a sweet beverage is made by combining powdered JT with water and sugar (Chaisiricharoenkul et al. 2011). The JT seed is used to make vinegar in Japan (Poddar et al. 2018). Sweet cocktails are also prepared using JT seeds (Biswas and Das 2022).

12.10.2 Food Industry

It is used in the production of various bakery products, including cookies (Indriati et al. 2021), American cookies made from JT (Lee et al. 2011a, b), and fermented gluten-free JT cookies (Marta et al. 2024). JT flour (30%) has also been blended with wheat flour for making bread and biscuits (Kutschera and Krasaekoopt 2012). The addition of JT is used in the form of powder or flakes in various products such as candies. In most places, it is consumed as pasta and dessert (Cainglet 2015). It's used to make, snacks, gruels (Khichdi), salads, soups, and sweets (Soni et al. 2023).

12.10.3 Fodder Industry

This crop's fodder can be fed to domestic animals in fresh, dried, or silage form due to its rapid regeneration capacity and high protein content. In India, the green leaves of wild species of JT are highly preferred by wild elephants. There is 8.5% of crude protein and 28% of crude fiber in JT fodder (Poddar et al. 2018). Sometimes molasses is added to it before feeding it to animals and poultry. The unshelled seeds after grinding can even be used to feed chickens. But when JT fodder is fresh, it tastes much better than when it is older (coarse plant pieces) (Soni et al. 2023). Figure 12.10 illustrates various uses of JT seeds in food industry.

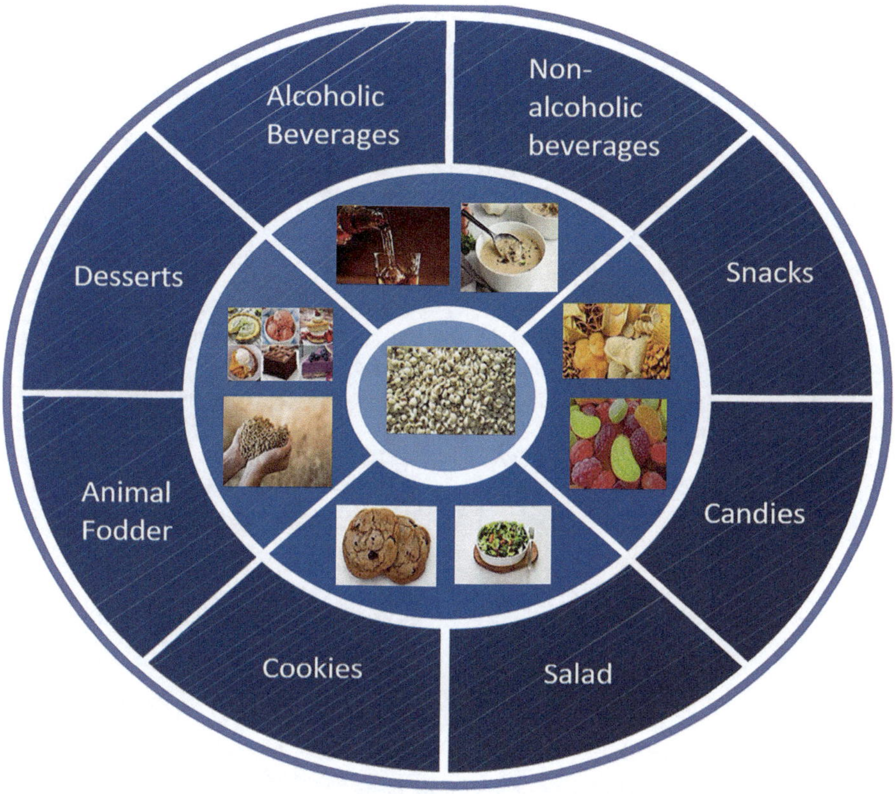

Fig. 12.10 Food applications of Job's tears seeds

12.10.4 Pharmaceutical Industry

In medicine, JT is frequently used. In traditional Indian Ayurveda and Chinese medicine, JT is well-known. Ayurvedic literature confirms that the leaves, roots, and seeds of this crop are all used to make medicine. As per Ayurvedic literature, JT can be applied in numerous ways to treat hair loss, warts, beriberi, bronchitis, dysuria, headache, hydrothorax, phthisis, pneumonia, smallpox, osteoporosis, hay fever, cholesterol levels, cancer, inflammation, hypertension, arthritis, asthma, appendicitis, immunological diseases, diabetes, rheumatism, menorrhagia, warts, urinary problems, puerperal fever, dysentery, intestinal parasites, tuberculosis, muscle and bone pains, swelling, pus accumulation, congestion, eczema, and diarrhea (Zeng et al. 2022; Feng et al. 2020; Hore and Rathi 2007).

Coix lacryma-jobi L., known as Gavedhukain Ayurveda, has attracted interest for its potential anti-obesity effects. A study examined the anti-obesity properties of JT using a network pharmacology approach, which integrates bioinformatics and computational tools to gain a deeper understanding of how its bioactive compounds interact with various targets and influence diseases like obesity (Pathak 2024). It

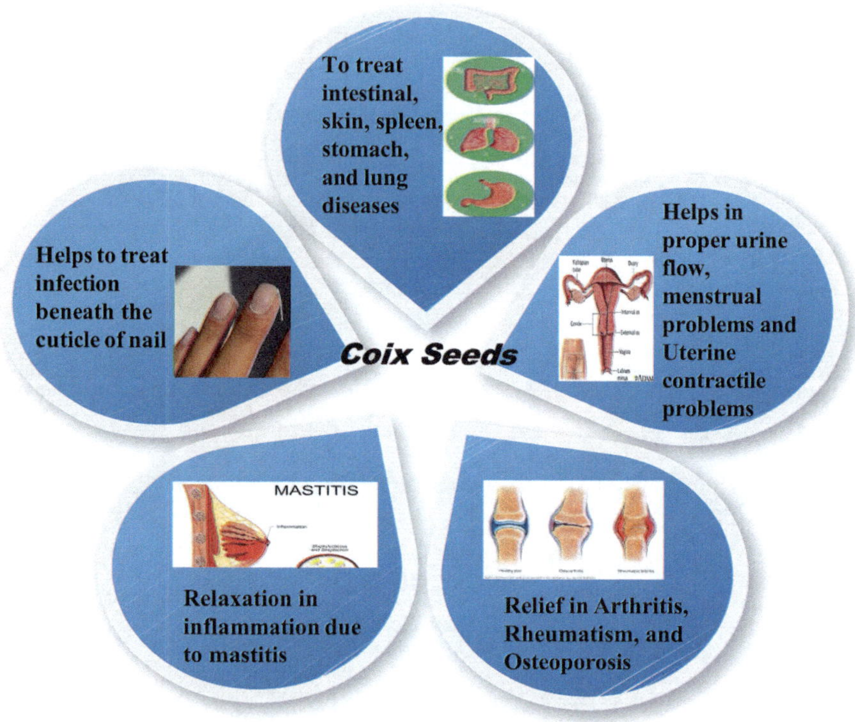

Fig. 12.11 Traditional health benefits of Job's tears seeds

contains chemicals that have antioxidant qualities that stop bacteria and other parasites from growing inside the body. It also has cooling diuretic properties. Tea prepared from the leaves or seeds of JT is traditionally believed to stimulate ovulation. The seeds are used as a decoction or tincture in places like China, the Philippines, India, and others (Biswas and Das 2022). JT is used in beauty industries. It is even used as bathing soap. Products from JT seeds are helpful in skin treatment (Igbokwe et al. 2021). JT is an allergen suppressant. It can suppress the serum level of IgE (Hsu et al. 2003). JT seeds reduce hyperpigmentation and help to moisturize the skin. It is a cooling herb and soothes skin ailments like rosacea, eczema (Cai et al. 2022; Suzuki and konaya 2021), and acne (Suzuki et al. 2019), prevents skin dryness and inflammation (Li et al. 2022), etc. JT seed extract helps in nourishment of skin, clarifies skin, removes impurities and dead skin cells, controls sebum production, enhances skin texture, and reduces skin pigmentation (Suzuki and konaya 2021). It is used to make an antibacterial facial mask (Lin et al. 2022). JT health benifits is represented in Fig. 12.11.

12.10.5 Other

JT seeds are used to make several musical instruments. It is used for preparing prayer beads, necklaces, rings, bracelets, anklets, and other ornaments. It is also used for making home decor such as decorative door and window curtains (Biswas and Das 2022). Moreover, JT stems are used to make matts (Shahrajabian et al. 2022).

12.11 Market Trends and Consumer Demand for JT Globally

The global market for JT is witnessing notable growth, driven by increasing consumer demand for healthy, nutrient-rich, gluten-free grains and plant-based natural products. This ancient grain, rich in protein, dietary fiber, and essential minerals, has gained significant popularity among health-conscious consumers seeking alternatives to conventional cereals (IMARC Group 2024). As a result, its versatility in culinary applications, ranging from soups and porridges to teas and desserts, has further expanded its adoption in both Asian and Western food markets. Moreover, JT is recognized for its medicinal properties, particularly its anti-inflammatory, antioxidant, and immunomodulatory effects, stimulating interest within the nutraceutical and functional food sectors (Fondazione Slow Food 2024). The grain is also increasingly utilized in skincare products and crafts, highlighting its diverse commercial potential. Additionally, the global shift toward sustainable and eco-friendly agriculture has also positioned JT as a viable crop for cultivation as it can be cultivated on marginal lands with minimal input, supporting agricultural resilience and biodiversity. As scientific research continues to substantiate the health benefits of JT and consumer preferences evolve toward holistic wellness, the global market for this crop is projected to expand significantly in the coming years (IMARC Group 2024; Fondazione Slow Food 2024).

12.12 Future Prospectives

Emerging research on JT emphasizes its significant potential in meeting nutritional demands, improving food security, and supporting sustainable agriculture. These crops are attracting growing interest due to their rich nutritional profiles, offering essential amino acids, proteins, vitamins, and minerals. Importantly, JT is naturally gluten-free, making it ideal for those with gluten intolerance or celiac disease. The rising consumer preference for healthier, minimally processed foods has encouraged the food industry to innovate and incorporate into a variety of new products, broadening their market trends. Research suggests that incorporating JT into the diet may help in reducing the risk of chronic conditions such as diabetes, cancer, and cardiovascular diseases. Their abundance of bioactive compounds, including powerful antioxidants like quercetin, rutin, and troxerutin, supports health by mitigating oxidative stress and inflammation. Hence, JT is not only a nutritious foods but it is a functional food with disease-preventive potential. Moreover, the extensive genetic diversity within JT enables them to thrive in diverse environmental conditions, making them valuable crops for climate change. These also contribute to agricultural biodiversity by growing on marginal lands unsuitable for conventional cereals, thereby enhancing land use and boosting food production in resource-limited regions. Cultivating these underutilized crops supports crop rotation and diversification practices, which help maintain soil health and reduce pest outbreaks. Such diversification is critical for developing resilient agricultural systems capable of adapting to climate variability. Additionally, ongoing research is dedicated to refining the processing and technological applications of JT to maximize their functionality in food systems. Advances in understanding their unique starch compositions and functional characteristics are paving the way for improved food products that meet the growing demand for gluten-free options. As scientific exploration continues, JT is expected to play an increasingly vital role in global food security and nutrition strategies aligned with sustainable development goals.

12.13 Conclusion

JT millet is a traditional cereal crop native to India, China, Burma, and Malaysia. This annual grass has been widely cultivated throughout Southeast Asia for many years and presents significant potential as a versatile and nutritious food source. This chapter focuses on the geographical distribution, cultivation, storage, appearance, morphology, and genetics of JT seeds. JT seed varies in size, appearance, and color among different species. The three most common forms of JT seeds are translucent, large, and small, providing a visual depiction of the several types of JT seed grains. Whole grain or JT extract consumption may be advantageous. Dehulled JT seed can be used as an ingredient in tea and fermented drinks, or it can be polished, cooked, and consumed like rice. In addition, it can be utilized whole or blended into flour to make pastries or other similar food items. Moreover, this chapter discussed the phytochemical composition, starch isolation, and its modification as well as the health benefits of JT seeds. A class of phytoconstituents and minerals such as polysaccharides, fats, vitamins, etc. are present in various parts (whole seed, hull, endosperm) of JT seeds. Modified JT starch shows various uses, from food formulations like cookies or noodles to pharmaceutical applications, including drug delivery systems and tablet manufacturing. In traditional medicine, JT seeds are used to cure arthritis, diarrhea, fever, and diuresis, aid in pus evacuation, and stimulate the spleen and lungs. However, ongoing breeding programs along with advanced storage and processing techniques are helping to overcome the challenges associated with its cultivation, enhancing JT productivity and market appeal. The chapter highlights the importance of further research to fully explore and expand JT applications, focusing the need for more studies on its chemical properties, nutritional benefits, and processing methods. Although the JT plant's seeds are known to have a wide range of advantages, researchers should also look into the advantages of the plant's other components and other JT family members to popularize the grain in the long run.

References

Acharya S, Gond H, Aslam S et al (2024) Job's tears (*Coix lacryma-jobi*): a medicinal plant of India. J Med Poaceae Ind 1:20. https://doi.org/10.5281/zenodo.10811091

Ahmad SR, Ghosh P (2022) A systematic investigation on flavonoids, catechin, β-sitosterol and lignin glycosides from Saracaasoca (ashoka) having anti-cancer & antioxidant properties with no side effect. J Indian Chem Soc 99(1):100293. https://doi.org/10.1016/j.jics.2021.100293

Ahmad R, Liaquat M, Sammi S et al (2024) Physicochemical and nutritional profiles of wild adlay (*Coix lacryma-jobi* Linn) accessions by GC, FTIR, and spectrophotometer. Food Chem 22:101418. https://doi.org/10.1016/j.fochx.2024.101418

Arora RK (1977) Job's-tears (*Coix lacryma-jobi*)—a minor food and fodder crop of Northeastern India. Econ Bot 31:358–366

Bai C, Zheng J, Zhao L et al (2018) Development of oral delivery systems with enhanced antioxidant and anticancer activity: coix seed oil and β-carotene coloaded liposomes. J Agric Food Chem 67(1):406–414

Bhandari SR, Park SK, Cho YC, Lee YS (2012) Evaluation of phytonutrients in Adlay (Coix lacrymajobi L.) seeds. Afr J Biotechnol 11(8):1872–1878. https://doi.org/10.5897/AJB11.2416

Biswas S, Das R (2022) Job's tears: a less known multipurpose crop. Just Agric 3(1):1–6

Bon SG, Antesco GP, Enicola EE (2023) Morphological characteristics and variability of traditional starch forming coix (*Coix lacryma-jobi* var. *ma-yuen*) populations from Mindanao Island, Philippines. Biodivers J 24(8):4382–4391

Cai Z, Liu H, Xie W, Jin W (2013) Research advances in Coix L. biology. J Field Crops Res 2:6–6

Cai X, Sun X, Liu L et al (2022) Efficacy and safety of Chinese herbal medicine for atopic dermatitis: evidence from eight high-quality randomized placebo-controlled trials. Front Pharmacol 13:927304. https://doi.org/10.3389/fphar.2022.927304

Cainglet KJA (2015) Spaghetti pasta from adlai flour (Coix lacryma-jobi L.) fortified with vitamin

Capule AB, Trinidad TP (2016) Isolation and characterization of native and modified starch from adlay (*Coix lacryma-jobi* L.). Int Food Res J 23(3):1199

Chaisiricharoenkul J, Tongta S, Intarapichet KO (2011) Structure and chemical and physicochemical properties of Job's tear (*Coix lacryma-jobi* L.) kernels and flours. Suranaree J Sci Technol 18(2):109–122

Chang HC, Huang YC, Hung WC (2003) Antiproliferative and chemopreventive effects of adlay seed on lung cancer in vitro and in vivo. J Agric Food Chem 51(12):3656–3660

Chang CC, Huang LH, Chiang W, Hsia SM (2018) Hexane fraction of adlay (Coix lachryma-jobi L.) testa ethanolic extract inhibits human uterine sarcoma cancer cells growth and chemosensitizes human uterine sarcoma cells to doxorubicin. PYTOEY 47:69–80. https://doi.org/10.1016/j.phymed.2018.03.056

Chang WC, Hu YT, Huang Q et al (2020) Development of a topical applied functional food formulation: Adlay bran oil nanoemulgel. LWT 117:108619. https://doi.org/10.1016/j.lwt.2019.108619

Chao G, Gao J, Liu R et al (2014) Starch physicochemical properties of waxy proso millet (*Panicum miliaceum* L.). Starch-Stärke 66(11–12):1005–1012. https://doi.org/10.1002/star.201400018

Chen HJ, Lo YC, Chiang W (2012) Inhibitory effects of adlay bran (*Coix lachryma-jobi* L. var. *ma-yuen Stapf*) on chemical mediator release and cytokine production in rat basophilic leukemia cells. J Ethnopharmacol 141(1):119–127

Chen Y, Qu D, Fu R et al (2018) A Tf-modified tripterine-loaded coix seed oil microemulsion enhances anti-cervical cancer treatment. Int J nanomed 7275–7287. https://doi.org/10.2147/IJN.S182475

Chiang Y, Chiang (2006) Effects of dehulled adlay on plasma glucose and lipid concentrations in streptozotocin-induced diabetic rats fed adiet enriched in cholesterol. Int J Vitam Nutr Res 76(5):299–305

Christopher J, Jacob B (1990) Cytology of a new Hexaploid cyto-type of *Coix lacryma-jobi* Linn. Cytologia 55(1):57–60

Chung CP, Hsia SM, Lee MY et al (2011) Gastroprotective activities of adlay (*Coix lachryma-jobi* L. var. *ma-yuen Stapf*) on the growth of the stomach cancer AGS cell line and indomethacin-induced gastric ulcers. J Agric Food Chem 59(11):6025–6033

Clayton WD (1980) Notes on the tribe Andropogoneae (Gramineae). Kew Bull 35(4):813–818

Corke H, Huang Y, Li JS (2016) Coix: overview. Encyclopedia Food Grains 1:184–189. https://doi.org/10.1016/8978-0-12-394437-5.00008-5

Devaraj RD, Jeepipalli SP, Xu B (2020) Phytochemistry and health promoting effects of Job's tears (*Coix lacryma-jobi*)—a critical review. Food Biosci 3:100537. https://doi.org/10.1016/j.fbio.2020.100537

Devi KS, Damayanti M, Singh DVNR (2015) Analysis of kidney stones by PXRD and evaluation of the antiurolithic potential of Coix lacryma jobi. Int J of Scientific and Res Publications 5(3):303–319

Diao X, Jia G (2017) Origin and domestication of foxtail millet. In: Genetics and genomics of setaria, pp 61–72. https://doi.org/10.1007/978-3-319-45105-3_4

Diningrat DS, Harahap NS, Risfandi M, Sari AN (2021) Antioxidant and antibacterial activities of Coix lacryma-jobi seed and root oil potential for meningitis treatment. Jordan J Biol Sci 14(5). https://doi.org/10.54319/jjbs/140501

Djaja N, Permadi I, Witjaksono F et al (2019) The effect of Job's tears-enriched yoghurt on GLP-1, calprotectin, blood glucose levels and weight of patients with type 2 diabetes mellitus. Mediterr J Nutr Metab 12(2):163–171. https://doi.org/10.3233/MNM-180258

Dwipa I, Martinsyah RH, Pamuji PAN et al (2022) Exploration and characterization of Hanjeli Nutfah Plasma (*Coix lacrima-jobi* L.) in West Sumatra Province. J Agronomi Tanaman Tropika 4(1):75–86

FAO (1995) Sorghum and millets in human nutrition. FAO food and nutrition series, no. 27. ISBN 92-5-103381-1. http://www.fao.org/docrep/T0818E/T0818E01.htm#Minormillets

FAO. *Coix lacryma-jobi* L. grassland species profiles. FAO/C IAT collaboration on tropical forages. http://www.fao.org/ag/AGP/AGPC/doc/Gbase/DATA/PF000205.HTM

Feng L, Zhao Y, Zhang Z et al (2020) Theedibleandmedicinal value of *Coix lacryma-jobi* and key cultivation techniques for high and stable yield. Nat Resour J 11(12):569. https://doi.org/10.4236/nr.2020.1112034

Ferreira RA, Borella J, Hüther C et al (2020) Drought-induced stress in leaves of *Coix lacryma-jobi* L. under exogenous application of proline and GABA amino acids. Rev Bras Bot 43(3):513–521. https://doi.org/10.1007/s40415-020-00637-0

Fondazione Slow Food (2024) Job's tears: traditional use and commercial potential. https://www.fondazioneslowfood.com

Fu YH, Yang C, Meng Q et al (2019) Genetic diversity and structure of *Coix lacryma-jobi* L. from its world secondary diversity center, Southwest China. Int J Genom 2019(1):9815697. https://doi.org/10.1155/2019/9815697

Guo M, Qu D, Qin Y et al (2019) Transferrin-functionalized microemulsions coloaded with coix seed oil and tripterine deeply penetrate to improve cervical cancer therapy. Mol Pharm 16(12):4826–4835

Harrisson T (1965) Borneo writing. Bijdragen tot de taal, land-en volkenkunde, 1ste Afl, pp 1–57

Hedayetullah M, Zaman, P (2022) Forage Crops of the World, 2-volume set: Volume I: Major Forage Crops; Volume II: Minor Forage Crops

Hong SS, Choi CW, Choi YH, Oh JS (2016) Coixlachryside A: A new lignan glycoside from the roots of Coix lachryma-jobi L. var. ma-yuen Stapf. Phytochem Lett 17:152–157. https://doi.org/10.1016/j.phytol.2016.07.004

Hore DK, Rathi RS (2007) Characterization of Job's tears germplasm in North-East India. Nat Prod Radiance 6(1):50–54

Hou JJ, Cao CM, Xu YW et al (2018) Exploring lipid markers of the quality of coix seeds with different geographical origins using supercritical fluid chromatography mass spectrometry and chemometrics. Phytomed 45:1–7. https://doi.org/10.1016/j.phymed.2018.03.010

Hsu HY, Lin BF, Lin JY et al (2003) Suppression of allergic reactions by dehulled adlay in association with the balance of TH1/TH2 cell responses. J Agric Food Chem 51(13):3763–3769

Hu J, Jin GL (2007) Nutritional and medicinal value of coix seeds. Food Nutr Chin 6:57–58

Hu X, Xu F, Li J et al (2022) Ultrasonic-assisted extraction of polysaccharides from coix seeds: Optimization, purification, and in vitro digestibility. Food Chem 374:131636. https://doi.org/10.1016/j.foodchem.2021.131636

Huang Y, Corke H, Li JS (2003) Coix. Enc Grain Sci. https://doi.org/10.1016/B0-12-765490-9/00035-5

Huang BW, Chiang MT, Yao HT, Chiang W (2005) The effect of adlay oil on plasma lipids, insulin and leptin in rat. Phytomedicine 12(6-7):433–439

Huang DW, Chung CP, Kuo YH et al (2009a) Identification of compounds in adlay (*Coix lachryma-jobi* L. var. *ma-yuen Stapf*) seed hull extracts that inhibit lipopolysaccharide-induced inflammation in RAW 264.7 macrophages. J Agric Food Chem 57(22):10651–10657

Huang DW, Kuo YH, Lin FY et al (2009b) Effect of adlay (*Coix lachryma-jobi* L. var. *ma-yuen Stapf*) Testa and its phenolic components on Cu2+-treated low-density lipoprotein (LDL) oxidation and lipopolysaccharide (LPS)-induced inflammation in RAW 264.7 macrophages. J Agric Food Chem 57(6):2259–2266

Huang Q, Xu M, Zhang H et al (2019) Transcriptome and proteome analyses of the molecular mechanisms associated with coix seed nutritional quality in the process of breeding. Food chem 272:549–558. https://doi.org/10.1016/j.foodchem.2018.07.116

Huang YJ, Chang CC, Wang YY et al (2021) Adlay testa (*Coix lachryma-jobi* L. var. *ma-yuen Stapf*) ethanolic extract and its active components exert anti-proliferative effects on endometrial cancer cells via cell cycle arrest. Molecules 26(7):1966. https://doi.org/10.3390/molecules26071966

Huda MN, Li X, Jahan T et al (2023) Acceleration of the genetic gain for nutraceutical improvement of adlay (Coix L.) through genomic approaches: Current status and future prospects. Food Rev Int 39(8):5377–5401. https://doi.org/10.1080/87559129.2022.2067175

Ibañez R (2024) Effects of storage methods and packaging materials on viability of Adlai (*Coix lacryma-jobi* L.). Seeds Diversitas J 9(2):560–573

Igbokwe CJ, Wei M, Feng Y et al (2021) Coix seed: a review of its physicochemical composition, bioactivity, processing, application, functionality, and safety aspects. Food Rev Int 38(1):921–939. https://doi.org/10.1080/87559129.2021.1892129

IMARC Group (2024) Job's tears market: global industry trends, share, size, growth, opportunity and forecast 2024–2029. https://www.imarcgroup.com/jobs-tears-market

Indriati A, Andriana Y, Mayasti NKI et al (2021) Techno-economic analysis on cookies production made from Adlay (*Coix lacryma-jobi*) flour that supplemented with moringa (*Moringa oleífera*) leaves powder. IOP Conf Ser Earth Environ Sci 672(1):012033. IOP Publishing

Jain SK, Banerjee DK (1974) Preliminary observations on the ethnobotany of the genus Coix. Econ Bot 28:38–42

Jinnouchi M, Miyahara T, Suzuki Y (2021) Coix seed consumption affects the gut microbiota and the peripheral lymphocyte subset profiles of healthy male adults. Nutrients. 13(11):4079. https://doi.org/10.3390/nu13114079

Kang JH, Song KB (2019) Characterization of Job's tears (*Coix lachryma-jobi* L.) starch films incorporated with clove bud essential oil and their antioxidant effects on pork belly during storage. Lwt 111:711–718

Khawas P, Deka SC (2017) Effect of modified resistant starch of culinary banana on physicochemical, functional, morphological, diffraction, and thermal properties. Int J Food Prop 20(1):133–150. https://doi.org/10.1080/10942912.2016.1147459

Khor HT, Chieng DY (1997) Effect of squalene, tocotrienols and α-tocopherol supplementations in the diet on serum and liver lipids in the hamster. Nutr Res 17(3):475–483. https://doi.org/10.1016/S0271-5317(97)00006-7

Kim Mulyono E, Kusuma A, Dewandari KT, Darniadi S (2019) Preliminary study of Hanjeli (*Coix lacryma-jobi* L) flour for food uses. J Earth Environ Sci 309(1):012057. https://doi.org/10.1088/1755-1315/309/1/012057

Kim HK, Cho DW, Hahm YT (2000) The effects of coix bran on lipid metabolism and glucose challenge in hyperlipidemic and diabetic rats. J Korean Soc Food Sci Nutr 29(2):140–146

Kim SO, Yun SJ, Lee EH (2007) The water extract of adlay seed (*Coix lachrymajobi* var. *mayuen*) exhibits anti-obesity effects through neuroendocrine modulation. Am J Chin Med 35(2):297–308

Kim MJ, Choi SJ, Shin SI et al (2008) Resistant glutarate starch from adlay: preparation and properties. Carbohydr Polym 74(4):787–796

Kim HW, Lee KY, Bae IY et al (2009) Physicochemical properties of Job's tears (*Coix lachryma-jobi* L.) starch modified with different levels of acid hydrolysis. Food Sci Biotechnol 18(5):1145–1149

Kim EJ, Kim HR, Choi SJ, Park CS, Moon TW (2016a) Low digestion property of amylosucrase-modified waxy adlay starch. Food Sci Biotechnol 25:457–460. https://doi.org/10.1007/s10068-016-0063-1

Kim SY, Choi CW, Hong SS (2016b) A new neolignan from *Coix lachryma-jobi* var. *mayuen*. Nat Prod Commun 11(2):1934578X1601100224. https://doi.org/10.1177/1934578X1601100224

Kristanti D, Ratnawati L (2022) Physicochemical and Functional Properties of Fermented Adlay (Coix lacryma jobi-L) Flour. Earth Environ Sci 1024(1):012026. https://doi.org/10.1088/1755-1315/1024/1/012026

Kumar YR (2021) Evaluation of germplasm lines for seed yield traits in Job's tears (*Coix lacryma-jobi* L). Int J Farm Sci 11(3):96–97

Kumar R, Kumawat N, Shukla UN (2014) Job's tears: an unexploited multipurpose shrub in North Eastern hilly region of India. Popular Kheti 2(4):118–122

Kuo CC, Chiang W, Liu GP (2002) 2, 2 'Diphenyl-1-picrylhydrazyl radical-scavenging active components from adlay (*Coix lachryma-jobi* L. Var. *ma-yuen Stapf*) hulls. J Agric Food Chem 50(21):5850–5855

Kuo CC, Chen HH, Chiang W (2012) Adlay ("soft-shelled Job's tears"); the seeds of (*Coix lachryma-jobi* L. var. *ma-yuen Stapf*) is a potential cancer chemopreventive agent toward multistage carcinogenesis processes. J Tradit Complement Med 2(4):267–275. https://doi.org/10.1016/s2225-4110(16)30112-2

Kutschera M, Krasaekoopt W (2012) The use of Job's tear (*Coix lacryma-jobi* L.) flour to substitute cake flour in butter cake. Aus J Technol 15(4):233–238

Lakkham K, Wangsomnuk P, Aromdee C (2009) Identification and quality of four varieties of adlay. Songklanakarin J Sci Technol 31(4):425–431

Laxmisha KM, Semwal DP, Gupta V et al (2022) Nutritional profiling and GIS-based grid mapping of Job's tears (*Coix lacryma-jobi* L.) germplasm. Appl Food Res 2(2):100169. https://doi.org/10.1016/j.afres.2022.100169

Laxmisha KM, Bhatt KC, Arya M et al (2023) Agro-morphological characterization and digitization of Job's tears (*Coix lacryma-jobi* L.) germplasm: a minor cereal crop of Northeastern India. Indian J Plant Genet Res 36(03):362–370

Lee MY, Lin HY, Cheng F et al (2008) Isolation and characterization of new lactam compounds that inhibit lung and colon cancer cells from adlay (*Coix lachryma-jobi* L. var. *ma-yuen Stapf*) bran. Food Chem Toxicol 46(6):933–1939

Lee HJ, Pak HO, Jang JS et al (2011a) Antioxidant activity and quality characteristics of American cookies prepared with Job's tears (*Coix lachryma-jobi* L.) chungkukjang powder and wheat bran powder. Korean J Food Nutr 24(1):85–93

Lee MJ, Kim SE, Lee SW et al (2011b) Quality characteristics of Korean and Chinese Job's tears (*Coix lachryma-jobi* L.). Food Ind Nutr 16(2):49–53

Li J, Corke H (1999) Physicochemical properties of normal and waxy job's tears (Coix lachryma-jobi L.) starch. Cereal Chem 76(3):413–416. http://dx.doi.org/10.1094/CCHEM.1999.76.3.413

Li J, Jin Z (2010) Potential allelopathic effects of *Mikania micrantha* on the seed germination and seedling growth of *Coix lacryma-jobi*. Weed Biol Manag 10(3):194–201. https://doi.org/10.1111/j.1445-6664.2010.00384.x

Li Z, Liu K (2012) Properties of Job's tears and high yield cultivation techniques in Liaoning province. Agric Sci Technol Equip 1:66–67

Li B, Qiao L, Li L et al (2017) Novel antihypertensive peptides derived from adlay (Coix larchryma-jobi L. var. ma-yuen Stapf) glutelin. Mol 22(1):123. https://doi.org/10.3390/molecules22010123

Li J, Han W, Zhang B et al (2018) Structure and physicochemical properties of resistant starch prepared by autoclaving-microwave. Starch-Stärke 70(9–10):1800060. https://doi.org/10.1002/star.201800060

Li Y, Tian X, Li S et al (2019) Total polysaccharides of adlay bran (Coix lachryma-jobi L.) improve TNF-α induced epithelial barrier dysfunction in Caco-2 cells via inhibition of the inflammatory response. Food Funct 10(5):2906–2913

Li F, Nakanishi Y, Murata K et al (2022) Coix seed extract prevents inflammation-mediated skin dryness induced by sodium dodecyl sulfate exposure in HR-1 hairless mice. Planta Med Int Open 9(01):e108–e115. https://doi.org/10.1055/a-1821-6332

Li XD, Lu XJ, Pan H, et al (2023) Component differentiation and functional analysis of flavonoids in coix seeds. J Shanxi Agric Sci 51(12):1410–1417. https://doi.org/10.3969/j.issn.1002-2481.2023.12.09

Lim TK (2012) Edible medicinal and non-medicinal plants, vol 1. Springer, Dordrecht, pp 285–292. https://doi.org/10.1007/978-94-007-5653-3-14

Lin LY, Liao YL, Chen MH et al (2022) Molecular action mechanism of Coixol from soft-shelled adlay on tyrosinase: the future of cosmetics. Molecules 27(14):4626. https://doi.org/10.3390/molecules27144626

Liu X, Rong YZ, Zhang X et al (2015) Rapid determination of total dietary fiber and minerals in Coix seed by near-infrared spectroscopy technology based on variable selection methods. Food Anal Methods 8:1607–1617. https://doi.org/10.1007/s12161-014-0037-y

Liu X, Wu JH, Xu JH et al (2016) The impact of heat-moisture treatment on the molecular structure and physicochemical properties of Coix seed starches. Starch-Stärke 68(7-8):662–674

Liu X, Zhang B, Xu JH et al (2017) Rapid determination of the crude starch content of Coix seed and comparing the pasting and textural properties of the starches. Starch-Stärke 69(1–2):1600115. https://doi.org/10.1002/star.201600115

Liu X, Yang YJ, Wang ZW (2018) Structure characteristics of Coix seeds prolamins and physicochemical and mechanical properties of their films. J Cereal Sci 79:233–239. https://doi.org/10.1016/j.jcs.2017.11.002

Liu S, Li F, Zhang X (2019) Structural modulation of gut microbiota reveals Coix seed contributes to weight loss in mice. Appl Microbiol Biotechnol 103(13):5311–5321. https://doi.org/10.1007/s00253-019-09786-z

Liu X, Li X, Zhou J et al (2025) The structure and digestibility of sequential enzyme modified starch from *Coix lachryma-jobi* L. Int J Biol Macromol 306:141462. https://doi.org/10.1016/j.ijbiomac.2025.141462

Lu P, Li YC (1996) The first discovery of Coix aquatica Roxnin China. Seed 1:54

Luithui Y, Meera MS (2019) Effect of heat processing on the physicochemical properties of Job's tears grain. J Food Meas Charact 13:874–882. https://doi.org/10.1007/s11694-018-0001-4

Luithui Y, Kamani MH, Sreerama YN, Meera MS (2021) Impact of hydrothermal processing on squalene, α-tocopherol, and fatty acid content in Job's tears grain milled fractions: evaluation of their storage stability. J Sci Food Agric 101(6):2319–2327

Luo C, Wang X, An C et al (2017) Molecular inhibition mechanisms of cell migration and invasion by coix polysaccharides in A549 NSCLC cells via targeting S100A4. Mol Med Rep 15(1):309–316

Luo XH, Liu C, Wang MY et al (2024) The hygroscopic properties and sorption isosteric heat of coix seeds. Food Sci Eng 5(2):255–270. https://doi.org/10.37256/fse.5220244583

Ma KH, Kim KH, Dixit A et al (2006) Newly developed polymorphic microsatellite markers in Job's tears (*Coix lacryma-jobi* L.). Mol Ecol Notes 6(3):689–691

Mandal S, Mandal M (2015) Coriander (*Coriandrum sativum* L.) essential oil: chemistry and biological activity. Asian Pac J Trop Biomed 5(6):421–428

Manosroi A, Sainakham M, Chankhampan C et al (2016) In vitro anti-cancer activities of Job's tears (*Coix lachryma-jobi* Linn.) extracts on human colon adenocarcinoma. Saudi J Biol Sci 23(2):248–256. https://doi.org/10.1016/j.sjbs.2015.03.008

Marta H, Firmansyah IN, Febiola C et al (2024) Physicochemical properties and sensory evaluation ofgluten-freecookies made from fermented adlay (*Coix lacryma-jobi* L.) flour using Rhizopus sp. Cogent Food Agric 10(1):2381608

Mayasti NKI, Iwansyah AC, Indriati A et al (2021) Nutritional and physical properties of three varieties of Hanjeli (*Coix lacryma-jobi*) flour for food diversification and application uses. IOP Conf Ser J Earth Environ Sci 733(1):012132

Meng X, Huang B, Zhou L et al (2010) Construction of a Coix BAC library and isolation of the 22 kDa α-coixin gene cluster. Genome 53(9):667–674

Moreda W, Pérez-Camino MDC, Cert A (2001) Gas and liquid chromatography of hydrocarbons in edible vegetable oils. J chromatogr A 936(1–2):159–171. https://doi.org/10.1016/S0021-9673(01)01222-5

Nurmala T (2011) Potensi dan Prospek Pengembangan Hanjeli (*Coix lacryma-jobi* L) sebagai Pangan Bergizi Kaya Lemak untuk Mendukung Diversifikasi Pangan Menuju Ketahanan Pangan Mandiri. Jurnal Pangan 20(1):41–48

Ochiai Y (2007) Conservation and creation of plant-using culture: a note on seed beads of Job's tears on the periphery of Myanmar. South East Asian Stud 45(3):382

Otsuka H, Hirai Y, Nagao T, Yamasaki K (1988) Anti-inflammatory activity of benzoxazinoids from roots of *Coix lachryma-jobi* var. *ma-yuen*. J Nat Prod 51(1):74–79

Ottoboni LM, Leite A, Yunes JA et al (1993) Sequence analysis of 22 kDa-like α-coixin genes and their comparison with homologous zein and kafirin genes reveals highly conserved protein structure and regulatory elements. Plant Mol Biol 21:765–778. https://doi.org/10.1007/bf00027110

Pang XF, Zhang SW, Qu ZC et al (1996) Growth characteristics and high yield cultivation techniques of *Coix lacryma-jobi*. Shandong Agric Sci 3:15–17

Patel B, Patel G, Shah S, Parmar S (2017) A review: *Coix lacryma-jobi* L. Res J Pharmacogn Phytochem 9(4):248–252. https://doi.org/10.5958/0975-4385.2017.00046.2

Pathak VK (2024) Network pharmacology-based anti-obesity activity of Gavedhuka (a millet): in silico study. J Drug Res Ayurvedic Sci 9(6):426–432. https://doi.org/10.4103/jdras.jdras_94_23

Pertanian K (2020) Buku Saku Bahan Pangan Potensialuntuk Anti Virus dan Imun Booster. Balai Besar Penelitian dan Pengembangan Pascapanen Pertanian

Pestana AC (1946) Cereal adlay. ALavoura 69:8–13

Poddar R, Jana K, Kundu CK, Das H (2018) Job's tears (coix). In: Forage crops of the world, volume I: major forage crops. Apple Academic Press, pp 111–120

Poddar R, Jana K, Kundu CK, Das H (2022) Job's tears (coix). In: Forage crops of the world, 2-volume set: volume I: major forage crops; volume II: minor forage crops, vol 29. Apple Academic Press, p 93

Puchongkavarin H, Varavinit S, Bergthaller W (2005) Comparative study of pilot scale rice starch production by an alkaline and an enzymatic process. Starch-Stärke 57(3–4):134–144

Qian MX (2016) Study on suitable row spacing and planting density of *Coix lacryma-jobi*. Peking Union Medical College and Chinese Academy of Medical Sciences, Beijing. https://doi.org/10.7666/d.Y3075222

Qian Y, Xiong Y, Feng D et al (2019) Coix seed extract enhances the anti-pancreatic cancer efficacy of gemcitabine through regulating ABCB1-and ABCG2-mediated drug efflux: a bioluminescent pharmacokinetic and pharmacodynamic study. Int J Mol Sci 20(21):5250

Quion KL, Bicoy JA, Alinsug MV et al (2024) Mapping the terrain: Baseline profiling of adlay (Coix lacyma-jobi L.) production in Bohol, Philippines. J Bio & Env Sci 2024(5):24–30

Ragasa CY, Caro JL, Lirio LG et al (2014) Chemical constituents of *Coix lacryma-jobi*. Res J Pharm, Biol Chem Sci 5(6):344–348

Ramadhan N, Hervani D, Martinsyah RH, Mutia YD (2023a) Evaluationofnutritioncontent of six local Job's tears (*Coix lacryma-jobi* L.) accessions in West Sumatera, Indonesia. IOP Conf Ser: Earth Environ Sci 1160(1):012016. https://doi.org/10.1088/1755-1315/1160/1/012016

Ramadhan N, Martinsyah RH, Muhsanati O et al (2023b) Article review: adlay (*Coix lacrima-jobi* L.) diversity in West Sumatra. Agroteknika 6(1):57–69. https://doi.org/10.55043/agroteknika.v6i1.193

Rao PN, Nirmala A (2010) Chromosomal basis of evolution in the genus Coix L. (Maydeae): a critical appraisal. Nucleus 53:13–24

Rick, Ellen B (2012) ECHO Asia Notes a regional supplement to ECHO development Notes. Issue 13

Rupitak Q, Yod-Arj J, Srisaikham S (2024) Job's tears (*Coix lacryma-jobi*) adaptability in Sa Kaeo Province, East of Thailand, and their chemical properties in an edible group for alternative ruminant animal forage and supplementary feed. Trends Sci 21(9):8060–8060. https://doi.org/10.48048/tis.2024.8060

Salami M, Tan H, Alizadeh B, Heidari B (2025) Photosynthetic performance, pigments and biochemicals influence seed yield in rapeseed under water deficit conditions: MGIDI index helps screening drought-tolerant genotypes. Field Crops Res 322:109733. https://doi.org/10.1016/j.fcr.2024.109733

Schaaffhausen RV (1952) Adlay or Job's tears—a cereal of potentially greater economic importance. Econ Bot 6(3):216–227

Seo GD (2019) Development of functional enzymes and diet supplement products through microbial cereal grain cultivation technology. Res J Pharm Technol 12(10):5042–5046. https://doi.org/10.5958/0974-360X.2019.00875.8

Shahrajabian MH, Sun W, Cheng Q, Khoshkharam M (2022) Exploring the quality of foods from ancient China based on traditional Chinese medicine. In: Functional foods and nutraceuticals in metabolic and non-communicable diseases. Academic, pp 87–105. https://doi.org/10.1016/B978-0-12-819815-5.00048-3

Shih CK, Chiang W, Kuo ML (2004) Effects of adlay on azoxymethane-induced colon carcinogenesis in rats. Food Chem Toxicol 42(8):1339–1347

Shouliang C, Phillips SM (2006) Coix. Flora China 22:648–649

Sira EEP, Amaiz ML (2004) A laboratory scale method for isolation of starch from pigmented sorghum. J Food Eng 64(4):515–519

Song ZQ, Ao MH, Shen G, Liu PF (2015) Effect of planting density, fertilizer level, and number of seedlings per hole onyield of *Coix lachryrma-jobi* L. Asian J Agric Res 7:69–73

Soni JK, Lalramhlimi B, Kumar A et al (2023) Coix: an underutilized functional food crop of Mizoram. Genet Resour Crop Evol 70(7):2143–2159. https://doi.org/10.1007/s10722-023-01587-8

Subroto E, Sitha N, Filianty F et al (2022) Freeze moisture treatment and ozonation of adlay starch (*Coix lacryma-jobi*): effect on functional, pasting, and physicochemical properties. Polym J 14(18):3854. https://doi.org/10.3390/polym14183854

Sui Y, Xu D, Sun X (2023) Identification of anti-hyperuricemic components from Coix seed. Food Biosci 52:102461

Sukkho T, Khanongnuch C, Lumyong S et al (2022) Local wisdom and diversity of medicinal plants in Cha Miang Forest in Mae Kampong Village, Chiang Mai, Thailand, and their potential for use as osteoprotective products. Plants 11(11):1492

Suzuki Y, Konaya Y (2021) Coix seed may affect human immune function. Nat Prod Commun 16(10):1934578X211048642. https://doi.org/10.1177/1934578X211048642

Suzuki N, Huo F, Kamibaba K, Hashimoto S (2019) Acne vulgaris successfully cured by intake of coix-seed reactive derivatives (CRD): a case-series study. Jpn Soc Complement Alternat Med 16(2):119–120. https://doi.org/10.1625/jcam.16.119

Tajalifar M (2018) Coix seeds: health and medicinal benefits. J Plant Genet Breed 2:2

Tensiska T, Cahyana Y, Nurhadi B et al (2021) In vitro digestibility and the characteristics of puffed products from different degrees of milling and moisture content of adlay grain (*Coix lacryma-jobi* L.). Food Res 5:371–378. https://doi.org/10.26656/fr.2017.5(2).548

Timilsena YP, Phosanam A, Stockmann R (2023) Perspectives on saponins: food functionality and applications. Int J Mol Sci 24(17):13538. https://doi.org/10.3390/ijms241713538

Van den Bergh MH, Iamsupasit N (1996) *Coix lacryma-jobi* L. In: Plant resources of South-East Asia, vol 10. Backhuys Publishers, Leiden, pp 84–87

Venkateswarlu J, Chaganti RSK (1973) Job's tears (*Coix lacryma-jobi* L.). Technical Bulletin 44, New Delhi

Wang L, Sun J, Yi Q, Wang X, Ju X (2012) Protective effect of polyphenols extract of adlay (Coix lachryma-jobi L. var. ma-yuen Stapf) on hypercholesterolemia-induced oxidative stress in rats. Mol 17(8):8886–8897. https://doi.org/10.3390/molecules17088886

Wang L, Chen C, Su A et al (2016) Structural characterization of phenolic compounds and antioxidant activity of the phenolic-rich fraction from defatted adlay (*Coix lachryma-jobi* L. var. *ma-yuen Stapf*) seed meal. Food Chem 196:509–517. https://doi.org/10.1016/j.foodchem.2015.09.083

Wang H, Ding J, Xiao N, Liu X, Zhang Y, Zhang H (2020) Insights into the hierarchical structure and digestibility of starch in heat-moisture treated adlay seeds. Food Chem 318:126489. https://doi.org/10.1016/j.foodchem.2020.126489

Weng WF, Peng Y, Pan X et al (2022) Adlay, an ancient functional plant with nutritional quality, improves human health. Front Nutr 9:1019375. https://doi.org/10.3389/fnut.2022.1019375

Wu SP (2010) Astudy on technological conditions for fermented glutinous wine made from seed of Job's tears. Cereal Feed Ind 2:26–37

Wu TT, Charles AL, Huang TC (2007) Determination of the contents of the main biochemical compounds of Adlay (*Coxi lachrymal-jobi*). Food Chem 104(4):1509–1515. https://doi.org/10.1016/j.foodchem.2007.02.027

Wu QH, Wei RC, Lin W (2014) Research progress of cultivation technology of *Coix lacryma-jobi*. Res Pract Chin Med 28:75–78

Wu TX, Yang ZT, Tang QL (2017) Method of yellow wine brewing with combined fermentation of coix seed and poria. Chinese Patent: 107189895A2017-09-22

Xi XJ, Zhu YG, Tong YP et al (2016) Assessment of the genetic diversity of different Job's tears (*Coix lacryma-jobi* L.) accessions and the active composition and anticancer effect of its seed oil. PLoS One 11(4):0153269. https://doi.org/10.1371/journal.pone.0153269

Xie J, Liu LZ, Lu MM, Wang Y (2016) Research progress on nutritional value and edible efficacy of coix seed. Grain Proc 41:50–52

Xu Y, Sun L, Gu Y et al (2023) Improving the emulsification performance of adlay seed starch by esterification combined with ultrasonication and enzymatic treatment. Int J Biol Macromol 242:124839. https://doi.org/10.1016/j.foodchem.2024.140105

Xu Y, Sun L, Zhuang Y et al (2024) Influence of gelatinized octenyl succinic anhydride-modified waxy adlay seed starch on the properties of astaxanthin-loaded emulsions: emulsion properties, stability and in vitro digestion properties. Food Chem 457:140105. https://doi.org/10.1016/j.foodchem.2024.140105

Xuan L, Xi X, Xu Z et al (2020) Genetic differences and variation in polysaccharide antioxidant activity found in germplasm resources for Job's tears (*Coix lacryma-jobi* L.). Botany 98(11):651–660

Yadav RK (2021) Evaluation of germplasm lines for seed yield traits in Job's tears (*Coix lacryma-jobi* L). Int J Farm Sci 11(3):96–97. https://doi.org/10.5958/2250-0499.2021.00044.6

Yang NW (2010) Formation and regulating of yield and edible quality of two *Coix lacryma-jobi* L. Peking Union Medical College and ChineseAcademy of Medical Sciences, Beijing

Yang SH, Peng J, Lui WB, Lin J (2008) Effects of adlay species and rice flour ratio on the physicochemical properties and texture characteristic of adlay-based extrudates. J Food Eng 84(3):489–494. https://doi.org/10.1016/j.jfoodeng.2007.06.010

Yang Q, Liu L, Li X et al (2021) Physicochemical characteristics of resistant starch prepared from Job's tears starch using autoclaving–cooling treatment. CyTA-J Food 19(1):316–325. https://doi.org/10.1080/19476337.2021.1897688

Yao FJ (2005) Effects of photoperiod and sowing date on growth and yield of coix. Sichuan Agricultural University, pp 16–19

Yin HM, Wang SN, Nie SP, Xie MY (2018) Coix polysaccharides: gut microbiota regulation and immunomodulatory. Bioact Carbohydr Diet Fibre 16:53–61. https://doi.org/10.1016/j.bcdf.2018.04.002

Yin Y, Yang W, Jiao Y et al (2019) Hypoglycaemic effect of coixan from Coix seeds on mice. EDP Sci 78:01009

Yu YT, Lu TJ, Chiang MT, Chiang W (2005) Physicochemical properties of water-soluble polysaccharide enriched fractions of adlay and their hypolipidemic effect in hamsters. J Food Drug Anal 13(4):7

Yu F, Li Y, Zhang J, Liu C (2015) *Coix lacryma-jobi* L. var. *ma-yuen* (Roman.) Stapf (Yiyiren, Job's tears). In: Dietary Chinese herbs: chemistry, pharmacology and clinical evidence. Springer, pp 339–346

Zeng Y, Yang J, Chen J et al (2022) Actional mechanisms of active ingredients in functional food adlay for human health. Molecules 27(15):4808

Zhang YM, Yang JS, Zhao YJ, Li XN (2002) Advances in chemical constituents and pharmacological activities of *Coix lacryma-jobi*. Chin Pharm J 37:8–11. https://doi.org/10.3321/j.issn:1001-2494.2002.01.004

Zhang C, Zhang W, Shi R et al (2019) *Coix lachryma-jobi* extract ameliorates inflammation and oxidative stress in a complete Freund's adjuvant-induced rheumatoid arthritis model. Pharm Biol 57(1):792–798

Zhang T, Xiao SY, Ding ZH, Song YT (2021) Effects of superfine grinding on physicochemical properties and morphological structure of coix seed powders. J Cereal Sci 102:103361. https://doi.org/10.1016/j.jcs.2021.103361

Zhao XH (2002) nutritional and medicalvalue of coix rice and development prospect of beverage making. Shanxi Food Ind 3:35–36. https://doi.org/10.3969/j.issn.1673-6044.2002.03.015

Zhao YJ, Liu ZH (1992) Study on the planting density of coix. Chin Med Mater 12:11–13

Zheng XL (2014) Problems and development strategies of coix production in China. Agric Dev Equip 5:40

Zheng HX, Zhang WM, Zhou HJ et al (2016) Effect of coixenolide on Foxp3+ CD4+ CD25+ regulatory T cells in collagen-induced arthritis mice. Chin J Integr Tradit West Med 36(3):348–350

Zhiyong, X (2018) Process Optimization and physicochemical properties of high-resistance coix seed flour prepared by ultrahigh pressure modification. Food and Machinery 34(1): 182–186. https://doi.org/10.13652/j.issn.1003-5788.2018.01.037

Zhou L, Huang B, Meng X et al (2010) The amplification and evolution of orthologous 22-kDa α-prolamin tandemly arrayed genes in coix, sorghum and maize genomes. Plant Mol Biol 74:631–643

Zhou S, Tang X, Hegyi F et al (2024) In vitro digestion and fermentation characteristics of soluble dietary fiber from adlay (Coix lacryma-jobi L. var. ma-yuen Staft) bran modified by steam explosion. Food Res Int. 192:114747. https://doi.org/10.1016/j.foodres.2024.114747

Zhu F (2017) Coix: chemical composition and health effects. Trends Food Sci Technol 61:160–175. https://doi.org/10.1016/j.tifs.2016.12.003

Zuloaga FO, Morrone O, Davidse G et al (2003) Catalogue of new world grasses (Poaceae): III. Subfamilies panicoideae, aristidoideae, arundinoideae, and danthonioideae. J Syst Evol 46:1–662

Chapter 13
Applicability of Pseudocereals (Buckwheat and Chaulai) in Food Industry

Priyanka Kumari, Suruchi Singh, Pankaj Bhatt, and Vivekanand Gaur

13.1 Introduction

Pseudocereals are non-grass plants that produce seeds or fruits used in a similar manner to true cereals, which belong to the grass family (*Poaceae*) (Martínez-Villaluenga et al. 2020; Poshadri et al. 2023). Unlike traditional cereals, pseudocereals are botanically distinct and include plants such as quinoa, amaranth, and buckwheat. These unique characteristics set them apart and contribute to their growing popularity in various culinary and health-conscious circles (Shahbaz et al. 2023). One of the most notable features of pseudocereals is their nutritional profile (Rodríguez et al. 2020). They are typically high in protein, essential amino acids, and dietary fiber, making them nutritionally valuable alternatives to traditional grains (Caeiro et al. 2022). For example, quinoa is particularly celebrated for its complete amino acid profile and high levels of essential fatty acids. This makes it an excellent choice for vegetarians and vegans looking to meet their protein needs (Upasana and Yadav 2022). Another significant advantage of pseudocereals is that they are naturally gluten-free. This characteristic has led to their increased popularity among individuals with celiac disease or gluten sensitivity (Estivi et al. 2022). As more people adopt gluten-free diets, pseudocereals provide a nutritious and

P. Kumari
Department of Zoology, Dhanauri PG College, Dhanauri-Haridwar, Uttarakhand, India

S. Singh (✉)
Accurate College of Pharmacy, Greater Noida, Uttar Pradesh, India

P. Bhatt
Lloyd Institute of Management and Technology, Greater Noida, Uttar Pradesh, India

V. Gaur
Columbus Industries Inc., Filtration Group, Dublin, OH, USA

© The Author(s), under exclusive license to Springer Nature Switzerland AG 2026

J. Singh, M. Kumar Malik (eds.), *Millets and Millet Starch: Chemical Composition, Functional Modification, and Applications*,
https://doi.org/10.1007/978-3-032-09502-2_13

versatile option that can easily replace wheat-based products in many recipes (Kaur 2023). Pseudocereals also exhibit remarkable adaptability when it comes to growing conditions (Torbica et al. 2022). Many of these plants can thrive in poor soil and are more resilient to environmental stresses compared to traditional cereal crops (Romano and Ferranti 2023). This adaptability allows them to be cultivated in regions where other grains may struggle, making them an important food source in diverse agricultural settings. Culinary uses for pseudocereals are varied and exciting. Their seeds can be processed similarly to true cereals; they can be ground into flour or cooked whole (Torbica et al. 2022; Peñalver et al. 2023). Common applications include porridge, baked goods, and as substitutes for rice or pasta in various dishes. For instance, buckwheat is often used to make pancakes and noodles, while amaranth can be incorporated into salads or served as a side dish (Vidaurre-Ruiz et al. 2023). Buckwheat and chaulai (amaranth) are both notable pseudocereals, valued for their nutritional benefits and versatility in culinary applications (Hitache et al. 2023).

13.1.1 *Buckwheat (*Fagopyrum esculentum*)*

Buckwheat, scientifically known as *Fagopyrum esculentum*, is a flowering plant belonging to the family *Polygonaceae* (Paucar-Menacho et al. 2023). It is cultivated primarily for its grain-like seeds, which are used in various food products, including flour and noodles. Despite its name, buckwheat is not related to wheat and is gluten-free, making it suitable for those with celiac disease or gluten intolerance (Graziano et al. 2022; Torbica et al. 2022). Originating around the 6th millennium BCE in what is now Yunnan Province, China, buckwheat has adapted well to diverse climatic conditions, particularly in hilly regions where soil fertility may be low. It thrives in acidic soils and can be grown quickly, typically maturing within 70–90 days (Bekkering and Tian 2019). This adaptability allows it to be cultivated as a late-season crop or as a cover crop to suppress weeds and improve soil quality. Nutritionally, buckwheat is rich in protein, fiber, and essential minerals such as manganese, copper, and magnesium (Habib et al. 2023; Hitache et al. 2023). It also contains bioactive compounds like rutin and quercetin, which have antioxidant properties and may support heart health by improving blood lipid profiles and reducing inflammation. Buckwheat's complex carbohydrates digest slowly, making it beneficial for blood sugar control and a good dietary option for diabetics (Agregán et al. 2022). Culinary uses of buckwheat include traditional dishes such as kasha (toasted buckwheat groats) and soba noodles (Tomás et al. 2022). It can also be incorporated into pancakes, porridge, and baked goods. The flowers of the buckwheat plant produce nectar that is harvested for honey, adding another dimension to its agricultural value (Peñalver et al. 2023).

13.1.2 Chaulai (Amaranth)

Chaulai, commonly known as amaranth, refers to several species within the *Amaranthus* genus. Like buckwheat, amaranth is classified as a pseudocereal due to its edible seeds that are used similarly to grains (Habib et al. 2023). It has been cultivated for thousands of years in various cultures around the world, particularly in Central and South America. Amaranth seeds are highly nutritious; they are an excellent source of protein and contain essential amino acids that are often lacking in other grains (Agregán et al. 2022). They are also rich in fiber, vitamins (such as vitamin A and vitamin C), and minerals like iron and calcium. The presence of antioxidants further enhances its health benefits (Paucar-Menacho et al. 2023). The cultivation of amaranth is notable for its resilience; it can thrive in poor soils and withstand drought conditions better than many traditional cereals. This makes it a sustainable crop option for regions facing climate challenges (Thakur et al. 2021). Culinary applications of amaranth include using the seeds as a grain substitute in dishes like porridge or salads (Thakur et al. 2021). The leaves of the amaranth plant are also edible and can be cooked similarly to spinach or used fresh in salads. In some cultures, amaranth is ground into flour for baking or used to make traditional snacks (Tomás et al. 2022; Vidaurre-Ruiz et al. 2023).

13.2 Importance of Pseudocereals in Global Food Systems

Pseudocereals play a significant role in global food systems, particularly as alternatives to traditional cereal grains like wheat, rice, and maize. Their importance is underscored by their nutritional benefits, adaptability to diverse growing conditions, and potential contributions to food security (Malik and Singh 2022; Romano and Ferranti 2023).

13.2.1 Nutritional Benefits

Pseudocereals such as quinoa, amaranth, and buckwheat are recognized for their high nutritional value (Botella-Martínez et al. 2021). They are rich in essential amino acids, vitamins, and minerals, making them valuable for addressing hidden hunger caused by micronutrient deficiencies that affect approximately 2 billion people globally (Malik and Singh 2022). Unlike major cereals, which often lack substantial amounts of micronutrients, pseudocereals are naturally nutrient-dense and can help improve dietary quality. For instance, buckwheat is noted for its high levels of antioxidants and dietary fiber, while quinoa offers a complete protein source (Kaur 2023).

13.2.2 Adaptability and Resilience

One of the key advantages of pseudocereals is their ability to thrive in poor soil conditions and under climatic stresses such as drought (Botella-Martínez et al. 2021). This resilience makes them particularly important in regions facing challenges related to climate change and food production (Morales et al. 2021). Pseudocereals can be cultivated in areas unsuitable for traditional cereals, thus enhancing agricultural biodiversity and providing alternative food sources where staple crops may fail (Caeiro et al. 2022).

13.2.3 Contribution to Food Security

As global populations grow and food security becomes an increasingly pressing issue, integrating pseudocereals into mainstream diets could play a crucial role (Morales et al. 2021). They can be utilized in various forms—such as whole grains, flours for baking, or as ingredients in processed foods—thereby expanding their market potential (He et al. 2022). The rising consumer interest in healthful eating has led to increased demand for gluten-free products, where pseudocereals serve as excellent substitutes due to their gluten-free nature (Shahbaz et al. 2023).

13.2.4 Economic Opportunities

The cultivation of pseudocereals also presents economic opportunities for farmers (He et al. 2022). With the growing global market for health foods, crops like quinoa have transitioned from traditional staples to globally traded commodities (Zhong et al. 2022). This shift not only supports local economies but also encourages sustainable agricultural practices that can benefit both producers and consumers alike (Poshadri et al. 2023).

13.3 Nutritional Profile

Nutritional profiling, also known as nutrient profiling, is the science of classifying or ranking foods based on their nutritional composition. This approach aims to promote health and prevent disease by providing a systematic way to evaluate the nutritional quality of foods. Nutritional profiling can be used in various contexts, including food labeling, dietary guidelines, and public health initiatives (Zhong et al. 2022). The primary purpose of nutritional profiling is to help consumers identify healthier food options. It involves analyzing the content of essential nutrients such as carbohydrates, proteins, fats, vitamins, and minerals in food items. By categorizing foods according to their nutrient density and overall health impact, nutritional profiling can guide consumers in making informed dietary choices (He et al. 2022). Different models of nutrient profiling have been developed by various organizations, including health authorities, academic institutions, and the food industry. These models may vary significantly in their criteria for classification, leading to different assessments of the same food item. For instance, some models might prioritize low sugar and fat content, while others may emphasize the presence of beneficial nutrients like fiber and vitamins (Morales et al. 2021). Overall, nutritional profiling serves as a valuable tool for public health nutrition interventions by identifying foods that can contribute to healthy eating patterns and those that may pose health risks if consumed excessively. It plays a critical role in shaping food policies aimed at reducing the prevalence of diet-related diseases and promoting overall well-being (Botella-Martínez et al. 2021).

13.3.1 Nutritional Analysis of Buckwheat

Buckwheat (*Fagopyrum esculentum*) is a highly nutritious pseudocereal that stands out for its unique health benefits and rich nutrient profile (Akın et al. 2023). In terms of macronutrients, buckwheat provides approximately 343 calories per 100 grams when raw, while cooked buckwheat groats yield about 92 calories per 100 grams (Malik and Singh 2022). It contains around 71.5 grams of carbohydrates per 100 grams, primarily in the form of starch, making it a good energy source (Dhondiram et al. 2023). Notably, buckwheat is rich in protein, offering about 13.25 grams per 100 grams, with a high-quality amino acid profile that includes essential amino acids such as lysine and arginine, which are often limited in other grains (Thakur et al. 2021). The fat content is relatively low at approximately 3.4 grams per 100 grams and consists mainly of healthy fatty acids like oleic and linoleic acids (Liu et al. 2021). Additionally, buckwheat is an excellent source of dietary fiber, providing about 10 grams per 100 grams, which aids in digestion and helps maintain stable blood sugar levels (Paucar-Menacho et al. 2023). In terms of vitamins and minerals, buckwheat is particularly rich in several essential nutrients. It contains approximately 57% of the Daily Value (DV) for manganese per 100 grams, which plays a crucial role in metabolism and antioxidant defenses. Additionally, it provides about 55% DV for magnesium, important for bone health and metabolic functions, and around 122% DV for copper, which supports cardiovascular health (Tomás et al. 2022). Buckwheat also offers significant levels of phosphorus (about 28% DV) and iron, which are vital for bone health and oxygen transport in the blood, respectively (Coţovanu and Mironeasa 2021). Furthermore, it contains various B vitamins such as riboflavin and niacin that are essential for energy metabolism. Buckwheat is also notable for its bioactive compounds, particularly flavonoids like rutin and quercetin (Habib et al. 2023). These compounds have antioxidant properties that can benefit cardiovascular health by strengthening blood vessels and reducing inflammation

(Aloo et al. 2021). The concentration of these bioactive compounds can vary significantly among different buckwheat cultivars, adding to its health-promoting potential (Agregán et al. 2022).

13.3.2 Nutritional Analysis of Chaulai (Amaranth)

Chaulai, commonly known as amaranth, refers to several species within the *Amaranthus* genus. Like buckwheat, amaranth is classified as a pseudocereal due to its edible seeds that are used similarly to grains (Živković et al. 2021). In terms of macronutrients, amaranth provides approximately 371 calories per 100 grams when cooked (Vidaurre-Ruiz et al. 2023). It consists of about 65–70 grams of carbohydrates per 100 grams, making it a significant energy source. Amaranth is exceptionally high in protein compared to many other grains, with about 13–14 grams per 100 grams (Coţovanu and Mironeasa 2021; Norbäck and Wieslander 2021). Its protein quality is high due to a balanced amino acid profile that is particularly rich in lysine, an essential amino acid often lacking in other grains (Romano and Ferranti 2023). The fat content ranges from 5 to 9 grams per 100 grams and consists predominantly of healthy unsaturated fats (Zhang et al. 2021). Additionally, amaranth provides around 7–8 grams of dietary fiber per 100 grams, contributing to digestive health and promoting satiety. Amaranth is also rich in several essential vitamins and minerals that enhance its nutritional profile. It contains significant levels of calcium compared to other grains, which is important for bone health (Kaur 2023; Akın et al. 2023). The grain provides a good amount of iron, aiding in the prevention of anemia, as well as magnesium that contributes to muscle function and energy production (Kreft et al. 2022). Furthermore, amaranth is a source of potassium, which is essential for heart health and regulating blood pressure (Shahbaz et al. 2023). It also contains B vitamins such as folate that play a crucial role in DNA synthesis and repair. In addition to its macronutrient composition, amaranth contains bioactive compounds that offer various health benefits. These antioxidants may help reduce oxidative stress and inflammation within the body (Cao et al. 2022). Research has also suggested that amaranth may possess potential anticancer properties due to its phytochemical content (Caeiro et al. 2022) (Table 13.1).

13.4 Comparison with Traditional Cereals

Pseudocereals, such as buckwheat and amaranth, present distinct advantages over traditional cereals like wheat and rice in several key areas, including nutritional composition, adaptability, and culinary applications (Akın et al. 2023).

Table 13.1 The nutritional composition of buckwheat and chaulai (amaranth) per 100 grams

Nutritional component	Buckwheat (common)	Chaulai (amaranth)	References
Calories	343 kcal	371 kcal	Poshadri et al. (2023)
Carbohydrates	71.5 g	65–70 g	Poshadri et al. (2023)
Protein	13.25 g	13–14 g	Shahbaz et al. (2023)
Fat	3.4 g	5–9 g	Caeiro et al. (2022)
Dietary fiber	10 g	7–8 g	Kaur (2023)
Ash	2.1 g	Not specified	Romano and Ferranti (2023)
Moisture	9.8 g	Not specified	Romano and Ferranti (2023)
Calcium	Not specified	Significant levels	Romano and Ferranti (2023)
Iron	Significant levels	Good amount	Vidaurre-Ruiz et al. (2023)
Magnesium	231 mg (55% DV)	High levels	Vidaurre-Ruiz et al. (2023)
Copper	1.1 mg (122% DV)	Present	Malik and Singh (2022)
Manganese	1.3 mg (57% DV)	Present	Malik and Singh (2022)
Phosphorus	347 mg (28% DV)	Present	Malik and Singh (2022)
Vitamins	B vitamins (e.g., niacin, riboflavin) present in good amounts; folate also available	Rich in vitamins, especially folate and vitamin A	Vidaurre-Ruiz et al. (2023)

13.4.1 Nutritional Comparison

One of the most significant differences between pseudocereals and traditional cereals is their nutritional profile (Živković et al. 2021). Pseudocereals tend to be higher in protein content compared to traditional grains. For instance, buckwheat contains approximately 13–14% protein, while quinoa can have up to 16% protein (Norbäck and Wieslander 2021). In contrast, wheat typically has around 10–12% protein. This higher protein content in pseudocereals is complemented by a more favorable amino acid profile, particularly in essential amino acids such as lysine, which is often limited in traditional grains like wheat and rice (Dhondiram et al. 2023). Moreover, pseudocereals are generally rich in dietary fiber, vitamins, and minerals (Zhang et al. 2021). For example, buckwheat is high in magnesium, manganese, and iron, while amaranth boasts significant levels of calcium and potassium. Traditional cereals often lack these micronutrients or contain them in lower amounts. Additionally, pseudocereals are gluten-free, making them suitable for individuals with celiac disease or gluten intolerance (Liu et al. 2021).

13.4.2 Adaptability and Cultivation

Pseudocereals also demonstrate greater adaptability to diverse growing conditions compared to traditional cereals (Bhatt et al. 2023). Many pseudocereals can thrive in poor soils and are more resilient to environmental stresses such as drought and high temperatures (Aloo et al. 2021). This resilience makes them valuable crops for regions facing climate change challenges where traditional cereals may struggle to grow (Kreft et al. 2022). In contrast, traditional cereals like wheat and rice often require specific soil conditions and ample water supply for optimal growth. This dependency on favorable agricultural conditions can make them more vulnerable to crop failures due to climate variability (Coţovanu and Mironeasa 2021).

13.4.3 Culinary Applications

From a culinary perspective, pseudocereals offer versatility similar to that of traditional cereals but with unique flavors and textures (Cao et al. 2022). Buckwheat can be used in various forms, including flour for pancakes and noodles or as whole groats in salads and porridge. Amaranth can be cooked similarly to rice or used as flour in baking (Norbäck and Wieslander 2021). While traditional cereals are staples in many diets worldwide providing the base for products like bread, pasta, and rice dishes pseudocereals are increasingly being incorporated into modern diets as health-conscious consumers seek alternatives that offer enhanced nutritional benefits (Zhong et al. 2022). The growing popularity of gluten-free diets has further propelled the use of pseudocereals in various food products such as breakfast cereals, snack bars, and baked goods (Živković et al. 2021).

13.4.4 Health Benefits Associated with Their Consumption

The consumption of pseudocereals like buckwheat and chaulai (amaranth) is associated with numerous health benefits (Fig. 13.1), making them valuable additions to a balanced diet. One of the most significant advantages of buckwheat is that it is naturally gluten-free, making it an excellent food option for individuals with celiac disease or those with gluten intolerance (He et al. 2022). Unlike many gluten-free products that may lack nutritional value, buckwheat is rich in fiber, vitamins, and minerals, providing a nutritious alternative to traditional grains like wheat and rice (Kreft et al. 2022). Buckwheat also promotes improved digestion due to its high

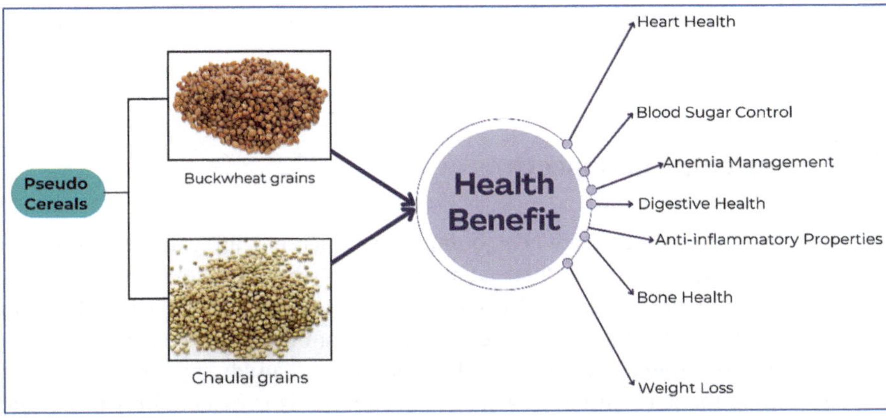

Fig. 13.1 Health benefits of buckwheat and chaulai

fiber content, which facilitates regular bowel movements and helps prevent constipation (Morales et al. 2021). The fiber in buckwheat supports the growth of beneficial gut bacteria, enhancing overall gut health. Additionally, buckwheat has a low to medium glycemic index, meaning it has a moderate effect on blood sugar levels. Its soluble carbohydrates, particularly D-chiro-inositol, help improve insulin sensitivity and may aid in managing type 2 diabetes by lowering post-meal blood sugar spikes (Cao et al. 2022).

Another notable health benefit of buckwheat is its positive impact on cardiovascular health. Regular consumption of buckwheat has been linked to lower cholesterol levels and improved heart health (Zhu 2021). The presence of antioxidants like rutin helps reduce inflammation, lower blood pressure, and prevent the formation of blood clots, thereby lowering the risk of cardiovascular diseases. Furthermore, buckwheat can contribute to weight management due to its high protein and fiber content, which promote feelings of fullness and satiety (Botella-Martínez et al. 2021). This effect can help reduce overall calorie intake and support weight loss efforts. In addition to these benefits, buckwheat is rich in antioxidants, including flavonoids such as rutin and quercetin. These compounds combat oxidative stress in the body, potentially reducing the risk of chronic diseases such as cancer and heart disease (Zhong et al. 2022). Similarly, chaulai (amaranth) offers a range of health benefits. It is nutrient-dense, being rich in protein, fiber, vitamins (especially B vitamins), and minerals like calcium and iron. Its high protein content makes amaranth an excellent choice for vegetarians and vegans seeking complete proteins (Malik and Singh 2022). Like buckwheat, amaranth is also gluten-free, making it suitable for those with gluten sensitivities or celiac disease (Botella-Martínez et al. 2021). Amaranth has a low glycemic index as well, which can help regulate blood sugar levels and may be beneficial for individuals with diabetes. Its fiber and antioxidant content contribute to cardiovascular health by lowering cholesterol levels and reducing inflammation. Additionally, the high fiber content in amaranth supports digestive health by promoting regular bowel movements and enhancing gut microbiota diversity (He et al. 2022).

13.5 Agricultural Aspects

Pseudocereals, including buckwheat, quinoa, and amaranth, are gaining recognition in agricultural systems due to their adaptability, nutritional benefits, and potential role in enhancing food security (Norbäck and Wieslander 2021). These plants are not true cereals; they belong to different botanical families and are characterized by their ability to produce starchy seeds that can be used similarly to grains. Understanding the agricultural aspects of pseudocereals is essential for promoting their cultivation and utilization in food systems (Morales et al. 2021).

13.5.1 Cultivation Practices for Buckwheat and Chaulai

13.5.1.1 Cultivation Practices for Buckwheat

Buckwheat (*Fagopyrum esculentum*) is a versatile and resilient crop that is relatively easy to cultivate (Akın et al. 2023). Known for its short growing season, buckwheat typically requires only 10–12 weeks from planting to harvest, making it particularly suitable for regions with short growing seasons or as a second crop following early-season crops. One of the primary advantages of buckwheat is its adaptability to various soil types (Tomasiak et al. 2022). It thrives in moderately fertile soils and can tolerate pH levels as low as 4.8. Buckwheat prefers light to medium-textured, well-drained soils such as sandy loams and loams; however, it does not perform well in heavy, wet soils or those high in limestone (Živković et al. 2021). Preparing a good seedbed is crucial; the soil should be worked several weeks in advance to ensure it is easy to manage and has adequate moisture for crop growth. Proper drainage is essential, as buckwheat seeds can rot if water collects around them (Zou et al. 2023). The ideal time to plant buckwheat varies by region but generally occurs in late spring or early summer in temperate climates. For example, in India, sowing typically takes place during the cooler months of autumn, particularly between October and November, to align with favorable climatic conditions, while controlled conditions (greenhouses) may allow for planting from February to October (Aloo et al. 2021). Buckwheat seeds should be sown at a depth of 2–4 cm with row spacing of 10–15 cm, and the recommended seed rate is about 20 kg per hectare. Throughout its growing season, buckwheat requires moderate moisture, particularly during critical stages such as flowering and seed formation. It is sensitive to extreme weather conditions; high temperatures during flowering can hinder seed set, while late spring or early fall frosts can severely damage the crop (Zhao et al. 2023). In terms of fertilization and pest management, buckwheat has modest nutrient requirements compared to other grains and often does not require significant fertilization. However, applying low rates of nitrogen can enhance growth, especially in nitrogen-depleted soils (Coţovanu and Mironeasa 2021). It is important to note that excessive nitrogen can lead to lodging, negatively impacting yield. Fortunately, buckwheat is generally resistant to pests and diseases, which reduces the need for chemical fertilizers; however, wildlife such as deer and turkeys may pose risks in some areas (Baillière et al. 2022).

13.5.2 Cultivation Practices for Chaulai (Amaranth)

Chaulai, commonly known as amaranth, encompasses several species within the *Amaranthus* genus that are cultivated for their nutritious seeds and leaves (Punia Bangar et al. 2022). Like buckwheat, amaranth is recognized for its adaptability to various growing conditions. Amaranth grows well in a range of soil types but prefers well-drained soils with moderate fertility (Liu et al. 2021). Sandy loams or loamy soils with good organic matter content are ideal for its cultivation. Prior to planting, land preparation involves tilling the soil to create a fine seedbed that ensures good seed-to-soil contact and drainage. Amaranth can be sown directly into the field or started in trays before transplanting (Paucar-Menacho et al. 2022a). The seeds are small and should be planted at a depth of about 1–2 cm with row spacing of 30–45 cm. The optimal planting time varies by region but typically occurs in late spring after the last frost when soil temperatures are warm enough for germination (Dhondiram et al. 2023). Amaranth requires consistent moisture during its growth period but is relatively drought-tolerant once established. It thrives in warm temperatures, ideally between 20 and 30 °C. When it comes to fertilization and pest management, amaranth has low fertilization requirements; however, applying organic matter such as compost can enhance soil fertility and improve yields (Akın et al. 2023). Regular weeding is essential during the early stages of growth since amaranth competes well against weeds once established. Pest management strategies often rely on cultural practices since amaranth is generally resistant to many pests. Crop rotation and intercropping can help reduce pest pressure while improving soil health (Dhondiram et al. 2023).

13.6 Challenges Faced in the Cultivation of These Crops

The cultivation of pseudocereals like buckwheat and chaulai (amaranth) presents several challenges that can impact their productivity and adoption among farmers. Understanding these challenges is crucial for developing strategies to enhance their cultivation and utilization (Liu et al. 2021).

13.6.1 Challenges in Buckwheat Cultivation

Buckwheat (*Fagopyrum esculentum*) faces a variety of challenges that hinder its widespread cultivation. One significant issue is its sensitivity to extreme weather conditions. Buckwheat thrives in cool, moist climates but is highly susceptible to freezing temperatures during fall and spring, which can lead to crop failure (Coţovanu and Mironeasa 2021). Additionally, high temperatures and dry weather during the flowering stage can prevent seed formation, resulting in reduced yields. This sensitivity limits the regions where buckwheat can be effectively grown, particularly as climate change leads to more unpredictable weather pattern (Paucar-Menacho et al. 2022a). Another challenge is the uneven ripening of buckwheat seeds, which complicates the harvesting process. Farmers often encounter difficulties in determining the optimal time for harvest, as seeds may not mature uniformly on the same plant (Aloo et al. 2021). This can lead to losses if harvesting occurs too early or too late. The lack of modern harvesting equipment specifically designed for buckwheat further exacerbates this issue, making it necessary for farmers to adopt traditional methods that may not be efficient. Additionally, buckwheat is often considered a marginal crop, leading to a decline in its cultivation over the years. Changing cropping patterns, improved economic conditions, and migration have contributed to reduced interest in growing buckwheat. As farmers shift toward more commercially viable crops with higher market demand, the area dedicated to buckwheat cultivation has diminished (Punia Bangar et al. 2022). This trend highlights the need for increased awareness and research into the benefits and potential of buckwheat as a nutritious food source (Živković et al. 2021).

13.6.2 Challenges in Chaulai (Amaranth) Cultivation

Chaulai (amaranth) also faces several challenges that can affect its cultivation. While amaranth is known for its adaptability to various soil types and climatic conditions, it still requires careful management to achieve optimal yields (Baillière et al. 2022). One challenge is the competition from weeds during the early stages of growth. Amaranth seedlings are relatively small and can struggle against aggressive weeds unless proper weed management practices are implemented. Failure to control weeds can lead to reduced yields and increased production costs (Norbäck and Wieslander 2021; Bhatt et al. 2024). Furthermore, while amaranth is generally resistant to many pests and diseases, certain pests such as aphids and leafhoppers can pose threats under specific conditions (Zhao et al. 2023). These pests can cause damage to young plants, impacting overall crop health and yield. Integrated pest management strategies are essential to mitigate these risks while minimizing chemical inputs. Market access and consumer awareness also present challenges for amaranth cultivation. Despite its nutritional benefits, amaranth remains underutilized in many regions due to limited consumer knowledge about its uses and health advantages (Zou et al. 2023). Farmers may hesitate to grow amaranth if they lack assurance of a market for their produce or if they face competition from more established crops (Norbäck and Wieslander 2021).

13.7 Processing Techniques

Pseudocereals, such as buckwheat, amaranth, and quinoa, are increasingly recognized for their nutritional benefits and versatility in food applications. Processing methods for these grains play a crucial role in enhancing their usability and nutritional value (Tomasiak et al. 2022). Various techniques are employed to transform pseudocereals from raw seeds into finished products suitable for consumption (Kreft et al. 2022).

13.7.1 Traditional Processing Methods

Traditional processing methods for pseudocereals include cooking, popping, roasting, and fermentation. Cooking is the most common method, often used to prepare porridge or soups (Shreeja et al. 2021). For instance, buckwheat can be cooked as groats or ground into flour for various baked goods. Amaranth can be boiled similarly to rice or used in porridge. Popping and roasting are techniques that enhance flavor and texture; popped amaranth is often used as a snack or ingredient in breakfast cereals (Suzuki et al. 2021). Fermentation is another traditional method that can improve the nutritional profile of pseudocereals. Fermentation processes can enhance the bioavailability of nutrients by breaking down anti-nutritional factors such as phytic acid, which inhibits mineral absorption. Sour dough fermentation is particularly beneficial for improving the digestibility of pseudocereal flours (Wang et al. 2021).

13.7.2 Modern Processing Techniques

Modern processing techniques have also been developed to optimize the use of pseudocereals in food production. These include extrusion cooking, which involves forcing a mixture through a die to create various shapes and textures (Kolarić et al. 2021). This method is commonly used to produce gluten-free pasta and snacks that incorporate pseudocereal flours. Extrusion can enhance the functional properties of pseudocereals, such as water-binding capacity and viscosity, making them suitable for a wider range of food applications. Milling is another critical processing step that allows pseudocereals to be transformed into flour or grits (Vollmannová et al. 2021). The milling process must consider the unique physical properties of pseudocereal seeds, which are generally smaller than those of traditional cereals. This difference in seed morphology requires specific adjustments in milling techniques to achieve the desired flour consistency and quality (Nešović et al. 2021).

13.7.3 Nutritional Enhancements

The processing of pseudocereals can also lead to nutritional enhancements. For example, roasting buckwheat not only improves its flavor but also increases its antioxidant content (Martínez-Villaluenga et al. 2020). Similarly, the processing methods employed can help retain essential nutrients while reducing anti-nutritional factors, making pseudocereals more appealing for health-conscious consumers (Ivankov et al. 2021). In addition to traditional and modern processing methods, there is ongoing research into innovative techniques that could further enhance the functional properties of pseudocereals. These innovations aim to improve their incorporation into various food products while maximizing their health benefits (Zhu 2021).

13.8 Culinary Applications

13.8.1 Traditional Uses of Buckwheat

Buckwheat (*Fagopyrum esculentum*) has a rich history of culinary use across various cultures, particularly in Eastern Europe, Asia, and parts of North America. In Russia and Ukraine, buckwheat is commonly prepared as "kasha," which consists of toasted buckwheat groats cooked with water or broth (De Bock et al. 2021). This dish is often served as a side or main course, paired with meats or vegetables. Additionally, buckwheat flour is used to make a variety of baked goods, including pancakes, bread, and blini—thin pancakes typically filled with savory or sweet ingredients (Tomasiak et al. 2022). In Japan, buckwheat is transformed into soba noodles, a staple food that can be served hot or cold and is often enjoyed with dipping sauces or in soups. The versatility of buckwheat extends to its use in various dishes across Europe; for instance, in Brittany, France, it is used to make galettes—savory buckwheat crepes filled with ingredients like cheese, ham, or eggs (Zou et al. 2023). Other European dishes include "ajdovi žganci," a Slovenian spoon bread made from buckwheat flour, and "kasha varnishkes," an Ashkenazi Jewish dish that combines buckwheat with noodles. In India, buckwheat flour is known as "kuttu ka atta" and is traditionally consumed during fasting periods associated with Hindu festivals (Zhao et al. 2023). It is used to prepare various dishes such as pancakes and breads. The leaves of the buckwheat plant are also utilized in Indian cuisine for making chutneys and vegetable curries. Furthermore, the seeds can be popped and consumed as a snack (Baillière et al. 2022).

13.8.2 Traditional Uses of Chaulai (Amaranth)

Chaulai, or amaranth, has been cultivated for thousands of years and holds significant cultural importance in many regions. In South America, particularly among indigenous communities, amaranth seeds are ground into flour for baking bread and making porridge. The leaves are also consumed as a green vegetable in salads or cooked dishes (Punia Bangar et al. 2022). In India, amaranth is known as "rajgira" and is widely used during fasting periods. The seeds are ground into flour and used to make traditional snacks like "rajgira ladoo" (sweet balls) and "rajgira puri" (fried bread). The leaves are often cooked similarly to spinach and served as a nutritious side dish (Paucar-Menacho et al. 2022a). Amaranth is also valued for its high protein content and is increasingly being recognized for its health benefits (Suzuki et al. 2021). In Mexico, amaranth is traditionally popped like popcorn to create "alegrías," a sweet treat made by mixing the popped grains with honey or syrup. This snack has historical roots dating back to the Aztec civilization, where amaranth was considered a sacred crop (Wang et al. 2021).

13.9 Modern Culinary Trends Incorporating These Grains

Modern culinary trends are increasingly incorporating pseudocereals like buckwheat and chaulai (amaranth) into various food products, reflecting a growing consumer interest in health-conscious eating, gluten-free options, and sustainable ingredients. Both grains are being utilized creatively across different cuisines and food applications, contributing to their resurgence in contemporary cooking (Wanniarachchi et al. 2022).

13.9.1 Buckwheat in Modern Culinary Trends

Buckwheat is experiencing a renaissance in the culinary world, recognized for its nutritional benefits and versatility. According to Whole Foods Market's trends report, buckwheat is highlighted as a key ingredient for 2024, appearing in diverse products ranging from pancake mixes to noodles and nondairy milk (Usman et al. 2022). This trend is partly driven by the increasing demand for gluten-free alternatives, as buckwheat is naturally gluten-free and offers a unique flavor profile that many consumers find appealing. Chefs and home cooks alike are experimenting with buckwheat flour in various baked goods. For instance, it is being used in pancakes, cookies, and artisanal breads, often combined with other flours to enhance texture and flavor (Cela et al. 2022). Notable examples include buckwheat pancakes served at popular diners and innovative recipes like buckwheat espresso cookies that utilize the grain's nutty flavor to complement chocolate and coffee. Additionally, buckwheat is being incorporated into breakfast cereals and snack foods, such as crackers and snack clusters, appealing to health-conscious consumers looking for nutritious options (Bez Batti et al. 2022). The versatility of buckwheat extends beyond baking; it can be cooked as groats or used in salads and stir-fries. Its ability to absorb flavors makes it a suitable base for various dishes, enhancing both taste and nutritional value (Kaur 2023). The growing popularity of plant-based diets has also led to increased interest in buckwheat as a protein source, further solidifying its place in modern culinary application (Vásquez-Ocmín et al. 2022).

13.10 Recipes Showcasing Their Versatility

13.10.1 Buckwheat Recipes

Buckwheat is a versatile ingredient that can be used in a variety of delicious recipes. One popular option is buckwheat pancakes, which are fluffy American-style pancakes made with buckwheat flour (Wang et al. 2021). These pancakes offer a nutty flavor and light texture, making them perfect for a weekend brunch when topped with fresh fruits, maple syrup, or yogurt. Another traditional dish is buckwheat khichdi, an Indian comfort food that combines soaked buckwheat with potatoes, peanuts, and spices (Kolarić et al. 2021). This dish is not only nutritious but also easy to prepare, making it a great choice for a wholesome meal. For those who enjoy risotto, a mushroom buckwheat risotto is an excellent alternative to traditional rice-based versions (Bez Batti et al. 2022). This creamy dish utilizes buckwheat groats sautéed with mushrooms and cooked in vegetable broth, resulting in a rich and earthy flavor profile. On cooler evenings, a buckwheat sweet potato chili can provide warmth and comfort (Cela et al. 2022). This hearty plant-based chili combines buckwheat groats with sweet potatoes and spices, creating a filling dish that is perfect for family dinners (Vollmannová et al. 2021). Breakfast enthusiasts can enjoy buckwheat porridge, which is made by cooking soaked buckwheat groats in water or milk. Topped with fruits, nuts, or sweeteners, this warm option provides a nutritious start to the day. Additionally, buckwheat noodle stir-fry showcases the versatility of buckwheat in Asian cuisine (Usman et al. 2022). Soba noodles (made from buckwheat) can be quickly stir-fried with vegetables and protein of choice, seasoned with soy sauce or teriyaki for a quick and nutritious meal. Finally, buckwheat granola offers a crunchy snack option made from buckwheat groats mixed with nuts, seeds, and natural sweeteners like honey or maple syrup (Nešović et al. 2021).

13.10.2 Chaulai (Amaranth) Recipes

Chaulai (amaranth) also shines in modern culinary applications due to its nutritional benefits and adaptability. A delightful way to start the day is with amaranth porridge, where cooked amaranth grains are simmered in milk or water and sweetened with honey or maple syrup (Huda et al. 2021). Topped with fresh fruits and nuts, this dish makes for a nutritious breakfast option (Mravlje et al. 2021). For those looking for gluten-free alternatives, amaranth flour pancakes are light and fluffy, ideal for breakfast or brunch when served with syrup or fruit compote. Another creative use of amaranth is in an amaranth salad, which features cooked amaranth grains mixed with chopped vegetables, herbs, and a tangy dressing (Yao et al. 2021). This refreshing salad is not only filling but also packed with nutrients. For a healthy snack, popped amaranth snacks can be made by popping amaranth grains like popcorn and mixing them with honey and nuts to create energy bites or snack bars. Incorporating amaranth leaves into cooking can lead to delicious results as well; an amaranth vegetable stir-fry involves sautéing the leaves with garlic and spices, served alongside cooked amaranth grains for a nutrient-rich side dish (Wu et al. 2021). For a comforting meal, an amaranth soup can be prepared by adding cooked amaranth to vegetable broth along with seasonal vegetables and spices for added flavor. Lastly, amaranth cookies made from amaranth flour combined with oats, nuts, and dried fruits offer a wholesome treat that satisfies sweet cravings while providing nutritional benefits (Li et al. 2021).

13.11 Functional Properties

Buckwheat and chaulai as pseudocereals provide diverse functional properties, enhancing their significance in contemporary dietary practices. Their unique nutritional profiles contribute to their appeal, particularly for those seeking gluten-free options, high fiber content, and other health benefits (Luthar et al. 2021b).

13.11.1 Buckwheat Functional Properties

One of the most notable functional properties of buckwheat is its gluten-free nature. Unlike traditional grains such as wheat, buckwheat does not contain gluten, making it an excellent dietary alternative for individuals with celiac disease or gluten intolerance (Lu et al. 2022). This characteristic allows buckwheat to be incorporated into a variety of foods, including flour for baking, noodles, and breakfast cereals, without the adverse effects associated with gluten consumption (Wen et al. 2022). Buckwheat is also rich in dietary fiber, which plays a crucial role in digestive health. It contains both soluble and insoluble fibers that aid in regular bowel movements and help maintain gut health (Kowalski et al. 2022). The soluble fiber found in buckwheat can help lower cholesterol levels and manage blood sugar levels by slowing down the digestion of carbohydrates. This makes buckwheat a beneficial food for individuals managing diabetes or those looking to maintain a healthy weight. Additionally, buckwheat is high in protein and contains essential amino acids, particularly lysine and arginine, which are often limited in other grains (Liu et al. 2022). The protein digestibility coefficient of buckwheat is approximately 80%, indicating that it provides a high-quality protein source. This high protein content contributes to increased satiety, helping individuals feel fuller for longer periods. Furthermore, buckwheat is rich in antioxidants such as rutin and quercetin (Wang et al. 2022). Rutin has been shown to possess anti-inflammatory and cardioprotective properties, contributing to heart health by improving blood vessel function and reducing blood pressure. The presence of these bioactive compounds enhances the overall health benefits of incorporating buckwheat into the diet (Sun et al. 2021).

13.11.2 Chaulai (Amaranth) Functional Properties

Chaulai (amaranth) also boasts several functional properties that make it an attractive addition to modern diets. Like buckwheat, amaranth is gluten-free, making it suitable for those with gluten sensitivities or celiac disease (Mesías et al. 2024). This quality allows amaranth to be utilized in various culinary applications, including flour for baking and as a grain substitute in salads and side dishes. Amaranth is particularly high in dietary fiber, which promotes digestive health by preventing constipation and supporting regular bowel movements. The fiber content also aids in weight management by promoting feelings of fullness (Nandan et al. 2024). Additionally, amaranth contains a higher protein content compared to many traditional grains, with a complete amino acid profile that includes essential amino acids such as lysine and methionine. This makes amaranth an excellent protein source for vegetarians and vegans. The nutritional profile of amaranth includes significant amounts of vitamins and minerals. It is rich in calcium, magnesium, iron, and potassium, which are essential for bone health, muscle function, and overall metabolic processes (Huda et al. 2021). The presence of these nutrients contributes to amaranth's role as a functional food that supports various aspects of health (Paucar-Menacho et al. 2022b). Moreover, amaranth contains bioactive compounds such as phytosterols and flavonoids that exhibit antioxidant properties. These compounds can help combat oxidative stress in the body and may reduce the risk of chronic diseases such as heart disease and cancer. The inclusion of amaranth in the diet can therefore provide additional health benefits beyond its basic nutritional value (Petrova and Petrov 2020) (Table 13.2).

Table 13.2 The functional properties of buckwheat and chaulai (amaranth)

Functional property	Buckwheat	Chaulai (amaranth)	References
Gluten-free	Naturally gluten-free, suitable for celiac disease	Naturally gluten-free, ideal for gluten-sensitive individuals	Graziano et al. (2022)
Protein content	High-quality protein (approx. 13–14%), rich in lysine and arginine	High protein content (approx. 13–14%), complete amino acid profile	Hitache et al. (2023)
Dietary fiber	Contains both soluble and insoluble fiber, promoting digestive health	High in dietary fiber, aiding in digestion and satiety	Graziano et al. (2022)
Antioxidants	Rich in antioxidants like rutin and quercetin, which have anti-inflammatory properties	Chaulai (amaranth) contains phytosterols (β-sitosterol, stigmasterol, campesterol) and flavonoids (quercetin, kaempferol) that provide antioxidant effects by neutralizing free radicals and reducing oxidative stress	Hitache et al. (2023)
Mineral content	Good source of magnesium, manganese, iron, and phosphorus	Rich in calcium, magnesium, iron, potassium, and zinc	Estivi et al. (2022)
Fat content	Low fat (approx. 3–4%), with healthy unsaturated fats	Moderate fat content (approx. 5–9%), primarily healthy unsaturated fats	Romano and Ferranti (2023)
Water absorption capacity	High water absorption capacity (137–154% depending on flour type)	Moderate water absorption capacity; effective in retaining moisture in baked goods	Vidaurre-Ruiz et al. (2023)
Oil absorption capacity	Higher oil absorption capacity than wheat flour, enhancing flavor retention	Good oil absorption capacity; useful in improving mouthfeel in foods	Vidaurre-Ruiz et al. (2023)
Swelling capacity	Moderate swelling capacity; contributes to texture in cooked dishes	High swelling capacity; enhances texture and volume in baked products	Hitache et al. (2023)
Foaming capacity	Moderate foaming capacity; useful in certain baked goods for aeration	Good foaming capacity; beneficial for creating light textures in baked items	Graziano et al. (2022)

13.12 Potential Applications of Buckwheat and Chaulai (Amaranth) in Food Formulation

Buckwheat and chaulai (amaranth) are gaining recognition for their versatility and nutritional benefits in various food formulations. Their unique properties make them suitable for a wide range of applications, particularly in bakery products, snacks, and other culinary creations (Graziano et al. 2022).

13.12.1 Buckwheat Applications

One of the most promising applications for buckwheat is in bakery products. Buckwheat flour is increasingly popular in gluten-free baking, allowing for the creation of a variety of baked goods, including breads, muffins, pancakes, and cookies (Rao and Poonia 2023). For instance, buckwheat flour can enhance the nutritional profile of bread by increasing its protein, fiber, and antioxidant content. Buckwheat's nutty flavor and versatile texture make it an excellent ingredient for a variety of recipes, appealing to health-conscious consumers. For instance, recipes like fermented buckwheat bread, often combined with seeds, highlight its ability to produce nutritious and flavorful loaves (Giménez-Bastida and Zieliński 2015). Additionally, buckwheat flour is widely used to create pancakes and crepes, valued for its light texture and distinctive taste, as seen in traditional dishes like French galettes (Lestido-Cardama et al. 2020). Dishes like Fluffy Buckwheat Chia Pancakes or Cherry Vanilla Crepes highlight the grain's versatility as a breakfast option or dessert (Paucar-Menacho et al. 2022c). These recipes not only provide a gluten-free alternative but also offer essential nutrients that appeal to a wide audience. Buckwheat can also be incorporated into granola mixes or snack bars, providing a crunchy texture and added nutritional value (Vemana and Jain 2010). For example, Date & Buckwheat Granola combines buckwheat with oats, nuts, and seeds to create a nutrient-dense snack option that appeals to consumers looking for healthy on-the-go choices. Beyond baked goods and snacks, buckwheat can be used in savory dishes such as risottos or pilafs (Singh et al. 2013). A dish like Buckwheat Risotto with Mushrooms and Peas demonstrates how buckwheat can serve as a hearty base for nutritious meals, showcasing its versatility across culinary traditions (Bahmanyar et al. 2022). Furthermore, buckwheat flour can be used in cookies and biscuits, providing health-promoting functional ingredients for gluten-intolerant individuals. Gluten-free cookies made from buckwheat have shown good consumer acceptability while enhancing nutritional value (Rollán et al. 2019).

13.12.2 Chaulai (Amaranth) Applications

Chaulai (amaranth) also offers exciting potential in food formulation. One notable application is in gluten-free breads, where amaranth flour is increasingly used due to its high protein content and complete amino acid profile (Bielecka et al. 2022). Combining amaranth flour with other gluten-free flours enhances the nutritional quality of the bread while maintaining desirable texture. Amaranth can also be utilized in breakfast cereals, either as whole grains or ground into flour. Its nutty flavor makes it an appealing addition to granola mixes or as a base for hot cereals, catering to health-conscious consumers seeking nutritious breakfast options (Langyan et al. 2023). Additionally, popped amaranth serves as a healthy snack alternative that can be enjoyed on its own or incorporated into snack bars and energy bites. The light texture and crunchy nature of popped amaranth make it an attractive ingredient for various snack formulations (Czerwonka and Białek 2023). Similar to buckwheat, amaranth flour can be used to make pancakes and waffles that are gluten-free yet rich in nutrients. Recipes utilizing amaranth flour provide an excellent way to incorporate this pseudocereal into breakfast menus while appealing to those with dietary restrictions. Furthermore, amaranth grains can be added to soups and stews for added nutrition and texture; their ability to absorb flavors makes them a valuable ingredient in hearty dishes that require a nutrient boost (Majzoobi et al. 2023).

13.13 Market Trends

13.13.1 Current Market Trends for Pseudocereals Globally

Current market trends for pseudocereals, such as buckwheat and chaulai (amaranth), reflect a growing global interest driven by health-conscious consumers, dietary diversity, and sustainability considerations (Wen et al. 2022). These grains are increasingly recognized for their nutritional profiles, which include high protein content, essential amino acids, and beneficial phytochemicals (Liu et al. 2022). As a result, they are being incorporated into a wide range of food products, from bakery items to snacks and breakfast cereals. One significant trend is the rising demand for gluten-free products. With the prevalence of gluten intolerance and celiac disease, pseudocereals like buckwheat and amaranth are becoming popular alternatives to traditional grains (Wang et al. 2022). Their gluten-free nature allows them to be used in various applications, including gluten-free breads, cookies, and pasta. This shift is supported by an expanding market for gluten-free foods projected to grow significantly over the coming years (Sun et al. 2021).

Additionally, the health benefits associated with pseudocereals are driving their popularity. Research highlights their role in managing conditions such as diabetes, obesity, and cardiovascular diseases due to their high fiber content and low glycemic index (Petrova and Petrov 2020). Products featuring these grains are increasingly marketed as functional foods that contribute to overall health and well-being (Mesías et al. 2024). For instance, buckwheat is often promoted for its antioxidant properties and potential to improve heart health. Pseudocereals are also gaining traction in the snack food sector. Popped amaranth and buckwheat-based granola bars are becoming popular choices among health-conscious consumers seeking nutritious on-the-go snacks. The versatility of these grains allows manufacturers to create innovative snack options that cater to diverse dietary preferences while maintaining high nutritional value (Nandan et al. 2024).

Sustainability is another critical factor influencing market trends. Pseudocereals are often cultivated in marginal lands where traditional crops may struggle, making them important for agricultural biodiversity and food security. Their resilience to climate change and ability to thrive in less fertile soils position them as viable crops for sustainable agriculture. This aspect appeals to environmentally conscious consumers and producers alike (Paucar-Menacho et al. 2022b).

Moreover, the global market for pseudocereals is expanding beyond traditional regions of cultivation. While quinoa has seen significant growth as a globally traded commodity, other pseudocereals like buckwheat and amaranth are beginning to enter international markets more prominently. This shift presents opportunities for farmers in various regions to diversify their crops and increase income potential (Bahmanyar et al. 2022).

13.13.2 Consumer Demand and Awareness Regarding Health Foods

Consumer demand and awareness regarding health foods have significantly evolved in recent years, driven by a growing emphasis on nutrition, health consciousness, and the impact of marketing strategies (Langyan et al. 2023). As consumers become more informed about the nutritional content of their food, they are increasingly willing to invest time and money in high-quality, healthy options. This shift is largely influenced by the popularization of health education and the dissemination of information regarding food choices (Paucar-Menacho et al. 2022c).

One key factor influencing consumer behavior is the increased focus on nutritional value. Research indicates that consumers are more inclined to compare and evaluate food products based on their health benefits, leading to a preference for items that align with their dietary needs (Czerwonka and Białek 2023). Advertising and publicity play a crucial role in shaping these preferences; food manufacturers utilize various channels to convey information about healthy ingredients and their associated benefits. The rise of social media has further amplified this effect, allowing consumers to share experiences and recommendations, which can significantly influence purchasing decisions. Price and convenience are also critical considerations for consumers when selecting healthy foods. Although the market for health foods has diversified, many products still come with a premium price tag (Hirdyani 2014). Consequently, consumers often face a trade-off between health benefits and affordability. Additionally, the convenience of purchasing healthy options is paramount in today's fast-paced lifestyle; busy consumers tend to favor accessible and easy-to-prepare foods that fit into their daily routines (Vasava et al. 2016).

The demand for transparency in food labeling has increased as well. Consumers are advocating for clearer information regarding food contents, including front-of-pack warning labels that highlight unhealthy ingredients (Baillière et al. 2022). This push for better labeling practices aims to empower consumers to make informed decisions about their dietary choices, particularly in light of rising concerns over noncommunicable diseases linked to poor nutrition (Diwan et al. 2017).

13.13.3 Economic Potential for Farmers and Producers

The economic potential for farmers and producers of pseudocereals, such as buckwheat and chaulai (amaranth), is significant, particularly in the context of rising global demand for healthy and sustainable food options (Zou et al. 2023). As consumer awareness of nutrition and health benefits increases, these underutilized crops are gaining traction in markets around the world (Yadav et al. 2018).

13.13.4 Agricultural Development and Economic Growth

Agriculture plays a crucial role in economic development, particularly in rural areas where a significant portion of the population relies on farming for their livelihoods (Nešović et al. 2021). The World Bank highlights that agricultural growth is two to four times more effective in raising incomes among the farmers compared to other sectors (Bez Batti et al. 2022). By promoting the cultivation of pseudocereals, governments and agricultural organizations can help improve food security, raise incomes, and reduce poverty among rural populations. Furthermore, pseudocereals are often well suited for cultivation in marginal lands where traditional crops may struggle (Aguiar et al. 2021). This adaptability allows farmers to utilize less fertile areas effectively, contributing to increased agricultural productivity without putting additional pressure on prime agricultural land. The resilience of these crops to climate change also positions them as valuable assets in sustainable agriculture initiatives (Mahangade et al. 2020).

13.13.5 Value Chains and Market Access

To fully realize the economic potential of pseudocereals, it is essential to strengthen value chains that connect farmers to consumers. Efficient market access is crucial for smallholder farmers who often face challenges related to logistics, storage, and distribution (Tiwari et al. 2016). By improving infrastructure and reducing barriers to entry in local and international markets, farmers can achieve better prices for their products and enhance their overall income. Innovative approaches such as e-commerce platforms can help bridge the gap between producers and consumers by allowing farmers to sell directly to buyers, thereby increasing their profit margins (Drub et al. 2021). For example, platforms that connect smallholder farmers with urban markets can facilitate direct sales, reducing reliance on intermediaries that often take a large share of profits (Chandrakar and Asati 2021).

13.13.6 Research and Development

Despite the promising economic potential of pseudocereals, challenges remain regarding research and development (Majzoobi et al. 2023). There is a need for increased investment in breeding programs focused on improving yield potential, pest resistance, and nutritional quality of these crops. The lack of scientific research compared to major staple grains has hindered the widespread adoption of pseudocereals in many regions (Czerwonka and Białek 2023). Addressing these gaps through targeted research initiatives can enhance productivity and marketability (Rao and Poonia 2023). Additionally, educational programs aimed at informing farmers about best practices for cultivating pseudocereals can further support their economic viability. Training on sustainable farming techniques, pest management, and postharvest processing will empower farmers to maximize their yields while maintaining environmental sustainability (Langyan et al. 2023).

13.14 Environmental Considerations

13.14.1 Sustainability Aspects of Cultivating Buckwheat and Chaulai

Sustainability in agriculture is increasingly important, and both buckwheat (*Fagopyrum esculentum*) and chaulai (*Chenopodium album*) exemplify sustainable practices in crop cultivation. Buckwheat is particularly noteworthy for its environmental benefits (Kumar Ratre et al. 2021). It has a short life cycle and low nutritional requirements, making it adaptable to various growing conditions, including poor soils (Bielecka et al. 2022). This crop produces significant biomass, which contributes positively to soil health and biodiversity. Additionally, buckwheat has allelopathic properties that help suppress weeds, reducing the need for chemical herbicides and promoting organic farming practices (Salunkhe and Kadam 1998). This makes it a valuable addition to sustainable diets and enhances food security in regions where it is cultivated. Chaulai, commonly known as lamb's quarters, also presents significant sustainability advantages. It thrives in diverse conditions and is often regarded as a weed; however, it has substantial potential as a cultivated crop due to its adaptability to poor soils and low water requirements (Rollán et al. 2019). This resilience allows chaulai to be an excellent option for sustainable agriculture, especially in areas facing climate challenges. Its rapid growth allows for multiple harvests within a single season, further contributing to food security. Ecologically, chaulai plays a vital role by preventing soil erosion and improving soil structure through its extensive root system (Cankurtaran Kömürcü 2023). It also supports local ecosystems by providing habitat for various species. As a cover crop, chaulai can naturally suppress weeds while enriching the soil with organic matter when tilled back into the ground (Paucar-Menacho et al. 2022c).

13.14.2 Benefits of Including Pseudocereals in Crop Rotation for Soil Health

Including pseudocereals in crop rotation offers numerous benefits for soil health, contributing to sustainable agricultural practices (Aguiar et al. 2021; Bahmanyar et al. 2022). Pseudocereals, such as buckwheat, quinoa, and amaranth, are known for their ability to improve soil structure and fertility. When integrated into crop rotations, these plants can enhance soil organic matter through their biomass, which, when decomposed, enriches the soil with essential nutrients (Cankurtaran Kömürcü 2023). This organic matter not only improves soil fertility but also enhances its water retention capacity, making it more resilient to drought conditions. One of the significant advantages of pseudocereals is their extensive root systems. These roots help break up compacted soil layers, promoting better aeration and water infiltration. Improved soil structure facilitates nutrient uptake by subsequent crops and supports the overall health of the soil ecosystem (Rollán et al. 2019). Additionally, pseudocereals can suppress weeds effectively due to their rapid growth and dense foliage. This natural weed suppression reduces the need for chemical herbicides, further benefiting soil health and promoting biodiversity. Pseudocereals also contribute to pest and disease management in crop rotations (Salunkhe and Kadam 1998). Certain pseudocereal species, such as buckwheat, release allelopathic compounds like phenolic acids that inhibit the growth of harmful soil pathogens, promoting a healthier microbial community (Achigan-Dako et al. 2014). This natural defense mechanism reduces the need for synthetic pesticides and supports sustainable farming practices. Incorporating pseudocereals like buckwheat and amaranth into crop rotation systems enhances ecosystem balance, improving resilience against pests and diseases. Additionally, pseudocereals are often drought-resistant and thrive in marginal soils where traditional cereals struggle, making them valuable for sustaining agricultural productivity amid climate change challenges (de Klerk et al. 2015). By diversifying crop rotations with pseudocereals, farmers can enhance the sustainability of their farming systems while improving soil health (Singh et al. 2013).

13.14.3 Contributions to Agricultural Biodiversity

The inclusion of pseudocereals in agricultural systems significantly contributes to agricultural biodiversity, enhancing ecosystem resilience and sustainability. Pseudocereals such as buckwheat, quinoa, and amaranth are not only nutritionally rich but also offer unique agronomic benefits that support diverse farming practices (Vemana and Jain 2010). Their cultivation can improve soil health, promote beneficial insects, and enhance crop rotation systems (Vemana and Jain 2010). One of the primary contributions of pseudocereals to agricultural biodiversity is their ability to thrive in various climatic and soil conditions. Unlike traditional monoculture crops, pseudocereals can grow in marginal lands and less fertile soils, making them valuable for regions facing environmental stressors such as drought or soil degradation (Singh et al. 2013). This adaptability allows farmers to diversify their cropping systems, reducing reliance on a limited number of staple crops and promoting a more resilient agricultural landscape. As they are often cultivated in rotation with other crops, pseudocereals help break pest and disease cycles, further enhancing biodiversity (Kumar Ratre et al. 2021). Pseudocereals also play a crucial role in supporting local ecosystems. For instance, buckwheat flowers are particularly attractive to pollinators like bees and butterflies, providing essential food sources that help maintain these critical populations. By incorporating pseudocereals into their farming practices, farmers can create habitats that support beneficial insects, thereby promoting overall ecosystem health and functionality (Salunkhe and Kadam 1998). This interaction not only aids in pollination but also contributes to the biological control of pests. Furthermore, the cultivation of pseudocereals can improve soil health through various mechanisms. Their extensive root systems help break up compacted soil, enhancing aeration and water infiltration (Bez Batti et al. 2022). Additionally, as cover crops, pseudocereals suppress weeds naturally and contribute organic matter to the soil when incorporated after harvest (Rollán et al. 2019). This organic matter enriches the soil with nutrients and improves its structure over time. The ability of certain pseudocereals to release bioactive compounds into the soil can also inhibit harmful pathogens, reducing the need for chemical inputs and fostering a healthier soil microbiome (Cankurtaran Kömürcü 2023).

13.14.4 Future Perspectives

Emerging research on pseudocereals highlights their potential benefits in addressing nutritional needs, enhancing food security, and promoting sustainable agricultural practices (Vásquez-Ocmín et al. 2022). These crops are particularly appealing because they are gluten-free, making them suitable for individuals with celiac disease or gluten intolerance (Aguiar et al. 2021). The increasing consumer demand for healthy and minimally processed foods has driven food producers to develop innovative products that incorporate pseudocereals, thereby expanding their market presence and usage in various food applications. Research indicates that incorporating pseudocereals into diets can help combat chronic diseases such as cancer, diabetes, and cardiovascular diseases (Kour et al. 2023). Their rich composition of bioactive compounds, including antioxidants like rutin and quercetin, contributes to health benefits by reducing oxidative stress and inflammation in the body (Drub et al. 2021). This makes pseudocereals not only a nutritious option but also a functional food that can play a role in disease prevention (Luthar et al. 2021a). Furthermore, the considerable genetic diversity of pseudocereals allows them to adapt to a wide range of environmental conditions, making them resilient crops for future agricultural systems facing climate change challenges. In addition to their health benefits, pseudocereals contribute to agricultural biodiversity

(Vásquez-Ocmín et al. 2022). They can be cultivated in marginal lands where traditional cereals may not thrive, thus improving land use efficiency and enhancing food production in areas with limited resources. The cultivation of these underutilized crops promotes crop rotation and diversification strategies that can improve soil health and reduce pest pressures (Huang et al. 2021). This diversification is crucial for building resilient agricultural systems capable of withstanding the pressures of climate variability. Moreover, ongoing research is focused on improving the processing and technological aspects of pseudocereals to enhance their usability in food products (Wanniarachchi et al. 2022). Understanding their unique starch structures and functional properties can lead to better food formulations that capitalize on their nutritional benefits while appealing to consumer preferences for gluten-free options (Usman et al. 2022). As research continues to uncover the potential of pseudocereals, they are likely to play an increasingly prominent role in global food production and nutrition strategies aimed at achieving sustainable development goals (Usman et al. 2022).

13.15 Conclusion

Buckwheat and chaulai represent promising avenues for enhancing the future of the food industry through their nutritional benefits, adaptability, and contributions to sustainable agricultural practices. As consumers increasingly seek healthier and more sustainable food options, these pseudocereals are well positioned to meet this demand. Buckwheat's rich nutritional profile and gluten-free status make it an ideal ingredient for a variety of health-focused products, while chaulai's resilience and nutrient density offer exciting opportunities for incorporation into diverse culinary applications. Both crops not only provide significant health benefits but also play crucial roles in promoting agricultural biodiversity. Their ability to thrive in marginal soils and contribute to soil health through crop rotation enhances ecosystem resilience, making them valuable assets in the face of climate change and food security challenges. Furthermore, ongoing research into their processing and culinary uses is likely to unlock new potential for these crops, expanding their presence in the market. As awareness of the importance of sustainable food systems grows, buckwheat and chaulai are expected to gain attraction among consumers and producers alike. Their integration into mainstream diets can foster a more diverse and resilient agricultural landscape, supporting both human health and environmental sustainability. Ultimately, the future role of buckwheat and chaulai in the food industry signifies a shift toward more inclusive, nutritious, and ecologically responsible food choices that benefit both people and the planet.

Conflict of Interest None.

References

Achigan-Dako EG, Sogbohossou OED, Maundu P (2014) Current knowledge on Amaranthus spp.: research avenues for improved nutritional value and yield in leafy amaranths in sub-Saharan Africa. Euphytica 197:303–317. https://doi.org/10.1007/S10681-014-1081-9/METRICS

Agregán R, Guzel N, Guzel M, Bangar SP, Zengin G, Kumar M, Lorenzo JM (2022) The effects of processing technologies on nutritional and anti-nutritional properties of pseudocereals and minor cereal. Food Bioprocess Technol 16:961–986. https://doi.org/10.1007/S11947-022-02936-8

Aguiar EV, Santos FG, Centeno ACLS, Capriles VD (2021) Influence of pseudocereals on gluten-free bread quality: a study integrating dough rheology, bread physical properties and acceptability. Food Res Int (Ottawa, Ont) 150. https://doi.org/10.1016/J.FOODRES.2021.110762

Akın M, Eyduran SP, Papageorgiou M, Bartkiene E, Rocha JM (2023) Bibliometric analysis on pseudocereals. Clean Circ Bioecon 6:100062. https://doi.org/10.1016/J.CLCB.2023.100062

Aloo SO, Ofosu FK, Oh DH (2021) Effect of germination on alfalfa and buckwheat: phytochemical profiling by UHPLC-ESI-QTOF-MS/MS, bioactive compounds, and in-vitro studies of their diabetes and obesity-related functions. Antioxidants (Basel, Switzerland) 10. https://doi.org/10.3390/ANTIOX10101613

Bahmanyar F, Hosseini SM, Shojaee-Aliabadi S, Mirmoghtadaie L (2022) Development of functional beef burgers with pseudocereals and study of their physicochemical and textural properties. Res Innov Food Sci Technol 11:247–256. https://doi.org/10.22101/JRIFST.2022.300264.1273

Baillière J, Laureys D, Vermeir P, Van Opstaele F, De Rouck G, De Cooman L, Vanderputten D, De Clippeleer J (2022) 10 unmalted alternative cereals and pseudocereals: a comparative analysis of their characteristics relevant to the brewing process. J Cereal Sci 106:103482. https://doi.org/10.1016/J.JCS.2022.103482

Bekkering CS, Tian L (2019) Thinking outside of the cereal box: breeding underutilized (pseudo)cereals for improved human nutrition. Front Genet 10. https://doi.org/10.3389/FGENE.2019.01289

Bez Batti ÉA, do Nascimento AB, Geraldo APG, Fernandes AC, Bernardo GL, Proença RP da C, Uggioni PL (2022) Use of the term whole grain on the label of processed and ultra-processed foods based on cereals and pseudocereals in Brazil. Front Nutr 9. https://doi.org/10.3389/FNUT.2022.875913

Bhatt D, Rasane P, Singh J, Kaur S, Fairos M, Kaur J, Gunjal M, Mahato DK, Mehta CM, Avinashe H, Sharma N (2023) Nutritional advantages of barnyard millet and opportunities for its processing as value-added foods. J Food Sci Technol 60:2748–2760. https://doi.org/10.1007/S13197-022-05602-1/METRICS

Bhatt P, Singh S, Garg S, Ashoka P, Mishra SK, Kaushik S, Deva V, Goel R, Kumar V (2024) Impact of cross-linking on the physicochemical and physiological characteristics of barnyard millet (*Echinochloa frumentacea*) grains starch. Starch Stärke 2300285. https://doi.org/10.1002/STAR.202300285

Bielecka J, Markiewicz-żukowska R, Puścion-Jakubik A, Grabia M, Nowakowski P, Soroczyńska J, Socha K (2022) Gluten-free cereals and pseudocereals as a potential source of exposure to toxic elements among polish residents. Nutrients 14. https://doi.org/10.3390/NU14112342

Botella-Martínez C, Pérez-álvarez JÁ, Sayas-Barberá E, Fernández-López J, Viuda-Martos M (2021) Assessment of chemical, physicochemical, and lipid stability properties of gelled emulsions elaborated with different oils chia (*Salvia hispanica* L.) or hemp (*Cannabis sativa* L.) and pseudocereals. Foods (Basel, Switzerland) 10. https://doi.org/10.3390/FOODS10071463

Caeiro C, Pragosa C, Cruz MC, Pereira CD, Pereira SG (2022) The role of pseudocereals in celiac disease: reducing nutritional deficiencies to improve well-being and health. J Nutr Metab 2022. https://doi.org/10.1155/2022/8502169

Cankurtaran Kömürcü T (2023) Impact of different ancient wheats and pseudocereals on physical, chemical, and functional properties of gevrek, cereal-based snack product. Cereal Chem 100:643–649. https://doi.org/10.1002/CCHE.10638

Cao P, Wu Y, Li Y, Xiang L, Cheng B, Hu Y, Jiang X, Wang Z, Wu S, Si L, Yang Q, Xu J, Huang J (2022) The important role of glycerophospholipid metabolism in the protective effects of polyphenol-enriched tartary buckwheat extract against alcoholic liver disease. Food Funct 13. https://doi.org/10.1039/D2FO01518H

Cela N, Galgano F, Perretti G, Di Cairano M, Tolve R, Condelli N (2022) Assessment of brewing attitude of unmalted cereals and pseudocereals for gluten free beer production. Food Chem 384. https://doi.org/10.1016/J.FOODCHEM.2022.132621

Chandrakar MK, Asati BS (2021) Genetic analysis for foliage yield attributes in vegetable red amaranth (*Amaranthus tricolor* L.). Int J Curr Microbiol App Sci 10:2143–2153. https://doi.org/10.20546/IJCMAS.2021.1002.255

Coţovanu I, Mironeasa S (2021) Buckwheat seeds: impact of milling fractions and addition level on wheat bread dough rheology. Appl Sci 11:1731. https://doi.org/10.3390/APP11041731

Czerwonka M, Białek A (2023) Fatty acid composition of pseudocereals and seeds used as functional food ingredients. Life (Basel, Switzerland) 13. https://doi.org/10.3390/LIFE13010217

De Bock P, Daelemans L, Selis L, Raes K, Vermeir P, Eeckhout M, Van Bockstaele F (2021) Comparison of the chemical and technological characteristics of wholemeal flours obtained from amaranth (Amaranthus sp.), quinoa (*Chenopodium quinoa*) and buckwheat (Fagopyrum sp.) seeds. Foods (Basel, Switzerland) 10. https://doi.org/10.3390/FOODS10030651

de Klerk P, Couwenberg J, Joosten H (2015) Pollen and macrofossils attributable to fagopyrum in western Eurasia prior to the late medieval: an intercontinental mystery. Palaeogeogr Palaeoclimatol Palaeoecol 440:1–21. https://doi.org/10.1016/j.palaeo.2015.08.024

Dhondiram DS, Sawant AA, Mamoon F, Aneja I, Duggal M (2023) Nutritional characterization of pseudocereals: a review. Pharm Innov J 12:109–118. https://doi.org/10.22271/TPI.2023.V12.I3B.19413

Diwan IS, Shukla N, Kurrey V (2017) Genetic studies in Amaranthus germplasm. Int J Curr Microbiol App Sci 6:2459–2470. https://doi.org/10.20546/IJCMAS.2017.608.291

Drub TF, Garcia dos Santos F, Ladeia Solera Centeno AC, Capriles VD (2021) Sorghum, millet and pseudocereals as ingredients for gluten-free whole-grain yeast rolls. Int J Gastron Food Sci 23:100293. https://doi.org/10.1016/J.IJGFS.2020.100293

Estivi L, Brandolini A, Condezo-Hoyos L, Hidalgo A (2022) Impact of low-frequency ultrasound technology on physical, chemical and technological properties of cereals and pseudocereals. Ultrason Sonochem 86. https://doi.org/10.1016/J.ULTSONCH.2022.106044

Giménez-Bastida JA, Zieliński H (2015) Buckwheat as a functional food and its effects on health. J Agric Food Chem 63:7896–7913. https://doi.org/10.1021/ACS.JAFC.5B02498/ASSET/IMAGES/MEDIUM/JF-2015-02498K_0004.GIF

Graziano S, Agrimonti C, Marmiroli N, Gullì M (2022) Utilisation and limitations of pseudocereals (quinoa, amaranth, and buckwheat) in food production: a review. Trends Food Sci Technol 125:154–165. https://doi.org/10.1016/J.TIFS.2022.04.007

Habib H, Singh J, Kumar A, Amin T, Bhat TA, Aziz N, Ercisli S (2023) Optimization of functional beverage using germinated pseudocereals. J Food Chem Nanotechnol 9:S108–S116. https://doi.org/10.17756/JFCN.2023-S1-015

He M, He Y, Zhang K, Lu X, Zhang X, Gao B, Fan Y, Zhao H, Jha R, Huda MN, Tang Y, Wang J, Yang W, Yan M, Cheng J, Ruan J, Dulloo E, Zhang Z, Georgiev MI, Chapman MA, Zhou M (2022) Comparison of buckwheat genomes reveals the genetic basis of metabolomic divergence and ecotype differentiation. New Phytol 235:1927–1943. https://doi.org/10.1111/NPH.18306

Hirdyani H (2014) Nutritional composition of chickpea (Cicerarietinum-L) and value added products—a review. Indian J Community Health 26:102–106

Hitache Z, Al-Dalali S, Pei H, Cao X (2023) Review of the health benefits of cereals and pseudocereals on human gut microbiota. Food Bioprocess Technol 16:2382–2399. https://doi.org/10.1007/S11947-023-03069-2

Huang J, Chen Q, Rong Y, Tang B, Zhu L, Ren R, Shi T, Chen Q (2021) Transcriptome analysis revealed gene regulatory network involved in PEG-induced drought stress in tartary buckwheat (*Fagopyrum tararicum*). PeerJ 9. https://doi.org/10.7717/PEERJ.11136

Huda MN, Lu S, Jahan T, Ding M, Jha R, Zhang K, Zhang W, Georgiev MI, Park SU, Zhou M (2021) Treasure from garden: bioactive compounds of buckwheat. Food Chem 335. https://doi.org/10.1016/J.FOODCHEM.2020.127653

Ivankov A, Naučienė Z, Degutytė-Fomins L, Žūkienė R, Januškaitienė I, Malakauskienė A, Jakštas V, Ivanauskas L, Romanovskaja D, Šlepetienė A, Filatova I, Lyushkevich V, Mildažienė V (2021) Changes in agricultural performance of common buckwheat induced by seed treatment with cold plasma and electromagnetic field. Appl Sci 11:4391. https://doi.org/10.3390/APP11104391

Kaur H (2023) Possible applications of certain pseudocereals within the food manufacturing sector. Pharm Innov J 12:1466–1469. https://doi.org/10.22271/TPI.2023.V12.I8SS.22405

Kolarić L, Popović V, Živanović L, Ljubičić N, Stevanović P, Todosijević LŠ, Simić D, Ikanović J (2021) Buckwheat yield traits response as influenced by row spacing, nitrogen, phosphorus, and potassium management. Agronomy 11:2371. https://doi.org/10.3390/AGRONOMY11122371

Kour DP, Sharma S, Gupta N, Gandotra A, Gupta S (2023) Utilization of pseudocereals into nutritious pancake premix. Plant Arch 23:411–415. https://doi.org/10.51470/PLANTARCHIVES.2023.v23.no2.070

Kowalski S, Mikulec A, Mickowska B, Buksa K (2022) Nutritional properties and amino acid profile of buckwheat bread. J Food Sci Technol 59:3020–3030. https://doi.org/10.1007/S13197-022-05518-W

Kreft I, Germ M, Golob A, Vombergar B, Bonafaccia F, Luthar Z (2022) Impact of rutin and other phenolic substances on the digestibility of buckwheat grain metabolites. Int J Mol Sci 23. https://doi.org/10.3390/IJMS23073923

Kumar Ratre A, Shah Jureshiya S, Kumar Joshi P (2021) Pharmacognostical & phytochemical evaluations of leaves of *Amaranthus spinousus* linn. in Chhattisgarh. https://doi.org/10.46607/iamj07p5062021

Langyan S, Khan FN, Kumar A (2023) Advancement in nutritional value, processing methods, and potential applications of pseudocereals in dietary food: a review. Food Bioprocess Technol 17:571–590. https://doi.org/10.1007/S11947-023-03109-X

Lestido-Cardama A, De Quirós ARB, Bustos J, Luisa Lomo M, Losada PP, Sendón R (2020) Estimation of dietary exposure to contaminants transferred from the packaging in fatty dry foods based on cereals. Foods 9:1038. https://doi.org/10.3390/FOODS9081038

Li H, Meng H, Sun X, Deng J, Shi T, Zhu L, Lv Q, Chen Q (2021) Integrated microRNA and transcriptome profiling reveal key miRNA-mRNA interaction pairs associated with seed development in Tartary buckwheat (*Fagopyrum tataricum*). BMC Plant Biol 21. https://doi.org/10.1186/S12870-021-02914-W

Liu J, Song Y, Zhao Q, Wang Y, Li C, Zou L, Hu Y (2021) Effects of tartary buckwheat protein on gut microbiome and plasma metabolite in rats with high-fat diet. Foods (Basel, Switzerland) 10. https://doi.org/10.3390/FOODS10102457

Liu Y, Tan ML, Zhu WJ, Cao YN, Peng LX, Yan ZY, Zhao G (2022) In vitro effects of tartary buckwheat-derived nanovesicles on gut microbiota. J Agric Food Chem 70:2616–2629. https://doi.org/10.1021/ACS.JAFC.1C07658

Lu Q, Zheng D, Li L, Liu F, Liu Y (2022) Research status of mechanized production technology and equipment of buckwheat in China/中国荞麦机械化生产技术与装备研究现状 67. https://doi.org/10.35633/inmateh-67-48

Luthar Z, Fabjan P, Mlinarič K (2021a) Biotechnological methods for buckwheat breeding. Plants (Basel, Switzerland) 10. https://doi.org/10.3390/PLANTS10081547

Luthar Z, Golob A, Germ M, Vombergar B, Kreft I (2021b) Tartary buckwheat in human nutrition. Plants (Basel, Switzerland) 10. https://doi.org/10.3390/PLANTS10040700

Mahangade PS, Mani I, Beaudry R, Müller N, Chopra S (2020) Using amaranth as a model plant for evaluating imperfect storages: assessment of solar-refrigerated and evaporatively-cooled structures in India. HortScience 55:1759–1765. https://doi.org/10.21273/HORTSCI15249-20

Majzoobi M, Wang Z, Teimouri S, Pematilleke N, Brennan CS, Farahnaky A (2023) Unlocking the potential of sprouted cereals, pseudocereals, and pulses in combating malnutrition. Foods (Basel, Switzerland) 12. https://doi.org/10.3390/FOODS12213901

Malik AM, Singh A (2022) Pseudocereals proteins—a comprehensive review on its isolation, composition and quality evaluation techniques. Food Chem Adv 1:100001. https://doi.org/10.1016/J.FOCHA.2021.100001

Martínez-Villaluenga C, Peñas E, Hernández-Ledesma B (2020) Pseudocereal grains: nutritional value, health benefits and current applications for the development of gluten-free foods. Food Chem Toxicol 137. https://doi.org/10.1016/J.FCT.2020.111178

Mesías M, Holgado F, Olombrada E, Morales FJ (2024) In vitro bioactivities of cereals, pseudocereals and seeds: assessment of antiglycative and carbonyl-trapping properties. Appl Sci 14:5684. https://doi.org/10.3390/APP14135684

Morales D, Miguel M, Garcés-Rimón M (2021) Pseudocereals: a novel source of biologically active peptides. Crit Rev Food Sci Nutr 61:1537–1544. https://doi.org/10.1080/10408398.2020.1761774

Mravlje J, Regvar M, Starič P, Mozetič M, Vogel-Mikuš K (2021) Cold plasma affects germination and fungal community structure of buckwheat seeds. Plants (Basel, Switzerland) 10. https://doi.org/10.3390/PLANTS10050851

Nandan A, Koirala P, Dutt Tripathi A, Vikranta U, Shah K, Gupta AJ, Agarwal A, Nirmal N (2024) Nutritional and functional perspectives of pseudocereals. Food Chem 448. https://doi.org/10.1016/J.FOODCHEM.2024.139072

Nešović M, Gašić U, Tosti T, Horvacki N, Nedić N, Sredojević M, Blagojević S, Ignjatović L, Tešić Ž (2021) Distribution of polyphenolic and sugar compounds in different buckwheat plant parts. RSC Adv 11:25816–25829. https://doi.org/10.1039/D1RA04250E

Norbäck D, Wieslander G (2021) A review on epidemiological and clinical studies on buckwheat allergy. Plants (Basel, Switzerland) 10. https://doi.org/10.3390/PLANTS10030607

Paucar-Menacho LM, Castillo-Martínez WE, Simpalo-Lopez WD, Verona-Ruiz A, Lavado-Cruz A, Martínez-Villaluenga C, Peñas E, Frias J, Schmiele M (2022a) Performance of thermoplastic extrusion, germination, fermentation, and hydrolysis techniques on phenolic compounds in cereals and pseudocereals. Foods (Basel, Switzerland) 11. https://doi.org/10.3390/FOODS11131957

Paucar-Menacho LM, Salvador-Reyes R, Castillo-Martinez WE, Símpalo-López WD, Verona-Ruiz A, Lavado-Cruz A, Quezada-Berrú S, López-Rodriguez W (2022b) Use of Andean pseudocereals in beer production. Sci Agropec 13:395–410. https://doi.org/10.17268/sci.agropecu.2022.036

Paucar-Menacho LM, Schmiele M, Lavado-Cruz AA, Verona-Ruiz AL, Mollá C, Peñas E, Frias J, Simpalo-Lopez WD, Castillo-Martínez WE, Martínez-Villaluenga C (2022c) Andean sprouted pseudocereals to produce healthier extrudates: impact in nutritional and physicochemical properties. Foods (Basel, Switzerland) 11. https://doi.org/10.3390/FOODS11203259

Paucar-Menacho LM, Vásquez Guzmán JC, Simpalo-Lopez WD, Castillo-Martínez WE, Martínez-Villaluenga C (2023) Enhancing nutritional profile of pasta: the impact of sprouted pseudocereals and cushuro on digestibility and health potential. Foods (Basel, Switzerland) 12. https://doi.org/10.3390/FOODS12244395

Peñalver R, Ros G, Nieto G (2023) Development of functional gluten-free sourdough bread with pseudocereals and enriched with *Moringa oleifera*. Foods (Basel, Switzerland) 12. https://doi.org/10.3390/FOODS12213920

Petrova P, Petrov K (2020) Lactic acid fermentation of cereals and pseudocereals: ancient nutritional biotechnologies with modern applications. Nutrients 12. https://doi.org/10.3390/NU12041118

Poshadri A, Deshpande HW, Machewad GM, Kshirsagar RB, Gadhe KS, Kadam SD (2023) Functional properties of selected composite gluten-free pseudocereals flour. Food Humanity 1:1200–1205. https://doi.org/10.1016/J.FOOHUM.2023.09.015

Punia Bangar S, Sharma N, Singh A, Phimolsiripol Y, Brennan CS (2022) Glycaemic response of pseudocereal-based gluten-free food products: a review. Int J Food Sci Technol 57:4936–4944. https://doi.org/10.1111/IJFS.15890

Rao V, Poonia A (2023) Protein characteristics, amino acid profile, health benefits and methods of extraction and isolation of proteins from some pseudocereals—a review. Food Prod Process Nutr 5:1–17. https://doi.org/10.1186/S43014-023-00154-Z

Rodríguez JP, Rahman H, Thushar S, Singh RK (2020) Healthy and resilient cereals and pseudocereals for marginal agriculture: molecular advances for improving nutrient bioavailability. Front Genet 11:510786. https://doi.org/10.3389/FGENE.2020.00049/BIBTEX

Rollán GC, Gerez CL, Leblanc JG (2019) Lactic fermentation as a strategy to improve the nutritional and functional values of pseudocereals. Front Nutr 6. https://doi.org/10.3389/FNUT.2019.00098

Romano A, Ferranti P (2023) Pseudocereals: quinoa (*Chenopodium quinoa* Willd.). Sustain Food Sci Comprehen Approach 1–4:141–149. https://doi.org/10.1016/B978-0-12-823960-5.00004-4

Salunkhe DK, Kadam SS (1998) Celery and other salad vegetables. In: Handbook of vegetable science and technology, pp 541–615. https://doi.org/10.1201/9781482269871-22

Sattar M, Saeed F, Afzaal M, Rasheed A, Asif A, Sharif S, Hussain M, Asad Ur Rehman H, Raza MA, Munir H, Al Jbawi E (2024) An overview of the nutritional and therapeutic properties of amaranth. Int J Food Prop 27:263–272. https://doi.org/10.1080/10942912.2024.2304266

Shahbaz M, Raza N, Islam M, Imran M, Ahmad I, Meyyazhagan A, Pushparaj K, Balasubramanian B, Park S, Rengasamy KRR, Gondal TA, El-Ghorab A, Abdelgawad MA, Ghoneim MM, Wan C (2023) The nutraceutical properties and health benefits of pseudocereals: a comprehensive treatise. Crit Rev Food Sci Nutr 63:10217–10229. https://doi.org/10.1080/10408398.2022.2071205

Shreeja K, Devi SS, Suneetha WJ, Prabhakar BN (2021) Effect of germination on nutritional composition of common buckwheat (*Fagopyrum esculentum* Moench). Int Res J Pure Appl Chem 22:1–7. https://doi.org/10.9734/IRJPAC/2021/V22I130350

Singh S, Agrawal M, Agrawal SB (2013) Differential sensitivity of spinach and amaranthus to enhanced UV-B at varying soil nutrient levels: association with gas exchange, UV-B-absorbing compounds and membrane damage. Photosynth Res 115:123–138. https://doi.org/10.1007/S11120-013-9841-2

Sun Z, Wang X, Liu R, Du W, Ma M, Han Y, Li H, Liu L, Hou S (2021) Comparative transcriptomic analysis reveals the regulatory mechanism of the gibberellic acid pathway of Tartary buckwheat (*Fagopyrum tataricum* (L.) Gaertn.) dwarf mutants. BMC Plant Biol 21. https://doi.org/10.1186/S12870-021-02978-8

Suzuki T, Morishita T, Noda T, Ishiguro K, Otsuka S, Katsu K (2021) Breeding of buckwheat to reduce bitterness and rutin hydrolysis. Plants (Basel, Switzerland) 10. https://doi.org/10.3390/PLANTS10040791

Thakur P, Kumar K, Dhaliwal HS (2021) Nutritional facts, bio-active components and processing aspects of pseudocereals: a comprehensive review. Food Biosci 42:101170. https://doi.org/10.1016/J.FBIO.2021.101170

Tiwari A, Verma O, Tripathi M (2016) Isolation and functional characterization of cadmium resistant strains RCP 1 using agronomical parameter of Chaulai (*Amaranthus tricolor* L.) AFBSAH-2016. Asian J Bio Sci 11:6–11. https://doi.org/10.15740/HAS/AJBS/11.1/6-11

Tomás MC, Muñoz L, Haros CM, Motta C, Castanheira I, Matos AS, Nascimento AC, Assunção R, Martins C, Alvito P (2022) Effect of cooking on the content and bioaccessibility of minerals in pseudocereals. Biol Life Sci Forum 17:17. https://doi.org/10.3390/BLSF2022017017

Tomasiak A, Zhou M, Betekhtin A (2022) Buckwheat in tissue culture research: current status and future perspectives. Int J Mol Sci 23. https://doi.org/10.3390/IJMS23042298

Torbica A, Radosavljević M, Belović M, Djukić N, Marković S (2022) Overview of nature, frequency and technological role of dietary fibre from cereals and pseudocereals from grain to bread. Carbohydr Polym 290. https://doi.org/10.1016/J.CARBPOL.2022.119470

Upasana, Yadav L (2022) Pseudocereals: a novel path towards healthy eating. Pseudocereals. https://doi.org/10.5772/INTECHOPEN.103708

Usman M, Patil PJ, Mehmood A, Rehman A, Shah H, Haider J, Xu K, Zhang C, Li X (2022) Comparative evaluation of pseudocereal peptides: a review of their nutritional contribution. Trends Food Sci Technol 122:287–313. https://doi.org/10.1016/J.TIFS.2022.02.009

Vasava H V, Chudasama VR, Rathva VD, Dalvaniya DG, Leua HN (2016) Performance of different varieties of amaranth (Amaranthus spp.) under net house and open field conditions. Copyright@ EM International 22

Vásquez-Ocmín PG, Marti G, Gadea A, Cabanac G, Vásquez-Briones JA, Casavilca-Zambrano S, Ponts N, Jargeat P, Haddad M, Bertani S, Molina L, Lima P (2022) Metabotyping of Andean pseudocereals and characterization of emerging mycotoxins. bioRxiv 2022.06.23.497323. https://doi.org/10.1101/2022.06.23.497323

Vemana K, Jain RK (2010) New experimental hosts of tobacco streak virus and absence of true seed transmission in leguminous hosts. Indian J Virol 21:117–127. https://doi.org/10.1007/S13337-010-0021-0

Vidaurre-Ruiz J, Bender D, Schönlechner R (2023) Exploiting pseudocereals as novel high protein grains. J Cereal Sci 114:103795. https://doi.org/10.1016/J.JCS.2023.103795

Vollmannová A, Musilová J, Lidiková J, Árvay J, Šnirc M, Tóth T, Bojňanská T, Čičová I, Kreft I, Germ M (2021) Concentrations of phenolic acids are differently genetically determined in leaves, flowers, and grain of common buckwheat (*Fagopyrum esculentum* Moench). Plants (Basel, Switzerland) 10. https://doi.org/10.3390/PLANTS10061142

Wang R, Li M, Wei Y, Guo B, Brennan M, Brennan CS (2021) Quality differences between fresh and dried buckwheat noodles associated with water status and inner structure. Foods (Basel, Switzerland) 10. https://doi.org/10.3390/FOODS10010187

Wang Y, Qi W, Guo X, Song G, Pang S, Fang W, Peng Z (2022) Effects of oats, tartary buckwheat, and foxtail millet supplementation on lipid metabolism, oxido-inflammatory responses, gut microbiota, and colonic SCFA composition in high-fat diet fed rats. Nutrients 14. https://doi.org/10.3390/NU14132760

Wanniarachchi PC, Pavalakumar D, Jayasinghe MA (2022) Pseudocereals: nutrition, health benefits, and potential applications in gluten-free food product developments. Curr Nutr Food Sci 19:377–385. https://doi.org/10.2174/1573401318666221010093914

Wen D, Wu L, Wang M, Yang W, Wang X, Ma W, Sun W, Chen S, Xiang L, Shi Y (2022) CRISPR/Cas9-mediated targeted mutagenesis of FtMYB45 promotes flavonoid biosynthesis in tartary buckwheat (*Fagopyrum tataricum*). Front Plant Sci 13. https://doi.org/10.3389/FPLS.2022.879390

Wu W, Li Z, Qin F, Qiu J (2021) Anti-diabetic effects of the soluble dietary fiber from tartary buckwheat bran in diabetic mice and their potential mechanisms. Food Nutr Res 65:1–14. https://doi.org/10.29219/FNR.V65.4998

Yadav RK, Tomar BS, Pachauri N, Jain V (2018) Studies of nutritional properties and antioxidant potential in green leafy vegetables. Int J Food Sci Agric 2. https://doi.org/10.26855/JSFA.2018.01.001

Yao Y, Sun L, Wu WJ, Wang S, Xiao X, Hu ML, Li CL, Zhao HX, Chen H, Wu Q (2021) Genome-wide investigation of major enzyme-encoding genes in the flavonoid metabolic pathway in tartary buckwheat (*Fagopyrum tataricum*). J Mol Evol 89:269–286. https://doi.org/10.1007/S00239-021-10004-6

Zhang K, He M, Fan Y, Zhao H, Gao B, Yang K, Li F, Tang Y, Gao Q, Lin T, Quinet M, Janovská D, Meglič V, Kwiatkowski J, Romanova O, Chrungoo N, Suzuki T, Luthar Z, Germ M, Woo SH, Georgiev MI, Zhou M (2021) Resequencing of global Tartary buckwheat accessions reveals multiple domestication events and key loci associated with agronomic traits. Genome Biol 22. https://doi.org/10.1186/S13059-020-02217-7

Zhao H, He Y, Zhang K, Li S, Chen Y, He M, He F, Gao B, Yang D, Fan Y, Zhu X, Yan M, Giglioli-Guivarc'h N, Hano C, Fernie AR, Georgiev MI, Janovská D, Meglič V, Zhou M (2023)

Rewiring of the seed metabolome during Tartary buckwheat domestication. Plant Biotechnol J 21:150–164. https://doi.org/10.1111/PBI.13932

Zhong L, Lin Y, Wang C, Niu B, Xu Y, Zhao G, Zhao J (2022) Chemical profile, antimicrobial and antioxidant activity assessment of the crude extract and its main flavonoids from tartary buckwheat sprouts. Molecules (Basel, Switzerland) 27. https://doi.org/10.3390/MOLECULES27020374

Zhu F (2021) Buckwheat proteins and peptides: biological functions and food applications. Trends Food Sci Technol 110:155–167. https://doi.org/10.1016/J.TIFS.2021.01.081

Živković A, Polak T, Cigić B, Požrl T (2021) Germinated buckwheat: effects of dehulling on phenolics profile and antioxidant activity of buckwheat seeds. Foods (Basel, Switzerland) 10. https://doi.org/10.3390/FOODS10040740

Zou L, Wu D, Ren G, Hu Y, Peng L, Zhao J, Garcia-Perez P, Carpena M, Prieto MA, Cao H, Cheng KW, Wang M, Simal-Gandara J, John OD, Rengasamy KRR, Zhao G, Xiao J (2023) Bioactive compounds, health benefits, and industrial applications of Tartary buckwheat (*Fagopyrum tataricum*). Crit Rev Food Sci Nutr 63:657–673. https://doi.org/10.1080/10408398.2021.1952161

Chapter 14
Nutri-cereals as a Source of Resistant Starch

Esha Vatsa, Mehak Aggarwal, and Nidhi Chaudhary

14.1 Introduction

Our body uses carbohydrates as its primary energy source before it consumes fat or protein. The primary type of carbohydrate found in the majority of foods is starch. Amylose and amylopectin are the two main ingredients that make up starch. Different plants and diets have varying amounts of these components. The ratio and arrangement of those two compounds inside the granules determines the starch's quality (Anju and Sarita 2010). When raw, starch is difficult to digest; however, when heated, it is easier to digest. After cooking, the majority of the starch becomes gelatinized and readily digestible, while some retrogrades and forms resistant starch. The portion of starch that is consumed that metabolizes in the large intestinal tract to produce short-chain fats (SCFA) in humans as opposed to being broken into smaller pieces or absorption in the small intestine is known as resistant starch (Annor et al. 2013).

As opposed to being broken apart by digestive enzymes, a kind of starch called resistant starch (RS) passes through the small intestine into the colon, where microorganisms ferment it (Aparicio-Saguilan et al. 2007). A portion of the starch that remained after enzymatic hydrolysis was discovered by Englyst and associates in 1982 when they created an in vitro assay for identifying non-starch polysaccharides (Ashwar et al. 2016). These starches may be used as a substrate for microbial fermentation, which generates hydrogen and carbon dioxide in addition to propionate,

E. Vatsa (✉) · N. Chaudhary
School of Pharmaceutical Sciences, Jigyasa University (Formerly Himgiri Zee University), Dehradun, Uttarakhand, India

M. Aggarwal
Department of Pharmacy, Himachal Institute of Pharmacy, Paonta Sahib, Himachal Pradesh, India

© The Author(s), under exclusive license to Springer Nature Switzerland AG 2026

J. Singh, M. Kumar Malik (eds.), *Millets and Millet Starch: Chemical Composition, Functional Modification, and Applications*,
https://doi.org/10.1007/978-3-032-09502-2_14

methane, butyrate, and acetate, according to further research (Asp 1992). A group of researchers funded by the European Union then determined the total amount of starch that is resistant to digestion in the small intestine of healthy people (Baixauli et al. 2008). The digesting fate of resistant starch depends on the physical structure of the grains and seeds, food processing methods, temperature, the size and shape of the starch granules, their quantity, and their interactions with other dietary components (Bednar et al. 2001). For example, higher amounts of amylose as well as resistant starch within the starch are linked to slower rates of digestion (Bello-Perez and Paredes-Lopez 2008).

The glucose polymers amylopectin and amylose combine to form starch, the main reserve carbohydrate found in plants (Bindels et al. 2017). Amylose is a linear structure that is hardly branched and joined by α1, 4 linkages. Within the endosperm of grain kernels, it is firmly enclosed within starch granules (Bojarczuk et al. 2022). The complicated arrangement of amylopectin, a branched monomer that has α1, 4 and α1, 6 connections, tends to occupy a large amount of space in the starch granules found in the grain endosperm (Bojarczuk et al. 2022). The salivary and pancreatic enzyme α-amylase hydrolyzes starch, which is an important supplier of digestible carbohydrates (Bora et al. 2017). Brush border enzymes within the small intestinal tract handle the tiny disaccharides and polysaccharides produced when α-amylase breaks down starch (Brown 1996). Disaccharides and polysaccharides are broken down by brush boundary enzymes into monosaccharides like fructose, fructose, galactose, and sugars which are subsequently absorbed by the bloodstream (Bora et al. 2017). The pancreas and other endocrine organs release insulin in response to these monosaccharides entering the bloodstream, which aids in the transportation of glucose to the cells and keeps plasma glucose levels within the normal range (Buttriss and Stokes 2008). Consuming refined carbohydrates can raise plasma glucose levels rapidly, which causes the pancreas to produce unnecessary amounts of insulin (Chandi and Annor 2016). Large boosts in glucose and insulin releases are among the several causes that can cause prolonged inflammation, which has been shown to play an important part in the formation of resistance to insulin or diabetes of the type 2 (Dahlqvist and Borgstrom 1962). The potential to prevent and/or manage the onset of diabetes through dietary interventions lies in their ability to reduce significant spikes in glucose and insulin levels. Additionally, such interventions contribute to enhancing insulin sensitivity and improving glucose uptake under these conditions (De Vadder et al. 2014).

Because resistant starch may enter the large intestinal tract and be digested by microorganisms there, there is interest in boosting its consumption. The short-chain acids such as butyrate, propionate, and acetate are byproduct of bacterial fermentation (Den Besten et al. 2013). Short-chain fatty acids production and release lowers the colon's pH, promoting the balance of beneficial bacteria and inhibiting the growth of pathogenic bacteria like *Enterobacteriaceae* and *Clostridium*. Apart from their synergistic effects, several short-chain fatty acids play important roles in many other pathways. Butyrate has anticancer properties in the colon and serves as the primary form of energy for colonic epithelial cells (Eerlingen and Delcour 1995). Propionate improves insulin-mediated glucose absorption into adipocytes, according to in vitro studies, which raises the possibility that it influences the control of human glucose homeostasis (Eerlingen et al. 1993). Acetate provides energy for peripheral tissues and aids in the production of cholesterol (Englyst and Cummings 1987). Additionally, resistant starch serves a prebiotic function by supporting the growth of numerous advantageous bacterial species, including Bifidobacterium (FAO 2019). Resistant starch has been demonstrated to improve insulin sensitivity, postprandial insulin reactivity, and postprandial glucose reactivity in addition to its functions as a fermentable fiber and prebiotic. The mechanisms underlying these effects have been hypothesized to be both independent of and reliant on the bacterial fermentation byproducts (Fausto et al. 1997; Fuentes-Zaragoza et al. 2011; Fuwa et al. 1977; Gallat et al. 1997).

14.2 Types of Resistant Starch

Resistant starch can be classified in five types, that is, RS1, RS2, RS3, RS4, and RS5 (Gallat et al. 1997; Giuberti et al. 2014; Gopalan et al. 2009; Grassini et al. 2013). These are described below:

Type I—Physically protected; because this form of starch is encased in whole or partially processed grains or seed and because these foods contain intact cell walls, they are physically indigestible. Because amylolytic and intestinal enzymes are inaccessible to this kind of RS, it avoids digestion and instead travels to the small intestine. The only way to get this RS to fully break down in the small intestine is via careful grinding or milling. Its heat-stable composition prevents it from disintegrating or opening up during typical cooking (Gupta et al. 2017).

Type II—Ungelatinized granules that are resistant with type B crystallinity and is hydrolyzed by alpha-amylase; RS2 are native starch granules that are structurally or conformationally protected from digestion. This kind of starch is present in raw potatoes and green bananas. Due to its crystalline and compact structure, RS2 is not readily accessible to amylases or digestive enzymes, making it less amenable to hydrolysis (Haberb et al. 1977).

Type III—Starches that have undergone physical alteration are referred to as retrograded starch (RS3). In essence, RS3 is a retrograded starch—more precisely, a retrograded and recrystallized amylose—that is created when gelatinized starch is heated and when starchy foods are prepared and stored at room temperature or low temperatures. It is produced when food is processed using wet heat. RS3 is used as an ingredient in a wide variety of traditional meals (Hernandez et al. 2008) because of its excellent thermostability (Han et al. 2014). The most common examples of RS3 are cooked and chilled potatoes and cornflakes. RS3 is an excellent form of RS that is essential for nutrition since it has a greater capacity to store water than granular starch (Hernandez et al. 2008). When gelatinized starch is stored at low temperatures, the amylose double helices realign and consolidate to create the exceptionally thermostable B-type crystalline form. RS3 is resistant to amylase because amylose retrogradation prevents amylase from accessing glucosidic linkages. Since these aggregates are thermostable, rehydrating them requires temperatures higher than 150 degrees Celsius. The production of RS3 is significantly influenced by temperature and storage duration. More RS3 is produced when starch paste is kept at a lower temperature as opposed to a higher one (Herrmann et al. 1995; ICCRISAT 2017). The crystallinity of starch is also influenced by the storage temperature. When gelatinized starch is stored below room temperature, it becomes B-type crystallinity; when it is stored above room temperature, it becomes A-type crystallinity (John 2017). Starch gels have both amorphous and crystalline fractions; the crystalline part is resistant while the amorphous fraction is degraded by amylolytic enzymes. The creation of amylose-lipid complexes inhibits the development of RS3, which reduces the amount of RS3 produced. Amylopectin also contributes to the production of partially crystalline gels in starch gels that are held at low temperatures. Nevertheless, in contrast to amylose crystallization structures, amylopectin crystallization is much slower and its structures are less thermostable, allowing them to be rehydrated at temperatures between 55 and 70 degrees Celsius (Kajuna 2001). The crystal structures of amylopectin are likewise resistant to the action of amylolytic enzymes. The process used to prepare it, the amount of amylose it contains, and the starch's botanical origin all affect the formation of RS3. The production of products resulting from amylopectin retrogradation is a time-consuming procedure that can be expedited by repeatedly heating and cooling starch paste (Karuppasamy 2015).

Type IV—Cross-linked starches that have undergone chemical modification; RS4 refers to chemically modified starches. RS4 starches undergo chemical processes including etherization, esterification, or cross-linking with other substances, which reduce their digestibility. Four subcategories of RS4 are further separated based on the analysis method and the solubility of RS4 in water (Kimura and Robyt 1995). The three primary processes in chemical modification are cross-linking, substitution, and conversion. The creation of atypical links during chemical modification prevents enzymes from accessing starch, reducing the amount of starch that hydrolytic enzymes can digest (Kossmann and Lloyd 2000). The degree of substitution is directly connected to the resistivity of acetylated and hydroxypropylated starches. Compared to native starches, acetylated and hydroxypropyldistarch phosphates are less vulnerable to enzyme hydrolysis. In RS4, the degree of resistance is closely proportional to the degree of chemical change (Krishnakumari and Thayumanavan 1995). Monostarch phosphate, for example, will be more resistant to the substitution of phosphoric acid. By changing the chemistry and structure of starch granules, chemical modification makes them more resistant to amylolytic enzymes (Kruger et al. 2014).

Type V—Amoxylase-lipid complexes; lipids found in starch derived from a variety of plant sources can pierce amylose strands. One kind of RS that results from amylose-lipid complexes is called RS5. These compounds are usually produced during food processing, although they can additionally be prepared under carefully monitored circumstances. The production of amylose-lipid complex is regulated by the amylose-amylopectin proportion of starch and botanical source, since these complexes were seen in high amylose starches. RS5 consists of water-insoluble linear poly α-1,4 glucan polysaccharides that are resistant to α-amylase degradation. These polysaccharides aid in the synthesis of SCFA, particularly butyrate, which is the most significant SCFA (Leszczynski 2004).

14.3 Millets

The word "millet" refers to all species of cereal with very small grains. They are particularly resilient, withstanding extreme heat and drought, and are primarily cultivated in developing nations. All things considered, millets' macro- and micronutrient content is similar to that of other cereals. Found in desert, tropical, as well as subtropical regions, millet is a group of perennial grass grain with tiny seeds. Pearl millet, finger millet, barnyard millet, proso millet, foxtail millet, kodo millet, and tiny millet are some of the main types. Millets provide numerous health advantages. They are high in phenolic compounds and free of gluten. Millets are used to make a wide variety of traditional and contemporary dishes and drinks all around the world. Food content is significantly impacted by the procedures of milling, the fermentation process, the malting process, and heat treatment, both positively and negatively. Research suggests that consuming foods high in flavonoid-type phenolics, such millet, on a daily basis may help prevent heart attacks, strokes, and type 2 diabetes (Liu et al. 2002) (Table 14.1).

Millets can tolerate longer heat cycles and can even flourish in droughty situations. Millets thrive in low rainfall conditions of 200–500 mm when grown without irrigation. They have the ability to grow even in the face of water stress. Large amounts of millets are an excellent source of stored nutrients. They consist of the

Table 14.1 Geographical distribution of millets (Lunn and Buttriss 2007)

Origin	Scientific name	Common name
India	*Paspalum scrobiculatum*	Kodo millet
Eastern Asia (China)	*Setaria italica*	Foxtail millet, Italian millet
Northeast region of Africa (Ethiopia-Sudan border)	*Sorghum bicolor*	Sorghum, great millet, jowar
Uganda or neighboring region	*Eleusine coracana*	Finger millet, African millet
Tropical West Africa	*Pennisetum glaucum*	Pearl millet, bajra
Japan	*Echinochloa crusgalli and Echinochloa coracana*	Barnyard millet, Sawa millet
Southeast Asia	*Panicum sumatrense*	Samai, little millet
Central and Eastern Asia	*Paniculam miliaceum*	Proso millet, hog millet

major and micronutrients that the body need. Thus, they can aid in resisting starvation (Martinez et al. 2010).

Like small millet, foxtail and barnyard millet were identified as having high levels of iron and zinc (3.7 as well as 280 mg per 100 g, respectively) as well as phosphorus (280 and 290 mg per 100 g, respectively) (Nowotny 1938). Additionally, proso millet contains magnesium, that is, 153 mg per 100 g. Additionally, finger millet and foxtail millet have substantial levels of carotenoids, that is, 32 and 42 mg per 100 g, respectively (Preetham et al. 2016). Millets are commonly used in gluten-free cuisine in place of wheat since they are high in carbohydrates (70%) and proteins (12%). Additionally, they are commonly consumed to prevent diseases brought on by vitamin deficits (Nowotny 1938). Specifically, finger millet has 450 mg/100 g of calcium, which is far more than other grains. It's also interesting to consider the possibility that finger millet calcium is more accessible since it's distributed in the endosperm, which has a lower concentration of phytic acid.

14.4 A Source of Resistant Starch: Millets

The primary source of starch is cereals. Global serial demand is expected to rise sharply to surpass 3036 million tons by 2028, according to the FAO research (Preetham et al. 2016). The starch industry will have challenges in managing grain storage, which is anticipated to occur in the future as a result of changes in the climate and rising temperatures. Millets will therefore be the most sustainable and environmentally friendly initial alternative for producing resistant starch. However, the existence of resistant starch in carbohydrate-rich food diet makes the diet suitable for health-conscious persons (Raigold et al. 2014). Unlike cereals, millets are prized for their nutritional value and characteristics that prevent diabetes (Raigond et al. 2014; Regina et al. 2015). Millets can be classified as major or minor according to their availability and truthfulness, as stated by the Semi-Arid Tropics International Crops Research Institute (AACC 2001).

14.5 Categorization of Millets

14.5.1 Finger Millet

The millet that is grown most frequently worldwide is finger millet. According to various research on Indian finger millet, ragi has contents of starch 55–65% weight by weight (Ridout et al. 2002; Roberts and Whelan 1960; Robertson et al. 2005; Roopa and Premavalli 2008). The examination of the starch profiles of 12 Indian ragi varieties revealed that 55–60.22% of the total starch was accessible, with 0.9–1.3% being resistance starch (Sajilata and Singhal 2005; Saleh et al. 2013).

14.5.2 Pearl Millet

Pearl millet makes for around 40% of the millet produced worldwide, which is extensively grown in Africa and India (Saha et al. 2011). Various Indian pearl millet cultivars' starch profiles have been the subject of various research, which have revealed starch contents that range from 53% to 68% w/w. A previous study indicated that the total amylose found in six Indian pearl millet cultivars ranged between 9.11% and 12.14% w/w (Sandhu and Siroha 2017). The estimated size of the starch globules was 10.9–20.45 micrometers. The percentage of resistant starch ranges from 6.03% to 11.06% w/w, according to the results of in vitro digestibility of starch experiments (Sanz et al. 2008).

14.5.3 Foxtail Millet

Italian millet, also known as foxtail millet, is a native of China and is ranked sixth in terms of grain output among the best-yielding crops (Sarikaya et al. 2000). Studies on the carbohydrate composition of foxtail millet have reported a starch concentration ranging from 56.2% to 73.1% w/w. The ratio of amylose to amylopectin is determined to be 25:75. Resistant starch recorded is 23.6–34.33% (Shamai et al. 2003).

14.5.4 Proso Millet

Common or proso millet is a hardy crop with a quick maturation rate. These millets range in starch content between 65.9% and 88.4% (w/w), while their contributions in amylose and amylopectin are 1.5–25.9% and 17.06–46.2%, respectively. Proso millet's starch digestibility varies depending on the type. As stated, the resistant starch content ranges from 5.44% to 12.74% (w/w) (Sharma et al. 2016; Shen et al. 2018).

14.5.5 Barnyard Millet

Sawa millet, also known as barnyard millet, matures in 6 weeks and grows quickly. Indian barnyard millet has a starch content between 54.77% and 68.41% and an amylose level between 5.82% and 12.40% (Siavoshian et al. 2000; Singh and Raghuvanshi 2012). In baked Indian barnyard millet flour, resistant starch makes up 29.90% of the content (Slavin 2004).

14.5.6 Little Millet

The little millet is a native to India and can reach heights of up to 2100 m (Sugimoto 1980). Due to its low yield, it is mostly grown in India and is not as important outside of that country. Starch content in small millet typically ranges from 47.73% to 65.11% and 13.58% amylose content. The value of resistant starch is 31.98% (Suma and Urooj 2015).

14.5.7 Kodo Millet

One tropical weed that is commonly grown in India is kodo millet. They may be grown on poor soil and can withstand drought. Usually, they have 17.39–19.89% amylose and 54.09–68.52% starch. They had resistant starch levels ranging from 25.14% to 34.33% (Tharanathan and Mahadevamma 2003).

14.6 Properties of Resistant Starch

Resistant starch has occupied significance due to the functional properties as well as its nutritional aspects. Due to its numerous qualities, which include increase in viscosity, swelling, formation of gel, and the water-binding capability, it can only be used in a variety of foods (Thippeswamy et al. 2016). Because of these qualities, it can efficiently substitute flour without changing the dough's rheology. Other characteristics of resistant starch include its insipid flavor, which does not change the flavor of other foods when added. It also has an extremely tiny particle size and appears whitish in hue. Due to its extremely tiny particle size, the food's taste, quality, and texture also don't change. Additionally, it has a low calorie content (1.6–2.8 Kcal g^{-1}) that enhances condensed fat and the production of food formulations with less sugar, giving it unique qualities. It also fortifies dietary fibers in foods that are high in fiber (Vali et al. 2018). A lower water-holding capacity also allows for better material handling throughout the food's processing phase, which enhances the final product's crispiness and best texture appearance.

Various other properties of resistant starch involve good extrusion capacity, higher temperature of gelatinization, and property of film formation. Incorporation of RS in food formulations mends in enhancement of the crispiness of the product via coating of RS and also improves the healthy life of breakfast cereals (Table 14.2).

Table 14.2 Benefits and functional characteristics of commercial suppliers of resistant starch (Veena et al. 2005)

S. no.	Functional properties and advantages of resistant starch
1.	Natural sources of ingredient
2.	Bland flavor constituent
3.	Cream and whitish in color
4.	Higher temperature of gelatinization
5.	Finest particle size
6.	Lower rate of water properties comparable to traditional fiber-rich available products
7.	Good property of extrusion
8.	Great film-forming qualities
9.	Lowering of the calorific value of functional foods
10.	Availability of functional food ingredients
11.	Enhanced coating crispness of the end products
12.	Improved health life of breakfast-consuming cereals
13.	Beneficial for both oral water retention formulations and large quantities laxative formulations for celiacs
14.	Aids in the creation of high-fiber and low-bulk products enhanced by the product's concealed structure and look
15.	Pleasant mouth feeling (viz., improved morphological qualities) in comparison to the traditional high-fiber content food products

14.7 Applications of Resistant Starch

Prospective physiological advantages and distinctive properties of RS have enticed the consideration of several potential nutritionists and food processors. As consumers' awareness of health and nutrient-dense food has grown over time, so too has their concern about the additional health benefits that come from regularly consuming resistant starch in addition to the conventional nutritional aspects of food consumption (Veena et al. 2005). Within this framework, food processors, researchers, academicians, and producers are conducting intensive research on the creation of higher-quality foods with slower digestion and additional health benefits. In order to improve the nutritional value and health benefits of food goods, resistant starch is used as a component for strengthening them. Resistant starch is available naturally in wide range of starch-oriented food products, which makes the food products very convenient in usage as a potential functional ingredient for the purpose of fortification (Vrinten and Nakamura 2000). The first commercially accessible resistant starch product was tested in the mid-1990s. However, more businesses are now producing resistant starch-rich powder formulations, using a variety of technologies in the process. Given the established excellent quality of the final product formulation enriched with resistant starch, traditional insoluble fiber, which is present in food products fortified with resistant starch, cannot take its place (Wolever et al. 1991). Food products that don't include moisture are made using RS.

Generally, cross-linked resistant starch is manufactured from maize starch, tapioca starch, and potato starch which are being used for the manufacturing of formulation which needed smoothness, flowability, pulpy texture, lower pH, and higher storage temperatures (Xu et al. 2003).

Imitation cheese often contains fats, but these days, RS has completely replaced them without negatively impacting the hardness or meltability. In those situations, RS offers two benefits: First, it lowers the fat content of a specific food product; second, it has health benefits related to RS itself. Today's market offers a wide variety of RS and fiber-fortified products, including high-fiber bread, cookies, and breakfast cereals.

RS auxiliate impenetrability among beverages. It is being utilized for thickening of opaque health drinks also under which insoluble fiber is very much desired. Resistant starch is the superior one in comparison to other kind of fibers, due to its bland flavor, which reduces the amount of grit in the mouth and also muffles flavor to a far lesser degree. On the other hand, other fibers contribute to a harsh texture, a strong flavor, and a bad, dry mouth sensation (Xu et al. 2003) (Table 14.3).

Table 14.3 Commercially prepared resistant starches usually utilized in various kinds of foods (Yang et al. 2018)

The commercial RS brand name	Kind	Content of RS/TDF%	Health and/or physiological advantages	Producer
Hi-maize	Type 2	30–60% TDF	Prebiotic qualities. Reduces the pH of feces. Raises SCFA levels, especially butyrate, which may lower the risk of cancer. Has a slight laxative effect and increases bowel motion. Enhances the beneficial microbiota associated with the bowel	United States: National Starch and Chemicals Co.
Crysta lean	Type 3	19.2–41% RS	Prebiotic impact. Raises a healthy amount of butyrate. It promotes cell growth in the colon's proximal region in rats. Lowers the glycemic index and offers solubility of fiber from food and prebiotic benefits	USA-based Opta Food Ingredients Inc.
Novelose240	Type 2	47% RS	As an alternative to wheat and other quickly absorbed carbs, it lowers glycemic response.	United States: National Starch and Chemicals Co.
C*Actistar	RS3	53% RS	Very well tolerated and gives potential health benefits, prebiotic impact, and butyrate's source. Aids in immune system support. Decreased glycemic reaction. Low calorie content. RS that ferments easily	Cargill Company

14.8 Health Benefits After Resistant Starch Consumption

Awareness regarding health improvement is rising rapidly among consumers. To meet the increasing demand from consumers for functional meals, the global food industry is searching for new ways to create food items with extra health benefits. Everybody's diet consists primarily of items high in carbohydrates, most of which are hyperglycemic. Therefore, the creation of foods high in carbohydrates and low in these nutrients is necessary in the modern era. Foods having a higher glycemic index have an impact on a person's health by preventing blood glucose levels from rising and ultimately causing insulin levels to rise. Products containing carbohydrates that have a lower glycemic index can help regulate diabetes and obesity, which in turn lowers the risk of cardiovascular diseases. All types of RS are not efficient in controlling the level of cholesterol (Yang et al. 2018).

In contrast to a low-fiber diet, resistant starch combines the physiological benefits of soluble fiber and improves colon health by increasing the pace at which cells are produced in the cellar and reducing colonal epithelial atrophy. However, as a result of shifting perspectives on nutritional value, in industrialized nations, food's nutritious value is expressed by its declining calorific content. Moderations in civilization also helped people consume more dietary fiber, which is essential for the body to operate properly. Dietary fiber contains RS with a lower calorific value, which has several health benefits, such as lowering blood glucose and cholesterol (Yang et al. 2018) (Table 14.4).

Table 14.4 Health benefits of resistant starch consumption

S. no.	Type of property	Functional benefits for health improvement
1.	Dietary fiber	Increase intestinal viscosity, impede intestinal transit, and postpone stomach emptying, all of which result in sluggish absorption of glucose and sterols
2.	Improves probiotic bacteria	Improve intestinal mucosal adhesion, strengthen the epithelial barrier, and simultaneously prevent pathogen adherence and competitive exclusion of harmful microbes
3.	Hypoglycemic in nature	Sluggish release of glucose and timing of glucose metabolism, inaccessibility of starch to digestive enzymes (e.g., pullulanase, isoamylase, and α-amylase), decreased insulin responsiveness, and postprandial blood glucose levels
4.	Hypocholesterolemic effect	Reduces blood cholesterol and triglycerides and increases the size, pool, and absorption of SCFA in the colon
5.	Reduction in gallstone formation	High intake of dietary fibers
6.	Enhanced minerals absorption	Improved fermentation process in the digestive tract's distal regions, which leads to improved absorption of copper, zinc, magnesium, and calcium

14.9 Conclusions

According to a thorough review of the literature, a number of reports have been developed and discussed in this manuscript that show that over the past few decades, the consumption of resistant starch has decreased, most likely as a result of consuming fast-food and lower-fiber foods. When it comes to fortifying ready-to-eat morning cereals, various snacks, noodles, baked goods, pasta products, and fried dishes, RS is said to be the perfect food material. When RS is added to products, the results are similar to those made with standard fibers in terms of improved crispness, mouthfeel, color, and flavor. Because of its physiological effects it can lower the risk of several illnesses, such as type 2 diabetes and colon cancer, as well as help regulate obesity, resistant starch has gained more attention.

References

AACC (2001) The definition of dietary fiber. Cereal Foods World 46(3):112–126

Anju T, Sarita S (2010) Suitability of foxtail millet (*Setaria italica*) and barnyard millet (*Echinochloa frumentacea*) for development of low glycaemic index biscuits. Malays J Nutr 16:361–368

Annor GA, Marcone M, Bertoft E et al (2013) In vitro starch digestibility and expected glycemic index in kodo millet (*Paspalum scrobiculatum*) as affected by starch protein lipid interactions. Cereal Chem 90:211–217. https://doi.org/10.1094/CCHEM-06-12-0074-R

Aparicio-Saguilan A, Sayagoayerdi S, Vargastorres A et al (2007) Slowly digestible cookies prepared from resistant starch-rich lintnerized banana starch. J Food Compos Anal 20(3–4):175–181

Ashwar BA, Gani A, Shah A et al (2016) Preparation, health benefits and applications of resistant starch—a review. Star 68(3–4):287–230. https://doi.org/10.1002/star.201500064

Asp NG (1992) Resistant starch. Proceedings from the second plenary meeting of EURESTA: European FLAIR concerted action, 11 on physiological implications of the consumption of resistant starch in man. Eur J Clin Nutr 46(Suppl 2):S1–S148

Baixauli R, Salvador A, Martinez-Cervera S et al (2008) Distinctive sensory features introduced by resistant starch in baked products. Food Sci Technol 41:1927–1933. https://doi.org/10.1016/j.lwt.2008.01.012

Bednar GE, Patil AR, Murray SM et al (2001) Starch and fiber fractions in selected food and feed ingredients affect their small intestinal digestibility and fermentability and their large bowel fermentability in vitro in a canine model. J Nutr 131:276–286. https://doi.org/10.1093/jn/131.2.276

Bello-Perez LA, Paredes-Lopez O (2008) Starches of some food crops, changes during processing and their nutraceutical potential. Food Eng Rev 1:50–65

Bindels LB, Munoz RRS, Gomes-Neto JC (2017) Resistant starch can improve insulin sensitivity independently of the gut microbiota. Microbiome 5:12. https://doi.org/10.1186/s40168-017-0230-5

Bojarczuk A, Skąpska S, Mousavi A et al (2022) Health benefits of resistant starch: a review of the literature. J Funct Foods 93:105094. https://doi.org/10.1016/j.jff.2022.105094

Bora P, Ragaee S, Marcone M (2017) Characterization of several types of millets as functional food ingredients. Int J Food Sci Nutr 70:714–724. https://doi.org/10.1080/09637486.2019.1570086

Brown I (1996) Complex carbohydrates and resistant starch. Nutr Rev 54:115–119. https://doi.org/10.1111/j.1753-4887.1996.tb03830.x

Buttriss JL, Stokes CS (2008) Dietary fiber and health: an overview. British nutrition foundation. Nutr Bull 33:186–200. https://doi.org/10.1111/j.1467-3010.2008.00705.x

Chandi GK, Annor GA (2016) Millet minor: overview. In: Encyclopedia of food grains, vol 2, pp 199–208. https://doi.org/10.1016/B978-0-12-394437-5.00012-7

Dahlqvist A, Borgstrom B (1962) Digestion and absorption of disaccharides in man. Nutr Rev 20:203–205. https://doi.org/10.1042/bj0810411

De Vadder F, Kovatcheva-Datchary P, Goncalves D (2014) Microbiota-generated metabolites promote metabolic benefits via gut-brain neural circuits. Cell 156:84–96. https://doi.org/10.1016/j.cell.2013.12.016

Den Besten G, Eunen K, Groen A (2013) The role of short-chain fatty acids in the interplay between diet, gut microbiota, and host energy metabolism. J Lipid Res 54:2325–2340. https://doi.org/10.1194/jlr.R036012

Eerlingen RC, Delcour A (1995) Formation, analysis, structure and properties of type III enzyme resistant starch. J Cereal Sci 22:129–138. https://doi.org/10.1016/0733-5210(95)90042-X

Eerlingen RC, Crombez M, Delcour JA (1993) Enzyme-resistant starch i. Qualitative and quantitative influence of incubation time and temperature of autoclaved starch on resistant starch formation. Cereal Chem 54:339–344

Englyst HN, Cummings JH (1987) Resistant starch a 'new' food component. A classification of starch for nutritional purposes. In: Morton ID (ed) Cereals in a European context. Ellis Horwood, Chichester, pp 221–233

FAO (2019) Food outlook—biannual report on global food markets. FAO

Fausto FD, Kacchi AI, Mehta D (1997) Starch products in confectionery. Bev Food World 24(4):4–16

Fuentes-Zaragoza E, Sanchez-Zapata E, Sendra E et al (2011) Resistant starch as prebiotic: a review. Starch-Starke 63:406–415. https://doi.org/10.1002/star.201000099

Fuwa H, Nakajima M, Hamada A et al (1977) Comparative susceptibility to amylases of starches from different plant species and several single endosperm mutant and their double-mutant combinations with opaque-2 inbred oh43 maize. Cereal Chem 54:230–237

Gallat DJ, Bouchet B, Baldwin PM (1997) Microscopy of starch: evidence of a new level of granule organization. Carbohydr Polym 32:177–191. https://doi.org/10.1016/S0144-8617(97)00008-8

Giuberti G, Gallo A, Masoero F et al (2014) Factors affecting starch utilization in large animal food production system: a review. Star 66(1–2):72–90. https://doi.org/10.1002/star.201300177

Gopalan C, Rama S, Balasubramanian SC (2009) Nutritive value of Indian foods. National Institute of Nutrition, Indian Council of Medical Research, Hyderabad, p 99

Grassini P, Eskridge KM, Cassmap KG (2013) Distinguishing between yield advances and yield plateaus in historical crop production trends. Nat Commun 4:2918. https://doi.org/10.1038/ncomms3918

Gupta SM, Arora S, Mirza N et al (2017) Finger millet: a certain crop for an uncertain future and a solution to food insecurity and hidden hunger under stressful environments. Front Plant Sci 8:1–11. https://doi.org/10.3389/fpls.2017.00643

Haberb GB, Heaton KW, Murphy D (1977) Depletion and disruption of dietary fibre. Effects on satiety, plasma-glucose, and serum-insulin. Lancet 2:679–682. https://doi.org/10.1016/s0140-6736(77)90494-9

Han J-H, Kim I-S, Jung S-H (2014) The effects of propionate and valerate on insulin responsiveness for glucose uptake in 3T3-L1 adipocytes and C2C12 myotubes via G protein-coupled receptor 41. PLoS One 9:95268. https://doi.org/10.1371/journal.pone.0095268

Hernandez O, Emaldi U, Tovar J (2008) In vitro digestibility of edible films from various starch sources. Carbohydr Polym 71:648–655

Herrmann C, Goke R, Richter G (1995) Glucagon-like peptide-1 and glucose-dependent insulin-releasing polypeptide plasma levels in response to nutrients. Digestion 56:117–126. https://doi.org/10.1159/000201231

ICCRISAT (2017) A short history of millets and how we are recognising their importance in the modern context-ICRISAT

John RN (2017) Chapter 4—millets: their unique nutritional and health-promoting attributes. In: Gluten-free ancient grains. Cereals, pseudocereals, and legumes: sustainable, nutritious, and health-promoting foods for the 21st century. Woodhead Publishing series in food science, technology and nutrition. Woodhead Publishing, pp 55–103. https://doi.org/10.1016/B978-0-08-100866-9.00004-2

Kajuna S (2001) MILLET post-harvest operation-post-harvest compendium. MILLET: Post-Harvest Operations

Karuppasamy P (2015) Overview on millets (nutri cereals). Tends Biosci 8:3269–3273

Kimura A, Robyt JF (1995) Reaction of enzymes with starch granules: kinetics and products of the reaction with glucoamylase. Carbohydr Res 277:87–107. https://doi.org/10.1016/0008-6215(95)00196-Z

Kossmann J, Lloyd J (2000) Understanding and influencing starch biochemistry. Crit Rev Plant Sci 19:171–226

Krishnakumari S, Thayumanavan B (1995) Content of starch and sugars and in vitro digestion of starch by alpha-amylase in five minor millets. Plant Foods Hum Nutr 48:327–333. https://doi.org/10.1007/bf01088491

Kruger J, Pineda-Vargas CA, Minnis-Ndimba R et al (2014) Visualization of the distribution of minerals in red non-tannin finger millet using PIXE microanalysis. J Cereal Sci 60:1–3

Leszczynski W (2004) Resistant starch-classification, structure, production. Pol J Food Nutr Sci 13:37–50

Liu S, Manson JE, Buring JE (2002) Relation between a diet with as high glycemic load and plasma concentrations of high-sensitivity C-reactive protein in middle-aged women. Am J Clin Nutr 75:492–498. https://doi.org/10.1093/ajcn/75.3.492

Lunn J, Buttriss JL (2007) Carbohydrates and dietary fibre. Nutr Bull 32(1):21–64. https://doi.org/10.1111/j.1467-3010.2007.00616.x

Martinez I, Kim J, Duffy PR (2010) Resistant starches types 2 and 4 have differential effects on the composition of the fecal microbiota in human subjects. PLoS One 5:15046. https://doi.org/10.1371/journal.pone.0015046

Nowotny E (1938) Effect of malt and barley amylase on raw non-gelatinised starch. Rocz Nauk Rol Leoen 45:1–38

Preetham KV, Dharmaraj U, Sakhare SD et al (2016) Flour functionally and nutritional characteristics of different roller milled streams of foxtail millet (*Setaria italica*). Lebensm Wiss Technol 73:274–279. https://doi.org/10.1016/j.lwt.2016.06.028

Raigold P, Ezekiel R, Kaundal B (2014) Starch fractions of cooked potatoes at low temperature. Potato J 41:58–67

Raigond P, Ezekiel R, Raigond B (2014) Resistant starch in food: a review. J Sci Food Agric 95:1968–1978. https://doi.org/10.1002/jsfa.6966

Regina A, Berbezy P, Kosar-Hashemi B (2015) A genetic strategy generating wheat with very high amylose content. Plant Biotechnol J13:1276–1286. https://doi.org/10.1111/pbi.12345

Ridout MJ, Gunning AP, Parker ML et al (2002) Using AFM to image the internal structure of starch granules. Carbohydr Polym 50:123–132. https://doi.org/10.1016/S0144-8617(02)00021-8

Roberts PJP, Whelan WJ (1960) The mechanism of carbohydrase action. 5. Action of human salivary α-amylase on amylopectin and glycogen. Biochem J 76:246–253. https://doi.org/10.1042/bj0760246

Robertson MD, Bickerton AS, Dennis AL (2005) Insulin-sensitizing effects of dietary resistant starch and effects on skeletal muscle and adipose tissue metabolism. Am J Clin Nutr 82:559–567

Roopa S, Premavalli KS (2008) Effect of processing on starch fractions in different varieties of finger millet. Food Chem 106:875–882

Saha S, Gupta A, Singh SRK et al (2011) Compositional and varietal influence of finger millet flour on rheological properties of dough and quality of biscuit. LWT—Food Sci Technol 44(3):616–621

Sajilata MG, Singhal RS (2005) Specialty starches for snack foods. Carbohydr Polym 59:131–151. https://doi.org/10.1016/j.carbpol.2004.08.012

Saleh AS, Zhang Q, Chen J et al (2013) Millet grains: nutritional quality, processing, and potential health benefits. Compr Rev Food Sci Food Saf 12:281–295. https://doi.org/10.1111/1541-4337.12012

Sandhu KS, Siroha AK (2017) Relationships between physicochemical, thermal, rheological, and in vitro digestibility properties of starches from pearl millet cultivars. LWT—Food Sci Technol 83:213–224. https://doi.org/10.1016/j.lwt.2017.05.015

Sanz T, Salvador A, Fiszman SM (2008) Resistant starch (RS) in battered fried products: functionally and high-fiber benefit. Food Hydrocoll 22:543–549. https://doi.org/10.1016/j.foodhyd.2007.01.018

Sarikaya E, Higasa T, Adachi M et al (2000) Comparison of degradation abilities of alpha-and beta-amylases on raw starch granules. Process Biochem 35:711–715. https://doi.org/10.1016/S0032-9592(99)00133-8

Shamai K, Blanco-Peled H, Shimoni E (2003) Polymorphism of resistant starch type III. Carbohydr Polym 27:363–369. https://doi.org/10.1016/S0144-8617(03)00192-9

Sharma M, Singh AK, Yadav DN et al (2016) Impact of octenylsuccinylation on rheological, pasting, thermal and physicochemical properties of pearl millet (*Pennisetum typhoides*) starch. Lebensm Wiss Technol 73:52–59. https://doi.org/10.1016/j.lwt.2016.05.034

Shen R, Ma Y, Jiang L et al (2018) Chemical composition, antioxidant and antiproliferative activities of nine Chinese proso millet varieties. Food Agric Immunol 29:625–637. https://doi.org/10.1080/09540105.2018.1428283

Siavoshian S, Segain JP, Kornprobst M (2000) Butyrate and trichostatin A effects on the proliferation/differentiation of human intestinal epithelial cells: induction of cyclin D3 and p21 expression. Gut 46:507–514. https://doi.org/10.1136/gut.46.4.507

Singh P, Raghuvanshi RS (2012) Finger millet for food and nutritional security. Afr J Food Sci 6:77–84. https://doi.org/10.5897/AJFSX10.010

Slavin J (2004) Whole grains and human health. Nutr Res Rev 17(1):99–110. https://doi.org/10.1079/nrr200374

Sugimoto Y (1980) Scanning electron microscopic observation of starch granules attacked by enzyme. J Jap Soc Starch Sci 27:28–40

Suma PF, Urooj A (2015) Isolation and characterization of starch from pearl millet (*Pennisetum typhoidium*) flours. Int J Food Prop 18:2675–2687. https://doi.org/10.1080/10942912.2014.981640

Tharanathan RN, Mahadevamma S (2003) Grain legumes: a boon to human nutrition. Trends Food Sci Technol 14:507–518. https://doi.org/10.1016/j.tifs.2003.07.002

Thippeswamy TG, Junna L, Shinde M (2016) Proximate composition, resistant starch and other phytochemical constituents of native finger millet cultivar. Int J Food Nutr Sci 5:67

Vali PK, Ratnavathi CV, Ajani J et al (2018) Proximate, mineral composition and antioxidant activity of traditional small millets cultivated and consumed in Rayalaseema region of South India. J Sci Food Agric 98:652–660. https://doi.org/10.1002/jsfa.8510

Veena S, Chimmad BV, Naik RK et al (2005) Physico-chemical and nutritional studies in barnyard millet. Karnataka J Agric Sci 18:101–105

Vrinten PL, Nakamura T (2000) Wheat granule-bound starch synthase I and II are encoded by separate genes that are expressed in different tissues. Plant Physiol 122:255–264. https://doi.org/10.1104/pp.122.1.255

Wolever TM, Spadafora P, Eshuis H (1991) Interaction between colonic acetate and propionate in humans. Am J Clin Nutr 53:681–687. https://doi.org/10.1093/ajcn/53.3.681

Xu H, Barnes GT, Yang Q (2003) Chronic inflammation in fat plays a crucial role in the development of obesity-related insulin resistance. J Clin Invest 112:1821–1830. https://doi.org/10.1172/JCI19451

Yang Q, Zhang P, Qu Y et al (2018) Comparison of physicochemical properties and cooking edibility of waxy proso millet (*Panicum miliaceum L.*). Food Chem 257:271–278. https://doi.org/10.1016/j.foodchem.2018.03.009

Chapter 15
Traditional Food Products of Little Millet, Kodo Millet, Pearl Millet, Proso Millets, and Sorghum Millet

Ajay, Vishakha Pundir, Sundeep Kumar, and Mahima Koushik

15.1 Introduction

Millets or Shri Dhanya are underappreciated crops grown in warm, arid, and humid areas with little likelihood of other supporting crops (Neelayathatchi et al. 2024). Foxtail millet (*Setaria italica*, kangani), finger millet (*Eleusine coracana*; ragi), little millet (*Panicum sumatrense*), proso millet (*Panicum miliaceum*; barri), barnyard millet (*Echinochloa utilis*; sanwa), and kodo millet (*Paspalum setaceum*; kodra) are recognized as small or minor millets. Major millets include sorghum (*Sorghum vulgare*; jawar) and pearl millet (*Pennisetum glaucam*; bajra). An old Kannada proverb in India emphasizes the importance of staple food that means "*A rice eater is always weightless like a bird; A jowar eater is strong like a wolf; A Ragi eater always remains nirogi (disease-free)*" (Saini et al. 2021). Millets are cultivated for thousands of years and one of the oldest food crop known to humans, according to both history and eating trends. Romans and Paulines were eating millets as porridges in the Middle Ages instead of rice, it's possible that millets were the first cereal ever utilized for food (Ambati and Sucharitha 2019).

Considering the oldest crops used for food purposes and to increase awareness among the communities for healthy diets, there is a captivating need in today's era to provide a detailed chapter focused on the traditional food products (TFs) developed from these future foods. This chapter summarized an overview of the numerous cuisines, traditional foods, and beverages with cultural significance developed from sorghum millet, kodo millet, pearl millet, proso millet, and little millet.

Ajay · V. Pundir · M. Koushik (✉)
Department of Chemistry, Gurukula Kangri (Deemed to Be University), Haridwar, India
e-mail: 23521003@gkv.ac.in

S. Kumar
Department of Applied Chemistry, M. J. P. Rohilkhand University, Bareilly, India

© The Author(s), under exclusive license to Springer Nature Switzerland AG 2026
J. Singh, M. Kumar Malik (eds.), *Millets and Millet Starch: Chemical Composition, Functional Modification, and Applications*,
https://doi.org/10.1007/978-3-032-09502-2_15

Cultures all around the world have a unique place for TFs, which are frequently derived from centuries-old culinary customs. These traditional dishes are significantly entwined in the rituals, customs, and histories of communities and tribes, which transmit their knowledge from one generation to another (Ghosh et al. 2023). Traditional cuisines are a veritable gold mine of culinary history because of their diversity, use of locally sourced ingredients, and distinctive cooking techniques. These foods often serve as a unique source of essential nutrients in addition to being a representation of culture. TFs are abundant in vitamins, minerals, dietary fibers, and antioxidants, which promote overall wellness and heath. Additionally, there is evidence linking TFs to reduced chance of developing chronic diseases including cardiac disease, diabetes, and obesity. TFs can be incorporated into contemporary diets to encourage improved eating habits and ward off health issues linked to diet (Cena and Calder 2020; Kim et al. 2016; Jung et al. 2014). A salient feature of these TFs of millets and their production processes have evolved mostly apart from those employed in the production of "Western" staple meals and drinks like malted barley beer, oat porridge, and wheat bread (Shahidi and Chandrasekara 2013). Skills to prepare such millet-based food products have evolved over generations passed down from mother to daughter and are enclosed with nutritional aspects resulting in incredible fabricating approaches still practiced in tribal and rural communities (Taylor and Duodu 2018).

In the last few decades, grains with better yields get more attention from farmers due to higher economic returns, while traditional cereals such as millets have low popularity because of low yield and other adverse conditions. As a result, people have forgotten the nutritional value and qualities of TFs and given more importance to less nutritious cereals which adversely impacts human health. The World Health Organization (WHO) evaluated that 80% of early deaths worldwide are caused by noncommunicable diseases, which include unhealthy lifestyle, cardiovascular disease, diabetes, cancer, and chronic lung and respiratory disorders (Amadou et al. 2013; Neelayathatchi et al. 2024; UNICEF 2019). To avoid all these consequences, people and researchers from around the globe are now focusing on a balance diet and TFs like millets to maintain their health and fitness. Also, the United Nations General Assembly (UNGA) proclaimed year 2023 to be the international year of millets (IYM) in March 5, 2021, on the proposal of the Indian government to raise awareness about millets in terms of nutritional benefits, ability to thrive in adverse climates, sustainable food source, and global food availability, and security, particularly in areas with adverse climate challenges (Panigrahy and Padhi 2023).

15.2 Processed and Traditional Food Products Made from Little Millet, Pearl Millet, Proso Millet, and Sorghum Millet

Millets are gluten-free cereals with rich nutritional index and consumed in numerous cuisines. Various traditional products are made from millets or millets mixed with other cereals or crops including traditional, processed, and non-processed foods with fermented and non-fermented beverages (Siroha and Bangar 2024).

Millets are an excellent source of energy, high-quality proteins, lipids, minerals, and vitamins in addition of bioactive phenolic compounds which promote health concern. Amino acid's profile of millets includes both nonessential and essential amino acids. Even millets also have amino acids with sulfur content (Seetha et al. 2019). Although their integration into the human food sector remains limited, they have numerous health benefits, including being gluten-free, possessing a low glycemic index, and offering high protein and dietary fiber content. Growing demand for gluten-free diets and consumer preference have revitalized interest in millet-based food products (Selladurai et al. 2023). This renewed focus has facilitated millets incorporation into various human food applications, including processed and non-processed food products, breakfast cereals, fermented foods, bakery products, and brewing industries (Saleem et al. 2023). Generally, millets are found in two types of varieties: naked and husked grains. Naked grains does not have the husk part and can be eaten after washing without any further processing. Finger millet, sorghum millet, and pearl millet are examples of naked millets, while kodo millet, proso millet, etc., are categorized as husked grains. Husked millets contain stiff and indigestible coats. Therefore, processing is required after harvesting, and husk must be removed before consumption (Sivarajasekar et al. 2023). Emerging technologies for processing of millets involved two steps: primary processing and secondary processing. Primary processing is used to clean husk and indigestible part of grains with dehulling, milling, and polishing. Secondary processing treats the primary processing products in single or multiple steps and transforms into consumable food forms. Secondary processing involves conventional and modern methods such as soaking, thermal or microwave cooking, roasting, puffing, extrusion, fortification, encapsulation, malting, fermentation, etc. (Mundassery et al. 2024). These processing techniques increase the digestibility, shelf life, product quality, and taste and cause the functional changes (Fig. 15.1). Cooking or heating reduces antioxidants and secondary metabolites but enhances the digestibility, while processing involved in formation of prebiotic and probiotic drinks with increase antioxidant property. In addition, overall nutritional and therapeutic qualities of millets improved by fermentation, but dietary fiber and micronutrients were decreased by extensive

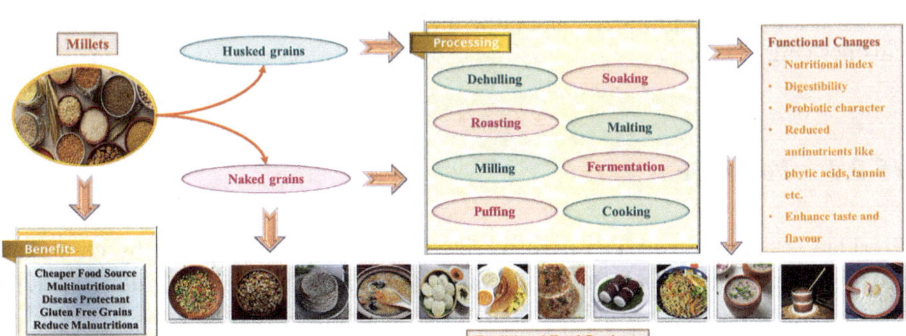

Fig. 15.1 Impact of processing on functional characteristics of traditional foods made from millets

Fig. 15.2 Traditional foods made from millet as a primary ingredient: porridge (**i**), flatbreads (**ii**), noodles (**iii**), dosa (**iv**), baked goods (**v**), non-alcoholic beverage (**vi**), alcoholic beverage (**vii**), and ragi puttu (**viii**)

dehulling, polishing, and milling (Budhwar et al. 2020). Despite these methods, germination is also a popular traditional method used to improve the nutritional index of millets (Gunenc et al. 2017). Consumable food recipes made from millets include millet porridge, flatbreads, sweet dishes like ragi puttu, ladoo, noodles, millet dosa, idli, baked products, alcoholic and non-alcoholic beverages in numerous cuisines (Fig. 15.2).

15.2.1 Porridge

Porridge is a staple and TFs used worldwide. Porridge is prepared by chopping, crushing, grounding, or boiling of starch-rich grains in water or milk. Porridge made with millets is an excellent preparation to get more nutrients with high-fiber and gluten-free properties (Zhang et al. 2022). Millet-based porridge is especially popular for its high energy content and easy digestibility. Nutritional contents of porridge made from various millets are summarized in Table 15.1.

Table 15.1 Nutritional index of porridge made from little millet, kodo millet, pearl millet, proso millet, and sorghum millet

Millets	Porridge	Nutritional content (per 100 g)									References
		Moisture (g)	Carbohydrate (g)	Protein (g)	Fiber (g)	Ash (g)	Fat (g)	Ca (mg)	Fe (mg)	Tannin (mg)	
Little millet	100 g	84.86	65.50	7.30	7.25	0.21	0.12	16.10	8.60	15.14	Karuppasamy and Malathi (2014)
Kodo millet	100 g	84.99	64.00	8.12	8.20	0.14	0.04	24.00	1.40	15.01	Karuppasamy and Malathi (2014)
Pearl millet	100 g	–	65.7	11.3	–	1.05	1.81	240.9	5.6	–	Meena et al. (2024)
Sorghum millet	100 g	77.24	19.08	2.03	0.75	0.85	0.97	3.78	0.204	–	Jackson et al. (2013)

15.2.2 Flatbreads

Flatbreads are made from a variety of cereals. Flour is the basic ingredient in a flatbread recipe as it affects the nutritional value and digestibility. Millet flour-based flatbreads are popular from several years in Asia and Africa with numerous names on the basis of preparation and local languages such as fulka, puri, roti, paratha, and chapati in India, Africa, and Nigeria (Renzetti et al. 2023). Flatbreads made from pearl millets, which is referred to as "*Bajra ki roti*" in Indian cuisine, is TFs from many centuries. Despite these, finger millets, pearl millets, sorghum millets, etc., are also used to make several types of flatbreads in various cuisines. Millet-based flour also affects the sensory properties and texture of flatbreads. Kumar et al. (2023) evaluated that chapatti prepared from millet enhanced the dough's rheological and pasting qualities as well as the nutritional value and texture of flatbreads. Primary ingredients that are used to make dough for flatbreads are flour, water, sodium chloride, and leaving agents (Mir et al. 2014). Multiple steps are carried out in order to mix these materials and develop (Fig. 15.3) dough with correct mechanical and sensory qualities that enable flatbreads to hold gas and produce a good expanded loaf of flatbread.

Other recipes of flatbread made from millets include millet dosa (pancake), puri, kachori, cheela, etc. Dosa is a famous South Indian dish rich in carbohydrate and protein. Generally, the main ingredients of dosa are rice and black gram pulse, but millet is also a good choice to prepare a value-added product. In recent years, a number of research have been published in support of nutritionally profuse millet-based dosa to meet the demands of healthy and tasty millet recipes. Vijayalaxmi et al. (2020) evaluated the nutritional component of kodo mix dosa and concluded that kodo millet dosa contains significantly high amount of carbohydrate (57.05 g/100 g), protein (12.22 g/100 g), and energy as well as fiber (15.40 g/100 g), calcium (48.08 mg/100 g), and iron (1.80 g/100 g) content in comparison of tradition dosa (Vijayalaxmi et al. 2020). Recently, Ayush systems of medicine in India emphasized the mixed millet-based dosa for its rich content of fiber and magnesium with gluten-free recipe (Fig. 15.4). Finger millet, sorghum millet, pearl millet, barnyard millet, foxtail millet, and little millet are the primary ingredients of mixed millet dosa with black gram, fenugreek seeds, and drumstick leaves as other major ingredients, which increase the overall nutritional index of dosa (Ayush 2020).

15.2.3 Millet Laddu

Ragi puttu, a traditional sweet dish from South India, is prepared using sorghum millet, pearl millet, foxtail millet, and finger millet, in different compositions. Ragi flour (finger millet), jaggery or sugar, almonds, and nuts are used in multistep procedure in preparation of ragi puttu as depicted in Fig. 15.5i.

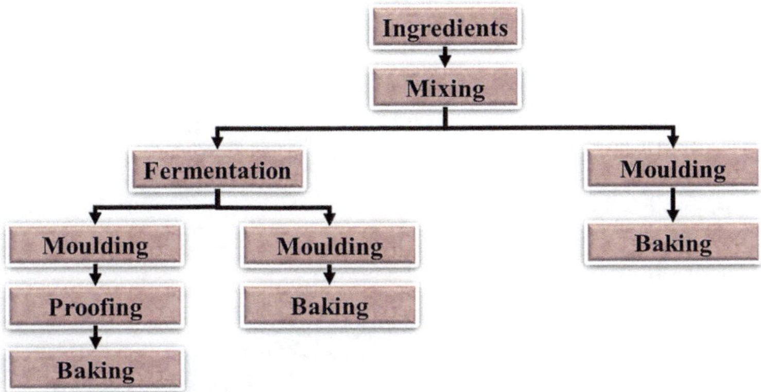

Fig. 15.3 The process involved in preparation of flatbreads

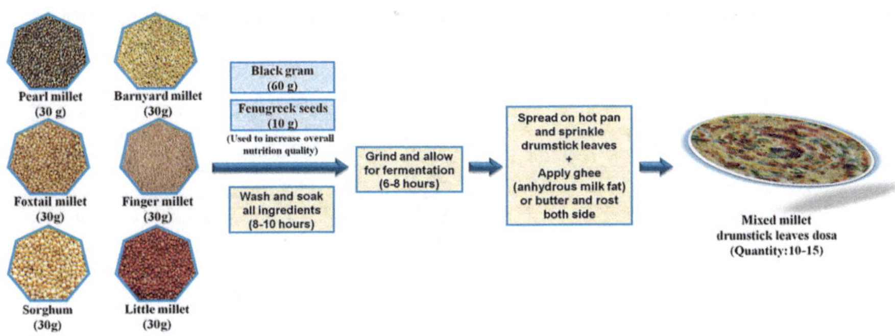

Fig. 15.4 The preparation method of mixed millets dosa

Traditional ragi puttu is recognized for having high calcium and iron content due to presence of finger millet and jaggery, which is iron rich. The nutritional value of ragi puttu has been analyzed by Tabassum et al. (2023) and represented in Fig. 15.5 ii.

Another famous preparation of sweet dish is mixed millet laddu. Pavitra et al. (2023) developed a highly recognized nutritionally dense composition of mixed millet laddu on industrial scale. Finger millet, pearl millet, and sorghum millet are the primary ingredients while ground nut, cardamom, flax seeds, poppy seeds, gingerly seeds, edible gum, desiccated coconut, jaggery, and ghee (anhydrous milk fat) are used to produce final product. This mixed millet laddu is extremely beneficial for bone health since it is a tasty recipe with high amount of calcium and energy. Composition and preparation of mixed millet laddu are shown in Fig. 15.6.

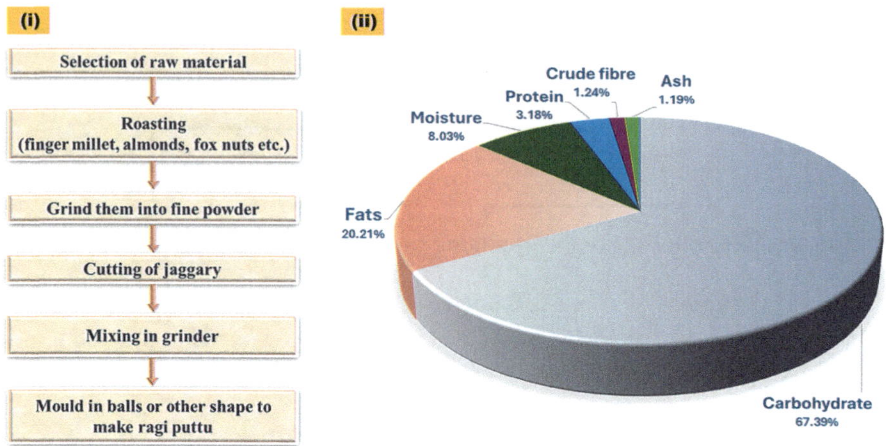

Fig. 15.5 The preparation process of ragi puttu (**i**) and nutritional profile of ragi puttu (**ii**)

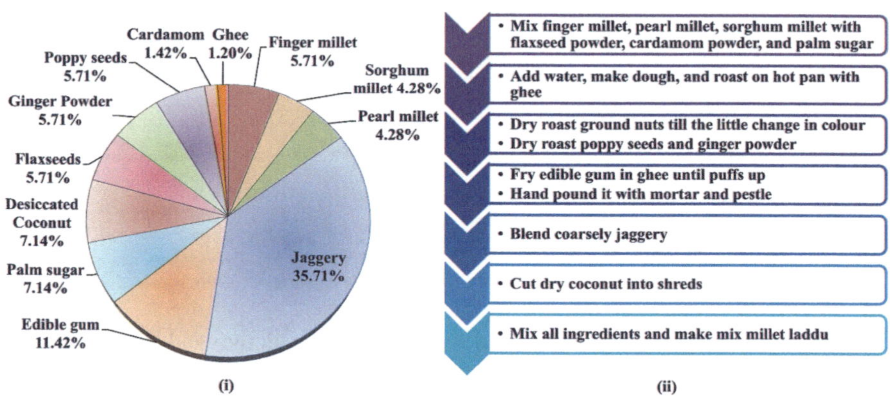

Fig. 15.6 The composition (**i**) and preparation method of mixed millet laddu (**ii**)

15.3 Fermented and Non-fermented Beverages Made from Millet

Millets are antioxidant-rich, healthy, and gluten-free grains that can be used to make plant-based beverages. Despite the traditional food, some beverages made from these nutri-cereals contribute to controlled blood pressure, blood sugar, cardiovascular disorders, help cure stomach disorder, nutritional deficiency, and are good for skin health, provides antimicrobial, anti-inflammatory, and antioxidant characteristics with the availability of alkaloids and bioactive compounds (Ragupathy et al. 2016; Tomar et al. 2025; Rao et al. 2017; Kanorwala et al. 2022). The culinary beverages made from millet in Asia and Africa signify the cultural importance of these drinks. Both non-alcoholic and alcoholic beverages are made from millets through fermentation and non-fermentation approaches. Non-fermented beverages made from millets maintain their natural composition of millets. In African continent, traditional drinks like Ogi (Onyema et al. 2021), Zoom koom (Séraphin et al. 2023), Togwa (Kitabatake et al. 2003), and Mahewa (Muyanja et al. 2010) are prepared by non-fermenting methods with unique flavors. Kunun-zaki (Jolayemi et al. 2023) in Nigeria, Malwa/Ajon (Muyanja et al. 2010) in Uganda, and Mangisi (Zvauya et al. 2024) in Zimbabwe are other non-alcoholic beverages, in which millets are used as primary constituent. Ragi Ambli (Shobana et al. 2013) is a famous non-alcoholic beverage in South India prepared via fermentation of ragi. Many other non-alcoholic drinks, geographical area, and their primary millet constituents are described in Table 15.2. Sometimes, millet beverages have low popularity among other plant beverages because of their characteristic millet smell and flavors (Adebiyi et al. 2016). Historically, fermentation and malting have been utilized for transforming millet into a variety of non-alcoholic and alcoholic beverages. Fermentation improves the digestibly, bioavailability, and probiotic properties of beverages through microbial transformation as well as taste and flavor of beverages (Budhwar et al. 2020). Mild alcoholic beverages such as Tongba (Majumder et al. 2022), Madua Apong (Shrivastava et al. 2012), Koozh (Shankar and Usha 2014), and Chhang (Bhardwaj et al. 2016) made through fermentation of millets have been commonly developed in Nepal and India. Various traditional and commercial alcoholic beverages developed from millets as primary constituent in different cuisines in Africa, Nigeria, China, India, and Nepal are described in Table 15.3.

15.4 Cultural Significance of Millet-Based Foods

Millets have contributed significantly to various cultures worldwide and play an enormous role in traditional ceremonies, rituals, preservation of culinary heritage, and contemporary dietary adaptation. The importance of millets is much more than just a grain in lots of cultures; they represent social bonds and community ties, nourishment, and tenacity of their rituals (Ankita and Seth 2025). Many researchers concur that profound association of millets with traditions and cultures has been crucial in their preservation and protection from local extinction (Rana 2022).

15.4.1 Millets in Traditional Ceremonies and Festivals

Millets have been crucial for various socio-culture fabric and contour the identity and cohesion of the community. These are used to symbolize connection, unity, sorrow, and joy in a variety of functions such as wedding ceremonies, worship rituals, and funeral rites, as well as folklore, art, and literature (Veerabadhran et al. 2023).

Table 15.2 Traditional and commercial non-alcoholic beverages made from millets as primary constituents

S. no.	Common name	Traditional/ commercial	Geographical region	Primary millet source	References
01	Kunun-zaki	Traditional	Nigeria	Sorghum, Pearl millet	Jolayemi et al. (2023)
02	Malwa/Ajon	Traditional	Uganda	Finger millet	Muyanja et al. (2010)
03	Mangisi	Traditional	Zimbabwe	Finger millet	Zvauya et al. (2024)
04	Masvusvu	Traditional	Zimbabwe	Finger millet	Zvauya et al. (2024)
04	Ogi	Traditional	West Africa	Fonio millet, Finger millet	Onyema et al. (2021)
05	Oshikundu	Traditional	Northen Namibia	Pearl millet, Sorghum millet	Embashu et al. (2013)
06	Togwa	Traditional	East Africa	Finger millet	Kitabatake et al. (2003)
07	Uji	Traditional	Kenya	Finger millet	Tomar et al. (2025)
08	Borde	Traditional	Ethiopia	Finger millet	Gebre and Emire (2024)
09	Zoom koom	Commercial	West Africa	Proso millet	Séraphin et al. (2023)
10	Ragi Ambli	Traditional	India (Karnataka, Tamil Nadu)	Finger millet	Shobana et al. (2013)
11	Bushera	Traditional	Uganda, Rwanda	Finger millet, Sorghum millet	Muyanja et al. (2003)
12	Boza	Traditional/ commercial	Turkey, Bulgaria, Central Asia	Proso millet, Maize	Pulatsu et al. (2023)
13	Mahewu	Traditional	Southern Africa (Zimbabwe, South Africa)	Pearl millet, Sorghum millet	Pswarayi and Gänzle (2019)
14	Fura da Nono, Fura da Yoghurt	Commercial	Nigeria	Pearl millet	Anyanwu (2019) and Jideani et al. (2001)
15	Koko	Traditional/ commercial	Africa	Pearl millet	Lei and Jakobsen (2004)

In some countries, at local scale, crop processing, harvesting celebrations, and other communal activity bring unity in people and strengthen the social bonds. In Indian state of Kerala, the Kuthiyottam festival is celebrated to signify a devotion to a goddess by a large community, to bring good luck and protection. This festival serves as harvest celebration associated with millets (Nair 2023). In North India, TFs prepared from millets like bajra roti, ragi dosa, and kangani ladoo have a vital role in Makar Sankranti, a harvest festival which expresses gratitude to nature, agricultural prosperity, and abundance. Another indigenous community, known as Gaddi,

Table 15.3 Traditional and commercial alcoholic beverages made from millets as primary constituents

S. no.	Common name	Traditional/ commercial	Geographical region	Primary millet source	References
01	Koozh	Traditional	South India	Finger millet, Pearl millet	Shankar and Usha (2014)
02	Madua Apong	Traditional/ commercial	North East India	Finger millets	Shrivastava et al. (2012)
03	Oti-oka	Traditional	Southwestern region of Nigeria	Pearl millet, Sorghum millet	Temitope and Taiyese (2012)
04	Shakparo Ale	Commercial	West Africa	Millet, Sorghum millet	Amadou (2019)
05	Tachapalo	Traditional	Côte d'Ivoire (Ivory Coast)	Finger millet, Sorghum millet	Aka et al. (2010)
06	Burukutu	Traditional	Northern part of Nigeria	Pearl millet	Ogbonna et al. (2016)
07	Busaa	Traditional	East Africa (Kenya, Uganda)	Finger millet, Sorghum millet	Nout (1980)
08	Tongba	Traditional	Himalayan region (Nepal, Sikkim, Darjeeling)	Brown Finger millet	Majumder et al. (2022)
09	Dolo	Traditional/ commercial	West Africa	Sorghum millet	Sawadogo-Lingani et al. (2021)
10	Chhang	Traditional	Tibet, Nepal, Bhutan	Finger millet, Barley	Bhardwaj et al. (2016)
11	Pito	Traditional	Western Africa Ghana and Nigeria	Pearl millet, Sorghum millet	Addison et al. (2024)
12	Kodo Ko Jaanr	Traditional	Eastern Himalayan regions, Nepal, Bhutan	Finger millet	Thapa and Tamang (2004)
13	Chibuku	Traditional/ commercial	Southern Africa (Zimbabwe, Zambia, Botswana)	Sorghum millet, Pearl millet	Ndondo and Kwaramba (2022)
14	Kaffir	Commercial/ traditional	Southern Africa	Sorghum millet	Jespersen (2003)
15	Doro	Traditional	Zimbabwe, Botswana	Sorghum millet	Jespersen (2003)

celebrates their festival for Lord Shiva, in which prasad is made from millet as their traditional practice. Additionally, pearl millet based during the harvest festival of Gangaur, celebrated in Rajasthan, are used by local people (Ankita and Seth 2025). Chepang community in Nepal offers food made from barnyard millets to a god of their community in Rishi Panchmi festival (Ragupathy et al. 2016). Similarly, The Yimchunger tribe celebrates the harvesting of millets with Metumniu festival in Nagaland (DIPR). Other harvest festivals in India such as Onam in Kerla, Gangaur in Rajsthan, Lohri in Panjab, Pongal in Tamil Nadu, etc., involve millet-based foods (Kumar 2023). The landraces of small millets are protected by the tribes of Kolli

Hill; their traits indicates the existence of new alleles for upcoming breeding initiatives (Venkatesan et al. 2015). Millets are also used in wedding rituals like offering millets as a gift to the groom, which is considered as good blessings in some states of India. In Madhya Pradesh, a paste made from millets mixed with turmeric is applied on the skin of the bride and groom in wedding rituals. Despite this, millets recipes are acceptable on fasting days in many religious ceremonies and festivals.

15.4.2 Cultural Heritage of Millets in Literature and Folklore

The cultural heritage of millets is described in various arts, literature, and folklore significantly. There, references in terms of grass (currently known as millets) are also mentioned in Vedas, and the great epics in India with their medicinal and nutritional value. Anu is a reference to proso millet in the Brhadaranyaka Upanishad. Rig Veda and Atharva Veda describe the various applications of balbaja (finger millet), and Sayamaka (barnyard millet) in fuel, baskets, handmade goods, and gift products, while many other millets are mentioned in Yajurveda and Ayurveda with the name shyaamaka (black finger millet), priyangava (foxtail millet), aanaya (barnyard millet), kodrava (kodo millet), Nartaki (finger millet), varaka (proso millet), and yavanaala (sorghum) (Kumar 2023). Sweets made from millets are also used in Terta Yuga to celebrate the return of Lord Ram to Ayodhya, and Sage Kanva poured (foxtail millet) at the time of bidding farewell to Sakuntala in Dushyant court was mentioned in literature Abhigyanshakuntalam written by Mahakavi Kalidasa. Despites these, Tamil literature in India such as Purananuru, Tholkappiyam, and Perumpatruppadai also interprets the role of finger millet, sorghum millet, foxtail millet, and kodo millet in food security. Sorghum is also mentioned as the great millet, and kodo millet is famoused for sweets (Veerabadhran et al. 2023). In a literature Ainkurunuru, a prayer song have these lines—"*ThaenumThinaiMaavum*" which means honey with foxtail millet flour (Senthamil et al. 2023). Another poem "Ragi thandheera" (Kannad language) in India demonstrated the importance of the ragi (finger millet) (Bhat et al. 2018). Fernao Nuniz, a Portuguese traveler, found sorghum as a main food grain consumed in Vijayanagara, India, in sixteenth century AD (Arora 2021). In Panjab, Rajasthan, Himachal Pradesh, Tamil Nadu, and Meghalaya, many folk songs express gratitude toward millets grains, and farming, such as famous folk song "Bajra Da Sitta" in Panjab, "Ngi is nom teia u krai" in Meghalaya, "Mulapari" with kummi dance, and other traditional dances (Senthamil et al. 2023).

15.5 Challenges

Traditional millet foods and their cultivation confront several challenges despite having various cultural importance, nutritional, and health benefits. Major part of millets, 97% are grown and consumed in underdeveloped nations by resource-poor farmers. Low market access and low demand of millets after cultivation decrease income and discourage the farmers from millet cultivation. Main limiting factors linked to decreased production and cultivation area of millets are a lack of agricultural aids and improved verities, adverse climate, and policy support (Meena et al. 2021). However, export and import of millets increased gently in twenty-first century but these milestones are less significant. By building infrastructure, well-developed marketing system, and offering logistical assistance, millet farmers may expand their market access and negotiate better returns. To overcome these challenges, it is essential to include millets into regular diets and agriculture practice. Although millets have highly nutritional index and health importance, lack of awareness in people is the main cause for its low popularity in health-conscious foods. To solve these issues, there is a need for national campaigns and educational initiatives to educate and increase awareness about millet-based foods. The use of millet in local cuisines and traditional knowledge of its cultivation methods help to promote millet foods. In addition, the incorporation of millet farming into everyday food distribution system facilitated by government incentives and assistance for millet cultivation promotes millet production. Food and Agriculture Organization (FAO) and India, has declared year 2018 the national year of millets (NYM) to encourage the production of millets and their nutritional as well as medicinal role. Further, to provide a steady market for the millets and their traditional food with numerous health benefits, the government of India purposed a revolutionary campaign globally, which had been supported by more than 70 countries, and UNGC proclaimed the year 2023 as IYM in March 2021 (Panigrahy and Padhi 2023). These initiatives for food and nutritional security through intensive millets export promotion program support a global market for millets with numerous recipes and value-added products in different cuisines. For their promotion and increased consumption, other major millet-growing nations need to implement similar national and worldwide multidisciplinary activities. On the whole, policy support, campaign for national and international initiative in concomitance with crop development, and improvement programs, market construction, and promoting knowledge about nutritional value will be beneficial to increase the cultivation and cultural significance of millets (Harish et al. 2024; Muskan et al. 2025).

15.6 Conclusion

Shri Dhanya or millets including proso millet, kodo millet, sorghum millet, and little millet have been a staple traditional and historical food significantly cultivated in Asia and Africa due to their cultural significance, nutritional and physiological advantages. However, the cultivation and consumption of traditional millet foods is less popular due to the prevalency of wheat, rice, and other cereals. Despite obstacles and challenges such as low consumer knowledge and awareness, weak supply chain and international market demand, complicated processing, and less availability of recipes, consumption of millets gains emphasis due to the high nutritional profile, gluten-free characteristic, climatic resistance, and huge potential of food security with availability of medically beneficial bioactive compounds. Additionally, market demand of millets recipes in different cuisines have been greatly developed by introduction of numerous value-added products at local as well as industrial level. Also, to guarantee their worldwide acceptance and promotion, future studies should concentrate on advancement in processing methods to boosting bioavailability, nutrition, and current demand of taste of flavor of various recipes of millets with incorporation into contemporary food systems. By leveraging scientific discoveries, legislative reforms, and innovative product development, millets might significantly contribute to rural economic empowerment, sustainable nutrition, and global food security. Hence, millet based products can be used as a substitute to rice and wheat-based food products.

References

Addison NN, Agbala MA, Abagale S, Sackey I (2024) Preliminary chemical evaluation of nutritional perspective of Pito, an indigenous alcoholic beverage brewed in the Upper East region of Ghana. Global J Pure Appl Chem Res 12(1):22–31. https://doi.org/10.37745/gjpacr.2023/vol12n12231

Adebiyi JA, Obadina AO, Mulaba-Bafubiandi AF et al (2016) Effect of fermentation and malting on the microstructure and selected physicochemical properties of pearl millet (*Pennisetum glaucum*) flour and biscuit. J Cereal Sci 70:132–139. https://doi.org/10.1016/j.jcs.2016.05.026

Aka S, Djeni N, N'guessan K et al (2010) Variabilité des propriétésphysico-chimiques et dénombrement de la florefermentaire du tchapalo, unebièretraditionnelle de sorghoen Côte d'Ivoire. Afr Sci Rev Int Sci Technol 4(2). https://doi.org/10.4314/afsci.v4i2.61688

Amadou I (2019) Millet based fermented beverages processing. In: Grumezescu AM, Holban AM (eds) Fermented beverages, vol 5. Elsevier, pp 433–472. https://doi.org/10.1016/B978-0-12-815271-3.00011-7

Amadou I, Gounga ME, Le GW (2013) Millets, nutritional composition, some health benefits and processing. J Agric Food Chem 25(7):501–508. https://doi.org/10.9755/ejfa.v25i7.12045

Ambati K, Sucharitha KV (2019) Millets—review on nutritional profiles and health benefits. Int J Recent Sci Res 10(7):33943–33948. https://doi.org/10.24327/ijrsr.2019.1007.3786

Ankita, Seth U (2025) Millets in India: exploring historical significance, cultural heritage and ethnic foods. J Ethn Food 12:2. https://doi.org/10.1186/s42779-024-00262-2

Anyanwu NCJ (2019) Microbiological and comparative analysis of indigenous and semi-industrial fermented milk drinks (Fura da Nono and Fura da Yoghurt) sold in Nigeria's capital. Int J Bioassays 8(1):5716–5723. https://doi.org/10.14303/ijbio.2019.8.1.5

Arora N (2021) Indians: a brief history of a civilization. Penguin Random House India Private Limited, pp 1–304

Ayushman Bharat Traditional Food Recipes from Ayush Systems of Medicines (2020) Delhi Ministry of Ayush (Ayush), Government of India, pp 1–44

Bhardwaj KN, Jain KK, Kumar S, Kuhad RC (2016) Microbiological analyses of traditional alcoholic beverage (Chhang) and its starter (Balma) prepared by Bhotiya Tribe of Uttarakhand, India. Indian J Microbiol 56(1):28–34. https://doi.org/10.1007/s12088-015-0560-6

Bhat BV, Rao BD, Vilas A (2018) The story of millets. Karnataka State Department of Agriculture and ICAR-Indian Institute of Millet Research. https://www.millets.res.in/pub/2018/The_Story_of_Millets.pdf. Accessed 11 Apr 2025

Budhwar S, Sethi K, Chakraborty M (2020) Efficacy of germination and probiotic fermentation on underutilized cereal and millet grains. Food Prod Process Nutr 2(1):12. https://doi.org/10.1186/s43014-020-00026-w

Cena H, Calder PC (2020) Defining a healthy diet: evidence for the role of contemporary dietary patterns in health and disease. J Nutr 12(2):334. https://doi.org/10.3390/nu12020334

Embashu W, Cheikhyoussef A, Kahaka GK, Lendelvo SM (2013) Processing methods of Oshikundu, a traditional beverage from sub-tribes within Aawambo culture in the Northern Namibia. J Stud Humanit Soc Sci 2:117–127

Festivals of Nagaland. Department of Information & Public Relations (DIPR), Nagaland. https://ipr.nagaland.gov.in/node/13707. Accessed 8 Apr 2025

Gebre TS, Emire SA (2024) Characterization of "borde": a traditional ethiopian cereal-based fermented beverage—microbial analysis, rheological profiling, and functional attributes. J Agric Food Res 18:101498. https://doi.org/10.1016/j.jafr.2024.101498

Ghosh S, Meyer-Rochow VB, Jung C (2023) Embracing tradition: the vital role of traditional foods in achieving nutrition security. J Food Res 12(23):4220. https://doi.org/10.3390/foods12234220

Gunenc A, Yeung MH, Lavergne C et al (2017) Enhancements of antioxidant activity and mineral solubility of germinated wrinkled lentils during fermentation in kefir. J Funct Foods 32:72–79. https://doi.org/10.1016/j.jff.2017.02.016

Harish MS, Bhuker A, Chauhan BS (2024) Millet production, challenges, and opportunities in the Asia-Pacific region: a comprehensive review. Front Sustain Food Syst 8:1386469. https://doi.org/10.3389/fsufs.2024.1386469

Jackson JC, Weatherspoon L, Nnyepi M et al (2013) Sorghum bean composite porridge nutritional quality and acceptability. Nutr Food Sci 43(5):453–461. https://doi.org/10.1108/NFS-03-2012-0024

Jespersen L (2003) Occurrence and taxonomic characteristics of strains of predominant in African indigenous fermented foods and beverages. FEMS Yeast Res 3(2):191–200. https://doi.org/10.1016/S1567-1356(02)00185-X

Jideani VA, Nkama I, Agbo EB, Jideani IA (2001) Survey of fura production in some northern states of Nigeria. Plant Foods Hum Nutr 56:23–36. https://doi.org/10.1023/A:1008116118051

Jolayemi OS, Ojo A, Ayo-Omogie HN, Enujiugha VN (2023) Multifactorial characterization of nutritional and quality profiles of organically-sweetened Kunun-zaki beverage. Meas Food 11:100106. https://doi.org/10.1016/j.meafoo.2023.100106

Jung SJ, Park SH, Choi EK et al (2014) Beneficial effects of Korean traditional diets in hypertensive and type 2 diabetic patients. J Med Food 17(1):161–171. https://doi.org/10.1089/jmf.2013.3042

Kanorwala B, Wadhawan N, Shekhawat P, Jain R (2022) Significance of finger millet consumption on nutritional status and management of diabetes mellitus: a review. Pharma Innov J 11(12S):977–979

Karuppasamy P, Malathi D (2014) Evaluation of quality characteristics of porridge from kodo and little millet. MAJ 101:201–205. https://doi.org/10.29321/MAJ.10.001182

Kim SH, Kim MS, Lee MS et al (2016) Korean diet: characteristics and historical background. J Ethn Foods 3(1):26–31. https://doi.org/10.1016/j.jef.2016.03.002

Kitabatake N, Gimbi DM, Oi Y (2003) Traditional non-alcoholic beverage, Togwa, in East Africa, produced from maize flour and germinated finger millet. Int J Food Sci Nutr 54(6):447–455. https://doi.org/10.1080/09637480120092053

Kumar S (2023) Millets an ingredient of celebrating harvest festivals and religion in Bharat. Akshara Multidiscip Res J 4(8):13–17

Kumar P, Kaur C, Rudra SG, Arora B (2023) Extruded finger millet improves rheology and quality of composite maize flatbread. Int J Gastron Food Sci 31(3):100638. https://doi.org/10.1016/j.ijgfs.2022.100638

Lei V, Jakobsen M (2004) Microbiological characterization and probiotic potential of koko and koko sour water, African spontaneously fermented millet porridge and drink. J Appl Microbiol 96(2):384–397. https://doi.org/10.1046/j.1365-2672.2004.02162.x

Majumder S, Chakraborty S, Ghosh A, Bhattacharya M (2022) The Himalayan ethnic beverage tongba with therapeutic properties in high-altitude illnesses and metabolomic similarities to Japanese sake. Acta Univ Sapientiae 15(1):67–83. https://doi.org/10.2478/ausal-2022-0006

Meena RP, Joshi D, Bisht JK, Kant L (2021) Global scenario of millets cultivation. In: Kumar A, Tripathi MK, Joshi D, Kumar V (eds) Millets and millet technology. Springer Singapore, pp 33–50. https://doi.org/10.1007/978-981-16-0676-2_2

Meena KK, Meena S, Joshi M, Dhotre AV (2024) Nutritional and functional exploration of pearl millet and its processing and utilization: an overview. Food Hum 3:100334. https://doi.org/10.1016/j.foohum.2024.100334

Mir SA, Naik HR, Shah MA et al (2014) Indian flat breads: a review. Food Nutr Sci 5(06):549–561. https://doi.org/10.4236/fns.2014.56065

Mundassery A, Ramaswamy J, Natarajan T et al (2024) Modern and conventional processing technologies and their impact on the quality of different millets. Food Sci Biotechnol 33:2441–2460. https://doi.org/10.1007/s10068-024-01579-z

Muskan F, Redhu M, Redhu S et al (2025) Millets in the global market: a critical review of challenges and opportunities. Food Prod Process Nutr 7:37. https://doi.org/10.1186/s43014-025-00316-1

Muyanja C, Kikafunda J, Narvhus JA et al (2003) Production methods and composition of Bushera: a Ugandan traditional fermented cereal beverage. Afr J Food Agric Nutr Dev 3(1):10–19. https://doi.org/10.4314/ajfand.v3i1.19108

Muyanja C, Birungi S, Ahimbisibwe M et al (2010) Traditional processing, microbial and physicochemical changes during fermentation of malwa. Afr J Food Agric Nutr Dev 10(10). https://doi.org/10.4314/ajfand.v10i10.62891

Nair AK (2023) Exploring the conjunction of childhood and rituals: an analytical study of kuthiyottam based on select temples in Kerala. Shodh Kosh J Vis Per Arts 4(1). https://doi.org/10.29121/shodhkosh.v4.i1.2023.1857

Ndondo JK, Kwaramba P (2022) The evolution of chibuku beer. IOSR J Humanit Social Sci 27(11):28–36

Neelayathatchi P, Neasri TM, Sanwal CS (2024) Ayurvedic perspective and clinical significance of millets. J Ayurveda Integr Med Sci 8(12):161–174. https://doi.org/10.21760/jaims.8.12.25

Nout MJR (1980) Process development and preservation of Busaa, a Kenyan traditional opaque maize beer. Chem Mikrobiol Technol Lebensm 11:51–55

Ogbonna A, Abuajah C, Umanah I (2016) Burukutu: healthy and superior indigenous African traditional opaque beverage. Am J Adv Food Sci Technol. https://doi.org/10.7726/ajafst.2016.1004

Onyema IEC, Ezeokoli OT, Sulyok M et al (2021) Metataxonomic analysis of bacterial communities and mycotoxin reduction during processing of three millet varieties into ogi, a fermented cereal beverage. Food Res Int 143:110241. https://doi.org/10.1016/j.foodres.2021.110241

Panigrahy AK, Padhi M (2023) The international year of millets 2023: a multi-dimensional opportunity for India. World J Bio Pharm Health Sci 15(2):122–127. https://doi.org/10.30574/wjbphs.2023.15.2.0331

Pavitra K, Usha Devi C, Navaneetha R (2023) Development of millet ladoo mix. Int J Sci Res 12(4):882–886. https://doi.org/10.21275/SR21622230210

Pswarayi F, Gänzle MG (2019) Composition and origin of the fermentation microbiota of Mahewu, a Zimbabwean fermented cereal beverage. Appl Environ Microbiol 85(11):e03130–e03118. https://doi.org/10.1128/AEM.03130-18

Pulatsu E, Malik S, Lin M et al (2023) Preparation and characterization of Boza enriched with non-fat dry milk and its impact on the fermentation process. Gels 10(1):22. https://doi.org/10.3390/gels10010022

Ragupathy S, Dhivya S, Patel K et al (2016) DNA record of some traditional small millet landraces in India and Nepal. 3 Biotech 6(2):133. https://doi.org/10.1007/s13205-016-0450-6

Rana NK (2022) The perpetual quest for 'authenticity' in Indian cuisine: an answer through history and folklore. Digest J Foodways Cult 9(1):Spring

Rao DB, Bhaskarachary K, Christina A et al (2017) Nutritional and health benefits of millets. ICAR-Indian Institute of Millets Research (IIMR) Rajendranagar, Hyderabad, p 2

Renzetti S, Aisala H, Ngadze RT et al (2023) Bread products from blends of african climate resilient crops: baking quality, sensory profile and consumers' perception. Foods 12(4):689. https://doi.org/10.3390/foods12040689

Saini S, Saxena S, Samtiya M et al (2021) Potential of underutilized millets as nutri-cereal: an overview. J Food Sci Technol 58(12):4465–4477. https://doi.org/10.1007/s13197-021-04985-x

Saleem S, Mushtaq NU, Shah WH et al (2023) Millets as smart future food with essential phytonutrients for promoting health. J Food Compos Anal 124:105669. https://doi.org/10.1016/j.jfca.2023.105669

Sawadogo-Lingani H, Owusu-Kwarteng J, Glover R et al (2021) Sustainable production of african traditional beers with focus on dolo, a West African sorghum-based alcoholic beverage. Front Sustain Food Syst 5:672410. https://doi.org/10.3389/fsufs.2021.672410

Seetha A, Mahalingam G, Kane-Potaka J (2019) Balanced amino acid and higher micronutrients in millets complements legumes for improved human dietary nutrition. Cereal Chem 97(1):74–84. https://doi.org/10.1002/cche.10227

Selladurai M, Pulivarthi MK, Raj AS et al (2023) Considerations for gluten free foods—pearl and finger millet processing and market demand. Grain Oil Sci Technol 6(2):59–70. https://doi.org/10.1016/j.gaost.2022.11.003

Senthamil E, Tamilmounika R, Kowsalya V (2023) Cultural heritage of millets. Farm Chronicle Agric Newslett 02:77–81

Séraphin WA, Gnamien TS, Judicaël AB et al (2023) Physicochemical and bacteriological characterization of an artisanal drink: the case of "zoom-koom" sold in the streets of Port-Bouët, Abidjan, Côte d'Ivoire. Int J Biosci 22(2):102–110. https://doi.org/10.12692/ijb/22.2.102-110

Shahidi F, Chandrasekara A (2013) Millet grain phenolics and their role in disease risk reduction and health promotion: a review. J Funct Foods 5(2):570–581. https://doi.org/10.1016/j.jff.2013.02.004

Shankar I, Usha A (2014) Assessment of the microbiological quality of koozh, a fermented millet beverage. Afr J Microbiol Res 8(3):308–312. https://doi.org/10.5897/AJMR2013.6482

Shobana S, Krishnaswamy K, Sudha V et al (2013) Finger millet (Ragi, *Eleusine coracana* L.): a review of its nutritional properties, processing, and plausible health benefits. Adv Food Nutr Res 69:1–39. https://doi.org/10.1016/B978-0-12-410540-9.00001-6

Shrivastava K, Greeshma AG, Srivastava B (2012) Biotechnology in tradition—a process technology of alcoholic beverages practiced by different tribes of Arunachal Pradesh, North East India. Indian J Tradit Knowl 11(1):81–89

Siroha AK, Bangar SP (2024) Millet-based food products: an overview. Curr Food Sci Tech Rep 2(2):213–220. https://doi.org/10.1007/s43555-024-00031-x

Sivarajasekar N, Nikitha D, Harsni V, Avantheka (2023) Millet seeds-production technology, harvesting and post harvesting harvest management, processing, and storage. In: Sharma S, Jain P (eds) Millets magical crops, vol 12. KD Publications, pp 236–250

Tabassum S, Bukya A, Harshitha NM (2023) Preparation and quality characterisation of ragi laddu. Int J Res Publ Rev 4(8):2115–2119. https://doi.org/10.55248/gengpi.4.823.51323

Taylor JRN, Duodu KG (2018) Traditional sorghum and millet food and beverage products and their technologies. In: Taylor JRN, Duodu KG (eds) Sorghum and millets: chemistry, technology and nutritional attributes, 2nd edn. Elsevier. https://doi.org/10.1016/B978-0-12-811527-5.00009-5

Temitope OS, Taiyese OB (2012) Quality assessment of 'oti-oka' like beverage produced from pearl millet. J Appl Biosci 51:3608–3617

Thapa S, Tamang JP (2004) Product characterization of kodo ko jaanr: fermented finger millet beverage of the Himalayas. Food Microbiol 21(5):617–622. https://doi.org/10.1016/j.fm.2004.01.004

Tomar T, Sachdeva A, Dutta J et al (2025) Fermentation dynamics of millet beverages: microbial interactions, nutritional enhancements, and health implications. Food Chem: X 25:102199. https://doi.org/10.1016/j.fochx.2025.102199

Unicef (2019) For every child, reimagine. UNICEF annual report

Veerabadhran V, Palakuru HR, Pratap GP et al (2023) A critical review of the history and practices of millets in India. J Drug Res Ayurvedic Sci 8(Suppl 1):S18–S25. https://doi.org/10.4103/jdras.jdras_113_23

Venkatesan P, Sundaramari M, Venkattakumar R (2015) Ethno taxonomical classification of little millet (*Panicum sumatrance*) by the tribal people in Tamilnadu, India. Eco Environ Cons 21(Dec Suppl):215–222

Vijayalaxmi KG, Baddi J, Umarji VK (2020) A study on development of instant kodo dosa mix and evaluation of its nutritional composition and shelf life. Int J Curr Microbiol App Sci 9(12):286–295. https://doi.org/10.20546/ijcmas.2020.912.038

Zhang J, Li P, Liu J et al (2022) Nutritional components of millet porridge cooked by different electric cookers based on principal component and cluster analyses. Foods 11(18):2823. https://doi.org/10.3390/foods11182823

Zvauya R, Mygochi T, Parawira W (2024) Microbial and biochemical changes occurring during production of Masvusvu and Mangisi, traditional Zimbabwean beverages. Plant Foods Hum Nutr 1997 51(1):43–51. https://doi.org/10.1023/A:1007972428849

Chapter 16
Role of Nutri-cereals to Control Obesity

Lata Arya, Shivani, and Ravendra Kumar

16.1 Introduction

16.1.1 Obesity: A Growing Health Concern

Obesity, characterized by an excessive accumulation of adipose tissue, is typically measured using body mass index (BMI), with values of 30 kg/m^2 or above indicating obesity (WHO 2020). The global prevalence of obesity has reached alarming levels in recent decades, with an estimated 650 million adults worldwide classified as obese (Global Health Observatory, WHO 2020). This increase has been particularly pronounced in high-income countries, but rising rates are also seen in low- and middle-income nations. The increasing prevalence of obesity is a consequence of multiple factors, including urbanization, sedentary lifestyles, unhealthy diets, and environmental influences (Ahmed and Mohammed 2025). Obesity is now recognized as a major preventable contributor to mortality worldwide, playing a significant role in the development of numerous health conditions such as heart disease, type 2 diabetes, and some forms of cancer (Gupta et al. 2023; WHO 2020). The growing burden of obesity on public health systems necessitates urgent action to address this issue.

L. Arya
Department of Home Science, S. N. Pant Govt. Degree College,
Bageshwar, Uttarakhand, India

Shivani
Department of Home Science, S. V. Govt. P.G. College, Champawat, Uttarakhand, India

R. Kumar (✉)
Department of Chemistry, College of Basic Sciences and Humanities, G.B. Pant University of Agriculture and Technology, Udham Singh Nagar, Uttarakhand, India

© The Author(s), under exclusive license to Springer Nature Switzerland AG 2026

J. Singh, M. Kumar Malik (eds.), *Millets and Millet Starch: Chemical Composition, Functional Modification, and Applications*,
https://doi.org/10.1007/978-3-032-09502-2_16

16.1.2 Health Risks Associated with Obesity

Obesity is recognized as a complex, multifactorial condition that significantly elevates the risk for a spectrum of serious health disorders, thereby adversely affecting both overall quality of life and life expectancy. Among the most critical health concerns associated with excess adiposity is the heightened risk of cardiovascular diseases (CVD), which encompasses coronary artery disease, hypertension, and cerebrovascular events such as stroke. The pathophysiological mechanisms underlying this association involve the accumulation of adipose tissue, which promotes a pro-inflammatory state and contributes to endothelial dysfunction and atherogenesis, ultimately increasing the likelihood of atherosclerotic plaque formation and vascular injury (Poirier et al. 2006). Beyond its association with cardiovascular disorders, excessive body weight particularly when classified as obesity has been extensively documented as a key predisposing factor in the onset and progression of type 2 diabetes mellitus. This connection arises from the metabolic disturbances caused by excess adiposity, including insulin resistance, chronic low-grade inflammation, and impaired glucose regulation, all of which significantly elevate the risk of developing this chronic endocrine condition. The presence of excess body fat, particularly visceral adiposity, impairs insulin signaling pathways and reduces tissue sensitivity to insulin, resulting in insulin resistance and subsequent disturbances in glucose metabolism (Kahn et al. 2006). Moreover, epidemiological and mechanistic studies have demonstrated that obesity is associated with an increased incidence of several malignancies, including but not limited to cancers of the breast, colon, and liver. This association is thought to be mediated by metabolic alterations linked to obesity, such as chronic hyperinsulinemia, elevated levels of insulin-like growth factors, and persistent low-grade inflammation, all of which can contribute to carcinogenesis (Luo et al. 2016). In addition to its well-known associations with serious health threats like cardiovascular disease and diabetes, obesity has been linked to a wide array of other medical and psychological conditions. These include respiratory disorders such as obstructive sleep apnea, metabolic dysfunctions like nonalcoholic fatty liver disease (NAFLD), and musculoskeletal issues including osteoarthritis due to increased mechanical stress on joints. Furthermore, a growing body of evidence highlights the strong correlation between excess body weight and mental health challenges, particularly mood disorders such as depression and anxiety, suggesting that obesity exerts both physiological and psychological burdens on affected individuals. Collectively, these obesity-related health challenges not only diminish individual well-being and functional capacity but also place a considerable financial strain on healthcare systems globally due to increased demand for medical care and management of chronic diseases.

16.1.3 Current Approaches to Obesity Management

Current approaches to obesity management are multifaceted, typically integrating lifestyle modification, pharmacological therapy, and, in select cases, surgical intervention. The foundation of obesity treatment continues to be comprehensive lifestyle changes, with a primary emphasis on dietary adjustments and increased physical activity. Dietary recommendations focus on achieving a caloric deficit while maintaining nutritional adequacy, often through the regular consumption of nutritionally balanced meals that emphasize on whole grains, fruits, vegetables, and lean sources of protein (Donnelly et al. 2009). Regular physical activity is equally critical as it not only facilitates energy expenditure and fat loss but also supports the preservation of lean body mass and enhances overall metabolic health. Despite the proven benefits of these lifestyle interventions, sustained adherence can be challenging, and long-term weight maintenance is often elusive for many individuals (Wing and Phelan 2005).

For those who do not achieve sufficient results through lifestyle modification alone, medical interventions may be considered. Prescription weight loss medications are available as adjuncts to diet and exercise, particularly for individuals with more severe obesity or obesity-related comorbidities. However, the use of pharmacotherapy is often limited by potential side effects and is generally reserved for patients with a higher-risk profile (Apovian 2016). For individuals with severe or treatment-resistant obesity, surgical intervention may be considered a medically appropriate option. Bariatric procedures most notably gastric bypass and sleeve gastrectomy have been shown to be effective in achieving significant and long-term reductions in body weight. In addition to promoting weight loss, these surgeries often lead to marked improvements in obesity-associated comorbidities, including type 2 diabetes, hypertension, and dyslipidemia, thereby contributing to enhanced overall health outcomes and quality of life. However, it is important to recognize that bariatric surgery is not without its inherent risks, as with any major surgical procedure. Potential complications can include infections, nutritional deficiencies, and long-term gastrointestinal issues. As a result, such interventions are generally reserved for patients with a body mass index (BMI) greater than 40 kg/m^2 or for those with a BMI over 35 kg/m^2 who also suffer from serious obesity-related health conditions, such as type 2 diabetes, cardiovascular disease, or obstructive sleep apnea. Clinical guidelines emphasize that candidates must undergo thorough medical evaluation and demonstrate a commitment to long-term lifestyle changes to ensure the procedure's success (Maggard et al. 2005). Overall, the selection of an appropriate obesity management strategy should be individualized, considering the severity of obesity, associated health risks, patient preferences, and the potential benefits and limitations of each intervention.

16.1.4 Potential of Millets in Obesity Control

Diet plays a central role in the development, progression, and clinical management of obesity, a condition characterized by excessive accumulation of adipose tissue and associated metabolic dysfunction. Emerging nutritional strategies increasingly emphasize the consumption of whole, minimally processed plant-based foods rich in bioactive compounds, fiber, and low-glycemic carbohydrates to mitigate obesity and its related comorbidities. Within this context, millets have garnered substantial scientific and public health interest as promising dietary components for obesity management, owing to their unique nutritional profile and long-standing role as staple grains in traditional diets. Millets are a diverse group of small-seeded cereal crops cultivated primarily in Asia and sub-Saharan Africa, historically valued for their resilience to harsh growing conditions and their significant contribution to food security (Sadh et al. 2024). Nutritionally, millets are abundant in complex carbohydrates, resistant starches, and both soluble and insoluble dietary fiber, which contribute to slower digestion and prolonged satiety mechanisms essential for energy intake regulation and body weight control.

In addition to their high fiber content, millets are notable for their considerable levels of plant-based protein, essential amino acids such as methionine and cysteine (often limited in other cereals), and a broad spectrum of micronutrients including magnesium, iron, zinc, and B-complex vitamins. One of the most clinically relevant features of millets is their low glycemic index (GI), which translates to reduced postprandial blood glucose and insulin levels, thereby improving insulin sensitivity and glycemic control over time. This attribute is particularly important for individuals with obesity and those at risk for type 2 diabetes mellitus. Furthermore, millets contain diverse phytochemicals, including phenolic compounds and flavonoids, which possess anti-inflammatory, antioxidant, and lipid-lowering properties factors increasingly recognized as key in the pathophysiology of obesity (Dhiman et al. 2025).

Several recent studies have underscored the potential of millet-based diets in reducing markers of adiposity and metabolic dysregulation. Experimental and clinical trials have reported favorable outcomes such as reductions in body mass index (BMI), waist circumference, serum lipid profiles, and fasting blood glucose following the incorporation of millet into daily meals (Popkin et al. 2012). Additionally, the high satiety index of millets may suppress appetite and reduce total energy intake, while their fermentable fibers contribute to gut health through the modulation of microbiota composition and short-chain fatty acid production. These physiological effects collectively suggest that millets can serve as functional grains in dietary interventions targeting obesity and metabolic syndrome. As millets are naturally gluten-free and ecologically sustainable, they also offer broader benefits for individuals with gluten sensitivities and for food systems seeking climate-resilient crops. Given these multifaceted advantages, millets represent a valuable addition to contemporary nutritional strategies designed to combat obesity and promote metabolic health (Manmohan et al. 2025).

The properties of these certain millets are so effective to the human body that it can treat these illnesses within roughly 2 years of usage (Rambabu 2018). Millets help with homeostasis or balance within the body by adjusting the energy or glucose levels. Most types of carbohydrates such as rice, bread, etc., tend to give you a burst of energy in the introduction and then you need to eat later again to refuel. These carbohydrates tend to use up all the energy as soon as the body metabolizes them which can be harmful to the body and cause imbalances such as storing the leftover unused glucose in fat or causes high sugar levels (Rao et al. 2017). These properties make millets an ideal choice for weight regulation as they can help to control hunger, support blood glucose regulation through reduced glycemic response, while increasing satiety via delayed gastric emptying and improved appetite control.

The elevated dietary fiber content inherent in millets plays a pivotal role in facilitating weight management by promoting early and sustained satiety, suppressing overall appetite, and contributing to improved glycemic regulation. These effects are largely mediated through delayed gastric emptying, reduced postprandial glucose absorption, and enhanced secretion of satiety-related gut hormones such as peptide YY and glucagon-like peptide-1 (GLP-1). As a result, the inclusion of millets in the diet may help lower overall caloric intake and stabilize blood glucose levels, factors that are particularly advantageous in the prevention and management of obesity and metabolic disorders (Sridevi et al. 2020). Furthermore, millets are characterized by a low glycemic index (GI), indicating that their carbohydrate content is digested and absorbed at a slower rate, resulting in a more gradual and sustained increase in postprandial blood glucose levels. This moderated glycemic response plays a crucial role in preventing rapid insulin surges, which are often implicated in promoting lipogenesis and fat storage. By minimizing glycemic fluctuations and improving insulin sensitivity, the regular consumption of low-GI foods such as millets may contribute to enhanced metabolic control and serve as an effective dietary strategy in the prevention and management of obesity and related metabolic disorders (Rao et al. 2019). The incorporation of millets into the daily diet, particularly within the framework of a balanced and nutrient-dense dietary pattern, has demonstrated beneficial effects on body weight regulation and overall metabolic health across both preclinical and clinical investigations. In animal studies, millet-based interventions have been associated with reduced adiposity, improved lipid profiles, enhanced insulin sensitivity, and lower markers of systemic inflammation. Parallel findings in human trials have shown that regular consumption of millet-containing meals can contribute to significant reductions in body mass index (BMI), waist circumference, fasting blood glucose, and serum cholesterol levels. These outcomes are attributed to the synergistic effects of millets' high dietary fiber, low glycemic index, and rich phytochemical content, which collectively influence appetite control, glycemic regulation, and lipid metabolism. Consequently, millets are increasingly recognized as functional grains with the potential to support weight loss and mitigate the risk of noncommunicable diseases such as obesity, type 2 diabetes, and cardiovascular disorders (Pandey et al. 2024). Additionally, millets are gluten-free, making them a suitable option for individuals with celiac disease or gluten intolerance. These characteristics make millets a promising food group for addressing the growing obesity epidemic. The Fig. 16.1 shows the compositional profile of two types of millets (Foxtail, and Finger) with its nutritional profile.

16.1.5 Importance of a Holistic Approach to Obesity Management

Effective management of obesity extends beyond dietary interventions, such as the incorporation of millets, and necessitates a comprehensive, multifaceted approach that addresses the diverse determinants of health. Sustainable obesity treatment requires simultaneous attention to nutritional intake, physical activity, psychological well-being, and behavioral modifications. Psychological support is particularly critical as individuals with obesity often encounter emotional and cognitive challenges related to eating behaviors and body image, which can undermine adherence to therapeutic regimens (Lillis et al. 2008). Additionally, environmental and socioeconomic factors significantly influence an individual's capacity to adopt and maintain healthier lifestyle choices. Access to nutritious foods, availability of safe and accessible spaces for exercise, and social support systems are pivotal in facilitating behavioral change. Policy-level interventions, including the regulation of unhealthy food marketing, implementation of nutrition education, and promotion of physical activity within educational institutions, serve as essential components of a broader public health strategy aimed at obesity prevention and management (Swinburn et al. 2019). Therefore, a personalized and integrative approach that considers biological, psychological, social, and environmental factors is imperative to effectively address the complex and multifactorial etiology of obesity and to achieve long-term health outcomes. In Fig. 16.2 the human figures are presenting a progression from slimmer to a larger body size.

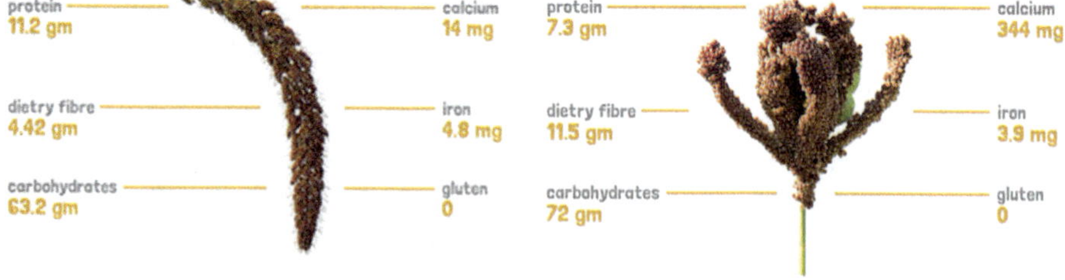

Fig. 16.1 Compositional profile of foxtail millet (left) and finger millet (right)

Fig. 16.2 Representation of fat gain in humans

16.1.6 Comparison of Millets with Other Whole Grains in Obesity Management

Millets offer unique advantages in obesity management when compared to other whole grains such as wheat, rice, and oats. While other whole grains are also rich in fiber and essential nutrients, millets tend to have a higher fiber content and lower glycemic index, which may provide superior benefits for weight management (Slavin 2013). For example, while oats are a well-known whole grain that is often recommended for weight control, they have a higher glycemic index compared to millets, which means they may not offer the same level of blood sugar regulation (Slavin 2013). Additionally, millets are more nutrient-dense, providing a greater concentration of minerals such as iron, magnesium, and phosphorus, which contribute to overall health and wellness (Sadh et al. 2017). Incorporating a variety of whole grains, including millets, into the diet may enhance the diversity of nutrients consumed, offering a more comprehensive approach to managing obesity and its associated health risks.

16.2 Nutri-cereals

Millets represent a diverse assemblage of small-seeded cereal crops that are taxonomically distributed across multiple genera within the grass family *Poaceae*, particularly within the subfamilies Panicoideae and Chloridoideae. This agronomically significant group includes both major and minor species that vary in botanical classification, ecological adaptability, and nutritional composition. Among the major millets, species such as *Sorghum bicolor* (sorghum), *Pennisetum glaucum* (pearl millet), and *Eleusine coracana* (finger millet) are widely cultivated for their high biomass yield and resilience to abiotic stress. The minor millets, which have gained renewed interest for their nutritional density and climate resilience, encompass *Setaria italica* (foxtail millet), *Paspalum scrobiculatum* (kodo millet), *Echinochloa frumentacea* (barnyard millet), *Panicum miliare* (little millet), and *Panicum miliaceum* (proso millet). In addition to true millets, certain pseudo-cereals such as *Amaranthus spp.* (amaranth) and *Fagopyrum esculentum* (buckwheat) are often grouped alongside them due to comparable nutritional profiles, including high-quality protein, dietary fiber, and micronutrient content (IIMR 2025). Together, these grains offer valuable potential for sustainable agriculture and improved nutritional security, especially in regions vulnerable to climatic variability and micronutrient deficiencies. Recognizing their superior nutrient density, characterized by higher levels of protein, dietary fiber, and essential minerals such as calcium, iron, zinc, and magnesium, low glycemic carbohydrates, and an abundance of phenolic antioxidants, the Government of India officially redesignated these grains as "nutri-cereals" in April 2018 to dispel the perception of inferiority attached to the former label "coarse cereals" and to mainstream their use in public nutrition program. India presently accounts for roughly 41% of global millet output, producing 11–12 million ton annually and cultivating the largest acreage worldwide, while Niger, China, Nigeria, and Mali jointly with India contribute about three-quarters of total world production. Despite this dominance in area and volume, India's mean productivity (≈1.24 t ha^{-1}) is modest relative to technologically advanced systems, situating the country around 31st globally in yield rankings, a gap attributed to limited input use, labor constraints, and varietal adoption. Agronomically, millets exhibit exceptional resilience thriving on marginal soils with minimal water and agrochemical inputs, and environmentally, they feature low carbon footprints, rendering them climate-smart alternatives capable of enhancing food and nutritional security, diversifying cereal-based diets, and mitigating micronutrient deficiencies in both developing and industrialized regions.

16.3 Nutrient Composition of Millet

Millets possess a rich nutrient profile, making them a vital component of diets in many developing regions and an emerging functional food in global markets. The nutrient composition of millets varies by species. However, millets are broadly characterized by their high concentrations of macronutrients including complex carbohydrates, dietary fiber, quality plant-based proteins, and lipids as well as a diverse array of micronutrients, such as essential vitamins (notably B complex) and minerals like iron, calcium, magnesium, and zinc. In addition to these foundational nutrients, millets also contain significant amounts of bioactive phytochemicals, including phenolic acids, flavonoids, tannins, phytosterols, and tocopherols, which are increasingly recognized for their antioxidant, anti-inflammatory, antidiabetic, and lipid-lowering properties. The synergistic effect of these nutritional and functional components contributes to millets' reputation as health-promoting grains, capable of supporting metabolic health, enhancing immune function, and reducing the risk of noncommunicable diseases such as obesity, cardiovascular disease, and type 2 diabetes. Their nutrient density and functional bioactives position millets as valuable candidates in the development of functional foods and nutraceuticals targeted at improving public health outcomes (Bhattacharya 2023).

Carbohydrates are the predominant macronutrient in millets, contributing approximately 60–75% of their dry weight. These carbohydrates are mainly present as starch, with a considerable proportion in the form of resistant starch and non-starch polysaccharides, which contribute to their low glycemic index and promote gut health (Gowda et al. 2022). Finger millet (*Eleusine coracana*) has been reported to have a high amylose-to-amylopectin ratio, which slows digestion and provides sustained energy release. In addition, millets contain a significant amount of dietary fiber, typically ranging between 7% and 12%, with both soluble and insoluble fractions. The fiber content aids in digestion, regulates blood glucose levels, and reduces the risk of cardiovascular diseases (Devi et al. 2014).

The protein content of millets is relatively high, varying between 7% and 14% depending on the species. The protein quality is influenced by the amino acid composition, with prolamins and globulins being the major storage proteins. Millets are notable not only for their complex carbohydrate and micronutrient content but also for their relatively high levels of essential amino acids, particularly methionine and cysteine, two sulfur-containing amino acids that are frequently limited in the protein profiles of conventional staple cereals such as rice and wheat. These amino acids play critical roles in numerous physiological processes, including protein synthesis, methylation reactions, antioxidant defense (via glutathione synthesis), and detoxification pathways. The presence of methionine and cysteine in appreciable quantities enhances the biological value of millet protein, making it a superior alternative in plant-based diets where amino acid completeness is often a nutritional concern (Anal et al. 2024). This attribute further reinforces the suitability of millets in addressing protein-energy malnutrition, especially in resource-limited settings where dietary diversity may be constrained. However, lysine remains a limiting amino acid in millet proteins. Nonetheless, their protein quality can be improved through complementary food formulations with legumes or processing techniques such as fermentation and germination, which enhance amino acid bioavailability.

In addition to their high fiber and complex carbohydrate content, millets also serve as a valuable source of dietary lipids, typically ranging between 2% and 6%, with a lipid profile that is notably rich in unsaturated fatty acids, including linoleic acid (omega-6) and oleic acid (omega-9). These fatty acids are recognized for their roles in maintaining cardiovascular health, modulating serum lipid profiles, and exerting anti-inflammatory effects (Singh et al. 2022). Beyond their fatty acid composition, the lipid fraction in millets contains a variety of bioactive compounds, such as phytosterols and tocopherols (vitamin E isoforms), which possess antioxidant properties and may contribute to the prevention of oxidative stress-related chronic diseases. Millets are also excellent sources of micronutrients, particularly iron, zinc, magnesium, and phosphorus, which are essential for numerous biochemical and physiological functions. Among these, finger millet (*Eleusine coracana*) is especially notable for its remarkably high calcium content, measured at approximately 344 mg per 100 g, making it a functional food for supporting bone mineralization and reducing the risk of osteoporosis, particularly in vulnerable populations such as postmenopausal women and older adults (Devi et al. 2014). However, the bioavailability of these mineral nutrients can be compromised by the presence of antinutritional factors such as phytates, tannins, and oxalates, which chelate minerals and hinder their absorption in the gastrointestinal tract. Nevertheless, a variety of traditional and modern processing techniques including soaking, fermentation, germination, and sprouting have been shown to significantly reduce these antinutrients, thereby enhancing mineral bio-accessibility and improving the overall nutritional quality of millet-based foods (Rao et al. 2017). These attributes collectively reinforce the nutritional and therapeutic relevance of millets in functional food systems aimed at promoting metabolic and skeletal health.

In addition to providing a rich supply of macronutrients and essential micronutrients, millets are a valuable source of bioactive phytochemicals, including phenolic acids, flavonoids, lignans, tannins, and phytosterols, which contribute significantly to their functional food properties. These compounds exhibit potent antioxidant, anti-inflammatory, antimicrobial, and antidiabetic activities, which have been increasingly documented in both in vitro and in vivo studies (Balyatanda et al. 2024). Their ability to scavenge free radicals, modulate inflammatory pathways, and improve glycemic regulation positions millets as promising candidates for nutraceutical development and the dietary prevention of noncommunicable diseases such as type 2 diabetes, cardiovascular conditions, and certain forms of cancer. Furthermore, the nutrient density, functional versatility, and stress-resilient nature of millets make them a strategic crop for enhancing food and nutrition security, particularly in low-resource settings where communities are vulnerable to climate variability, water scarcity, and soil nutrient depletion. Their adaptability to arid and semiarid environments, combined with low input requirements and resistance to pests and diseases, reinforces their value not only as a dietary solution to undernutrition but also as a sustainable agricultural resource capable of supporting resilient food systems in the face of global environmental challenges. Table 16.1 illustrates the macronutrients level and fiber compositions of selected grain millets.

Table 16.1 Macronutrient and fiber composition of selected whole grain millets

Millet type	Carbohydrate level	Protein level	Fat level	Mineral residue	Fiber level
Pearl millet	Moderate (60–80%)	Moderate (8–20%)	Low (1–7%)	Low (3–4%)	Low–moderate (1–11%)
Finger millet	Very high (80–90%)	Low–moderate (7–11%)	Low (1–2%)	Moderate (3–6%)	Moderate (3–4%)
Foxtail millet	Moderate (60–75%)	Moderate (11–13%)	Moderate (3–4%)	Moderate (3–3.5%)	Moderate (4–8%)
Little millet	Moderate (70–80%)	Moderate (10–14%)	Moderate (4%)	Moderate (3–4%)	Moderate (4–9%)
Kodo millet	Moderate (65–73%)	Low–moderate (6–13%)	Moderate (3–5%)	Moderate (3–4%)	High (8–11%)
Barnyard millet	Low (50–65%)	Moderate (11–13%)	Low–moderate (2–6%)	Moderate (4–5%)	Very high (14%)
Proso millet	Moderate (66–79%)	Moderate (11–15%)	Low (2–4%)	Low–moderate (2–4%)	Low–moderate (2–9%)

Adapted from Chauhan et al. (2018), Vali Pasha et al. (2018), Nithiyanantham et al. (2019), Embashu and Nantanga (2019), Jayawardana et al. (2019) and Serna-Saldivar and Espinosa-Ramírez (2019)

16.4 Mineral Composition of Millet

Millets are increasingly acknowledged for their diverse and nutritionally dense mineral profiles, which distinguish them from conventional staple cereals and support their inclusion in micronutrient-rich diets, especially in resource-limited settings. These grains contain notably higher concentrations of essential minerals including iron, calcium, magnesium, zinc, and phosphorus when compared to commonly consumed cereals such as rice and wheat, positioning them as key contributors to addressing widespread micronutrient deficiencies in populations with limited dietary diversity. The mineral composition of millets is influenced by a combination of genotypic variation, soil fertility, and agroecological conditions, thereby offering opportunities for biofortification and selective breeding to enhance their nutritional value (Chandra et al. 2016; Serna-Saldivar and Espinosa-Ramírez 2019). Among the microminerals, potassium and phosphorus are abundantly present across all millet species. Finger millet (*Eleusine coracana*) stands out for its exceptionally high calcium content (300–350 mg/100 g) nearly tenfold higher than that of wheat, making it especially beneficial for bone health and osteoporosis prevention, particularly where dairy intake is low. It also provides significant levels of magnesium and phosphorus, vital for cellular energy metabolism and neuromuscular functions. Pearl millet (*Pennisetum glaucum*), another nutritionally prominent variety, is particularly rich in iron (5–6.5 mg/100 g) and zinc, supporting hematological and immune functions, alongside appreciable quantities of magnesium, copper, and manganese. Additionally, minor millets such as kodo millet (*Paspalum scrobiculatum*) and little millet (*Panicum miliare*) are notable for their high concentrations of trace elements, with kodo millet showing elevated levels of iron and little millet exhibiting the highest zinc content among the group. Despite natural variability due to environmental and varietal influences, millets consistently provide nutritionally relevant amounts of calcium, iron, zinc, magnesium, potassium, phosphorus, copper, and manganese, which are essential for maintaining skeletal integrity, enzymatic function, oxygen transport, and immune defense mechanisms. As a result, millets hold considerable promise as dietary interventions in combatting hidden hunger and micronutrient malnutrition, especially in regions impacted by soil degradation, climate stress, and food insecurity (Kulthe et al. 2016; Vali Pasha et al. 2018; Serna-Saldivar and Espinosa-Ramírez 2019).

16.5 Vitamins Profile

Millets, comprising a diverse group of small-seeded cereal crops primarily cultivated in arid and semiarid agroecological zones, have attracted growing scientific interest due to their remarkable adaptability and rich nutritional profile. These grains are particularly noted for their high concentrations of essential micronutrients and vitamins, including iron, calcium, zinc, magnesium, and B-complex vitamins, which are critical for physiological functions such as hematopoiesis, immune response, and enzymatic activity. This nutrient density, combined with their natural resilience to harsh climatic conditions, positions millets as a strategic food resource for enhancing dietary quality in resource-constrained and nutritionally vulnerable populations, where conventional staples often fall short in micronutrient content. Furthermore, their potential to address hidden hunger and diet-related chronic diseases, alongside their agronomic suitability for low-input farming systems, underscores their role in promoting nutrition-sensitive agriculture and long-term food and nutritional security in the face of global environmental and public health challenges. Among the suite of vitamins present in millets, the B-complex group is notably abundant, except for vitamin B12, a cobalamin that is characteristically absent in plant-derived foods and is primarily sourced from animal-based products, thereby necessitating dietary diversification for populations reliant on vegetarian diets to meet their B12 requirements (Jacob et al. 2024). Detailed compositional analyses have demonstrated that millets are an excellent source of B-complex vitamins, particularly thiamine (vitamin B1) and riboflavin (vitamin B2), which play critical roles in energy metabolism, neural function, and cellular enzymatic activity. Quantitative assessments commonly report thiamine concentrations in millets ranging from 0.15 to 0.60 mg per 100 grams, while riboflavin levels typically fall between 0.09 and 0.28 mg per 100 grams. These values frequently exceed those observed in conventional staple cereals such as rice and wheat, highlighting the comparative nutritional advantage of millets in meeting micronutrient needs. Among the different millet varieties, pearl millet (*Pennisetum glaucum*), one of the most widely consumed species, contains thiamine and riboflavin at approximately 0.25 mg/100 g and 0.20 mg/100 g, respectively. Other millets, such as finger millet (*Eleusine coracana*), foxtail millet (*Setaria italica*), and proso millet (*Panicum miliaceum*), exhibit comparable or even higher levels of these vitamins, reflecting interspecies variability and emphasizing the broader nutritional potential of millets in supporting vitamin sufficiency. Given the essential role of B vitamins in preventing deficiencies such as beriberi and ariboflavinosis, and their involvement in numerous metabolic pathways, the regular consumption of millet-based foods may offer a practical and sustainable means of improving micronutrient intake, particularly in regions where vitamin deficiencies are prevalent and dietary diversification is limited (Jacob et al. 2024). The B vitamins in millets are predominantly localized in the bran and germ layers, which implies that processing techniques such as decortication, commonly employed to enhance edibility, may inadvertently reduce the vitamin content, thereby highlighting the importance of whole grain consumption for optimal nutrient retention. In addition to being rich sources of thiamine (vitamin B1) and riboflavin (vitamin B2), millets also contribute appreciable quantities of other B-complex vitamins, including niacin (vitamin B3), pantothenic acid (vitamin B5), pyridoxine (vitamin B6), and folic acid (vitamin B9). These micronutrients are essential for a wide range of biological processes, such as DNA synthesis, red blood cell formation, neurotransmitter function, and energy metabolism. The concentration and distribution of these vitamins vary across millet species, with certain varieties particularly foxtail millet (*Setaria italica*) and proso millet (*Panicum miliaceum*) demonstrating comparatively higher levels, thus enhancing their overall micronutrient profile. This variability offers opportunities for both targeted dietary planning and crop improvement programs. The presence of these additional B vitamins further underscores millets' role as nutritionally dense grains capable of supporting cognitive health, metabolic regulation, and fetal development, particularly in populations with limited access to diversified diets. As such, the broader spectrum of B-complex vitamins found in millets reinforces their utility in combatting micronutrient malnutrition and promoting nutritional adequacy through sustainable and culturally adaptable food systems. Notably, the presence of carotenoids, specifically β-carotene, has been documented in pearl millet, with reported values around 5.4 mg/kg, contributing provitamin A activity and enhancing the antioxidant capacity of the grain, which is a crucial attribute for combating micronutrient deficiencies and oxidative stress in vulnerable populations (McDonough et al. 2000). The comprehensive micronutrient profile of millets extends beyond vitamins, encompassing significant levels of minerals such as iron, calcium, zinc, and phosphorus, as well as dietary fiber and an array of phytochemicals, including phenolic acids, flavonoids, and tannins, all of which confer additional health-promoting properties such as improved glycemic control, enhanced satiety, and reduced risk of chronic diseases. The unique combination of high micronutrient density, resilience to environmental stresses, and adaptability to marginal agroecological conditions renders millets not only a "smart food" in the context of sustainable agriculture and public health nutrition but also an indispensable dietary staple for economically disadvantaged populations, earning them the moniker of the "poor man's best food" due to their affordability, accessibility, and capacity to address multiple dimensions of food and nutrition security in both developing and developed regions. The mineral composition of millets is listed in Table 16.2.

Table 16.2 Minerals composition of millets (Jacob et al. 2024)

	Minerals/micronutrients (g/kg)				
Millet variety	Calcium	Phosphorus	Potassium	Sodium	Magnesium
Pearl millet	0.28–0.44	2.35–3.85	4.05–4.65	0.09–0.13	1.28–1.40
Finger millet	1.05–3.30	3.00–5.80	4.15–10.80	0.13–0.69	1.40–3.65
Foxtail millet	0.18–0.33	2.80–7.00	3.70–8.95	0.04–0.66	1.39–2.95
Little millet	0.16–0.25	2.30–7.10	1.22–4.80	0.08–0.70	2.25–3.58
Kodo millet	0.21–0.36	1.85–4.60	1.52–6.30	0.60–0.66	2.16–2.98
Barnyard millet	0.21–0.24	2.65–6.35	7.12–7.85	0.67–0.71	2.50–3.11
Proso millet	0.14–0.21	2.15–5.45	2.05–5.20	0.58–0.61	1.89–2.90

16.6 Healthy Benefits

Millets are nutrient-dense grains that have garnered significant scientific attention due to their high content of essential vitamins (such as B-complex vitamins), minerals (including iron, magnesium, phosphorus, and potassium), complex carbohydrates, and dietary fiber, rendering them an important class of food commonly referred to as "nutri-cereals" (Devi et al. 2011; Saleh et al. 2013). These nutritional attributes make millets highly beneficial for addressing food insecurity and malnutrition, particularly in underdeveloped regions, while also contributing to the management of noncommunicable diseases prevalent in urban populations with sedentary and unhealthy lifestyles (Kumar et al. 2018). Furthermore, millets are rich in bioactive compounds such as polyphenols, flavonoids, and other antioxidants that play a pivotal role in neutralizing free radicals, thereby reducing oxidative stress and potentially lowering the risk of chronic conditions such as cancer and cardiovascular diseases. Additionally, the presence of unsaturated fatty acids enhances their medicinal value by promoting cardiovascular health. A notable functional property of millets is their low glycemic index compared to conventional cereals, which contributes to better glycemic control and is particularly advantageous for individuals aiming to manage obesity and prevent type 2 diabetes (Shobana et al. 2013). Moreover, by aiding in the detoxification process and reducing the accumulation of toxic metabolites, the regular inclusion of millets in the diet supports overall metabolic well-being.

Obesity develops when an individual consistently consumes more energy, derived from dietary fats, carbohydrates, sugars, and proteins, than is expended through metabolic processes and physical activity. The surplus energy is converted and stored as triglycerides within adipose tissue, leading to an increase in body fat mass. Therefore, dietary regulation is a primary strategy for the prevention and management of obesity, as controlling caloric intake directly influences energy balance. Obesity is clinically defined using the body mass index (BMI), a metric calculated from an individual's weight and height, and sometimes adjusted for age. According to the classification system established by the National Health Service (NHS), body mass index (BMI) serves as a widely accepted anthropometric indicator for assessing weight status in adults. A BMI ranging from 25.0 to 29.9 kg/m^2 is categorized as overweight, indicating an increased risk of developing weight-related health conditions. Values between 30.0 and 39.9 kg/m^2 fall within the obese category, reflecting a significantly elevated risk for metabolic and cardiovascular disorders, including type 2 diabetes, hypertension, and coronary artery disease. This BMI-based classification system forms a fundamental component of clinical and public health strategies aimed at identifying individuals at risk and implementing appropriate lifestyle or medical interventions. It is important to note that BMI values and the prevalence of obesity can vary significantly across populations due to differences in dietary habits, cultural practices, and environmental factors. Thus, effective obesity management requires a multifaceted approach that considers individual lifestyle, food culture, and environmental influences in addition to quantitative dietary control (Lee et al. 2000; Safaei et al. 2021).

Excessive accumulation of body fat is a well-documented contributor to a wide range of cardiovascular conditions, highlighting the urgent need for effective dietary approaches to manage and prevent obesity. In this regard, millets have gained attention as a valuable nutritional tool, offering health benefits due to their rich content of dietary fiber, essential minerals, and low glycemic index carbohydrates. These properties support better blood sugar regulation, appetite control, and lipid balance, positioning millets as a practical dietary option for mitigating obesity-related cardiovascular risk. A principal attribute of millets is their low glycemic index (GI), which translates to a gradual and sustained elevation of postprandial blood glucose levels, thereby minimizing rapid glycemic excursions that are typically implicated in the pathogenesis of obesity, insulin resistance, and type 2 diabetes mellitus (Jenipher et al. 2024). The hypoglycemic potential of millets has been recognized for several years and is primarily attributed to the structural characteristics of millet starch, including the presence of relatively large starch granules that present a reduced surface area for enzymatic hydrolysis, as well as a less porous starch matrix that impedes the penetration of digestive enzymes into the inner regions of the granule, collectively resulting in a slower rate of starch degradation and glucose release into the bloodstream. This attenuated glycemic response is particularly advantageous for individuals with impaired glucose tolerance or diabetes as it facilitates improved glycemic control and may contribute to the amelioration of hyperglycemia. Moreover, the lower glucose content inherent in millets reduces the total energy available for immediate metabolic use, thereby diminishing the likelihood of surplus caloric accumulation and subsequent adiposity. Millets are also characterized by a substantial dietary fiber content, ranging from 20% to 30%, which exerts multiple anti-obesogenic effects: The fiber induces a sensation of satiety and gastric distension, thereby curbing excessive food intake and promoting adherence to a lower-calorie diet (Venkatesh et al. 2005). Notably, dietary fiber is resistant to enzymatic digestion in the human gastrointestinal tract, but it can be fermented by specific gut microbiota, yielding short-chain fatty acids and other metabolites that further modulate energy balance and metabolic health. In terms of macronutrient composition, millets possess carbohydrate and fat contents comparable to those of conventional cereals, yet they are distinguished by a higher proportion of slowly digestible starch (SDS), which ensures a prolonged and steady release of glucose, thereby sustaining energy supply over an extended period and reducing the propensity for fat deposition. The naturally low-fat content of millets adds to their appeal as a staple food in diets aimed at preventing and managing obesity. In addition to being low in lipids, millets are also distinguished by their comparatively high protein content, with an average value of approximately 11.6%, which exceeds that of many conventional cereal grains such as rice and maize. This elevated protein level contributes to enhanced satiety, preservation of lean body mass, and improved metabolic outcomes, further supporting the role of millets in weight regulation and overall nutritional adequacy (Jenipher et al. 2024). Millet proteins are characterized by a well-balanced amino acid profile, containing key essential amino acids such as lysine, threonine, methionine, and cysteine, with a composition that closely resembles that of maize. Among the various species, pearl millet stands out for its particularly favorable protein quality, making it a valuable source of plant-based protein. This amino acid richness plays a vital role in supporting muscle maintenance, tissue repair, and lean mass development, all of which are essential for sustaining a healthy body composition and basal metabolic rate. Furthermore, the digestibility of millet proteins is relatively high compared to other cereals such as rice and corn. In vitro protein digestibility values for millets typically range between 72% and 75%, indicating efficient absorption and utilization. These values can be significantly improved through processing techniques such as germination, which enhances enzymatic activity, reduces antinutritional factors, and increases protein bioavailability further reinforcing the nutritional efficacy of millets in plant-based and protein-conscious diets. Collectively, these attributes of low glycemic index, high dietary fiber content, slow-release carbohydrates, low fat, and superior protein quality underscore the multifaceted role of millets in obesity control and cardiovascular risk reduction, positioning them as a strategic component of dietary regimens aimed at promoting metabolic health and preventing the sequelae of excessive adiposity (Venkatesh et al. 2005).

Millets are notable for containing a diverse array of anthocyanins and related polyphenolic compounds, including condensed tannins, which play a multifaceted role in human nutrition and health. These condensed tannins, a subclass of proanthocyanidins, are recognized for their potential to modulate the bioavailability of macronutrients and micronutrients within the gastrointestinal tract. Specifically, the presence of these bioactive compounds in millet grains has been associated with enhanced absorption of proteins, beneficial lipids, and essential minerals. This improvement in nutrient bioavailability is thought to occur through the modulation of digestive enzyme activity and the alteration of nutrient-binding properties in the gut environment. Furthermore, the regular consumption of millets has been shown to promote the proliferation of beneficial gut microbiota, which not only facilitates more efficient digestion and nutrient assimilation but also contributes to the regulation of metabolic processes, thereby reducing the risk of excessive nutrient accumulation and associated metabolic disorders. Collectively, the anthocyanins and condensed tannins in millets contribute to their functional food properties by supporting improved nutrient utilization and maintaining gastrointestinal health (Jenipher et al. 2024).

16.7 Case Studies on Role of Nutri-cereals to Control Obesity

A synthesis of findings from eight independent studies reveals an average 7.0% reduction in body mass index (BMI) among participants initially classified as overweight or obese, with mean BMI decreasing from 28.5 ± 2.4 kg/m^2 to 26.7 ± 1.8 kg/m^2, indicating the potential for individuals to approach or attain the clinically defined normal BMI threshold (<25 kg/m^2). Among these studies, six were randomized controlled trials (RCTs) with intervention durations ranging from 3 to 4 months, while two were cross-sectional investigations conducted over 1–2 years. The RCTs demonstrated no detectable heterogeneity ($I^2 = 0.0\%$), likely due to the limited number of trials, modest sample sizes, and the geographic clustering of the research, which was predominantly conducted in Asian populations, especially in India. Despite the demographic and methodological similarities, the consistency in BMI reduction across RCTs strengthens the reliability of the findings. Additionally, the standardized mean difference (SMD) for both fixed-effect and random-effect models was −0.18, reflecting a uniform effect size and supporting the reproducibility of the intervention's efficacy across different statistical frameworks (Jalilzadeh and Goharinezhad 2025).

The cumulative findings from multiple studies provide statistically significant support, under both fixed-effect and random-effect meta-analysis models, for the efficacy of millet-based dietary interventions in reducing overweight and obesity, particularly among individuals with a body mass index (BMI) exceeding 30 kg/m^2. This dual-model statistical significance lends robustness to the observed outcomes and reduces the likelihood of model-dependent bias, indicating consistency across different assumptions of study heterogeneity. The inclusion of millet-derived foods, particularly those from low-glycemic index varieties such as foxtail millet, appears to exert favorable effects on body weight and fat mass, as evidenced in several controlled interventions. In a 12-week open-label, self-controlled clinical trial conducted by Ren et al. (2018), participants who incorporated foxtail millet into their regular diet exhibited a statistically significant reduction in BMI, alongside a measurable decrease in body fat mass, which declined from 22.1 ± 7.1 kg to 21.1 ± 7.2 kg over the intervention period (Lee and Lee 2021). Importantly, this decline in adiposity was not accompanied by a significant loss in skeletal muscle mass, suggesting that the weight reduction was primarily attributable to fat loss rather than generalized or nonspecific weight depletion. This selective targeting of fat mass has important implications for body composition optimization as preservation of lean mass is essential for sustaining basal metabolic rate, insulin sensitivity, and long-term metabolic health.

Further reinforcing these findings, Surekha et al. (2013) reported that consumption of millet-based foods was associated with enhanced satiety, which likely contributes to reduced caloric intake and better appetite regulation. These satiating effects may be explained by the high dietary fiber content, slow digestibility, and prebiotic activity of millets, all of which influence gastric emptying and gut hormone secretion. In support of this, Ren et al. (2018) also observed a significant rise in circulating leptin levels, from 8.3 ± 6.4 to 9.6 ± 7.0 ng/mL, following the same 12-week millet intervention. Leptin, an adipocyte-derived hormone, plays a central role in energy homeostasis by acting on the hypothalamus to suppress appetite and enhance energy expenditure. The observed increase in leptin suggests that millet intake may positively modulate neuroendocrine pathways involved in hunger signaling and fat metabolism, potentially creating a physiological environment conducive to sustained weight loss and prevention of weight regain.

Collectively, these findings emphasize that millet-based diets offer multifactorial benefits for obesity management, acting through both physiological mechanisms (such as fat mass reduction and muscle mass preservation) and hormonal pathways (including enhanced leptin-mediated satiety). The low glycemic response, favorable macronutrient composition, and bioactive compounds in millets likely synergize to support these outcomes. Consequently, millets represent a promising functional food component within holistic dietary strategies aimed at reducing the prevalence, severity, and long-term risks associated with obesity and related metabolic disorders (Bhattacharya 2023).

16.8 Conclusion

In conclusion, the inclusion of millets in routine dietary regimens represents a scientifically validated and contextually practical approach for addressing the global burden of obesity and enhancing broader health outcomes. Millets, as ancient cereal grains, are nutritionally superior staples, abundant in dietary fiber; essential micronutrients such as magnesium, iron, zinc, and phosphorus; and B-complex vitamins that collectively contribute to their classification as functional foods. A defining characteristic of millets is their low glycemic index (GI), which facilitates a slower and more controlled digestion of carbohydrates, resulting in attenuated postprandial glucose excursions and reduced insulin surges, factors directly linked to diminished lipogenesis and improved insulin sensitivity. Their substantial fiber content aids in prolonging gastric emptying and modulating gut hormone release, including glucagon-like peptide-1 (GLP-1) and peptide YY, which play central roles in appetite suppression and satiety enhancement. These mechanisms jointly support lower energy intake, improved glycemic regulation, and a reduced propensity for weight gain, thereby positioning millets as valuable components in dietary strategies targeting both the prevention and management of obesity and its associated metabolic disturbances. In addition to their metabolic benefits, millets' gluten-free nature, adaptability to diverse agroecological conditions, and suitability in various culinary traditions further reinforce their relevance in contemporary nutrition. Their consistent consumption has been associated with reduced risk factors for noncommunicable diseases, including type 2 diabetes, cardiovascular disorders, and metabolic syndrome, especially in populations undergoing rapid nutritional transitions. Despite the promising data, there is a growing need for mechanistic and longitudinal research to better elucidate the pathways through which millets modulate energy balance, glucose metabolism, and lipid homeostasis. Moreover, further investigation is required to optimize their integration into culturally specific diets, exploring preparation techniques, bioavailability of nutrients, and synergistic effects with other dietary components. Ultimately, leveraging the health-promoting properties of millets within balanced dietary frameworks offers an evidence-based, sustainable, and culturally adaptable strategy for curbing obesity and improving long-term health across diverse populations.

References

Ahmed SK, Mohammed RA (2025) Metabolic effects of diet-based therapies in obesity. Metab Open 27:100375. https://doi.org/10.1016/j.metop.2025.100375

Anal AK, Singh R, Rice D et al (2024) Millets as sustainable food grains: nutritional, processing and therapeutic perspectives. Sustain Food Technol 2(4):908–925. https://doi.org/10.1039/D4FB00047A

Apovian CM (2016) The clinical and economic consequences of obesity. Obes Res Clin Pract 10(2):100–107. https://doi.org/10.1210/jc.2014-3415

Balyatanda BB, Gowda NAN, Subbiah J et al (2024) Sustainable strategies for millet value chains: integrating processing, nutrition, and food security. Foods 13(22):3684. https://doi.org/10.3390/foods13223684

Bhattacharya S (2023) Recent advances in millet research for nutrition and health security. Front Nutr 10:1190111. https://doi.org/10.3389/fnut.2023.1190111

Chandra D, Chandra S, Sharma AK (2016) Review of finger millet (*Eleusine coracana* (L.) Gaertn): a power house of health benefiting nutrients. Food Sci Human Wellness 5(3):149–155. https://doi.org/10.1016/j.fshw.2016.05.004

Chauhan M, Sonawane SK, Arya SS (2018) Nutritional and nutraceutical properties of millets: a review. Clin J Nutr Diet 1(1):1–10

Devi KN, Singh MS, Singh NG et al (2011) Effect of integrated nutrient management on growth and yield of wheat (*Triticum aestivum* L.). J Crop Weed 7(2):23–27

Devi PB, Vijayabharathi R, Sathyabama S et al (2014) Health benefits of finger millet (*Eleusine coracana* L.) polyphenols and dietary fiber: a review. J Food Sci Technol 51(6):1021–1040. https://doi.org/10.1007/s13197-011-0584-9

Dhiman S, Kumar K, Jan T et al (2025) Millets as climate-smart crops: nutritional composition and health promoting phytochemicals. J Appl Biol Biotechnol 13(3):4–14. https://doi.org/10.7324/JABB.2025.190911

Donnelly JE, Blair SN, Jakicic JM (2009) Appropriate physical activity intervention strategies for weight loss and prevention of weight regain for adults. Am J Clin Nutr 89(4):1130–1135. https://doi.org/10.1249/MSS.0b013e3181949333

Embashu W, Nantanga KK (2019) Malts: Quality and phenolic content of pearl millet and sorghum varieties for brewing nonalcoholic beverages and opaque beers. Cereal Chem 96(4):765–774. https://doi.org/10.1002/cche.10178

Gowda NAN, Siliveru K, Prasad PVV et al (2022) Processing strategies for enhancing shelf-life and nutrition of millets. Foods 11(4):499. https://doi.org/10.3390/foods11040499

Gupta M, Asfaha DM, Ponnaiah G (2023) An overview of millet research in improving food security and nutrition. Cureus 15:e47769. https://doi.org/10.7759/cureus.47769

Indian Institute of Millets Research (IIMR) (2025) Official repository on millet research and innovation. Retrieved July 31, 2025, from https://millets.res.in

Jacob J, Krishnan V, Antony C et al (2024) Millets: the miracle grains in ancient medicine and modern nutrition. Front Nutr 11:1346869. https://doi.org/10.3389/fnut.2024.1346869

Jalilzadeh M, Goharinezhad S (2025) Natural environmental determinants of childhood obesity: a systematic review and meta-analysis. Front Public Health 13:1540756. https://doi.org/10.3389/fpubh.2025.1540756

Jayawardana SAS, Samarasekera JKRR, Hettiarachchi, GHCM, et al (2019) Dietary fibers, starch fractions and nutritional composition of finger millet varieties cultivated in Sri Lanka. J Food Compos Anal 82:103249. https://doi.org/10.1016/j.jfca.2019.103249

Jenipher C, Santhi VP, Amalraj S et al (2024) Nutraceutical potential and ethnobotany of millets in India. S Afr J Bot 170:10–22

Kahn SE, Prigeon RL, Boyko EJ (2006) The relationship between beta-cell function and the development of type 2 diabetes in the Pima Indians. Diabetologia 49(8):1775–1781. https://doi.org/10.2337/dc08-1478

Kulthe AA, Thorat SS, Lande SB (2016) Health benefits and utilization of small millets: an overview. Adv Life Sci 5:4672–4675

Kumar A, Tomer V, Kaur A, Kumar V, Gupta K (2018) Millets: a solution to agrarian and nutritional challenges. Agric Food Secur 7(1):31. https://doi.org/10.1186/s40066-018-0183-3

Lee A, Cardel M, Donahoo WT (2000) Social and environmental factors influencing obesity. In: Feingold KR, Ahmed SF, Anawalt B et al (eds) Endotext. MDText.com, Inc. https://www.ncbi.nlm.nih.gov/books/NBK278977/

Lee HS, Lee J (2021) The association between nutrition labeling and consumer behaviors: a systematic review. Int J Environ Res Public Health 18(5):2635. https://doi.org/10.3390/ijerph18052635

Lillis J, Thomas JG, Wing RR (2008) Acceptance-based behavioral treatment for weight control: a review and future directions. J Consult Clin Psychol 76(6):1081–1094. https://doi.org/10.1037/a0012986

Luo J, Ruan Z, Li S (2016) Obesity and risk of liver cancer: a meta-analysis. Eur J Cancer 52:134–141. https://doi.org/10.1097/CEJ.0000000000000128

Maggard MA, Shugarman LR, Suttorp M (2005) Meta-analysis: surgical treatment of obesity. Ann Intern Med 142(7):547–559. https://doi.org/10.7326/0003-4819-142-7-200504050-00013

McDonough CM, Lloyd WR, Serna-Saldivar SO (2000) The millets: handbook of cereal science and technology. CRC Press, Boca Raton. https://doi.org/10.1201/9781420027228

Mohanan MM, Vijayakumar A, Bang-Berthelsen CH et al (2025) Bioactive metabolites in finger millet and their potential therapeutic roles. Foods 14(10):1733. https://doi.org/10.3390/foods14101733

Nithiyanantham S, Kalaiselvi P, Mahomoodally MF, et al (2019) Nutritional and functional roles of millets—A review. J food biochem, 43(7):e12859. https://doi.org/10.1111/jfbc.12859

Pandey M, Tripathi R, Mishra et al (2024) Millets for food and nutritional security: a review. Int J Plant Soil Sci 36:238–247

Poirier P, Giles TD, Bray GA (2006) Obesity and cardiovascular disease: pathophysiology, evaluation, and effect of weight loss. Arterioscler Thromb Vasc Biol 26(5):968–976. https://doi.org/10.1161/01.ATV.0000216787.85457.f3

Popkin BM, Adair LS, Ng SW (2012) Global nutrition transition and the pandemic of obesity in developing countries. Nutr Rev 70(1):3–21. https://doi.org/10.1111/j.1753-4887.2011.00456.x

Rambabu P (2018) Siridhanyalu (millets): food that heals. Telangana Retired Engineers' Association (TREA), Hyderabad

Rao B, Kandlakunta B, Christina GD, et al (2017) Nutritional and health benefits of millets. Retrieved March 3, 2024

Rao PS, Kumar A, Reddy BV (2019) Nutritional profile and health benefits of minor millets: a review. Int J Food Sci Nutr 70(6):725–736

Ren X, Yin R, Hou D et al (2018) The comparative study of sorghum and millet starches: a nutritional perspective. Nutrients. https://doi.org/10.3390/nu10101509

Sadh PK, Chawla P, Bhandari L, et al (2017) *In vitro* assessment of bio-augmented minerals from peanut oil cakes fermented by *Aspergillus oryzae* through Caco-2 cells. J Food Sci Technol 54(11):3640–3649. https://doi.org/10.1007/s13197-017-2825-z

Sadh PK, Kamboj A, Kumar S et al (2024) Millet-based functional foods for lifestyle diseases: recent advancements. Discov Food 4(1):187. https://doi.org/10.1007/s44187-024-00266-6

Safaei M, Sundararajan EA, Driss M et al (2021) A machine learning approach for early detection of obesity using clinical data. Comput Biol Med 136:104754. https://doi.org/10.1016/j.compbiomed.2021.104754

Saleh ASM, Zhang Q, Chen J, Shen Q (2013) Milling of cereals: impact on nutritional properties and health implications. Compr Rev Food Sci Food Saf 12(3):281–295. https://doi.org/10.1111/1541-4337.12012

Serna-Saldivar SO, Espinosa-Ramírez J (2019) Sorghum and millets. Elsevier, Duxford

Shobana S, Krishnaswamy K, Sudha V et al (2013) Finger millet (Ragi, *Eleusine coracana*): a review of its health benefits and cultural significance. Adv Food Nutr Res 69:1–39. https://doi.org/10.1016/B978-0-12-410540-9.00001-6

Singh RB, Watanabe S, Li D et al (2022) Functional foods and nutraceuticals for prevention of non-communicable diseases. In: Singh RB, Watanabe S, Isaza AA (eds) Functional foods and nutraceuticals in metabolic and non-communicable diseases. Academic, pp 165–185. https://doi.org/10.1016/B978-0-12-819815-5.00011-2

Slavin JL (2013) Whole grains and human health. Forum Nutr 29(4):478–482. https://doi.org/10.1016/j.nut.2004.08.018

Sridevi V, Subramanian V, Ramesh S (2020) Role of dietary intervention in reducing obesity: a study among South Indian adults. Int J Obes 44(5):1182–1189

Surekha N, Chimmad BV, Naik SR (2013) Nutritional qualities of minor millets and their importance. Asian J Home Sci 8:383–387

Swinburn BA, Sacks G, Hall KD (2019) The global obesity pandemic: shaped by global drivers and local environments. Lancet 378(9793):804–814. https://doi.org/10.1016/S0140-6736(11)60813-1

Vali Pasha K, Ratnavathi CV, Ajani J et al (2018) Nutritional and functional characteristics of sorghum and millets: implications for food processing industries. J Sci Food Agric 98:652–660. https://doi.org/10.1002/jsfa.8510

Venkatesh T, Ashokkumar K, Gopalakrishnan G et al (2005) Nutritional interventions and long-term weight maintenance. Am J Clin Nutr 82(1):222S–225S. https://doi.org/10.1093/ajcn/82.1.222S

Wing RR, Phelan S (2005) Long-term weight loss maintenance. The Am J Clin Nutr, 82(1):222S-225S. https://doi.org/10.1093/ajcn/82.1.222S

World Health Organization (WHO) (2020) Obesity and overweight: key facts. Retrieved from https://www.who.int/news-room/fact-sheets/detail/obesity-and-overweight

Chapter 17
Nutri-cereal Starch in Native and Modified Form—A Myriad of Transformation of Opportunities for Novel (Functional) Food and Non-food Applications

Manila Kashni and Neha

17.1 Introduction

The high nutritional content of millet in contrast to conventional grains such as rice and wheat has led to their growing recognition as "nutri-cereals" (Hassan et al. 2021). Proteins, vitamins, minerals, dietary fibers, carbs, and vital fatty acids are all abundant in these small-seeded grains. Millet starch, which makes up 55–61% of the grain, is essential to its nutritional profile and functionality (Prasad et al. 2023a, b). Millets can help to preclude lifestyle-related health issues like high blood pressure, diabetes, and cardiovascular disorders, among other health benefits. Due to inconvenience in food preparation, processing technologies, and a lack of awareness, millet consumption is still low despite its nutritional benefits (Porwal et al. 2023a, b). But in an effort to combat malnutrition and improve nutritional value, there is increasing interest in adding millets to food products (Prasad et al. 2023a, b).

The distinct physicochemical characteristics and adaptability of millet starch in its natural and modified forms make it extremely promising for industrial applications (Mahajan et al. 2021a, b). The functional characteristics of starch that is extracted from millet grains allow it for utilization in non-food and food sectors (Annor et al. 2014). Millets are nutrient-dense, climate-resilient grains that are gaining popularity for their contribution to sustainable agriculture and food security. Modification of native starch through physical, chemical, or enzymatic treatment improves starch functionality , and is prized for its diverse applications (Sultana et al. 2024a, b). These modifications increase the starch's mechanical strength, water retention, and thermal stability, increasing its use in novel food products, drug

M. Kashni (✉) · Neha
Department of Chemistry, Gurukula Kangri (Deemed to be University), Haridwar, Uttarakhand, India
e-mail: manila@gkv.ac.in

© The Author(s), under exclusive license to Springer Nature Switzerland AG 2026

J. Singh, M. Kumar Malik (eds.), *Millets and Millet Starch: Chemical Composition, Functional Modification, and Applications*,
https://doi.org/10.1007/978-3-032-09502-2_17

delivery, biodegradable packaging, and other industrial uses. This chapter examines the revolutionary possibilities that millet starch offers, highlighting how it might spur the creation of innovative functional food solutions and ecologically friendly non-food uses.

17.2 Millet Starch: Structure and Characteristics

A key factor in assessing the caliber of millet-based products is millet starch (Zhu 2014a, b; Mahajan et al. 2021a, b). Its structure and properties vary among millet varieties, with range of apparent amylose content from 12.3% to 27.4% (Shi et al. 2023a, b). The granules of starch show diverse sizes and shapes, influencing their physicochemical characteristics (Shi et al. 2023a, b). Native millet starch often lacks specific properties required for various applications, necessitating modifications to enhance its functionality (Dimri and Singh 2022a, b). The ordered molecular structure of low-amylose millet starch affects its swelling and viscosity properties, while higher amylose content correlates with lower rapidly digestible starch levels (Shi et al. 2023a, b). Despite growing interest in millet utilization because of its health advantages and food security potential, gaps in knowledge about millet starch hinder its broader application in food product development.

17.2.1 Physicochemical Properties of Millet Starch

Millet starches exhibit diverse physicochemical properties depending on their source and processing. Amylose content, water absorption, solubility, swelling power, and gel strength are key functional properties of native starches (Solarte-Montufar et al. 2019). Amylose content varies among millet types, with non-waxy varieties containing higher amounts (Kim et al. 1987b). Millet starch granules generally have polygonal round shapes and show A-type crystallinity (Bangar et al. 2023). Waxy millet starches exhibit greater enthalpies and gelatinization temperatures than non-waxy ones (Yang et al. 2019a, b). Pasting properties differ significantly among millet starches, with variations in breakdown, peak viscosity, and final viscosity. Malting affects starch properties, resulting in smaller granules, higher gelatinization temperatures, and increased solubility (Malleshi et al. 1986a, b). The solubility and swelling power of millet starches vary among different varieties. These diverse properties make millet starches suitable alternatives to conventional starch sources (Li et al. 2020).

- *Starch Granule Morphology and Size*—Granule morphology and size are crucial factors in understanding the properties and behavior of various materials. The crystalline structure of millet starch granules adds to their functional diversity. Most millet starches have A-type crystallinity, which is often seen in cereal starches. However, some small-granule millets have a mix of A- and B-type structures. This variation in starch structure impacts gelatinization, water absorption, and thermal stability. A-type starches usually hydrate more easily and have lower gelatinization temperatures. In contrast, mixed polymorphs show greater thermal resistance, which is helpful for high-temperature processing (Liu et al. 2020). Granule diameters range from 0.3 to 17 µm, and the underutilized form of proso millet has A-type polymorphism patterns (Kumar et al. 2023a, b). Millet starch granules usually possess well-defined hilum and growth rings, which have important implications during hydration. A more compact central hilum, like that of pearl millet, results in a higher granule integrity and reduced susceptibility to enzymatic attack. Millets with porous surfaces of the starch granules have increased susceptibility to gelatinization and enzymatic hydrolysis, hence are considered suitable for producing malted or fermented products (Obilana et al. 2021)..

- *Amylose-Amylopectin Ratio*—The amylose-amylopectin ratio (Am:Ap) in starch significantly influences its properties and applications. Higher Am:Ap ratios in meals result in lower initial glucose and insulin responses, increased satiety, and altered free glycerol and fatty acid concentrations in blood (Amelsvoort and Weststrate 1992). Acid hydrolysis of starch decreases the Am:Ap ratio (Wuttisela et al. 2008). The retrogradation behaviour of millet starch gels is similarly influenced by variations in the Am:Ap. Stronger retrogradation is shown by high-amylose millet starch, resulting in stiffer gels when cooled. Applications including starch-based noodles, biodegradable films, and resistant starch development benefit from this characteristic. On the other hand, millets high in amylopectin show slower retrogradation, which enhances freeze-thaw stability and decreases syneresis in frozen goods (Choi & Lim 2021). The Am:Ap ratio varies among different millet, with finger millet, foxtail millet, and barnyard millet-starches having ratios of 35.10:64.9, 31.33:68.67, and 33.33:66.67, respectively (Verma et al. 2018) The nutritional profile of millet species is also associated with variations in amylose concentration. Increased dietary resistant starch, which promotes the creation of low-glycemic meals targeted at diabetes control, is a result of high-amylose millet starches. According to several studies, compared to main grains like rice and wheat, finger millet and barnyard millet have starch portions that resist quick enzymatic hydrolysis, causing a slow release of glucose (Kumari and Thakur 2022). In high-amylose maize starch gels, the Am:Ap ratio and amylopectin nature affect gelation behavior, with amylose addition accelerating gel formation and increasing retrogradation enthalpy in mixtures with waxy or acid-hydrolyzed waxy starches (Klucinec and Thompson 2002). These findings highlight the importance of Am:Ap ratio in determining millet starch functionality across various applications.
- *Swelling Power and Solubility*—The naturally occurring lipid–amylose complexes in millet grain starches also affect the swelling behaviour of millet starch. A higher presence of such complexes reduces granule growth by reinforcing internal structural stability. According to Patel et al. (2020), pearl millet starch has comparatively larger concentrations of these complexes, which contribute to reduced swelling power but enhanced heat stability qualities beneficial for

extrusion and baking applications. Due to the disruption of hydrogen bonds and crystalline domains caused by high-pressure treatments or alkaline conditions, millet starch becomes much more soluble. Millet starch can be added to soups, drinks, instant foods, and emulsified systems when quick hydration is needed because this alteration improves solubility and paste clarity (Ramaswamy & Tharanathan 2021). The swelling power and solubility depend on granule size, the crystalline arrangement of millet starch, and amylose content. Millet starches with lower amylose generally exhibit higher swelling capacity as a result of the highly branched amylopectin that causes high water absorption. In contrast, millet starches having high amylose contents hinder the swelling of granules, hence resulting in a low level of solubility. Further modification in the swelling behaviour can be achieved through HMT, malting, and ultrasonication due to the breakdown of crystalline regions and improving the absorption of water. It tailors the functional properties of the millet starch, making them suitable for their use as a thickener, stabiliser, and biodegradable packaging material (Zhao et al. 2021; Kaur & Gill 2020).).

17.2.2 Rheological Properties

Native starches from different botanical sources have shown diverse functional and rheological properties. Similar to those reported for corn, rice, and tapioca starches (Kim et al. 2012a, b),millet starches also exhibit distinct pasting and gelatinization behavior. Millet starches in general show moderate viscosity with species-specific differences. The gelatinization temperatures of millet starch are within the range reported for other cereal starches (Ai and Jane 2015;Kim et al. 1987a). However, it is relatively higher in finger millet due to a more compact granule structure and higher amylose content.Waxy millet varieties have also been reported to have increased swelling power and clearer pastes than normal millet starch reported by (Lewandowicz et al. 2024a, b)and such functional and rheological properties make millet starch a suitable ingredient for food formulations that require moderate viscosity, thermal stability, and desirable gel texture...

17.2.3 Variability Among Different Millet Species

Millets are diverse, nutrient-rich crops valued for their resilience to environmental stresses and adaptability to various climates (Goron and Raizada 2015). The Indian subcontinent and Africa are major centers of millet cultivation, with several endemic species domesticated in India (Chandra and Koppar 1990). Different millet species, including finger millet, sorghum, foxtail millet, and pearl millet, exhibit significant genetic and proteomic diversity (Ravichandran et al. 2023a, b, c). This diversity is reflected in their varied traits, for instance, drought tolerance, nutritional content, and growth patterns. Foxtail millet, for instance, shows high variability in characteristics like panicle length, grain yield, and maturity time (Anuradha and Patro 2020a, b). The preservation of millet biodiversity in remote regions provides valuable genetic resources for crop improvement and adaptation to changing climatic conditions (Goron and Raizada 2015).

17.3 Modification of Millet Starch: Techniques and Approaches

The nutritional and functional qualities of millet starch have drawn interest (Thakur et al. 2022). To improve its functionality, variety of modification techniques are used, such as physical, chemical, enzymatic, and combination methods (Dey and Sit 2017). Solubility, freeze-thaw stability, swelling power, viscosity, and gel texture can all be changed by these modifications To enhance functional qualities, nonthermal physical techniques such as high-pressure processing, pulsed electric fields, and ultrasonication are also employed (Heena et al. 2024). The food and non-food industries may use modified millet starches as binding agents, thickeners, and stabilizers. To fully explore the potential of millet starches in a variety of applications, more research is required.

17.3.1 Physical Modifications

Physical modifications of millet starch can significantly change its characteristics and expand its applications. Annealing, heat moisture treatment, ultrasound, and microwave treatments have been demonstrated to modify starch characteristics (Dey and Sit 2017). These modifications generally decrease solubility, swelling power, and viscosity while increasing gelatinization temperatures and improving freeze-thaw stability (Amarnath et al. 2023a, b). Physically modified millet starches exhibit altered amylose content, water absorption capacity, and crystallinity. These changes can impact the starch's functionality in various applications, such as film formation (Bangar et al. 2023). The growing interest in millet starch as a novel plant-based ingredient has led to increased research on extraction methods and modification techniques to improve its practical qualities for usage in the food and non-food sectors (Heena et al. 2024). These modifications offer potential for developing value-added millet-based products with improved characteristics.

- *Heat Moisture Treatment (HMT)*—The HMT is an effective method for modifying millet starch properties. HMT generally reduces swelling power, starch solubility, and pasting viscosities while increasing thermal stability (Widyastuti 2021; Sandhu et al. 2020a, b). The treatment typically involves conditioning starch to 25% moisture content and heating at temperatures between 100 and

150°C for 3–7 h. HMT can increase amylose and resistant starch content, enhancing the nutritional value of millet starch (Zheng et al. 2020). The modification also affects starch crystallinity, with some studies reporting decreased relative crystallinity (Zheng et al. 2020) while others reported no change in the A-type crystalline structure (Wang et al 2023). HMT can increase slowly digestible starch content, potentially leading to healthier starch-derived food items (Wang and Zhang 2021). Pearl millet starch treated with HMT produced more stable starch paste profiles by reducing solubility and swelling power (Widyastuti 2021). Likewise, HMT-altered foxtail millet starch demonstrated enhanced freeze-thaw stability and decreased both solubility and swelling power (Dey and Sit 2017;(Zavareze and Dias 2011). HMT modification can increase water-binding capacity and peak viscosity (Balasubramanian et al. 2011). HMT was found to enhance thermal stability and nutritional quality of pearl millet starches, enhancing resistant starch, and slowly digestible starch content and reducing rapidly digestible starch content (Sandhu et al. 2020a, b).

- *Annealing*—The alteration of millet starch by annealing and other physical processes has been investigated in recent research. White finger millet starch annealing changed the gel texture and amylose content while raising the gelatinization temperature and relative crystallinity (Amarnath et al. 2023a, b). Annealing of proso millet starch reduced peak viscosity and crystallinity while raising the gelatinization temperature and slowly digested starch content. Dual modification of foxtail millet starch by ultrasonication and annealing produced improved acid and shear resistance, freeze-thaw stability, and gel texture (Babu et al. 2019a, b; Fonseca et al. 2021). HMT and annealing both improved white millet starch's resilience to mechanical and thermal operations. In particular, annealing reduced viscosity characteristics while increasing solubility, water absorption, gel strength, and gelatinization temperature (Marta et al. 2016). These changes show promise for increasing the use of millet starch in culinary items (Zavareze and Dias 2011).
- *Non-Thermal Modifications*—Several modifications such as cold plasma treatment, ultrasonication, irradiation, and ultra-high pressure are also explored by researchers to tailor the millet starch characteristics (Li et al. 2018; Sujatha et al. 2018; Yadav et al. 2021; Heena et al. 2024). Ultrasonication, can improve millet starch functional, pasting, thermal, and structural properties (Kumari et al. 2024). Depending on the modification technique, starch alterations generally reduced swelling power and viscosity across millet cultivars, with differing impacts on solubility. These changes provide the opportunity to customize the qualities of millet starch for particular culinary uses (Heena et al. 2024a, b). Recent advancements include the use of unconventional media like ionic liquids and supercritical CO_2, as well as technologies such as microwaves and ultrasounds for starch modification (Gilet et al. 2018). Cold plasma treatment of Kodo millet starch reduced relative crystallinity from 23.06 to 21.63% and improves the oil absorption capacities from 1.01 to 1.34 g/g (Sonkar et al. 2023). In an another study, pulse electric field treatment of browntop millet starch significantly enhanced the powder flowability, water and oil absorption capacity of browntop millet

starch and reduced cohesion index, confirming better dispersibility, handling properties and industrial applicability (Kaushik et al 2025).

17.3.2 Chemical Modifications

The chemical modification of starch is introducing functional group such as hydroxypropyl, carboxyl, acetyl, amine, amide, or any other to the starch polymer. The novel functional groups provide starches specialized features to increase its functionality and nutritional attributes that customize starch to specific industrial use. The choice of modification method and reaction conditions significantly impacts the resulting starch properties (Siroha et al. 2020). Physical modifications are preferred for producing more natural food components, while chemical modifications are used to overcome native starch deficiencies like retrogradation and syneresis (Kapri et al. 2017a, b). Overall, starch modification techniques aim to refine the functionality and stability of millet starches for various industrial applications.

- *Acid Hydrolysis*—Acid hydrolysis is a common method for modifying millet starches, often combined with other techniques like heat moisture treatment, enzymatic modification, and ultrasound treatment. These modifications generally decrease swelling power and viscosity while increasing solubility (Balasubramanian et al. 2014; Sanchez de la Concha et al. 2018). Acid-modified starches typically show reduced protein and ash content, better freeze-thaw stability, and increased paste clarity. The type of acid used affects the extent of modification, with HNO_3 and HCl producing the highest alkali fluidity numbers, then by H_3PO_4 and H_2SO_4 (Singh and Ali 2008a, b). Chemical modifications, like stearic acid treatment of pearl millet starch, have shown increased thermal stability and potential for food packaging applications (Sultana et al. 2024a, b). Acid modification also impacts molecular weight, intrinsic viscosity, and iodine binding capacity of starches, with the degree of substitutions varying among different starch sources (Singh and Ali 2008a, b). These modifications alter the functional characteristics of millet starches, potentially expanding their applications in food and non-food industries.
- *Cross-Linking*—Cross-linking of millet starches has been investigated by researchers as a way to improve their functional qualities. Epichlorohydrin cross-linking of pearl millet starch changed its rheological characteristics and increased its resistant starch content; the effects were stronger at higher concentrations (Siroha et al. 2021). The cross-linking of barnyard millet starch with sodium trimetaphosphate enhanced film properties while reducing swelling power and amylose concentration (Sharma et al. 2021a, b). The digestibility and physicochemical properties of cereal pearl millet were impacted by the degree of cross-linking that differed between cultivars (Siroha and Sandhu 2018). Adding amino acids, especially lysine, to kutki millet starch at higher pH values enhanced its textural qualities and gel formation (Mahajan et al. 2021a, b). Across various

millet starches, these changes typically led to changed thermal characteristics, higher pasting temperatures, and decreased breakdown viscosity (Sharma et al. 2021a, b). Based on the findings, cross-linking can modify the characteristics of millet starch for a range of uses, such as hydrogels and edible films.

- *Etherification and Esterification*—Etherification and esterification of starch have drawn a lot of attention because of their numerous uses in different sectors (Farkhani et al. 2024). These processes alter the physicochemical properties of starch, enhancing its functionality for use in adsorption, adhesives, pharmaceuticals, and nanoparticle synthesis (He et al. 2023). 2018 Organic acids and their derivatives have gained particular interest in starch modification for food applications due to their natural occurrence and GRAS status (Ačkar et al. 2015a, b). However, these modification processes may result in the partial degradation of starch molecular structure (Gilet et al. 2018). Pearl millet starch modifications, including acetylation, succinylation, oxidation, and octenylsuccinylation, have been shown to enhance its functional properties. These modifications can improve swelling power, paste clarity, solubility, and cold-storage stability while reducing retrogradation (Shaikh et al. 2017a, b). Acetylated finger millet starch based tablets shows better tensile and crushing strength (Afolabi et al. 2011). The esterification of kodo millet starch using octenyl succinic anhydride enhanced its compatibility with proteins, leading to better emulsifying characteristics and more stable emulsions (Mahanure et al. 2025). Octenyl succinic anhydride modification of pearl millet starch has demonstrated changes in pasting properties, rheological behavior, and in vitro digestibility, with potential applications in film and coating development (Siroha et al. 2022)

17.3.3 Biological/Enzymatic Modifications

Millet starches, including those from foxtail, finger, pearl, and proso millets, can be modified through enzymatic treatment to enhance their functional properties (Dey and Sit 2017; Kumar et al. 2023a, b). These modifications generally result in decreased swelling power and viscosity, while solubility tends to increase (Balasubramanian et al. 2011). Freeze-thaw stability and paste clarity are improved in some modified starches, particularly those treated enzymatically or with acid (Balasubramanian et al. 2014).. Enzyme modified millet starches have changed properties that make them appropriate for a range of food and non-food applications, with the potential for increased use in the food sector (Kumar et al. 2023a, b).

- *Amylase Treatment*—Millet starch, which constitutes 70% of millet grain, is responsible for characterizing the quality of millet products (Zhu 2014a, b).. These changes extend the scope of uses of millet starch in the food market to meet the increasing demand for millet utilization based on its nutritional value and its contribution to food security (Mahajan et al. 2021a, b). Reduced peak viscosity and pseudoplastic flow behavior were seen in proso millet starch treated with α-amylase, followed by hydroxypropylation and cross-linking (Mathew et al. 2023).

- *Fermentation-Based Changes*—The alteration of millet starch through fermentation has demonstrated encouraging outcomes in terms of changing its functional characteristics. Solid-state fermentation using fungi like *Aspergillus sojae* and *Rhizopus azygosporus* can create pores on starch granules, reduce particle size, and increase amylose content and solubility (Purewal et al. 2024a, b). Natural fermentation of pearl and finger millet flour for 24 h significantly impacts starch functionality without altering granule structure (Gupta and Gaur 2023).

17.3.4 Nanoscale Modifications

Millet starch, comprising 60–70% of the grain, has gained attention for its nutritional and functional properties (Joye 2019). Native millet starch exhibits A-type polymorphic patterns and granule sizes of 0.3–17 μm (Kumar et al. 2023a, b). However, its limited functionality necessitates modification through physical, chemical, enzymatic, or combined methods. These modifications can alter solubility, swelling power, freeze-thaw stability, and viscosity (Dey and Sit 2017). Nonthermal physical processes like ultrasonication, high-pressure processing, and irradiation have been employed to enhance millet starch's functional properties (Heena et al. 2024). Modified millet starches find applications in food and non-food industries as thickeners, binders, and stabilizers. The growing interest in millet starch highlights the need for further research to expand its utilization in various sectors (Thakur et al. 2022).

- *Starch Nanoparticles/Nanocrystals*—Recent studies have explored the modification of millet starch to produce nanoparticles with enhanced properties. Starch nanocrystals can be surface-modified using various compounds, maintaining their crystalline structure while improving individualization (Thielemans et al. 2006). Proso millet starch nanoparticles, prepared through enzymolysis and recrystallization, exhibit increased crystallinity and altered crystal patterns (Sun et al. 2014). Acid hydrolysis of pearl millet starch results in nanoparticles with improved oil and absorption capacities, along with altered pasting properties (Chavan et al. 2023). Nano-reduction of pearl and proso millet starches using collision ball milling produces stable nanoparticles with reduced crystallinity and particle size, as well as enhanced antioxidant potential (Jhan et al. 2020a). These modifications to millet starch nanoparticles offer potential applications in a wide range of industries, including food and pharmaceuticals, due to their improved functional properties and nutraceutical potential.

17.4 Food and Non-food Applications of Nutri-cereal Starch

Cereal starch is also widely employed in the manufacturing of food items, where it gives gummies and jellies their structure and acts as a mold release agent; in dairy products, such as yoghurt and custards, it improves viscosity and prevents syneresis, ensuring a smooth, creamy texture; in coating applications, such as batter and breading for fried foods, it enhances crispness and reduces oil absorption; it is an essential ingredient in gluten-free formulations, providing structure and binding properties that mimic gluten in baked goods; and, in addition, cereal starch helps to flavor encapsulation in seasonings and powdered food products, protecting sensitive ingredients and extending shelf life. Its versatility and affordability make it an essential tool in contemporary food processing and innovation (Kurichh et al. 2025). Millet grains possess favorable functional properties, including good bulk density, oil and water absorption capacities, and pasting properties, enabling their use in various food applications such as beverages, gruel, and traditional dishes (Abah et al. 2020).

17.4.1 Food Packaging Film

Millet-based nutri-cereal starch has gained significant attention as a sustainable material for food packaging films due to its biodegradability, film-forming ability, and functional properties. The starch extracted from millets such as pearl millet, finger millet, and sorghum can be modified to enhance its mechanical strength, water resistance, and barrier properties, making it suitable for eco-friendly packaging solutions (Khatkar et al. 2009). Millets offer a nutrient-rich, gluten-free alternative to plastic packaging, with edible films prolonging shelf life and reducing environmental impact (Anitha and Priya 2022). Studies on barnyard millet starch films demonstrate varying properties based on additive concentrations, showing promise for packaging applications (Soujanya and Hema 2021). Pearl millet starch nanoparticle-based films exhibit improved mechanical characteristics, reduced water permeability, and increased biodegradability, making them suitable for specialized food packaging needs (Chavan et al. 2024; Tomar et al. 2022).

These starch-based films are often blended with natural plasticizers, biopolymers, or antimicrobial agents to improve their flexibility and shelf-life extension capabilities, reducing the reliance on synthetic plastics (Adewale et al. 2022). Additionally, millet starch films can be used in active packaging systems, where they help in preserving food quality by controlling moisture and oxygen permeability, thereby enhancing food safety and storage stability. Beyond food applications, these biodegradable films have potential uses in non-food sectors, including pharmaceutical packaging, biodegradable carry bags, and agricultural mulch films, contributing to environmental sustainability(Malik et al. 2023; Kaimal et al. 2021).

The utilization of millet starch in packaging aligns with global efforts to reduce plastic waste while promoting the use of renewable and natural resources.

17.4.2 Textile Industry

After petroleum, the textile industry is the second most polluting sector of the global economy. A key strategy for a sustainable textile industry is the use of sustainable chemicals for sizing, printing, and finishing textiles because wet textile processing is very polluting (Singh and Sahu 2019). However, native starch has poor adhesion properties due to its brittleness. Numerous techniques have been used to modify starch, such as cationization (Hebeish et al. 2005), double etherification (Li et al. 2018), and grafting (Kovancevic et al. 2020). Before weaving, sizing is done in which a protective adhesive coating to the yarn's surface in textiles is applied to minimize rupturing of fibers during the weaving process (Shen et al. 2021).

17.4.3 Papermaking Industry

The papermaking industry, with its developed and commercially viable technologies, continues to be one of the major users of starch in non-food applications. Starch must be chemically modified before it can be utilized in binding or sizing applications, but the cost of the modifications is needed to be balanced against the expected improvement in performance. There has been growing interest in using bio-latex (chemically cross-linked starch particles) to replace petroleum-based synthetic latexes such as styrene butadiene and polyvinyl acetate latex as binders in coating colors (Du and Hsieh 2017). For example, the cross-linking of cassava starch with sodium hexametaphosphate to produce a bio-latex showed good adhesion performance when 25% of the petroleum-based latex was substituted with the bio-latex (Liu et al. 2020). Currently, Eco-sphere biopolymers by Ecosynthetix are one commercial starch bio-latex for the papermaking industry. The use of millet starch in paper making industry has gained growing attention as a sustainable and biodegradable alternative to conventional synthetic additives. Owing to its excellent film- forming,adhesive, and binding properties, the millet starch can be employed in various stages of paper production, including surface sizing, coating and strengthening of paper sheets. When applied during the manufacturing process, it enhances paper stiffness, smoothness, printability and resistance to water absorption. Additionally, the natural viscosity and amylose- amylopectin ration of millet starch contribute to improved fiber bonding and better surface finish, making it suitable for high- quality printing and packaging papers. Unlike chemical binders, millet starch is non- toxic, renewable and cost effective, aligning with the current industrial trend toward eco-friendly and sustainable materials. Its use also helps in reducing waste and dependence on petroleum-based products, offering an environmentally responsible solution for the modern paper industry.

17.4.4 Bioplastics

Petroleum feedstocks are used to produce the majority of polymeric materials. According to their origin and biodegradability, alternatives to petroleum-based, nonbiodegradable plastic materials fall into one of three categories: (1) biobased, nonbiodegradable bioplastics (like polyamide and polyethylene terephthalate (PET)); (2) biobased, biodegradable bioplastics (like polyhydroxyalkanoates (PHA), or starch); or (3) fossil-based, biodegradable bioplastics (like polycaprolactone (PCL))(Tai et al. 2021a). Babu et al. 2024 utilized foxtail millet husk to develop bio-based plastic and concluded that addition of foxtail millet husk significantly enhanced the thermal stability of the bioplastic composite.

17.4.5 Pharmaceutical Industry

Millet starch's favourable swelling behaviour, biodegradability, and non-toxicity have drawn interest as a possible medicinal excipient. It can work well in tablet formulations as a filler, disintegrant, and binder. Modified millet starches are appropriate for sustained-release medication delivery systems because of their improved compressibility and regulated swelling (Malik et al. 2023). Millet starch's intrinsic stability and abundance provide a sustainable substitute for traditional excipients used in oral dosage forms (Adejumo & Samuel 2022; Patel et al. 2021). Native millet starch showed promising efficiency as a disintegrant due to its good water absorption capacity and swelling within the compressed matrix, which is very useful for fast-dissolving tablets. According to studies, millet starch exhibited almost similar performances to maize starch and microcrystalline cellulose in disintegration and mechanical strength tests (Shah & Bhatt 2021). Moreover, pre-gelatinized millet starch exhibits an improved binding ability that enables uniform granulation in wet granulation processes. Due to its smooth particle morphology, it is more compressible and less friable into tablets. These attributes underpin millet starch as one of the sustainable and cost-effective excipients for conventional and novel drug delivery systems (Mishra et al. 2022). Millet starch is also used as a stabilizer to increase the consistency and shelf life of liquid medications and suspensions. Modified starch derivatives are utilized to improve controlled release mechanisms in biodegradable drug delivery systems, and cereal starch is a natural source, non-toxic, and biocompatible, resulting it for a perfect option for different pharmaceutical uses (Malik et al. 2024; Kunle 2019).

17.4.6 Nanotechnology

When compared to native starch, millet starch-based nanomaterials, such as starch nanoparticles, nanocapsules, and nanocrystals, show enhanced mechanical strength, thermal stability, and film-forming capacity. These nanostructures provide higher surface area and better compatibility with biopolymers, enabling their use in biodegradable packaging, edible coatings, and controlled-release matrices. It has been demonstrated that adding millet starch nanocrystals to polymer blends improves barrier qualities and lowers moisture permeability, which supports their use in environmentally friendly food packaging technologies (Nazir et al. 2024; Jhan et al. 2020a,b). A lot of research has been undertaken on nanoscale starch as a possible filler for biocomposites to add functionality and strengthen reinforcement.Many chemical, enzymatic, and physical techniques can be used to synthesize starch nanomaterials, producing a range of morphologies and sizes (Le Corre et al. 2010). A recent review examined the application of starch non-isocyanate and isocyanate polyurethanes in their production (Tai et al. 2021b). Pearl millet nanoparticles were used to create biodegradable films and showed enhanced thermal stability (Bhardwaj et al. 2025; Chavan et al. 2024).

17.5 Conclusion

Nutri-cereal starch is derived from different types of millet. Numerous industrial applications exist for nutri-cereal starch as a binding, texture-modifying, structural, and viscosity-regulating agent. Because millet starches digest more slowly than other starches, they are a better alternative for diabetics.. Various physical, chemical, and biological methods have been reported for modifying millet starches. As starch requires modification before a specific use and has a low functionality, it is rarely used in its native form. Every modification treatment makes a unique contribution to the structural alterations in starch that may affect the starch's resistance to digestive enzymes. In addition to this, the potential culinary uses of several millet varieties, including barnyard, small millets, teff, white and black fonio, and kodo needs significant attention to achieve food security... Additional investigations into the nutritional value of cereal starch will aid in the creation of novel food products, and generating revenue for farmers. However, the future research should be conducted to examine the impacts of diverse physical, chemical, biological and mechanical procedures on the functional and structural aspects of millet starch for wide applications.

References

Abah A, Agbaji B, Eze I (2020) Functional properties and utilization of millet grains. Food Sci Nutr Technol 5(1):1–6. https://doi.org/10.23880/fsnt-16000211

Ačkar D, Babić J, Šubarić D et al (2015a) Modification of starch with organic acids and their derivatives. Carbohydr Polym 132:366–375. https://doi.org/10.1016/j.carbpol.2015.06.074

Ačkar D, Šubarić D, Miličević B et al (2015b) Isolation of starch from two wheat varieties and their modification with ascorbic and citric acids. Carbohydr Polym 119:141–151. https://doi.org/10.1016/j.carbpol.2014.11.033

Adewale P, Shokrolllahi M, Lam E (2022) Starch modification for non-food, industrial applications: market intelligence and critical review. Carbohydr Polym 291:119590. https://doi.org/10.1016/j.carbpol.2022.119590

Ai Y, Jane JL (2015) Gelatinization and rheological properties of starch. In: BeMiller JN, Whistler RL (eds) Starch: chemistry and technology. 3rd ed. Academic, pp 379–429. doi:https://doi.org/10.1016/B978-0-12-800140-6.00009-4

Amarnath L, Sethi S, Rana S, Sharma S (2023a) Influence of annealing on techno-functional and nutritional properties of millet starches. Int J Biol Macromol 242:124671. https://doi.org/10.1016/j.ijbiomac.2023.124671

Amarnath D, Sit N, Dey A (2023b) Physical modification of starch from white finger millet: effect on structural and functional properties. Int J Biol Macromol 228:1187–1195. https://doi.org/10.1016/j.ijbiomac.2023.05.017

Anitha S, Priya M (2022) Millet-based edible films for sustainable food packaging. Mater Today Proc 65:1002–1006. https://doi.org/10.1016/j.matpr.2022.05.314

Annor A, Marcone M, Bertoft E et al (2014) Physical and molecular characterization of millet starches. Cereal Chem 91:286–292. https://doi.org/10.1094/CCHEM-08-13-0155-R

Anuradha N, Patro TS (2020a) Genetic variability in foxtail millet for yield and maturity traits. J Pharmacogn Phytochem 9(5):2361–2365

Anuradha N, Patro TS (2020b) Agronomic diversity in foxtail millet (Setaria italica) genotypes. Int J Curr Microbiol App Sci 9(4):1030–1038. https://doi.org/10.20546/ijcmas.2020.904.123

Babu A, Dey A, Sit N (2019a) Dual modification of foxtail millet starch: effect on physicochemical and functional properties. Carbohydr Polym 216:41–48. https://doi.org/10.1016/j.carbpol.2019.03.013

Babu AS, Reddy CK, Haripriya S (2019b) Dual modification of foxtail millet starch by ultrasonication and annealing. Int J Biol Macromol 135:962–969. https://doi.org/10.1016/j.ijbiomac.2019.05.235

Balasubramanian S, Singh KK, Kumar R (2011) Effect of hydrothermal modification on millet starch properties. J Food Sci Technol 48(5):587–591. https://doi.org/10.1007/s13197-011-0204-y

Balasubramanian S, Borah A, Padmavathi T (2014) Effect of modification techniques on the properties of millet starch. J Food Sci Technol 51(10):2570–2576. https://doi.org/10.1007/s13197-012-0784-y

Bangar SP, Dhull SP, Manzoor M et al (2023) Functionality and applications of non-conventional starches from different sources. Starch-Stärke 76:2300073. https://doi.org/10.1002/star.202300073

Chandra S, Koppar MN (1990) Domestication and cultivation of millets in India. Econ Bot 44(3):333–340. https://doi.org/10.1007/BF02861439

Chavan UD, Patil JV, Kadam PS et al (2023) Development and characterization of acid hydrolyzed millet starch nanoparticles. Starch-Stärke 75(5–6):2200276. https://doi.org/10.1002/star.202200276

Chavan UD, Kadam PS, Pawar NJ et al (2024) Development of biodegradable films using pearl millet starch nanoparticles. Int J Biol Macromol 251:127365. https://doi.org/10.1016/j.ijbiomac.2024.127365

Chen Y, Yang Q, Xu X et al (2017) Structural changes of waxy and normal maize starches modified by heat moisture treatment and their relationship with starch digestibility. Carbohydr Polym 177:232–240

Dey A, Sit N (2017) Modification of foxtail millet starch by using physical and enzymatic methods. J Food Sci Technol 54(9):2804–2812. https://doi.org/10.1007/s13197-017-2717-7

Dimri S, Singh S (2022a) A brief review on millet starch. Bhartiya Krishi Anusandhan Patrika 37(2):126–132. https://doi.org/10.18805/BKAP489

Dimri U, Singh AK (2022b) Millet starch: composition, structure and applications—a review. J Food Process Preserv 46(8):e16785. https://doi.org/10.1111/jfpp.16785

Du J, Hsieh YL (2017) Bio-latex from crosslinked starch particles for paper coating. Ind Crop Prod 95:535–543. https://doi.org/10.1016/j.indcrop.2016.11.007

Farkhani M, Ahmadzadeh S, Sheykhsaran E et al (2024) Etherified and esterified starches: advances in modification and applications. Carbohydr Polym 311:120797. https://doi.org/10.1016/j.carbpol.2023.120797

Fonseca LM, El Halal SLM, Dias ARG, Zavareze ER (2021) Physical modification of starch by heat-moisture treatment and annealing and their applications: a review. Carbohydr Polym 274:118665. https://doi.org/10.1016/j.carbpol.2021.118665

Gilet A, Aussenac T, Merah O (2018) Structural alteration of starch by dual modification: ionic liquids and ultrasound. Int J Biol Macromol 117:722–729. https://doi.org/10.1016/j.ijbiomac.2018.05.033

Goron TL, Raizada MN (2015) Genetic diversity and genomic resources available for the small millet crops to accelerate a new green revolution. Front Plant Sci 6:157. https://doi.org/10.3389/fpls.2015.00157

Gupta S, Gaur V (2023) Effect of fermentation on functional characteristics of millet flour starch. J Cereal Sci 108:103540. https://doi.org/10.1016/j.jcs.2023.103540

Hassan ZM, Sebola NA, Mabelebele M (2021) The nutritional use of millet grain for food and feed: a review. Agric Food Secur 10:16. https://doi.org/10.1186/s40066-020-00282-6

He B, Zeng X, Liu W et al (2023) Green synthesis and applications of modified starch nanoparticles. Carbohydr Polym 303:120389. https://doi.org/10.1016/j.carbpol.2023.120389

Heena, Panesar PS, Bera MB (2024) Functional and structural properties of modified millet starches: a review. Int J Biol Macromol 251:129584. https://doi.org/10.1016/j.ijbiomac.2024.129584

Heena A, Dey A, Sharma S (2024a) Emerging non-thermal technologies for starch modification. Int J Biol Macromol 242:124689. https://doi.org/10.1016/j.ijbiomac.2023.124689

Heena, Kumar N, Singh R et al (2024b) Application and functional properties of millet starch: wet milling extraction process and different modification approaches. Heliyon 10(3):e25330. https://doi.org/10.1016/j.heliyon.2024.e25330

Jane J, Kasemsuwan T, Leas S et al (1994) Anthology of starch granule morphology by scanning electron microscopy. Starch-Stärke 46(4):121–129. https://doi.org/10.1002/star.19940460402

Jhan FH, Lin HY, Chang YC et al (2020a) Preparation of millet starch nanoparticles and their antioxidant activity. Food Chem 315:126238. https://doi.org/10.1016/j.foodchem.2020.126238

Joye IJ (2019) Cereal biopolymers for nano- and microtechnology: a myriad of opportunities for novel (functional) food applications. Trends Food Sci Technol 83:1–11. https://doi.org/10.1016/j.tifs.2018.10.009

Kaimal A, Mujumdar A, Thorat B (2021) Resistant starch from millets: recent developments and applications in food industries. Trends Food Sci Technol 111. https://doi.org/10.1016/j.tifs.2021.02.074

Kapri A, Bhardwaj A, Sharma S (2017a) Chemical modification of starch for industrial applications. Int J Biol Macromol 104:1380–1391. https://doi.org/10.1016/j.ijbiomac.2017.06.090

Kapri A, Bhatt R, Bisht TS et al (2017b) Recent advances in starch modification for food and non-food applications. J Food Sci Technol 54(7):1747–1760. https://doi.org/10.1007/s13197-017-2667-2

Khatkar BS, Panghal A, Singh U (2009) Application of cereal starches in food processing. Indian Food Ind 28:37

Kim SK, Kim SH, Kim HW et al (1987a) Physicochemical properties of millet starches. Korean J Food Sci Technol 19(3):368–372

Kim SK, Wiesenborn D, Orr PH, Grant LA (1987b) Screening of corn starch for novel properties. Cereal Chem 64:516–519

Kim Y, Wiesenborn DP, Orr PH (2012a) Rheological properties of cross-linked waxy, normal, and high amylose maize starches. Starch-Stärke 64(5):367–377. https://doi.org/10.1002/star.201100101

Kim YS, Wiesenborn DP, Orr PH, Grant LA (2012b) Pasting and thermal properties of potato and bean starches. Cereal Chem 69(5):551–555. https://doi.org/10.1094/CCHEM.1987.64.4.365

Klucinec JD, Thompson DB (2002) Structure of maize starch hydrolysates and relationships to gelation and retrogradation. Cereal Chem 79(1):24–30. https://doi.org/10.1094/CCHEM.2002.79.1.24

Kovancevic D, Bilic M, Subaric D et al (2020) Functionalisation of starch for textile applications. Text Res J 90(7–8):807–822. https://doi.org/10.1177/0040517519865684

Kumar S, Sharma S, Kumari P (2023a) Starch polymorphism and properties of underutilised proso millet. Int J Food Prop 26(1):1772–1785. https://doi.org/10.1080/10942912.2023.2261735

Kumar P, Yadav R, Sharma N, Bhardwaj A (2023b) Functional characteristics and bioactive potential of proso millet (*Panicum miliaceum* L.) starch. Food Chem 405:134877. https://doi.org/10.1016/j.foodchem.2022.134877

Kumari P, Rani S, Joshi S et al (2024) Functional improvements in millet starches via ultrasonication. Carbohydr Polym 312:121038. https://doi.org/10.1016/j.carbpol.2024.121038

Kunle OF (2019) Starch as an excipient in pharmaceutical formulations. Starch-Stärke 71(9–10):1800243. https://doi.org/10.1002/star.201800243

Le Corre D, Bras J, Dufresne A (2010) Starch nanoparticles: a review. Biomacromolecules 11(5):1139–1153. https://doi.org/10.1021/bm901428y

Lewandowicz G, Fornal E, Urbaniak M, Walkowiak R (2024a) Physicochemical properties and rheology of native and modified starches: a comparative study. Carbohydr Polym 320:121012. https://doi.org/10.1016/j.carbpol.2024.121012

Lewandowicz G, Fornal Ł, Walkowiak-Tomczak D (2024b) Comparison of waxy and non-waxy starches from cereals and tubers. Food Chem 436:136891. https://doi.org/10.1016/j.foodchem.2024.136891

Li Z, Liu X, Liu W et al (2018) Double etherification of starch for improved textile performance. Carbohydr Polym 181:651–659. https://doi.org/10.1016/j.carbpol.2017.11.076

Liu X, Wang Y, Zhang L et al (2020) Biolatex from crosslinked cassava starch for paper coating. Carbohydr Polym 236:116051. https://doi.org/10.1016/j.carbpol.2020.116051

Mahajan P, Bera MB, Panesar PS, Chauhan A (2021a) Millet starch: a review. Int J Biol Macromol 180:61–79. https://doi.org/10.1016/j.ijbiomac.2021.03.063

Mahajan R, Thakur D, Guleria P (2021b) Modification and potential applications of millet starch: a review. J Food Sci Technol 58(7):2509–2522. https://doi.org/10.1007/s13197-020-04764-2

Malleshi NG, Hadimani NA, Chinnaswamy R, Klopfenstein CF (1986a) Studies on raw and malted finger millet. J Food Sci Technol 23(3):170–173

Malleshi NG, Klopfenstein CF, Rooney LW (1986b) Malting of pearl millet. Cereal Chem 63(4):255–259

Marta H, Lewandowicz G, Fornal Ł (2016) Thermal and rheological properties of starches subjected to annealing. Starch 68(5–6):556–562. https://doi.org/10.1002/star.201500159

Mathew A, Bera MB, Panesar PS et al (2023) Enzymatic modification of millet starches and evaluation of their functional properties. Food Chem 414:135722. https://doi.org/10.1016/j.foodchem.2023.135722

Osman MA, Ahmed SI, Gad AY (2020) Applications of starch in pharmaceutical formulations. J Pharm Sci Res 12(4):498–505

Porwal A, Bhagwat G, Sawarkar J, Kamble P, Rode M (2023a) An overview of millets—the nutricereals: its nutritional profile, potential health benefits and sustainable cultivation approach. Int J Sci Res Arch 10:841–859. https://doi.org/10.30574/ijsra.2023.10.1.0828

Porwal M, Jain D, Saxena J et al (2023b) Challenges and perspectives in millet consumption in India. Curr Nutr Rep 12:221–231. https://doi.org/10.1007/s13668-023-00476-1

Prasad P, Sahu JK, Srivastava AK (2023a) Millet starch—a potential functional ingredient. Int J Plant Soil Sci 35(21):1060–1065. https://doi.org/10.9734/ijpss/2023/v35i214077

Prasad R, Sharma S, Thakur D (2023b) Nutritional evaluation and starch content of millets. J Cereal Sci 110:103519. https://doi.org/10.1016/j.jcs.2023.103519

Purewal SS, Sharma HK, Kumar N et al (2024a) Functional and structural characteristics of fermented millet starches. Int J Biol Macromol 250:129020. https://doi.org/10.1016/j.ijbiomac.2024.129020

Purewal SS, Singh A, Kaur M et al (2024b) Starch fermentation for enhanced functionality in millet. Carbohydr Polym 315:120087. https://doi.org/10.1016/j.carbpol.2024.120087

Ravichandran M, Chandra R, Lal M, Singh AK (2023a) Comparative proteomics and genomics of millets: a review. Plant Gene 35:100413. https://doi.org/10.1016/j.plgene.2023.100413

Ravichandran G, Pandey R, Yadav AK et al (2023b) Proteomic insights into millet species: diversity and drought tolerance. J Cereal Sci 108:103621. https://doi.org/10.1016/j.jcs.2022.103621

Ravichandran G, Parveen A, Mahalingam PU et al (2023c) Diversity and genomics of millet crops. Genet Resour Crop Evol 70:1–14. https://doi.org/10.1007/s10722-023-01533-3

Sanchez de la Concha BB, Agama-Acevedo E et al (2018) Acid hydrolysis of waxy starches with different granule size for nanocrystal production. J Cereal Sci 79:193–200. https://doi.org/10.1016/j.jcs.2017.10.018

Sandhu KS, Singh N, Lim ST (2020a) Effect of heat moisture treatment on physicochemical and functional properties of pearl millet starches. Food Hydrocoll 103:105654. https://doi.org/10.1016/j.foodhyd.2020.105654

Sandhu KS, Siroha AK, Kaur M (2020b) Effect of heat moisture treatment on pearl millet starch. J Food Sci Technol 57(4):1417–1423. https://doi.org/10.1007/s13197-019-04195-w

Shaikh AF, Patil MB, Deshpande HW (2017a) Physicochemical properties of chemically modified pearl millet starch. Int J Chem Stud 5(6):1867–1871

Shaikh F, Sit N, Dey A (2017b) Effect of acetylation on properties of pearl millet starch. Starch 69(5–6):1600040. https://doi.org/10.1002/star.201600040

Sharma B, Singh B, Panesar PS (2021a) Modification of barnyard millet starch and its application in biodegradable films. Int J Biol Macromol 184:970–977. https://doi.org/10.1016/j.ijbiomac.2021.06.077

Sharma S, Thakur D, Mahajan R (2021b) Modification of barnyard millet starch for edible film applications. J Appl Polym Sci 138(8):49978. https://doi.org/10.1002/app.49978

Shen Y, Zhang H, Zhu Z et al (2021) Sizing in textiles using modified starch. Text Res J 91(1):44–56. https://doi.org/10.1177/0040517520939326

Shi Y, Liu L, He R et al (2023a) Structure and functionality of millet starch. Carbohydr Polym 310:120008. https://doi.org/10.1016/j.carbpol.2023.120008

Shi M, Wang F, Zhang M, Jin Z (2023b) Physicochemical properties and digestive characteristics of starch from proso millet. Int J Biol Macromol 242:124753. https://doi.org/10.1016/j.ijbiomac.2023.124753

Singh G, Ali SZ (2008a) Comparative acid modification of starches from pearl millet, maize, and sorghum. J Food Sci Technol 45(4):339–343

Singh G, Ali SZ (2008b) Comparative study on acid hydrolysis of millet starches. Starch-Stärke 60(10):579–586. https://doi.org/10.1002/star.200800004

Singh R, Sahu V (2019) Textile pollution: sustainable approaches and use of starch in textile. J Clean Prod 238:117877. https://doi.org/10.1016/j.jclepro.2019.117877

Siroha AK, Sandhu KS (2018) Impact of cross-linking on physicochemical properties of pearl millet starch. Int J Food Prop 21(1):1780–1792. https://doi.org/10.1080/10942912.2018.150333

Siroha AK, Sandhu KS, Kaur M (2020) Structural, morphological, and functional properties of foxtail millet starch. Int J Food Prop 23(1):887–900. https://doi.org/10.1080/10942912.2020.1770188

Solarte-Montufar I, Agama-Acevedo E, Islas-Hernández JJ, Bello-Pérez LA (2019) Physicochemical properties of native potato starches from Solanum phureja varieties. Starch-Stärke 71(1–2):1800108. https://doi.org/10.1002/star.201800108

Solarte-Montufar JE, Restrepo A, Serna JA (2022) Functional properties of high-amylose starches. Starch-Stärke 74(3–4):2100164. https://doi.org/10.1002/star.202100164

Soujanya N, Hema S (2021) Biodegradable starch film from barnyard millet. Int J Agric Sci 13(3):453–457

Sultana H, Kumar N, Singh R et al (2024a) Application and functional properties of millet starch: wet milling extraction process and different modification approaches. Heliyon 10(3):e25330. https://doi.org/10.1016/j.heliyon.2024.e25330

Sultana S, Prabhasankar P, John A (2024b) Stearic acid modification of pearl millet starch for edible films. Food Packag Shelf Life 39:101056. https://doi.org/10.1016/j.fpsl.2023.101056

Sun QJ, Xu LM, Xing XH et al (2014) Preparation of proso millet starch nanoparticles. Food Sci Technol Res 20(5):897–902. https://doi.org/10.3136/fstr.20.897

Tai KW, Wu TL, Chiang YC et al (2021a) Non-isocyanate polyurethane from starch. Carbohydr Polym 258:117636. https://doi.org/10.1016/j.carbpol.2021.117636

Tai NL, Ghasemlou M, Adhikari R, Adhikari B (2021b) Starch-based isocyanate- and non-isocyanate polyurethane hybrids: a review on synthesis, performance and biodegradation. Carbohydr Polym 265:118029. https://doi.org/10.1016/j.carbpol.2021.118029

Thakur K, Sharma S, Sharma R (2022) Morphological and functional properties of millet starches as influenced by different modification techniques: a review. Starch 75:2200184. https://doi.org/10.1002/star.202200184

Thielemans W, Belgacem MN, Gandini A (2006) Surface modification of starch nanocrystals. Langmuir 22(12):4804–4810. https://doi.org/10.1021/la053221x

Tomar M, Bhardwaj R, Verma SPS, Dahuja A, Krishnan V, Kansal R, Yadav VK, Praveen S, Sachdev A (2022) Interactome of millet-based food matrices: a review. Food Chem 385:132636. https://doi.org/10.1016/j.foodchem.2022.132636

Wang Y, Zhang G (2021) The preparation of modified nano-starch and its application in food industry. Food Res Int 140:110009. https://doi.org/10.1016/j.foodres.2020.110009

Wang Z, Xiao H, Wang G et al (1998) Giant magnetoresistance in Cu–Co alloys. J Appl Phys 83(11):7273–7277

Widyastuti Y (2021) Changes in physicochemical properties of pearl millet starch by heat-moisture treatment. J Food Process Preserv 45(10):e15781. https://doi.org/10.1111/jfpp.15781

Wuttisela K, Rungsardthong U, Jitcharoen J et al (2008) Enzymatic hydrolysis and properties of acid-thinned starch. Carbohydr Polym 71:186–194

Yang W, Ding S, Dong Q et al (2019a) Gelatinisation properties of waxy and non-waxy millet starches. Int J Biol Macromol 130:778–784. https://doi.org/10.1016/j.ijbiomac.2019.02.191

Yang Z, Gu Q, Li et al (2019b) Thermal properties of waxy and non-waxy proso millet starches. J Cereal Sci 88:17–23

Zavareze ER, Dias ARG (2011) Impact of heat-moisture treatment and annealing in starches: a review. Carbohydr Polym 83(2):317–328. https://doi.org/10.1016/j.carbpol.2010.08.064

Zheng B, Wang L, Zeng H et al (2020) Effect of heat moisture treatment on structure and digestibility of high amylose starch. Int J Biol Macromol 154:193–201. https://doi.org/10.1016/j.ijbiomac.2020.03.120

Zhu F (2014a) Composition, structure, and physicochemical properties of starches from diverse botanical sources. Int J Biol Macromol 64:205–212. https://doi.org/10.1016/j.ijbiomac.2013.11.019

Zhu F (2014b) Structure, physicochemical properties, and uses of millet starch. Food Res Int 64:200–211. https://doi.org/10.1016/j.foodres.2014.06.030

Siroha AK, Bangar SP, Rose PK, Kumar A (2022) Novel Approaches to Improve Functional Potential of Cereals. InFunctional Cereals and Cereal Foods: Properties, Functionality and Applications (pp. 27-49). Cham: Springer International Publishing.

Hebeish A, Abdel-Rahman A, El-Hilw Z, Hashem M (2005) Cationized Starch Derived From Pre-Oxidized Starch For Textile Sizing and Printing. Starch/Stärke, 57: 616-623. https://doi.org/10.1002/star.200400388

Liu H, Yu L, Chen L (2020) Structure and morphology of cereal starch granules. Food Chemistry, 331, 127280.

Obilana AB, & Manyasa E (2021) Granular structure and digestibility of millet starches. Journal of Food Science and Technology, 58, 2492–2501.

Choi S, Lim S (2021) Effect of amylose content on retrogradation and functional properties of cereal starches. Carbohydrate Polymers, 258, 117670.

Kumari M, Thakur S (2022) Glycemic response and resistant starch content of selected millets. Journal of Cereal Science, 105, 103451.

Chandra D, Singh S (2022) Structural diversity and functional implications of amylose–amylopectin ratio in millet starches. Starch/Stärke, 74(3–4), 210030.

Saleh AS, Zhang Q, Chen J, Shen Q (2013) Millet grains: nutritional quality, processing, and potential health benefits. Comprehensive Reviews in Food Science and Food Safety, 12(3), 281–295.

Patel A, Singh J, Bajaj R (2020) Lipid–starch interactions in cereal starches: Implications for functional properties. Food Hydrocolloids, 101, 105470.

Ramaswamy B, Tharanathan R (2021) Modifying solubility characteristics of cereal starches using physical treatments. Food Research International, 146, 110419.

Adejumo B, Samuel T (2022) Evaluation of millet starch as a pharmaceutical excipient: physicochemical and functional characteristics. Journal of Drug Delivery Science and Technology, 72, 103421.

Patel H, Shah N, Chauhan B (2021) Application of cereal and millet starches in oral drug delivery: A comparative assessment. International Journal of Pharmaceutics, 603, 120705.

Shah R, Bhatt S (2021) Millet starch as a natural super-disintegrant: A comparative tablet formulation study. International Journal of Biological Macromolecules, 170, 329–337.

Mishra G, Verma N, Tiwari A (2022) Role of cereal starches in pharmaceutical formulation development. Journal of Drug Delivery Science and Technology, 75, 103597.

Bangar SP, Kaushik N(2022) Functional cereals and cereal foods. Springer International Publishing.https://doi.org/10.1007/978-3-031-05611-6_2

Verma VC, Kumar A, Zaidi MG et al (2018) Starch isolation from different cereals with variable amylose/amylopectin ratio and its morphological study using SEM and FT-IR. Int J Curr Microbiol App Sci 7(10):211-28.

Wang J, Ye Y, Chai Z et al (2023) Physicochemical properties and in vitro digestibility of proso millet starch modified by heat-moisture treatment and annealing processing. Int J Bio Macromol 235:123829.

Zheng MZ, Xiao Y, Yang S et al (2020) Effects of heat–moisture, autoclaving, and microwave treatments on physicochemical properties of proso millet starch. Food Science & Nutrition 8(2):735-43.

Yadav S, Mishra S, Pradhan RC (2021) Ultrasound-assisted hydration of finger millet (Eleusine Coracana) and its effects on starch isolates and antinutrients. Ultrasonics sonochemistry 73:105542.

Li W, Gao J, Saleh AS et al (2018) The modifications in physicochemical and functional properties of proso millet starch after Ultra-High pressure (UHP) process. Starch-Stärke 70(5-6):1700235.

Sujatha M, Hymavathi T, Devi KU et al (2017) Effect of heat treatment and gamma irradiation on in vitro protein digestibility of selected Millet Grains. International Journal of Current Microbiology and Applied Sciences 6(11):1014-20.

Sonkar S, Jaddu S, Pradhan RC et al (2023) Effect of atmospheric cold plasma (pin type) on hydration and structure properties of kodo-millet starch. Lwt 182:114889.

Kaushik A, Saxena DC, Singh S (2025) Modification of browntop millet (Brachiaria ramosa) starch: Effects of pulse electric field and gamma-irradiation on morphological, thermal, rheological, and powder flow properties. Int J Biol Macromol 309:143132.

Siroha AK, Bangar SP, Sandhu KS et al (2021) Effect of cross-linking modification on structural and film-forming characteristics of pearl millet (Pennisetum glaucum L.) starch. Coatings 11(10):1163.

Afolabi TA, Olu-Owolabi BI, Adebowale KO et al (2012) Functional and tableting properties of acetylated and oxidised finger millet (Eleusine coracana) starch. Starch-Stärke 64(4):326-37.

Mahanure K, Bist Y, Kumar Y et al (2025) Impact of octenyl succinic anhydride esterification on Kodo millet starch-commercial protein blends functionality and Pickering emulsion properties. Int J Bio Macromol 308:142734.

Kurichh R, Kumar Y, Bist Y et al (2025) Enhancing Kodo millet starch-based hybrid gel for fat replacement in cookies: Dual modification (heat treatment-octenyl succinic anhydride) for improved properties. Food Chemistry 478:143709.

Malik MK, Kumar V, Singh J et al (2023) Efficiency of phosphorylated mandua starch in matrix tablet for targeted release of mesalamine in colon. Journal of Drug Delivery Science and Technology 81:104251.

Babu A, Kumar AR, Amrutha NR et al (2024) Utilizing foxtail millet husk waste for sustainable new bioplastic composites with enhanced thermal stability and biodegradability. Int J Biol Macromol 282:137283.

Malik MK, Kumar V, Kumarasamy V et al (2024) Film Coating of Phosphorylated Mandua Starch on Matrix Tablets for pH-Sensitive Release of Mesalamine. Molecules 29(13):3208.

Nazir M, Jhan F, Gani A (2024) Fabrication of millet starch nanocapsules loaded with beta carotene using acid hydrolysis and ultrasonication: Characterisation, release behaviour and bioactivity retention. Ultrasonics Sonochemistry 111:107112.

Jhan F, Shah A, Gani A et al (2020b) Nano-reduction of starch from underutilised millets: Effect on structural, thermal, morphological and nutraceutical properties. Int J Biol Macromol 159:1113-21.

Bhardwaj M, Bist Y, Saxena, DC (2025) Effect of Amylose Content and Starch Nanocrystals on the Structure–Function Properties of Pearl Millet Starch Nanocomposite Films. Food Biophysics 20,169. https://doi.org/10.1007/s11483-025-10058-9

Zhao Y, Zhang B, Chen H (2021) Swelling behaviour and water solubility of starches from minor cereals: influence of structural variations. Food Chemistry, 340, 128160.

Kaur N, Gill B (2020) Functional properties of starches from pearl and finger millet: Influence of processing. International Journal of Biological Macromolecules, 164, 4203–4211.

Chapter 18
Therapeutics of Indian Cereals

Lalit Raj Singh and Karan Singh

Abbreviations

Anti-TPO	Antithyroid peroxidase
COX-2	Cyclooxygenase-2
CVDs	Cardiovascular disorders
GABA	Gamma-aminobutyric acid
GI	Glycemic index
GNA	Galanthus nivalis agglutinin
GRAS	Generally recognized as safe
HIV	Human immunodeficiency viruses
HT-29	Human colon cancer cells
IBS	Irritable bowel syndrome
LAB	Lactic acid bacteria
MCF-7	Michigan cancer foundation-7
NSPS	Non-starch polysaccharides
PUFA	Polyunsaturated fatty acid
QPS	Qualified presumption of safety
RS	Resistant starch
SCFA	Short-chain fatty acids
T2DM	Type 2 diabetes mellitus
TI	Therapeutic index
TNF	Tumor necrosis factor
TPC	Total phenolic content
XOS	Xylooligosaccharides

L. R. Singh (✉)
National Innovation Foundation, Gandhinagar, Gujarat, India

K. Singh
Department of Medicinal Plants Sciences, Dev Sanskriti University, Haridwar, Uttarakhand, India

© The Author(s), under exclusive license to Springer Nature Switzerland AG 2026
J. Singh, M. Kumar Malik (eds.), *Millets and Millet Starch: Chemical Composition, Functional Modification, and Applications*,
https://doi.org/10.1007/978-3-032-09502-2_18

18.1 Rationale of the Study

The rapidly escalating lifestyle disorders have become a cause of concern for the generation. The "*food as medicine*" fortifies the body and rejuvenates immunity through the scientific justification of the cereals for therapeutic purposes. The extraction of nutrients, processing methodology, and changes in metabolic rate due to changes in the chemical properties of molecules have significant effects on our health. The chemical composition of cereals needs to be understood to harness their optimum health benefits.

18.2 Introduction

Over the last few decades, lifestyle-related disorders escalated rapidly including diabetes, high cholesterol level, and cardiovascular disorders. Dietary therapy having cereals is one of the safest resorts for the management of lifestyle disorders as evidenced by scientific studies over time (Singhal and Kaushik 2016). Cereals are any grass cultivated for its edible larger size of grains and composed of an endosperm, a germ, and a bran classified as cereals including wheat, rice, rye, oats, barley, and maize. Millets are also *Poaceae* family members with smaller size of grains (Kumar et al. 1992). Edible grains from other plant families are called pseudocereals such as buckwheat, quinoa, and chia (Table 18.1). Cereals are the most important staple foods grown in approximately 60% of the cultivated area globally and an imperative source of energy food for humans and animals. It provides major human nutrients such as proteins, lipids, carbohydrates, minerals, and vitamins. Whole grains embrace the notable quantity of dietary fibers (soluble and insoluble), phytochemicals, antioxidants, and minerals, a panacea of bioactive components essential for comprehensive health benefits (Ahmad et al. 2015). The bioactive components are higher in less processed or unprocessed cereals resulting in the prevention and control of cardiovascular disorders (CVDs), cancer, diabetes, and hypertension by facilitating healthy gut biota and detoxification of the body (Cho et al. 2013).

Probiotics are stated as "unique amalgamation of healthy microorganisms that when administered in rational amount exhibit a beneficial effect on the host's health." Probiotics can be administered as part of fermented foods or as dietary supplements depending upon the availability. The fermented cereal products act as live probiotics that balance the microbiota of intestine, which consequently has improved health benefits (Yasar et al. 2018). The lactic acid bacteria (LAB) Lactobacillus and Bifidobacterium species have been qualified and classified as "Generally Recognized as Safe" (GRAS) and "Qualified Presumption of Safety" (QPS) status, as their consumption does not present risks for the host's health but also supplements the body with indispensable minerals (Colautti et al. 2022). A healthy diet is essential to keep a healthy microbiota of the gut that implicit an important role in the vital functioning of the body and protects it from several diseases. Therefore, a healthy diet comprises whole grains of cereals.

Table 18.1 Classification of cereals

Cereals (class 1)	Millets (class 2)	Pseudocereals (class 3)
Barley	Foxtail	Amaranth
Wheat	Finger millet	Buckwheat
Rice	Barnyard millet	Quinoa
Maize	Japanese barnyard millet	Chia
Oat	Pearl millet	
	Little millet	
	Kodo millet	
	Teff millet	
	Proso millet	
	Sorghum	
	Black fonio	
	White fonio	
	Job's tear	

Cereals are a crucial part of our daily diet and principally the source of carbohydrates. The adequate amount of nutrients such as vitamins, trace elements, minerals, and essential amino acids are highly required to lead the quality of life. Some therapeutically valuable cereals are enumerated in Fig. 18.1.

Class I These cereals are most commonly used as staple food. The grain size of cereals is larger than millets and belongs to the *Poaceae* family. The bioavailability and nutrient extraction efficacy can be enhanced by processing which results in specific modifications in active constituents of the plants and enhances the assimilation rate. The therapeutic index (TI) and nutraceutical potency of the less processed products are enhanced through the elimination of phytotoxic alkaloids and tannins (Bravo et al. 2013).

(i) Barley (*Hordeum vulgare* L.): It is considered an ancient cultivated plant of human civilization. It is a food with a low glycemic index (GI) and is beneficial to gluten-sensitive people. It is a high-fiber cereal grain that contains elevated levels of β-glucan (BG) relative to other cereal grains that lower cholesterol (Abumweis et al. 2016). It is a recommended food for type 2 diabetes or prediabetes and irritable bowel syndrome (IBS)-afflicted people (Balmus et al. 2022). The composition of grain shows 60–65% starch, 11–16% protein, 5–10% β-glucans, 1.5–2.5% minerals, and 2–3% lipids. The cereal possesses essential minerals (iron, calcium, magnesium, manganese, molybdenum, lithium, germanium, zinc, copper), vitamins (E, C, B1, B2, B6, B12), bioflavonoids, polysaccharides, and polypeptides (Obadi et al. 2021). The juice (Swaras) of fresh leaves is reported in ancient *Ayurvedic* texts as general body rejuvenator and considered as green blood. Several researchers in recent times have established anticancer activity of its fresh juice (Czerwonka et al. 2017).

(ii) Wheat (*Triticum aestivum* L.): It is recognized as commonly used food grain across the globe. It rejuvenates innate immunity and detoxifies the body while administered in fresh conditions. Five percent wheat

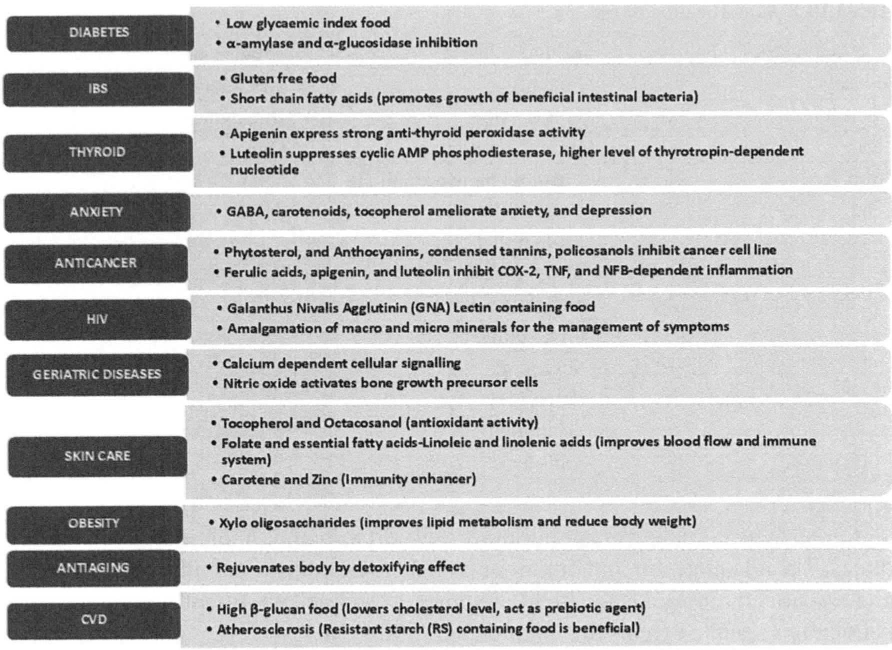

Fig. 18.1 Therapeutic effect of cereals in remediating diseases

bran xylooligosaccharides (XOS) improve blood lipid metabolism, reduce body weight, improve blood glucose level, and have antioxidant activity (Wang et al. 2011a, b). Whole grains possess the high content of magnesium and act as cofactors inducing insulin secretion and resulting in reduced risk of type 2 diabetes. Wheat germ oil is rich in octacosanol (28-carbon long-chain alcohol), tocopherol, folate, essential fatty acids-linolenic acid, linoleic acid, beta-carotene, and zinc that are good for the skin due to the presence of antioxidants that improve blood flow, strengthen the immune system, and restore overall health (Ghafoor et al. 2017).

(iii) Rice (*Oryza sativa* L.): Paste of fresh paddy applied with butter to cure skin infections, eruptions, itching, eczema, and flaring. *Herpes zoster* marks are remediated by applying paste with coconut oil (Jemimah et al. 2019). It is used after soaking in water to cure leukorrhea problems by the presence of starch and natural supplements. These therapeutic effects are expressed by the plethora of compounds including vitamin B complex (thiamine, niacin, and riboflavin), amino acids, minerals (Ca, Mg, P, Fe, Cu, Zn, and Mn), and carbs (amylose and amylopectin) (Wang et al. 2011a, b). Nowadays, in Asia Pacific countries, biofortified rice-fermented beverages are in huge demand that combat vitamin A and micronutrients like Fe and Zn deficiencies easily. The beverages fermented at 37°C for 4 days consecutively in anaerobic conditions increase the antioxidant and enzymatic activity leading to the production of glucoamylase, α-amylase, and phytase (therefore, an increase of phytate bioavailability) which results in higher mineral availability, that is, Fe, Ca, Mg, Na, and Mn. Fermentation of rice increases the hydrosoluble vitamin s.a. folic acid, thiamine, ascorbic acid, and pyridoxine as compared with the unfermented process (Chavan et al. 1989).

(iv) Maize (*Zea mays* L.): It is one of the highly consumed cereals that contains approx. 72% starch, 10% proteins, and 4% fiber, along with vitamin B and essential minerals. The beverages prepared from it scored 5 out of 9 points for general acceptance, with lactic acid as a major organic acid collected during the fermentation cycle, along with a total of 70 identified volatile compounds. Its phytochemicals such as galanthus nivalis agglutinin (GNA) lectin and resistant starch (RS) are used respectively for controlling HIV, cecal cancer, obesity, and atherosclerotic complications (Rouf Shah et al. 2016). The decoction of maize silk, roots, cob, and leaves ameliorates kidney dysfunction, urinary bladder, and urolithiasis conditions of the patient. However, leaves, maize silk, and cobs are thrown by people in the light of ignorance of their therapeutic benefits (Ab Razak et al. 2022).

(v) Oat (*Avena sativa* L.): It is a potential functional cereal, due to its rich content of proteins, soluble fiber, and antioxidants along with β-glucan as a prebiotic agent. It generates short-chain fatty acids (SCFAs) which promote the growth of beneficial intestinal bacteria (*Enterobacteria*) and inhibit harmful bacteria including anaerobes and clostridia (Drzikova et al. 2005). The oat beverages fermented by *Lactobacillus* strains exhibited high antioxidant activity due to the presence of rich polyphenol content. The fermentation increases the β-glucan content, probiotic viability, and antimicrobial activity simultaneously while reducing the hydrolysis index of starch and autoaggregation capacity (the process of reducing/blocking the adhesion of pathogenic bacteria to epithelial cells) (Gyawali et al. 2015). The rational processing technologies change the chemical structures, distribution, and bioavailability that ameliorate the therapeutic benefits of the cereals. The α-amylase (reversible anticompetitive inhibition by ginsenoside Re) and α-glucosidase (reversible linear mixed inhibition by soyasaponin Ba) are inhibited by ginsenoside Re and soyasaponin Ba released during digestion of cereals in the gastrointestinal tract (Wei et al. 2021). The α-amylase inhibitors check the hydrolysis of starch by blocking the active centers of the enzymes, while α-glucosidase inhibitors reduce sugar-binding sites and shorter oligomers by occupying enzymes, which results in delayed intestinal absorption of glucose (Zhang et al. 2021).

Class II Millets are small-sized food grains belonging to *Poaceae* family that include sorghum (*Sorghum bicolor* L. Moench), foxtail millet (*Setaria italica (L.) P. Beauv.*), proso millet (*Panicum miliaceum L.*), pearl millet (*Pennisetum glaucum (L.) R. Br.*), and finger millet (*Eleusine coracana* (L.) Gaertn.). Minor millets includes little millet (*Panicum sumatrense* L.), kodo millet (*Paspalum scrobiculatum* L.), and barnyard millet (*Echinochloa colonum* (L.) Link.).

(i) Sorghum (*Sorghum bicolor* L. Moench): Sorghum is a drought-tolerant millet and a rich source of copper, magnesium, phosphorus, selenium, iron, zinc, and B1, B3, B5, and B6 vitamins. It has the highest concentration of phenolic compounds that exhibit antioxidant effects. It is consumed as a functional food to overcome protein and amino acids deficiency. Several multifunctional food additives and colorants (apigeninidin) are derivatives of 3-deoxyanthocyanidins, phenolic compounds, and condensed tannins (Akogou et al. 2018). Phytosterols and anthocyanins from bran inhibit Michigan Cancer Foundation-7 (MCF-7) breast cancer cell lines, phenols of bran extract inhibit human colon cancer cells (HT-29), and tannins, phytosterols, and policosanols attributed to anticancer activity (Duodu and Awika 2019). Sorghum bran fiber possesses non-starch polysaccharides (NSPs), which are implicated to decreased fat absorption and improved intestinal health. Ferulic acid, apigenin, and luteolin inhibit cyclooxygenase-2 (COX-2), tumor necrosis factor (TNF), and nuclear factor kappa B (NF-kB)-dependent inflammatory pathways (Meena et al. 2022). 100 g powder approximately comprises 63 g of carbohydrates, 10.2 g of protein, 3.4 g of fat, and 11 g of dietary fiber that liberates 345 kcal of energy (Mohammed et al. 2011).

(ii) Foxtail (*Setaria italica (L.) P. Beauv.*): Foxtail millet is a rich source of vitamins B1, B3, B5 and B6, and Ca, Cu, Mg, and P. It is a source of iron, niacin, vitamin B_6, and zinc. 100 g powder gives 67 g of carbohydrates, 9.7 g of proteins, 4.4 g of fat, and 4.6 g of dietary fiber and it generates 356 kcal energy. The cellular signaling involves calcium as a stimulant for mature bone cells by facilitating the production of nitric oxide, triggering the activation of bone growth precursor cells, and resulting in the regeneration of bone tissue. Geriatric degenerative bone diseases involving decreased bone density, hormonal changes, and metabolic disequilibrium, causing osteopenia, osteoarthritis, and osteoporosis are controlled by their usage. However, a specific mode of action needs to be revealed (Nithya et al. 2023).

(iii) Proso millet (*Panicum miliaceum L.*): The high content of B1, B2, B3, B5 and B6, Cu, P, Mg, Zn, Fe, and Se. 100 g powder gives 66 g of carbohydrates, 10.5 g of proteins, 4 g of fat, and 6.5 g of dietary fiber that liberates 350 kcal of energy. The sulfur-containing amino acids (methionine and cysteine), high crude fiber, and total phenolic contents (TPC) exhibit antioxidant activity (Li et al. 2022a, b).

(iv) Pearl millet (*Pennisetum glaucum (L.) R. Br.*): It is a heat- and drought-enduring crop rich in Cu, Fe, Mg, P, Se, and Zn. It is also a source of vitamins B1 and vitamin B3. 100 g powder comprises 62.5 g of carbohydrates, 10 g of proteins, 6.8 g of fat, and 11 g of dietary fiber that liberates 366 kcal of energy. It's gluten-free, is rich in dietary fiber and essential amino acids (leucine, isoleucine, and lysine) containing cereal, is low in glycemic index, controls weight, and reduces the risk of chronic diseases like diabetes. It also ameliorates geriatric diseases and is reported to be less susceptible to aflatoxins after storage (Dias Martins et al. 2018).

(v) Finger millet (*Eleusine coracana* (L.) Gaertn.): It is a potential source of B1, Cu, Mg, P, Se, and Fe. 100 g powder comprises 68 g of carbohydrates, 6.8 g of proteins, 2 g of fat, and 11.2 g of dietary fiber that generates 336 kcal of energy. Bioaccessibility and its potential health benefits are increased by the malting or fermentation process. The extraction efficiency of bioactive constituents (ferulic acids, feraxans, and quercetin) increases that exhibits higher healing effects such as antioxidative, antimicrobial, and anti-inflammatory (Udeh et al. 2017).

(vi) Little millet (*Panicum sumatrense* L.): Little millet is a powerhouse of Cu, Mg, Se, B1, P, and Zn minerals. 100 g powder comprises 66.2 g of carbohydrates, 9.4 g of proteins, 4 g of fat, and 7.5 g of dietary fiber that liberates 353 kcal of energy. The malting process increases the therapeutic potential of its bioactive constituents such as gamma-aminobutyric acid (GABA), carotenoids, tocopherol, and phenolic compounds. GABA is a neurotransmitter that ameliorates anxiety, depression, and geriatric diseases (Guha et al. 2015).

(vii) Kodo millet (*Paspalum scrobiculatum* L.): Kodo millet is a rich source of Mg, Se, B1, B2, Cu, and Zn. 100 g powder comprises 69 g of carbohydrates, 8.3 g of protein, 2.6 g of fat, and 6.4 g of dietary fiber giving 336 kcal of energy. It is a good source of polyphenols, flavonoids, and antioxidants. Phytates exhibit anticancerous, weight-lowering, and antiarthritic activity, and linoleic acid revealed antitumor capabilities (Shukla et al. 2024).

(viii) Barnyard millet (*Echinochloa colonum* (L.) Link): It is a rich source of B1, Cu, Ca, Mg, P, Zn, Mn, and Fe. 100 g powder comprises 48 g of carbohydrates, 5.8 g of proteins, 2.9 g of fat, and 23 g of dietary fiber that generates 315 kcal of energy. It relieves nutritional deficiency disorders (anemia, irregular menses), skin affliction, IBS, and hysteroepilepsy. It strengthens intestinal motility and remediates digestive disorders. The processing (malting/fermentation) reduces tricin, one of the phytotoxic compounds (Singh 2019).

(ix) Japanese barnyard millet (*Echinochloa esculenta* (A. Braun) H. Scholz.): It is a rich source of B1, B5, P, Zn, Cu, and Mg. The 100 g powder gives 69.6 g of carbohydrates, 8.8 g of proteins, 3.6 g of fat, and 4.3 g of dietary fiber that liberates 351 kcal of energy. It revealed antimutagenicity and antioxidant activity (scavenging H_2O_2 radicals) due to the presence of higher polyphenolic compounds (Renganathan et al. 2020).

(x) Teff millet (*Eragrostis tef* (Zucc.) Trotter): Teff is rich in vitamins B1, B2, B3, B5, and B6, Cu, Fe, Mg, and P. 100 g powder comprises 66.3 g of carbohydrates, 12.4 g of proteins, 2.4 g of fat, and 8.0 g of dietary fiber that gives 351 kcal of energy. It is preferred by farmers due to its high resilience in waterlogged and wasteland conditions and by consumers due to gluten-free food. It is a potential source of iron and lysine amino acids that play a crucial role in generating energy by fatty acid metabolism and lowering bad cholesterol levels (Mengesha et al. 2022).

(xi) Black fonio (*Digitaria iburua* Stapf): It is an ample source of micro minerals and vitamins. 100 g powder comprises 70.4 g of carbohydrates, 8 g of proteins, 4 g of fat, and 4 g of dietary fiber that generates 354 kcal of energy. The malting process improves phytoextraction efficiency and removes anti-nutritional factors like phytic acid, saponins, and tannins that result in enhanced bioavailability. The low gluten content and glycemic index are helpful in the management of IBS and diabetes (Mengesha et al. 2022).

(xii) White fonio (*Digitaria exilis* (Kippist) Stapf): It is an ample source of Cu, Mg, P, Zn, and folate. 100 g powder comprises 77.4 g of carbohydrates, 7.2 g of proteins, 1.8 g of fat, and 2.4 g of dietary fiber that gives 356 kcal of energy. Apigenin and luteolin manifest strong antithyroid peroxidase (anti-TPO) activities, resulting in a significant reduction of the hormonogenic capacity of this enzyme. In addition, luteolin suppresses cyclic AMP phosphodiesterase, implying a concomitant higher production of the thyrotropin-dependent nucleotide that helps in equilibration of thyroid hormone levels (Bassey et al. 2023).

(xiii) Job's tears (*Coix lacryma-jobi* L.): Job's tears is a potential source of Cu, Mg, P, Fe, Zn, and vit B1. 100 g powder liberates 67.7 g of carbohydrates, 13.5 g of protein, 3 g of fat, and 3.2 g of dietary fiber which gives 357 kcal of energy (FAO 2023). Its active constituents (coixdines A-B) show cytotoxicity against CT-26 (colon cancer) and BxPC-3(human pancreatic cancer) cell lines. The compound 24-ethylcholesta-4E22-dien-3Kt exhibited pro-apoptotic effects on three types of tumor cells (HeLa, HepG2, and SGC-7901) by expressing an inhibitory effect on DNA topoisomerase I (Yin et al. 2023; Yu et al. 2021).

Class III Pseudocereals: These are non-grasses that are used in the same way as cereals and provide nutrition by the presence of proteins, carbohydrates, fatty acids, and minerals.

(i) Amaranth (*Amaranthus species*): Some of the species of *Amaranthaceae* family including *A. caudatus*, *A. cruentus*, *A. hypochondriacus*, *A. gangeticus*, *A. tricolor*, *A. mangostanus,* etc., are used during ritual and religious ceremonies. These are consumed as dietary supplements and functional foods. These species generally contain saponins, proto alkaloids, betacyanins, iron, and minerals that exhibit antidiabetic, antimicrobial, antioxidant, cardiac, and digestive stimulant effects (Baraniak and Kania-Dobrowolska 2022).

(ii) Buckwheat (*Fagopyrum esculentum* Moench): It is a genus of the *Polygonaceae* family and possesses D-chiro-inositol, proteins, flavonoids, iron, calcium, and minerals that express anticancer, antidiabetic, anticholesteremic, antihypertensive, anti-inflammatory, and neuroprotective effect (Antonio Gimenez-Bastida and Zielinski 2015).

(iii) Quinoa (*Chenopodium quinoa* L.): It is a genus of the *Amaranthaceae* family and principally contains protein, lipids, fibers, vitamins, and minerals that are beneficial for lactose intolerants, anemic, obese, elderly, malnutrition, and celiac disease patients (Ghafoor et al. 2017). This pseudocereal is commonly used as in our food matrix. It is the only plant-based food that has all the essential amino acids (lysine, methionine, and threonine), trace elements, vitamins, and proteins that match milk. It also decreases the risk of type 2 diabetes and cardiovascular diseases. Furthermore, it is gluten-free and does not cause allergic problems in celiac people after consumption. Therefore, quinoa-based fermented beverages are ample source of proteins, fiber, vitamins, and minerals, simultaneously with probiotic effect that exerts health benefits by improving microbiota of the human gastrointestinal tract (Valero-Cases et al. 2020).

(iv) Chia (*Salvia hispanica* L.): It is a member of the *Lamiaceae* family, which provides macro- and micronutrients, consequently a wonderful source of phenolic compounds like caffeic acid (hypoglycemic and antiepileptic), rosmarinic acid (antimicrobial, antioxidant, anti-inflammatory, and immune regulatory), quercetin (antiaging), myricetin (anticarcinogenic), and protocatechuic acid (antioxidant). Chia seeds have protein, polyunsaturated fatty acid (PUFA), fiber, vitamins, and minerals that remediate diabetes, hypertension, dyslipidemia, anxiety, depression, constipation, and autoimmune disorders (Biswas et al. 2023).

18.3 Chemical Structures and *In Vivo* Activities of Cereals

The structure-activity relationship elucidates that the number of hydroxyl and methoxy groups present in the aromatic ring of phenolic acids exhibits the inhibitory activity. The enzyme inhibiting potency of phenolic compounds is highly correlated with their structures. It could be mixed, uncompetitive, and competitive types depending upon the binding free energy. Further, the glucoside substitutions and position of their substitutions escalate the enzyme inhibition activities due to an increased number of total aromatic hydroxyl groups. For example, the hydroxy (–OH) groups at the C-3 position of ring C, the C-3′ and C-4′ position of ring B, and the glucoside substitutions at the C-3 position of ring C were crucial for the enzyme inhibition activities of flavanols. Polyphenols with higher polymerization also have inhibition effects on the enzymes due to increased number of aromatic hydroxyl groups (Yahfoufi et al. 2018). Pigmented cereals are highly demanded due to their health benefits which is associated with adequate anthocyanins. For example, traditional red rice varieties s.a. Masuran and Sudu Heenati showed significant anti-amylase activity in a dose-dependent manner (Thushara et al. 2022). Some of the important classes of the compounds are given as under for specific understanding:

I. Phenolic compounds: The enzyme inhibition is expressed prominently by phenolic compounds via hydrogen bonding, hydrophobic interactions, and van der Waals interactions. Phenolic acids such as caffeic, ferulic, and sinapic acids are found in soluble fractions that exhibit hydrogen bonding for enzyme inhibition, while ferulic and p-coumaric acids are abundantly found in bound fractions that exhibit hydrophobic and van der Waals interactions. Naringenin, kaempferol, luteolin glycoside, apigenin, (+)-catechin/(−)-epicatechin, daidzein, caffeic acid, ferulic acid, and syringic acid are some of phenolic compounds found from cereals and considered as the enzyme inhibitors (Naresh and Nidhi 2019).

II. Peptides: These are hydrolysates of different proteins s.a. albumin, globulin, prolamin, and glutelin obtained from therapeutically valuable cereals s.a. rice bran proteins subjected to different protease hydrolysis (Alcalase, Neutrase, Flavourzyme, and Protamax) that resulted in different metabolic activities. The highest inhibition activities are found in Protamax and Alcalase-catalyzed hydrolysates of albumin and glutelin, respectively. The potential hydrolysates s.a. IQAEGGLT and DKKYPK were identified from proteins of quinoa and that inhibit α-glucosidase activity via hydrophobic interaction. The hydrolysates released from quinoa showed the highest inhibitory effects, reaching an IC_{50} value of 0.19 mg protein/ml for α-amylase inhibition and 1.75 mg protein/ml for α-glucosidase inhibition, respectively (He et al. 2022).

III. Non-starch polysaccharides: Barley, buckwheat, and wheat exhibit α-glucosidase and α-amylase inhibitory activities due to the presence of non-starch polysaccharides. Barley polysaccharide exhibited a noncompetitive inhibition of α-glucosidase with IC_{50} at 22.49 mg/ml. The sulfation process significantly raises the enzyme inhibitory activity. Buckwheat exhibits α-glucosidase inhibition activity due to presence of neutral polysaccharide (TBP-II, 26 kDa). Wheat exhibits α-amylase inhibition activity due to presence of arabinoxylan and β-glucan that impair diffusion of α-amylase due to similarity in size (competitive inhibition), which implicit starch hydrolysis in the small intestine due to poor diffusion and less availability of α-amylase (Qian et al. 2015).

IV. Lipids: The wheat germ oil comprises unsaturated fatty acids and phosphate group in the glycerides s.a. 1,2-dilinoleylglycerol-3-phosphate and 1-palmitoyl-2-linoleoylglycerol-3-phosphate that exhibit the α-glucosidase inhibitory activity. Alkylresorcinols are another important component found in wheat bran lipid that exhibit the α-glucosidase inhibitory activity. It directly regulates enzyme activity, such as inhibition of acetylcholinesterase, and the inhibitory effect is associated with the length of the alkyl side chain (Li et al. 2022a, b).

V. α-glucosidase and α-amylase inhibited by 70% ethanolic extracts of different types and varieties of cereals s.a. sorghum, foxtail millet, and proso millet (Kim et al. 2011). The therapeutic effect of cereals on some diseases is given in Fig. 18.1.

Carbohydrate digestibility is regulated by hydrolysing enzymes (α-amylase and α-glucosidase), and to control postprandial hyperglycemia, the inhibition of hydrolysing enzymes is vital. Cereals are a viable and potential treatment for prophylactic management of type 2 diabetes mellitus (T2DM). Cereals possess a rich number of phenolic compounds, peptides, non-starch polysaccharides, lipids, and minerals that inhibit several hydrolysing enzymes. For example, the number and location of hydroxyl groups of phenolic acids, the molecular weight of polysaccharides, acetylation, methylation, hydrophilic, hydrophobic, and van der Waals interactions, all these physicochemical activities affect the enzyme inhibition activities (Gong et al. 2020). In developing countries, the scarcity of animal products is most commonly reported wherein people faces malnutrition challenges. Hence, the potential sources of nutrients could be materialized by the consumption of alternate and easily available powerhouse of energy, that is, cereals. The cereals are the major crops, and due to their immense nutritional potential, these have become an integral part of people's diets for millennia. Today, they are important not only to ensure food security but also to combat malnutrition, reduce poverty, improve human health, and improve agricultural sustainability (Shavanov 2021). In alignment with Hippocrates' aphorisms, "*Let food be your medicine and medicine be your food*" is now been proven and testified on scientific merits that for healthy life, one should include cereals in their diets to enhance immunity, for amelioration of minerals, and for sustaining enriched microbiota of intestine. The fermentation process facilitates the delivery of probiotic bacteria into the human gut and detoxifies (removal of aflatoxins effect) cereals. The fermented process not only reduces the toxins of raw materials such as aflatoxins that are causal elements for suppressing the human immune system by affecting nutrient absorption, accumulation of free radicals, and inducing liver cancer but also promotes probiotics (Aslam et al. 2021). Good quality probiotics are the fulcrum of our living system, that is, soil. Our soil is essential for higher crop productivity and nutritional value of the agriproducts. Soil health is directly connected with healthy food and assuring good human health (Frac et al. 2018).

18.4 Conclusions

The scarcity of resources and challenges for a balanced diet are escalating across the globe. The role of cereals in the management of lifestyle disorders via providing a balanced diet is crucial. The chapter showcases the therapeutic benefits of cereals for the mitigation of several lifestyle disorders. The famous and testified quote of Hippocrates, "*Let food be thy medicine and medicine be thy food,*" should be emphasized for sustainable holistic health through cereals (Wegener 2014). The rational application of cereals for nutrition and therapy needs to be met by understanding their composition, processing, mode of action, regulation pathways, signaling hormones, and antioxidant and antimicrobial activity. It helps in rejuvenation of immunity through detoxifying vital organs. Processed food products enhance the bioavailability of active constituents and the removal of phytotoxins. The fermented foods of cereals act as probiotics that maintain healthy gut biota. Fermentation process escalates antioxidant capacity, reduces glucoamylase and α-amylase production, increases phytase activity, and increases hydrosoluble vitamins and antibiotic susceptibility. The rational fermentation processing exhibited an increase in lactic, succinic, and acetic acid, high probiotic viability, antimicrobial activities, and an increased bioavailability of minerals. These solutions could be sustainable, effective, and better if good quality cereals are available. Quality cereals are the fulcrum of a living soil ecosystem. The productivity and nutritional value of cereals are outcome of soil health. Soil health needs to be promoted through emphasizing organic cultivation. Hence, it can be deduced that for realizing sustainable therapeutic benefits of cereals, the bioavailability of active constituents is imperative and may be ensured through good soil health.

Acknowledgments The authors extend their appreciation to faculty members who provide data and access to resources up to the extent possible.

No Conflict-of-Interest Statement There was no conflict of interest in this study.

References

Ab Razak SN, Amrang NZ, Halid NA et al (2022) Corn silk tea: artificial chemical free. Multidiscip Appl Res Innov 3:154–159

AbuMweis S, Thandapilly SJ, Storsley J, Ames N (2016) Effect of barley β-glucan on postprandial glycaemic response in the healthy human population: a meta-analysis of randomized controlled trials. J Funct Foods 27:329–342

Ahmad IM, Samuel E, Makama SA, Kiresur VR (2015) Trend of area, production and productivity of major cereals: India and Nigeria scenario. Res J Agric For Sci 3(2):10–15

Akogou FUG, Kayodé APP, den Besten HMW et al (2018) Extraction methods and food uses of a natural red colorant from dye sorghum. J Sci Food Agric 98:361–368

Antonio Giménez-Bastida J, Zieliński H (2015) Buckwheat as a functional food and its effects on health. J Agric Food Chem 63:7896–7913

Aslam M, Aslam A, Sheraz M et al (2021) Lead toxicity in cereals: mechanistic insight into toxicity, mode of action, and management. Front Plant Sci 11:587785

Balmus IM, Copolovici D, Copolovici L, Ciobica A, Gorgan DL (2022) Biomolecules from plant wastes potentially relevant in the management of irritable bowel syndrome and co-occurring symptomatology. Molecules 27:2403

Baraniak J, Kania-Dobrowolska M (2022) The dual nature of amaranth—functional food and potential medicine. Foods 11(4):618

Bassey SO, Chinma CE, Ezeocha VC et al (2023) Nutritional and physicochemical changes in two varieties of fonio (*Digitaria exilis* and *Digitaria iburua*) during germination. Heliyon 9(6):e17452

Biswas S, Islam F, Imran A et al (2023) Phytochemical profile, nutritional composition, and therapeutic potentials of chia seeds: a concise review. Cogent Food Agric 9(1):2220516

Bravo HR, Copaja SV, Lamborot M (2013) Phytotoxicity of phenolic acids from cereals. In: Herbicides-advances in research. InTech, p 37

Chavan JK, Kadam SS, Beuchat LR (1989) Nutritional improvement of cereals by fermentation. Crit Rev Food Sci Nutr 28:349–400

Cho Susan S, Qi L, Fahey GC Jr, Klurfeld DM (2013) Consumption of cereal fiber, mixtures of whole grains and bran, and whole grains and risk reduction in type 2 diabetes, obesity, and cardiovascular disease. Am J Clin Nutr 98(2):594–619

Colautti A, Arnoldi M, Comi G, Iacumin L (2022) Antibiotic resistance and virulence factors in lactobacilli: something to carefully consider. Food Microbiol 103:103934. https://doi.org/10.1016/j.fm.2021.103934

Czerwonka A, Kawka K, Cykier K et al (2017) Evaluation of anticancer activity of water and juice extracts of young Hordeum vulgare in human cancer cell lines HT-29 and A549. Ann Agric Environ Med 24:345–349

Dias-Martins AM, Pessanha KLF, Pacheco S et al (2018) Potential use of pearl millet (*Pennisetum glaucum* (L.) R. Br.) in Brazil: food security, processing, health benefits and nutritional products. Food Res Int 109:175–186

Drzikova B, Dongowski G, Gebhardt E (2005) Dietary fibre-rich oat-based products affect serum lipids, microbiota, formation of short-chain fatty acids and steroids in rats. Br J Nutr 94:1012–1025

Duodu KG, Awika JM (2019) Phytochemical-related health-promoting attributes of sorghum and millets. In: Sorghum and millets. AACC International Press, pp 225–258

FAO (2023) Millets recipe book. Rome, Italy. https://openknowledge.fao.org/handle/20.500.14283/cc8019en https://doi.org/10.4060/cc8019en

Frac M, Hannula SE, Belka M et al (2018) Fungal biodiversity and their role in soil health. Front Microbiol 9(707):2018. https://doi.org/10.3389/fmicb.2018.00707

Ghafoor K, Özcan MM, AL-Juhaımı F et al (2017) Nutritional composition, extraction, and utilization of wheat germ oil: a review. Eur J Lipid Sci Technol 119(7). https://doi.org/10.1002/ejlt.201600160

Gong L, Feng D, Wang T et al (2020) Inhibitors of α-amylase and α-glucosidase: potential linkage for whole cereal foods on prevention of hyperglycemia. Food Sci Nutr 8(12):6320–6337

Guha M, Sreerama YN, Malleshi NG (2015) Influence of processing on nutraceuticals of little millet (*Panicum sumatrense*). In: Preedy V (ed) Processing and impact on active components in food. Academic, pp 353–360. https://doi.org/10.1016/B978-0-12-404699-3.00042-1

Gyawali R, Minor RC, Donovan B et al (2015) Inclusion of oat in feeding can increase the potential probiotic bifidobacteria in sow milk. Animals 5:610–623

He R, Liu M, Zou Z et al (2022) Anti-inflammatory activity of peptides derived from millet bran in vitro and in vivo. Food Funct 13(4):1881–1889

Jemimah G, Reddy TS, Keerthana M et al (2019) Medical plants used in the treatment of skin diseases. World J Pharm Res 8(13):496–510

Kim JS, Hyun TK, Kim MJ (2011) The inhibitory effects of ethanol extracts from sorghum, foxtail millet and proso millet on α-glucosidase and α-amylase activities. Food Chem 124(4):1647–1651

Kumar LS, Sivaraman L, Ranjekar PK (1992) Genome turnover in great millet and related millets (Poaceae). Plant Syst Evol 179:155–165

Li M, Chang L, Ren J, Jiang F et al (2022a) Nutritional, physical, functional properties and antioxidant potential of different colors of proso millet husks and brans. LWT 171:114092

Li X, Bai Y, Jin Z et al (2022b) Food-derived non-phenolic α-amylase and α-glucosidase inhibitors for controlling starch digestion rate and guiding diabetes-friendly recipes. Lwt 153. https://doi.org/10.1016/j.lwt.2021.112455

Meena K, Visarada KBRS, Meena DK (2022) *Sorghum bicolor* (L.) Moench a multifarious crop-fodder to therapeutic potential and biotechnological applications: a future food for the millennium. Future Foods. https://doi.org/10.1016/j.fufo.2022.100188

Mengesha Y, Tebeje A, Tilahun BA (2022) Review on factors influencing the fermentation process of teff (*Eragrostis teff*) and other cereal-based Ethiopian injera. Int J Food Sci 2022:4419955

Mohammed NA, Ahmed IAM, Babiker EE (2011) Nutritional evaluation of sorghum flour (*Sorghum bicolor* L. Moench) during processing of injera. World Acad Sci Eng Technol 51:72–76

Naresh K, Nidhi G (2019) Phenolic acids: natural versatile molecules with promising therapeutic applications. Biotechnol Rep. https://doi.org/10.1016/j.btre.2019.e00370

Nithya VS, Athul TP, Deogade M et al (2023) Role of foxtail millet (*Setaria italica* (L.) P. Beauvois) in senile bone degenerative diseases: a conceptual review. J Drug Res Ayurvedic Sci 8:S76–S81

Obadi M, Sun J, Xu B (2021) Highland barley: chemical composition, bioactive compounds, health effects, and applications. Food Res Int 140:110065

Qian JY, Bai YY, Tang J et al (2015) Antioxidation and α-glucosidase inhibitory activities of barley polysaccharides modified with sulfation. LWT-Food Sci Technol 64(1):104–111. https://doi.org/10.1016/j.lwt.2015.05.034

Renganathan VG, Vanniarajan C, Karthikeyan A et al (2020) Barnyard millet for food and nutritional security: current status and future research direction. Front Genet 11:500

Rouf Shah T, Prasad K, Kumar P et al (2016) A potential source of human nutrition and health: a review. Cogent Food Agric 2:1166995

Shavanov MV (2021) The role of food crops within the Poaceae and Fabaceae families as nutritional plants. IOP Conf Ser Earth Environ Sci 624(1) IOP Publishing

Shukla V, Srivastava S, Singh S et al (2024) Unveiling the intricacies of phytate antinutrients in millets and their therapeutic implications in breast cancer. Intell Pharm 2:516–527

Singh LR (2019) Pharmacognostic evaluation of organically cultivated herb *Echinochloa colonum* (L.) Link. World J Pharm Res 8(12):1279–1286

Singhal P, Kaushik G (2016) Therapeutic effect of cereal grains: a review. Crit Rev Food Sci Nutr 56(5):748–759

Thushara N, Malawiarachchi P, Samarakoon SR et al (2022) Anti-diabetic and anti-cancer related health food properties of selected Sri Lankan traditional rice-based porridges. J Food Sci Technol 59:3745–3753. https://doi.org/10.1007/s13197-021-05341-9

Udeh HO, Duodu KG, Jideani AIO (2017) Finger millet bioactive compounds, bioaccessibility, and potential health effects—a review. Czech J Food Sci 35(1):7–17

Valero-Cases E, Cerdá-Bernad D, Pastor JJ (2020) Non-dairy fermented beverages as potential carriers to ensure probiotics, prebiotics, and bioactive compounds arrival to the gut and their health benefits. Nutrients 12(6):1666

Wang J, Cao Y, Wang C et al (2011a) Wheat bran xylooligosaccharides improve blood lipid metabolism and antioxidant status in rats fed a high-fat diet. Carbohydr Polym 86(3):1192–1197

Wang KM, Wu JG, Li G, Yang ZW et al (2011b) Distribution of phytic acid and mineral elements in three indica rice (*Oryza sativa* L.) cultivars. J Cereal Sci 54(1):116–121

Wegener G (2014) ‘Let food be thy medicine, and medicine be thy food’: hippocrates revisited. Acta Neuropsychiatr 26(1):1–3

Wei M, Tang M, Wang L et al (2021) Endogenous bioactive compounds of naked oats (*Avena nuda* L.) inhibit α-amylase and α-glucosidase activity. Lwt 149:111902

Yahfoufi N, Alsadi N, Jambi M, Matar C (2018) The immunomodulatory and anti-inflammatory role of polyphenols. Nutrients 10(11):1618. https://doi.org/10.3390/nu10111618

Yasar S, Boselli E, Rossetti F, Gok MS (2018) Effect of fermented cereals, probiotics, and phytase on the sensory quality of poultry meat. Sci Agric Bohem 49:225–235

Yin X, Zhu WQ, Zhou YQ et al (2023) Two new amides from the seeds of *Coixlacryma-jobi* var. lacryma-jobi. Nat Prod Res 37(20):3499–3504

Yu Q, Ye G, Lei Z et al (2021) An isolated compound from stems and leaves of *Coixlacryma-jobi* L. and its anticancer effect. Food Biosci 42:101047

Zhang K, Dong R, Hu X et al (2021) Oat-based foods: chemical constituents, glycemic index, and the effect of processing. Foods 10:1304

Chapter 19
Millet-Based Fermented Beverages Processing

Pankaj Bhatt, Pushpendra Kumar Jain, Puneet Nirmal, and Surya Pratap Singh

19.1 Introduction

Millets, a varied group of small-seeded grasses from the *Poaceae* family, are widely grown for their grains, which form a fundamental part of the diet in various regions, especially across Asia and Africa. These ancient grains have held historical importance for millennia, with records of their cultivation tracing back to approximately 3500 B.C (Sudha et al. 2016). In regions like the Korean peninsula and ancient India, they were mentioned in sacred texts such as the Yajurveda (Sudha et al. 2016). These grains are particularly valued for their resilience; they thrive in marginal soils and are capable of growing under harsh climatic conditions, including drought and high temperatures (Neeru et al. 2024). This adaptability makes millets a vital crop for food security, especially in developing countries where conventional cereals may fail to produce adequate yields (Neeru et al. 2024). Across the diverse range of millet species, pearl millet, finger millet, foxtail millet, and proso millet are the most widely grown varieties. These millets provide distinct nutritional advantages; for example, finger millet stands out for its exceptional calcium content, while pearl millet is noted for being a great source of iron and protein. Additionally, millets are naturally gluten-free, making them an ideal food option for those who suffer from celiac disease or have gluten intolerance (Joseph et al. 2024). Millets offer numerous health benefits beyond their nutritional value, being abundant in dietary fiber, essential vitamins, and minerals like magnesium and phosphorus. These grains also

P. Bhatt (✉) · P. Nirmal
Lloyd Institute of Management and Technology, Greater Noida, Uttar Pradesh, India

P. K. Jain
IIMT College of Pharmacy, Greater Noida, Uttar Pradesh, India

S. P. Singh
MRU-Uttar Pradesh University of Medical Sciences, Etawah, Uttar Pradesh, India

© The Author(s), under exclusive license to Springer Nature Switzerland AG 2026
J. Singh, M. Kumar Malik (eds.), *Millets and Millet Starch: Chemical Composition, Functional Modification, and Applications*,
https://doi.org/10.1007/978-3-032-09502-2_19

possess bioactive substances that provide advantages such as antioxidant effects and a low glycemic index, which can help regulate blood sugar levels effectively (Amadou 2019). Despite their numerous advantages, millets have been somewhat overshadowed by more popular grains like wheat and rice due to changing agricultural practices and consumer preferences influenced by the Green Revolution (Semwal et al. 2021). This shift has led to a decline in millet production over the past several decades; however, there is a renewed interest in these "nutri-cereals" as global awareness of health and sustainability grows. Millets are now being promoted as climate-resilient crops that can play a crucial role in ensuring food security amid changing environmental conditions (Chamarthy and Tonapi 2021). Their short growing season (typically 75–120 days) allows them to be harvested quickly compared to other grains, making them an attractive option for farmers facing unpredictable weather patterns. Furthermore, millets require minimal inputs in terms of fertilizers and pesticides due to their natural resistance to pests and diseases (Srivastava et al. 2024). This characteristic not only reduces the environmental impact of agriculture but also lowers production costs for farmers. The versatility of millets extends beyond traditional uses; they can be processed into various forms such as flour for baking or fermented beverages, showcasing their adaptability in modern culinary applications(Bhattacharjee et al. 2021). As interest in sustainable diets grows, millets are being recognized not just as food for rural populations but as valuable components of global food systems that can contribute to healthier diets and sustainable agriculture practices. Leading millet-producing countries in the world are shown in the Fig. 19.1.

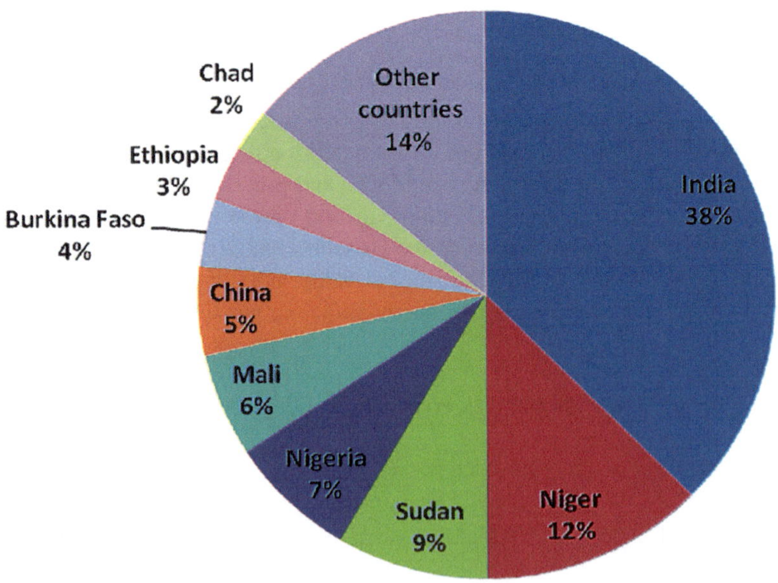

Fig. 19.1 Top millet-producing nations globally

19.1.1 Importance of Fermented Beverages

Fermented beverages made from millets play a crucial role in enhancing both the nutritional profile and shelf life of these products, making them increasingly significant in contemporary diets (Shrivastava et al. 2012). Fermentation is a metabolic process in which microorganisms like bacteria, yeast, and molds break down sugars into acids, gases, or alcohol. This technique has been employed for centuries to enhance food preservation and improve digestibility. When applied to millet-based beverages, fermentation not only prolongs their storage life but also greatly enhances their nutritional profile (Cuamatzin-garcía et al. 2022). One of the primary benefits of fermentation is the increase in protein levels; studies have shown that fermentation can elevate protein content by as much as 15.2% in finger millet porridge. This enhancement occurs because microbes produce hydrolytic enzymes that break down complex proteins into simpler forms, thereby increasing the availability of essential amino acids such as lysine and methionine (Adakole et al. 2021). Additionally, fermented millet products tend to be more easily digestible than non-fermented versions, making them an ideal choice for individuals with dietary limitations or conditions such as celiac disease. Fermentation also decreases the presence of anti-nutritional elements like phytic acid and tannins, which can interfere with the body's ability to absorb minerals efficiently (Amadou 2019). Fermentation reduces these substances, increasing the bioavailability of key minerals such as calcium, iron, and zinc, thus transforming millet-based drinks into a valuable source of essential nutrients. For example, the fermenting process in finger millet has shown a marked improvement in the extractability of both iron and calcium, greatly enhancing its overall nutritional worth. Moreover, fermentation promotes the production of bioactive compounds with antioxidant capabilities, amplifying the health advantages linked to millet consumption (Makwana and Hati 2019). The increased levels of phenolic compounds and vitamins such as riboflavin and niacin during fermentation contribute to the overall antioxidant capacity of these beverages. Moreover, fermented millet products are often rich in dietary fiber due to the breakdown of complex carbohydrates during fermentation. This increase in fiber not only aids digestion but also promotes gut health by fostering a balanced microbiome. The presence of probiotics in fermented beverages can help reduce gastrointestinal issues and improve overall digestive health (Estruch and Lamuela-Raventós 2023). Fermentation serves not only to boost the nutritional value of millet-based products but also plays a key role in extending their shelf life. By producing an acidic environment, fermentation helps prevent the growth of harmful microorganisms and spoilage agents, thus maintaining the product's freshness for a longer period (Leeuwendaal et al. 2022). This is particularly important in regions where refrigeration may not be readily available, allowing consumers to enjoy these nutritious beverages over extended periods without compromising safety or quality (Day and McSweeney 2016). The sensory attributes of fermented millet beverages also improve during fermentation; they often develop unique flavors and aromas that enhance consumer appeal. Traditional fermentation processes vary by region and culture, leading to a diverse array of flavors and textures that cater to different palates. This cultural significance adds an additional layer of value to fermented millet products as they connect consumers with their culinary heritage while offering modern health benefits (Makwana and Hati 2019). As global interest in healthful eating continues to rise, fermented millet beverages present an attractive option for those seeking nutritious alternatives to conventional drinks. They align with current trends favoring functional foods that offer both health benefits and sustainability. By utilizing locally sourced millets and traditional fermentation methods, these beverages support local agriculture while promoting dietary diversity.

19.1.2 Significance of Studying Millets-based Fermented Beverages

Studying millet-based fermented beverages is significant for several compelling reasons, particularly in cultural heritage, sustainability, and health benefits. These beverages are deeply rooted in the traditions of many cultures, especially in Africa and Asia, where millets have been cultivated for thousands of years. For example, Ambali, a traditional Indian drink made from fermented millet, showcases not only unique flavors but also the nutritional wisdom passed down through generations. By exploring such beverages, we can preserve valuable cultural narratives that are at risk of being lost in modern dietary trends (Makwana and Hati 2019).

From a sustainability perspective, millets are considered "climate-smart" crops due to their resilience to adverse conditions like drought and poor soil quality. They require significantly less water compared to staple grains such as rice and wheat, making them ideal for sustainable agriculture in water-scarce regions (Makwana and Hati 2019; Rios-Corripio et al. 2021; Vila-Real et al. 2022).

Health benefits associated with these beverages provide another compelling reason for their study. Fermentation improves the digestibility of millets and enhances their nutritional profile. It increases the bioavailability of essential nutrients like iron, calcium, and various B vitamins while reducing anti-nutritional factors that inhibit mineral absorption. Research has shown that fermented finger millet porridge can alleviate gastrointestinal issues and malnutrition among children, highlighting its potential as a therapeutic food for vulnerable populations (Leeuwendaal et al. 2022). Additionally, the probiotics present in these beverages support gut health and boost immunity. Millet-based fermented beverages also serve as functional foods that contribute to overall well-being. Rich in dietary fiber, they help regulate blood sugar levels—crucial for managing conditions like diabetes (Cuamatzin-garcía et al. 2022).

19.2 Millets as a Raw Material for Fermented Beverages

Millets are an ideal ingredient for producing fermented drinks, thanks to their high nutritional content, gluten-free properties, and versatility in accommodating different fermentation methods (Hassan et al. 2021). Their high fiber content and essential nutrients enhance the health benefits of the final products. Additionally, millets' resilience in diverse climates makes them a sustainable choice for producing traditional and innovative fermented drinks worldwide (Anitha et al. 2024).

19.2.1 Nutritional Composition of Millets

Millets are abundant in macronutrients, as illustrated in Table 19.1, containing carbohydrates, proteins, and fats, along with a diverse range of vitamins and minerals. The specific nutrient profile can vary widely across different types of millets, including pearl millet, finger millet, foxtail millet, proso millet, and others (Dimidi et al. 2019).

Carbohydrates Millets are predominantly composed of carbohydrates, positioning them as a superb source of energy. The carbohydrate levels generally fall between 65% and 75%. For instance, pearl millet has approximately 67% carbohydrates, whereas finger millet features a marginally higher carbohydrate content, reaching around 75% (Anitha et al. 2024). The carbohydrate composition mainly includes starches, which are made up of amylose and amylopectin. This starch structure influences the digestibility and glycemic response of millets (Hassan et al. 2021).

Proteins Millets are also notable for their protein content, which ranges from 7% to 12% depending on the variety. Pearl millet typically has about 11.6–11.8% protein, while finger millet contains around 7.7% (Hiremath et al. 2018). Millets are particularly notable for their elevated levels of essential amino acids relative to numerous other grains. For example, finger millet stands out for its abundant lysine and methionine content two amino acids that are frequently deficient in cereals such as wheat and rice. This characteristic makes millets an important source of protein for those following vegetarian diets (Jacob et al. 2024).

Fats The fat content in millets is relatively low but varies among types. Pearl millet has the highest fat content at about 4.8–5.0%, while finger millet contains around 1.8% (Agrawal et al. 2023). The fats present in millets are primarily unsaturated fatty acids, such as linoleic and oleic acid, which contribute to their health benefits (Jacob et al. 2024).

Dietary Fiber A key characteristic of millets is their significant dietary fiber content, which varies between 8% and 15%. Notably, finger millet can have fiber levels

Table 19.1 Nutritional profile of millets

Nutrient (per 100 g)	Pearl millet	Finger millet	Foxtail millet	Barnyard millet	Little millet	References
Moisture (%)	12.4	7.15–13.1	10.0	11.0	10.0	Anitha et al. (2024)
Protein (g)	11.6–11.8	7.7	11.79–13.2	11.0	7.0	Anitha et al. (2024)
Fat/lipids (g)	4.8–5.0	1.8	4.0	2.0	3.5	Dayakar et al. (2017)
Dietary fiber (g)	11.3	15–22	8.0	9.0	10.0	Dimidi et al. (2019)
Carbohydrates (g)	67–67.5	75.0–83.3	72.0	73.5	75.0	Vila-Real et al. (2022)
Gross energy (MJ/kg)	17.0	15.8	–	–	–	Puranik et al. (2017)
Phosphorus (mg)	296	130–250	280	200	180	Hiremath et al. (2018)
Potassium (mg)	307	430–490	350	400	370	Dimidi et al. (2019)
Magnesium (mg)	137	78–201	90	95	110	Hiremath et al. (2018)
Calcium (mg)	42	398	30	15	20	Vila-Real et al. (2022)
Zinc (mg)	3.1	2.3	1.2	1.5	1.4	Shrivastava et al. (2012)
Iron (mg)	5–6.5	2.2–17.7	2–3	3–4	2–3	Dayakar et al. (2017)

reaching between 15% and 22%, positioning it as an exceptional option for enhancing digestive health (Vila-Real et al. 2022). The fiber in millets consists of both soluble and insoluble types, which help regulate blood sugar levels and improve gut health by fostering a balanced microbiome (Dayakar et al. 2017).

19.2.2 Vitamins and Minerals

Millets are rich in essential vitamins and minerals that contribute to overall health.

Calcium Finger millet is particularly notable for its calcium content, providing between 162 and 487 mg per 100 g, making it one of the richest vegetarian sources of this mineral (Anitha et al. 2024).

Iron Pearl millet contains significant amounts of iron (up to 17 mg per 100 g), while finger millet also provides a good source at approximately 3.9 mg per 100 g (Dayakar et al. 2017).

Magnesium Millets are rich in magnesium, with values ranging from 78 mg to over 500 mg per 100 g, depending on the variety (Hiremath et al. 2018).

Phosphorus Millets provide substantial phosphorus levels (around 189–296 mg per 100 g), essential for bone health and energy metabolism (Agrawal et al. 2023).

Zinc The zinc content varies but generally falls between 2 and 3 mg per 100 g, contributing to immune function (Puranik et al. 2017).

19.2.3 Antioxidants

Millets are rich in various antioxidants that play a crucial role in alleviating oxidative stress within the body. They are particularly high in phenolic compounds like ferulic acid and catechins, which are associated with multiple health advantages, such as anti-inflammatory effects and enhanced wound healing (Saleem et al. 2021).

19.3 Benefits of Using Millets in Fermentation

Incorporating millets into fermentation provides numerous advantages that enrich both the nutritional content and health benefits of the final products. This natural fermentation process has been employed for ages to enhance food preservation, digestibility, and overall nutritional quality (Debnath et al. 2024).

19.3.1 Enhanced Gut Health

A major advantage of fermenting millet is its ability to enhance gut health. The fermentation process encourages the proliferation of probiotics—beneficial bacteria that are essential for maintaining a balanced gut microbiome. These probiotics are crucial for digestion and nutrient absorption and also aid in bolstering immune health (Singh et al. 2023). For instance, traditional Indian beverages made from fermented millet, such as Ambali, have long been enjoyed for their health-promoting effects. The probiotics found in these drinks help to eliminate harmful pathogens and enhance overall digestive performance, alleviating problems such as bloating and constipation (Cuamatzin-garcía et al. 2022).

19.3.2 Improved Nutrient Absorption

Fermentation improves the absorption of vital nutrients present in millets. This process breaks down complex carbohydrates and proteins into simpler components, facilitating the body's ability to absorb essential vitamins and minerals (Shafiq et al. 2007; Rathore 2016). Research indicates that fermented finger millet contains greater amounts of accessible calcium, iron, phosphorus, and zinc than its unfermented version. This enhanced nutrient availability is especially crucial for groups susceptible to deficiencies. Additionally, fermentation boosts the concentrations of B vitamins and omega-3 fatty acids, both of which are essential for numerous metabolic functions (Makokha et al. 2002).

19.3.3 Increased Dietary Fiber

Millets are already known for their high dietary fiber content, which ranges from 8% to 15%. Fermentation can further enhance this fiber content by breaking down hemicellulose and cellulose into more digestible forms. This increase in dietary fiber contributes to improved bowel movement and overall digestive health (Lei and Jakobsen 2004). A diet rich in fiber is linked to lower risks of chronic conditions, including heart disease, diabetes, and some forms of cancer. Including fermented millet products in one's diet allows individuals to take advantage of both their fiber content and the probiotics generated through the fermentation process (Lei et al. 2006).

19.3.4 Functional Food Development

Fermenting millets creates opportunities for the development of functional foods—items that offer health advantages that extend beyond mere nutritional value (Devi et al. 2014). Fermented millet porridges and beverages can be fortified with additional probiotics and prebiotics, making them excellent candidates for symbiotic foods that support gut health. These products can be tailored to meet specific dietary needs or preferences, such as gluten-free options for those with gluten sensitivities (Di Stefano et al. 2017).

19.3.5 Weight Management

Fermented millet products may also aid in weight management efforts. The elevated fiber levels contribute to a sense of fullness and satisfaction, aiding in appetite regulation. Additionally, a balanced gut microbiome has been linked to better weight regulation (Mishra and Panda 2022). By incorporating fermented millets into their diets, individuals may find it easier to maintain a healthy weight (Adebiyi et al. 2018).

19.3.6 Cultural Significance

Fermented millet products have cultural significance in many regions around the world. They are often part of traditional diets and rituals, contributing not only to nutrition but also to community bonding (Semwal et al. 2021). By promoting these traditional practices through modern research and culinary innovation, we can preserve cultural heritage while enhancing public awareness about the nutritional benefits of millets (Hema et al. 2022).

19.4 Comparison with Other Grains Used in Fermentation

Millets are increasingly recognized for their potential in fermentation processes, offering unique benefits compared to other grains like wheat and barley. This comparison focuses on the nutritional enhancements, fermentation efficiency, and health benefits of millets, particularly emphasizing their gluten-free nature and superior mineral bioavailability (Newlove et al. 2022).

19.4.1 Nutritional Enhancements

Millets offer a robust nutritional profile, further improved through fermentation, which enhances nutrient absorption (Nkhata et al. 2018). This is critical in regions with prevalent micronutrient deficiencies. Compared to other grains, millets stand out: finger millet contains ~7.7% protein, pearl millet ~11.6%, surpassing wheat (~7.5%) and barley (~9.9%) in essential amino acid content (Ali et al. 2003). Fermentation boosts protein availability, producing oligopeptides with health benefits absent in unfermented millets (Chavan et al. 1988). It also reduces antinutritional factors like phytic acid and tannins, which hinder mineral absorption (Duodu et al. 2003). For instance, fermentation significantly lowers phytic acid, enhancing mineral bioavailability. In contrast, grains like rice (~6.8% protein) and maize (~9.4%) have lower protein content and less pronounced fermentation benefits compared to millets (Dordević et al. 2010).

19.4.2 Antinutrient Reduction

Fermentation plays a crucial role in reducing antinutritional factors in millets more effectively than in wheat or barley. Antinutrients such as phytic acid and tannins can inhibit mineral absorption and negatively impact nutritional quality (Mohapatra et al. 2024). Research has shown that fermentation significantly decreases phytic acid levels in millets, enhancing mineral bioavailability. For instance, Jan et al. (2022) reported that fermentation reduced phytic acid levels in finger millet, thereby improving the availability of essential minerals like iron and zinc. In contrast, while wheat and barley also undergo reductions in antinutritional factors during fermentation, the extent of these reductions is often less pronounced compared to millets. Studies indicate that the reduction of phytic acid in wheat during fermentation can vary but is generally not as significant as that observed in millets (Jan et al. 2022).

19.4.3 Rapid Fermentation

Millets demonstrate swift fermentation properties because of their elevated starch levels and reduced viscosity when compared to barley and wheat (Nkhata et al. 2018). The glucose content in millet wort facilitates quicker yeast activity, leading to shorter fermentation times for millet-based beverages. A study found that beers brewed with 100% malted proso millet completed fermentation effectively within 2 days, demonstrating a rapid fermentation rate. This contrasts with barley, which typically requires longer fermentation periods due to its complex starch structures and higher viscosity. The average extract content from millet ranges from 43% to 63%, while barley averages between 76% and 85%. This difference in extract potential influences the efficiency of fermentation; however, the rapid fermentability of millets can offset this by reducing overall production time (Mohapatra et al. 2024).

19.4.4 Gluten-Free Options

Millets offer a considerable benefit compared to wheat and barley because they are naturally gluten-free. This trait is especially advantageous for those who suffer from celiac disease or have sensitivities to gluten (Rao and Poonia 2023). Wheat and barley contain gluten proteins that can trigger adverse reactions in sensitive individuals, limiting their consumption options (Zhu 2021). By utilizing millets in fermented products, manufacturers can create gluten-free alternatives that cater to a growing market segment seeking safe dietary options (Dymchenko et al. 2024). Fermented millet beverages such as Ambali or Kunu not only meet gluten-free requirements but also offer nutritional benefits without compromising taste or quality (Dhondiram et al. 2023). One of the most significant advantages of using millets in fermentation is their gluten-free nature. This characteristic makes them suitable for individuals with celiac disease or gluten sensitivity—conditions that affect a substantial portion of the population (Kaur 2023). In contrast, wheat and barley contain gluten proteins that can trigger adverse reactions in sensitive individuals. By utilizing millets in fermented beverages, producers can cater to a growing market seeking gluten-free alternatives without sacrificing flavor or quality (Dymchenko et al. 2024). Fermented millet beverages such as Ambali and Kunu not only provide nutritional benefits but also preserve cultural heritage associated with millet consumption in various regions. The gluten-free aspect allows these beverages to be marketed as healthful options for those avoiding gluten-containing grains (Newlove et al. 2022). Fermented millet beverages, such as Ambali (a traditional South Indian drink), Kunu (a Nigerian millet-based beverage), Boza (a fermented millet drink from the Balkans and Turkey), and Bhaati Jaanr (a Himalayan fermented millet beverage), not only deliver nutritional benefits but also preserve cultural heritage tied to millet consumption across diverse regions. These beverages leverage the gluten-free nature of millets, making them suitable for broader markets seeking healthful, gluten-free alternatives. For example, Kunu, made from pearl millet, is rich in B vitamins and minerals, while Boza offers probiotic benefits due to fermentation (Amadou 2019). Similarly, Bhaati Jaanr, prepared from finger millet, is valued for its high iron content and probiotic properties (Singh et al. 2018). By incorporating millets into these traditional fermented beverages, producers can meet the rising global demand for gluten-free products while honoring cultural traditions and enhancing nutritional accessibility (Lorri and Svanberg 1993; Amadou 2019).

19.4.5 Nutritional Profile

When fermented, millets typically provide a more advantageous nutritional profile than wheat and barley. They are abundant in crucial minerals, including calcium, iron, magnesium, and phosphorus. For instance, studies indicate that finger millet contains around 344 mg of calcium for every 100 grams, making it one of the richest vegetarian sources of this mineral. In contrast, wheat contains about 15 mg of calcium per 100 g. Furthermore, studies have demonstrated that fermented millets exhibit enhanced bioavailability of these minerals post-fermentation (Drub et al. 2021). A study indicated that fermented finger millet showed improved extractability rates for iron and calcium after fermentation. This enhanced mineral availability is crucial for populations at risk of micronutrient deficiencies (Majzoobi et al. 2023).

Additionally, millets provide higher levels of dietary fiber than both wheat and barley. The fiber content in millets ranges from 8% to 15%, which aids digestion and contributes to overall gut health. Fermentation can further increase the fiber content by breaking down complex carbohydrates into simpler forms that are more easily digestible (Dev et al. 2024).

19.4.6 Probiotic Potential

The ability of fermented millets to support probiotic growth enhances their health benefits beyond what is typically observed in wheat or barley-based products (Di Stefano et al. 2017). Probiotics are beneficial microorganisms that contribute positively to gut health by promoting a balanced microbiome. Fermented millet products have been shown to reach viable counts greater than 10^6 CFU/mL, indicating substantial probiotic presence (Budhwar et al. 2020). In comparison, while wheat and barley can also be fermented to produce probiotic-rich products (such as certain types of beer), they do not offer the same level of safety for individuals with gluten sensitivities. Furthermore, research has indicated that fermented millet products can contain various bioactive compounds that contribute to their health benefits (Di Stefano et al. 2017). For instance, millet fermented with lactic acid bacteria has been shown to enhance antioxidant capacity while increasing phenolic compound levels. The existence of these bioactive compounds not only promotes digestive well-being but may also offer additional health advantages, including decreased inflammation and enhanced metabolic function (Atter et al. 2024).

19.4.7 Fermentation Efficiency

Millets demonstrate remarkable fermentation efficiency due to their high starch content, which is readily fermentable by yeast (Balli et al. 2023). A study involving the fermentation of 100% malted proso millet found that fermentation was effectively completed within 2 days, showcasing millets' rapid fermentability. The quick fermentation observed in millet is primarily due to its elevated glucose levels in the wort, which enables yeast to ferment swiftly (Sheethal et al. 2022). In comparison, barley usually necessitates extended fermentation times because of its intricate starch composition. Although barley possesses a greater diastatic power—the capacity to transform starches into fermentable sugars—millets can achieve favorable outcomes when adequately malted (Agu and Palmer 2013). The average extract content from millet is reported to be between 43% and 63%, compared to 76–85% for barley. This lower extract potential necessitates careful consideration during the brewing process, particularly regarding mashing efficiency (Torello Pianale et al. 2022).

19.5 Processing of Millets-based Fermented Beverages

The preparation of fermented beverages made from millet includes several important stages: initially, it is essential to choose and thoroughly clean the millets, then soak them to promote germination (Balli et al. 2023). The soaked grains are then steamed or roasted to develop flavor and reduce anti-nutrients. Next, they are ground into a paste or flour, mixed with water, and inoculated with starter cultures (Sheethal et al. 2022). Fermentation occurs at controlled temperatures for several hours to days, resulting in a nutritious and flavorful beverage ready for consumption (Mehrotra 2016) (Table 19.2 and Fig. 19.2).

19.5.1 Selection and Preparation of Millets

The process of preparing millet grains for fermentation is complex and plays a crucial role in determining the quality and nutritional benefits of the end product. This preparation includes meticulous selection of millet varieties, followed by processes such as soaking, malting, and germination. Each step is crucial for optimizing the enzymatic activity and nutrient availability in the grains, ultimately enhancing the fermentation process (Hassan et al. 2021).

19.5.2 Selection of Millet Varieties

The initial stage in the fermentation process of millets involves choosing the right type. Popular varieties consist of pearl millet (*Pennisetum glaucum*), finger millet (*Eleusine coracana*), foxtail millet (*Setaria italica*), and proso millet (*Panicum miliaceum*) (Gupta and Gaur 2024). The selected variety can greatly affect the fermentation dynamics and the nutritional qualities of the end product (Gupta and Gaur 2024).

On the Basis of Nutritional Composition Various types of millet display differing concentrations of crucial nutrients. For example, finger millet stands out for its significant calcium content, approximately 344 mg per 100 grams, and iron levels around 3.9 mg per 100 grams. In contrast, pearl millet boasts higher amounts of protein, reaching up to 11.6%, as well as dietary fiber, which is about 12%. Selecting a variety with a favorable nutritional profile can enhance the health benefits of the fermented product (Sheethal et al. 2022).

On the Basis of Adaptability The choice of millet should also consider local agro-climatic conditions. Certain varieties are better suited for specific environments; for example, pearl millet thrives in arid regions due to its drought resistance, while finger millet can adapt to diverse soil types (Hassan et al. 2021).

On the Basis of Breeding Programs Innovations in breeding methods have resulted in the creation of varieties that produce higher yields and resist diseases. Initiatives emphasizing mass selection and hybrid breeding have enhanced characteristics like grain productivity, resilience to diseases, and drought resistance.

Table 19.2 Production of fermented beverages from millet

Millet type	Product name	Main ingredients	Processing method	Fermentation duration	Nutritional benefits	References
Pearl millet	Rabadi	Pearl millet, buttermilk, deionized water	Mixed and fermented for 4 h at 37 °C, then cooked for 7–8 min	4 h	High in antioxidants and phenolic compounds	Sheethal et al. (2022)
Finger millet	Uji	Finger millet, water	Soaked, ground, boiled, and allowed to ferment overnight	Overnight	Rich in probiotics and dietary fiber	(Palanisamy and Thiyagamoorthy (2025)
Pearl millet	Mangisi	Pearl millet, water	Malting, boiling, cooling, and spontaneous fermentation	Several hours	Contains alcohol and beneficial microbes	Jideani et al. (2021)
Finger millet	Masvusvu	Finger millet, water	Soaked, boiled, cooled, and allowed to ferment	Overnight	Nutrient-rich tonic	Zvauya et al. (1997)
Mixed millets	Flavored ready-to-drink	Sorghum, finger millet, jaggery, coconut	Sprouted seeds dried, powdered, mixed with flavors; optimized for sensory evaluation	Varies	High total soluble solids	Chamarthy and Tonapi (2021)
Pearl millet	Probiotic fermented beverage	Millet flour, milk/water	Pre-treatment followed by fermentation with bacterial cultures	12 h	Enhanced digestibility and probiotic content	Neeru et al. (2024)

Millet type	Product name	Main ingredients	Processing method	Fermentation duration	Nutritional benefits	References
Finger millet	EU	Finger millet, water	The process of making Eu involves boiling finger millet, mixing it with a rice-based starter culture (Moaa), and fermenting the mixture for 5–8 days before consumption	5–8 days	Probiotic properties for gut health Increased antioxidant capacity Low alcohol content Enhanced nutrientbioavailability Traditional medicinal uses Hydration support	Bhattacharjee et al. (2021)

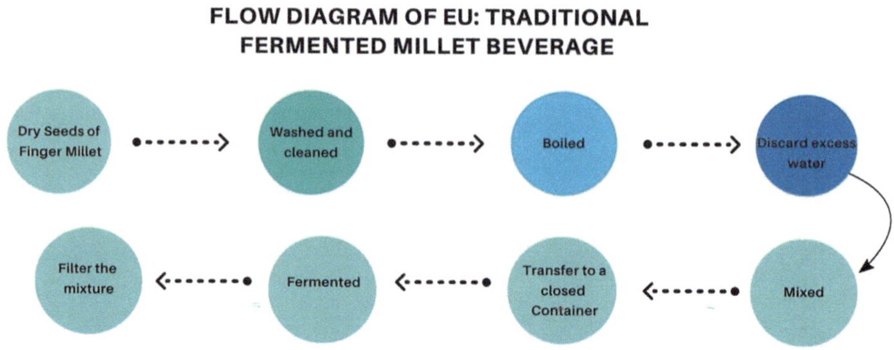

Fig. 19.2 Flow diagram of Eu: traditional fermented millet beverage

Selecting improved cultivars can significantly impact fermentation efficiency and product quality (Balli et al. 2023).

19.5.3 Preparation of Millet Grains

Once the appropriate variety is selected, proper preparation of millets is crucial for successful fermentation. This preparation involves several steps:

1. *Cleaning*

   The initial stage involves meticulously washing the millets to eliminate any impurities, including stones, dirt, and other unwanted substances. This can be accomplished by employing screens or air classifiers to guarantee that only premium grains are selected. Proper cleaning is essential to prevent contamination during fermentation (Balli et al. 2023).
2. *Soaking*

   Soaking millets in water activates enzymes responsible for breaking down starches into fermentable sugars during fermentation. The soaking process typically lasts between 4 and 12 h, depending on the variety and desired fermentation characteristics. This process also helps reduce antinutritional factors like phytic acid. Research indicates that soaking can lead to a reduction in phytic acid levels by approximately 50%, thereby enhancing mineral bioavailability (Nkhata et al. 2018).
3. *Malting*

   Malting is a key process that enhances the nutritional profile of millets by promoting enzymatic activity necessary for starch conversion during fermentation. The malting procedure involves three primary phases: soaking, sprouting, and drying (Sheethal et al. 2022).

   Steeping: The cleaned millets are steeped in water at a ratio of approximately 1:4 (millet to water) for about 20 h at a temperature of 25–30 °C. This process hydrates the grains and initiates metabolic activities.

Germination: After steeping, the water is drained, and the grains are spread out in a controlled environment for germination at temperatures between 25 and 30 °C for 72 h. During this period, enzymes such as amylase are activated, which break down starches into simpler sugars that are more easily fermentable (Chauhan et al. 2024).
Drying: Once germination is complete, the sprouted grains are dried at temperatures around 50 °C for approximately 24 h to halt enzymatic activity while preserving beneficial enzymes produced during germination. Proper drying is crucial to prevent spoilage and maintain quality (Balli et al. 2023).

4. *Grinding*

   After drying, the malted millets are ground into a fine powder using an electric grinder. The grinding process increases surface area, facilitating better interaction with fermentative microorganisms during fermentation. The particle size is typically standardized to around 80 mesh to ensure uniformity (2018).

19.5.4 Germination Process

The germination process of millets is a critical phase that significantly enhances their nutritional composition and functional properties, making them more suitable for fermentation. This process involves several biochemical changes, including enzyme activation, nutrient enhancement, and protein modification, all of which contribute to the overall quality of the final fermented product (Budhwar et al. 2020).

(a) *Enzyme Activation*

During germination, various enzymes are activated that play essential roles in breaking down complex carbohydrates into simpler sugars. One of the key enzymes involved is amylase, which catalyzes the hydrolysis of starch into maltose and glucose (Aparicio-García et al. 2021). This enzymatic activity is crucial for fermentation, as yeast can efficiently ferment these simpler sugars into alcohol and carbon dioxide. Studies have shown that the activity of α-amylase increases significantly during germination, peaking after approximately 48 h. The increased amylase activity correlates with a rise in fermentable sugars, which can enhance the efficiency of fermentation processes (Kumari et al. 2024).

(b) *Nutrient Enhancement*

Germination also leads to substantial improvements in the bioavailability of nutrients. One significant change is the reduction of antinutritional factors such as phytic acid and tannins, which inhibit mineral absorption (Samtiya et al. 2021). Research indicates that soaking and germinating millets can reduce phytic acid levels by up to 50%, thereby enhancing mineral bioavailability. For example, studies have demonstrated that extractability rates for iron and calcium in finger millet increase from 18.1% to 37.3% and from 76.9% to 90.2%, respectively, following germination. This increase in mineral availability is crucial for populations at risk of micronutrient deficiencies (Budhwar et al. 2020).

Furthermore, germination has been found to elevate the concentrations of bioactive substances like polyphenols and γ-aminobutyric acid (GABA). These compounds are linked to a range of health advantages, including enhanced antioxidant properties and better metabolic function. The activation of specific enzymes during germination facilitates the release of these beneficial compounds from their bound forms within the grain matrix (Joshi 2018).

(c) *Protein Modification*

The germination process also induces significant changes in protein content and composition. During germination, storage proteins are broken down into free amino acids and peptides through proteolytic enzymes such as proteases (Chethan Kumar et al. 2022). This process not only increases the overall protein solubility but also enhances the nutritional value by providing essential amino acids that are often limiting in cereal grains (Ali et al. 2017; Joshi 2018).

For instance, studies have reported a 34% increase in protein solubility in germinated foxtail millet compared to ungerminated samples. The breakdown of proteins during germination leads to an increase in free α-amino nitrogen levels, which can enhance flavor and improve the sensory attributes of fermented products. Furthermore, the synthesis of new proteins during germination contributes to an overall increase in nutritional quality (Yu et al. 2021).

19.5.5 Fermentation Process

The fermentation process is a complex biochemical reaction that transforms carbohydrates into various products, primarily through the action of microorganisms (Chauhan et al. 2024). This procedure is vital for creating a diverse array of fermented foods and drinks, particularly those made from millets (Vila-Real et al. 2022). Grasping the different microbial cultures that play a role in fermentation, especially lactic acid bacteria (LAB) and yeasts, is key to enhancing the fermentation process of millet-derived items (Versari et al. 2016).

19.6 Types of Microbial Cultures Used in Fermentation

19.6.1 Lactic Acid Bacteria (LAB)

Lactic acid bacteria (LAB) are a category of Gram-positive, non-spore-forming microorganisms that are essential in the fermentation process of millets and various grains. Prominent genera in this group are *Lactobacillus, Leuconostoc, Streptococcus, Enterococcus,* and *Pediococcus* (Aguirre-Garcia et al. 2024). These bacteria primarily convert sugars into lactic acid, which effectively reduces the pH of the environment, thereby preventing the growth of spoilage organisms.

Homofermentative LAB These bacteria convert glucose predominantly into lactic acid. For instance, Lactococcus lactis is widely used in dairy fermentations, while *Lactobacillus plantarum* is utilized in millet-based products like idlis and fermented beverages. In this pathway, glucose undergoes glycolysis to form pyruvate, which is then reduced to lactic acid by lactate dehydrogenase, regenerating NAD^+ necessary for glycolysis (Kondrotiene et al. 2023).

Heterofermentative LAB These bacteria produce multiple end products, including lactic acid, ethanol, and carbon dioxide. Leuconostoc mesenteroides is an example that contributes to the fermentation of finger millet products like bhakari (Shin and Han 2015). In heterofermentation, glucose is metabolized to pyruvate, which can then be converted into lactic acid or further fermented into ethanol and CO_2 through mixed pathways (Li et al. 2020).

19.6.2 Yeasts

Yeasts, which are single-celled fungi, are crucial participants in the fermentation process (Semwal et al. 2021). The yeast species most frequently utilized in millet fermentation is Saccharomyces cerevisiae. These microorganisms primarily convert sugars into alcohol and carbon dioxide via alcoholic fermentation. During this process, pyruvate generated from glycolysis undergoes decarboxylation to form acetaldehyde, resulting in the release of carbon dioxide (Zhang et al. 2023). Afterward, acetaldehyde is transformed into ethanol through the action of alcohol dehydrogenase. This process not only contributes to the alcoholic content of beverages but also enhances flavor complexity (Walker and Stewart 2016).

19.6.3 Molds

Certain molds are also utilized in mixed-culture fermentations. For example, Aspergillus oryzae is employed in traditional Japanese koji-making processes where it breaks down starches into sugars through enzymatic activity. This mold can be combined with LAB and yeasts to produce complex flavors and enhance nutrient availability (Sun et al. 2023).

19.7 Mechanisms of Fermentation

The fermentation process consists of several key stages that facilitate the conversion of sugars into various end products (Sarsan et al. 2021). Initially, glycolysis occurs, where glucose is broken down into pyruvate, resulting in a net gain of two ATP molecules per glucose molecule. Following this, in the absence of oxygen, pyruvate undergoes different transformations depending on the microbial culture present. In homofermentative lactic acid bacteria (LAB), pyruvate is converted directly to lactic acid, while in heterofermentative LAB, pyruvate can yield lactic acid alongside ethanol and carbon dioxide. Yeasts, on the other hand, convert pyruvate into ethanol and carbon dioxide (Stokes and Gillings 2011). A crucial aspect of fermentation is the regeneration of NAD+, which is essential for glycolysis to continue. In LAB, this regeneration occurs as NADH donates electrons during the conversion of pyruvate to lactic acid; in yeast, it happens during the conversion of acetaldehyde to ethanol. This intricate interplay of biochemical reactions ensures a continuous supply of energy and metabolic intermediates necessary for microbial growth and fermentation (Semwal et al. 2021).

19.7.1 Mixed-Culture Fermentation

Mixed-culture fermentation is a process that involves the simultaneous use of multiple microbial species to enhance the fermentation of substrates, such as millets. This approach leverages the synergistic interactions among different microorganisms, which can lead to improved flavor profiles, increased stability of the final product, and more efficient utilization of available nutrients. The mechanisms underlying mixed-culture fermentation are complex and involve various biochemical and physiological processes (Yaqoob et al. 2024).

19.7.2 Synergistic Interactions

In fermentations involving multiple cultures, various microbial species can enhance each other's metabolic functions. For instance, in millet-based beverages such as togwa or mahewu, lactic acid bacteria (LAB), and yeasts coexist, collaborating effectively to create appealing flavors. This partnership also helps prevent spoilage organisms from thriving by outcompeting them (Sarsan et al. 2021). LAB, such as *Lactobacillus plantarum*, produce lactic acid as a primary fermentation product, which lowers the pH of the medium. This acidic environment not only enhances flavor but also creates conditions unfavorable for pathogenic microbes, thus improving food safety.

The metabolic pathways of LAB and yeasts can be interconnected (Stokes and Gillings 2011). For instance, while LAB ferment sugars primarily into lactic acid, yeasts like Saccharomyces cerevisiae convert sugars into ethanol and carbon dioxide. The presence of ethanol can inhibit the growth of undesirable bacteria and molds, further stabilizing the fermentation process. Additionally, the production of organic acids by LAB can enhance the flavor complexity of fermented beverages, contributing to a more palatable product (Sieuwerts et al. 2008).

19.7.3 Nutrient Utilization

Mixed cultures exhibit a broader range of enzymatic capabilities compared to single-species cultures. This diversity allows them to utilize a wider variety of substrates efficiently. For example, LAB can break down lactose and other carbohydrates that yeasts may not effectively ferment. Conversely, yeasts can metabolize sugars that are not readily accessible to LAB due to their complex structure (Devi et al. 2014).

The ability of mixed cultures to degrade complex carbohydrates is particularly beneficial in millet fermentation. Millets contain various polysaccharides that may be difficult for a single organism to metabolize fully. However, when LAB and yeasts are combined, they can collaboratively break down these compounds into simpler sugars that are readily fermentable. This cooperation not only enhances substrate utilization but also improves the overall yield of fermentation products (Zhao et al. 2020).

19.8 Factors Affecting the Fermentation Process

Several crucial elements impact the fermentation process, playing a significant role in the growth and metabolic functions of the microorganisms that participate in fermentation. Recognizing these factors is vital for enhancing the quality and efficiency of fermented products, especially those made from millets. The primary factors include temperature, pH, and fermentation duration (Shrivastava et al. 2012).

19.8.1 Temperature

Temperature plays a crucial role in the fermentation process, significantly impacting the metabolic rates of microorganisms, enzyme functions, and the overall dynamics of fermentation. Lactic acid bacteria (LAB) and yeasts used in millet fermentation thrive at an optimal temperature range of 25–30 °C. Within this range, the metabolic efficiency of LAB, such as Lactobacillus plantarum, is maximized, leading to efficient lactic acid production (Guzzon et al. 2020; Maleke et al. 2022).

At temperatures below 25 °C, the rate of fermentation slows down due to reduced enzymatic activity, while temperatures above 30 °C can lead to thermal stress and denaturation of enzymes, adversely affecting microbial viability. Additionally, high temperatures may lead to the production of off-flavors or undesirable byproducts due to increased metabolic byproducts like acetic acid. Therefore, maintaining a stable temperature within the optimal range is crucial for achieving consistent fermentation results (Guzzon et al. 2020).

19.8.2 pH Levels

The pH of the fermentation medium significantly impacts microbial growth and metabolic activity. For effective millet fermentation, maintaining pH levels between 4.5 and 5.5 is essential. This acidic environment not only promotes the growth of LAB but also inhibits the growth of spoilage organisms and pathogens. LAB can thrive in acidic conditions due to their ability to ferment sugars into lactic acid, further lowering the pH and creating a self-preserving environment (Gänzle 2015).

Research indicates that LAB have diverse mechanisms for energy generation that are influenced by pH. For example, at lower pH levels, LAB can utilize proton motive force across their plasma membranes to generate energy anaerobically. Conversely, if the pH rises above 5.5, it may hinder the growth of LAB while allowing undesirable microorganisms to proliferate. Therefore, regular monitoring and adjustment of pH during fermentation are critical for maintaining an optimal environment (Petrova and Petrov 2020).

19.8.3 Fermentation Duration

The duration of fermentation plays a vital role in determining the flavor profile, texture, and overall quality of the final product. A typical fermentation period for millet-based products ranges from 24 to 48 h, depending on various factors such as temperature and microbial strains used. During this time, microorganisms metabolize sugars and produce organic acids, alcohols, and gases that contribute to flavor development (Semwal et al. 2021).

Shorter fermentation durations may result in incomplete sugar conversion and insufficient flavor development, while excessively long fermentation can lead to over-acidification or spoilage. The balance between these extremes is crucial; thus, monitoring sensory attributes such as taste and aroma during fermentation can help determine the optimal duration (Zhao et al. 2020).

19.8.4 Nutrient Availability

In addition to temperature, pH, and duration, nutrient availability also significantly affects fermentation outcomes. Microorganisms require a balanced supply of carbon sources (sugars), nitrogen sources (amino acids), vitamins, and minerals for optimal growth (Adebo et al. 2022). The composition of the substrate used in millet fermentation can influence nutrient availability; for instance, millets naturally contain various carbohydrates that serve as fermentable substrates. Research has shown that varying the carbon-to-nitrogen ratio can impact microbial growth rates and metabolite production. For example, an adequate supply of nitrogen enhances protein synthesis in LAB, which can improve flavor complexity in fermented products (Adebo et al. 2022).

19.9 Health Benefits of Millets-based Fermented Beverages

19.9.1 Probiotic Content and Its Impact on Gut Health

Fermented drinks made from millet, like Ambali and Togwa, provide a variety of health advantages mainly due to their high probiotic content. Probiotics are live microorganisms that deliver health benefits when ingested in sufficient quantities, especially for digestive wellness. The fermentation process not only improves the nutritional value of millets but also promotes the growth of beneficial bacteria that are essential for sustaining a balanced gut microbiome (Jacob et al. 2024).

19.9.2 Probiotic Content

The process of fermenting millets usually incorporates LAB and yeasts, which enhance the probiotic levels in the resulting product (Oloo et al. 2014). Certain LAB strains, including *Lactobacillus plantarum* and *Lactobacillus helveticus*, are notably efficient at converting the sugars found in millets into lactic acid. This conversion reduces the pH, creating favorable conditions for their proliferation. Research has indicated that fermented millet products can have viable probiotic counts surpassing 10^6 CFU/mL, a level deemed advantageous for digestive health (Rivera-Espinoza and Gallardo-Navarro 2010).

19.9.3 Impact on Gut Health and Other Health Benefits

Enhanced Digestion The inclusion of probiotics in fermented millet drinks supports the breakdown of complex carbohydrates and proteins. Through fermentation, these larger molecules are converted into simpler forms, facilitating easier absorption by the body (Nkhata et al. 2018). For example, fermenting millets can improve the availability of critical nutrients, including iron and calcium, by minimizing the presence of anti-nutritional components such as phytic acid. Research indicates that extractability rates for iron and calcium increase significantly during fermentation, from 18.1% to 37.3% for iron and from 76.9% to 90.2% for calcium (Kiczorowski et al. 2022).

Prebiotic Effects The dietary fiber found in millets serves as a prebiotic, encouraging the proliferation of helpful bacteria in the gut. This fiber comprises both soluble and insoluble varieties, which support digestive health by increasing stool bulk and aiding in regular bowel movements. Prebiotics foster the development of probiotics, thus enhancing the diversity of gut flora and optimizing overall gut function (Cuamatzin-garcía et al. 2022).

Immune System Regulation The functionality of the immune system is closely connected to the health of the gut microbiome. Fermented millet drinks contain probiotics that aid in fighting off harmful pathogens and regulating immune reactions. During fermentation, short-chain fatty acids (SCFAs) are produced, providing energy for colonocytes, the cells that line the colon, and demonstrating anti-inflammatory properties. This regulation is vital for preserving gut health and preventing disorders such as leaky gut syndrome (Ahmed et al. 2016).

Weight Management Fermented foods have been associated with increased satiety due to their fiber content and probiotic activity. Studies suggest that a balanced gut microbiome can influence weight regulation by affecting metabolism and energy extraction from food. The consumption of millet-based fermented beverages may therefore aid in weight management efforts (Geetha et al. 2020).

Potential Therapeutic Benefits Fermented millets also exhibit potential therapeutic benefits due to their antioxidant properties. The fermentation process enhances the concentration of bioactive substances, including phenolics and flavonoids, recognized for their antioxidant properties. These compounds play a crucial role in shielding cells from oxidative stress, thereby supporting overall health (Kiczorowska et al. 2019).

19.10 Mechanisms of Antioxidant Activity

19.10.1 Free Radical Scavenging

The primary mechanism by which phenolic compounds in fermented millet beverages exert their antioxidant effects is through free radical scavenging (Gull et al. 2016). Millets are abundant in a range of bioactive substances, especially phenolic compounds recognized for their antioxidant capabilities (Steinkraus 1997). Free radicals are unstable molecules that can inflict oxidative harm on cellular structures, potentially resulting in numerous health issues such as cancer and heart disease. Phenolic compounds, which include flavonoids and phenolic acids, can counteract these free radicals by donating electrons, thereby stabilizing them and preventing further cellular injury (Mokoena et al. 2016). A recent study demonstrated that the antioxidant capacity of germinated millet flours is considerably higher than that of dehusked varieties (Charalampopoulos et al. 2002). For example, studies indicate that germinated millet flours exhibit antioxidant activity ranging from 165.60 to 315.58 μg AEAC/g (Ascorbic Acid Equivalent Antioxidant Capacity per gram), while dehusked millet flours range from 132.27 to 193.57 μg AEAC/g. Fermentation enhances the release of bioactive compounds, such as phenolics, increasing the free radical scavenging capacity by up to 250% in certain millet varieties (Pushparaj and Urooj 2014).

19.10.2 Metal Chelation

Another significant mechanism of antioxidant activity is metal chelation. Phenolic compounds can bind metal ions such as iron and copper, which act as catalysts for oxidative reactions. By chelating these metal ions, phenolics inhibit the Fenton reaction—a process that generates highly reactive hydroxyl radicals from hydrogen peroxide. This action reduces oxidative stress within cells (Borresen et al. 2012).

For instance, phenolic compounds in fermented millets can effectively bind iron ions, preventing them from participating in harmful redox reactions that generate free radicals. The ability of phenolics to chelate metals not only helps mitigate oxidative damage but also enhances the overall antioxidant capacity of fermented products (Tou et al. 2007).

19.10.3 Modulation of Antioxidant Enzymes

Fermented millet drinks may bolster the body's natural antioxidant mechanisms by increasing the production of essential antioxidant enzymes such as superoxide dismutase (SOD), catalase (CAT), and glutathione peroxidase (GPx) (Fasreen et al. 2017). These enzymes are vital for neutralizing reactive oxygen species (ROS) and safeguarding cells from oxidative harm. Studies suggest that consuming fermented foods can enhance the activity of these protective enzymes (Khorshidian et al. 2020). For instance, research on lactic acid fermentation demonstrated a notable rise in SOD and CAT activity in human cell lines subjected to oxidative stress. The elevated levels of these enzymes contribute to strengthening the body's inherent defenses against oxidative stress, ultimately promoting better cellular health (Muyanja et al. 2003).

19.10.4 Potential Role of Millet-Based Fermented Beverages in Preventing Chronic Diseases

Fermented drinks made from millet, including Ambali and Togwa, have attracted interest due to their possible health advantages, especially in reducing the risk of chronic illnesses. These beverages are abundant in bioactive substances, probiotics, and antioxidants, all of which play a role in their beneficial effects on health (Panghal et al. 2018).

1. *Cardiovascular Diseases*
   Cardiovascular diseases (CVDs) are among the leading causes of mortality worldwide. The consumption of millet-based fermented beverages may reduce the risk of CVDs through several mechanisms:

   Cholesterol Reduction: Millets are high in soluble fiber, which has been shown to lower LDL cholesterol levels (Singh et al. 2018). A study indicated that dietary fiber can reduce total cholesterol by approximately 10% to 15%. The fermentation process enhances fiber solubility, making it more effective in cholesterol management (Sarita and Singh 2016).
   Antioxidant Activity: The antioxidant properties of fermented millets help combat oxidative stress, a significant contributor to cardiovascular diseases. Phenolic compounds in millets exhibit free radical scavenging activity, which can reduce lipid peroxidation and improve endothelial function (Khorshidian et al. 2020).

2. *Diabetes Management*
   Diabetes is a chronic condition characterized by elevated blood glucose levels. Millet-based fermented beverages may play a role in managing diabetes through:

   Glycemic Control: Fermented millets have a low glycemic index (GI), which helps regulate blood sugar levels. Studies show that consuming low-GI foods can lead to better glycemic control and reduced insulin resistance (Fasreen et al. 2017). The fermentation process further breaks down complex carbohydrates into simpler sugars, making them easier to digest without causing spikes in blood glucose (Bajaj et al. 2012).
   Probiotic Influence: Probiotics present in fermented millet beverages can improve gut health and enhance insulin sensitivity. Research indicates that certain strains of Lactobacillus can modulate glucose metabolism and improve insulin sensitivity (Di Stefano et al. 2017).

3. *Cancer Prevention*
   The potential anticancer properties of millet-based fermented beverages are attributed to their high antioxidant content and the presence of bioactive compounds:

   Phenolic Compounds: Fermented millets contain abundant phenolic compounds known for their anti-cancer properties. Research indicates that these compounds can prevent the growth of cancer cells and trigger programmed cell death in different types of cancer cell lines (Gupta et al. 2023). Specifically, ferulic acid has demonstrated the ability to halt the proliferation of colon cancer cells by causing cell cycle arrest (Byresh et al. 2022).
   Gut Health: Probiotics found in fermented millets play a significant role in maintaining a healthy gut microbiome, which is gaining recognition for its potential in cancer prevention. A well-balanced gut microbiome helps to avoid dysbiosis, a state associated with a heightened risk of colorectal cancer (Budhwar et al. 2020).

4. Digestive Disorders

   Beverages made from fermented millet may also help in reducing the occurrence of digestive issues like irritable bowel syndrome (IBS) and constipation.

   Enhanced Digestion: During fermentation, complex carbohydrates and proteins are transformed into simpler compounds, enhancing their digestibility. The presence of probiotics supports a balanced gut microbiome, facilitating both digestion and the absorption of nutrients (Saleh et al. 2013).

   Prebiotic Effects: The dietary fiber found in millets serves as a prebiotic, fostering the proliferation of helpful bacteria within the gastrointestinal tract. Maintaining this balance is essential for promoting digestive wellness and mitigating issues such as irritable bowel syndrome (IBS) (Xie and Liu 2004).

19.11 Challenges and Future Directions in Millet-Based Fermented Beverages

Millet-based fermented beverages are emerging as functional foods with significant health benefits, particularly in improving gut health and preventing chronic diseases. However, several challenges must be addressed to fully realize their potential. These challenges include the reduction of anti-nutritional factors, standardization of fermentation processes, consumer acceptance, and the need for innovative processing technologies (Adakole et al. 2021).

19.11.1 Challenges

Reduction of Anti-nutritional Factors Although fermentation can greatly diminish anti-nutritional elements like phytic acid and tannins, the degree of this reduction is influenced by the specific fermentation conditions and the types of microorganisms involved (Annor et al. 2014). For example, studies have shown that lactic acid fermentation can reduce phytic acid levels by up to 50%, but achieving consistent results across different batches remains a challenge. Therefore, optimizing fermentation parameters such as temperature, pH, and duration is crucial for maximizing the reduction of these compounds (Di Stefano et al. 2017; Kumar Siroha et al. 2024).

Standardization of Fermentation Processes The fermentation process is often influenced by local practices and traditional methods, leading to variability in product quality. Standardizing the fermentation process using specific starter cultures can help achieve consistent flavor profiles and nutritional benefits (Choi et al. 2004). Research indicates that using elite strains of *Lactobacillus* can enhance the texture and viscosity of millet beverages while improving their probiotic content. However, further studies are needed to determine the optimal strains and conditions for industrial-scale production (Malleshi et al. 1986).

Consumer Acceptance Despite their health benefits, millet-based products often face challenges related to consumer acceptance. The perception of millets as "traditional" or "less desirable" compared to mainstream grains like wheat and rice can hinder their adoption. To overcome this barrier, it is essential to educate consumers about the nutritional advantages of millets and develop appealing products that cater to modern tastes. Innovative processing (Siroha and Bangar 2024) techniques that enhance flavor and texture can also play a crucial role in increasing consumer acceptance.

Nutritional Profiling While millet-based fermented beverages are recognized for their health benefits, comprehensive nutritional profiling is often lacking. Detailed analyses of micronutrient content, bioactive compounds, and their effects on human health are necessary to substantiate health claims (Jacob et al. 2024). Research indicates that fermented millet drinks may improve the bioavailability of minerals like iron and calcium. Nevertheless, further large-scale clinical studies are required to validate these results in human subjects (Patimah et al. 2019).

19.11.2 Future Directions

Innovative Processing Technologies Progress in food technology has the potential to greatly improve both the nutritional value and sensory qualities of products made from millet. Methods like extrusion cooking, malting, and nonthermal processing can enhance digestibility and make nutrients more accessible (Amadou 2021). For instance, combining malting with fermentation has been shown to increase protein digestibility by up to 30% while reducing antinutrients (Dayakar et al. 2017).

Functional Food Development Cutting-edge processing technologies are increasingly sought after as consumers show a heightened interest in functional foods that provide health advantages beyond mere nutritional value (Mehta et al. 2024). Future research should focus on developing novel millet-based products fortified with additional nutrients or bioactive compounds. For example, incorporating microgreens from millets could enhance the antioxidant profile and overall nutritional value of fermented beverages (Saleem et al. 2023).

Exploration of Diverse Strains Investigating a broader range of microbial strains for fermentation can lead to improved product quality (Jacob et al. 2024). The use of mixed cultures comprising LAB and yeasts can enhance flavor complexity while improving probiotic content. Identifying specific strains that maximize health benefits while minimizing undesirable flavors will be crucial for product development (Shukla and Bhise 2023).

Sustainability Considerations As global food systems face challenges related to climate change and food security, millets offer a sustainable alternative due to their resilience in adverse conditions (Kumar Siroha et al. 2024). Research should focus on promoting millet cultivation as part of sustainable agricultural practices while developing value-added products that leverage local resources (Puranik et al. 2017). This approach aligns with various United Nations Sustainable Development Goals (SDGs), including Zero Hunger (SDG 2) and Responsible Consumption and Production (SDG 12) (Hassan et al. 2021).

19.12 Conclusion

Millet-based fermented beverages offer significant potential to enhance public health due to their nutrient-rich profile, including high protein (7.7–11.6%), essential amino acids, and bioactive compounds. Fermentation enhances nutrient bioavailability, reduces antinutritional factors like phytic acid, and introduces probiotics, supporting gut health and reducing chronic disease risks, such as diabetes and heart disease. To maximize benefits, challenges like inconsistent fermentation methods, sensory appeal, and consumer acceptance must be addressed. Future research should focus on optimizing microbial strains (*Lactobacillus plantarum*), improving sensory attributes, and validating health benefits through clinical trials. Promoting sustainable millet farming aligns with global food security goals. Comprehensive nutritional analyses are needed to integrate these gluten-free, functional beverages into modern diets. By leveraging interdisciplinary expertise from food scientists, nutritionists, and agricultural specialists, millet-based fermented drinks can become a sustainable, nutritious option, blending traditional wisdom with contemporary health demands, and gaining global market acceptance.

References

Adakole MI, Ogori AF, Ikya JKH et al (2021) Fermented millet "Ibyer" beverage enhanced with ginger powder: an assessment of microbiological, pasting, proximate, and sensorial properties. Appl Sci 11:3151. https://doi.org/10.3390/APP11073151

Adebiyi JA, Obadina AO, Adebo OA, Kayitesi E (2018) Fermented and malted millet products in Africa: expedition from traditional/ethnic foods to industrial value-added products. Crit Rev Food Sci Nutr 58:463–474. https://doi.org/10.1080/10408398.2016.1188056

Adebo JA, Njobeh PB, Gbashi S et al (2022) Fermentation of cereals and legumes: impact on nutritional constituents and nutrient bioavailability. Fermentation 8:63. https://doi.org/10.3390/FERMENTATION8020063

Agrawal P, Singh BR, Gajbe U et al (2023) Managing diabetes mellitus with millets: a new solution. Cureus 15:e44908. https://doi.org/10.7759/CUREUS.44908

Agu RC, Palmer GH (2013) Evaluation of the potentials of millet, sorghum and barley with similar nitrogen contents malted at their optimum germination temperatures for use in brewing. J Inst Brew 119:258–264. https://doi.org/10.1002/JIB.91

Aguirre-Garcia YL, Nery-Flores SD, Campos-Muzquiz LG et al (2024) Lactic acid fermentation in the food industry and bio-preservation of food. Fermentation 10:168. https://doi.org/10.3390/FERMENTATION10030168

Ahmed ST, Mun HS, Islam MM et al (2016) Effects of dietary natural and fermented herb combination on growth performance, carcass traits and meat quality in grower-finisher pigs. Meat Sci 122:7–15. https://doi.org/10.1016/j.meatsci.2016.07.016

Ali MAM, El Tinay AH, Abdalla AH (2003) Effect of fermentation on the in vitro protein digestibility of pearl millet. Food Chem 80:51–54. https://doi.org/10.1016/S0308-8146(02)00234-0

Ali AS, Elozeiri AA, Ali AS, Elozeiri AA (2017) Metabolic processes during seed germination. Adv Seed Biol. https://doi.org/10.5772/INTECHOPEN.70653

Amadou I (2019) Millet based fermented beverages processing. In: Fermented beverages: volume 5 the science of beverages, pp 433–472. https://doi.org/10.1016/B978-0-12-815271-3.00011-7

Amadou I (2021) Millet based functional foods: bio-chemical and bio-functional properties. Funct Foods:303–329. https://doi.org/10.1002/9781119776345.CH9

Anitha S, Rajendran A, Botha R et al (2024) Variation in the nutrient content of different genotypes and varieties of millets, studied globally: a systematic review. Front Sustain Food Syst 8:1324046. https://doi.org/10.3389/FSUFS.2024.1324046/BIBTEX

Annor GA, Marcone M, Bertoft E, Seetharaman K (2014) Physical and molecular characterization of millet starches. Cereal Chem 91:286–292. https://doi.org/10.1094/CCHEM-08-13-0155-R

Aparicio-García N, Martínez-Villaluenga C, Frias J, Peñas E (2021) Production and characterization of a novel gluten-free fermented beverage based on sprouted oat flour. Foods 10:139. https://doi.org/10.3390/foods10010139

Atter A, Diaz M, Tano-Debrah K et al (2024) The predominant lactic acid bacteria and yeasts involved in the spontaneous fermentation of millet during the production of the traditional porridge Hausa koko in Ghana. BMC Microbiol 24:1–17. https://doi.org/10.1186/S12866-024-03317-1/TABLES/4

Bajaj A, Rao MRP, Pardeshi A, Sali D (2012) Nanocrystallization by evaporative antisolvent technique for solubility and bioavailability enhancement of telmisartan. AAPS PharmSciTech 13:1331–1340. https://doi.org/10.1208/S12249-012-9860-X/METRICS

Balli D, Cecchi L, Pieraccini G et al (2023) Millet fermented by different combinations of yeasts and lactobacilli: effects on phenolic composition, starch, mineral content and prebiotic activity. Foods 12:748. https://doi.org/10.3390/FOODS12040748

Bhattacharjee S, Kar P, Sarkar I et al (2021) Biochemical and microbial profiling establish "Eu" (a traditional fermented beverage of Toto people) as a probiotic health drink. J Ethn Foods 8:1–13. https://doi.org/10.1186/S42779-021-00093-5/FIGURES/9

Borresen EC, Henderson AJ, Kumar A et al (2012) Fermented foods: patented approaches and formulations for nutritional supplementation and health promotion. Rec Patent Food Nutr Agric 4:134–140. https://doi.org/10.2174/2212798411204020134

Budhwar S, Sethi K, Chakraborty M (2020) Efficacy of germination and probiotic fermentation on underutilized cereal and millet grains. Food Prod Process Nutr 2. https://doi.org/10.1186/S43014-020-00026-W

Byresh TS, Malini B, Meena L et al (2022) Effect of addition of pineapple peel powder on white finger millet vegan probiotic beverage. J Food Process Preserv 46. https://doi.org/10.1111/JFPP.16905

Chamarthy VR, Tonapi VA (2021) Millet-based traditional processed food beverages. In: Millets and millet technology, pp 305–320. https://doi.org/10.1007/978-981-16-0676-2_15

Charalampopoulos D, Wang R, Pandiella SS, Webb C (2002) Application of cereals and cereal components in functional foods: a review. Int J Food Microbiol 79:131–141. https://doi.org/10.1016/S0168-1605(02)00187-3

Chauhan D, Gujral HS, Kaler RSS (2024) Modification of the bioactivity, digestibility and rheological behaviour of different millets by lactic acid fermentation. Int J Food Sci Technol 59:4336–4347. https://doi.org/10.1111/IJFS.17009

Chavan UD, Chavan JK, Kadam SS (1988) Effect of fermentation on soluble proteins and in vitro protein digestibility of sorghum, green gram and sorghum-green gram blends. J Food Sci 53:1574–1575. https://doi.org/10.1111/J.1365-2621.1988.TB09329.X

Chethan Kumar P, Amutha S, Oberoi HS et al (2022) Germination induced changes in bioactive compounds and nutritional components of millets. J Food Sci Technol 59:4244–4252. https://doi.org/10.1007/S13197-022-05485-2/METRICS

Choi H, Kim W, Shin M (2004) Properties of korean amaranth starch compared to waxy millet and waxy sorghum starches. Starch/Staerke 56:469–477. https://doi.org/10.1002/STAR.200300273

Cuamatzin-garcía L, Rodríguez-rugarcía P, El-kassis EG et al (2022) Traditional fermented foods and beverages from around the world and their health benefits. Microorganisms 10:1151. https://doi.org/10.3390/MICROORGANISMS10061151

Day L, McSweeney PLH (2016) Beverages. In: Reference module in food science. https://doi.org/10.1016/B978-0-08-100596-5.03409-0

Dayakar B, Bhaskarachary RK, Arlene GD et al (2017) Nutritional and health benefits of millets. ICAR_Indian Institute of Millets Research (IIMR) Rajendranagar, Hyderabad

Debnath B, Manna K, Singh WS, Goswami S (2024) Formulation and evaluation of new herbal nutraceutical tablet for the treatment of diabetes mellitus. J Drug Res Ayurvedic Sci 9:29–39. https://doi.org/10.4103/JDRAS.JDRAS_48_23

Dev R, Bhatt S, Gupta M (2024) Effect of lactic acid fermentation on the physico-chemical, functional, and antioxidant properties, and in vitro protein digestibility of malted ragi (*Eleusine coracana* L.). Sustain Food Technol 2:1128–1138. https://doi.org/10.1039/D4FB00134F

Devi PB, Vijayabharathi R, Sathyabama S et al (2014) Health benefits of finger millet (*Eleusine coracana* L.) polyphenols and dietary fiber: a review. J Food Sci Technol 51:1021–1040. https://doi.org/10.1007/S13197-011-0584-9

Dhondiram DS, Sawant AA, Mamoon F et al (2023) Nutritional characterization of pseudocereals: a review. Pharma Innov J 12:109–118. https://doi.org/10.22271/TPI.2023.V12.I3B.19413

Di Stefano E, White J, Seney S et al (2017) A novel millet-based probiotic fermented food for the developing world. Nutrients 9:529. https://doi.org/10.3390/NU9050529

Dimidi E, Cox SR, Rossi M, Whelan K (2019) Fermented foods: definitions and characteristics, impact on the gut microbiota and effects on gastrointestinal health and disease. Nutrients 11:1806. https://doi.org/10.3390/NU11081806

Dordević TM, Šiler-Marinković SS, Dimitrijević-Branković SI (2010) Effect of fermentation on antioxidant properties of some cereals and pseudo cereals. Food Chem 119:957–963. https://doi.org/10.1016/J.FOODCHEM.2009.07.049

Drub TF, Garcia dos Santos F, Ladeia Solera Centeno AC, Capriles VD (2021) Sorghum, millet and pseudocereals as ingredients for gluten-free whole-grain yeast rolls. Int J Gastron Food Sci 23:100293. https://doi.org/10.1016/J.IJGFS.2020.100293

Duodu KG, Taylor JRN, Belton PS, Hamaker BR (2003) Factors affecting sorghum protein digestibility. J Cereal Sci 38:117–131. https://doi.org/10.1016/S0733-5210(03)00016-X

Dymchenko A, Kirilenko Y, Ochkolyas O, Bal-Prylypko L (2024) Gluten-free beer with unmalted millet. Eur Food Res Technol 250:999–1005. https://doi.org/10.1007/S00217-023-04405-8/TABLES/4

Estruch R, Lamuela-Raventós RM (2023) Cardiovascular benefits of fermented foods and beverages: still up for debate. Nat Rev Cardiol 20:789–790. https://doi.org/10.1038/s41569-023-00938-3

Fasreen MMF, Perera ODAN, Weerahewa HLD (2017) Development of finger millet based probiotic beverage using *Lactobacillus casei* 431®. OUSL J 12:128. https://doi.org/10.4038/OUSLJ.V12I1.7384

Gänzle MG (2015) Lactic metabolism revisited: metabolism of lactic acid bacteria in food fermentations and food spoilage. Curr Opin Food Sci 2:106–117. https://doi.org/10.1016/J.COFS.2015.03.001

Geetha K, Yankanchi GM, Hulamani S, Hiremath N (2020) Glycemic index of millet based food mix and its effect on pre diabetic subjects. J Food Sci Technol 57:2732–2738. https://doi.org/10.1007/S13197-020-04309-5

Gull A, Prasad K, Kumar P (2016) Evaluation of functional, antinutritional, pasting and microstructural properties of millet flours. J Food Meas Charact 10:96–102. https://doi.org/10.1007/S11694-015-9281-0

Gupta M, Asfaha DM, Ponnaiah G (2023) Millets: A nutritional powerhouse with anti-cancer potential. Cureus, 15(10):e47769. https://doi.org/10.7759/cureus.47769

Gupta R, Gaur S (2024) Investigating the effect of natural fermentation in modifying the physico-functional, structural and thermal characteristics of pearl and finger millet starch. J Sci Food Agric 104:2440–2448. https://doi.org/10.1002/JSFA.13129

Guzzon R, Roman T, Larcher R (2020) Impact of different temperature profiles on simultaneous yeast and bacteria fermentation. Ann Microbiol 70:1–14. https://doi.org/10.1186/S13213-020-01565-W/FIGURES/4

Hassan ZM, Sebola NA, Mabelebele M (2021) The nutritional use of millet grain for food and feed: a review. Agric Food Secur 10:16. https://doi.org/10.1186/S40066-020-00282-6

Hema V, Ramaprabha M, Saraswathi R et al (2022) Millet food products. In: Handbook of millets—processing, quality, and nutrition status, pp 265–299. https://doi.org/10.1007/978-981-16-7224-8_12

Hiremath N, Geetha K, Vikram SR et al (2018) Minerals content in finger millet [*Eleusine coracana* (L.) Gaertn]: a future grain for nutritional security. Int J Curr Microbiol App Sci 2018:3448–3455

Institute of Food Science and Technology (UK) (2018) Food and drink—good manufacturing practice. A guide to its responsible management. https://doi.org/10.1002/9781119388494

Jacob J, Krishnan V, Antony C et al (2024) The nutrition and therapeutic potential of millets: an updated narrative review. Front Nutr 11:1346869. https://doi.org/10.3389/FNUT.2024.1346869

Jan S, Kumar K, Yadav AN et al (2022) Effect of diverse fermentation treatments on nutritional composition, bioactive components, and anti-nutritional factors of finger millet (*Eleusine coracana* L.). J Appl Biol Biotechnol 10:46–52. https://doi.org/10.7324/JABB.2022.10S107

Jideani VA, Ratau MA, Okudoh VI (2021) Non-alcoholic pearl millet beverage innovation with own bioburden: *Leuconostoc mesenteroides*, *Pediococcus pentosaceus* and *Enterococcus gallinarum*. Foods 10:1447. https://doi.org/10.3390/FOODS10071447

Joseph B, Bhavadharani M, Lavanya M et al (2024) Comparative analysis of LAB and non-LAB fermented millet drinks fortified with Chlorella sp. Food Bioeng 3:352–364. https://doi.org/10.1002/FBE2.12102

Joshi R (2018) Role of enzymes in seed germination. Int J Creative Res Thought 6:2320–2882

Kaur H (2023) Possible applications of certain pseudocereals within the food manufacturing sector. Pharma Innov J 12:1466–1469. https://doi.org/10.22271/TPI.2023.V12.I8SS.22405

Khorshidian N, Yousefi M, Mortazavian AM (2020) Fermented milk: the most popular probiotic food carrier. Adv Food Nutr Res 94:91–114. https://doi.org/10.1016/BS.AFNR.2020.06.007

Kiczorowska B, Samolińska W, Andrejko D et al (2019) Comparative analysis of selected bioactive components (fatty acids, tocopherols, xanthophyll, lycopene, phenols) and basic nutrients in raw and thermally processed camelina, sunflower, and flax seeds (*Camelina sativa* L. Crantz, Helianthus L., and Linum L.). J Food Sci Technol 56:4296–4310. https://doi.org/10.1007/s13197-019-03899-z

Kiczorowski P, Kiczorowska B, Samolińska W et al (2022) Effect of fermentation of chosen vegetables on the nutrient, mineral, and biocomponent profile in human and animal nutrition. Sci Rep 12:1–13. https://doi.org/10.1038/s41598-022-17782-z

Kondrotiene K, Zavistanaviciute P, Aksomaitiene J et al (2023) *Lactococcus lactis* in dairy fermentation—health-promoting and probiotic properties. Fermentation 10:16. https://doi.org/10.3390/FERMENTATION10010016

Kumari K, Kashyap P, Chakrabarti P (2024) Germination and probiotic fermentation: a way to enhance nutritional and biochemical properties of cereals and millets. Food Sci Biotechnol 33:505–518. https://doi.org/10.1007/S10068-023-01401-2/METRICS

Leeuwendaal NK, Stanton C, O'toole PW, Beresford TP (2022) Fermented foods, health and the gut microbiome. Nutrients 14:1527. https://doi.org/10.3390/NU14071527

Lei V, Jakobsen M (2004) Microbiological characterization and probiotic potential of koko and koko sour water, African spontaneously fermented millet porridge and drink. J Appl Microbiol 96:384–397. https://doi.org/10.1046/J.1365-2672.2004.02162.X

Lei V, Friis H, Michaelsen KF (2006) Spontaneously fermented millet product as a natural probiotic treatment for diarrhoea in young children: an intervention study in Northern Ghana. Int J Food Microbiol 110:246–253. https://doi.org/10.1016/J.IJFOODMICRO.2006.04.022

Li L, Yan Y, Ding W et al (2020) Improvement in the quality of kimchi by fermentation with Leuconostoc mesenteroides ATCC 8293 as starter culture. Microbiol Biotechnol Lett 48:533–538. https://doi.org/10.48022/MBL.2008.08008

Lorri W, Svanberg U (1993) Lactic-fermented cereal gruels with improved in vitro protein digestibility. Int J Food Sci Nutr 44:29–36. https://doi.org/10.3109/09637489309017420

Majzoobi M, Wang Z, Teimouri S et al (2023) Unlocking the potential of sprouted cereals, pseudocereals, and pulses in combating malnutrition. Foods (Basel, Switzerland) 12. https://doi.org/10.3390/FOODS12213901

Makokha AO, Oniang'o RK, Njoroge SM, Kamar OK (2002) Effect of traditional fermentation and malting on phytic acid and mineral availability from sorghum (*Sorghum bicolor*) and finger millet (*Eleusine coracana*) grain varieties grown in Kenya. Food Nutr Bull 23:241–245. https://doi.org/10.1177/15648265020233S147

Makwana M, Hati S (2019) Fermented beverages and their health benefits. In: Fermented beverages: volume 5 the science of beverages, pp 1–29. https://doi.org/10.1016/B978-0-12-815271-3.00001-4

Maleke M, Doorsamy W, Abrahams AM et al (2022) Influence of fermentation conditions (temperature and time) on the physicochemical properties and bacteria microbiota of amasi. Fermentation 8:57. https://doi.org/10.3390/FERMENTATION8020057

Malleshi NG, Desikachar HSR, Tharanathan RN (1986) Physico-chemical properties of native and malted finger millet, pearl millet and foxtail millet starches. Starch-Stärke 38:202–205. https://doi.org/10.1002/STAR.19860380608

Mehrotra P (2016) Biosensors and their applications—a review. J Oral Biol Craniofac Res 6:153. https://doi.org/10.1016/J.JOBCR.2015.12.002

Mehta D, Vyas S, Dudhagara D et al (2024) Significance of Indian millets in enhancing global food security: a comprehensive review. Trends Food Sci Technol 149:104527. https://doi.org/10.1016/J.TIFS.2024.104527

MILLETS: the future food | India Science, Technology & Innovation—ISTI Portal. https://www.indiascienceandtechnology.gov.in/listingpage/millets-future-food. Accessed 25 Oct 2024

Mishra G, Panda BK (2022) Cereal based fermented foods and non-alcohol beverages. In: Functional cereals and cereal foods: properties, functionality and applications, pp 189–213. https://doi.org/10.1007/978-3-031-05611-6_8

Mohapatra D, Nickhil C, Kar A et al (2024) Impact of LAB fermentation on the nutrient content, amino acid profile, and estimated glycemic index of sorghum, pearl millet, and kodo millet. Front Biosci Elite 16:18. https://doi.org/10.31083/J.FBE1602018/1945-0508-16-2-018/FIG1.JPG

Mokoena MP, Mutanda T, Olaniran AO (2016) Perspectives on the probiotic potential of lactic acid bacteria from African traditional fermented foods and beverages. Food Sci Nutr 60. https://doi.org/10.3402/FNR.V60.29630

Muyanja CMBK, Narvhus JA, Treimo J, Langsrud T (2003) Isolation, characterisation and identification of lactic acid bacteria from bushera: a Ugandan traditional fermented beverage. Int J Food Microbiol 80:201–210. https://doi.org/10.1016/S0168-1605(02)00148-4

Neeru, Choudhary P, Dhewa T, Kumar A (2024) Process optimization of pearl millet-based fermented beverage components using response surface methodology and evaluation of its physicochemical parameters. J Food Sci Technol 61:1557–1568. https://doi.org/10.1007/S13197-024-05926-0/METRICS

Newlove AA, Mahunu GK, Tahir HE (2022) Fermented millet for porridge production: a model for improved gastrointestinal health. In: African fermented food products—new trends, pp 445–458. https://doi.org/10.1007/978-3-030-82902-5_29

Nkhata SG, Ayua E, Kamau EH, Shingiro JB (2018) Fermentation and germination improve nutritional value of cereals and legumes through activation of endogenous enzymes. Food Sci Nutr 6:2446–2458. https://doi.org/10.1002/FSN3.846

Oloo BO, Shitandi A, Mahungu S et al (2014) Effects of lactic acid fermentation on the retention of beta-carotene content in orange fleshed sweet potatoes. Int J Food Stud 3:13–33. https://doi.org/10.7455/ijfs/3.1.2014.a2

Panghal A, Janghu S, Virkar K et al (2018) Potential non-dairy probiotic products—a healthy approach. Food Biosci 21:80–89. https://doi.org/10.1016/J.FBIO.2017.12.003

Palanisamy S, Thiyagamoorthy UM (2025) Probiotication of finger millet porridge for enhancing its function and safety. Bangladesh J Bot 54(1):75–81. https://doi.org/10.3329/bjb.v54i1.78226

Patimah S, Arundhana AI, Mursaha A, Syam A (2019) Development of foxtail millet and flying fish flour-based cookies as functional food. Curr Res Nutr Food Sci 7:504–516. https://doi.org/10.12944/CRNFSJ.7.2.20

Petrova P, Petrov K (2020) Lactic acid fermentation of cereals and pseudocereals: ancient nutritional biotechnologies with modern applications. Nutrients 12. https://doi.org/10.3390/NU12041118

Puranik S, Kam J, Sahu PP et al (2017) Harnessing finger millet to combat calcium deficiency in humans: challenges and prospects. Front Plant Sci 8:1311. https://doi.org/10.3389/FPLS.2017.01311

Pushparaj FS, Urooj A (2014) Antioxidant activity in two pearl millet (*Pennisetum typhoideum*) cultivars as influenced by processing. Antioxidants 3:55. https://doi.org/10.3390/ANTIOX3010055

Rao V, Poonia A (2023) Protein characteristics, amino acid profile, health benefits and methods of extraction and isolation of proteins from some pseudocereals—a review. Food Prod Process Nutr 5:1–17. https://doi.org/10.1186/S43014-023-00154-Z

Rathore S (2016) Millet grain processing, utilization and its role in health promotion: a review. Int J Nutr Food Sci 5:318. https://doi.org/10.11648/J.IJNFS.20160505.12

Rios-Corripio G, Hernández-Carranza P, Ruiz-López I et al (2021) Health benefits of fermented alcoholic beverages. In: Bioactive compounds in fermented foods, pp 265–299. https://doi.org/10.1201/9780429027413-14

Rivera-Espinoza Y, Gallardo-Navarro Y (2010) Non-dairy probiotic products. Food Microbiol 27:1–11. https://doi.org/10.1016/j.fm.2008.06.008

Saleem S, Ul Mushtaq N, Shah WH et al (2021) Morpho-physiological, biochemical and molecular adaptation of millets to abiotic stresses: a review. Phyton Int J Exp Bot 90:1363–1385. https://doi.org/10.32604/PHYTON.2021.014826

Saleem S, Mushtaq NU, Shah WH et al (2023) Millets as smart future food with essential phytonutrients for promoting health. J Food Compos Anal 124:105669. https://doi.org/10.1016/J.JFCA.2023.105669

Saleh ASM, Zhang Q, Chen J, Shen Q (2013) Millet grains: nutritional quality, processing, and potential health benefits. Compr Rev Food Sci Food Saf 12:281–295. https://doi.org/10.1111/1541-4337.12012

Samtiya M, Aluko RE, Puniya AK, Dhewa T (2021) Enhancing micronutrients bioavailability through fermentation of plant-based foods: a concise review. Fermentation 7:63. https://doi.org/10.3390/FERMENTATION7020063

Sarita, Singh E (2016) Potential of millets: nutrients composition and health benefits. J Sci Innov Res 5:46–50. https://doi.org/10.31254/JSIR.2016.5204

Sarsan S, Pandiyan A, Rodhe AV, Jagavati S (2021) Synergistic interactions among microbial communities. In: Microbes in microbial communities: ecological and applied perspectives, pp 1–37. https://doi.org/10.1007/978-981-16-5617-0_1

Semwal J, Kamani MH, Meera MS (2021) Fermented millet technology and products. In: Millets and millet technology, pp 255–271. https://doi.org/10.1007/978-981-16-0676-2_12

Shafiq S, Shakeel F, Talegaonkar S et al (2007) Development and bioavailability assessment of ramipril nanoemulsion formulation. Eur J Pharm Biopharm 66:227–243. https://doi.org/10.1016/j.ejpb.2006.10.014

Sheethal HV, Baruah C, Subhash K et al (2022) Insights of nutritional and anti-nutritional retention in traditionally processed millets. Front Sustain Food Syst 5:735356. https://doi.org/10.3389/FSUFS.2021.735356/BIBTEX

Shin S-Y, Han NS (2015) Leuconostoc spp. as starters and their beneficial roles in fermented foods. In: Beneficial microorganisms in food and nutraceuticals, pp 111–132. https://doi.org/10.1007/978-3-319-23177-8_5

Shrivastava K, Greeshma A, Srivastava B (2012) Biotechnology in tradition—a process technology of alcoholic beverages practiced by different tribes of Arunachal Pradesh, North East India. Indian J Tradit Knowl 11:81–89

Shukla P, Bhise S (2023) Mighty millets: bespoke for multi nutrients. Curr J Appl Sci Technol 42:54–62. https://doi.org/10.9734/cjast/2023/v42i324220

Sieuwerts S, De Bok FAM, Hugenholtz J, Van Hylckama Vlieg JET (2008) Unraveling microbial interactions in food fermentations: from classical to genomics approaches. Appl Environ Microbiol 74:4997–5007. https://doi.org/10.1128/AEM.00113-08/ASSET/07083657-0C5E-4F99-B04E-C12B55CF0280/ASSETS/GRAPHIC/ZAM0160890970001.JPEG

Simple fermented millet (GF, wild-fermented)—fermenting for foodies. https://www.fermentingforfoodies.com/simple-fermented-millet/. Accessed 26 Oct 2024

Singh RB, Khan S, Chauhan AK et al (2018) Millets as functional food, a gift from Asia to Western world. In: The role of functional food security in global health, pp 457–468. https://doi.org/10.1016/B978-0-12-813148-0.00027-X

Singh SB, Meena AK, Sisodia BS et al (2023) From farm to gut: unraveling the role of millets in promoting metabolic well-being via gut microbiota. J Drug Res Ayurvedic Sci 8:S50–S54. https://doi.org/10.4103/JDRAS.JDRAS_192_23

Siroha AK, Bangar P (2024) Millet-based food products: an overview. Curr Food Sci Technol Rep 2:213–220. https://doi.org/10.1007/S43555-024-00031-X

Srivastava U, Singh A, Mazia et al (2024) Functional and physicochemical characterization of a novel pearl millet—soy milk-based synbiotic beverage. Food Safety Health. https://doi.org/10.1002/FSH3.12076

Steinkraus KH (1997) Classification of fermented foods: worldwide review of household fermentation techniques. Food Control 8:311–317. https://doi.org/10.1016/S0956-7135(97)00050-9

Stokes HW, Gillings MR (2011) Gene flow, mobile genetic elements and the recruitment of antibiotic resistance genes into Gram-negative pathogens. FEMS Microbiol Rev 35:790–819. https://doi.org/10.1111/J.1574-6976.2011.00273.X

Sudha A, Priyenka Devi KS, Sangeetha V, Sangeetha A (2016) Development of fermented millet sprout milk beverage based on physicochemical property studies and consumer acceptability data. J Sci Ind Res 75:239–243

Sun Z, Bin, Zhang X, Yan Y et al (2023) The effect of buckwheat resistant starch on intestinal physiological function. Foods (Basel, Switzerland) 12. https://doi.org/10.3390/FOODS12102069

Torello Pianale L, Rugbjerg P, Olsson L (2022) Real-time monitoring of the yeast intracellular state during bioprocesses with a toolbox of biosensors. Front Microbiol 12:802169. https://doi.org/10.3389/FMICB.2021.802169/BIBTEX

Tou EH, Mouquet-Rivier C, Picq C et al (2007) Improving the nutritional quality of ben-saalga, a traditional fermented millet-based gruel, by co-fermenting millet with groundnut and modifying the processing method. LWT 40:1561–1569. https://doi.org/10.1016/J.LWT.2006.12.001

Versari A, Patrizi C, Parpinello GP et al (2016) Effect of co-inoculation with yeast and bacteria on chemical and sensory characteristics of commercial Cabernet Franc red wine from Switzerland. J Chem Technol Biotechnol 91:876–882. https://doi.org/10.1002/JCTB.4652

Vila-Real C, Pimenta-Martins A, Mbugua S et al (2022) Novel synbiotic fermented finger millet-based yoghurt-like beverage: nutritional, physicochemical, and sensory characterization. J Funct Foods 99:105324. https://doi.org/10.1016/J.JFF.2022.105324

Walker GM, Stewart GG (2016) Saccharomyces cerevisiae in the production of fermented beverages. Beverages 2:30. https://doi.org/10.3390/BEVERAGES2040030

Xie X, Liu Q (2004) Development and physicochemical characterization of new resistant citrate starch from different corn starches. Starch/Staerke 56:364–370. https://doi.org/10.1002/STAR.200300261

Yaqoob S, Imtiaz A, Awan KA et al (2024) Impact of fermentation through synergistic effect of different lactic acid bacteria (mono and co-cultures) on metabolic and sensorial profile of mulberry juice. J Food Meas Charact 18:9364–9384. https://doi.org/10.1007/S11694-024-02885-0/METRICS

Yu F, Li M, He D, Yang P (2021) Advances on post-translational modifications involved in seed germination. Front Plant Sci 12:642979. https://doi.org/10.3389/FPLS.2021.642979/BIBTEX

Zhang H, Zhang P, Wu T, Ruan H (2023) Bioethanol production based on *Saccharomyces cerevisiae*: opportunities and challenges. Fermentation 9:709. https://doi.org/10.3390/FERMENTATION9080709

Zhao H, Carrau F, Tao Y et al (2020) Impact of nutrient availability on the fermentation and production of aroma compounds under sequential inoculation with *M. pulcherrima* and *S. cerevisiae*. Front Microbiol 11:518930. https://doi.org/10.3389/FMICB.2020.00305

Zhu F (2021) Buckwheat proteins and peptides: biological functions and food applications. Trends Food Sci Technol 110:155–167. https://doi.org/10.1016/J.TIFS.2021.01.081

Zvauya R, Mygochi T, Parawira W (1997) Microbial and biochemical changes occurring during production of masvusvu and mangisi, traditional Zimbabwean beverages. Plant Foods Hum Nutr 51:43–51. https://doi.org/10.1023/A:1007972428849/METRICS

Chapter 20
Molecular Biology with Nutritional Attributes of Sorghum

Nazia Tarannum, Km Pooja, and Ranu Agrawal

20.1 Introduction

The *Gramineae* family of crops includes the carbohydrate-rich crop Sorghum millet, also known as *Sorghum bicolor L.*, which is native to Africa (de Morais Cardoso et al. 2017). After wheat, maize, rice, and barley, it was domesticated 3000–5000 years ago and is now ranked as the 5th largest crop in the world, which is rich in carbohydrates. Sorghum millet serves as a critical food crop in dealing with hunger and ensuring food security, particularly for vulnerable populations living in fragile agro-ecological regions across the globe. Apart from being one of the most significant crops rich in carbohydrates and a crucial staple crop that provides nutrition for millions of people in tropical places, it is also referred as "King of Millets" and is popularly grown in China, Africa, Mexico, the United States, and India (Khaton et al. 2016). Sorghum millet is grown in around 100 nations worldwide and has become more important, especially in Africa, Asia, and other semi-arid regions. Sorghum millet is a staple food for people even though it is primarily cultivated as a significant feed grain for animal nutrition in many western countries, such as Australia, the United States, and Brazil. Nevertheless, a lot of research has been done on Sorghum for human consumption because of its enormous nutritional and physiological potential (Rakshit et al. 2014).

Examining nutritional qualities of sorghum millet aims to give a thorough grasp of the grain dietary importance and potential health effects. The purpose of this study is to inform readers about the particular nutrients found in sorghum, their

N. Tarannum (✉) · K. Pooja
Department of Chemistry, Chaudhary Charan Singh University, Meerut, Uttar Pradesh, India

R. Agrawal
Department of Applied Science, SCRIET, Chaudhary Charan Singh University, Meerut, Uttar Pradesh, India

© The Author(s), under exclusive license to Springer Nature Switzerland AG 2026

J. Singh, M. Kumar Malik (eds.), *Millets and Millet Starch: Chemical Composition, Functional Modification, and Applications*,
https://doi.org/10.1007/978-3-032-09502-2_20

possible health advantages, and the ways in which sorghum millet in diets can improve general health (Awika and Rooney 2004). This chapter explores the diverse nutritional attributes of sorghum, emphasizing its significance as a functional food in both traditional and modern diets. A detailed analysis of macro and micronutrient of sorghum millet composition, including dietary fiber, essential vitamins and minerals, proteins, lipids, and carbohydrates, is provided to highlight how these components contribute to its overall nutritional value. The chapter also investigates the bioavailability of these nutrients, identifying factors that enhance or hinder their absorption within the human body. Special attention is given to the contribution of dietary fiber to maintaining digestive wellness with discussions on how high fiber content of sorghum millet can aid in digestion and offer broader health benefits. Furthermore, the potential of sorghum millet in disease prevention is examined, particularly its contributions to blood glucose regulation, cardiovascular health, and the mitigation of certain chronic illnesses.

A comparative analysis is conducted to evaluate nutritional profile against other commonly consumed grains, emphasizing the unique properties that make it a suitable addition to a balanced healthy diet. The chapter also presents the culinary versatility of sorghum millet, offering practical guidance on incorporating it into meals along with examples of nutritious recipes. Considerations for individuals with specific dietary needs like those requiring gluten-free options or managing particular medical conditions are also addressed. Lastly, the global importance of sorghum millet is discussed, underscoring its role as a staple crop in various regions and cultures, as well as its contribution to sustainable agriculture and global food security.

20.1.1 Background and Overview

A significant crop used as a cereal in Africa, Asia, and other semi-arid parts of world is "Sorghum millet" (Afify Ael et al. 2011). The 5th-most-significant cereal produced worldwide is sorghum millet (Serna-Saldivar et al. 1988). It may thrive in tropical climates with higher temperatures and yield fruit. The most common cereal grain in Asia and Africa is sorghum. It is a great source of nutrients and bioactive compounds (like protein, carbohydrates, and others) for humans (de Morais Cardoso et al. 2017). Cookies, cereal, cakes, bread, flakes, couscous, pancakes, and drinks (alcoholic or nonalcoholic), all contain sorghum millet (Yousif, et al. 2012). The demand of sorghum millet as an important food is increasing regularly worldwide. This increase can be attributed to both the growing population that consumes sorghum millet and the growing industrial use of sorghum millet (Akintayo and Sedgo 2001). An indigenous crop of Africa, sorghum millet, is still a mainstay diet in many rural areas. In South Africa, sorghum millet is grown mostly in dry regions in average temperatures of around 25 °C, mainly on shallow, heavy clay soil. In a year, sorghum millet yield ranges between 100,000 tonnes (130,000 ha) and 180,000 tonnes (150,000 ha). The production of sorghum millet was shifted from the drier portion of western areas of production to the wet portions of eastern areas in recent years. Because of this shift in the production area, the cultivars that can withstand lower temperatures were developed after identification (Devi et al. 2014). It folds its leaves during dry spells to minimize water loss from evaporation. It does not die if the drought persists; instead, it goes dormant. A waxy cuticle covers the leaves to prevent evapotranspiration.

It is employed in the cooking of adult and children's meals. Baby food is created in tropical climates from gruels of sorghum millet and maize, sweetened with sugar. The low concentration of vital amino acids in sorghum millet grain including tryptophan, lysine, and threonine results in low-quality protein (Badi et al. 1990). The quality of protein in cereals is improved by malting as it causes an increase in lysine. Although sorghum millet exhibits limited digestibility in infants, its nutritional value as a weaning food can be enhanced through supplementation with lysine-rich foods, thereby making it a more suitable dietary option during early childhood (MacLean et al. 1981). The sorghum millet proteins of sorghum millet become less digestible when cooked. Like many other cereals, sorghum millet is a great source of vitamins B, including niacin, biotin, thiamin, and riboflavin (Hegedus et al. 1985). The composition of minerals in sorghum millet is quite similar with that of millet. Potassium and phosphorus are the key minerals that occur in sorghum millet grain, and amount of calcium is low (Khalil 1984). Condensed tannins are a class of polyphenolic chemicals found in sorghum millet that have anti-nutritional properties. Because condensed tannins can bind to minerals like iron, dietary proteins, digestive enzymes, and vitamins B like vitamin B6 and thiamin, they reduce the value of nutrition present in sorghum millet grains (Wang and Kies 1991). These compounds are absent in both colored and white sorghum millet varieties lacking a pigmented testa but are present in sorghum millet varieties that possess a pigmented testa (Anglani 1998).

20.2 Nutritional Quality

A highly prized grain crop, sorghum millet is known for its excellent nutritious content. It is grown in many African nations, the United States, and India (Tanwar et al. 2023; Kazungu et al. 2023). The United States Food and Drug Administration (USFDA) has provided the following nutritional data for 1/2 cup (96 g) of uncooked grains of sorghum. It is essential to note that the nutritional values provided refer to sorghum millet in its uncooked, dry form. Upon cooking, 1/2 cup of dry sorghum millet typically yields approximately 1.5 cups of the cooked grain due to water absorption during the cooking process. Consequently, since most individuals are likely to consume only 1/2–1 cup of the cooked product per serving, the actual intake of calories and carbohydrates per serving will be lower than the values listed for the uncooked grain. Figure 20.1 represents nutritional qualities of sorghum.

Nutritional makeup of sorghum millet varies slightly based on variety and growth environment. However, here is a general overview of the key nutritional components found in sorghum millet. Like other cereal and pseudocereal grains, sorghum

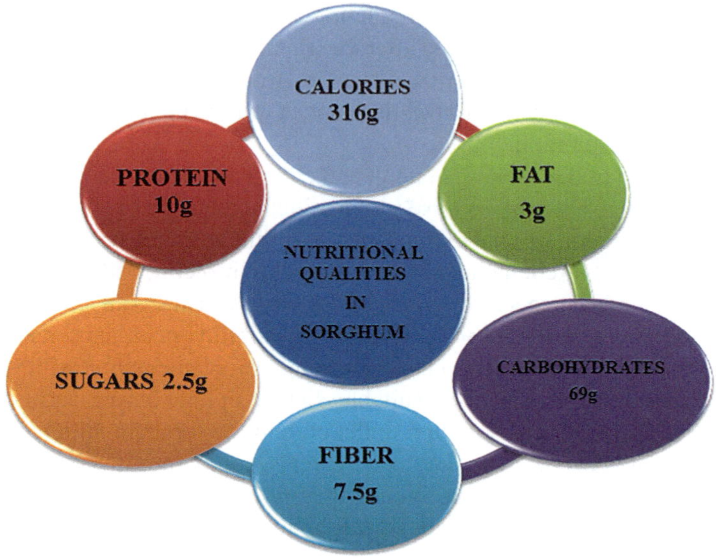

Fig. 20.1 Nutritional qualities of sorghum

millet is mostly composed of starch, which makes up roughly 65–75% of the entire grain and 85–90% of the endosperm when dried (Dicko et al. 2006). Unlike many other kinds of cereal grains, several sorghum millet varieties contain starch not only in the endosperm but also within the pericarp layer. Starch is primarily found in the endosperm. Because of this, grains that are genetically categorized as tannin-sorghums and have a colored testa may seem white in color. Sorghum millet starch particles ranges in size from 4 to 25 μm and can be spherical to angular depending on where it is located in the endosperm, much like maize starch (Beta et al. 2000). Sorghum millet starch has a high tendency to have a higher temperature range of gelatinization (68–75 °C) (Ratnavathi and Patil 2013) than maize (62–72 °C) (Tester et al. 2004), and many other cereals (like wheat, 51–72 °C; and rice, 61–74 °C) (Zhou et al. 2002). This is a crucial characteristic that must be taken into account while creating ideal processing conditions for the grain because it could take more energy to cook sorghum millet to the same degree as other cereals. Over 90% of the dietary fiber in sorghum millet is insoluble and is mostly produced from cellulosic carbohydrates found in the plant cell walls (Taylor and Emmambux 2008a, b).

20.2.1 Macronutrients and Micronutrients

A versatile cereal grain that is abundant in macro and micronutrients, a variety of nutrients is present in sorghum. The nutrient content can change based on the type of sorghum, the growth location (site), and the state of the soil. Sorghum millet nutritional makeup can change on the basis of numerous variables, including the variety, growing environment, and processing techniques. Here is a more detailed breakdown of the macro and micronutrients found in sorghum:

A. *Macronutrients:* The macronutrient composition of sorghum millet includes carbohydrates, proteins, fats, and dietary fiber.

   1. *Carbohydrates*
      Sorghum millet is primarily composed of carbohydrates, including starch and dietary fiber. The carbohydrate content can vary, but on an average, sorghum millet contains around 72–75% carbohydrates. It gives a good source of energy (Tasie and Gebreyes 2020). Table 20.1 shows the amount of carbohydrates in sorghum millet.
   2. *Protein*
      Sorghum millet is a very good protein source, with an average protein content ranging from 9% to 14%. The protein in sorghum millet is relatively high in essential amino acids, which makes it an extremely valuable source of protein based on plants (Harden et al. 1976). Table 20.2 represents the content of amino acid and protein in sorghum millet.
   3. *Dietary Fiber*
      Sorghum millet is a valuable source of dietary fiber, which includes both insoluble and soluble fibers. The fiber content can range from 6% to 15% contributing to digestive health and aiding in the regulation of blood sugar levels (Rao et al. 2018).
   4. *Fat*
      Sorghum millet is low in fat with an average fat content of about 2–4%. The fat is primarily unsaturated including both monounsaturated and polyunsaturated fats. Table 20.3 represents the content of fat and fatty acids (Pontieri et al. 2022).
   5. *Calories*
      Sorghum millet is a calorie-dense food providing energy in the form of carbohydrates and fats. On average, 1 cup of cooked sorghum millet may provide around 220–240 calories.

B. *Micronutrients:* Sorghum millet is a nutrient-dense cereal grain that contains many micronutrients, while the particular levels might vary based on factors such as variety, soil conditions, and growth environment.

Table 20.1 Carbohydrate content of sorghum millet (Palavecino et al. 2016)

Carbohydrates	
Nutrient	Amount (g)
Carbohydrate	83.45 %
Fiber	6.7 g
Sugars	4.87 g
Ash content	0.68 g

Table 20.2 Essential amino acids content of sorghum millet grain

Nutrient	*Protein*	*Alanine*	*Arginine*	*Aspartic acid*	*Cystine*
Amount (g)	20.39	1.983	0.682	1.427	0.244
Nutrient	*Glutamic acid*	*Glycine*	*Histidine*	*Isoleucine*	*Leucine*
Amount (g)	4.683	0.664	0.472	0.831	2.863
Nutrient	*Lysine*	*Methionine*	*Phenylalanine*	*Proline*	*Serine*
Amount (g)	0.440	0.324	1.048	1.636	0.887
Nutrient	*Threonine*	*Tryptophan*	*Tyrosine*	*Valine*	*Phenylalanine + tyrosine*
Amount (g)	0.664	0.238	0.616	1.077	1.664

1. *Vitamins*

   Vitamins B: Sorghum millet contains various vitamins B, such as thiamine (B1), niacin (B3), riboflavin (B2), pantothenic acid (B5), folate (B9), and pyridoxine (B6).
   Vitamin E: Sorghum millet contains small amounts of vitamin E, which has antioxidant properties. Table 20.4 represents the vitamins available in sorghum.

2. *Minerals*

   Iron: Sorghum millet is an Iron source. It is a necessary mineral for transportation of oxygen in the blood.
   Phosphorus: Important for energy metabolism, bone health, and deoxyribonucleic acid (DNA) synthesis.
   Magnesium: Contributes to nerve system function and muscle, as well as bone health.
   Zinc: Important for function of immune system and healing of wound.
   Potassium: Necessary for fluid balance, nerve system function, and muscle contractions. Table 20.5 represents the minerals content in sorghum millet (Motlhaodi et al. 2018).

3. *Antioxidants*

   Sorghum millet is considered a rich source of natural antioxidants, particularly due to its high content of polyphenolic compounds; these antioxidants might help to shield cells from inflammation and oxidative damage (Oboh and Amusan 2009). Here are some key antioxidant compounds present in sorghum:

   (i) *Phenolic Acids*

   Mainly caffeic acid, ferulic acid, sinapic acid, and p-coumaric acid present in sorghum. It is occurring in both free and bound forms (bound to cell

Table 20.3 Content of fat and fatty acids in sorghum millet (Pontieri et al. 2022)

Nutrient	*Fat*	*Saturated fatty acids*	*Docosanoic acid*	*Dodecanoic acid*	*Heptadecanoic acid*
Amount (g)	6.643	1.171	0.008	0.002	0.004
Nutrient	*Hexadecanoic acid*	*Octadecanoic acid*	*Pentadecanoic acid*	*Tetracosanoic acid*	*Tetradecanoic acid*
Amount (g)	1.043	1.100	0.002	0.015	0.004
Nutrient	*Monounsaturated fatty acids*	*EicosenoicAcid*	*Hexadecenoic acid*	*Octadecenoic acid*	*Tetracosenoiccis acid*
Amount (g)	2.172	0.021	0.031	2.114	0.006
Nutrient	*Trans-hexadecenoic acid*	*Polyunsatura ted fatty acids*	*Octadecadienoic acid*	*Octadecatrienoic acid*	*Fatty acids, total trans*
Amount (g)	0.004	2.991	2.867	0.1	0.1

Table 20.4 Vitamins content in sorghum millet (https://fdc.nal.usda.gov/food-details/169716/nutrients)

Nutrient	*Vitamin A*	*Retinol*	*Thiamin*	*Riboflavin*	*Niacin*
Amount (mg)	0.00	0.00	2.927	0.142	7.081
Nutrient	*Pantothenic acid*	*Vitamin B 6*	*Vitamin B 12*	*Folate*	
Amount (mg)	0.705	0.590	0.00	0.02	
Nutrient	*Folic acid*	*Vitamin C*	*Vitamin D*	*Vitamin E*	*Tocopherol, alpha*
Amount (mg)	0.00	0.0	0.00	0.810	0.5

Table 20.5 Minerals content in sorghum millet (Motlhaodi et al. 2018)

Nutrient	*Calcium*	*Copper*	*Iron*	*Magnesium*	*Manganese*
Amount (mg)	24.96	0.55	6.45	316.80	3.082
Nutrient	*Phosphorus*	*Potassium*	*Selenium*	*Sodium*	*Zinc*
Amount (mg)	554.88	696.96	23.42	3.84	3.21

wall structures like lignin and hemicellulose). The most abundant and a major contributor to antioxidant activity is Ferulic acid (Dicko et al. 2005).

(ii) *Flavonoids*

It includes luteolin, apigenin, and their glycosides. Pigmented sorghum millet varieties (especially red, brown, and black) have higher flavonoid content (Dykes and Rooney 2006).

(iii) *Tannins (Condensed Polyphenols)*

It is found in high-tannin sorghum millet varieties (usually with pigmented pericarp). It has strong radical-scavenging and metal-chelating properties. It may reduce nutrient bioavailability but has beneficial effects at moderate levels (Awika et al. 2003).

(iv) *3-Deoxyanthocyanidins*

It has unique class of anthocyanin-like pigments (e.g., luteolinidin and apigeninidin) found almost exclusively in sorghum. It possesses high free radical scavenging activity and is heat stable, useful in functional foods (Yang et al. 2018).

20.3 Typical Composition of Sorghum Millet Relative to Other Cereals

Sorghum millet grain composition and its other constituent parts are largely identical to maize with the only exception of reduced oil content. The grain has 2–6% fiber, 65–76% starch, and 8–12% protein. Even though cellulose and hemicellulose make up the majority of bran, a sizable amount of starch is deposited in the mesocarp tissue of this fraction. Wax, not oil, makes up the majority of the lipids in bran. Various sources of sorghum millet grain may have different compositions due to a variety of factors (Hegde and Chandra 2005). The grains are bigger, more densely packed with carbohydrates, and contain less protein as a result of the more recent hybrids and extensive irrigation practices. A thorough analysis has been performed on the nutritional characteristics and proximate sorghum millet grain composition. Lu et al. investigated that the content of protein in grain of sorghum millet ranges between 4.4% and 21.1% with an average of 11.4%. Compared to other food grains, the grain of sorghum millet is renowned because of its hardness. Grain hardness is caused by a greater concentration of prolamin protein (3.6–5.1%) (Mohamed et al. 2009). There is a range of 1.06–3.64% lysine content. According to studies on protein fractionation in sorghum, the distribution of total nitrogen is roughly 15%, 26%, and 44% for albumin-globulin, prolamin, and glutelin, respectively. The total concentration of soluble sugar in sorghum millet grain ranges from 0.7% to 4.2%, while its reducing sugar content ranged between 0.05% and 0.53%. The fat content of sorghum millet grain ranges from 2.1% to 7.6% while its crude fiber content ranges between 1.0% and 3.4% and its ash content ranges between 1.3% and 3.3% (Rathore et al. 2019). There was significant variance in the protein (7.99–17.8%), starch (51.88–85%), lipids (2.52–4.76%), and amylose (12.30–28.38%) content of another study that examined the physico-chemical characteristics of sorghum millet accessions. Oleic acid (18:1) and linoleic acid (18:2) were the two main fatty acid components of sorghum millet lipids. The majority of nutrients retained in the testa were frequently used up with the grain. Sorghum millet is a rich source of minerals due to its wide variation in trace element and mineral; the mineral composition is affected by environmental factors. Table 20.6 represents the typical composition of sorghum millet relative to other cereals (Maciel et al. 2024; Widowati and Luna 2022).

20.4 Molecular Biology of Sorghum

20.4.1 Genomic Structure and Organization

The genetic structure and organization of sorghum millet refer to the arrangement of genetic material within the species. Sorghum millet genome is organized into chromosomes. The exact number of chromosomes in sorghum millet is 10. The chromosomes contain both coding regions (genes) and noncoding regions. Sorghum millet shares genomic synteny with other grass species including rice, maize, and millet. Synteny refers to the conservation of content and gene order across various genome species. Genomics comparison helps in understanding the evolutionary relationships between different plant species. The sorghum millet has diploid genome and has haploid genome size of approximately 730 Mega base pair (Mbp), which is much larger in comparison of Rice and Arabidopsis (510 Mbp and 155 Mbp, respectively). Although these three plants belong same number of gene families (Paterson et al. 2009). In the 90s, genome mapping of sorghum millet was started, and at present several genetic maps are available for genome. Sorghum

Table 20.6 Typical composition of sorghum millet relative to other cereals (Maciel et al. 2024; Widowati and Luna 2022)

S. no.	Cereals	Protein (g)	Fiber (g)	Fat (g)	Carbohydrates (g)	Energy (kcal)
1	Wheat	13.3	10.7	2	71	339
2	Emmer	13	5	4%	34	360
3	Einkorn	4	2	1%	20	100
4	Rice	6.81	2.8	1.9	81.68	370
5	White basmati rice	3	0	0	36	160
6	Millet	9.9	3.2	2.9	72.9	378
7	Pearl millet	11.8	2.3	4.8	67	363
8	Barley	9.9	15.6	2.3	77.7	352
9	Maize	8.9	2	3.9	72.2	360
10	Oats	7	5	3	29	180
11	Durum	7	7	Greater than 1%	40	198
12	Spelt	3	7.6	Greater than 1.7%	23	110
13	Buckweat	13	10	3%	71	343
14	Brown rice cooked	2.6	0.4	1.8	23	111
15	Jasmine rice	4	2	1.5	38	180
16	Finger millet	7.3	3.6	1.3	72	336
17	Sorghum	10.62	6.7	3.3	72.09	329
18	Quinoa	14.12	7	6	64.16	368
19	Psyllium	2	6.7	14.6	8.5	160
20	Flaxseed	20.3	37.1	24	28.9	530

millet consists of ten diploid chromosomes and has fever possibilities to be duplicated genetically, which make it an exceptional model for the advancement of functional genomics studies. Assumption was made that sorghum millet originated from tetraploid because of the presence of more amount of complementary gene loci, and this was also supported by the phenomenon of fluorescence in situ hybridization (FISH). Later using other techniques and other different genomic resources, which included genomic clones associated with large inserts in the bacterial artificial chromosomes (BAC), recognized the ten haploid chromosomes (Kim et al. 2002; Gomez et al. 1997). After this, Paterson et al. found the homologies and identities between various linkage groups in metaphase state and determined sorghum millet diploid ($2n = 20$) chromosomes and genome size of 730 Mega base pair. The genomic map was made by using restriction fragment length polymorphisms (RFLPs), amplified fragment length polymorphism (AFLP) or using recombinant inbred lines (RIL) population and various other tests. Kong et al. (2020) plotted a RIL population with 31 simple sequence repeats (SSRs) polymorphic loci using isolated clones from sorghum. A map with high genetic density was built with the help of specific length amplified fragments markers (SLAFs). The genetic map was marked with 2246 markers covering 10 chromosomes having an overall length of 2158.10 cm and between adjacent markers, the average distance was of 0.98 cm. Sequencing was done using 52,928 SLAFs and in this 12% were found polymorphic. In sorghum, about 125,000 ESTs, which were grouped in 22,000 unigenes and shown in more than 20 diverse libraries of different genotypes, indicate the spatial structure of the genes (Pratt et al. 2005). Sorghum millet genome contains a diverse set of genes that encode proteins and noncoding RNAs. Sorghum, like many plant genomes, contains transposable elements (TEs). These are the DNA sequences that are capable of moving within the genome and influence genetic diversity and evolution. Genes are responsible for various biological functions, including growth, metabolism, development, and responses to environmental stimuli. Sorghum millet has a moderate gene density, and genes are distributed across the chromosomes. Advances in genomics play a significant function in expediting breeding initiatives and unraveling the genetic makeup of sorghum millet and for this advantageous cereal crop.

20.4.2 Genetic Diversity and Breeding Advances

Sorghum millet includes three subspecies: bicolor, arundinaceum, and drummondii. On the basis of morphological classification, all cultivated sorghum millet species are grouped in feasible varieties, which are characterized into five races: caudatum, bicolor, guinea durra, and kafire. These races mostly differ generally in their panicle morphology, yield potential, and grain size. Durra primarily originated in Ethiopia, and now it became spread and popular in Nigeria and some part of South Africa. Kafir developed in southern and eastern Africa. Guinea race grows well in West and Central Africa, while the bicolor race developed in East Africa but not important for production in Africa. Sorghum millet is generally a grass and belongs to graminae family, it grows annually. Sorghum millet is generally classified into four major groups as: sweet, grain, grass, and broom sorghum. Grain sorghum millet is the most cultivated type in the world and a chief staple food in semi-arid tropical area of Africa and Asia. The food is made in the form of boiled porridge, popped, dumplings, fermented beers, and nonalcoholics beverages. The sweet sorghum millet is produced mainly for sugar production as sweet syrup. Grain sorghum, millet, and sweet sorghum millet differ phenotypically in several aspects: higher plant length, sugar-rich juicy stem, higher production of biomass, and fewer amounts of grains in the spike (Rooney et al. 2007). The genetic variation cannot explain the differences between two varieties, which suggest epigenetic mechanisms and DNA methylation are key factors to explain those. International Crops Research Institute for the Semi-Arid Tropics (ICRISAT) is a member of the Consultative Group on International Agricultural Research (CGIAR), located at Patancheru, India. It was established in 1972 and considered one of its five main crops. Total 36,774 accessions have been submitted from 90 different countries and arranged in a gene accession bank (Reddy et al. 2008). These accession collections reveal 80% of the genetic diversity found in the crop. Different varieties and accessions of sorghum millet may exhibit genetic variation. This variation can include differences in gene sequences, copy number variations, and other structural variations. Genetic variation is crucial for breeding programs focused at improving traits like disease resistance yield and stress tolerance (Mengistu et al. 2018). ICRISAT is mostly engaged in diversification of breeding populations of sorghum. Now it has incorporated abiotic and biotic resistance traits, which were not inserted earlier in improvement programs. The traits which are presently being emphasized globally include tolerance toward drought, Aluminum toxicity, heat, salt, stem pests, and grain molds (Tari et al. 2013). Fast growth to obtain high quantity grain, biomass yield, and tillering capacity are being highlighted. Genetic diversity in sorghum millet can be measured using molecular markers like single nucleotide polymorphisms (SNPs) or SSRs, and map traits of interest in sorghum millet (Mace et al. 2008; Weerasooriya et al. 2016). Genetic markers assist in the selection and identification of desirable traits in breeding programs. Genomic tools, such as genetic maps and genomic databases, are employed for studying the genetic structure and organization of sorghum. Genetic basis of important agronomic traits can be understood via these tools.

A comprehensive understanding of the sorghum millet genome is provided by genome sequencing, facilitating the recognition of genes that are responsible for important traits (Paterson et al. 2009). It acts as a foundation for various molecular breeding techniques. Molecular breeding in sorghum millet involves molecular biology techniques application to accelerate the breeding process, enabling the development of sorghum millet varieties with desirable traits. Marker-assisted selection (MAS) includes the utilization of molecular markers that are linked to specific traits to assist in selecting the plants during the breeding process. Molecular markers such as SNPs and SSRs are used to detect and select plants having target traits, like drought tolerance, disease resistance, or high yield (Guo et al. 2019). Genomic selection uses information from the entire genome to identify the breeding value of an individual, even in the absence of detailed phenotypic data. Genomic selection has been employed in sorghum millet breeding to improve the accuracy of predicting complex traits. It utilizes dense marker information to determine the genetic value of individuals. Quantitative trait locus (QTL) mapping recognizes the genomic regions linked with quantitative traits. QTL mapping in sorghum millet helps pinpoint regions of the genome that influence traits like yield, grain quality, and stress tolerance (Mwamahonje et al. 2024). This information guides the breeding process. Precise modification of specific genes can be edited by using CRISPR-Cas9, which is a revolutionary genome editing technology. CRISPR-Cas9 has been used to create sorghum millet varieties with targeted genetic modifications leading to traits like improved disease resistance or enhanced nutritional content. High-throughput phenotyping consists of the automated and rapid measurement of plant traits on a large scale. High-throughput phenotyping technologies help quantify plant traits efficiently, enabling breeders to pick plants with desirable characteristics (Juliana et al. 2019). Integration of different omics technologies, including transcriptomics, genomics, and metabolomics, provides a holistic acknowledgement of traits. Molecular breeding in sorghum millet has significantly accelerated the evolution of improved varieties with increased resilience, productivity, and nutritional quality (Semere et al. 2023). Continuous advancements in molecular techniques contribute to the efficiency and precision of sorghum millet breeding programs.

20.4.3 Transcriptomics and Gene Expression Profiling

The transcriptomics of sorghum millet involves the research of the complete set of ribonucleic acid (RNA) transcripts generated by the genome under particular conditions or at different developmental stages. Transcriptomics provides insights into gene expression patterns, alternative splicing, and other aspects of RNA biology. The identification of differential genes has been parallel to the progress in transcriptomes expressing in response to abiotic and biotic factors, and also the damage created by dehydration, insects, concentration of high salt, salicylic acid, abscisic acid (Buchanan et al. 2005), methyl-jasmonate, and amino-cyclopropane carboxylic acid (Salzman et al. 2005; McKinley et al. 2016). The genes (Rio, R9188, and BTx406) are biosynthetic for cellulose-like structures and pectin cellulose (Li et al. 2019). New insights into growth phenomenon of stem vegetative and development of intercalary meristem are provided by comparing stem areas within the same genotype and between different genotypes of sorghum millet (Kebrom et al. 2017; Yu et al. 2022).

There are various key findings and trends in sorghum millet transcriptomics like abiotic stress responses when sorghum millet is exposed to abiotic stresses, like high temperatures and drought. The molecular mechanisms of sorghum millet responses to these stress environments were studied in 2016 by Hamza and coworkers in which 8 deregulated miRNAs by abiotic stress were studied in 11 elite varieties of sorghum millet under drought and low availability of water. miRNAs intervene in a number of biological processes, like stress responses, metabolism, and identification and development of organs. This research revealed that the miRNAs, miR393, miR396, miR397-5p, miR167, miR166, and miR168 have a significant range of deregulation, being sbi-miRNA398 and sbi-miR396 the ones with higher overexpression for all the genotypes. The similar effect of salinity and drought in the miRNAs profiles developed in S. bicolor by the other research group. This confirmed that the miRNAs expression profile is connected to the dosage of stress applied to the plants. However, each miRNA reacted in an exclusive manner in each of the six genotypes.

Transcriptomics has been explored for gene expression toward sorghum millet responses to biotic stresses, having pathogens-pest defense-related genes and pathways, which are crucial for breeding programs focused at developing resistant varieties (Hao et al. 2021). Studies revealed gene expression changes during different developmental stages of sorghum, from seed germination to flowering and grain development. Sorghum millet is a bioenergy crop that has the expression of genes incriminated in biomass production, biofuel traits, and responses to bioenergy-related stress (Mace et al. 2013).

Understanding the transcriptomics of sorghum can provide insights into the expression of genes related to nutritional content including those that are engaged in the synthesis of nutrients like vitamins and amino acids. Yu and co-worker (2015) proposed mir-271 as a particular miRNA of the Rio sweet sorghum millet variety related to the sugar accumulation and cellulose synthesis. Transcriptomics also plays a key role in identifying candidate genes for crop improvement. A complete list with most of the relevant miRNAs for the genetic enhancement of sorghum millet in the production of biofuels was studied. Researchers use differential expression analysis to compare gene expression between different conditions or genotypes. This helps identify genes that are upregulated or downregulated under specific circumstances. Identifying transcription factors and unraveling regulatory networks is a key aspect of sorghum millet transcriptomics. Understanding the regulatory machinery is essential for manipulating gene expression in a controlled manner. Advances in technologies of high-throughput sequencing, such as RNA-seq, have revolutionized transcriptomic studies in sorghum. These technologies provide a comprehensive aspect of the entire transcriptome.

20.5 Chemistry of Starch and Proteins in Sorghum

Sorghum millet is a cereal grain that contains starch and proteins. Both of which are essential components contributing to its nutritional and functional properties.

20.5.1 Composition and Structure

20.5.1.1 Starch

20.5.1.1.1 Composition

Starch is the primary storage form of carbohydrate in sorghum millet grains. The starch in sorghum millet is the complex carbohydrate that is the primary energy storage molecule in the grain. It is made up of two types of polymers: amylopectin and amylose. Amylose is one of the components of starch in sorghum. It is a linear polymer made up of α-1,4-glycosidic linkages connecting glucose molecules. Amylose content can vary among sorghum millet varieties and is an essential factor influencing the properties of sorghum millet starch. Amylopectin is another major component of sorghum millet starch. It is a highly branched polymer of glucose with α-1,6-glycosidic bonds forming branches and α-1,4-glycosidic bonds forming the backbone. Amylopectin contributes to the branching and structure of starch granules. Sorghum millet starch can include a portion of resistant starch, which is not digested properly in the small intestine and reaches the colon intact. Resistant starch has potential health benefits. The metabolic analysis of sorghum millet juice and stem internodes explained that it also contains fructose, glucose and sucrose type nonstructural carbohydrates in which the main abundant soluble sugar in most sorghum millet genotypes is sucrose (Li et al. 2018; Sekhon et al. 2016).

20.5.1.1.2 Structural Characteristics

The starch granules in sorghum millet are organized in the endosperm of the grain. Granules exhibit a characteristic bimodal size distribution, containing both large and small granules. Starch is stored in amyloplasts, which are specialized organelles within plant cells (Hao et al. 2021). The size and organization of amyloplasts can influence the functional properties of the starch. Starch granules have both amorphous and crystalline regions. Crystalline regions made by organized amylopectin chains contribute to the granule structure and properties during gelatinization. The gelatinization properties of sorghum millet starch are essential for food processing (Ubwa et al. 2012). Gelatinization is the phenomenon during which granules of starch absorb water and swell, leading to the loss of their semi-crystalline nature. The gelatinization temperature and viscosity of sorghum millet starch can affect the texture of products prepared from sorghum millet flour (Subramanian et al. 1982).

20.5.1.2 Protein

20.5.1.2.1 Composition

Protein composition of sorghum millet includes various classes of proteins. Sorghum millet contains both storage proteins and structural proteins. The major storage proteins in sorghum millet are kafirins, which make up the majority of the protein fraction being the predominant storage proteins. Kafirins are prolamin proteins unique to sorghum. Kafirins are a group of prolamin proteins unique to sorghum millet (Proietti et al. 2015). They are divided into three major types: α-kafirin, β-kafirin, and γ-kafirin. Each type has different solubility properties in alcohol, with α-kafirin being soluble, β-kafirin partially soluble, and γ-kafirin insoluble.

Sorghum millet also contains non-storage proteins such as albumins and globulins. These proteins are water-soluble and play roles in various metabolic processes (Tanwar et al. 2023). Sorghum millet grains may also contain enzymes involved in processes such as germination, including amylases, and proteases (Kayisoglu et al. 2024).

20.5.1.2.2 Structural Characteristics

Kafirins are rich in glutamine and proline contributing to their unique structural properties. The primary structures of proteins are characterized with a large content of repetitive sequences including pentapeptide repeats and hexapeptide repeats. The secondary structure of kafirin exhibits a high proportion of β-sheet, which contributes to their unique solubility characteristics. The β-sheet structure is important to form the protein bodies in sorghum millet endosperm. This solubility property influences the extraction and separation of kafirins during processing.

20.5.2 Functionality and Source of Energy

Sorghum millet starch is considered more slowly digestible compared to some other cereals, which can be beneficial for managing blood glucose levels. The structure and composition of sorghum millet starch can influence its digestibility and glycemic response (Barros et al. 2014). Sorghum millet proteins having low digestibility make it promising food candidate for people with diabetes and contribute to the texture and structure of various food products. The unique properties of kafirins influence the dough rheology and affect the quality of sorghum-based products (Belton et al. 2006).

20.5.3 Nutritive and Anti-nutritive Properties

20.5.3.1 Nutritive Properties

Sorghum millet is a versatile and nutritious cereal grain with several key nutritive properties. Its nutritional composition makes it an important food source, especially in regions with arid and semiarid climates. Sorghum millet is a nutritious cereal grain that provides a range of essential nutrients. The nutritional composition can vary based on factors such as the growing conditions, variety, and processing methods. The following are some of the nutritive properties of sorghum:

(a) *Dietary Fiber:* Sorghum millet is a rich source of dietary fiber including both insoluble fiber (75–90%) and soluble fibers (10–25%). These fibers are present in the endosperm cell wall and pericarp, and the concentration is about 6–15 g per 100 g of grain (Martinol et al. 2012). Dietary fiber contributes to digestive health, helps to maintain bowel regularity, and may aid in weight management.
(b) *Macromolecules:* Sorghum millet is a great source of complex carbohydrates including starch. It provides a steady release of energy. While sorghum millet proteins may be lower in lysine compared to some other grains. Data reveals that 100 g of the sorghum millet cereals possess on average, 72.1 g carbohydrate, 10.6 g protein, 3.5 g lipids, 6.7 gm fibers, and 12.4 g water and gives about 1377 kj energy sorghum millet is rich in proline, glutamic acid, and leucine and contribute to the overall protein content in the diet (Xiong et al. 2019).
(c) *Minerals:* Sorghum millet is a rich source of minerals like iron, phosphorus, magnesium, and zinc. These minerals play a crucial role in energy metabolism, immune function, and bone health. The bioavailability of these minerals can be affected by factors such as the presence of antinutritional factors, processing methods, and dietary context. Zinc, manganese, and copper are principal trace minerals that help to scavenge free radicals and lower harmful effects (Wierzbicka and Gromadzka 2014).

(d) *Vitamins:* Sorghum millet contains various vitamins possessing vitamins B like niacin (B3), thiamine (B1), and riboflavin (B2). The vitamin content can vary among sorghum millet varieties and is influenced by factors such as growing conditions (Afify et al. 2012).
(e) *Antioxidants:* Sorghum millet contains antioxidants carrying phenolic compounds and flavonoids. Antioxidants help in protecting cells from oxidative stress, and they might also have potential health benefits (Xu et al. 2012).
(f) *Resistant Starch:* Sorghum millet contains resistant starch, which is not completely digested in the small intestine. Resistant starch reaches the colon intact, promoting gut health (Mahasukhonthachat et al. 2010).
(g) *Low Glycemic Index:* Sorghum millets have low glycemic index in comparison to some other grains. Glycemic index food can help to manage blood sugar levels (Chinenye et al. 2016).
(h) *Gluten-Free:* Sorghum millet is naturally gluten-free, which makes it appropriate for individuals with celiac disease or gluten sensitivity. It can be utilized as an alternative to wheat in gluten-free products.
(i) *Phytochemicals:* Sorghum millet contains various phytochemicals including polyphenols, which have potential health-promoting effects. These compounds may have antioxidant, anti-inflammatory, and other bioactive properties (Awika et al. 2003).
(j) *Energy Density:* Sorghum millet provides a moderate amount of calories, making it a valuable energy source in the diet. When considering its biomass biofuel, it contains 15.07–19.7 MJ/Kg of dry matter energy density (Morales et al. 2024). It serves as a staple food in many regions, contributing to the daily caloric intake.

While sorghum millet offers many nutritional benefits, it's essential to consider the specific nutritional needs of individuals and populations. The nutritional value of sorghum millet can be affected by factors like cooking methods and the inclusion of other foods in the diet (Tanwar et al. 2023). The nutritional composition can vary among sorghum millet varieties, and the impact of anti-nutritional factors may need to be addressed through processing methods. Additionally, incorporating a diverse range of foods into the diet ensures a well-rounded nutritional intake.

20.5.3.2 Anti-nutritive Properties

Sorghum millet is a nutritious cereal grain but like many plant-based foods, it can contain certain antinutritional factors that cause interference with the utilization or absorption of nutrients (Gunawan et al. 2022). There are some anti-nutritive properties associated with sorghum millet contents as tannins, phytates, protease inhibitor, amylase inhibitor, trypsin inhibitor, and oxalates. Some sorghum millet varieties contain tannins, which are polyphenolic compounds. Tannins can bind to proteins and other nutrients and form complexes that are less digestible. Tannins may also interfere in the minerals absorption such as iron and reduce protein digestibility

(Sharma et al. 2021). Sorghum millet like many grains contains phytates, the salt form of phytic acid. Phytates have the ability to bind with minerals like magnesium, calcium, iron, and zinc forming insoluble complexes. These complexes decrease the bioavailability of minerals, potentially leading to mineral deficiencies especially in diets where sorghum millet is a main source of nutrients (Makkar 2003). Protease inhibitors inhibit the digestive enzymes activity (proteases) in the gastrointestinal tract. Inhibition of proteases can affect protein digestion and absorption (Kumar et al. 1978). Some varieties of sorghum millet may contain amylase inhibitors, which interfere with the amylase enzymes activity necessary for starch digestion. Amylase inhibitors can impact the breakdown of carbohydrates in the digestive system (Mulimani and Supriya 1993). Trypsin inhibitors can interfere with the trypsin activity involved in protein digestion. Inhibition of trypsin can affect the digestion and utilization of dietary proteins. Oxalates can form insoluble salts with minerals like calcium. These salts can lead to the kidney stones formation in some individuals. It is necessary to note that the presence and levels of anti-nutritional factors in sorghum millet can vary among different varieties. Additionally, many traditional processing methods, such as fermentation, heat treatments, and soaking, can cause the reduction of levels of antinutrients and enhance the nutritional quality of sorghum. These methods are often employed in traditional food preparation practices to improve the bioavailability and digestibility of nutrients (Donfrancesco et al. 2018).

20.6 Health Promoting Phytochemicals and Antioxidants in Sorghum

In addition to carbohydrates, protein, and fiber, sorghum millet contains various phytochemicals, which include alkaloids, flavanoids, tannins, polyphenol, saponins, anthocyanins, terpenoids, phytosterols, and policosanols (Awika and Rooney 2004). These are bioactive compounds found in plants that can have potential health benefits. The phytochemical composition of sorghum millet can vary among different varieties and is influenced by factors like growing conditions and processing methods.

20.6.1 Alkaloids

Alkaloids are defined as pharmacologically active compounds having specific physiological activities. Sorghum millet contains certain alkaloids though their concentrations are generally lower compared to other alkaloid-containing plants. In sorghum, the primary alkaloids are dhurrin and hordenine (Wang et al. 2024). Dhurrin is a cyanogenic glycoside found in sorghum. It is a precursor to hydrogen cyanide (HCN), a toxic compound. It is present in sorghum millet leaves, stems, and roots, but the concentration is usually higher in the leaves. Cyanogenic glycosides serve as a defense mechanism in plants, releasing toxic compounds when the plant tissue is damaged (Busk and Møller 2002). Another alkaloid hordenine is a naturally occurring phenylethylamine alkaloid with potential pharmacological effects (Smith 1977).

20.6.2 Flavonoids

Sorghum millet possesses more concentration of phenolic compounds like ferulic acid, gallic acids, and vanillic acids, which have been proven to show antioxidant and anti-inflammatory characteristics having several other health benefits. The phenolic chemicals are produced via the shikimate pathway (Ahmad et al. 2015). It can prevent blockages in blood vessels, improve blood circulation throughout, decrease accumulation of fat in blood vessel walls, so can reduce cholesterol amount and reduce the risk of coronary heart disease. It may help to reduce pain or swelling if bleeding occurs (Janicsák 1999). A study shows a highly significant correlation between the polyphenols content and flavonoids with antioxidant activity and free radicals. Generally, flavonoids further grouped into subgroups viz., flavanone (example: naringenin), flavones (example: luteolin), flavonol (e.g., anthocyanin, kaempferol, and chalcone), and condensed tannins (also called proanthocyanins) are produced from the flavonoids present in the sorghum millet cereals (Tanwar et al. 2023). An anthocyanin cyanidine is responsible for the red or purple color in pigmented or red sorghum millet varieties. Anthocyanins are potent antioxidants, chemically 3-deoxy anthicyanidines, holds up to 79% of the total flavonoids in the plant (Taleon et al. 2012). Other flavonoids are Quercetin, Kaempferol, Apigenin with anti-inflammatory and antioxidant effects which is present in certain varieties of sorghum. Sorghum millet also contains lignans, which are polyphenolic compounds with antioxidant and potential health-promoting effects.

20.6.2.1 Polyphenols

Sorghum millet is rich in polyphenolic compounds, which are a diverse group of naturally occurring plant chemicals with potential health benefits. The polyphenol content in sorghum millet can vary among different varieties, processing methods, and growing conditions. Polyphenols have many phenolic groups in their molecule. It plays an important role providing color to the plant leaves (Meena et al. 2024). Sorghum millet Polyphenol plays an important role as defensive against herbivores, competitors, and pathogen. Polyphenols act as antioxidants, which is good for health. It helps in reducing oxidative stress, neutralizing free radicals in the body, and potential damage to cells (Tanwar et al. 2023). It can lower the risk of diseases related to blood vessels and heart, Alzheimer's diseases, or cancer (Cox et al. 2019).

20.6.2.2 Tannins

Tannins are a type of polyphenolic compounds having molecular weight ranging from 500 to 20,000 daltons present in sorghum. They are considered anti-nutritional due to their potential impact on protein digestibility and decreasing nutritional value of protein; they also contribute to the overall polyphenol content in sorghum. Tannin prevents proteins from microbial degradation and enzymatic digestion. Although soghum tannin is anti-nutritional, it also functions as biological antioxidants. The free radical formation in the chain reaction is hindered via the antioxidants, and it can stabilize the formed free radicals by complementing their electron deficiency. Tannin also exhibits other properties as antibacterial and anti-diarrhea as well (Palacios et al. 2021).

20.6.2.3 Saponins

Saponins have characteristics like a soap to form leather when they are shaken with water. It is a glycoside compound which consists of glycons (containing sugar moiety) and aglycones (sapogenin). Sapogenins are composed of two groups triterpenoid saponins and steroidal saponins (Kumar et al. 2009). It shows amphiphilic characteristics and is found in the outer layers of sorghum millet grains. Saponin compounds have the ability to cure several types of diseases (Rachmawati 2009). Saponins were studied for their antibacterial, antiviral, increase vitality, immunomodulatory effects, reduce sugar levels, and reduce the risk of blood clotting.

20.6.3 Phytosterols

Sorghum millet contains phytosterols, which are plant-derived compounds with a structure similar to cholesterol. Steroids belong to secondary metabolites and are considered to have antidiabetic, bioinsecticide, antibacterial, and antifungal activities (Hidayah et al. 2016). Steroids are significant molecules that are bioactive with a fundamental framework of 17 carbon atoms consisting of combination of 4 rings, 3 of which are cyclopentane and cyclohexane. Steroid compounds are needle-shaped crystals with characteristics containing methyl groups, OH groups, methyl groups, and double bonds (unconjugated). Sorghum millet grains contain phytosterols including various types such as stigmasterol, campesterol, and beta-sitosterol. Structure of phytosterols is similar to cholesterol but differs in the side-chain structure. Beta-sitosterol is one of the most common phytosterols found in sorghum. This steroid is very necessary in the field of medical. More than 150 types of steroids have been registered as drugs. Steroids may be used as contraception and drugs, for example, estrogens can develop female sexual organs, whereas androgens named steroid hormones develop male sexual organs, adrenocorticoids hormones can protect from rheumatism and inflammation. Stigmasterol can reduce blood cholesterol by inhibiting the absorption of dietary cholesterol through intestine, resultantly inhibiting the progress of colon cancer, and restraining liver cholesterol. On the basis of their chemical composition, steroids can be utilized as secondary metabolites. Additionally, it can also be utilized as an insecticide and pesticide, for example, rotenone and rotenoid (Jones et al. 2000). Phytosterols are often included in functional foods and dietary supplements designed to promote heart health. Consuming whole grains, including sorghum, can contribute to the intake of these beneficial compounds. Sorghum, being a whole grain, can be an important component of a heart-healthy diet, providing not only phytosterols but also a range of other essential nutrients and bioactive compounds.

20.6.4 Terpenoids

Terpenoids are a diverse class of natural compounds that include terpenes and their oxygenated derivatives. Terpenoids found in nature are mostly essential oil components. The essential oil compounds are known to have a remedial effect in substitute medicine (Kabera et al. 2014). These compounds give taste, smell, and color to the plants and usually occur in the fruits and leaves of higher plants. Sorghum millet contains various terpenoids contributing to the plant's overall chemical composition. There are some types of terpenoids found in sorghum millet as mono terpenes, diterpenes, sesquiterpenes, and sorgoleone (Takemura et al. 2017). The specific terpenoid profile of sorghum millet can vary among different varieties, environmental conditions, and stages of plant development. Sorgoleone is a lipophilic compound classified as a terpenoid. It is produced by sorghum millet roots and serves as an allelopathic compound inhibiting the growth of competing plants by affecting their root systems (Jesudas et al. 2014). Terpenoids play an essential role in plant defense mechanisms and also in interactions with other organisms. Terpenoids often serve ecological functions, like defense against pathogens, competing plants, and herbivores (Meena et al. 2022).

20.7 Opportunities for Using Sorghum Millet to Promote Human Health and Well Being

Sorghum millet presents various opportunities for promoting human health and well-being due to its nutritional composition and potential health benefits. In previous section, the nutritional properties of sorghum millet have been discussed. Sorghum millet adds nutritional value to promote human health as it is an excellent source of essential nutrients including carbohydrates, dietary fiber, proteins, and various vitamins and minerals. The nutritional content of sorghum millet makes it a valuable component of a balanced diet providing energy and essential micronutrients (Tanwar et al. 2023). It may be Gluten-Free alternative, which makes it a good alternative for individuals with celiac disease or gluten sensitivity. Sorghum millet flour and products can be used in gluten-free recipes offering diversity in dietary choices. Sorghum millet contains antioxidants including flavonoids and polyphenols, which help in neutralizing free radicals in the body (Ghinea et al. 2021). Antioxidants play a role in reducing inflammation and oxidative stress contributing to overall health. The presence of phytosterols in sorghum millet may contribute to lowering cholesterol levels supporting heart health. The fiber content in sorghum millet can aid in maintaining healthy cholesterol levels and promoting cardiovascular well-being. Sorghum millet has a low glycemic index compared to various other grains, which may be beneficial for individuals managing blood sugar levels. The presence of resistant starch and dietary fiber contributes to slower digestion and absorption of carbohydrates and acts as a prebiotic supporting the growth of beneficial gut bacteria (Taylor and Duodu 2023). Its use in diet can help maintain weight by reducing overall calorie intake. Sorghum millet contains various bioactive compounds, such as phytochemicals, which may have anti-inflammatory and other health-promoting effects. Sorghum millet is a hardy crop that can thrive in challenging environmental conditions contributing to food security and sustainable agriculture (Xiong et al. 2019). Its resilience to drought makes it an important crop in regions with water scarcity. As with any dietary component, it's essential to consume sorghum millet as part of a varied and balanced diet. Consulting with registered dietitians or healthcare professionals can provide personalized advice based on individual health conditions and dietary needs.

20.8 Challenges and Future Perspectives

Molecular biology has revolutionized our understanding of physiology, biochemistry, and genetics and provided us with deep insights into the complex biological processes that underlie living things. When it comes to sorghum millet nutritional qualities, it provides challenges as well as prospects:

Challenges

A. *Genetic Diversity:* Sorghum millet exhibits a high degree of genetic variability, which makes it difficult to identify the precise genes causing desired nutritional characteristics. Because of this diversity, methods like quantitative trait locus (QTL) mapping and genome-wide association studies (GWAS) are essential but complex (Upadhyaya et al. 2013).
B. *Gene Regulation:* It is difficult to understand how the genes associated with nutritional qualities are regulated in sorghum. With the use of advanced molecular techniques like transcriptomics and epigenetics, it is possible to unravel the intricate networks that control characteristics like grain protein content, micronutrient levels, and digestibility (Wang et al. 2021).

C. *Implementation of Research into Varietal Enhancement:* Although molecular biology makes it easier to identify the genes linked to desired nutritional qualities, there are still many obstacles to overcome before new sorghum millet varieties can be produced through breeding programs. It requires precise genetic engineering techniques and effective gene editing tools such as CRISPR/Cas9 (Li et al. 2021).
D. *Environmental Influence:* The nutritional characteristics of sorghum millet can be influenced by environmental factors, including soil quality, climate, and farming practices. Developing robust sorghum millet cultivars with consistent nutritional profiles requires an understanding of how genetics and environment interact through molecular techniques.

Future Prospects

A. *Nutritional Improvement:* Molecular biology offers ways to modify the expression of genes involved in the metabolism and production of nutrients, therefore improving the nutritional value of sorghum. This includes increasing the amount of protein, fine-tuning the composition of amino acids, increasing levels of micronutrients (iron, zinc, etc.), and reducing anti-nutritional components (Borrell et al. 2014).
B. *Biofortification:* Molecular methods can support biofortification efforts to increase the concentrations of important nutrients in sorghum millet grains. Specific gene editing has the potential to increase the uptake of vitamins, minerals, and antioxidants, so mitigating micronutrient deficits in people who rely on diets high in sorghum.
C. *Resistance to Disease and Pests:* The production of sorghum millet variants resistant to pests and illnesses is made possible by molecular biology. By using genetic engineering and marker-assisted selection to find and add resistance genes, researchers can improve the nutritional value of sorghum millet grown under biotic stress conditions while reducing yield losses.
D. *Precision Breeding:* With the aid of molecular tools, breeding tactics can be precisely adapted to meet particular nutritional goals and environmental constraints. Through the combination of marker-assisted breeding, genome editing, and genomic selection, breeders can accelerate the production of sorghum millet varieties with optimized agronomic attributes and nutritional profiles.
E. *Functional Genomics:* The functional genomics of sorghum millet has been clarified by advances in molecular biology, providing insight into the biological mechanisms underlying nutritional features. The integration of omics technologies, including transcriptomics, proteomics, metabolomics, and genomes, enables a thorough comprehension of the biology and nutritional value of sorghum. To sum up, molecular biology presents enormous promise for addressing the difficulties related to improving the nutritional qualities of sorghum. To satisfy the changing needs of food security and nutrition, researchers can overcome genetic obstacles, improve nutritional quality, and grow resilient sorghum millet cultivars by utilizing molecular approaches (Kumar et al. 2022).

20.9 Sorghum Millet in Global Food Security

Poverty is an additional problem that makes it difficult to access food. The potential obstacles in the future will make the situation even direr. Climate variability is probably considered one of the major obstacles to food security (Giannini et al. 2017; Shaheen 2012).

Research production was primarily driven by new opportunities created by the sequencing of the sorghum millet genome, the crop's evolution as a model for drought tolerance, and the increasing awareness of sorghum millet as a promising crop for food and energy security in response to climate change. The literature data shows that in comparison to major cereal crops like rice and maize, sorghum millet yield and nutritional quality are more resilient to climate change. End-use patterns indicate that sorghum millet is now more competitive in the food market than it was in the feed and bioenergy sectors, which previously dominated its production. Compared to other major cereal crops like maize, sorghum millet typically has lower production costs (such as those associated with fertilizer, seed, pesticides, land rent, etc.) and yields higher net profits (Staggenborg et al. 2008). Sorghum millet performed better than maize in terms of water use efficiency, aboveground biomass, and harvest index in an experimental test of relative water inadequacy between maize and sorghum. This indicates that sorghum millet yields higher profits than maize in semi-arid and arid regions of the world. According to a meta-analysis published by Zereyesus and Dalton (2017), the collective results of research expenditures in millets and sorghum millet are great and progressive, coming in at about 20% annually. Sorghum millet plays a pivotal role in enhancing global food security, particularly in regions characterized by harsh agro-climatic conditions such as drought, heat, and nutrient-poor soils. As a highly resilient cereal crop, sorghum millet demonstrates remarkable adaptability to semi-arid and arid environments where other staple cereals such as rice, wheat, and maize often fail to thrive. It is notably efficient in water use, requiring as little as 450–650 mm of water per growing season, which is significantly lower than the water requirements of maize and rice. This characteristic makes sorghum millet a vital crop in the context of increasing climate variability and water scarcity, enabling sustainable food production in vulnerable regions (FAO 2018).

Nutritionally, sorghum millet offers a well-balanced profile of macronutrients and micronutrients that contribute to dietary diversity and nutritional security. It is a rich source of complex carbohydrates, dietary fiber, and proteins and contains essential micronutrients such as iron, zinc, phosphorus, and B-complex vitamins. Moreover, sorghum millet is inherently gluten-free, making it an ideal food option for individuals with celiac disease or gluten intolerance. Sorghum millet also contains significant levels of bioactive compounds, including phenolic acids and flavonoids, which exhibit antioxidant, anti-inflammatory, and antidiabetic properties. These attributes collectively enhance the nutritional value of sorghum, particularly in addressing micronutrient deficiencies prevalent in low-income populations. From an agronomic perspective, sorghum millet contributes to food security by ensuring yield stability under marginal land conditions. It exhibits resistance to biotic and abiotic stresses and maintains productivity with minimal inputs, thereby reducing dependency on fertilizers and pesticides. Improved sorghum millet varieties developed through breeding programs have shown yield increases of 20–40% in drought-prone areas, particularly in Sub-Saharan Africa and South Asia. Furthermore, its compatibility with intercropping systems, especially with legumes, aids in improving soil health and nitrogen fixation, thereby promoting sustainable agricultural practices (Taylor and Emmambux 2008a, b).

The versatility of sorghum millet in food processing further enhances its contribution to food security. It can be utilized in various forms such as flour, whole grain, popped grain, and syrup and is commonly incorporated into porridges, flatbreads, snack products, and fermented beverages. In addition to direct human consumption, sorghum millet by-products such as bran and stalks are valuable in animal feed, contributing to livestock productivity and food chain sustainability. Its incorporation into school feeding programs and food relief efforts in developing countries underscores its role in combating hunger and malnutrition. Moreover, sorghum millet contributes to sustainable agriculture and rural livelihoods. Its low input requirements and adaptability make it an economically viable crop for smallholder farmers. By reducing reliance on major cereals and supporting agro biodiversity, sorghum millet plays a strategic role in enhancing national and household food security. Given its climate resilience, nutritional richness, and multifunctional use, sorghum millet is increasingly being recognized as a critical crop for achieving the United Nations Sustainable Development Goals, particularly those related to zero hunger, good health, and climate action (Dykes and Rooney 2006). In summary, sorghum millet has great potential as a sustainable and profitable crop-growing system that is remarkably adaptable to the environmental stresses brought on by climate variation.

20.10 Conclusion

A significant crop used as a cereal in Africa, Asia, and the other semiarid parts of world is "Sorghum". The fifth-most-significant cereal produced worldwide is sorghum. It is a great source of nutrients and bioactive compounds (like protein, carbohydrates, and others) for humans. Cookies, cereal, cakes, bread, flakes, couscous, pancakes, and drinks (alcoholic or nonalcoholic), all contain sorghum. The popularity of sorghum millet as a functional food is increasing continuously worldwide. Nutritional qualities of sorghum millet aim to give a thorough grasp of the grain dietary importance and potential health effects. The purpose of this study is to inform readers about the particular nutrients found in sorghum, their possible health advantages, and the ways in which sorghum millet in diets can improve general health.

The investigation of sorghum millet yields an intricate description about its molecular, chemical, and nutritional characteristics. The introduction establishes the context by defining the goal and environment and highlighting the importance of sorghum millet in many scenarios. Its potential as a dietary staple is revealed by comparing its macronutrient and micronutrient profile to other cereals and delving into its nutritional quality. The section on molecular biology clarifies the complex genomic structure and genetic variety of Sorghum, which is important for improving breeding practices and its possible uses. Subsequently, the starch and protein chemistry of sorghum millet is revealed, revealing its structure, uses, and nutritional qualities as well as any possible anti-nutritive elements. Furthermore, the potential of sorghum millet to improve human health and wellbeing is highlighted by the presence of antioxidants and phytochemicals that support good health. Finally, its contribution to global food security highlights its importance beyond personal health and resonates with wider societal issues. In conclusion, research on sorghum millet spans a number of fields including nutrition and molecular biology providing a thorough picture of the crop potential as a useful and adaptable resource for advancing human health and tackling issues related to global food security.

Conflict of Interest The authors do not have any conflict of interest.

References

Afify Ael M, El-Beltagi HS, El-Salam SM, Omran AA (2011) Bioavailability of iron, zinc, phytate and phytase activity during soaking and germination of white sorghum varieties. PLoS One 6(10):e25512. https://doi.org/10.1371/journal.pone.0025512

Afify AE, El-Beltagi HS, El-Salam SM et al (2012) Biochemical changes in phenols, flavonoids, tannins, vitamin E, β-carotene and antioxidant activity during soaking and germination of white sorghum varieties. Asian Pac J Trop Biomed 2(3):203–209. https://doi.org/10.1016/S2221-1691(12)60045-8

Ahmad A, Kaleem M, Ahmed Z, Shafiq H (2015) Therapeutic potential of flavonoids and their mechanism of action against microbial and viral infections—a review. Food Res Int 77:221–235. https://doi.org/10.1016/j.foodres.2015.06.021

Akintayo I, Sedgo J (2001) Towards sustainable sorghum production, utilization, and commercialization in West and Central Africa. In: Akintayo I, Sedgo J (eds) Proceedings of a technical workshop of the West and Central Africa sorghum research network. ICRISAT. https://oar.icrisat.org/357/

Anglani C (1998) Sorghum for human food—a review. Plant Foods Hum Nutr 52(1):85–95. https://doi.org/10.1023/A:1008065519820

Awika JM, Rooney LW (2004) Sorghum phytochemicals and their potential impact on human health. Phytochemistry 65(9):1199–1221. https://doi.org/10.1016/j.phytochem.2004.04.001

Awika JM, Rooney LW, Wu X et al (2003) Screening methods to measure antioxidant activity of sorghum (*Sorghum bicolor*) and sorghum products. J Agric Food Chem 51(23):6657–6662

Badi S, Pedersen B, Monowar L, Eggum BO (1990) The nutritive value of new and traditional sorghum and millet foods from Sudan. Plant Foods Hum Nutr 40(1):5–19. https://doi.org/10.1007/BF02193775

Barros F, Awika J, Rooney LW (2014) Effect of molecular weight profile of sorghum proanthocyanidins on resistant starch formation. J Sci Food Agric 94(6):1212–1217

Belton PS, Delgadillo I, Halford NG, Shewry PR (2006) Kafirin structure and functionality. J Cereal Sci 44(3):272–286. https://doi.org/10.1016/j.jcs.2006.05.004

Beta T, Corke H, Rooney LW, Taylor JR (2000) Starch properties as affected by sorghum grain chemistry. J Sci Food Agric 81(2):245–251. https://doi.org/10.1002/1097-0010(20010115)81:2<245::AID-JSFA805>3.0.CO;2-S

Borrell AK, van Oosterom EJ, Mullet JE et al (2014) Stay-green alleles individually enhance grain yield in sorghum under drought by modifying canopy development and water uptake patterns. New Phytol 203(3):817–830. https://doi.org/10.1111/nph.12869

Buchanan CD, Lim S, Salzman RA et al (2005) Sorghum bicolor's transcriptome response to dehydration, high salinity and ABA. Plant Mol Biol 58(5):699–720. https://doi.org/10.1007/s11103-005-7876-2

Busk PK, Møller BL (2002) Dhurrin synthesis in sorghum is regulated at the transcriptional level and induced by nitrogen fertilization in older plants. Plant Physiol 7:1222–1231

Chinenye JN, Uchechukwu I, Olatunde GO (2016) Glycemic index and glycemic load of selected staple foods in Nigeria. Niger J Nutr Sci 37(1):65–72

Cox S, Noronha L, Herald T et al (2019) Evaluation of ethanol-based extraction conditions of sorghum bran bioactive compounds with downstream anti-proliferative properties in human cancer cells. Heliyon 5(5):e01589. https://doi.org/10.1016/j.heliyon.2019.e01589

de Morais Cardoso L, Pinheiro SS, Martino HS, Pinheiro-Sant'Ana HM (2017) Sorghum (*Sorghum bicolor* L.): nutrients, bioactive compounds, and potential impact on human health. Crit Rev Food Sci Nutr 57(2):372–390. https://doi.org/10.1080/10408398.2014.887057

Devi PB, Vijayabharathi R, Sathyabama S, Malleshi NG, Priyadarisini VB (2014) Health benefits of finger millet (*Eleusine coracana* L.) polyphenols and dietary fiber: a review. J Food Sci Technol 51(6):1021–1040. https://doi.org/10.1007/s13197-011-0584-9

Di Donfrancesco B, Koppel K, Aldrich CG (2018) Pet and owner acceptance of dry dog foods manufactured with sorghum and sorghum fractions. J Cereal Sci 83:42–48

Dicko MH, Gruppen H, Traore AS, Voragen AGJ, van Berkel WJH (2006) Sorghum grain as human food in Africa: relevance of content of starch and amylase activities. Afr J Biotechnol 5(5):384–395. https://edepot.wur.nl/40432

Dicko MH, Gruppen H, Barro C, Traore AS, Van Berkel WJ, Voragen AG (2005) Impact of phenolic compounds and related enzymes in sorghum varieties for resistance and susceptibility to biotic and abiotic stresses. J Chem Ecol, 31(11):2671–2688. https://doi.org/10.1007/s10886-005-7619-5

Dykes L, Rooney LW (2006) Sorghum and millet phenols and antioxidants. J Cereal Sci 44(3):236–251

FAO (2018) Sorghum and millets in human nutrition. Food and Agriculture Organization United Nations

Ghinea IO, Mihaila MD, Blaga GV et al (2021) HPLC-DAD polyphenolic profiling and antioxidant activities of Sorghum bicolor during germination. Agronomy 11(3):417. https://doi.org/10.3390/agronomy11030417

Giannini A, Krishnamurthy PK, Cousin R, Labidi N, Choularton RJ (2017) Climate risk and food security in Mali: a historical perspective on adaptation. Earth's Future 5(2):144–157. https://doi.org/10.1002/2016ef000404

Gómez MI, Islam-Faridi MN, Woo SS et al (1997) FISH of a maize sh2-selected sorghum BAC to chromosomes of *Sorghum bicolor*. Genome 40(4):475–478. https://doi.org/10.1139/g97-063

Gunawan S, Dwitasari I, Rahmawati N et al (2022) Effect of process production on antinutritional, nutrition, and physicochemical properties of modified sorghum flour. Arab J Chem 15(10):104134. https://doi.org/10.1016/j.arabjc.2022.104134

Guo T, Yu X, Li X et al (2019) Optimal designs for genomic selection in hybrid crops. Mol Plant 12(3):390–401. https://doi.org/10.1016/j.molp.2018.12.022

Hao H, Li Z, Leng C et al (2021) Sorghum breeding in the genomic era: opportunities and challenges. Theor Appl Genet 134:1899–1924. https://doi.org/10.1007/s00122-021-03789-z

Harden ML, Stanaland R, Briley M, Yang SP (1976) The nutritional quality of proteins in sorghum. J Food Sci 41(5):1082–1085. https://doi.org/10.1111/j.1365-2621.1976.tb14392.x

Hegde PS, Chandra TS (2005) ESR spectroscopic study reveals higher free radical quenching potential in kodo millet (*Paspalum scrobiculatum*) compared to other millets. Food Chem 92(1):177–182. https://doi.org/10.1016/j.foodchem.2004.08.002

Hegedôs M, Pedersen B, Eggum BO (1985) The influence of milling on the nutritive value of flour from cereal grains. 7. Vitamins and tryptophan. Qual Plant Plant Foods Hum Nutr 35(2):175–180. https://doi.org/10.1007/bf01092134

Hidayah WW, Kusrini D, Fachriyah E (2016) Isolasi, identifikasi senyawa steroid dari daun getih-getihan (*Rivina humilis* L.) dan uji aktivitas sebagai antibakteri. J Kimia Sains Dan Apl 19(1):32–37. https://doi.org/10.14710/jksa.19.1.32-37

Janicsák G (1999) Comparative studies of the rosmarinic and caffeic acid contents of Lamiaceae species. Biochem Syst Ecol 27(6):733–738. https://doi.org/10.1016/S0305-1978(99)00007-1

Jesudas AP, Kingsley JS, Ignacimuthu S (2014) Sorgolone from *Sorghum bicolor* as a potent bioherbicide. Res J Recent Sci 3:32–36

Jones PJ, Raeini-Sarjaz M, Ntanios FY et al (2000) Modulation of plasma lipid levels and cholesterol kinetics by phytosterol versus phytostanol esters. J Lipid Res 41(5):697–705

Juliana P, Montesinos-López OA, Crossa J et al (2019) Integrating genomic-enabled prediction and high-throughput phenotyping in breeding for climate-resilient bread wheat. Theor Appl Genet 132(1):177–194. https://doi.org/10.1007/s00122-018-3206-3

Kabera JN, Semana E, Mussa AR, He X (2014) Plant secondary metabolites: biosynthesis, classification, function and pharmacological properties. J Pharm Pharmacol 2:377–392

Kayisoglu C, Altikardes E, Guzel N, Uzel S (2024) Germination: a powerful way to improve the nutritional, functional, and molecular properties of white- and red-colored sorghum grains. Foods 13(5):662. https://doi.org/10.3390/foods13050662

Kazungu FK, Muindi EM, Mulinge JM (2023) Overview of sorghum (*Sorghum bicolor* L.), its economic importance, ecological requirements and production constraints in Kenya. Int J Plant Soil Sci 35(1):62–71. https://doi.org/10.9734/ijpss/2023/v35i12744

Kebrom TH, McKinley B, Mullet JE (2017) Dynamics of gene expression during development and expansion of vegetative stem internodes of bioenergy sorghum. Biotechnol Biofuels 10(1). https://doi.org/10.1186/s13068-017-0848-3

Khalil J (1984) Chemical composition and nutritional quality of sorghum flour and bread. Plant Foods Hum Nutr. https://doi.org/10.1007/BF01094842

Khaton A, Sagar A, Tajkia J, Mahmud MS et al (2016) Effect of moisture stress on morphological and yield attributes of four sorghum varieties. Progress Agric 27(3):265–271. https://doi.org/10.3329/pa.v27i3.30806

Kim JS, Childs KL, Islam-Faridi MN et al (2002) Integrated karyotyping of sorghum by in situ hybridization of landed BACs. Genome 45(2):402–412. https://doi.org/10.1139/g01-141

Kong W, Jin H, Goff VH et al (2020) Genetic analysis of stem diameter and water contents to improve sorghum bioenergy efficiency. G3 (Bethesda) 10(11):3991–4000. https://doi.org/10.1534/g3.120.401608

Kumar PM, Virupaksha TK, Vithayathil PJ (1978) Sorghum proteinase inhibitors: purification and some biochemical properties. Int J Pept Protein Res 12(4):185–196

Kumar A, Chanda I, Singh A, Kopal (2009) Extraction and evaluation of pharmacological activity of saponin extract of *Plumeria rubra* leaves. Pharmacology 1:969–974

Kumar R, Yadav N, Kumar S et al (2022) Nutritional improvement of cereals by genome editing: a review. Front Genet 13:858730. https://doi.org/10.3389/fgene.2022.858730

Li Y, Wu S, He Y et al (2018) A new high-throughput assay for determining soluble sugar in sorghum internode-extracted juice. Planta 248(4):785–793. https://doi.org/10.1007/s00425-018-2932-8

Li Y, Wang W, Feng Y et al (2019) Transcriptome and metabolome reveal distinct carbon allocation patterns during internode sugar accumulation in different sorghum genotypes. Plant Biotechnol J 17(4):614–626. https://doi.org/10.1111/pbi.13006

Li C, Wang X, Dong N et al (2021) CRISPR/Cas9-mediated gene editing in sorghum: progress and perspectives. Front Genome Edit 3:646229. https://doi.org/10.3389/fgeed.2021.646229

Mace ES, Xia L, Jordan DR et al (2008) DArT markers: diversity analyses and mapping in *Sorghum bicolor*. BMC Genomics 9:26. https://doi.org/10.1186/1471-2164-9-26

Mace ES, Hunt CH, Jordan DR (2013) Supermodels: sorghum and maize provide mutual insight into the genetics of flowering time. Theor Appl Genet 126:1377–1395. https://doi.org/10.1007/s00122-013-2059-z

Maciel LS, Malenica D, Herodes K et al (2024) Detection and quantification of free amino compounds in selected agricultural by-products with derivatization followed by liquid chromatography–tandem mass spectrometry analysis. Food Chem Adv 4:100629. https://doi.org/10.1016/j.focha.2024.100629

Maclean WC, Lopez de Romaña G, Placko RP et al (1981) Protein quality and digestibility of sorghum in preschool children: balance studies and plasma free amino acids. J Nutr 111(11):1928–1936. https://doi.org/10.1093/jn/111.11.1928

Mahasukhonthachat K, Sopade PA, Gidley MJ (2010) Kinetics of starch digestion and functional properties of twin-screw extruded sorghum. J Cereal Sci 51(3):392–401. https://doi.org/10.1016/j.jcs.2010.01.005

Makkar HPS (2003) Effects and fate of tannins in ruminant animals, adaptation to tannins, and strategies to overcome detrimental effects of feeding tannin-rich feeds. Small Rumin Res 49(3):241–256

Martinol HSD, Tomaz PA, Moraes EA et al (2012) Chemical characterization and size distribution of sorghum genotypes for human consumption. Rev Inst Adolfo Lutz 71(2):337–344

McKinley B, Rooney W, Wilkerson C et al (2016) Dynamics of biomass partitioning, stem gene expression, cell wall biosynthesis, and sucrose accumulation during development of *Sorghum bicolor*. Plant J 88(4):662–680. https://doi.org/10.1111/tpj.13269

Meena K, Visarada KBRS, Meena DK (2022) *Sorghum bicolor* (L.) Moench—a multifarious crop: fodder to therapeutic potential and biotechnological applications. Future Foods 6:100188. https://doi.org/10.1016/j.fufo.2022.100188

Meena K, Meena DK, Jacob J et al (2024) Unveiling the antioxidant arsenal of colored sorghum: a path to functional food development. Agriculture 14(4):566. https://doi.org/10.3390/agriculture14040566

Mengistu G, Shimelis H, Laing M, Lule D (2018) Breeding for anthracnose (*Colletotrichum sublineolum* Henn.) resistance in sorghum: challenges and opportunities. Aust J Crop Sci 12:1911–1920

Mohamed TK, Zhu K, Issoufou A et al (2009) Functionality, in vitro digestibility and physicochemical properties of two varieties of defatted foxtail millet protein concentrates. Int J Mol Sci 10(12):5224–5238. https://doi.org/10.3390/ijms10125224

Morales MM, Hoshide AK, Carvalho LMP et al (2024) Sorghum biomass as an alternative source for bioenergy. Biomass 4(3):1017–1030

Motlhaodi T, Bryngelsson T, Chite S et al (2018) Nutritional variation in sorghum (*Sorghum bicolor* (L.) Moench) accessions from Southern Africa revealed by protein and mineral composition. J Cereal Sci. https://doi.org/10.1016/j.jcs.2018.08.010

Mulimani VH, Supriya D (1993) Alpha amylase inhibitors in sorghum (*Sorghum bicolor*). Plant Foods Hum Nutr 44:261–266. https://doi.org/10.1007/BF01088321

Mwamahonje A, Mdindikasi Z, Mchau D et al (2024) Advances in sorghum improvement for climate resilience in the global arid and semi-arid tropics: a review. Agronomy 14(12):3025. https://doi.org/10.3390/agronomy14123025

Oboh G, Amusan TV (2009) Nutritive value and antioxidant properties of cereal gruels produced from fermented maize and sorghum. Food Biotechnol 23(1):17–31. https://doi.org/10.1080/08905430802671899

Palacios CE, Nagai A, Torres P et al (2021) Contents of tannins of cultivars of sorghum cultivated in Brazil, as determined by four quantification methods. Food Chem 337:127970

Palavecino PM, Penci MC, Calderón-Domínguez G et al (2016) Chemical Composition and Physical Properties of Sorghum Flour Prepared from Different Sorghum Hybrids Grown in Argentina. Starch-Stärke 68(11–12):1055–1064. https://doi.org/10.1002/star.201600111

Paterson AH, Bowers JE, Bruggmann R et al (2009) The *Sorghum bicolor* genome and the diversification of grasses. Nature 457(7229):551–556. https://doi.org/10.1038/nature07723

Pontieri P, Troisi J, Calcagnile M, Miazzi M, Milano F, Del Giudice L et al (2022) Chemical composition, fatty acid and mineral content of food-grade white, red and black sorghum varieties grown in the Mediterranean environment. Foods 11(3):436. https://doi.org/10.3390/foods11030436

Pratt LH, Liang C, Shah T, Sun Q, Wang J, Thomas TL et al (2005) Sorghum expressed sequence tags identify signature genes for drought, pathogenesis, and skotomorphogenesis from a milestone set of 16,801 unique transcripts. Plant Physiol 139(2):869–884. https://doi.org/10.1104/pp.105.066134

Proietti I, Frazzoli C, Mantovani A (2015) Exploiting nutritional value of staple foods in the world's semi-arid areas: risks, benefits, challenges and opportunities of sorghum. Healthcare 3(2):172–193. https://doi.org/10.3390/healthcare3020172

Rachmawati S (2009) Studi makroskopi mikroskopi, dan skrining fitokimia daun *Anredera cordifolia* (Ten.) Steenis. Skripsi, Universitas Airlangga. http://lib.unair.ac.id

Rakshit S, Hariprasanna K, Gomashe S, Ganapathy KN, Rathore A, Panwar S et al (2014) Changes in area, yield gains, and yield stability of sorghum in major sorghum-producing countries, 1970 to 2009. Crop Sci 54(4):1571–1584. https://doi.org/10.2135/cropsci2012.12.0697

Rao BD, Kulkarni DB, Kavitha C (2018) Study on evaluation of starch, dietary fiber and mineral composition of cookies developed from 12 sorghum cultivars. Food Chem 238:82–86. https://doi.org/10.1016/j.foodchem.2016.12.069

Rathore T, Khare P, Jain S, Jain D, Kumar R (2019) Review on finger millet: processing and value addition. Pharma Innov J 8(4):283–291. https://www.thepharmajournal.com/archives/?year=2019&vol=8&issue=4&ArticleId=3247

Ratnavathi CV, Patil JV (2013) Sorghum utilization as food. J Nutr Food Sci 3(1):1000208

Reddy BVS, Ramesh S, Ashok KA, Ramaiah B, Panwar S (2008) Bio-fuel crops research for energy security and rural development in developing countries. BioEnergy Res 1(3–4):248–258. https://doi.org/10.1007/s12155-008-9022-x

Rooney W, Blumenthal J, Bean B, Mullet J (2007) Designing sorghum as a dedicated bioenergy feedstock. Biofuels Bioprod. Biorefining, 1:147–157. https://doi.org/10.1002/bbb.15

Salzman RA, Brady JA, Finlayson SA, Buchanan CD, Summer EJ, Sun F, Klein PE, Klein RR, Pratt LH, Mullet JE (2005) Transcriptional profiling of sorghum induced by methyl jasmonate, salicylic acid, and aminocyclopropane carboxylic acid reveals cooperative regulation and novel gene responses. Plant Physiol 138(1):352–368. https://doi.org/10.1104/pp.104.058206

Sekhon RS, Childs KL, Santoro N, Foster CE, Buell CR, de Leon N, Kaeppler SM (2016) Stover composition in maize and sorghum reveals remarkable genetic variation and plasticity for carbohydrate accumulation. Front Plant Sci 7:822. https://doi.org/10.3389/fpls.2016.00822

Semere T, Tsehaye Y, Tareke L, Araya T (2023) Nutritional and antinutritional potentials of sorghum: a comparative study among different sorghum landraces of Tigray, Northern Ethiopia. Agriculture 13(7):1753. https://doi.org/10.3390/agriculture13071753

Serna-Saldivar SO, Gomez MH, Rooney LW, Ortega R (1988) Production of tortilla chips from sorghum and maize. J Cereal Sci 8(3):275–284. https://doi.org/10.1016/S0733-5210(88)80039-0

Shaheen N (2012) National situation of food and nutrition security in Bangladesh. (No publication details provided—consider updating this with source or citation format if it's a report)

Sharma K, Kumar V, Kaur J, Kaur R (2021) Health effects, sources, utilization and safety of tannins: a critical review. Toxins 13(4):432–444. https://doi.org/10.3390/toxins13040432

Smith TA (1977) Phenethylamine and related compounds in plants. Phytochemistry 16(1):9–18. https://doi.org/10.1016/0031-9422(77)83004-5

Staggenborg S, Dhuyvetter KC, Gordon WB, Maddux LD, Marsh BH, Fjell DL (2008) Grain sorghum and corn comparisons: yield, economic, and environmental responses. Agron J 100(6):1600–1604. https://doi.org/10.2134/agronj2008.0129

Subramanian V, Murty DS, Jambunathan R, Borse BB (1982) Boiled sorghum characteristics and their relationship to starch properties. In: Rooney LW, Murty DS (eds) Proceedings of the international symposium on sorghum grain quality, Hyderabad, India, 28–31 October 1981. ICRISAT, Patancheru, pp 103–109

Takemura M, Tanaka R, Misawa N (2017) Pathway engineering for the production of β-amyrin and cycloartenol in Escherichia coli—a method to biosynthesize plant-derived triterpene skeletons in *E. coli*. Appl Microbiol Biotechnol 101:6615–6625. https://doi.org/10.1007/s00253-017-8409-z

Taleon V, Dykes L, Rooney WL, Rooney LW (2012) Effect of genotype and environment on flavonoid concentration and profile of black sorghum grains. J Cereal Sci 56(2):470–475. https://doi.org/10.1016/j.jcs.2012.06.008

Tanwar R, Panghal A, Chaudhary G, Yadav N, Kaur M (2023) Nutritional, phytochemical and functional potential of sorghum: a review. Food Chem Adv 3:100501. https://doi.org/10.1016/j.focha.2023.100501

Tari I, Laskay G, Takács Z, Poór P (2013) Response of sorghum to abiotic stresses: a review. J Agron Crop Sci 199(4):264–274. https://doi.org/10.1111/jac.12017

Tasie MM, Gebreyes BG (2020) Characterization of nutritional, antinutritional, and mineral contents of thirty-five sorghum varieties grown in Ethiopia. Int J Food Sci 2020:1–11. https://doi.org/10.1155/2020/8243617

Taylor JRN, Duodu KG (2023) Resistant-type starch in sorghum foods—factors involved and health implications. Starch 75:2100296. https://doi.org/10.1002/star.202100296

Taylor JRN, Emmambux MN (2008a) Nutritional aspects of sorghum and the millets. In: Cereal grains for the food and beverage industries. https://doi.org/10.1016/B978-012373739-7.00006-9

Taylor JRN, Emmambux MN (2008b) Gluten-free foods and beverages from millets and sorghum. In: Gluten-free cereal products and beverages, pp 119–148. https://doi.org/10.1016/B978-012373739-7.00006-9

Tester RF, Karkalas J, Qi X (2004) Starch—composition, fine structure and architecture. J Cereal Sci 39(2):151–165

Ubwa ST, Abah J, Asemave K, Abah E (2012) Studies on the gelatinization temperature of some cereal starches. Int J Chem 4(1):22–28

Upadhyaya HD, Wang YH, Gowda CLL, Sharma S, Sharma R, Singh S (2013) Association mapping of maturity and plant height using SNP markers with the sorghum mini core collection. Theor Appl Genet 126(8):2003–2015. https://doi.org/10.1007/s00122-013-2104-3

Wang RS, Kies C (1991) Niacin status of humans as affected by eating decorticated and whole-ground sorghum (*Sorghum gramineae*) grain, ready-to-eat breakfast cereals. Plant Food Hum Nutr 41(4):355–369. https://doi.org/10.1007/BF02310629

Wang N, Wang Y, Tian F, Wang Y, Li Y (2021) Transcriptome and epigenome-wide association studies reveal complex gene regulatory networks underlying agronomic traits in crops. Trends Plant Sci 26(11):1121–1134. https://doi.org/10.1016/j.tplants.2021.05.009

Wang B, Xiong W, Guo Y (2024) Dhurrin in sorghum: biosynthesis, regulation, biological function and challenges for animal production. Plants 13(16):2291

Weerasooriya DK, Maulana FR, Bandara AY, Menz MA, Little CR (2016) Genetic diversity and population structure among sorghum (*Sorghum bicolor* L.) germplasm collections from Western Ethiopia. Afr J Biotechnol 15(23):1147–1158

Widowati S, Luna P (2022) Nutritional and functional properties of sorghum (*Sorghum bicolor* (L.) Moench)-based products and potential valorisation of sorghum bran. IOP Conf Ser Earth Environ Sci 1024(1):012031. https://doi.org/10.1088/1755-1315/1024/1/012031

Wierzbicka D, Gromadzka G (2014) Ceruloplasmin, hephaestin and zyklopen: the three multicopper oxidases important for human iron metabolism. Postepy Hig Med Dosw 68:912–924

Xiong Y, Zhang P, Warner RD, Fang Z (2019) Sorghum grain: from genotype, nutrition, and phenolic profile to its health benefits and food applications. Compr Rev Food Sci Food Saf 18(6):2025–2046. https://doi.org/10.1111/1541-4337.12506

Xu J, Huang W, Chen X, Wang Z, Yan B (2012) Changes in phenolics and antioxidant activity of whole grain sorghum during germination. Food Chem 134(3):1679–1685. https://doi.org/10.1016/j.foodchem.2012.03.091

Yang L, Wen KS, Ruan X, Zhao YX, Wei F, Wang Q (2018) Response of plant secondary metabolites to environmental factors. Molecules 23(4):762. https://doi.org/10.3390/molecules23040762

Yousif A, Ismail B, Fayed A, Abdelrahman A (2012) Influence of sorghum flour addition on flat bread in vitro starch digestibility, antioxidant capacity and consumer acceptability. Food Chem 134(2):880–887. https://doi.org/10.1016/j.foodchem.2012.02.199

Yu H, Xie W, Wang D, Zhang H, Guo X (2015) Identification of differentially expressed microRNA in the stems and leaves during sugar accumulation in sweet sorghum. Gene 571(2):221–230. https://doi.org/10.1016/j.gene.2015.06.056

Yu KMJ, Xu C, Zhang M, Liu R, Zhou Y, Yu Q, Chuck G (2022) Bioenergy sorghum stem growth regulation: intercalary meristem localization, development, and gene regulatory network analysis. Plant J 112(2):476–492. https://doi.org/10.1111/tpj.15960

Zereyesus YA, Dalton TJ (2017) Rates of return to sorghum and millet research investments: a meta-analysis. PLoS One 12(7):e0180414. https://doi.org/10.1371/journal.pone.0180414

Zhou Z, Robards K, Helliwell S, Blanchard C (2002) Composition and functional properties of rice. Int J Food Sci Technol 37(8):849–868. https://doi.org/10.1046/j.1365-2621.2002.00611.x

Chapter 21
Millets for Food Security

Om Prakash Nautiyal, Anjali, and R. D. Kaushik

Abbreviations

CEE	Carbon Equivalent Potential
EBT	Electronic Benefits Transfer
FAO	Food and Agriculture Organization
FDPIR	Food Distribution Program on Indian Reservations
GFSI	Global Food Security Index
GST	Goods and Service Tax
GWP	Global Warming Potential
ICAR	Indian Council of Agricultural Research
IFAD	International Fund for Agricultural Development
IIMR	Indian Institute of Millets Research
PDS	Public distribution system
SBP	Survivor Benefit Plan
SNAP	Supplemental Nutrition Assistance Program
TEFAP	The Emergency Food Assistance Program
UNICEF	United Nations International Children's Emergency Fund
WFP	World Food Programme
WHO	World Health Organization

O. P. Nautiyal
Information Technology, Good Governance and Science Technology, Government of Uttarakhand, Uttarakhand Council of Science and Technology, Dehradun, Uttarakhand, India

Anjali (✉) · R. D. Kaushik
Department of Chemistry, Gurukula Kangri (Deemed to be University), Haridwar, Uttarakhand, India

© The Author(s), under exclusive license to Springer Nature Switzerland AG 2026

J. Singh, M. Kumar Malik (eds.), *Millets and Millet Starch: Chemical Composition, Functional Modification, and Applications*,
https://doi.org/10.1007/978-3-032-09502-2_21

21.1 Introduction

Millets (Family: Poaceae) are generally group of cereal grains that are widely grown around the world and commonly known as poor man crop. Millets are rich in nutritional contents (Jain et al. 2024). It is said that millets may be the first cultivated crop among other cultivated crops that were grown in "hoe age," earlier the "plow age". These can be easily grown in challenging climate and location due to their good capability of drought tolerance (Yadav et al. 2023). Millets have great potential in research and development as well as in the food mercenary system.

Millets are commonly planted, small-seeded, highly nutritious crops that can resist drought and require very little irrigation to produce. Millets are high in dietary fiber and minerals. These are high in vitamins, protein, minerals, and micronutrients. They contain 2–6% fat, 15–20% dietary fiber, 65–75% carbohydrates, and 7–12% protein. They also contain sufficient amount of critical amino acids than other cereals (Kumar et al. 2023a). Millets consist of Finger Millet, Barnyard Millet, Pearl Millet, Foxtail Millet, Kodo Millet, Proso Millet, and Little Millet. All of these millets are high in phytochemicals, antioxidants, and insoluble dietary fiber with antioxidant characteristics, as well as of various minerals (mostly micronutrients like zinc, magnesium, and iron), dietary energy, and vitamins. As millets can be stored under normal condition for a long duration, they are often known as "famine reserves". So, millets are most suited for intercropping, allowing farmers to manage resources sustainably while also providing food and livelihood stability. Generally, millets are cultivated for both food and fodder. This improves the economic condition of the farmers (Tiwari et al. 2022). Millets play an important role in traditional diets. The proof of the existence of millets can be seen in the documents of ancient works such as Ain-i-Akbari, Kautilya's Arthashastra, and others (Pal et al. 2023). In this chapter, we will discuss about the food security, definition of food security, and some major steps taken by the government to resolve this problem. Furthermore, this chapter also describes how millets can be the best alternative of staple food and the role of millets in food security. This chapter also contains the nutritional value and health benefits of consuming millets, and what are the advantages and challenges of growing millets with their expected solutions.

21.2 Food Security

Food security is a flexible concept. Many attempts have been taken to define food security in research and policy usage. The concept of food security originated around 50 years ago in the early 1970s at a time of global food crises (Peng and Berry 2018). The Food and Agriculture Organization's (FAO) annual report on food security provided the current widely accepted concept of food security. The FAO came up with the definition of food security that is "To ensure the availability of basic food stuffs at all time to avoid acute food shortages in the case of crop failure or natural disasters or any other pandemic, and to sustain a steady expansion of production as well as consumption, also decreases the fluctuation in production and prices" (FAO 2002). The last amendment to this definition occurred at the 2009 World Summit on Food Security (FAO 2009).

The backbone of any developing country like India is agriculture undoubtedly. Directly or indirectly every living being depends on agriculture. Many small and large-scale industries get raw materials such as flower, fruit, fuel, food, fodder, fiber, and timber as a primary source from agriculture produce (Jamal et al. 2023). In the modern world, the two major challenges are to tackle hunger and provide nutritional food (Carolan 2013). The global agriculture production according to the United Nations (UN) will need to be atleast 60% higher in 2050 than the levels of 2007. As the population increases day by day, the demand for food also increases. This raises the issue of food security (Saxena et al. 2018). Various studies estimated that only 1.5 billion of additional hectares of lands are available to be brought under cultivation (Schade and Pimentel 2010; Balmford et al. 2005). However, the land rights must be resolved, market and infrastructure must be in place, and credit must be available before land brought in production. The annual growth rate will slow down from 0.30% between 1961 and 2005 to 0.9% between 2005 and 2050. This means that the annual net increase of cultivable land is around 2.75 million hectare per year between the year 2005 and the year 2050 (Ray et al. 2013; Sjauw 2010). On the other hand, in developed and transitional countries, the cultivable land will continue to decline with an estimation of 0.23% per year between the year 2005 and 2050 (Bruinsma 2009). So, in order to tackle hunger and provide nutritional food, major staple crops are not enough. In this context, millets play a vital role in food security of a nation. The food security of a nation is to ensure that the healthy and nutritious food must be enough for everyone and at every place so that no one will remain hungry, meet with their daily nutritious needs, and can live a healthy life. The reason is the shortage in the production of foods and deficiency in the supply of nutrients (Jamal et al. 2023).

To resolve this problem, the annual report of global hunger index of 2023 gives some recommendations that states must formalize the right to food in real laws and regulations, accompanied by open monitoring and strong accountability systems. Government should invest in nutrition and health status of youth, especially young women, and also need to provide nutritional education and employ tax and regulatory policies (Global Hunger Index 2023).

The United States also runs some nutrition-related policies and programs with the largest reach to ensure nutrition security in the United States. Some of them are Supplemental Nutrition Assistance Program (SNAP) in which citizens get monthly cash value on Electronic Benefits Transfer (EBT) card that can be exchanged for food at SNAP-authorized retailer. Also, Food Distribution Program on Indian Reservations (FDPIR), The Emergency Food Assistance Program (TEFAP), Charitable Food System, etc., are also run by the United States government for adults, family, and children and Special Supplemental Nutrition Program for Women, Infants, and Children. Survivor Benefit Plan (SBP), Head Start Program, etc. are also run by the US government for Children and pregnant or postpartum women (Thorndike et al. 2022). In India, the public distribution system (PDS) was created in order to manage the food security in the year 1940s, which became the most efficient policy to fight against the food security after the independence in India. In spite of various improvements in the production of food and distribution of food, appropriate nutrition and enough food remain evasive for every people at every times. Hunger and nutrition is still a major challenge in the world that is also continued in India (George and McKay 2019).

Various countries introduced various programs and policies to tackle hunger and provide nutritional food. The increasing population, changing climatic conditions, and increasing nutritional requirement arise a need of new solution or policies in order to achieve food and nutritional security. Food security contains majorly four major approaches (Pandey et al. 2024) described in Fig. 21.1.

In India, crop diversification and climatic change receive little attention among the area of ensuring food and nutrition security. To achieve food and nutrition security, researchers identify limited water and land availability, climatic changes, labor shortage, and dwindling natural sources as challenges (Sharma and Sharma 2017). One of the major challenge is change in climate, which is a natural phenomenon that affects the production of food. The National Intelligence Council report on India anticipates the rise in temperature by 2–4 °C by the end of the century (Islam and Manaloor 2021). This rising temperature results in drought and heavy rainfall,

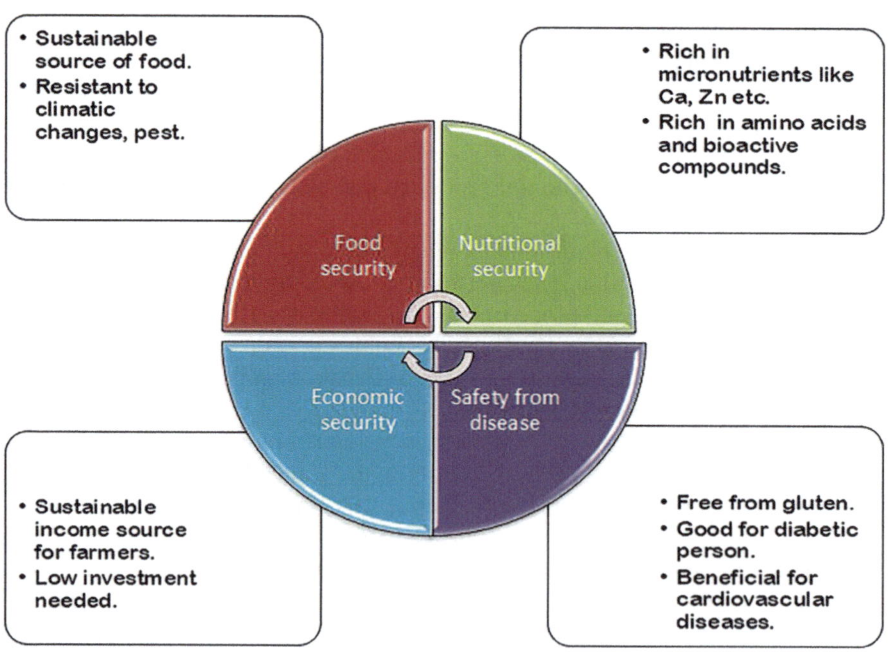

Fig. 21.1 Role of millets in food, nutritional and economic security

which will severely affect the water availability and agriculture. Generally, India's major agriculture areas are drought-prone, and the changing climate makes the condition worse. So, to improve food and nutrition security, crop diversification can be a solutions that need attention. Nutritional security can be improved by including crops that are drought resistant into the cropping system. Millets, a group of cereal grains, can play a significant role in food and nutritional security (Singh et al. 2022).

21.3 Role of Millets in Food Security

Millets can play a vital role in food security. On comparing small millets with major cereals, no doubt that millets are more capable to fulfill the need of food security. Millets are one of the important drought-resistant crops and the sixth cereal crop in term of world agriculture production. They can be eaten as a major cereal or in place of major cereals and are a strong source of protein and calories. Millets are not easily affected by climatic changes, rainfall, drought, etc. (Saxena et al. 2018). This makes less yield loss and ensures income to the population, although millets grow comparatively in a small area compared to the major cereals. Also, millets can grow in under agricultural conditions where other cereals are not harvested properly (Tripathi and Vyas 2023). So, better gross return can be made by millets. Millets cultivation ensures sustainability of agriculture as the dependency of synthetic fertilizers such as pesticides, weedicides, and insecticides became negligible (Mohod et al. 2023). Also, the carbon footprint of the millets could be less than major cereals, although the exact value are not available to date. However, the global warming potential (GWP) and carbon equivalent potential (CEE) of rice and wheat are the highest among the cereals. The GWP of wheat is around 4 tons and the GWP of rice is 3.4 tons, whereas the CEE of wheat and rice is 1000 tons and 956 tons, respectively (Naser et al. 2019).

Millets can add to the variety of food. Only 7000 of the 10,000 plant species that are known to exist on Earth are edible, and of these 7000, 20% provide 90% of the world's food needs (Garn and Leonard 1989; Chivenge et al. 2015). Wheat, maize, and rice make up about 60% (Rodriguez et al. 2020) of all the staple crops which resulting in a boring diet. This lack of variety in food may make production disruptions more likely. To raise the awareness among the community and raise the understanding of importance of millets for food security, India has designated 2023 as millet year. This declaration also aims to make people aware to build up their high-quality, environmental friendly production (Kumar et al. 2023b).

21.4 Specific Major Nutrients of Millets and Their Role in Health

Millets are high in nutritional contents and also have many health benefits. The scientific names of millets are Pearl Millet (Bajra; *Pennisetumglaucum*), Finger Millet (Ragi; *Eleusinecoracana*), Proso Millet (Cheena; *Panicummiliaceum*), Barnyard Millet (Sanwa; *Echinochloautilis*), Foxtail Millet (Kauni; *Setariaitalica*), Little Millet (Kutki; *Panicumsumatrense*), and Kodo Millet (Arikalu; *Paspalumsetaceum*). All the above-mentioned millets are rich in phytochemicals and also high in insoluble dietary fiber, vitamins, and dietary energy. The dietary fiber, moisture, fat, protein, carbohydrate and mineral content of some millets are discussed in Figs. 21.2 and 21.3. Millets are not only high in antioxidant characteristics but also contain various minerals, majorly micronutrients such as magnesium, zinc, and iron (Singh and Chauhan 2019; Ambati and Sucharitha 2019; Gull et al. 2014; Awadelkareem et al. 2015) (Table 21.1).

21.4.1 Finger Millet

Finger millet is also known as ragi. Ragi is commonly consumed by people all over the world. Ragi is the highest in calcium and mineral concentration in comparison to all whole grains. It is known as "wonder grain." Ragi is the powerhouse of protein and amino acids. It has health benefits to treat anemia and ulcers. It is also used for

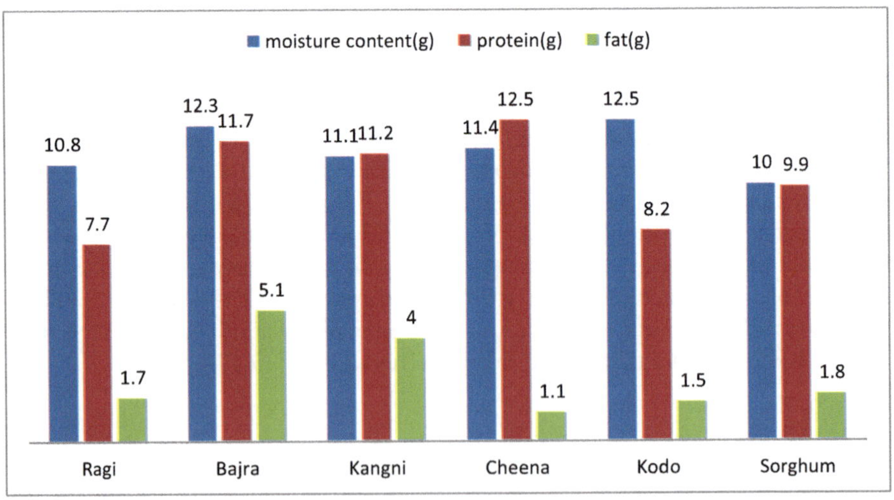

Fig. 21.2 Comparision between moisture content, protein, and fat value of millets (Dayakar et al. 2017; Shahidi and Chandrasekara 2013; Singh and Chauhan 2019; Annor 2015; Paschapur et al. 2021)

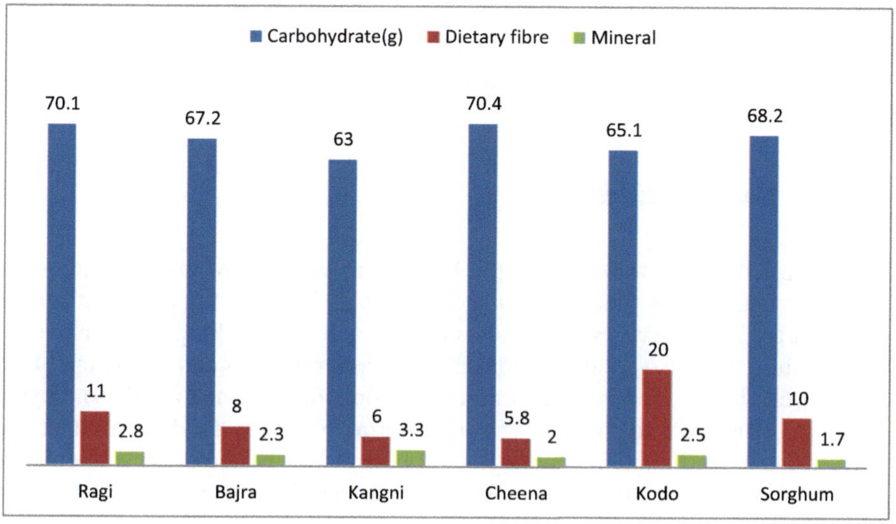

Fig. 21.3 Comparision between carbohydrate, dietary fiber, and mineral value of millets (Dayakar et al. 2017; Shahidi and Chandrasekara 2013; Singh and Chauhan 2019; Annor 2015; Paschapur et al. 2021)

Table 21.1 Minerals and trace elements in millets (mg/100g)

Millets	Minerals							References
	Calcium (mg)	Iron (mg)	Magnesium (mg)	Phosphorus (mg)	Zinc (mg)	Sodium (mg)	Potassium (mg)	
Ragi (finger millet)	344	3.9	137	283	2.4	11	408	Chandi and Annor (2015)
Bajra (pearl millet)	40.1	7.9	134	298	3.0	11.0	305	Paschapur et al. (2021)
Kangni (foxtail millet)	31	2.8	81	297	2.5	4.5	251	Annor (2015)
Cheena (proso millet)	30.0	2.2	152	205	1.5	–	–	Pandey et al. (2024)
Kodon (kodo millet)	28.0	0.5	1.47	189	0.8	4.5	145	Annor (2015)
Jowar (sorghum millet)	13.1	3.35	166	223	1.8	2.1	365	Gopalan et al. (2014)

decreasing the body heat. Ragi is an excellent substitute for wheat and rice especially for diabetics (Gull et al. 2014). It is one of the most digestible, nonacid forming and least allergic warming grains so it is helpful to heat the body in cold and rainy seasons. Finger millet is highly nutritious, richer in fat, protein mainly iron and calcium compared to the rice. So, finger millet is an excellent cereals for food security and developing countries (Singh and Raghuvanshi 2012).

21.4.2 Foxtail Millet

Foxtail millet is also known as kangni. Kangni can easily grow in low-fertility and drought-prone regions. They have a short life span. Foxtail millets have plenty of fibers, carbs, and minerals such as iron and copper. It helps to keep the immune and body healthy. It is also good for controlling cholesterol levels and regulating blood sugar levels. It was used in ancient time very commonly (Jain et al. 2024). Foxtail millet can contribute to food security by offering a healthy food supply that can tolerate harsh conditions. It is also gluten-free and may be the greatest alternative grain for those who are intolerant to gluten (Kalsi and Bhasin 2023).

21.4.3 Kodo Millet

Kodo millets are the coarsest millet among all the minor millets. Kodo is primarily a digestible millet. Kodo millet is rich in phytochemicals and phytate, which have the capability to lower cancer risk. It is beneficial for an obese person as it helps in reducing body weight. It is also utilized to cure and manage irregular menstrual periods in women. Kodo millet alleviates neurological diseases particularly those affecting the eyes, also aids to cure diabetes (Jain et al. 2024). Kodo millet, can aid with food security because it is nutritious, drought-resistant, and can be kept for long periods of time (Chandrakar and Shahi 2023).

21.4.4 Little Millet

Little millet is generally known as kutki or saamai. It is rich in fiber content, which prevents fat deposition in body. Little millet help to treat any gastrointestinal disorders and to prevent constipation. It is also helpful in irregular menstrual periods (Patra et al. 2023). They can be used to complement carbohydrate-rich meals with micronutrients. Kutki millet can be stored for long periods of time without being damaged by insects, providing households with food security and diet diversification during shortages. Also, it is one of the millets that can help combat malnutrition and hidden hunger (Patra et al. 2023).

21.4.5 Proso Millet

Proso millet is also known as cheena. The life span of proso millet is short. It has good capability to grown in drought and heat areas in western India (Ambati and Sucharitha 2019). Proso millet can be rapidly sown as a backup cash crop in the event that other crops do not succeed. Comparatively, proso millets are rich in vitamins, minerals, proteins, and other micronutrients than other staple cereals. Thus, it has the capability to ensure both food and nutritional security (Bhat et al. 2019).

21.4.6 Pearl Millet

Another regional name of Pearl millet is Bajra. Pearl millet is another versatile cereal generally cultivated for food, specifically in Asian and African countries (Nambiar et al. 2011). Pearl millet has the great potential to grow in hot and drought regions. It can easily survive in those regions where maize, wheat, and other cereal crops fail to persist. India is the largest producer of pearl millet among other developing countries (Basavaraj et al. 2010), which covers an area of 9.8 million hectare out of the production of total world (Rathore et al. 2016). Pearl millet is rich in protein, fiber, fat, and ash content (Sade 2009) in comparison of major cereal crops like wheat (Kavitha and Parimalavalli 2014), rice (Ahmed et al. 2014), and sorghum (Awadelkareem et al. 2015). Thus, the pearl millet has the potential to increase income and food security of farming communities generally in arid regions (Mehta et al. 2024).

21.5 Millets in Human Health

According to the global nutritional report 2021, whole world is dealing with various health challenges, including chronic diseases. There are around 149.2 million underdeveloped children's under the age of five, 45.4 million are wasting children, and 38.9 million are overweight children. More than 40% of the population suffers from severe overweight, obesity, and malnutrition (Pandey et al. 2024). The root cause of all these disorders is nutrient-imbalanced diet. The FAO estimates that 795 million people, or about 10% of the world's population and 15% of India's total population are malnourished (FAO 2022). Millets are ranked sixth in terms of total cereal grain production in the world (UNICEF 2022). In many regions of the world, they are still staple food source. The best substitute for regular food intake is millets (Jamal et al. 2023). Some of the health benefits of millets are summarized in Fig. 21.4.

Millets have a lower glycemic index. Millets are also free from gluten. Millets contain high amount of dietary fiber, proteins with all essential amino acids,

MILLETS IN HEALTH
❖ Detoxify the body
❖ Lower bad cholesterol level
❖ Effective in reducing BP
❖ Prevent onset of breast cancer
❖ Beneficial in heart disease
❖ Effective in curing respiratory conditions
❖ Helps in optimize liver, kidney and immune health system
❖ Aids in curing gastrointestinal condition like gastric ulcer
❖ Regulate Blood Glucose Levels

Fig. 21.4 Millets in human health

vitamins, and minerals, which help in the stabilization of blood sugar (Jamal et al. 2023). So, millets can be added to the diet of diabetic person, which helps to minimize blood sugar levels and also promote insulin sensitivity (Pandey et al. 2024).

21.5.1 Millets in Cancer

Millets are the great source of antinutrients such as phenolic acids, tannins, and phytates. These antinutrients have the capability to minimize the risk of breast cancer and colon cancer. The balanced diet of fruits and millets, which is rich in fiber, may be very effective in protecting the menopausal women from breast cancer (Jain et al. 2024). Butrytes, which are the small chain of fatty acids which is release due to the resistant starch, are the necessary metabolites that are effective in colonic cell proliferation stabilization. It works as a prevention strategy for colon cancer. Though some experts are unsure about this, they suspect that high level of magnesium and phytate compound in millets might be responsible for such effects, as it was reported that magnesium ion deficiency is closely associated with tumor development (Pandey et al. 2024; Huang et al. 2024; Dayakar et al. 2017).

21.5.2 Millets in Cardiovascular Disease

Nowadays poor diet is a main reason of health problems such as heart attack, diabetes, obesity, and strokes. The study shows that the consumption of finger millet, proso millet significantly reduced the triglycerides in hyperlipidemic rats. Pandey et al. also represent the inhibition effect of various millets (Pandey et al. 2024).

21.5.3 Promotes Digestion

The fiber-rich diet is best for good digestive system. Millets are rich in dietary fiber, which is effective in boosting the human digestive system. It is beneficial to cure constipation, bloating, and cramping. It is also good for liver and kidney to keep them healthy (Ambati and Sucharitha 2019).

21.5.4 Millets in Ageing

Millets are capable of fighting against the aging due to the presence of high content of antioxidant and phenolic content. It was reported that bioactive compounds found in the methanolic extract of finger millet and kodo millet have antiaging characteristics due to the inhibition of crosslinking of collagen and glycation (Singh and Chauhan 2019).

21.5.5 Millets in Gallstones

It is found that eating meals rich in dietary fiber both soluble and unsoluble is good for women in avoiding galls stones. According to the study conducted by some researchers, women who ate high fiber foods such as millets and fruits had a lower risk of getting gall stones by 13%. The extra bile production in our intestine can lead to gall stone. Pearl millet is one of the best millet in preventing gall stone as it contains insoluble fiber, which decreases the production of extra bile in our system (Tsai et al. 2004; Pandey et al. 2024).

21.5.6 Aids in Sleep

Pearl millet is a rich source of tryptophan, an amino acid that increases the production of serotonin and melatonin in body. Both serotonin and melatonin are known as feel-good hormone as they reduce stress and regulate sleep patterns and promote better sleep (Kautz et al. 2024). Also, millets ensure the gradual release of energy that promotes consistent blood sugar levels. This stability prevents disturbance in sleep that is caused by fluctuations in blood sugar, fostering a more stable and restorative sleep cycle (Jamal et al. 2023).

21.5.7 Millet in Diabetes

Millets are very beneficial for diabetic patients. Normal grain products digest quickly and release glucose instantly. Millets are rich in fiber and digest slowly that leads to slow production of glucose and mix in blood at a slow rate. This helps in maintaining blood sugar level (Dayakar et al. 2017).

21.6 Current Food Security Issues

Food security is the major concern in current scenario in all over the world. India ranked 68th out of 113 major countries in global food security index (GFSI 2021). In India, nearly 195 million people are undernourished. Also, it is roughly estimated that around 40% of children are chronically undernourished in India (Menon et al. 2018). Although the daily food requirements meet 100% current nutritional standards but India still lags far behind by 20% quality protein intake. It is found by the human right measurement initiative that India is utilizing only 56.8% of its capacity in order to fulfill the need of food security for citizens (Haytowitz et al. 2019). Around 60% of the population of India depends directly upon agriculture for their livelihoods. Both for the economy of India and food security, the agricultural sector is critical across the country. During COVID pandemic, community has faced various threats and challenges to food security globally. Due to the corona virus disease, lots of serious issues arise during this time such as disturbance in food supply chain, shortage of essential goods, rise in prices, and threat to human life. Food security and economic recession are one of the major risks in them (Jain et al. 2024). However, there are various challenges in India's agriculture sector including lower agriculture production due to climatic changes and lack of farmland due to the rapidly growing population. The rapidly growing population not only affects the farmland but also affects the economic growth that strains land availability and natural sources (Saxena et al. 2018). So, it is one of the major causes of millions of malnourished children around India. The crops are also affected by the climatic changes, which is also a major issue to achieve food and nutrition security. Although some reports showed that climatic changes are beneficial in some regions but it will still not be sufficient to feed the higher no of population globally (Kang et al. 2009).

21.7 Growing Millets for Food Security

Millets are the best alternative to address this issue and can be an effective way to achieve food security. Millets is not only a important sources of macro and micronutrients but also helps in reducing global warming due to low GWP in comparision to rice and wheat. Millets can be grown easily under challenging climatic conditions (Saxena et al. 2018). Millets were the backbone of India's food and farming system, but for a variety of reasons, both production and consumption have drastically decreased, and farmers have neglected millions of hectares of dry land. These dry land areas, if converted to agricultural use, might create employments to most undeveloped districts such as Rajasthan in India, Africa, and many other countries (Jamal et al. 2023). Furthermore, both food sovereignty and food security will be preserved. The Government of India is executing a number of efforts to provide grains to low-income households at a reduced cost, but the food security bill is expected to be the first to include millet in the Public Distribution System (PDS) (Islam and Manaloor 2021). As previously stated, millets have a higher mineral content than other grains; nevertheless, due to antinutrient factors, their bioavailability is reduced. According to scientific study, soaking, malting, popping, puffing, germination, and fermentation are simple and inexpensive home processing processes that have been demonstrated to boost nutrient availability from millets (Kumar et al. 2021). As a result, if receivers of food security bills are aware on these simple approaches, nutrient security will be addressed to some extent within the restrictions of available resources (Tiwari et al. 2023). The ability of millets to adapt in harsh climatic conditions requires nominal inputs. Also, due to the presence of adequate nutrients makes millets more valuable among other cereals.

21.8 Benefits of Growing Millet for Farmers

Millets are also good in the perspective of farmer's economy. Millets had a great potential to generate livelihoods, ensure food and nutritional security, and increase farmers' income (Singh and Chauhan 2019). As millets require low input compared to the others cereals, this offers a key advantage to small-scale farmers, who often lack expensive agricultural inputs and extensive irrigation systems. Since millets can flourish in nutrient-deprived soil and under restricted irrigation, farmers have the potential to lower their production expenses while maximizing their profits (Mohod et al. 2023).

Table 21.2 Climate resilient traits and life span of millets

Sorghum	Duration: 100–125 days Drought tolerant, adaptive to soils, altitudes and temperatures, excellent recovery mechanism
Pearl millet	Duration: 80–95 days Adaptable in hot and dry conditions, easily cultivated in poor soils
Finger millet	Duration: 90–130 days Adaptable across wide altitude, moderately resistant to heat, humidity and drought
Foxtail millet	Duration: 70–120 days Adaptable to low rainfall and high altitude conditions
Proso millet	Duration: 60–95 days Adaptable to low rainfall and high altitude condition, short-duration crop
Little millet	Duration: 70–110 days Easily grown in poor soil and required less water comparatively
Kodo millet	Duration: 100–150 days Tough crop with long duration, adaptable to poor soils and low rainfall
Barnyard millet	Duration: 45–60 days Short-duration crop, adaptable to high altitudes range and less rainfall conditions

By including millets as a crop in cropping system small-scale farmer not only earn their livelihoods but also help in building climate resilience against the uncertainties posed by climatic changes. The wide variety of millets ensures greater biodiversity on-farm, reduces pests and climate risks, and improves farmers' overall resilience. The past few decades are the witnessed of increasing global demand of nutritional rich and gluten-free food, which also adds benefits in human health (Sharma and Sandal 2024). Millet flour is one of the gluten-free and nutrient-dense products that make it a new market trend. This opens up a new direction for small-scale farmers as well as for business to diversify their income by taking part in various activities that add the value such as packaging, processing, and marketing of millet-based products (Mohod et al. 2023).

Therefore, agriculture organizations and policymakers may support the development of small-scale farmers and improve food and nutritional security by encouraging millet cultivation and creating easy access to markets.

So, by promoting millet cultivation and making smooth paths to access the markets, agriculture organization and policymakers can help in empowering small-scale farmer's development and enhancing food and nutritional security. Also, the life span of millet cultivation is much less than other crops and can be grown in a challenging climate (Bhat et al. 2018). The climate resilient traits and life span of millets are shown in Table 21.2.

21.9 Challenges and Solutions in Increasing Millet Production

There is no doubt that millets have various advantages in sustainable agriculture practice, but there are several challenges that require attention before millets can be a practical option. Millets need more labor than other crops as well as require specialized farming techniques (Yadav et al. 2012). Also, millets are not commonly available as any other crop and farmers may not have the necessary resources to grow them. A number of actions must be taken and implemented properly in order to make millets as an alternative for sustainable agriculture. So, various initiatives should be taken to raise the awareness about millets and their benefits and increase the availability of millets (Sher et al. 2019). Also, in order to develop better technologies and techniques for millet cultivation, research must be promoted. Additionally, various steps should be taken to increase the millet-based diet and market value of millets. Some challenges and strategies to overcome these problems are given below.

21.9.1 Lack of Adequate storage Facilities

Another challenge that millet producers face is inadequate storage facilities. Inadequate storage facilities might cause crops to rot too soon and lower farmers' profits (Kheya et al. 2023).

21.9.1.1 Improve Storage Facilities

To address the challenge of poor storage, proper storage facilities should be developed to preserve the millet crops and to reduce the postharvest loss such as moisture proof containers, hermetic bag, metal silos, etc. (Patra et al. 2023). Additionally, by developing decentralized storage facilities at the village or community level, farmers can be empowered to maintain crop quality and increase their economic stability. These technologies help farmers to reduce postharvest losses by protecting millets from pests, humidity, and microbial spoilage.

21.9.2 Limited Availability of Quality Seed

The availability of quality seeds is one of the major problems which is faced by farmers who produce millets. Generally, the seeds available in the market are either of low quality or affected by diseases or weeds. This causes a lack of desired crop yields for the farmers (Kheya et al. 2023).

21.9.2.1 Availability of Quality Seeds

To cultivate good crops, quality seeds are important for farmers. So, farmers should be aware and promote to use certified disease-free seeds for agriculture to ensure high-quality inputs. Also, it is essential to strengthen seed production and distribution systems through public-private partnerships and promote community-based seed banks. This helps the farmer to get good income as well as good crops (Annor 2015).

21.9.3 Soil Degradation

Soil degradation is one of another reasons that affects the yield of millets by lowering soil fertility and water retention. In millet crops, degraded soils, particularly those undergoing erosion or nutrient depletion, can lead to stunted growth, reduced yields, and heightened vulnerability to pests and diseases (Nishigaki et al. 2025). So, sometimes it is hard for farmers to get the desired price for their hard work (Das and Rakshit 2016).

21.9.3.1 Improve Soil and Water Management

To reduce soil degradation and increase millet yield, use of sustainable water management techniques and conservation tillage technique is essential. This may include the use of organic fertilizers, green manures, composts, proper irrigation system along with crop rotation to prevent soil erosion. Also, promoting agroforestry and integrated nutrient management supports soil health (Sukanya et al. 2024).

21.9.4 High Cost of Production

Millet cultivation, although climate-resilient, often requires lots of hard work such as manual weeding, sowing, and harvesting. Also, the rising prices of agricultural inputs—such as certified seeds, organic or chemical fertilizers, pest control measures, and irrigation facilities—further increase production costs. Millets receive less assistance and subsidies than more commercial crops like wheat or rice, making input procurement more expensive, thereby reducing the net profitability and discouraging widespread millet cultivation (Yamuna et al. 2024).

21.9.4.1 Sustainable Agriculture Practices

To reduce production costs, the agriculture practice should be low input and sustainable by adopting organic farming, use of farmyard manure, and biofertilizers. Also, mechanization support for small tools, subsidies on inputs, and promotion of Farmer

Producer Organizations (FPOs) can help in reducing input costs and improve millet production and make millet cultivation more profitable (Gayathri and Manimozhi 2024).

21.9.5 Poor Production

Poor production of millet is one of the challenges faced by the millet farmers. The lesser the production lesser will be the profits as profit depends upon production costs, quantity produced, and market demand. The revenue will automatically decline if the production is less, unless prices sharply increase, which typically won't be sufficient to make up for the decline in production (Das and Rakshit 2016).

21.9.5.1 Promote Intercropping

To address low production and to increase the profit and productivity of the millet production, intercropping should be promoted. Also, farmers should adopt high-yielding millet varieties, pest management, healthy seeds, and integrated nutrient. This will not only improve the fertility of the soil but also reduce the crop failure risk and can significantly (Sukanya et al. 2024).

21.9.6 Poor Market Access and Infrastructure

To access the market, infrastructure is the main factor, but poor infrastructure can make it difficult to access market for farmers. Infrastructure refers to the basic physical and organizational facilities needed for farmers to transport, store, and sell their produce in the market such as good quality of rural roads, availability of vehicles, designated areas or facilities where farmers can bring their produce for sale, auction, or distribution (Mandis, etc.). This may lead to the poor income source for farmers (Carolan 2017).

21.9.6.1 Market Access

To improve the profit in millet production, markets should be accessible for farmers as well as for buyers. For this, new markets should be created, and infrastructure should be developed and maintained regularly. Also, by setting mandis, transport subsidies, and rural market linkages through FPOs can improve market opportunities (Chandra et al. 2021).

21.9.7 *Low Prices*

Farmers may find it challenging to turn a profit from their millet crops due to the low prices that buyers are willing to pay. This may result in a loss of interest for millet production (Sukanya et al. 2024).

21.9.7.1 Encourage Millet Products

Millets' product should be encouraged by the government. People should be aware of the benefits of millet-based products and they should be easily available in the market. Government can give relaxation on GST of millet-based based products in order to promote them. Additionally, millet-based meals might increase customer demand by being marketed and branded as wellness and health items (Jain et al. 2024).

21.9.8 *Unfavorable Government Policies*

The government policies including subsidies and taxes give the negative effect on the production of millets. Unlike rice and wheat, millets often don't have strong government procurement policies or minimum support prices (MSP). Farmers feel insecure about earning a good price for their millets. This discourages millet cultivation. Many government subsidy programs (e.g., for wheat, rice, or sugarcane) provide cheaper seeds, fertilizers, electricity for irrigation, and guaranteed MSP for these crops so, farmers are encouraged to grow these subsidized crops rather than millets, because they seem more profitable or less risky. This resulted in decline in millet production as farmers switched to these supported crops (Jain et al. 2024).

21.9.8.1 Encourage Research

The government should fund research based on millets to create disease-free, more resistant, and more effective millets in order to promote millet as a profitable crop. As a result, farmers can achieve better income and better yield (Gayathri and Manimozhi 2024).

21.9.9 *Lack of Extension Services*

Latest advices and information of millets are important for farmers so extension services are essential for farmers, but these services are not enough. This may lead to poor income and crop yield (Patra et al. 2023).

21.9.9.1 Aware and Educate Farmers

To address these problems, proper training and awareness program should be organized for farmers to promote millet production. Additionally, awareness program should include knowledge about best farming practice, disease management, climate-resilient techniques, and market opportunities. Also, working together with local agricultural colleges, Krishi Vigyan Kendras (KVKs) helps guarantee that farmers remain knowledgeable and equipped to implement better millet farming techniques (Gayathri and Manimozhi 2024).

21.10 Conclusion

Food security is one of the biggest issues faced by the world. In the modern world, the two major challenges are to tackle hunger and provide nutritional food. In conclusion, millets consumption and utilization for food and fooder purposes are the greatest way to ensure food security. Millets are not only a staple source of food but also rich in nutrition and can help defeat malnutrition and other major problems. They have the potential to grow even in drought conditions. However, a number of obstacles must be overcome before millets may be considered a practical choice. Millets are a good source of nutrients and also have lots of health benefits like beneficial in the prevention of cancer, sleep disorders, cardiovascular disease, digestion, etc. Millets also provide economic security to farmers as by giving minimum inputs farmers received the maximum output. However, there are some challenges in growing millet like the availability of proper healthy seeds, soil and irrigation system, infrastructure, etc. But by overcoming these challenges, millets can be profitable crops for farmers. Also, government should makes various steps to promote these millets and continuously invests and works in this direction. So, it's time to include our traditional food crop millets in our daily diet. Millets have the capacity to fight against hunger and provide nutritional food to the nations of the world.

Conflict of Interest The authors declare that there is no conflict of interest.

References

Ahmed K, Shoaib M, Akhtar MN et al (2014) Chemical analysis of different cereals to access nutritional components vital for human health. Int J Chem Biochem Sci 6:61–67. https://doi.org/10.13140/RG.2.2.30783.43684

Ambati K, Sucharitha KV (2019) Millets-review on nutritional profiles and health benefits. Int J Rec Sci Res 10(7):33943–33948. https://doi.org/10.24327/ijrsr.2019.1007.3786

Annor GA (2015) Millet minor: overview GK Chandi, University of Guelph, Guelph, ON, Canada GA Annor, University of Ghana, Accra, Ghana© 2016 Elsevier Ltd. All rights reserved. Encyclopedia of food grains, pp 199. https://doi.org/10.1016/B978-0-12-394437-5.00012-7

Awadelkareem AM, Hassan EG, Fageer AS et al (2015) The nutritive value of two sorghum cultivar. Int J Food Nutr Sci 4(1):1

Balmford A, Green RE, Scharlemann JP (2005) Sparing land for nature: exploring the potential impact of changes in agricultural yield on the area needed for crop production. Glob Chang Biol 11(10):1594–1605. https://doi.org/10.1111/j.1365-2486.2005.001035.x

Basavaraj G, Rao PP, Bhagavatula S et al (2010) Availability and utilization of pearl millet in India. SAT ejournal 8. http://oar.icrisat.org/id/eprint/96

Bhat BV, Rao DB, Tonapi VA (2018) The story of millets Karnataka state department of Agriculture in association with ICAR. Indian Institute of Millet Research, Hyderabad, pp 57–59

Bhat S, Nandini C, Srinathareddy S et al (2019) Proso millet (*Panicum miliaceum* L.)—a climate resilient crop for food and nutritional security: a review. Environ Conserv J 20(3):113–124. https://doi.org/10.36953/ECJ.2019.20315

Bruinsma J (2009) The resource outlook to 2050: by how much do land, water and crop yields need to increase by 2050? Expert meeting on how to feed the world in 2050. https://openknowledge.fao.org/handle/20.500.14283/ak971e

Carolan MS (2013) Reclaiming food security. Routledge. https://doi.org/10.4324/9780203387931

Carolan M (2017) Food security and policy. In: Sustainable protein sources. Academic, pp 391–407. https://doi.org/10.1016/B978-0-12-802778-3.00024-X

Chandra AK, Chandora R, Sood S et al (2021) Global production, demand, and supply. In: Millets and pseudo cereals. Woodhead Publishing, pp 7–18. https://doi.org/10.1016/B978-0-12-820089-6.00002-1

Chandi GK, Annor GA (2015) Millet Minor: Overview. In The World Of Food Grains. 1-4. Elsevier Inc. 199–208

Chandrakar L, Shahi S (2023) A systematic review on kodo. J Adv Zool 44(5). https://doi.org/10.53555/jaz.v44i5.3242

Chivenge P, Mabhaudhi T, Modi AT et al (2015) The potential role of neglected and underutilised crop species as future crops under water scarce conditions in Sub-Saharan Africa. Int J Environ Res Public Health 12(6):5685–5711. https://doi.org/10.3390/ijerph120605685

Das IK, Rakshit S (2016) Millets, their importance, and production constraints. In: Biotic stress resistance in millets. Academic, pp 3–19. https://doi.org/10.1016/B978-0-12-804549-7.00001-9

Dayakar Rao B, Bhaskarachary K et al (2017) Nutritional and health benefits of millets. ICAR Indian Institute of Millets Research (IIMR) Rajendranagar, Hyderabad, p 2

FAO (2002) The state of food insecurity in the world 2001. FAO, Rome

FAO (2009) Declaration of the world food summit on food security. FAO, Rome

FAO FOOD R, AFFORDABLE HD (2022) Food security and nutrition in the World. https://doi.org/10.4060/cc0639en

Garn SM, Leonard WR (1989) What did our ancestors eat? Nutr Rev 47(11):337–345. https://doi.org/10.1111/j.1753-4887.1989.tb02765.x

Gayathri M, Manimozhi K (2024) Bridging the knowledge gap: enhancing awareness among farmers about millet cultivation. Curr Agric Res J 12(1). https://doi.org/10.12944/CARJ.12.1.23

George NA, McKay FH (2019) The public distribution system and food security in India. Int J Environ Res Public Health 16(17):3221. https://doi.org/10.3390/ijerph16173221

GFSI (2021) Global food security index (GFSI). London and IMF Staff Calculations

Global Hunger Index (2023). https://www.globalhungerindex.org/policy-recommendations-2023

Gopalan C, Rama Sastri BV et al (2014) Nutritive value of Indian foods. National Institute of Nutrition, Hyderabad

Gull A, Jan R, Nayik GA et al (2014) Significance of finger millet in nutrition, health and value added products: a review. Magnesium 130(32):120

Haytowitz DB, Ahuja JK, Wu X et al (2019) USDA national nutrient database for standard reference, legacy release. https://figshare.com/articles/dataset/USDA_National_Nutrient_Database_for_Standard_Reference_Legacy_Release/24661818

Huang L, Lin R, Chen J et al (2024) Magnesium Ion: A New Switch in Tumor Treatment. Biomedicines 12(8):1717 https://doi.org/10.3390/biomedicines12081717

Islam S, Manaloor V (2021) Millets for food and nutrition security in India: determinants and policy implications. J Nutr Food Secur. https://doi.org/10.18502/jnfs.v6i2.6074

Jain V, Agarwal N, Bhatia V (2024) Promoting millets: charting a journey from food security to health. Indian J Community Med 49(1):5–10. https://doi.org/10.4103/ijcm.ijcm_107_23

Jamal T, Ghosh L, Maity D et al (2023) Millets: the nourishing solution to food security challenges over cereal-based cropping systems. Int J Environ Clim Chang 13(12):417–426. https://doi.org/10.9734/ijecc/2023/v13i123698

Kalsi R, Bhasin JK (2023) Nutritional exploration of foxtail millet (*Setaria italica*) in addressing food security and its utilization trends in food system. Food 4(5):111

Kang Y, Khan S, Ma X (2009) Climate change impacts on crop yield, crop water productivity and food security—a review. Prog Nat Sci 19(12):1665–1674. https://doi.org/10.1016/j.pnsc.2009.08.001

Kautz A, Meng Y, Yeh KL et al (2024) Dietary intake of nutrients involved in serotonin and melatonin synthesis and prenatal maternal sleep quality and affective symptoms. J Nutr Metab 2024(1):6611169. https://doi.org/10.1155/2024/6611169

Kavitha S, Parimalavalli R (2014) Effect of processing methods on proximate composition of cereal and legume flours. J Hum Nutr Food Sci 2(4):1051

Kheya SA, Talukder SK, Datta P et al (2023) Millets: the future crops for the tropics-status, challenges and future prospects. Heliyon. https://doi.org/10.1016/j.heliyon.2023.e22123

Kumar A, Tripathi MK, Joshi D et al (2021) Millets and millet technology. Springer, Singapore, p 438. https://doi.org/10.1007/978-981-16-0676-2

Kumar S, Singh S, Yadav R et al (2023a) Millets: an ancient food and its nutritional importance for human health. Agrosphere e-Newslett 4(7):1–4

Kumar S, Sridhar R, Monika S et al (2023b) A comprehensive review on millets: a potential source of energy and nutrients for health. Int J Environ Clim Chang 13(9):2531–2538. https://doi.org/10.9734/ijecc/2023/v13i92487

Mehta D, Vyas S, Dudhagara D et al (2024) Significance of Indian millets in enhancing global food security: a comprehensive review. Trends Food Sci Technol 149:104527

Menon P, Headey D, Avula R et al (2018) Understanding the geographical burden of stunting in India: A regression-decomposition analysis of district-level data from 2015–16. Maternal & child nutrition 14(4):e12620

Mohod NB, Borah A, Goswami P et al (2023) The international year of millet 2023: a global initiative for sustainable food security and nutrition. Int J Plant Soil Sci 35(19):1204–1211. https://doi.org/10.9734/ijpss/2023/v35i193659

Nambiar VS, Dhaduk JJ, Sareen N et al (2011) Potential functional implications of pearl millet (*Pennisetum glaucum*) in health and disease. J Appl Pharm Sci 1(10):62–67

Naser HM, Nagata O, Sultana S et al (2019) Carbon sequestration and contribution of CO2, CH4 and N2O fluxes to global warming potential from paddy-fallow fields on mineral soil beneath peat in Central Hokkaido, Japan. Agriculture 10(1):6. https://doi.org/10.3390/agriculture10010006

Nishigaki T, Ikazaki K, Shinjo H et al (2025) Pearl millet yield reduction by soil erosion and its recovery potential through fertilizer application on an Arenosol in the Sahel. Soil Tillage Res 246:106324. https://doi.org/10.1016/j.still.2024.106324

Pal B, Mukhopadhyay A, Kishore A et al (2023) Revitalizing the potential of minor millets: Agrarian constraints, possible solutions and future roadmap. Int J Environ Clim Chang 13(7):328–338. https://doi.org/10.9734/ijecc/2023/v13i71884

Pandey M, Singh S, Tripathi R et al (2024) Millets for food and nutrition security: a review. Int J Plant Soil Sci 36(2):238–247. https://doi.org/10.9734/ijpss/2024/v36i24385

Paschapur AU, Joshi D, Mishra KK (2021) Millets for life: a brief introduction. In: Millets and millet technology, pp 1–32. https://doi.org/10.1007/978-981-16-0676-2_1

Patra A, Singh RP, Kundu M (2023) Millet production in India: challenges and opportunities. Bio Res Today 5(3):238–241

Peng W, Berry EM (2018) The concept of food security. In: Ferranti P, Berry EM, Anderson JR (eds) Encyclopedia of food security and sustainability. Elsevier, Amsterdam

Rathore S, Singh K, Kumar V (2016) Millet grain processing, utilization and its role in health promotion: a review. Int J Food Sci Nutr 5(5):318–329. http://www.sciencepublishinggroup.com/j/ijnfs

Ray DK, Mueller ND, West PC et al (2013) Yield trends are insufficient to double global crop production by 2050. PLoS One 8(6):6428. https://doi.org/10.1371/journal.pone.0066428

Rodriguez JP, Rahman H, Thushar S et al (2020) Healthy and resilient cereals and pseudo-cereals for marginal agriculture: molecular advances for improving nutrient bioavailability. Front Genet 11:49. https://doi.org/10.3389/fgene.2020.00049

Sade FO (2009) Proximate, antinutritional factors and functional properties of processed pearl millet (*Pennisetum glaucum*). J Food Technol 7(3):92–97

Saxena R, Vanga SK, Wang J et al (2018) Millets for food security in the context of climate change: a review. Sustainability 10(7):2228. https://doi.org/10.3390/su10072228

Schade C, Pimentel D (2010) Population crash: prospects for famine in the twenty-first century. Environ Dev Sustain 12:245–262. https://doi.org/10.1007/s10668-009-9192-5

Shahidi F, Chandrasekara A (2013) Millet grain phenolics and their role in disease risk reduction and health promotion: a review. J Funct Foods 5(2):570–581. https://doi.org/10.1016/j.jff.2013.02.004

Sharma A, Sandal A (2024) Millets for food and health security: a review. Emerg Life Sci Res:1–9. https://doi.org/10.31783/elsr.2024.1010109

Sharma S, Sharma P (2017) Agricultural production, marketing and food security in India: a peep into progress. Productivity 58(2):155

Sher A, Nawaz A, Sarfraz M et al (2019) Advanced production technologies of millets. In: Agronomic crops: volume 1: production technologies, pp 273–296. https://doi.org/10.1007/978-981-32-9151-5_15

Singh S, Chauhan ES (2019) Role of underutilized millets and their nutraceuticals importance in the new era. A review. Int J Sci Res 8(2):2844–2857

Singh P, Raghuvanshi RS (2012) Finger millet for food and nutritional security. Afr J Food Sci 6(4):77–84

Singh RP, Qidwai S, Singh O et al (2022) Millets for food and nutritional security in the context of climate resilient agriculture: a review. Int J Plant Soil Sci:939–953. https://doi.org/10.9734/ijpss/2022/v34i232504

Sjauw-Koen-Fa A (2010) Sustainability and security of the global food supply chain. Rabobank Group, Utrecht

Sukanya TS, Kumar A, Sathya K et al (2024) Millet based cropping systems for enhanced productivity. In: Mishra S, Kumar S, Srivastava RC (eds) Genetic improvement of small millets, pp 63–86. https://doi.org/10.1007/978-981-99-7232-6_4

Thorndike AN, Gardner CD, Kendrick KB et al (2022) Strengthening US food policies and programs to promote equity in nutrition security: a policy statement from the American Heart Association. Circulation 145(24):1077–1093.a

Tsai CJ, Leitzmann MF, Willett WC et al (2004) Long-term intake of dietary fiber and decreased risk of cholecystectomy in women. Am J Gastroenterol 99(7):1364–70

Tiwari H, Naresh RK, Kumar L et al (2022) Millets for food and nutritional security for small and marginal farmers of North West India in the context of climate change: a review. Int J Plant Soil Sci 34(23):1694–1705. https://doi.org/10.9734/ijpss/2022/v34i232594

Tiwari H, Naresh RK, Bhatt R et al (2023) Underutilized nutrient rich millets: challenges and solutions for India's food and nutritional security: a review. Int J Plant Soil Sci 35(2):45–56. https://doi.org/10.9734/ijpss/2023/v35i22758

Tripathi T, Vyas S (2023) From ancient grains to modern solutions: a history of millets and their significance in agriculture and food security. Int J Home Sci 9(2):72–78

UNICEF Data (2022) The state of food security and nutrition in the world UNICEF Data: monitoring the situation of women and children. https://data.unicef.org/resources/sofi-2022/. Accessed 6 July 2022

Yadav OP, Rai KN, Rajpurohit BS et al (2012) Twenty-five years of pearl millet improvement in India. ICAR. http://oar.icrisat.org/id/eprint/10887

Yadav R, Singh LB, Patel A et al (2023) A review on yield gap analysis of millets in India. Int J Plant Soil Sci 35(18):1800–1804. https://doi.org/10.9734/ijpss/2023/v35i183462

Yamuna N, Kolekar DV, Rakesh N (2024) Trend and instability analysis of area, production, productivity of millets in India. J Sci Res Rep 30(8):686–699. https://doi.org/10.9734/jsrr/2024/v30i82290

Chapter 22
Medicinal Value of Little Millet, Kodo Millet, and Pearl Millet

Vishakha Pundir, Naveen Kumar, and Mani Govindasamy

Abbreviations

BM	Barnyard Millet
CVD	Cardiovascular Disease
DPPH	2,2-diphenyl-1-picrylhydrazyl
FBG	Fasting Blood Glucose
FLM	Foxtail Millet
FM	Finger Millet
GAE	Gallic Acid Equivalent
GI	Glycemic Index
IC50	Inhibitory Concentration 50
KM	Kodo Millet
LDL	Low-Density Lipoprotein
LM	Little Millet
PM	Pearl Millet
PSM	Proso Millet
SM	Sorghum Millet
TFC	Total Flavonoid Content
TPC	Total Phenolic Content

V. Pundir (✉)
Department of Chemistry, Faculty of Sciences, Gurukula Kangri (Deemed to be University), Haridwar, Uttarakhand, India

N. Kumar
Department of Chemistry, K.L.D.A.V. (P.G.) College Roorkee, Haridwar, Uttarakhand, India

M. Govindasamy
Department of Materials Engineering, Ming Chi University of Technology, Taishan City, New Taipei City, Taiwan

© The Author(s), under exclusive license to Springer Nature Switzerland AG 2026

J. Singh, M. Kumar Malik (eds.), *Millets and Millet Starch: Chemical Composition, Functional Modification, and Applications*,
https://doi.org/10.1007/978-3-032-09502-2_22

22.1 Introduction

The term "grass family" refers to a widely diversified group of grasses with small seeds from the *Poaceae* family called millets (Michaelraj and Shanmugam 2013; Vetriventhan et al. 2020). They are nurtured as grains, or cereal crops, primarily for human nourishment and animal feed. Archaeological documents indicates that millet has grown in India for 1000 years (Choudhary et al. 2023). Millets are coarse grains, like rice and wheat, which were later grown in the Indian subcontinent (Azarbad et al. 2019). There are many different types of millets, each with unique properties and uses in cooking, including foxtail millet (FLM) (kangni), barnyard millet (BM) (jhangora), proso millet (PSM) (barri), sorghum millet (SM) (jowar), kodo millet (KM) (kodra), little millet (LM) (kutki), finger millet (FM) (Mandua), pearl millet (PM)) (bajra) and commonly grown in India (Vetriventhan et al. 2020). Rich in antioxidants, dietary fiber, minerals, and vitamins, millets are an excellent addition to a well-balanced diet because they don't contain gluten.

The semi-arid and dry areas of Karnataka, Andhra Pradesh, Rajasthan, Maharashtra, and Tamil Nadu are among the dryland locations where these millets are mainly grown (Michaelraj and Shanmugam 2013). Due to their numerous culinary applications, environmental sustainability, and high nutritional content, these ancient grains have become increasingly well-liked (Rathore et al. 2016). They have been a staple food source in many parts of the world, especially Asia and Africa (Shahidi and Chandrasekara 2013). Millets can also be used as puffed grains, flakes, flour, or whole grains and are useful in a variety of culinary applications. They are delightfully chewy and have a nutty flavor. Millets are used to prepare pancakes, porridge, breads and alcoholic drinks in many traditional cuisines. Millets can be used as a nutritious alternative to refined grains in baked dishes, soups, and other ready-to-eat foodstuffs (Bisht et al. 2022; Malleshi 1989).

Millets are nutritional powerhouses that offer several essential elements vital to human health. While most industrialized countries have mostly limited their use to animal feed, most developing countries consume nearly all millet for human use. Millets are often rich in minerals, vitamins, fiber, and anti-nutritional components, such as phytates and tannins, which have an impact on the minerals' bioavailability (Rao and Deosthale 1983). Additionally, millets are a great source of phytochemicals that promote health, such as lignans, polyphenols, phytosterols, phytoestrogens, and phytocyanin (Kavitha et al. 2024). They serve as guardians against age-related diseases like cancer, diabetes, and cardiovascular ailments (CVD) by functioning as immunological enhancers, antioxidants, and purifying agents (Rao et al. 2011). Their phytochemicals' high dietary fiber content and prebiotic properties have been demonstrated to improve gut health. Dietary fiber from millet prevents constipation and helps in regular bowel movements and overall digestive health (Saleh et al. 2013). The prebiotic effect of millet fibers promotes the growth of *Lactobacillus* and *Bifidobacterium*, two kinds of good gut bacteria, resulting in a balanced gut microbiota (Chethan et al. 2018). A diverse and robust gut microbiota is associated with many health benefits, including improved micro-nutrients absorption, a reduced inflammatory response, and a stronger immunological response. Millets are also rich in antioxidants and other cancer-preventive compounds (Amadou et al. 2011). Overall, this chapter explores the distinctive nutritional profiles and health advantages of little, kodo, and pearl millet, with the objective of synthesizing traditional knowledge and scientific understanding to facilitate their incorporation into contemporary dietary practices and healthcare systems.

22.2 Health Benefits of Millets

Millets are excellent energy sources. Some of the common health benefits of the millets have been represented in Fig. 22.1. These offered dietary fiber, polyphenols, minerals, vitamins, fatty acids, and protein. A remarkable abundance of vital amino acids, especially those rich in sulfur (methionine and cysteine), can be found in typical millet protein (Nithiyanantham et al. 2019). According to epidemiological data derived from various studies, millets have the potential to be beneficial to one's health. Eating millet can help in detoxification of the body, minimize the chance of cardiovascular, metabolic, and malignant disorders, boost respiratory immunity,

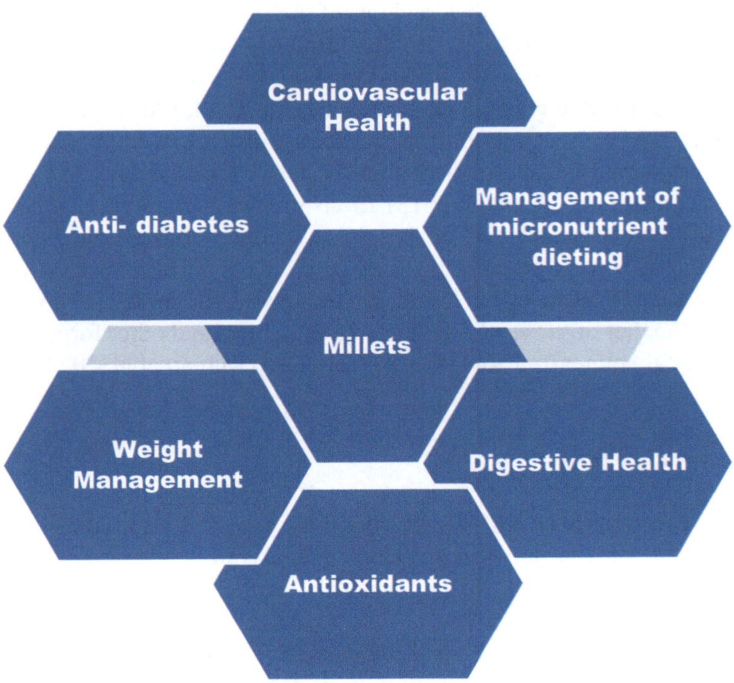

Fig. 22.1 Schematic of the health benefits of millets

improve muscular and neural systems, and shield against various degenerative ailments including metabolic disorders and Parkinson's illness (Scalbert et al. 2005; Chandrasekara and Shahidi 2012; Manach et al. 2005). Millets include essential ingredients such as resistant starch, oligosaccharides, lipids, cancer prevention agents such as avenanthramides and phenolic acids lignans, phytosterols, and flavonoids. These nutrients are thought to have numerous health advantages (Edge et al. 2005) (Table 22.1).

22.2.1 General Overview of Little Millet, Kodo Millet, and Pearl Millet

22.2.1.1 Little Millet

Little millet (*Panicum sumatrense*) is commonly referred to as samai (Tamil), kutki (Hindi), and sum aloo (Telugu) (Kumari et al. 2022). "Cool food" is a different name for LM due to the refreshing effect on the body when ingested during the summer (Pradeep and Sreerama 2018).

In India, LM is grown as a traditional crop. In 2024, the total production of LM was 1.44 lac tons in India. In Gujarat state, it is cultivated across 10,634 hectares, yielding 9526 tons with a productivity of 896 kg/ha (Rajput et al. 2022). This crop is primarily grown in the tribal regions of Andhra Pradesh, Chhattisgarh, and Madhya Pradesh (Padulosi et al. 2015). Despite its diminutive size, LM is more adaptable and nutritious than other grains, making it a good choice for a diet (Maitra and Shankar 2019; Rao et al. 2017; Srilekha et al. 2019). LM is a better substitute for other foods because it is high in phytochemicals and antioxidants. It helps against lifestyle and metabolic diseases like diabetes, and obesity (Chauhan et al. 2018; Dey et al. 2022; Saleh et al. 2013). LM is well-known among health-conscious people due to bioactive nutraceuticals such as tocopherols, phenolic contents, low glycemic index, and carotenoids. It serves as a rich source of phosphorus and fiber, facilitating the reduction of fat levels in the body (Indirani 2021). LM is a great addition to the diet of those who practice cardio, yoga, and other forms of exercise because they are gluten-free and do not create acid (Kumari et al. 2022).

22.2.1.2 Kodo Millet

The ancient, underutilized KM (*Paspalum scrobiculatum L.*) from the *Poaceae* family is a staple grain consumed by the native Himalayan Indian population (Sharma et al. 2017a). Figure 22.2 shows the botanical classification and appearance of KM grains. It is widely cultivated in several states of India like Andhra Pradesh, Karnataka, and Tamil Nadu. It's also cultivated in Vietnam, Indonesia, Thailand, the Philippines, and West Africa. In the west of Africa, it grows wild as a perennial and is consumed during the famine (Singh et al. 2023; Tripathi and Vyas 2023;

Table 22.1 Comparison of the nutritional value of different millets with rice, wheat, and corn (Saleh et al. 2013; Muthamilarasan et al. 2016: Sheng et al. 2018)

Nutrient	Pearl millet (Bajra)	Finger millet (Ragi)	Foxtail millet	Sorghum (Jowar)	Barnyard millet	Little millet	Kodo millet	Rice (Brown)	Wheat (Whole)	Corn (Maize)
Calories (kcal)	363	336	351	329	342	207	309	362	348	365
Protein (g)	11.6	7.6	12.3	10.4	6.2	7.7	8.3	2.6	13.2	9.4
Fat (g)	4.2	1.5	4.3	3.1	2.2	4.7	3.6	0.9	2.5	4.7
Carbohydrates (g)	67.0	72.6	60.9	72.6	65.5	67.0	66.6	23.5	71.2	74.3
Fiber (g)	1.2	3.6	8.0	6.7	9.8	7.6	9.0	1.8	12.2	7.3
Calcium (mg)	42	344	31	25	20	17	27	10	34	7
Iron (mg)	8.0	3.9	2.8	4.1	15.2	9.3	0.5	0.4	3.5	2.7
Magnesium (mg)	137	137	81	166	82	114	77	43	138	127
Phosphorus (mg)	296	283	290	289	280	220	188	115	346	210
Potassium (mg)	430	408	150	305	280	220	188	43	363	287
Zinc (mg)	3.1	2.7	2.2	1.7	2.3	3.4	2.6	0.5	2.6	2.2
Thiamin (mg)	0.38	0.42	0.59	0.33	0.33	0.30	0.15	0.07	0.47	0.39
Riboflavin (mg)	0.29	0.19	0.10	0.14	0.10	0.09	0.09	0.03	0.17	0.20
Niacin (mg)	2.8	1.1	3.2	2.1	4.0	3.2	2.0	1.6	5.5	3.6

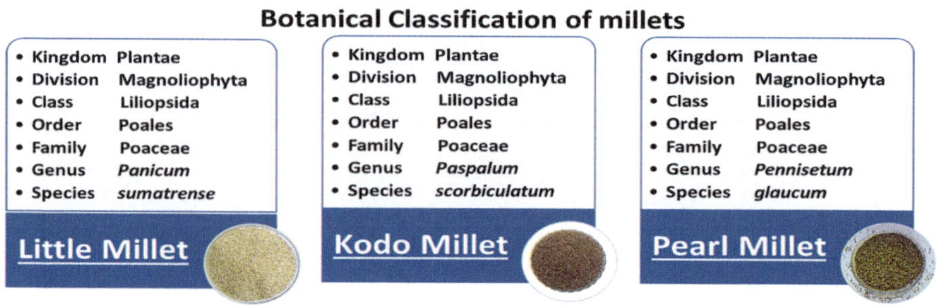

Fig. 22.2 The botanical classification of Little millet, Kodo millet, and Pearl millet

Deshpande et al. 2015). In most of India, it is planted as a minor food crop. As a pasture crop, KM is also cultivated in hot, desert areas of Asia, the United States, and New Zealand. Its frequently grows like a weed in rice fields, but farmers do not pull it out since they can use it as a backup crop in case their main crop fails (Saxena et al. 2018; Bhat et al. 2018). KM is also known as Araka, is believed to have been tamed by humans around 3000 years in the past (Kumar et al. 2016; Arendt and Dal 2011). KM is valued for its nutritional content, and adaptability is receiving more attention in modern diets and sustainable agricultural methods (Sarabhai et al. 2021; Yadav et al. 2013). KM is an excellent choice for individuals suffering from celiac disease or other wheat allergies or intolerances because they don't contain gluten. In addition, they can greatly improve the nutritional security of a sizable portion of the population because of their high fiber, polyphenol, and protein composition (Desai et al. 2010).

22.2.1.3 Pearl Millet (PM)

As a substitute for the primary cereals, pearl millet (PM) *(Pennisetum glaucum),* commonly referred to as bajra (Fig. 22.2), serves as a substantial reservoir of nutritional and health-enhancing constituents. Grown mostly in nations like Nigeria, India, and Sudan, it serves as a fundamental food source for millions in these regions (Merga 2020). Phytate, oxalic acid, and tannins are anti-nutritional components that cause low sustaining consistency and poor mineral availability because of increased lipase activity (Padulosi et al. 2015). Farmers and consumers benefit greatly from this versatile grain (Tripathi and Vyas 2023). Some of the harsh agricultural conditions, such as increases in temperatures, little rainfall, and low soil fertility, are favorable to PM growth. The grain also contains antioxidants, amino acids, and vitamins, all of which contribute to a well-rounded, healthful diet. It is a valuable dietary staple because of its nutritional composition, particularly in places where access to a wide range of nutrient-rich foods is limited (Saleh et al. 2013). Its deep-rooted nature allows it to collect water from deeper soil layers, making it suitable for rain-fed agriculture and drought-resistant (Merga 2020). Other cereals might not be able to survive in sandy or acidic soils, but the crop may flourish in a range of soil types. PM has very short growing cycle—it matures in 70–120 days—making it a reliable crop for farmers dealing with erratic weather patterns (Rao et al. 1986). Further highlighting its importance for sustainable farming practices is its ability to provide decent yields with minimal input.

22.2.2 Phytoactive Constituents of Little Millet, Kodo Millet, Pearl Millet Along with Their Beneficial Effects on Human Health

22.2.2.1 Helps to Fight Against Diabetes

Research findings indicate that individuals who consume millet exhibit a lower prevalence of diabetes (Shobana et al. 2009). The elevated fiber content inherent in millet is responsible for the attenuation of carbohydrate digestion, thereby categorizing it as a food with a low glycemic index (GI) or low-sugar content food. A prolonged blood glucose transport mechanism effectively sustains blood glucose concentrations (Indirani 2021). Cultivars of LM exhibited enhanced suppression of α-amylase and α-glucosidase activity in comparison to those of FLM. The inhibitory capacity of alpha-glucosidase was evaluated in a concentration range of 30–120 μg FAE/ml (bound) and 10–50 μg FAE/ml (soluble) phenolic extracts for both FLM and LM varieties. The bound and soluble fractions of FLM varieties (IC50, 57.26 and 22.37 μg/ml) and LM varieties (IC50, 55.69 and 18.97 μg/ml) exhibited significant suppression of α-glucosidase action. The reduction of α-glucosidase activity observed in these millet cultivars suggests that they may effectively impede glucose uptake in the small intestine, thereby providing a potential therapeutic advantage for diabetes management by mitigating postprandial hyperglycemia (Pradeep and Sreerama 2018). Saini and Verma (2018) executed an investigation aimed at assessing the wound healing efficacy of ethyl acetate and ethanolic extracts obtained from the shade-dried roots of KM in Wistar albino rats suffered streptozocin-induced diabetes and ulcers. The ethyl acetate and ethanol extracts were given in doses of 100 mg/kg, 200 mg/kg, and 400 mg/kg over 21 days. The study concluded a substantial decrease in the level of blood glucose and wound dimensions, suggesting the potential of these extracts as a promising adjunct in the management of diabetic patients with impaired wound healing (Saini and Verma 2018). In another study, Jain et al. (2010) assessed the antidiabetic properties of aqueous and ethanolic extracts derived from the grains of KM, which were administered at dosages of 250 mg and 500 mg/kg body weight to male Wistar rats. Alloxan monohydrate was used to induce diabetes in rats. A significant decrease in fasting blood glucose (FBG) levels (35.14%) was observed in the rats treated with the ethanolic extracts when given a dosage of 500 mg/kg. Additionally, a marked increase in serum insulin levels was recorded, concomitant with a restoration of serum lipid levels toward values approaching normal. The administration of FM or KM (55% by weight of the tested animal) for 28 days resulted in the reversal of weight reduction level in diabetic rats to approximating normality. As anticipated, an elevation in blood glucose and cholesterol levels was documented in the alloxan-induced diabetic rats. Conversely, the dietary inclusion of KM was associated with a reduction in glucose and cholesterol levels. Notably, the KM dietary regimen resulted in a more pronounced decrease in blood glucose and cholesterol levels (42% and 27%, respectively) compared to the FM dietary intervention (36% and 13% decrease). Moreover, blood protein levels were observed to be restored in diabetic rats that were fed with these millet varieties (Hegde et al. 2005). Furthermore, a study involving 40 healthy male rats, categorized into four groups of ten each, was conducted. In this study, groups (A) and (B) served as normal control rats, whereas groups (C) and (D) were classified as diabetic rats, demonstrating the efficacy of pearl millet-based feed in lowering blood glucose levels. Diabetes was induced via intraperitoneal administration of 40 mg/kg streptozocin and was validated by elevated blood glucose levels, which marked day zero. Groups (A and C) received equivalent doses of pearl millet, with parameters assessed from days 0 to 28, revealing a significant decrease in insulin ($P < 0.05$), adiponectin ($P < 0.001$), and blood glucose ($P < 0.001$) levels in diabetic rats within group C. Conversely, groups (B and D) were administered 20% PM, with parameters measured from days 0 to 28, indicating significant reductions in blood glucose and LDL levels ($P < 0.05$), alongside increases in adiponectin and HDL levels in the rats of groups B and D (Sukar et al. 2020).

LM is characterized by a substantial concentration of soluble polyphenols (12.55 μg/g) and insoluble bound polyphenols (8.19 μg/g), which synergistically function alongside apigenin (Pradeep and Sreerama 2018; Dey et al. 2022). Apigenin has been associated with numerous beneficial health outcomes, including its antioxidant capacity, efficacy in combating inflammation and cancer, as well as its antidiabetic attributes. LM serves as a substantial reservoir of apigenin in conjunction with other flavonoids. Research conducted on both animal models and human subjects has demonstrated that LM positively influences lipid and glucose metabolism, inflammatory processes, and oxidative stress (Shah et al. 2021).

22.2.2.2 Millets as Anti-Aging Food

The biochemical relationship between the aldehyde reduction group in sugars and the amino group in proteins, referred to as nonenzymatic glycosylation, is predominantly implicated in the pathophysiology of diabetes and the aging process. Hegde et al. (2005) confirmed that KM represents one among numerous millets characterized by elevated levels of antioxidants and phenolic compounds, including phytates, tannins, and phenols, which may exert a substantial antioxidant influence on aging, metabolic disorders, and overall well-being.

A study by Patil et al. (2020) reported that KM contains 143 ± 0.5 mg of polyphenols (expressed as catechol equivalents) per 100 g and exhibits an antioxidant activity of 5.04 ± 0.01% (equivalent to 50.4 mg tannic acid equivalents per 100 g). These antioxidants can neutralize free radicals, potentially mitigating aging-related oxidative stress.

22.2.2.3 Combating Cancer with Millets

Millets contain a lot of tannins, phenolic acids, and phytes, which are "anti-nutrients." On the other hand, these antinutrients reduce the risk of breast and colon cancer in animals. Chandrasekara and Shahidi in (2010) claimed that millet's phenolics can effectively prevent cancer from growing in both in vitro and in-vivo. Small millet is regarded as a nutritional powerhouse, providing an exceptional source of protein, B vitamins, dietary fiber, zinc, phosphorus, iron, and magnesium. The processes of chemotherapy and radiation therapy can deplete the body's energy reserves, necessitating the provision of adequate vitality to endure each therapeutic intervention. LM contains many antioxidant and phenolic compounds that helps in reversing the adverse effects of therapy (Indirani 2021). Patients who have a genetic susceptibility to celiac disease have immune-mediated enteropathy when they eat gluten (Taylor 2006). Millets present considerable opportunities in the realm of food and beverage production, particularly for individuals who have celiac disease, owing to their gluten-free nature (Ezeogu et al. 2008). As a result, millet grains and their uses may be beneficial for both the manufacturing of celiac food products and the prevention of cancer.

KM (*Paspalum scrobiculatum*) has exhibited significant anticancer capabilities, primarily attributed to its abundant phenolic composition (Gupta et al. 2023). A study conducted by Chandrasekara and Shahidi (2011b) explored the impact of phenolic extracts derived from KM on the proliferation of HT-29 colon cancer cells. The results indicated that at 0.1 mg/ml concentration, hulled KM phenolic extracts achieved a complete inhibition of cellular proliferation. Furthermore, at 0.5 mg/ml concentration, both whole and dehulled KM grain phenolic extracts demonstrated inhibition rates of 100% and 76%, respectively, against HT-29 cells. These findings underscore the potential of phenolic compounds from KM in the realms of cancer prevention and treatment.

Remarkably, a low concentration (0.1 mg/ml) of phenolic extracts derived from dehulled PM grain exhibited 1.3 times greater anticarcinogenic efficacy compared to a higher concentration (0.5 mg/ml). Moreover, it was noted that phenolic compounds obtained from whole PM grain (52.7%) and dehulled grain (43.8%) exhibited more significant inhibitory effects on the HT-29 colon cancer cell line when compared to their KM counterparts (Chandrasekara and Shahidi 2011b).

22.2.2.4 Improves Heart Health

Hundreds of metabolic events that take place in the body depend on the mineral magnesium. Millets can be a boon for maintaining heart rate and blood pressure due to presence of high magnesium content (Srilekha et al. 2019).

High concentrations of iron (8 mg/100 mg) and zinc (3.1 mg/100 g) could contribute to an increase in hemoglobin level (Tako et al. 2015).

Factors such as obesity, tobacco use, poor dietary habits, and insufficient physical activity significantly elevate the risk of myocardial infarction and cerebrovascular accidents. The prevalence of cardiovascular disease remains markedly high and continues to escalate in numerous countries worldwide. Millets demonstrate enhanced free radical scavenging capabilities, which could reduce the chance of cardiovascular illnesses (Hegde and Chandra 2005).

Khare et al. (2020) investigated the polyphenol extracts from FM and KM for their potential to ameliorate metabolic disturbances caused by a high-fat diet and found the presence of catechin in KM with 1.10 ppm concentration, which showed anti-obesity effects.

22.2.2.5 Role in Cardiovascular Diseases

The susceptibility to cardiovascular disease is augmented by factors such as obesity, tobacco consumption, and sedentary lifestyles. LM contains ample quantities of essential minerals such as calcium, magnesium, and potassium, which function as vasodilators and contribute in managing hypertension, atherosclerotic lesions, and the overall risk of stroke associated with cardiovascular diseases (CVDs). The research demonstrated that phenolic extracts (0.05 mg/mL) obtained from LM inhibited the oxidation of low-density lipoprotein (LDL) cholesterol by an extent ranging from 1% to 41% (Dey et al. 2022). Consequently, LM grains and their extracts provide a viable intervention against CVDs.

Kaempferol, classified as a flavonoid, is a naturally occurring compound that is widely distributed across various plant genera (Bangar et al. 2023). It has been proposed that the flavonoid kaempferol possesses antidiabetic, anti-inflammatory, anticancer, and antiaging characteristics (Dey et al. 2022; Shah et al. 2021). According to a scholarly investigation conducted by Dey et al. (2022), LM is characterized by a concentration of 0.6 mg/100 g of kaempferol along with various other flavonoids. These flavonoids possess the capacity to modulate signaling pathways that govern cellular proliferation, inflammatory responses, programmed cell death, and oxidative stress.

The investigation utilized a combination of optimization, gas chromatography-mass spectrometry, and analytical methodologies to elucidate the impact of germination on the phenolic compounds, dietary fibers, mineral content, and gamma-aminobutyric acid levels in KM, revealing that the concentration of campesterol in both raw and germinated KM flour was quantified at 0.31% and 2.60%, respectively, which is associated with cholesterol-lowering properties (Sharma et al. 2017b).

22.2.2.6 Antioxidant Properties of Millets

Millets serve as substantial reservoirs of antioxidants, encompassing polyphenols, phenolic compounds, flavonoids, and tannins. These play a pivotal role in the prevention of lifestyle-related diseases, including diabetes, cardiovascular conditions, obesity, cataracts, cancers, inflammatory disorders, and gastrointestinal complications—all of which represent significant public health challenges globally. Polyphenols are the largest class of phytochemicals with their ability to act as antioxidants, metal chelators, and reducers. Small millet's health benefits include reducing metabolic syndrome, delaying aging, and enhancing immunity. Neuroprotective, antiviral, anticarcinogenic, and anti-inflammatory effects are among the bioactive characteristics of polyphenols (Bhat et al. 2018).

In 2018, Revankar et al. determined that PM exhibits the highest total phenolic and flavonoid content, measuring 106.86 ± 2.57 mg GAE/100 g flour and 40.22 ± 2.41 mg/100 g flour, respectively. The DPPH radical scavenging activity was found to be maximal for PM at 70.39 ± 2.57% and minimal for LM at 39.87 ± 2.0%. PM varieties CO (CU) 9 and ICMV 221 demonstrated the highest reducing capacities, recorded at 0.513 and 0.560 mg AAE/g, respectively. Moreover, PM ICMV 221 exhibited the greatest metal-chelating activity at 88.45% in comparison to other millet varieties. There was a moderate connection with lowering power ($R = 0.55456$), and a strong correlation was observed in DPPH radical scavenging activity ($R = 0.97618$) and TFC. The TPC also displayed a moderate correlation with DPPH radical scavenging activity. These results indicate that both PM and LM serve as rich sources of phenolic antioxidants, suggesting their potential utility in mitigating diseases associated with free radical activity. Another study demonstrated that varying concentrations of 10% v/v, 20% v/v, 30% v/v, 40% v/v, and 50% v/v correspond to an increase in the reducing capacity of the extract derived from pearl millet. The findings indicate a progressive enhancement in the ability of PM extract to reduce Fe^{3+} to Fe^{2+}. The TPC for PM at the specified concentrations was recorded as 0.19, 0.46, 0.85, 1.10, and 1.50, evidencing an increase in phenolic content concomitant with elevated extract concentrations from 10% v/v to 50% v/v. Elevated concentrations suggest a positive correlation with higher phenolic contents within the pearl millet extracts. This study also shows how PM extracts scavenges free radicals. The PM extracts' capacity to scavenge free radicals rises from 33.33 to 93.33 as the concentration rises from 10% v/v to 50% v/v (Odusola et al. 2013). Plants with antioxidant properties have free radical scavenging activity, which contributes to severe diseases like cancer, renal failure, diabetes, liver diseases, and degenerative disorders. (Aiyegoro and Okoh 2009; Das and Pereira 1990). This study further elucidates the free radical scavenging capabilities of PM extracts, highlighting their pronounced activity and potential role as antioxidants within dietary regimens.

22.2.2.7 Anti-microbial Properties of Millets

The study isolated 19 endophytic *Bacillus* strains from pearl millet that exhibit antagonistic properties against pathogens such as *Rhizoctonia solani, Fusarium solani,* and *Sclerotium rolfsii*. Eight antagonistic endophytes were identified, including *B. amyloliquefaciens, Bacillus cereus, and Bacillus subtilis subsp. subtilis*. The isolates demonstrated variability in their production of indole acetic acid and siderophores, with their antifungal characteristics substantiated by the presence of three antimicrobial peptide genes (ituD, srfA, and bmyC). Additionally, under greenhouse circumstances, a singular application of three promising strains provided protection ranging from 35.68% to 45.74%. Nonetheless, the microbial consortium significantly increased plant biomass while offering optimal protection (71.96%) against root rot and wilt disease. Overall, the findings imply that pearl millet plants harbor diverse endophytic *Bacillus* species that possess notable biocontrol and plant growth promotion capabilities (Kushwaha et al. 2020).

22.3 Traditional Uses of Millets

Millets are abundant in essential vitamins, phytochemicals, minerals, and sulfur-containing amino acids, thus gained the designation of "Nutri cereals." Additionally, millets are advocated for those with celiac disease because of their gluten-free nature (Chandrasekara and Shahidi 2010). Cereals, particularly those derived from millet, enjoy widespread popularity globally and maintain their status as a dietary staple in numerous African nations (Amadou et al. 2011; Obilana and Manyasa 2002). KM and its flour are predominantly utilized in the preparation of porridge combined with milk and various beverages. Traditional culinary practices involving KM encompass the preparation of dosa, breads, idlis, chaklis, dosa, papads, porridges, infant formulations, and various snacks (Chandrasekara and Shahidi 2011a). In regions such as Madhya Pradesh and Chhattisgarh, KM is regarded as a staple grain, frequently consumed as a substitute for rice (Kaushik et al. 2022). In Haryana, the dish "Kodo ki Roti," crafted from KM flour, is a celebrated culinary offering typically served with ghee, yogurt, or vegetables (Siroha and Bangar 2024). In the Himalayan foothills, millet is utilized as a cereal, incorporated into soups, and transformed into dense, whole-grain bread known as Chapatti. A traditional alcoholic beverage, referred to as "Kodo ko jaanr," is produced in the Darjeeling hills and Sikkim regions of India, utilizing the dry seeds of KM. Koozh is a fermented beverage produced using flour derived from PM or FM combined with rice, commonly consumed by ethnic groups in Tamil Nadu (Ilango and Antony 2014). Additionally, porridge, samai dosa, paysam, and paddu are traditional dishes prepared from LM across different millet-cultivating regions in India (Jaybhaye et al. 2014). Despite the capability to prepare numerous traditional dishes within domestic settings, the limited commercialization of millet serves to dissuade farmers from cultivating these grains (Subramanian and Viswanathan 2003). Consequently, there exists a pressing necessity for the global community to develop industrial applications utilizing millets.

22.4 Incorporating Millets into Everyone's Daily Diet

Incorporating millets into everyone's daily diet can enhance everyone's nutrition, provide essential nutrients like fiber, protein, vitamins, and minerals while offering delightful taste and tecture to everyone's meals.

Several ways are there to incorporate millets into daily diet as shown in Fig. 22.3 (Vyer 2024).

22.5 Conclusion

Historically, millets have been integral as a crucial cereal source due to their nutritional and health-promoting attributes. The varieties of millets discussed in this chapter, including, LM, KM and PM, present significant potential for application as raw materials within the food and pharmaceutical industries. Although compared with other popular cereals like rice, wheat, and maize, millets have more health advantages and nutritional value. The existing technological advancements are insufficient in enhancing their acceptance among consumers regarding quality, nutritional value, and functional attributes. The food processing industry now offers several options to produce millet-based, high-quality products that enhance the nutritional value of everyone's meal and provide a delightful taste to food. Recent progress in post-harvest and value-added processes has augmented consumer acceptance of millet-based food products. Enhanced nutritional awareness is compelling the food industry to innovate new food products with unique characteristics that can contribute positively to public health. The dietary constituents present in these millet varieties, particularly polyphenols, exhibit a spectrum of health-promoting effects, including antimicrobial, antioxidant, antidiabetic, and cholesterol-lowering properties, thereby offering protective measures against diet-related diseases,

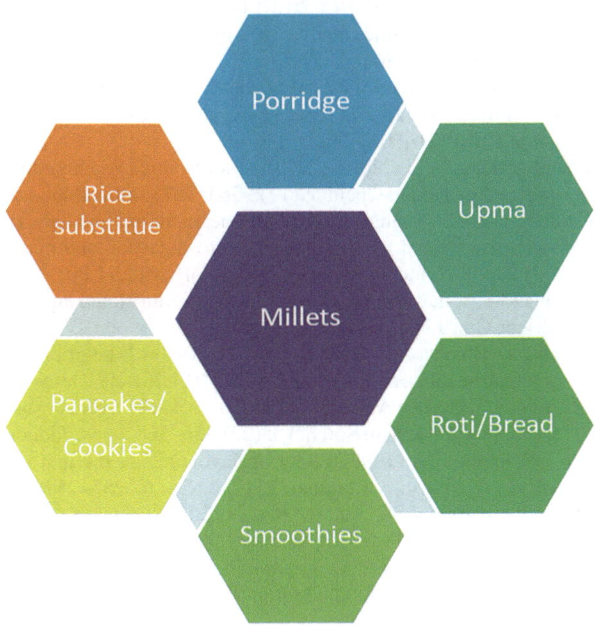

Fig. 22.3 The different ways to incorporate millets into daily diet

particularly within rural communities. Thus, due to the abundance of various bioactive compounds, millets are adeptly utilized in the creation of nutraceuticals, functional foods, and pharmaceutical products, which can be consumed safely without any health concerns. This chapter also provides an overview of the medicinal values of LM, KM, and PM.

References

Aiyegoro OA, Okoh AI (2009) Phytochemical screening and polyphenolic antioxidant activity of aqueous crude leaf extract of *Helichrysum pedunculatum*. Int J Mol Sci 10:4990–5001. https://doi.org/10.3390/ijms10114990

Amadou I, Gbadamos OS, Le GW (2011) Millet based traditional processed foods and beverages—a review. Cereal Foods World 56(3):115–121. https://doi.org/10.1094/CFW-56-3-0115

Arendt E, Dal Bello F (eds) (2011) Gluten-free cereal products and beverages. Elsevier

Azarbad HR, Mazaheri TM, Rashidi H (2019) Optimization of gluten-free bread formulation using sorghum, rice, and millet flour by D-optimal mixture design approach. J Agric Sci Technol 21(1):101–115

Bangar SP, Chaudhary V, Sharma N et al (2023) Kaempferol: a flavonoid with wider biological activities and its applications. Crit Rev Food Sci Nutr 63(28):9580–9604. https://doi.org/10.1080/10408398.2022.2067121

Bhat S, Nandini C, Tippeswamy V, Prabhakar P (2018) Significance of Small Millets In Nutrition And Health -A Review. Asian J Dairy Food Res 37(1):35–40. https://doi.org/10.18805/ajdfr.DR-1306

Bisht A, Joshi S, Srivastava S (2022) Small millets: An overview. In: Srivastava Sarita (ed) Small Millet Grains: The Superfoods in Human Diet. Springer, Singapore, p 1–3. https://doi.org/10.1007/978-981-16-9306-9

Chandrasekara A, Shahidi F (2010) Content of insoluble bound phenolics in millets and their contribution to antioxidant capacity. J Agric Food Chem 58(11):6706–6714. https://doi.org/10.1021/jf100868b

Chandrasekara A, Shahidi F (2011a) Inhibitory activities of soluble and bound millet seed phenolics on free radicals and reactive oxygen species. J Agric Food Chem 59(1):428–436. https://doi.org/10.1021/jf103896z

Chandrasekara A, Shahidi F (2011b) Bioactivities and antiradical properties of millet grains and hulls. J Agric Food Chem 59(17):9563–9571. https://doi.org/10.1021/jf201849d

Chandrasekara A, Shahidi F (2012) Bioaccessibility and antioxidant potential of millet grain phenolics as affected by simulated in vitro digestion and microbial fermentation. J Funct Foods 4(1):226–237. https://doi.org/10.1016/j.jff.2011.11.001

Chauhan M, Sonawane SK, Arya SS (2018) Nutritional and nutraceutical properties of millets: a review. Clin J Nutr Diets 1(1):1–10

Chethan S, Bhavya ML, Sowbhagya HB (2018) Millet grains: nutritional quality, processing, and potential health benefits. Food Rev Int 34(5):461–481. https://doi.org/10.1111/1541-4337.12012

Choudhary S, Boruah A, Ram N et al (2023) Millet's role in sustainable agriculture: a comprehensive review. Int J Plant Soil Sci 35(22):556–568. https://doi.org/10.9734/ijpss/2023/v35i224165

Das NP, Pereira TA (1990) Effect of flavonoids on thermal auto oxidation of palm oil structure activity relationship. J Animal Oil Chem Soc 67:255–258. https://doi.org/10.1007/BF02540652

Desai AD, Kulkarni SS, Sahoo AK et al (2010) Effect of supplementation of malted ragi flour on the nutritional and sensorial quality characteristics of cake. Adv J Food Sci Technol 2(1):67–71

Deshpande SS, Mohapatra D, Tripathi MK, Sadvatha RH (2015) Kodo millet-nutritional value and utilization in Indian foods. J Grain Process Storage 2(2):16–23

Dey S, Saxena A, Kumar Y et al (2022) Understanding the antinutritional factors and bioactive compounds of kodo millet (*Paspalum scrobiculatum*) and little millet (*Panicum sumatrense*). J Food Qual 2022(1):1578448. https://doi.org/10.1155/2022/1578448

Edge MS, Jones JM, Marquart L (2005) A new life for whole grains. J Am Diet Assoc 105(12):1856–1860. https://doi.org/10.1016/j.jada.2005.10.022

Ezeogu LI, Duodu KG, Emmambux MN, Taylor JR (2008) Influence of cooking conditions on the protein matrix of sorghum and maize endosperm flours. Cereal Chem 85(3):397–402. https://doi.org/10.1094/CCHEM-85-3-0397

Gupta M, Asfaha DM, Ponnaiah G (2023) Millets: a nutritional powerhouse with anti-cancer potential. Cureus 15(10). https://doi.org/10.7759/cureus.47769

Hegde PS, Chandra TS (2005) ESR spectroscopic study reveals higher free radical quenching potential in Kodo millet (*Paspalum scrobiculatum*) compared to other millets. Food Chem 92(1):177–182. https://doi.org/10.1016/j.foodchem.2004.08.002

Hegde PS, Rajasekaran NS, Chandra TS (2005) Effects of the antioxidant properties of millet species on oxidative stress and glycemic status in alloxan-induced rats. Nutr Res 25(12):1109–1120. https://doi.org/10.1016/j.nutres.2005.09.020

Ilango S, Antony U (2014) Assessment of the microbiological quality of koozh, a fermented millet beverage. Afr J Microbiol Res 8(3):308–312. https://doi.org/10.5897/AJMR2013.6482

Indirani K (2021) Review on nutritional profiles and health benefits of little millets–India. Int J Res Eng Sci 9(11):07–11

Jain S, Bhatia G, Barik R et al (2010) Antidiabetic activity of *Paspalum scrobiculatum* Linn. in alloxan-induced diabetic rats. J Ethnopharmacol 127:325–328. https://doi.org/10.1016/j.jep.2009.10.038

Jaybhaye RV, Pardeshi IL, Vengaiah PC, Srivastav PP (2014) Processing and technology for millet based food products: a review. J Ready Eat food 1(2):32–48

Kaushik NI, Chauhan KO, Aggarwal MA, Kumar RA (2022) State-of-the-art of knowledge on underutilized millets: Kodo and kutki, grown in tribal areas of India. Int J Agric Sci Res 12(1):35–56

Kavitha U, Nihar SS, Kader MM (2024) The therapeutic potential of millets: a comprehensive review. J Res Siddha Med 7(Suppl 1):S29–S35. https://doi.org/10.4103/jrsm.jrsm_15_23

Khare P, Maurya R, Bhatia R et al (2020) Polyphenol rich extracts of finger millet and kodo millet ameliorate high fat diet induced metabolic alterations. Food Funct 11(11):9833–9847. https://doi.org/10.1039/D0FO01643H

Kumar D, Patel S, Naik RK, Mishra NK (2016) Study on physical properties of indira kodo-I (*Paspalum scrobiculatum* L.) millet. Int J Eng Res Technol 5(1):39–45. https://doi.org/10.17577/IJERTV5IS010092

Kumari P, Singh YV, Yadav R et al (2022) Analytical study of little millets and therapeutic effect of little millets based on its components in type 2 diabetes mellitus. Int J Herb Med 10(5):11–14. https://doi.org/10.22271/flora.2022.v10.i5a.832

Kushwaha P, Kashyap PL, Srivastava AK, Tiwari RK (2020) Plant growth promoting and antifungal activity in endophytic Bacillus strains from pearl millet (*Pennisetum glaucum*). Braz J Microbiol 51(1):229–241. https://doi.org/10.1007/s42770-019-00172-5

Maitra S, Shankar T (2019) Agronomic management in little millet (*Panicum sumatrense* L.) for enhancement of productivity and sustainability. Int J Bioresour Sci 6(2):91–96. https://doi.org/10.30954/2347-9655.02.2019.9

Manach C, Mazur A, Scalbert A (2005) Polyphenols and prevention of cardiovascular diseases. Curr Opin Lipidol 16(1):77–84. https://doi.org/10.1097/00041433-200502000-00013

Malleshi NG (1989) Small millets for food and industrial uses. In: Seetharam A, Riley KW, Harinarayana G (ed) Proceedings of the First International Small Millets Workshop, Bangalore India, November 1986.Oxford & IBH Publishing Co. Pvt. Ltd., India, p 325–39

Merga D (2020) Pearl millet (*Pennisetum glaucum* L.) breeding for adaptation and performance under drought condition. Markers (Tanksley, 1993 and Mohan et al., 1997). J Environ Earth Sci 10(4):1–10

Michaelraj PSJ, Shanmugam A (2013) A study on millets-based cultivation and consumption in India. Int J Marketing Financ Serv Manag Res 2(4):49–58

Muthamilarasan M, Dhaka A, Yadav R, Prasad M (2016) Exploration of millet models for developing nutrient rich graminaceous crops. J Plant Sci 242:89–97

Nithiyanantham S, Kalaiselvi P, Mahomoodally MF et al (2019) Nutritional and functional roles of millets—a review. J Food Biochem 43(7):e12859. https://doi.org/10.1111/jfbc.12859

Obilana AB, Manyasa E (2002) Millets. In: Pseudocereals and less common cereals. Springer, Berlin/Heidelberg, pp 177–217

Odusola KB, Ilesanmi FF, Akinloye OA (2013) Assessment of nutritional composition and antioxidant ability of pearl millet (*Pennisetum glaucum*). Am J Res Commun 1(6):262–272

Padulosi S, Mal B, King OI, Gotor E (2015) Minor millets as a central element for sustainably enhanced incomes, empowerment, and nutrition in rural India. Sustainable 7(7):8904–8933. https://doi.org/10.3390/su7078904

Patil RB, Vijayalakshmi KG, Vijayalakshmi D (2020) Physical, functional, nutritional, phytochemical and antioxidant properties of kodo millet (*Paspalum scrobiculatum*). J Pharmacogn Phytochem 9(5):2390–2393

Pradeep PM, Sreerama YN (2018) Phenolic antioxidants of foxtail and little millet cultivars and their inhibitory effects on α-amylase and α-glucosidase activities. Food Chem 247:46–55. https://doi.org/10.1016/j.foodchem.2017.11.103

Rajput JS, Thakur AK, Nag NK et al (2022) Effect of nano fertilizer in relation to growth, yield and economics of little millet (*Panicum sumatrense* roth) under rainfed conditions. Pharm Innov 11(7):153–156

Rao DS, Deosthale YG (1983) Mineral composition, ionisable iron and soluble zine in malted grains of pearl millet and ragi. Food Chem 11(3):217–223. https://doi.org/10.1016/0308-8146(83)90104-8

Rao SA, Mengesha MH, Sibale PK, Reddy R (1986) Collection and evaluation of pearl millet (Pennisetum) germplasm from Malawi. Econ Bot 40:27–37. https://doi.org/10.1007/BF02858943

Rao BR, Nagasampige MH, Ravikiran M (2011) Evaluation of nutraceutical properties of selected small millets. J Pharm Bioallied Sci 3(2):277–279. https://doi.org/10.4103/0975-7406.80775

Rao B, Bhaskarachary K, Arlene Christina GD et al (2017) Nutritional and health benefits of millets. ICAR IIMR Rajendranagar, Hyderabad

Rathore S, Singh K, Kumar V (2016) Millet grain processing, utilization and its role in health promotion: a review. Int J Nutr Food Sci 5(5):318–329. https://doi.org/10.11648/ijnfs.20160505.12

Revankar J, Ankita Shamnani DK, Podili K (2018) Antioxidant activities of Pearl millet (*Pennisetum glaucum*) and Little millet (*Panicum sumatrense*) in different in vitro models. Int J Bioassays 7:5595–5601. https://doi.org/10.21746/ijbio.2018.7.2.5

Saini P, Verma PK (2018) The wound healing potential of *Paspalum scrobiculate* Linn. in streptozotocin-induced diabetic rats. Int J Pharm Biol Arch 9(4):237–241. https://doi.org/10.22377/ijpba.v9i04.1717

Saleh AS, Zhang Q, Chen J, Shen Q (2013) Millet grains: nutritional quality, processing, and potential health benefits. Compr Rev Food Sci Food Saf 12(3):281–295. https://doi.org/10.1111/1541-4337.12012

Sarabhai S, Tamilselvan T, Prabhasankar P (2021) Role of enzymes for improvement in gluten-free foxtail millet bread: it's effect on quality, textural, rheological and pasting properties. LWT 137:110365. https://doi.org/10.1016/j.lwt.2020.110365

Saxena R, Vanga SK, Wang J et al (2018) Millets for food security in the context of climate change: a review. Sustainable 10:1–31. https://doi.org/10.3390/su10072228

Scalbert A, Manach C, Morand C et al (2005) Dietary polyphenols and the prevention of diseases. Crit Rev Food Sci Nutr 45:287–306. https://doi.org/10.1080/1040869059096

Shah P, Kumar A, Kumar V, Tripathi MK (2021) Millets, phytochemicals, and their health attributes. In: Kumar A, Tripathi MK, Joshi D, Kumar V (eds) Millets and millet technology. Springer, Singapore. https://doi.org/10.1007/978-981-16-0676-2_9

Shahidi F, Chandrasekara A (2013) Millet grain phenolics and their role in disease risk reduction and health promotion: a review. J Funct Foods 5(2):570–581. https://doi.org/10.1016/j.jff.2013.02.004

Sharma S, Sharma N, Handa S, Pathania S (2017a) Evaluation of health potential of nutritionally enriched Kodo millet (*Eleusine coracana*) grown in Himachal Pradesh, India. Food Chem 214:162–168. https://doi.org/10.1016/j.foodchem.2016.07.086

Sharma S, Saxena DC, Riar CS (2017b) Using combined optimization, GC–MS and analytical technique to analyze the germination effect on phenolics, dietary fibers, minerals and GABA contents of Kodo millet (*Paspalum scrobiculatum*). Food Chem 233:20–28. https://doi.org/10.1016/j.foodchem.2017.04.099

Sheng S, Li T, Liu R (2018) Corn phytochemicals and their health benefits. Food Sci Hum Wellness 7(3):185–195

Shobana S, Sreerama YN, Malleshi NG (2009) Composition and enzyme inhibitory properties of finger millet (*Eleusine coracana* L.) seed coat phenolics: mode of inhibition of a-glucosidase and pancreatic amylase. Food Chem 115(4):1268–1273. https://doi.org/10.1016/j.foodchem.2009.01.042

Singh N, Kumari P, Bassi JG (2023) Nutrient-rich Kodo millet, importance and value addition: an overview. Pharma Innov J 12:3713–3719

Siroha AK, Bangar SP (2024) Millet-based food products: an overview. Curr Food Sci Technol Rep 2(2):213–220. https://doi.org/10.1007/s43555-024-00031-x

Srilekha K, Kamalaja T, Uma Maheswari K, Neela Rani R (2019) Nutritional composition of little millet flour. Int Res J Pure App Chem 20(4):1–4. https://doi.org/10.9734/irjpac/2019/v20i430140

Subramanian S, Viswanathan R (2003) Thermal properties of minor millet grains and flours. Biosyst Eng 84(3):289–296. https://doi.org/10.1016/S1537-5110(02)00222-2

Sukar KA, Abdalla RI, Humeda HS, Alameen AO, Mubarak EI (2020) Effect of pearl millet on glycaemic control and lipid profile in streptozocin induced diabetic wistar rat model. Asian J Med Health 18:40–51. https://doi.org/10.9734/ajmah/2020/v18i330193

Tako E, Reed SM, Budiman J et al (2015) Higher iron pearl millet (*Pennisetum glaucum* L.) provides more absorbable iron that is limited by increased polyphenolic content. Nutr J 14:1–9. https://doi.org/10.1186/1475-2891-14-11

Taylor SE (2006) Tend and befriend: biobehavioral bases of affiliation n under stress. Curr Dir Psychol Sci 15(6):273–277. https://doi.org/10.1111/j.1467-8721.2006.00451.x

Tripathi T, Vyas S (2023) From ancient grains to modern solutions: a history of millets and their significance in agriculture and food security. Int J Home Sci 9:72–78

Vetriventhan M, Azevedo VC, Upadhyaya HD et al (2020) Genetic and genomic resources, and breeding for accelerating improvement of small millets: current status and future interventions. Nucleus 63:217–239. https://doi.org/10.1007/s13237-020-00322-3

Vyer (2024) How to Incorporate Millets into Your Daily Diet? Available via DIALOG. https://www.rootsveyr.com/blogs/how-to-incorporate-millets-into-your-dailydiet/?srsltid=AfmBOoqRXcCjW2JO5nPzgYwAVW4Emzmbl14xoYQjtug6YppGl1Tl1EF. Accessed 31 July 2024

Yadav N, Chaudhary K, Singh A, Gupta A (2013) Evaluation of hypoglycemic properties of kodo millet-based food products in healthy subjects. IOSR J Pharm 3(2):14–20. https://doi.org/10.9790/3013-32201420

Chapter 23
Medicinal Value of Finger Millet, Sorghum Millet, and Foxtail Millet

Sadaf Hashmi, Prabhakar Vishvakarma, Gazala Parveen, Nawal Kishor, and Neeru Sharma

23.1 Introduction

Millets, including finger millet (*Eleusine coracana*), sorghum millet (*Sorghum bicolor*), and foxtail millet (*Setaria italica*), are ancient grains renowned for their exceptional medicinal and nutritional value (Nithya et al. 2023; Jacob et al. 2024). These crops have been staples in arid and semi-arid regions of Asia and Africa for centuries, offering a rich source of essential nutrients, phytochemicals, and therapeutic properties that address various health challenges (Jacob et al. 2024).

Finger millet stands out for its high calcium content, making it ideal for bone health (Sahaya Rani et al. 2021). It is also rich in dietary fibers, proteins, and polyphenols, which contribute to its antioxidant properties (Banerjee et al. 2012). These antioxidants protect against oxidative stress-related diseases such as diabetes, cardiovascular disorders, neurodegenerative conditions, and certain types of cancer (Jayawardana et al. 2021). Finger millet has demonstrated potential in managing diabetes by improving antioxidant enzyme activity and reducing glycation complications (Agrawal et al. 2023). Additionally, its insoluble fiber helps prevent

S. Hashmi
Axis Institute of Pharmacy, Kanpur, Uttar Pradesh, India

P. Vishvakarma
Department of Pharmaceutics, Sharda School of Pharmacy, Sharda University Agra, Agra, Uttar Pradesh, India

G. Parveen
College of Pharmacy, Knowledge University, Erbil, Kurdistan Region, Iraq

N. Kishor
Department of Chemistry, Gurukula Kangri (Deemed to be University), Haridwar, India

N. Sharma (✉)
Government of Haryana, Chandigarh, Haryana, India

© The Author(s), under exclusive license to Springer Nature Switzerland AG 2026

J. Singh, M. Kumar Malik (eds.), *Millets and Millet Starch: Chemical Composition, Functional Modification, and Applications*,
https://doi.org/10.1007/978-3-032-09502-2_23

gastrointestinal disorders and colon cancer, while its slowly digested carbohydrates help regulate blood sugar levels (Gupta et al. 2023).

Sorghum millet is recognized for its gluten-free nature and significant bioactive compounds like phenolics and flavonoids (Kulamarva et al. 2009). These compounds exhibit anti-inflammatory and antioxidant effects, reducing the risk of chronic diseases such as cancer and cardiovascular issues (Kumar and Kotwal 2023). Sorghum's α-kafirin has shown promise in lowering blood pressure by inhibiting angiotensin-converting enzymes (Khalid et al. 2022). Its dietary fiber supports gut health by promoting beneficial microbiota and preventing metabolic disorders like obesity and diabetes (Stefoska-Needham 2024).

Foxtail millet is particularly valued for its role in managing hypertension and diabetes (Agrawal et al. 2023). Its prolamins inhibit α-amylase activity, aiding in blood sugar regulation (Zhang et al. 2024). Foxtail millet's bran-derived polyphenols exhibit anticancer properties by inducing apoptosis in colorectal cancer cells (Shan et al. 2021). Furthermore, it helps alleviate nonalcoholic fatty liver disease (NAFLD) through improved lipid metabolism and gut microbiota modulation (Ghani et al. 2024). The grain's antioxidant capacity also contributes to wound healing, antiaging effects, and protection against gastric ulcers (Abedin et al. 2022).

Millets collectively offer a low glycemic index, making them suitable for individuals with diabetes or those seeking weight management solutions (Agrawal et al. 2023). Their nutrient density, rich in vitamins (thiamine, niacin) and minerals (iron, phosphorus), addresses deficiencies like anemia and calcium depletion, especially in vulnerable populations such as pregnant women and children (Kumar et al. 2016). As climate-resilient crops, millets also provide sustainable solutions to global food security challenges while promoting health through their nutraceutical potential (Jacob et al. 2024). Table 23.1 presents the total, soluble, and insoluble dietary fiber content (g/100 g) of various millet varieties. Finger millet shows the highest fiber content, while foxtail and little millet are rich in soluble fiber. These values highlight the nutritional diversity of millets as functional foods. The global rise of

Table 23.1 Dietary fiber content and composition of soluble and insoluble dietary fiber in different millet varieties (Jacob et al. 2024)

Millet variety	Fiber content (g/100 g)	Fiber type and composition (approx. g/100 g)	
		Soluble DF*	Insoluble DF**
Foxtail millet	8.00	5.87	11.06
Pearl millet	8.00	1.91	9.85
Barnyard millet	6.00	4.15	8.19
Kodo millet	9.00	3.85	13.13
Little millet	7.00	5.65	8.57
Proso millet	3.00	2.05	9.48
Finger millet	15.00	1.40	15.70
Brown top millet	9.00	–	–

* = soluble dietary fiber.
** = insoluble dietary fiber

lifestyle diseases, particularly malnutrition, diabetes, and obesity, has become a pressing public health crisis (Dhanushkodi et al. 2023). Noncommunicable diseases (NCDs) account for 74% of global deaths, with diabetes affecting nearly 10% of adults (Varghese et al. 2025). In 2021, NCDs caused 43 million deaths, highlighting the prevalence of conditions linked to poor dietary habits (Li et al. 2025a, b).

Malnutrition presents a dual challenge: over 2 billion people face micronutrient deficiencies while 1.9 billion adults are overweight or obese. The shift toward energy-dense, nutrient-poor diets, driven by the accessibility of processed foods, exacerbates this issue (Passarelli et al. 2024). Poor dietary choices lead to obesity, increasing the risk of type 2 diabetes due to insulin resistance and chronic inflammation. Micronutrient deficiencies further impair metabolic health (Li et al. 2025a, b).

Addressing these interconnected issues requires promoting functional foods that offer health benefits beyond basic nutrition (Sahu et al. 2024). Foods rich in bioactive compounds can help prevent lifestyle diseases, including whole grains, fruits, vegetables, nuts, and fatty fish (Dixit et al. 2023). The Mediterranean diet exemplifies how to incorporate functional foods effectively (Bavaro et al. 2024).

However, barriers to adopting functional foods exist, particularly in low- and middle-income countries where access to fresh produce is limited (Ng et al. 2022). Economic factors and aggressive marketing of unhealthy foods overshadow healthier options. To overcome these challenges, governments and healthcare organizations must implement policies that promote food security and dietary diversity while educating communities about nutrition and functional foods (Muthayya et al. 2013). Collaborative efforts across public health sectors and food industries are essential for fostering healthier eating habits (Bavaro et al. 2024; Mozaffarian 2016).

The chapter will focus on three specific millets: finger millet, sorghum, and foxtail millet, each of which holds significant nutritional and medicinal potential that warrants exploration in the context of modern dietary needs and health challenges (Fig. 23.1).

Fig. 23.1 Inflorescence of finger millet, sorghum millet and foxtail millet

23.2 Finger Millet (*Eleusine coracana*): A Powerhouse of Health

Finger millet, scientifically known as *Eleusine coracana*, is emerging as a powerhouse of health due to its impressive nutritional profile and numerous health benefits (Kumar et al. 2016). This ancient grain has been a staple food in various regions, particularly in India and Africa, for centuries (Tripathi and Vyas 2023). Its rich composition includes essential nutrients such as calcium, iron, protein, and dietary fiber, making it an ideal choice for individuals seeking to improve their overall health and manage specific health conditions (Devi et al. 2014).

One of the standout features of finger millet is its exceptionally high calcium content, which can be up to three times more than that found in milk (Anitha et al. 2021a, b). This makes it particularly beneficial for populations at risk of osteoporosis, including postmenopausal women and children who require adequate calcium for bone development (Puranik et al. 2017; Sahaya Rani et al. 2021). Regular consumption of finger millet can significantly contribute to bone health and prevent conditions associated with calcium deficiency (Sahaya Rani et al. 2021; Warhade et al. 2023).

In addition to its mineral wealth, finger millet is rich in dietary fiber, which plays a crucial role in digestion (Gaikwad et al. 2024). The fiber content in finger millet consists of both soluble and insoluble forms, aiding bowel regularity and preventing gastrointestinal disorders (Devi et al. 2014; Arachchige et al. 2019). Soluble fiber helps lower cholesterol levels by trapping fats in the digestive tract, while insoluble fiber promotes satiety and aids in weight management by slowing digestion and reducing hunger pangs (Ahamed et al. 2024). This dual action makes finger millet an excellent food choice for individuals looking to control their weight or manage diabetes, as it has a low glycemic index that helps regulate blood sugar levels (Dhanushkodi et al. 2023).

Furthermore, finger millet is packed with antioxidants, particularly polyphenols that exhibit significant health benefits (Devi et al. 2014). These compounds are known to combat oxidative stress caused by free radicals, which can lead to chronic diseases such as cancer and cardiovascular disorders (Jayawardana et al. 2021; Kumar et al. 2021). Research has shown that the phenolic content in finger millet is considerably higher than that in other grains like rice and wheat, providing enhanced protective effects against various degenerative diseases (Kumari et al. 2016).

The nutritional benefits of finger millet extend beyond its macro- and micronutrient content (Devi et al. 2014; Sirin 2023). It is also gluten-free, making it an excellent alternative for individuals with coeliac disease or gluten intolerance (Mudau et al. 2022). This characteristic allows a broader audience to incorporate finger millet into their diets without adverse effects (Sood et al. 2017). Moreover, finger millet's versatility in culinary applications enhances its appeal as a health food. It can be prepared in various forms such as porridge, rotis (flatbreads), dosas (fermented pancakes), and even baked goods making it easy to incorporate into daily meals.

This adaptability not only preserves its nutritional benefits but also brings diverse flavors and textures to diets (Nayak 2024; Sood et al. 2017).

23.2.1 Nutritional Profile of Finger Millet

Finger millet (*Eleusine coracana*) is recognized for its exceptional nutritional profile, making it a valuable addition to diets aimed at improving health and preventing chronic diseases (Devi et al. 2014). This ancient grain is particularly noteworthy for its high content of calcium, iron, fiber, and essential amino acids, which contribute to its status as a nutritional powerhouse (Gaikwad et al. 2024).

23.2.1.1 Calcium

Finger millet is the richest source of calcium among cereal grains, containing approximately 344 mg per 100 grams (Abioye et al. 2022). This concentration is significantly higher than other grains like rice and wheat and about three times more than milk (Anitha et al. 2021a, b). The high calcium content is crucial for bone health, making it an excellent choice for populations at risk of osteoporosis, particularly children and women (Puranik et al. 2017). Adequate calcium intake is essential for maintaining strong bones and teeth, as well as proper muscle function and nerve transmission (Anitha et al. 2024).

23.2.1.2 Iron

In addition to its calcium content, finger millet is rich in iron, providing around 87% of the daily value (DV) per 100 grams (Anitha et al. 2021a, b). This makes it an important food source for preventing iron deficiency anemia, which is prevalent among women and children (Parikh et al. 2024). The bioavailability of iron from finger millet can be enhanced through proper cooking methods and dietary practices, making it a practical option for improving iron status in vulnerable populations (Sood et al. 2017).

23.2.1.3 Fiber

Finger millet is an excellent source of dietary fiber; it contains approximately 22.6 grams per 100 grams (Devi et al. 2014). This high fiber content promotes digestive health by encouraging regular bowel movements and preventing constipation (Kumar et al. 2016). Additionally, soluble fiber in finger millet helps regulate blood sugar levels by slowing glucose absorption, which benefits individuals with diabetes

(Agrawal et al. 2023; Ahamed et al. 2024). Fiber also aids in weight management by increasing satiety and reducing overall calorie intake (Krishnan 2024).

23.2.1.4 Amino Acids

Finger millet boasts a balanced profile of essential amino acids, including lysine, methionine, isoleucine, leucine, and phenylalanine (Kumar et al. 2016). These amino acids are crucial for various bodily functions, such as protein synthesis, immunity, and hormone production (Sood and Mohiuddin Bhat 2023; Verma 2024). Unlike many cereals that lack lysine, finger millet offers a more complete amino acid profile, making it an excellent option for those on plant-based diets (Devi et al. 2014).

23.2.2 Key Bioactive Compounds in Finger Millet

Finger millet (*Eleusine coracana*) is increasingly recognized for its nutritional benefits and a rich array of bioactive compounds that enhance its health-promoting properties (Pandey et al. 2023). Compounds such as polyphenols, flavonoids, phytosterols, and antioxidants play pivotal roles in making finger millet a valuable addition to diets aimed at improving health and preventing chronic diseases (Devi et al. 2014).

23.2.2.1 Polyphenols

Polyphenols are abundant in finger millet, significantly contributing to its antioxidant capacity (Kumari et al. 2016). These compounds neutralize free radicals, reducing oxidative stress and lowering the risk of noncommunicable diseases like cardiovascular diseases, diabetes, and certain cancers (Jayawardana et al. 2021). The ~10.2 mg/g polyphenol content in finger millet is linked to antidiabetic effects by improving insulin sensitivity and modulating glucose metabolism (Singh et al. 2022). The gut microbiota enhances the bioavailability of these polyphenols, further supporting their health benefits. Additionally, polyphenolic compounds from finger millet exhibit anti-inflammatory properties, helping mitigate chronic inflammation associated with metabolic disorders (Dhanushkodi et al. 2023).

23.2.2.2 Flavonoids

Flavonoids, a subclass of polyphenols found in finger millet, include compounds like quercetin and catechins known for their strong antioxidant properties (Devi et al. 2014). With 2.5–3.8 mg/g flavonoids, they help reduce blood pressure and cholesterol levels while providing anti-inflammatory effects (Owheruo et al. 2018). The presence of flavonoids in finger millet is associated with improved cardiovascular health and enhanced immune function. Regular consumption of flavonoid-rich foods can lower the risk of chronic diseases, reinforcing the importance of incorporating finger millet into diets (Kumar et al. 2016).

23.2.2.3 Phytosterols

Phytosterols are plant-derived compounds structurally similar to cholesterol that can help lower cholesterol levels in the body (Venkateswaran and Vijayalakshmi 2010). Although specific data on phytosterol content in finger millet is limited compared to other grains, studies estimate 0.6–1.2 mg/g phytosterols in finger millet (Owheruo et al. 2018). These compounds work by competing with cholesterol for absorption in the intestines, thereby promoting heart health (Devi et al. 2014; Owheruo et al. 2018).

23.2.2.4 Antioxidants

The antioxidant activity of finger millet primarily stems from its rich content of polyphenols and flavonoids, which work together to protect cells from oxidative damage (Kumar et al. 2021). With an oxygen radical absorbance capacity (ORAC) of ~25–35 μmol TE/g, finger millet's antioxidants are crucial for combating oxidative stress linked to ageing and diseases like cancer and neurodegenerative disorders (Jayawardana et al. 2021). Consuming antioxidant-rich foods, like finger millet, enhances the body's defense mechanisms against oxidative damage, promoting overall health and longevity (Venkateswaran and Vijayalakshmi 2010).

23.2.3 Medicinal Properties

Finger millet (*Eleusine coracana*) is gaining recognition not only for its nutritional value but also for its significant medicinal properties, particularly in the realms of diabetes management, bone health, and anti-inflammatory and antimicrobial activities (Agrawal et al. 2023; Nayak 2024).

23.2.3.1 Antidiabetic Effects (α-Glucosidase Inhibition)

One of the most interesting health benefits of finger millet is that it may help people with diabetes. This is because it can stop an enzyme called α-glucosidase from working (Agrawal et al. 2023). This enzyme is what breaks down carbohydrates into glucose in the intestines. By inhibiting this enzyme, finger millet slows down the digestion and absorption of carbohydrates, leading to a more gradual increase in blood sugar levels after meals (Agrawal et al. 2023). This mechanism is crucial for managing type 2 diabetes, as it helps maintain glycemic control and reduces postprandial hyperglycemia (Ahamed et al. 2024). Studies have indicated that the consumption of finger millet can lead to improved insulin sensitivity and better overall glucose metabolism, making it a valuable dietary component for individuals with diabetes or those at risk of developing the condition.

23.2.3.2 Bone Health (Calcium Bioavailability)

Finger millet is regarded as one of the richest sources of calcium among cereals, containing approximately 344 mg of calcium per 100 grams—three times more than milk (Anitha et al. 2021a). This high calcium content is particularly beneficial for bone health, especially in populations at risk of osteoporosis and other calcium deficiency-related conditions (Sahaya Rani et al. 2021). Research indicates that finger millet not only provides a substantial amount of calcium but also has high bioavailability, meaning that a significant portion of the calcium consumed is effectively absorbed by the body (Anitha et al. 2021a; Warhade et al. 2023). For instance, studies have reported that boiled finger millet retains about 28% of its calcium in a bioavailable form, which is higher than many other staple grains like rice and maize (Anitha et al. 2021a). This characteristic makes finger millet an excellent dietary choice for enhancing bone density and preventing osteoporosis, particularly among vulnerable groups such as children, pregnant women, and the elderly (Puranik et al. 2017).

23.2.3.3 Anti-inflammatory and Antimicrobial Activities

In addition to its benefits for diabetes and bone health, finger millet exhibits notable anti-inflammatory and antimicrobial properties (Jayawardana et al. 2021). The presence of various bioactive compounds, including polyphenols and flavonoids, contributes to its ability to combat inflammation (Sirin 2023). These compounds help reduce oxidative stress in the body, which is linked to chronic inflammatory conditions such as arthritis and cardiovascular diseases (Devi et al. 2014). Furthermore, finger millet has demonstrated antimicrobial activity against various pathogens, making it a potential natural remedy for preventing infections. The grain's phytochemical composition supports immune function and may enhance the body's ability to fight off infections and diseases (Singh et al. 2015).

23.2.4 Traditional Uses: Ayurvedic and African Ethnomedicine

Finger millet (*Eleusine coracana*), also known as ragi, has a rich history of traditional uses in both Ayurvedic and African ethnomedicine, where it is valued not only as a staple food but also for its medicinal properties (Pandey et al. 2023). This ancient grain has been integrated into various cultural practices and dietary regimens, reflecting its significance in promoting health and well-being (Veerabadhran et al. 2023).

23.2.4.1 Ayurvedic Uses

In Ayurveda, finger millet is regarded as a highly nutritious grain that supports various aspects of health. It is believed to possess cooling properties, making it beneficial for balancing pitta dosha, which is associated with heat and inflammation in the body (Shisode 2024). Finger millet is often recommended for individuals suffering from conditions like hypertension, migraines, and digestive issues because of its high fiber content and ability to regulate blood sugar levels (Park et al. 2021).

Moreover, finger millet is used as a weaning food for infants because of its rich calcium and protein content, which are essential for growth and development (Devi et al. 2014). The grain is commonly prepared as porridge or malted into beverages that are easy to digest for young children (Shisode 2024). In Ayurvedic practices, finger millet is also used to strengthen bones and teeth, making it particularly valuable for pregnant women and the elderly who require additional calcium for bone health (Veerabadhran et al. 2023; Shisode 2024).

23.2.4.2 African Ethnomedicine

In many African cultures, finger millet holds a prominent place in traditional medicine and dietary practices (Maharajan et al. 2021). It is often used to prepare local alcoholic beverages such as *pombe* in Tanzania and *chaang* in Nepal, which are not only culturally significant but also believed to have health benefits (Kubo 2016). The fermentation process involved in brewing these drinks enhances the bioavailability of nutrients, making them more beneficial for health (Shisode 2024).

Additionally, finger millet is employed in traditional healing practices to address various ailments. Its high antioxidant content helps combat oxidative stress, while its anti-inflammatory properties are utilized to manage conditions such as arthritis and respiratory issues (Kubo 2016). In some African communities, finger millet porridge is prescribed for individuals recovering from illness or surgery due to its nourishing qualities (Veerabadhran et al. 2023).

The grain's versatility allows it to be incorporated into various dishes, including flatbreads and porridge, which are staples in many households. This adaptability not only preserves the nutritional benefits of finger millet but also reinforces its cultural significance as a source of sustenance.

23.3 Sorghum Millet (*Sorghum bicolor*)

Sorghum millet (*Sorghum bicolor*), often referred to as the "gluten-free healer," has garnered attention in recent years for its impressive nutritional profile and medicinal properties (Khalid et al. 2022). This ancient grain is not only a staple food in many cultures but also serves as a functional food with significant health benefits, particularly in managing diabetes and anemia (Ahamed et al. 2024). Recent research, including clinical trials, has highlighted the efficacy of sorghum in glycemic control and improving iron status, making it a valuable addition to dietary interventions (Kulamarva et al. 2009).

23.3.1 Nutritional Profile

Sorghum millet (*Sorghum bicolor*), often celebrated as a gluten-free grain, boasts an impressive nutritional profile that positions it as a valuable food source for health-conscious individuals and those with dietary restrictions (Reddy et al. 2022). This ancient grain is particularly notable for its high protein content, B vitamins, and resistant starch, making it a powerhouse of nutrition that can contribute to overall health and well-being (Hymavathi et al. 2020).

23.3.1.1 High Protein Content

Sorghum millet contains approximately 7–12% protein, depending on the variety and processing methods. This protein content is comparable to other grains, making sorghum an excellent source of plant-based protein for vegetarians and vegans (Sood and Mohiuddin Bhat 2023). The protein in sorghum is also rich in essential amino acids, particularly lysine, methionine, and threonine, which are often lacking in other starchy staples (Mosse et al. 1988). These amino acids are crucial for various bodily functions, including tissue repair, muscle growth, and immune system support (Stefoska-Needham 2024). The presence of a well-balanced amino acid profile in sorghum makes it a valuable dietary option for those looking to enhance their protein intake without relying on animal products (Reddy et al. 2022).

23.3.1.2 B-Vitamins

Sorghum is a significant source of several B vitamins, including thiamine (B1), riboflavin (B2), niacin (B3), and folate (B9). These vitamins play essential roles in energy metabolism, red blood cell formation, and the maintenance of healthy skin and nerves (Frankowski et al. 2022). For instance, thiamine is vital for converting carbohydrates into energy, while riboflavin is important for cellular function and growth. Niacin aids in DNA repair and the production of stress hormones in the adrenal glands (Oduor et al. 2010). Folate is crucial during pregnancy for fetal development and helps prevent neural tube defects (Surjana et al. 2010). The presence of these B vitamins in sorghum makes it an excellent choice for promoting overall health and preventing deficiencies that can lead to various health issues.

23.3.1.3 Resistant Starch

One of the standout features of sorghum millet is its high content of resistant starch, which is a type of carbohydrate that resists digestion in the small intestine (Taylor and Duodu 2023). Resistant starch acts similarly to dietary fiber in that it promotes gut health and improves digestive function (Niu et al. 2025). It serves as a prebiotic, providing nourishment for beneficial gut bacteria and supporting a healthy microbiome (Topping et al. 2003). Regular consumption of resistant starch has been linked to improved insulin sensitivity, reduced blood sugar spikes after meals, and enhanced feelings of fullness, making it beneficial for weight management (Johnston et al. 2010).

Research has shown that incorporating resistant starch into the diet can lead to various health benefits. For example, studies indicate that diets rich in resistant starch can help lower cholesterol levels and improve metabolic markers associated with type 2 diabetes. Furthermore, resistant starch has been associated with improved colon health by promoting regular bowel movements and reducing the risk of colorectal cancer.

23.3.2 *Bioactive Compounds*

Sorghum millet (*Sorghum bicolor*) is not only a staple food in many cultures but also a rich source of bioactive compounds that contribute to its numerous health benefits (de Morais Cardoso et al. 2017). Among these compounds, tannins, phenolic acids, and 3-deoxyanthocyanins stand out for their significant roles in promoting health and preventing chronic diseases (Taylor and Duodu 2015).

23.3.2.1 Tannins

Tannins are polyphenolic compounds found abundantly in sorghum, particularly in the outer layers of the grain (Wu et al. 2012). These compounds are known for their astringent properties and have been linked to various health benefits. Tannins possess antioxidant activity, which helps neutralize free radicals in the body, thereby reducing oxidative stress and lowering the risk of chronic diseases, such as cancer and cardiovascular disorders (Stefoska-Needham 2024; Xiong et al. 2019). Additionally, tannins have been shown to exhibit antimicrobial properties, making them effective against a range of pathogens (Kauther et al. 2020). Research indicates that the presence of tannins in sorghum can also inhibit digestive enzymes, which may help regulate blood sugar levels by slowing carbohydrate digestion (Kauther et al. 2020; Sonawane 2024). This characteristic makes sorghum a valuable dietary option for individuals managing diabetes (Park et al. 2012).

23.3.2.2 Phenolic Acids

Phenolic acids are another significant group of bioactive compounds present in sorghum (Xiong et al. 2019). They are classified into two main categories: benzoic acids and cinnamic acids. Key phenolic acids found in sorghum include gallic acid, p-coumaric acid, ferulic acid, caffeic acid, and sinapic acid (Shen et al. 2018). These compounds are renowned for their potent antioxidant properties, which contribute to reducing inflammation and protecting against oxidative damage (Xu et al. 2021). Studies have shown that phenolic acids can improve gut health by promoting beneficial gut microbiota and enhancing digestive function (Sleem et al. 2024). Furthermore, they play a role in regulating glucose metabolism and may help lower cholesterol levels, contributing to cardiovascular health (Shen et al. 2018; Xiong et al. 2019).

The antioxidant capacity of phenolic acids is particularly noteworthy; sorghum has been found to contain higher levels of these compounds compared to other major cereal grains. This high concentration of phenolic acids enhances sorghum's potential as a functional food that can support overall health and well-being (Pontieri et al. 2021).

23.3.2.3 3-Deoxyanthocyanins

One of the unique features of sorghum millet is its content of 3-deoxyanthocyanins (3-DXAs), which are a type of flavonoid pigment not commonly found in other grains (Li et al. 2022). These compounds impart color to certain varieties of sorghum, particularly the red and purple ones (Li et al. 2022). 3-DXAs have garnered attention for their strong antioxidant properties, which can help combat oxidative stress and inflammation in the body (Yang et al. 2009). Research indicates that these pigments may also possess anticancer properties by inhibiting tumor growth and promoting apoptosis (programmed cell death) in cancer cells (Yang et al. 2009).

3-Deoxyanthocyanins have a special structure that lets them interact with many biological pathways. This could improve metabolic health and lower the risk factors for chronic diseases. Their presence in sorghum adds to the grain's appeal as a functional food that can contribute to a healthy diet (Yang et al. 2009).

23.3.3 Medicinal Properties

Sorghum millet (*Sorghum bicolor*), often recognized for its gluten-free properties, possesses a range of medicinal benefits that contribute to its reputation as a health-promoting food (Benson et al. 2013). Recent research has highlighted its potential in three key areas: anticancer effects, cardiovascular benefits, and antioxidant capacity (Lee et al. 2020). These properties are largely attributed to the bioactive compounds present in sorghum, particularly phenolic compounds and tannins (de Morais Cardoso et al. 2017).

23.3.3.1 Anticancer Potential

Sorghum millet has demonstrated significant anticancer potential, particularly in inhibiting the proliferation of colon cancer cells (Gupta et al. 2023). Studies have shown that high phenolic sorghum bran extracts can effectively suppress the growth of human colon cancer cells through various mechanisms (Lee et al. 2020). For example, one study showed that treating colon cancer cells with these extracts decreased cell growth in a way that depended on the dose. This happened in several lines of cells, such as HCT15 and SW480. The extracts induced apoptosis (programmed cell death) and caused S phase growth arrest, which inhibited the migration and invasion of cancer cells (Xu et al. 2021). Furthermore, the extracts modulated key signaling pathways involved in cancer progression, such as downregulating β-catenin and inhibiting the IGF-1-stimulated PI3K/AKT pathway (Lee et al. 2021). These findings suggest that the bioactive compounds in sorghum millet may play a crucial role in preventing colon cancer and could be employed as part of dietary strategies for cancer management (Gupta et al. 2023).

23.3.3.2 Cardiovascular Benefits

Sorghum millet also offers cardiovascular benefits, particularly through its ability to reduce cholesterol levels (Jacob et al. 2024). The presence of soluble fiber and bioactive compounds, such as tannins, contributes to improved lipid profiles by lowering low-density lipoprotein (LDL) cholesterol while increasing high-density lipoprotein (HDL) cholesterol (Shisode 2024). Research indicates that regular consumption of sorghum can lead to significant reductions in total cholesterol and triglyceride levels. One study reported that participants who included sorghum-based products in their diet experienced an average reduction of 10–15% in total cholesterol levels over several weeks (Lee et al. 2020). This cholesterol-lowering effect is essential for reducing the risk of heart disease and stroke, making sorghum a heart-healthy grain that can be easily incorporated into various diets.

23.3.3.3 Antioxidant Capacity

The antioxidant capacity of sorghum millet is another vital aspect of its medicinal properties. Sorghum is rich in phenolic compounds and tannins, which exhibit strong free radical scavenging activity (Mawouma et al. 2022). These antioxidants help neutralize harmful free radicals in the body, thereby reducing oxidative stress a key factor in the development of chronic diseases such as cancer, cardiovascular disease, and neurodegenerative disorders (Shaikh et al. 2014). The antioxidant activity of sorghum has been shown to be higher than that of many other grains, making it an excellent choice for enhancing overall health (Pontieri et al. 2021). Research has demonstrated that the phenolic extracts from sorghum can significantly inhibit oxidative damage to cells, further supporting its role as a functional food with protective health benefits (Chen et al. 2021; Lee et al. 2021).

23.3.4 Traditional Uses of Sorghum Millet

Sorghum millet (*Sorghum bicolor*), a staple grain in both African and Asian cultures, has been utilized for centuries not only as a food source but also in traditional medicine, particularly for addressing digestive disorders (Benson et al. 2013). This ancient grain is celebrated for its nutritional benefits and its role in various folk remedies aimed at promoting digestive health (Adebo 2020; Benson et al. 2013).

23.3.4.1 Traditional Uses in African Folk Remedies

In many African communities, sorghum is revered for its therapeutic properties, especially concerning digestive health (Adebo 2020). The grain is often ground into flour and used to prepare traditional dishes such as porridge, which is easily digestible and helps soothe gastrointestinal discomfort (Stark et al. 2013). Fermented sorghum products, such as *ogi* and *nasha*, are commonly consumed for their probiotic benefits (Rapoo et al. 2023). These fermented foods are believed to promote gut health by increasing the balance of beneficial bacteria in the intestines, which can alleviate issues such as bloating, diarrhea, and constipation (Adebo 2020).

Moreover, the high fiber content of sorghum aids digestion by promoting regular bowel movements and preventing constipation (Adebo 2020). In some cultures, a decoction made from sorghum grains is consumed to relieve stomach aches and improve overall digestive function (Elmaghraby et al. 2023). The grain's natural tannins may also contribute to its ability to soothe the digestive tract by reducing inflammation and irritation (Shaikh et al. 2014).

23.3.4.2 Traditional Uses in Asian Folk Remedies

In Asia, particularly in India, sorghum (known as *jowar*) has long been used in folk remedies for digestive disorders. Traditional preparations often include sorghum flour mixed with water or milk to create a soothing porridge that is easy on the stomach (Elmaghraby et al. 2023). This dish is especially popular among individuals recovering from illness or those with sensitive digestive systems (Gupta et al. 2023).

Additionally, sorghum is often included in dietary regimens aimed at managing conditions like irritable bowel syndrome (IBS) because of its high fiber content, which helps regulate bowel function (Bijkerk et al. 2004). In some regions, the grain is also used in herbal concoctions that combine it with other medicinal plants known for their digestive benefits. These mixtures are believed to enhance digestion and alleviate symptoms of indigestion (El-Salhy et al. 2017; Niu et al. 2025).

23.3.4.3 Nutritional Benefits Supporting Digestive Health

The medicinal properties of sorghum are supported by its rich nutritional profile. Sorghum is high in dietary fiber, which not only aids digestion but also plays a crucial role in maintaining gut health (Ashley et al. 2019). The presence of resistant starch in sorghum further enhances its digestibility and promotes the growth of beneficial gut bacteria. This prebiotic effect is vital for improving overall gut health and preventing digestive disorders (Stark et al. 2013).

Moreover, sorghum's gluten-free nature makes it an excellent alternative for individuals with coeliac disease or gluten intolerance who may experience digestive issues when consuming gluten-containing grains. By incorporating sorghum into their diets, these individuals can enjoy a nutritious food source that supports their digestive health without adverse effects (Adebo 2020; Xiong et al. 2019).

23.4 Foxtail Millet (*Setaria italica*)

Foxtail millet (*Setaria italica*), often referred to as a "metabolic regulator," has garnered attention in recent research for its significant role in promoting metabolic health and adaptability to various environmental conditions (Ren et al. 2021). This ancient grain is not only a staple food in many cultures but also exhibits numerous health benefits that make it a valuable addition to modern diets, particularly in the context of metabolic regulation (Adebo 2020).

Foxtail millet is rich in essential nutrients, including high-quality protein, dietary fiber, B-vitamins, and minerals such as iron and magnesium (Kalsi and Bhasin 2023). Its protein content is approximately 11–14%, making it an excellent source of plant-based protein (Arora et al. 2023). The presence of dietary fiber aids digestion and helps regulate blood sugar levels, which contributes to its role as a metabolic regulator (Cui et al. 2023). The low glycemic index of foxtail millet makes it particularly beneficial for individuals managing diabetes, as it leads to slower absorption of carbohydrates and more stable blood sugar levels (Agrawal et al. 2023; Ahamed et al. 2024).

Recent studies have highlighted the metabolic pathways involved in foxtail millet's health benefits (Kalsi and Bhasin 2023). For instance, research utilizing multiomics approaches has revealed insights into the carbon fixation processes and metabolic regulation within the plant (Aggarwal et al. 2022). The integration of transcriptomic and metabolomic data has shown that foxtail millet employs mixed decarboxylation modes for C4 photosynthesis, which enhances its efficiency in carbon utilization (De Oliveira Dal'Molin et al. 2016). This efficiency not only supports the plant's growth under various environmental stresses but also contributes to the nutritional quality of the grain (Aggarwal et al. 2022; Arora et al. 2023).

Foxtail millet is also known for its antioxidant properties, attributed to its rich content of phenolic compounds and flavonoids (Sharma et al. 2015). These bioactive compounds play a crucial role in scavenging free radicals, thereby reducing oxidative stress and inflammation in the body (Suma and Urooj 2011). Recent research has demonstrated that the consumption of foxtail millet can enhance gut microbiota diversity, promoting the growth of beneficial bacteria that contribute to improved digestive health. The prebiotic effects of the dietary fiber found in foxtail millet support a healthy gut environment, which is essential for overall metabolic function (Chen et al. 2022).

One of the remarkable aspects of foxtail millet is its strong tolerance to environmental stresses such as drought and low nitrogen conditions (Nematpour et al. 2019). Research has shown that foxtail millet exhibits physiological and molecular adaptations that enable it to thrive under these challenging conditions. For example, studies have identified key genes involved in systemic nitrogen signaling that enhance plants' growth and productivity even when nitrogen availability is limited. This adaptability makes foxtail millet a resilient crop and highlights its potential as a sustainable food source in regions facing climate change challenges (Zhao et al. 2023).

The implications of incorporating foxtail millet into diets extend beyond basic nutrition; they encompass significant clinical benefits for individuals at risk of metabolic disorders (Jacob et al. 2024). By improving glycemic control, enhancing gut health, and providing essential nutrients, foxtail millet can serve as a functional food that supports overall health (Chen et al. 2022).

Future research should focus on further elucidating the mechanisms behind foxtail millet's health benefits, particularly its effects on metabolic pathways related to obesity and diabetes management. Additionally, exploring genetic strategies to enhance nutrient content and stress tolerance will be crucial for maximizing the potential of foxtail millet as a staple food crop (Cui et al. 2023).

23.4.1 Nutritional Profile

Foxtail millet (*Setaria italica*), often hailed as a nutritional powerhouse, boasts a remarkable profile that makes it an excellent choice for health-conscious individuals (Nithya et al. 2023). This ancient grain is particularly notable for its low glycemic index (GI), high magnesium content, and zinc levels, which collectively contribute to its status as a beneficial food for metabolic regulation and overall health (Kalsi and Bhasin 2023).

23.4.1.1 Low Glycemic Index

Glycemic Index: ~50 (Front. Nutr., 2023) | Dietary Fiber: ~6–8 g/100 g. One of the standout features of foxtail millet is its low glycemic index (GI), which is approximately 50-significantly lower than refined grains like white rice (GI 73) or wheat (GI 70). This makes it ideal for individuals managing diabetes or aiming to stabilize blood sugar levels. The slow digestion and gradual glucose release are attributed to its high fiber content (~6–8 g per 100 g), which also promotes satiety and aids in weight management. Studies highlight that replacing high-GI grains with foxtail millet can reduce postprandial blood glucose spikes by ~20–25%, enhancing metabolic health.

23.4.1.2 High Magnesium Content

Foxtail millet is rich in magnesium, providing 81 mg per 100 g equivalent to 20% of the recommended daily intake (RDI) for adults (Abedin et al. 2022). Magnesium is critical for over 300 enzymatic reactions, including energy production, nerve function, and glucose metabolism (Jahnen-Dechent and Ketteler 2012). Research indicates that diets rich in magnesium (≥350 mg/day) can lower the risk of type 2 diabetes by ~15–20% and reduce hypertension incidence by ~10%. Its role in improving insulin sensitivity and regulating blood pressure underscores foxtail millet's value in combating metabolic disorders (Arora et al. 2023).

23.4.1.3 Zinc Levels

Foxtail millet provides 2.4–3.0 mg of zinc per 100 g, covering 22–27% of the daily requirement for adults (Abedin et al. 2022). Zinc is vital for immune function, DNA synthesis, and wound healing. Deficiency in zinc (affecting ~17% of the global population) compromises immunity and increases susceptibility to infections. Regular consumption of zinc-rich foxtail millet can help bridge this gap, supporting immune resilience and cellular repair processes (Jacob et al. 2024).

23.4.2 Key Bioactive Compounds

Foxtail millet (*Setaria italica*) is increasingly recognized for its rich array of bioactive compounds, which contribute to its numerous health benefits (Kalsi and Bhasin 2023). Among these compounds, dietary fiber, γ-aminobutyric acid (GABA), and lignans play significant roles in promoting overall health and well-being (Kumari et al. 2025).

23.4.2.1 Dietary Fiber

Foxtail millet is an excellent source of dietary fiber, containing approximately 8–10% of its weight in fiber (Arora et al. 2023). This high fiber content is beneficial for digestive health because it aids in regular bowel movements and helps prevent constipation (Chen et al. 2022). The soluble fiber found in foxtail millet can also help regulate blood sugar levels by slowing glucose absorption into the bloodstream, making it a valuable food for individuals managing diabetes (Ren et al. 2018). Moreover, dietary fiber contributes to satiety, which can assist in weight management by reducing total calorie intake (Howarth et al. 2001). The presence of both soluble and insoluble fiber in foxtail millet promotes a healthy gut microbiome by supporting the growth of beneficial bacteria that play a crucial role in digestion and overall metabolic health (Li et al. 2025b).

23.4.2.2 γ-Aminobutyric Acid (GABA)

GABA is a naturally occurring amino acid that functions as a neurotransmitter in the brain. It is known for its calming effects on the nervous system and has been associated with various health benefits, including reduced anxiety and improved sleep quality (Wu et al. 2023). Recent studies have shown that foxtail millet contains significant amounts of GABA, particularly when fermented or germinated. The presence of GABA in foxtail millet may contribute to its potential as a functional food that supports mental well-being and stress reduction. Additionally,

GABA has been linked to improved metabolic health, as it may enhance insulin sensitivity and help regulate blood sugar levels (Yu et al. 2024).

23.4.2.3 Lignans

Lignans are a group of phytoestrogens found in various plant foods, including foxtail millet (Arora et al. 2023). These compounds have garnered attention for their potential health benefits, particularly in hormone regulation and cancer prevention (Peterson et al. 2010). Lignans exhibit antioxidant properties that help combat oxidative stress and inflammation in the body (Jang et al. 2022). Research suggests that lignans may play a role in reducing the risk of hormone-related cancers, such as breast and prostate cancers, by modulating estrogen metabolism. The presence of lignans in foxtail millet adds to its appeal as a functional food that can support women's health and overall hormonal balance (Lowcock et al. 2013).

23.4.3 Medicinal Properties

Foxtail millet (*Setaria italica*) has emerged as a potent functional food with diverse medicinal properties, including anti-obesity effects, neuroprotective potential, and immune modulation (Lakhani et al. 2024). These benefits are driven by its unique bioactive compounds and nutritional profile, making it a valuable dietary intervention for modern health challenges (Kalsi and Bhasin 2023).

23.4.3.1 Anti-obesity Effects

Foxtail millet exhibits anti-obesity properties through its ability to regulate lipid metabolism and promote satiety. Its high dietary fiber content (8–10%) delays gastric emptying and prolongs feelings of fullness, reducing overall calorie intake (Lakhani et al. 2024). Studies indicate that regular consumption of foxtail millet reduces adiposity by increasing leptin secretion, a hormone critical for energy balance and appetite suppression (Lakhani et al. 2024; Radha et al. 2024). In diabetic rat models, foxtail millet supplementation significantly lowered triglyceride (TG) and LDL cholesterol levels while improving glucose metabolism (Ren et al. 2021). The grain's unsaturated fatty acids, such as linolenic (66.68%) and oleic acid (16.11%), further enhance lipid regulation by reducing cholesterol oxidation and inflammation (Kalsi and Bhasin 2023). Additionally, foxtail millet's resistant starch modulates gut microbiota, promoting beneficial bacteria like *Lactobacillus*, which improve bile acid metabolism and inhibit fat accumulation. These mechanisms collectively position foxtail millet as an effective tool in combating obesity and metabolic syndrome (Chen et al. 2022).

23.4.3.2 Neuroprotective Potential

Foxtail millet's neuroprotective properties are linked to its γ-aminobutyric acid (GABA) content, a neurotransmitter known for reducing anxiety and improving stress resilience (Yu et al. 2024). Germinated foxtail millet contains up to 336.52 mg/100 g of GABA29.5 times higher than ungerminated grains (Wu et al. 2023). Clinical studies demonstrate that GABA-enriched diets alleviate anxiety and improve sleep quality by modulating the brain's neuroendocrine pathways (Linseman et al. 2024). For instance, GABA administration in stressed individuals reduced anxiety and induced relaxation by inhibiting sympathetic nervous system activity (Nuss 2015). Foxtail millet's GABA also enhances insulin sensitivity and regulates energy metabolism, indirectly supporting cognitive health (Hepsomali et al. 2020). These effects highlight its potential as a natural remedy for stress-related disorders and neurodegenerative conditions.

23.4.3.3 Immune Modulation

Foxtail millet strengthens immune function through its rich antioxidant profile and micronutrients (Zhang et al. 2021). It contains polyphenols, lignans, and vitamins that neutralize free radicals, reducing the oxidative stress and inflammation linked to chronic diseases (Rodríguez-García et al. 2019). The grain's iron content (~3.5 mg/100 g) supports red blood cell production, ensuring optimal oxygen transport and immune cell activity (Jacob et al. 2024). Additionally, its dietary fiber acts as a prebiotic, fostering a healthy gut microbiome, which is crucial for 70% of immune functions. Animal studies reveal that foxtail millet alleviates gut dysbiosis in diabetic mice, enhancing populations of beneficial bacteria that regulate immune responses (Rinninella et al. 2019). Its anti-inflammatory compounds, such as *p*-coumaric acid, further mitigate systemic inflammation, reducing the risk of autoimmune conditions and infections (Shi et al. 2017).

23.4.4 Traditional Uses of Foxtail Millet

Foxtail millet (*Setaria italica*) has long been valued in both Chinese and Indian traditional medicine for its health benefits, particularly in promoting longevity and aiding digestion (Jacob et al. 2024). This ancient grain is not only a staple food but also plays a significant role in holistic health practices, reflecting its deep-rooted cultural significance (Tripathi and Vyas 2023).

23.4.4.1 Traditional Uses in Indian Medicine

In Ayurveda, the traditional system of medicine in India, foxtail millet is revered for its ability to balance the body's doshasVata, Pitta, and Kapha (Balkrishna et al. 2023). Ayurvedic texts describe foxtail millet as having a sweet taste and astringent properties, making it beneficial for digestive health (Agrawal et al. 2023). Its high fiber content promotes regular bowel movements, prevents constipation, and supports overall digestive efficiency. The grain is often recommended for individuals with sensitive stomachs or digestive disorders due to its easily digestible nature (Kalsi and Bhasin 2023).

Foxtail millet is also recognized for its ability to enhance longevity and vitality (Jacob et al. 2024). It is considered a "sattvic" food, which means it promotes clarity and calmness in the mind while providing sustained energy (Abedin et al. 2022). This aligns with Ayurvedic principles, which emphasize the importance of wholesome foods for maintaining health and preventing disease. The grain's rich nutritional profile, including essential vitamins and minerals like iron and magnesium, supports various bodily functions and contributes to overall well-being (Suma and Urooj 2011; Abedin et al. 2022; Arora et al. 2023).

Moreover, foxtail millet is often incorporated into Ayurvedic recipes such as kitchari (a nourishing porridge) and can be paired with spices like turmeric and cumin to enhance its therapeutic properties. These combinations improve flavor and amplify the health benefits associated with the grain.

23.4.4.2 Traditional Uses in Chinese Medicine

In traditional Chinese medicine (TCM), foxtail millet is celebrated as a sacred crop with cooling properties that support kidney health and strengthen "Yin" fluids in the body (Jacob et al. 2024). It is believed to be particularly beneficial during the summer months when the body can become overheated. The grain is used to prepare nourishing congee or porridge, which is easy on the digestive system and helps replenish energy (Chen et al. 2022).

Chinese herbalists have historically utilized foxtail millet to address digestive issues such as stomach reflux and abdominal distention (Jacob et al. 2024). Its gentle nature makes it suitable for individuals recovering from illness or those with weakened digestive systems (Jacob et al. 2024). The cooling effects of foxtail millet are thought to alleviate heat-related conditions, making it an ideal food during hot weather or for individuals experiencing internal heat (Table 23.2 and Fig. 23.2).

Table 23.2 Nutrient composition comparison of millets, rice, and wheat (per 100 g) (Abioye et al. 2022; Muthamilarasan et al. 2016)

Nutrient	Finger millet	Foxtail millet	Pearl millet	Barnyard millet	Kodo millet	Little millet	Proso millet	Rice	Wheat
Calories (Kcal)	336	329	378	360	353	329	354	362	348
Protein (g)	7.3	12.3	11.8	11.2	8.3	7.7	12.5	6.8	11.8
Carbohydrate (g)	72	60.2	67	72	65.9	67	71	78.2	71.2
Fat (g)	1.3	4.3	4.8	2.5	1.4	4.7	1.1	0.5	1.5
Fiber (g)	3.6	8	8.5	10.1	5.2	7.6	2.3	1	2
Calcium (mg)	344	31	42	22	35	17	8	33	30
Iron (mg)	3.9	2.8	2.2	15	0.6	9.3	2.9	1	3.5

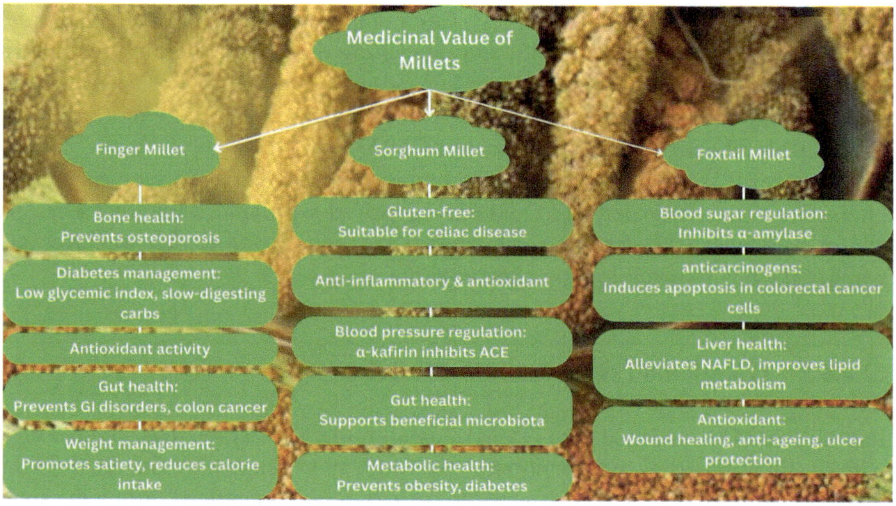

Fig. 23.2 Medicinal value of millet

23.5 Challenges and Future Directions

The promotion of millets as a sustainable food source faces multifaceted challenges across agricultural, economic, and social dimensions (Harish et al. 2024). Despite their rich nutritional profile and potential to combat food security and malnutrition, millets have seen a significant decline in cultivation, dropping from 35 million hectares to 15 million due to low yields, labor-intensive processing, and minimal marketing efforts (Chandra et al. 2021; Harish et al. 2024). This decline threatens their sustainability as a food source (Chandra et al. 2021).

One of the primary challenges is low productivity; many millet varieties, such as sorghum and pearl millet, have stagnated or declined in yield over the past decade (Chandra et al. 2021; Harish et al. 2024). Factors contributing to such an outcome include inadequate rainfall distribution, poor crop management practices, and limited access to high-yielding varieties (Silungwe et al. 2019). Additionally, there is a lack of public awareness regarding the health benefits of millets compared to more popular grains like rice and wheat, leading to low consumer demand (Lakra et al. 2024).

Economic barriers also hinder millet adoption; they often come with a higher price tag than traditional cereals, limiting accessibility for low-income consumers (Banerjee et al. 2025). Moreover, millets are not widely available in retail markets or e-commerce platforms, making it difficult for consumers to purchase these nutritious grains (Lakra et al. 2024; Banerjee et al. 2025). Cultural preferences further complicate matters, as the taste and texture of millets may not appeal to all consumers accustomed to rice and wheat (Alekhya and Shravanthi 2019).

To address these issues, several strategic directions can be pursued. Enhanced research and development are crucial for creating high-yielding millet varieties through genomics and marker-assisted breeding. Public awareness campaigns can educate consumers about the health benefits of millets and promote their culinary versatility (Silungwe et al. 2019). Stronger government policies that support millet farmers through subsidies and incentives can bolster production efforts.

Additionally, integrating millets into national food fortification programs can enhance their nutritional profiles in baby foods and dietary supplements. Establishing robust market linkages can improve profitability for millet producers while ensuring better availability for consumers (Chandra et al. 2021). Promoting sustainable agricultural practices will enhance resilience against climate change impacts. Collaboration across sectors government agencies, NGOs, academic institutions, and private stakeholders will facilitate the knowledge sharing and resource mobilization necessary for advancing millet production and consumption (Banerjee et al. 2025). By overcoming these challenges through strategic initiatives, millets can play a vital role in creating a sustainable food future (Alekhya and Shravanthi 2019; Lakra et al. 2024).

23.6 Conclusion

Finger millet (*Eleusine coracana*), sorghum millet (*Sorghum bicolor*), and foxtail millet (*Setaria italica*) are nutrient-dense, climate-resilient grains with significant medicinal potential. Finger millet stands out for its exceptional calcium content (344 mg/100 g), threefold higher than milk, making it pivotal for bone health and osteoporosis prevention. It's high dietary fiber (15 g/100 g) and polyphenols (10.2 mg/g) aid in glycemic control, antioxidant defense, and gastrointestinal health. Sorghum millet, rich in tannins, phenolic acids, and 3-deoxyanthocyanins, exhibits potent anticancer properties by inducing apoptosis in colon cancer cells and lowering LDL cholesterol through soluble fiber and bioactive compounds. Its gluten-free nature and resistant starch content further support metabolic and gut health. Foxtail millet, with a low glycemic index (~50) and high magnesium (81 mg/100 g), regulates blood sugar and hypertension. Its γ-aminobutyric acid (GABA) content

(336.52 mg/100 g in germinated forms) alleviates anxiety, while lignans and dietary fiber enhance immune modulation and lipid metabolism. Collectively, these millets address global health challenges such as diabetes, cardiovascular diseases, and malnutrition through their high fiber, antioxidants, and essential nutrients. Their adaptability to arid environments positions them as sustainable solutions for food security. However, challenges like low yield, consumer awareness gaps, and market accessibility must be addressed via genomic research, policy support, and public campaigns. Integrating millets into diets as functional foods offers a holistic approach to combating lifestyle diseases while promoting environmental resilience.

References

Abedin MJ, Abdullah ATM, Satter MA, Farzana T (2022) Physical, functional, nutritional and antioxidant properties of foxtail millet in Bangladesh. Heliyon 8:e11186. https://doi.org/10.1016/J.HELIYON.2022.E11186

Abioye VF, Babarinde GO, Ogunlakin GO, Adejuyitan JA, Olatunde SJ, Abioye AO (2022) Varietal and processing influence on nutritional and phytochemical properties of finger millet: a review. Heliyon 8:e12310. https://doi.org/10.1016/J.HELIYON.2022.E12310

Adebo OA (2020) African sorghum-based fermented foods: past, current and future prospects. Nutrients 12:1111. https://doi.org/10.3390/NU12041111

Aggarwal PR, Pramitha L, Choudhary P, Singh RK, Shukla P, Prasad M, Muthamilarasan M (2022) Multi-omics intervention in Setaria to dissect climate-resilient traits: progress and prospects. Front Plant Sci 13:892736. https://doi.org/10.3389/FPLS.2022.892736/FULL

Agrawal P, Singh BR, Gajbe U, Kalambe MA, Bankar M (2023) Managing diabetes mellitus with millets: a new solution. Cureus 15:e44908. https://doi.org/10.7759/CUREUS.44908

Ahamed VPA, Joshi A, Mudey A, Choudhari S, Raut J, Ahmed S (2024) Unlocking the potential: millets and their impact on diabetes management. Cureus 16:e59283. https://doi.org/10.7759/CUREUS.59283

Alekhya P, Shravanthi AR (2019) Buying behaviour of consumers towards millet based food products in Hyderabad District of Telangana, India. Int J Curr Microbiol App Sci 8:223–236. https://doi.org/10.20546/ijcmas.2019.810.023

Anitha S, Givens DI, Botha R, Kane-Potaka J, Sulaiman NLB, Tsusaka TW, Subramaniam K, Rajendran A, Parasannanavar DJ, Bhandari RK (2021a) Calcium from finger millet—a systematic review and meta-analysis on calcium retention, bone resorption, and in vitro bioavailability. Sustainability 13:8677. https://doi.org/10.3390/SU13168677

Anitha S, Kane-Potaka J, Botha R, Givens DI, Sulaiman NLB, Upadhyay S, Vetriventhan M, Tsusaka TW, Parasannanavar DJ, Longvah T, Rajendran A, Subramaniam K, Bhandari RK (2021b) Millets can have a major impact on improving iron status, hemoglobin level, and in reducing iron deficiency anemia—a systematic review and meta-analysis. Front Nutr 8:725529. https://doi.org/10.3389/FNUT.2021.725529/FULL

Anitha S, Arjun P, Palli NC, Sreekanth N, Miruthika Devi SA, Pandey S, Krishnan S, Prasad S, Sharma S, Chidambara Murthy KN, Botha R, Upadhyay S, Kane-Potaka J (2024) Sensory and nutritional evaluation of nine types of millet substituted for polished white rice in select Indian meal preparations. Front Sustain Food Syst 8:1331260. https://doi.org/10.3389/FSUFS.2024.1331260/BIBTEX

Arachchige S, Jayawardana S, Koralalage J, Samarasekera RR, Chamari GH, Hettiarachchi M, Gooneratne J, Datta Mazumdar S, Banerjee R (2019) Dietary fibers, starch fractions and nutritional composition of finger millet varieties cultivated in Sri Lanka. J Food Compos Anal 82(103249):1–6. https://doi.org/10.1016/j.jfca.2019.103249

Arora L, Aggarwal R, Dhaliwal I, Gupta OP, Kaushik P (2023) Assessment of sensory and nutritional attributes of foxtail millet-based food products. Front Nutr 10:1146545. https://doi.org/10.3389/FNUT.2023.1146545

Ashley D, Marasini D, Brownmiller C, Lee JA, Carbonero F, Lee SO (2019) Impact of grain sorghum polyphenols on microbiota of normal weight and overweight/obese subjects during in vitro fecal fermentation. Nutrients 11:217. https://doi.org/10.3390/NU11020217

Balkrishna A, Shankar R, Prajapati UB, Joshi RA, Srivastava A, Awasthi C (2023) Exploring the common millets of India: a comprehensive review. J Drug Res Ayurvedic Sci 8:S26–S33. https://doi.org/10.4103/JDRAS.JDRAS_132_23

Banerjee S, Sanjay KR, Chethan S, Malleshi NG (2012) Finger millet (*Eleusine coracana*) polyphenols: investigation of their antioxidant capacity and antimicrobial activity. Afr J Food Sci 6:362–374. https://doi.org/10.5897/AJFS12.031

Banerjee R, Bharti B, Das P, Barman S, Devi S, Ankita A (2025) Comprehensive analysis of millets in India: area, production, cost of production and export statistics. Curr Agric Res J 12:1181–1192. https://doi.org/10.12944/CARJ.12.3.13

Bavaro AR, Tarantini A, Bruno A, Logrieco AF, Gallo A, Mita G, Valerio F, Bleve G, Cardinali A (2024) Functional foods in Mediterranean diet: exploring the functional features of vegetable case-studies obtained also by biotechnological approaches. Aging Clin Exp Res 36:208. https://doi.org/10.1007/S40520-024-02860-1

Benson KF, Beaman JL, Ou B, Okubena A, Okubena O, Jensen GS (2013) West African sorghum bicolor leaf sheaths have anti-inflammatory and immune-modulating properties in vitro. J Med Food 16:230. https://doi.org/10.1089/JMF.2012.0214

Bijkerk CJ, Muris JWM, Knottnerus JA, Hoes AW, De Wit NJ (2004) Systematic review: the role of different types of fibre in the treatment of irritable bowel syndrome. Aliment Pharmacol Ther 19:245–251. https://doi.org/10.1111/J.0269-2813.2004.01862.X

Chandra AK, Chandora R, Sood S, Malhotra N (2021) Global production, demand, and supply. In: Millets and pseudo cereals, pp 7–18. https://doi.org/10.1016/B978-0-12-820089-6.00002-1

Chen X, Shen J, Xu J, Herald T, Smolensky D, Perumal R, Wang W (2021) Sorghum phenolic compounds are associated with cell growth inhibition through cell cycle arrest and apoptosis in human hepatocarcinoma and colorectal adenocarcinoma cells. Foods 10:993. https://doi.org/10.3390/FOODS10050993

Chen Y, Zhang R, Xu J, Ren Q (2022) Alteration of intestinal microflora by the intake of millet porridge improves gastrointestinal motility. Front Nutr 9:965687. https://doi.org/10.3389/FNUT.2022.965687/BIBTEX

Cui X, Wang B, Chen Z, Guo J, Zhang T, Zhang W, Shi L (2023) Comprehensive physiological, transcriptomic, and metabolomic analysis of the key metabolic pathways in millet seedling adaptation to drought stress. Physiol Plant 175:e14122. https://doi.org/10.1111/PPL.14122

de Morais Cardoso L, Pinheiro SS, Martino HSD, Pinheiro-Sant'Ana HM (2017) Sorghum (*Sorghum bicolor* L.): nutrients, bioactive compounds, and potential impact on human health. Crit Rev Food Sci Nutr 57:372–390. https://doi.org/10.1080/10408398.2014.887057

De Oliveira Dal'Molin CG, Orellana C, Gebbie L, Steen J, Hodson MP, Chrysanthopoulos P, Plan MR, McQualter R, Palfreyman RW, Nielsen LK (2016) Metabolic reconstruction of *Setaria italica*: a systems biology approach for integrating tissue-specific omics and pathway analysis of bioenergy grasses. Front Plant Sci 7. https://doi.org/10.3389/FPLS.2016.01138

Devi PB, Vijayabharathi R, Sathyabama S, Malleshi NG, Priyadarisini VB (2014) Health benefits of finger millet (*Eleusine coracana* L.) polyphenols and dietary fiber: a review. J Food Sci Technol 51:1021–1040. https://doi.org/10.1007/S13197-011-0584-9/METRICS

Dhanushkodi V, Hemavathy AT, Shenbagavalli S, Sangeetha S, Anitha R, UmaMaheshwari T (2023) A review on nutritional properties and health benefits of finger millet. Int J Plant Soil Sci 35:753–761. https://doi.org/10.9734/IJPSS/2023/V35I183342

Dixit V, Joseph Kamal SW, Bajrang Chole P, Dayal D, Chaubey KK, Pal AK, Xavier J, Manjunath BT, Bachheti RK (2023) Functional foods: exploring the health benefits of bioactive compounds from plant and animal sources. J Food Qual 2023:5546753. https://doi.org/10.1155/2023/5546753

Elmaghraby DA, Alsalman GA, Alawadh LH, Al-Abdulqader SA, Alaithan MM, Alnuwaysir BI (2023) Integrated traditional herbal medicine in the treatment of gastrointestinal disorder: the pattern of use and the knowledge of safety among the Eastern Region Saudi population. BMC Complement Med Ther 23:1–12. https://doi.org/10.1186/S12906-023-04197-0/PEER-REVIEW

El-Salhy M, Ystad SO, Mazzawi T, Gundersen D (2017) Dietary fiber in irritable bowel syndrome (review). Int J Mol Med 40:607. https://doi.org/10.3892/IJMM.2017.3072

Frankowski J, Przybylska-Balcerek A, Stuper-Szablewska K (2022) Concentration of pro-health compound of sorghum grain-based foods. Foods 11:216. https://doi.org/10.3390/FOODS11020216

Gaikwad V, Kaur J, Rasane P, Kaur S, Singh J, Kumar A, Kumar A, Sharma N, Mohan Mehta C, Singh Patel A (2024) Nutritional significance of finger millet and its potential for using in functional products. Foods Raw Mater 12:110–123. https://doi.org/10.21603/2308-4057-2024-1-593

Ghani I, An Y, Qiao Q, He S, Li Z (2024) Polyphenols from foxtail millet improve non-alcoholic fatty liver disease by regulating intestinal microbiome in mice. Foods 13. https://doi.org/10.3390/FOODS13111683

Gupta M, Asfaha DM, Ponnaiah G (2023) Millets: a nutritional powerhouse with anti-cancer potential. Cureus 15:e47769. https://doi.org/10.7759/CUREUS.47769

Harish MS, Bhuker A, Chauhan BS (2024) Millet production, challenges, and opportunities in the Asia-Pacific region: a comprehensive review. Front Sustain Food Syst 8:1386469. https://doi.org/10.3389/FSUFS.2024.1386469/BIBTEX

Hepsomali P, Groeger JA, Nishihira J, Scholey A (2020) Effects of oral gamma-aminobutyric acid (GABA) administration on stress and sleep in humans: a systematic review. Front Neurosci 14:923. https://doi.org/10.3389/FNINS.2020.00923

Howarth NC, Saltzman E, Roberts SB (2001) Dietary fiber and weight regulation. Nutr Rev 59:129–139. https://doi.org/10.1111/J.1753-4887.2001.TB07001.X

Hymavathi TV, Jyothsna E, Robert TP, Sri VT (2020) Effect of resistant starch (RS) rich sorghum food consumption on lipids and glucose levels of diabetic subjects. J Pharm Res Int:86–92. https://doi.org/10.9734/JPRI/2020/V32I3030909

Jacob J, Krishnan V, Antony C, Bhavyasri M, Aruna C, Mishra K, Nepolean T, Satyavathi CT, Visarada KBRS (2024) The nutrition and therapeutic potential of millets: an updated narrative review. Front Nutr 11:1346869. https://doi.org/10.3389/FNUT.2024.1346869

Jahnen-Dechent W, Ketteler M (2012) Magnesium basics. Clin Kidney J 5:i3. https://doi.org/10.1093/NDTPLUS/SFR163

Jang WY, Kim MY, Cho JY (2022) Antioxidant, anti-inflammatory, anti-menopausal, and anti-cancer effects of lignans and their metabolites. Int J Mol Sci 23. https://doi.org/10.3390/IJMS232415482

Jayawardana SAS, Samarasekera JKRR, Hettiarachchi GHCM, Gooneratne J, Choudhary MI, Jabeen A (2021) Anti-inflammatory and antioxidant properties of finger millet (*Eleusine coracana* (L.) Gaertn.) varieties cultivated in Sri Lanka. Biomed Res Int 2021:7744961. https://doi.org/10.1155/2021/7744961

Johnston KL, Thomas EL, Bell JD, Frost GS, Robertson MD (2010) Resistant starch improves insulin sensitivity in metabolic syndrome. Diabet Med 27:391–397. https://doi.org/10.1111/J.1464-5491.2010.02923.X

Kalsi R, Bhasin JK (2023) Nutritional exploration of foxtail millet (*Setaria italica*) in addressing food security and its utilization trends in food system. eFood 4:e111. https://doi.org/10.1002/EFD2.111

Kauther A, Azza ME, Hind A, Huda O, Hassan A, Nada BH, Magdoleen O, Hussien MD (2020) Phytochemical screening and antibacterial activities of Sorghum bicolor leaves derived from in vitro culture. GSC Biol Pharm Sci 2020:65–072. https://doi.org/10.30574/gscbps.2020.10.1.0241

Khalid W, Ali A, Arshad MS, Afzal F, Akram R, Siddeeg A, Kousar S, Rahim MA, Aziz A, Maqbool Z, Saeed A (2022) Nutrients and bioactive compounds of Sorghum bicolor L. used to prepare functional foods: a review on the efficacy against different chronic disorders. Int J Food Prop 25:1045–1062. https://doi.org/10.1080/10942912.2022.2071293

Krishnan V (2024) Editorial: millets: superfood of the century—challenges and opportunities. Front Nutr 11:1462567. https://doi.org/10.3389/FNUT.2024.1462567

Kubo R (2016) The reason for the preferential use of finger millet (*Eleusine coracana*) in Eastern African brewing. J Inst Brew 122:175–180. https://doi.org/10.1002/JIB.309

Kulamarva AG, Sosle VR, Raghavan GSV (2009) Nutritional and rheological properties of sorghum. Int J Food Prop 12:55–69. https://doi.org/10.1080/10942910802252148

Kumar S, Kotwal N (2023) Millets (Shrianna) and lifestyle diseases: a healing touch. Med J Armed Forces India 79:249. https://doi.org/10.1016/J.MJAFI.2023.04.001

Kumar A, Metwal M, Kaur S, Gupta AK, Puranik S, Singh S, Singh M, Gupta S, Babu BK, Sood S, Yadav R (2016) Nutraceutical value of finger millet [*Eleusine coracana* (L.) Gaertn.], and their improvement using omics approaches. Front Plant Sci 7:198656. https://doi.org/10.3389/FPLS.2016.00934/XML/NLM

Kumar A, Rani M, Mani S, Shah P, Singh DB, Kudapa H, Varshney RK (2021) Nutritional significance and antioxidant-mediated antiaging effects of finger millet: molecular insights and prospects. Front Sustain Food Syst 5:684318. https://doi.org/10.3389/FSUFS.2021.684318/XML/NLM

Kumari D, Madhujith T, Chandrasekara A (2016) Comparison of phenolic content and antioxidant activities of millet varieties grown in different locations in Sri Lanka. Food Sci Nutr 5:474. https://doi.org/10.1002/FSN3.415

Kumari A, Sadh PK, Kamboj A, Yadav B, Kumar A, Sivakumar S, Surekha, Saharan BS, Brar B, Goyal C, Dhull SB, Duhan JS (2025) Exploring the benefits of nutrition of little millet: unveiling the effect of processing methods on bioactive properties. J Food Biochem 2025:2488816. https://doi.org/10.1155/JFBC/2488816

Lakhani KG, Hamid R, Gupta S, Prajapati P, Prabha R, Patel S, Suthar KP (2024) Exploring the therapeutic mechanisms of millet in obesity through molecular docking, pharmacokinetics, and dynamic simulation. Front Nutr 11:1453819. https://doi.org/10.3389/FNUT.2024.1453819/FULL

Lakra P, Nashier Gahlawat I, Hussain A, Author C (2024) Millet mania: exploring awareness and consumption patterns among college students. Mysore J Agric Sci 58:232–240

Lee SH, Lee J, Herald T, Cox S, Noronha L, Perumal R, Lee HS, Smolensky D (2020) Anticancer activity of a novel high phenolic sorghum bran in human colon cancer cells. Oxidative Med Cell Longev 2020:2890536. https://doi.org/10.1155/2020/2890536

Lee SH, Lee HS, Lee J, Amarakoon D, Lou Z, Noronha LE, Herald TJ, Perumal R, Smolensky D (2021) Polyphenol containing sorghum brans exhibit an anti-cancer effect in apc min/+ mice treated with dextran sodium sulfate. Int J Mol Sci 22. https://doi.org/10.3390/IJMS22158286

Li Z, Zhao X, Zhang X, Liu H (2022) The effects of processing on bioactive compounds and biological activities of sorghum grains. Molecules 27:3246. https://doi.org/10.3390/MOLECULES27103246

Li J, Pandian V, Davidson PM, Song Y, Chen N, Fong DYT (2025a) Burden and attributable risk factors of non-communicable diseases and subtypes in 204 countries and territories, 1990-2021: a systematic analysis for the global burden of disease study 2021. Int J Surg 111:2385–2397. https://doi.org/10.1097/JS9.0000000000002260

Li Q, Wang X, Ma C, Onyango S, Wu W, Gao H, Li Q (2025b) Foxtail millet bran dietary fibres foster in vitro beneficial gut microbes and metabolites while suppressing pathobionts. Food Chem 464. https://doi.org/10.1016/j.foodchem.2024.141933

Linseman D, Marini HR, Almutairi S, Sivadas A, Kwakowsky A (2024) The effect of oral GABA on the nervous system: potential for therapeutic intervention. Nutraceuticals 4:241–259. https://doi.org/10.3390/NUTRACEUTICALS4020015

Lowcock EC, Cotterchio M, Boucher BA (2013) Consumption of flaxseed, a rich source of lignans, is associated with reduced breast cancer risk. Cancer Causes Control 24:813–816. https://doi.org/10.1007/S10552-013-0155-7

Maharajan T, Antony Ceasar S, Ajeesh Krishna TP, Ignacimuthu S (2021) Finger millet [*Eleusine coracana* (L.) Gaertn]: an orphan crop with a potential to alleviate the calcium deficiency in the semi-arid tropics of Asia and Africa. Front Sustain Food Syst 5:684447. https://doi.org/10.3389/FSUFS.2021.684447/BIBTEX

Mawouma S, Condurache NN, Turturică M, Constantin OE, Croitoru C, Rapeanu G (2022) Chemical composition and antioxidant profile of sorghum (*Sorghum bicolor* (L.) Moench) and pearl millet (*Pennisetum glaucum* (L.) R.Br.) grains cultivated in the Far-North Region of Cameroon. Foods 11:2026. https://doi.org/10.3390/FOODS11142026

Millet consumption can reduce total cholesterol, triacylglycerols and BMI. https://www.news-medical.net/news/20210818/Millet-consumption-can-reduce-total-cholesterol-triacylglycerols-and-BMI.aspx. Accessed 10 May 2025

Mosse J, Huet J-C, Baudet J (1988) The amino acid composition of whole sorghum grain in relation to its nitrogen content. Cereal Chem 65(4):271–277

Mozaffarian D (2016) Dietary and policy priorities for cardiovascular disease, diabetes, and obesity. Circulation 133:187–225. https://doi.org/10.1161/CIRCULATIONAHA.115.018585

Mudau M, Mashau ME, Ramashia SE (2022) Nutritional quality, antioxidant, microstructural and sensory properties of spontaneously fermented gluten-free finger millet biscuits. Foods 11:1265. https://doi.org/10.3390/FOODS11091265/S1

Muthamilarasan M, Dhaka A, Yadav R et al (2016) Exploration of millet models for developing nutrient rich graminaceous crops. Plant Sci 242:89–97

Muthayya S, Rah JH, Sugimoto JD, Roos FF, Kraemer K, Black RE (2013) The global hidden hunger indices and maps: an advocacy tool for action. PLoS One 8. https://doi.org/10.1371/JOURNAL.PONE.0067860

Nayak RR (2024) The superfood finger millet (*Eleusine coracana*) and its therapeutic activities—a review. Int J Res Rev 11:5. https://doi.org/10.52403/ijrr.20240543

Nematpour A, Eshghizadeh HR, Zahedi M (2019) Drought-tolerance mechanisms in foxtail millet (*Setaria italica*) and proso millet (*Panicum miliaceum*) under different nitrogen supply and sowing dates. Crop Pasture Sci 70:442–452. https://doi.org/10.1071/CP18501

Ng SH, Yeatman H, Kelly B, Sankaranarayanan S, Karupaiah T (2022) Identifying barriers and facilitators in the development and implementation of government-led food environment policies: a systematic review. Nutr Rev 80:1896. https://doi.org/10.1093/NUTRIT/NUAC016

Nithya VS, Athul TP, Deogade M, Nesari TM (2023) Role of foxtail millet (*Setaria italica* (L.) P. Beauvois) in senile bone degenerative diseases: a conceptual review. J Drug Res Ayurvedic Sci 8:S76–S81. https://doi.org/10.4103/JDRAS.JDRAS_218_23

Niu Y, Wang L, Gong H, Jia S, Guan Q, Li L, Cheng H (2025) Nutrition and gut health: preparation and efficacy of resistant starch. Foods 14:471. https://doi.org/10.3390/FOODS14030471

Nuss P (2015) Anxiety disorders and GABA neurotransmission: a disturbance of modulation. Neuropsychiatr Dis Treat 11:165. https://doi.org/10.2147/NDT.S58841

Oduor S, Christine Akoth O, Alfred Mwanjallah M, Joy Kagwiria O, Francis Mutiso M (2010) Effects of malting and fermentation treatments on group B-vitamins of red sorghum, white sorghum and pearl millets in Kenya. J Appl Biosci 34:2128–2134

Owheruo JO, Ifesan BOT, Kolawole AO (2018) Physicochemical properties of malted finger millet (*Eleusine coracana*) and pearl millet (*Pennisetum glaucum*). Food Sci Nutr 7:476. https://doi.org/10.1002/FSN3.816

Pandey SN, Mittal P, Singh G (2023) Phyto-pharmacological importance with nutritional potential of *Eleusine coracana* Linn.: a review. Int J Pharm Investig 13:688–697. https://doi.org/10.5530/IJPI.13.4.086

Parikh T, Bansal C, Shukla U, Zainab S (2024) Preventive potential of finger millet (ragi *Eleusine coracana* L.) in nutritional anemia and lifestyle diseases: a comprehensive review. Int J Res Ayurveda Pharm 15:2024. https://doi.org/10.7897/2277-4343.156197

Park JH, Lee SH, Chung IM, Park Y (2012) Sorghum extract exerts an anti-diabetic effect by improving insulin sensitivity via PPAR-γ in mice fed a high-fat diet. Nutr Res Pract 6:322. https://doi.org/10.4162/NRP.2012.6.4.322

Park SY, Jeong EW, Yang YS, Kim HJ, Go GW, Lee HG (2021) Finger millet ethanol extracts prevent hypertension by inhibiting the angiotensin-converting enzyme level and enhancing the antioxidant capacity in spontaneously hypertensive rats. Antioxidants 10:1766. https://doi.org/10.3390/ANTIOX10111766

Passarelli S, Free CM, Shepon A, Beal T, Batis C, Golden CD (2024) Global estimation of dietary micronutrient inadequacies: a modelling analysis. Lancet Glob Health 12:e1590–e1599. https://doi.org/10.1016/S2214-109X(24)00276-6

Peterson J, Dwyer J, Adlercreutz H, Scalbert A, Jacques P, McCullough ML (2010) Dietary lignans: physiology and potential for cardiovascular disease risk reduction. Nutr Rev 68:571. https://doi.org/10.1111/J.1753-4887.2010.00319.X

Pontieri P, Pepe G, Campiglia P, Merciai F, Basilicata MG, Smolensky D, Calcagnile M, Troisi J, Romano R, Del Giudice F, Aletta M, Guida M, Alifano P, Del Giudice L (2021) Comparison of content in phenolic compounds and antioxidant capacity in grains of white, red, and black sorghum varieties grown in the Mediterranean area. ACS Food Sci Technol 1:1109–1119. https://doi.org/10.1021/ACSFOODSCITECH.1C00115

Puranik S, Kam J, Sahu PP, Yadav R, Srivastava RK, Ojulong H, Yadav R (2017) Harnessing finger millet to combat calcium deficiency in humans: challenges and prospects. Front Plant Sci 8:1311. https://doi.org/10.3389/FPLS.2017.01311

Radha S, Mekala A, Premkumar K, Bhanu S, Samraj K, Thilagavathi R (2024) Therapeutic potential of millets in the management of metabolic syndrome—a review. J Res Siddha Med 7:36. https://doi.org/10.4103/JRSM.JRSM_18_23

Rapoo SM, Budeli P, Thaoge ML (2023) Recovery of potential starter cultures and probiotics from fermented sorghum (ting) slurries. Microorganisms 11:715. https://doi.org/10.3390/MICROORGANISMS11030715/S1

Reddy KJ, Bindu GSM, Hussain SA, Devi MU, Maheshwaramma S (2022) Nutritional composition of yellow pericarp sorghum (*Sorghum bicolor* L. Moenech): a review. Int J Environ Clim Chang:1732–1738. https://doi.org/10.9734/IJECC/2022/V12I121618

Ren X, Yin R, Hou D, Xue Y, Zhang M, Diao X, Zhang Y, Wu J, Hu J, Hu X, Shen Q (2018) The glucose-lowering effect of foxtail millet in subjects with impaired glucose tolerance: a self-controlled clinical trial. Nutrients 10:1509. https://doi.org/10.3390/NU10101509

Ren X, Wang L, Chen Z, Hou D, Xue Y, Diao X, Shen Q (2021) Foxtail millet improves blood glucose metabolism in diabetic rats through pi3k/akt and nf-κb signaling pathways mediated by gut microbiota. Nutrients 13:1837. https://doi.org/10.3390/NU13061837/S1

Rinninella E, Cintoni M, Raoul P, Lopetuso LR, Scaldaferri F, Pulcini G, Miggiano GAD, Gasbarrini A, Mele MC (2019) Food components and dietary habits: keys for a healthy gut microbiota composition. Nutrients 11:2393. https://doi.org/10.3390/NU11102393

Rodríguez-García C, Sánchez-Quesada C, Toledo E, Delgado-Rodríguez M, Gaforio JJ (2019) Naturally lignan-rich foods: a dietary tool for health promotion? Molecules 24:917. https://doi.org/10.3390/MOLECULES24050917

Sahaya Rani G, Swaminathan A, Vijayaraghavan R (2021) Effectiveness of physical activity and finger millet-based food supplement on biochemical parameters and bone mineral density among premenopausal women. Evid Based Complement Alternat Med 2021:4757991. https://doi.org/10.1155/2021/4757991

Sahu S, Kumar S, Nagtode NR, Sahu M (2024) "The burden of lifestyle diseases and their impact on health service in India"—a narrative review. J Family Med Prim Care 13:1612–1619. https://doi.org/10.4103/JFMPC.JFMPC_693_23

Shaikh R, Pund M, Dawane A, Iliyas S (2014) Evaluation of anticancer, antioxidant, and possible anti-inflammatory properties of selected medicinal plants used in indian traditional medication. J Tradit Complement Med 4:253–257. https://doi.org/10.4103/2225-4110.128904

Shan S, Lu Y, Zhang X, Shi J, Li H, Li Z (2021) Inhibitory effect of bound polyphenol from foxtail millet bran on miR-149 methylation increases the chemosensitivity of human colorectal cancer HCT-8/Fu cells. Mol Cell Biochem 476:513–523. https://doi.org/10.1007/S11010-020-03906-4

Sharma S, Saxena DC, Riar CS (2015) Antioxidant activity, total phenolics, flavonoids and antinutritional characteristics of germinated foxtail millet (*Setaria italica*). Cogent Food Agric 1. https://doi.org/10.1080/23311932.2015.1081728

Shen S, Huang R, Li C, Wu W, Chen H, Shi J, Chen S, Ye X (2018) Phenolic compositions and antioxidant activities differ significantly among sorghum grains with different applications. Molecules 23:1203. https://doi.org/10.3390/MOLECULES23051203

Shi J, Shan S, Li H, Song G, Li Z (2017) Anti-inflammatory effects of millet bran derived-bound polyphenols in LPS-induced HT-29 cell via ROS/miR-149/Akt/NF-κB signaling pathway. Oncotarget 8:74582. https://doi.org/10.18632/ONCOTARGET.20216

Shisode N (2024) Millets: a review of its properties as per ayurveda. Ann Geriatr Educ Med Sci 11:3–10. https://doi.org/10.18231/J.AGEMS.2024.002

Silungwe FR, Graef F, Bellingrath-Kimura SD, Tumbo SD, Kahimba FC, Lana MA (2019) Analysis of intra and interseasonal rainfall variability and its effects on pearl millet yield in a semiarid agroclimate: significance of scattered fields and tied ridges. Water 11:578. https://doi.org/10.3390/W11030578

Singh N, Sekhar A, Abraham J (2015) Evaluation of antimicrobial and anticancer properties of finger millet (*Eleusine coracana*) and pearl millet (*Pennisetum glaucum*) extracts. Pharma Innov J 3(11):82–86

Singh V, Lee GD, Son HW, Amani S, Baunthiyal M, Shin JH (2022) Anti-diabetic prospects of dietary bio-actives of millets and the significance of the gut microbiota: a case of finger millet. Front Nutr 9. https://doi.org/10.3389/FNUT.2022.1056445

Sirin F (2023) A review on finger millet: its nutritional value, bioactive compounds, processing and value-addition. Pharma Innov J 12:2193–2199

Sleem I, Smolensky D, Dia V (2024) Gastrointestinal health benefits of sorghum phenolics. Plant Foods Hum Nutr 79. https://doi.org/10.1007/S11130-024-01230-6

Sonawane G (2024) Antioxidant and antimicrobial activities of methanolic extract of sorghum bicolor seeds. Int J Pharm Sci 2. https://doi.org/10.5281/ZENODO.14563308

Sood S, Mohiuddin Bhat F (2023) Evaluation of chemical composition protein quality and amino acid scoring WHO/FAO standards of functional cereals oat pearl millet sorghum and finger millet. Int J Pharmacogn Chinese Med 7(2):000241. https://doi.org/10.23880/ipcm-16000241

Sood S, Kant L, Pattanayak A (2017) Finger millet [*Eleusine coracana* (L.) Gaertn.]: a minor crop for sustainable food and nutritional security: a mini review. Asian J Chem 29:707–710. https://doi.org/10.14233/AJCHEM.2017.20284

Stark TD, Mtui DJ, Balemba OB (2013) Ethnopharmacological survey of plants used in the traditional treatment of gastrointestinal pain, inflammation and diarrhea in Africa: future perspectives for integration into modern medicine. Animals 3:158–227. https://doi.org/10.3390/ANI3010158

Stefoska-Needham A (2024) Sorghum and health: an overview of potential protective health effects. J Food Sci 89:A30. https://doi.org/10.1111/1750-3841.16978

Suma PF, Urooj A (2011) Antioxidant activity of extracts from foxtail millet (*Setaria italica*). J Food Sci Technol 49:500. https://doi.org/10.1007/S13197-011-0300-9

Surjana D, Halliday GM, Damian DL (2010) Role of nicotinamide in DNA damage, mutagenesis, and DNA repair. J Nucleic Acids 2010:157591. https://doi.org/10.4061/2010/157591

Taylor JR, Duodu KG (2015) Effects of processing sorghum and millets on their phenolic phytochemicals and the implications of this to the health-enhancing properties of sorghum and millet food and beverage products. J Sci Food Agric 95:225–237. https://doi.org/10.1002/JSFA.6713

Taylor JRN, Duodu KG (2023) Resistant-type starch in sorghum foods—factors involved and health implications. Starch-Stärke 75:2100296. https://doi.org/10.1002/STAR.202100296

Topping DL, Fukushima M, Bird AR (2003) Resistant starch as a prebiotic and synbiotic: state of the art. Proc Nutr Soc 62:171–176. https://doi.org/10.1079/PNS2002224

Tripathi T, Vyas S (2023) From ancient grains to modern solutions: a history of millets and their significance in agriculture and food security. Int J Home Sci 9(2):72–78. https://doi.org/10.22271/23957476.2023.v9.i2b.1466

Varghese C, Prem A, Nongkynrih B, Chattu VK, Mikkelsen B (2025) Fourth UNHLM on noncommunicable diseases 2025: an opportunity to bridge the transcending priorities for impact in global south. PLOS Glob Public Health 5:e0004287. https://doi.org/10.1371/JOURNAL.PGPH.0004287

Veerabadhran V, Palakuru HR, Pratap GP, Rajeshwari B, Prasad GP, Srikanth N, Acharya R (2023) A critical review of the history and practices of millets in India. J Drug Res Ayurvedic Sci 8:S18–S25. https://doi.org/10.4103/JDRAS.JDRAS_113_23

Venkateswaran V, Vijayalakshmi G (2010) Finger millet (*Eleusine coracana*)—an economically viable source for antihypercholesterolemic metabolites production by *Monascus purpureus*. J Food Sci Technol 47:426. https://doi.org/10.1007/S13197-010-0070-9

Verma A (2024) Organochemistry of amino acids in finger millets—a brief review. J Sci Technol Res 5:1–13. https://doi.org/10.51514/JSTR.5.3.2023.1-13

Warhade S, Atram P, Sawarkar N (2023) Finger millet (ragi): a nutritional boon for osteoporosis. Int J Res Appl Sci Eng Technol 11:562–564. https://doi.org/10.22214/IJRASET.2023.54700

Wu Y, Li X, Xiang W, Zhu C, Lin Z, Wu Y, Li J, Pandravada S, Ridder DD, Bai G, Wang ML, Trick HN, Beane SR, Tuinstra MR, Tesso TT, Yu J (2012) Presence of tannins in sorghum grains is conditioned by different natural alleles of Tannin1. Proc Natl Acad Sci USA 109:10281–10286. https://doi.org/10.1073/PNAS.1201700109/SUPPL_FILE/SAPP.PDF

Wu T, Li H, Li J, Hao J (2023) Nutrient composition of germinated foxtail millet flour treated with mixed salt solution and slightly acidic electrolyzed water. Foods 12:75. https://doi.org/10.3390/FOODS12010075/S1

Xiong Y, Zhang P, Warner RD, Fang Z (2019) Sorghum grain: from genotype, nutrition, and phenolic profile to its health benefits and food applications. Compr Rev Food Sci Food Saf 18:2025–2046. https://doi.org/10.1111/1541-4337.12506

Xu J, Wang W, Zhao Y (2021) Phenolic compounds in whole grain sorghum and their health benefits. Foods 10:1921. https://doi.org/10.3390/FOODS10081921

Yang L, Browning JD, Awika JM (2009) Sorghum 3-deoxyanthocyanins possess strong phase II enzyme inducer activity and cancer cell growth inhibition properties. J Agric Food Chem 57:1797–1804. https://doi.org/10.1021/JF8035066

Yu S, Li C, Wang X, Herrera-Balandrano DD, Johnson JB, Xiang J (2024) Optimization of germination conditions for enriched γ-aminobutyric acid and phenolic compounds of foxtail millet sprouts by response surface methodology. Foods 13:3340. https://doi.org/10.3390/FOODS13203340

Zhang Y, Liu W, Zhang D, Yang Y, Wang X, Li L (2021) Fermented and germinated processing improved the protective effects of foxtail millet whole grain against dextran sulfate sodium-induced acute ulcerative colitis and gut microbiota dysbiosis in C57BL/6 mice. Front Nutr 8:694936. https://doi.org/10.3389/FNUT.2021.694936/BIBTEX

Zhang W, Zhang G, Liang W, Tian J, Sun S, Zhang X, Lv X, Guo P, Qu A, Wu Z (2024) Structure, functional properties, and applications of foxtail millet prolamin: a review. Biomolecules 14:913. https://doi.org/10.3390/BIOM14080913

Zhao X, Huang M, Huang X, Liu E (2023) Evaluation of drought resistance and index screening of foxtail millet cultivars. J Water Clim Change 14:2384–2396. https://doi.org/10.2166/WCC.2023.086

Index

© The Editor(s) (if applicable) and The Author(s), under exclusive license to Springer Nature Switzerland AG 2026
J. Singh, M. Kumar Malik (eds.), *Millets and Millet Starch: Chemical Composition, Functional Modification, and Applications*,
https://doi.org/10.1007/978-3-032-09502-2

GPSR Compliance

The European Union's (EU) General Product Safety Regulation (GPSR) is a set of rules that requires consumer products to be safe and our obligations to ensure this.

If you have any concerns about our products, you can contact us on ProductSafety@springernature.com

In case Publisher is established outside the EU, the EU authorized representative is:

Springer Nature Customer Service Center GmbH
Europaplatz 3
69115 Heidelberg, Germany

Batch number: 09598733

Printed by Printforce, the Netherlands